Routledge Handbook of Science, Technology, and Society

Over the last decade or so, the field of science and technology studies (STS) has become an intellectually dynamic interdisciplinary arena. Concepts, methods, and theoretical perspectives are being drawn from both long-established and relatively young disciplines. From its origins in philosophical and political debates about the creation and use of scientific knowledge, STS has become a wide and deep space for the consideration of the place of science and technology in the world, past and present.

The *Routledge Handbook of Science, Technology, and Society* seeks to capture the dynamism and breadth of the field by presenting work that pushes the reader to think about science and technology and their intersections with social life in new ways. The inter-disciplinary contributions by international experts in this handbook are organized around six topic areas:

- Embodiment
- Consuming technoscience
- Digitization
- Environments
- Technoscience as work
- Rules and standards

This volume highlights a range of theoretical and empirical approaches to some of the persistent – and new – questions in the field. It will be useful for students and scholars throughout the social sciences and humanities, including science and technology studies, history, geography, critical race studies, sociology, communications, women's and gender studies, anthropology, and political science.

Daniel Lee Kleinman is Associate Dean for Social Studies at the Graduate School and Professor in the Department of Community and Environmental Sociology at the University of Wisconsin–Madison.

Kelly Moore is Associate Professor of Sociology at Loyola University Chicago.

'This Handbook shows how power relations are both exercised and disguised through apparently neutral expertise or artefacts, as well as how such linkages are disrupted by subaltern groups. The articles offer STS methods for critical analysis to learn from struggles for social justice and to inform them.'

Les Levidow, Editor, Science as Culture

'This timely set of essays results in much more than a summary of a field; it is an incisive and forward-looking collection, offering a substantive journey into new directions in STS scholarship today. The book will be widely read for its diversity of approaches, yet coherence of chapters that together challenge a rethinking of sociotechnical processes as they unfold in the major areas of contemporary public debate.'

Laura Mamo, Author of Queering Reproduction

Routledge Handbook of Science, Technology, and Society

Edited by Daniel Lee Kleinman and Kelly Moore

LONDON AND NEW YORK

First published 2014 by Routledge

2 Park Square, Milton Park, Abingdon, Oxfordshire OX14 4RN
52 Vanderbilt Avenue, New York, NY 10017

Routledge is an imprint of the Taylor & Francis Group, an informa business.

First issued in paperback 2019

British Library Cataloguing in Publication Data
A catalogue record for this book is available from the British Library.

Library of Congress Cataloging in Publication Data
Routledge handbook of science, technology and society / edited by Daniel Lee Kleinman
and Kelly Moore.
pages cm
Includes bibliographical references and index.
1. Science—Social aspects. 2. Technology—Social aspects. 3. Science and civilization.
4. Technology and civilization. I. Kleinman, Daniel Lee. II. Moore, Kelly.
Q175.5.R68 2014
303.48'3—dc23
2013039839

ISBN: 978-0-415-53152-8 (hbk)
ISBN: 978-0-367-33593-9 (pbk)

Typeset in Bembo
by FiSH Books Ltd, Enfield

To the memory of my parents, Barbara and Gerald Kleinman.
– Daniel Lee Kleinman

To EB, KKB, ESC, and RHW for their ongoing intellectual support and inspiration.
– Kelly Moore

Contents

Contents

Contents

List of Illustrations

Figures

Tables

Contributors

Samer Alatout is Associate Professor of Community and Environmental Sociology, Environmental Studies, and Geography at the University of Wisconsin-Madison. He is also affiliated with the Holtz Center for Science and Technology Studies and the Center for Culture, History, and Environment. He has published extensively on water history, policy, and politics in the Middle East, focusing on the relationship between notions of water scarcity/abundance and political identity, state-building and governance. His work has appeared in, among others, Social Studies of Science, Political Geography, Environment and Planning D, and the Annals of the Association of American Geographers. He is conceptually interested in biopolitics, border studies, and the sociology and history of science and technology. At present, Alatout is writing a book manuscript on water politics in historic Palestine tentatively titled *A Palestinian Century: Water Politics from Empire to Globalization*.

Mathieu Albert is Associate Professor in the Department of Psychiatry and the Wilson Centre for Research in Education, University of Toronto. His current work primarily focuses on the multidisciplinary relationship between academics and the struggle for scientific authority. He has published in a wide range of disciplinary and interdisciplinary journals in social science and in medicine, including articles on symbolic boundaries between scientific groups (*Minerva* and *Social Science and Medicine*), science policy-making process (*Science, Technology & Human Values*), academic assessment criteria (*Higher Education*) and funding agencies (*Canadian Journal of Higher Education*). He received the 2011 American Sociological Section on Science, Knowledge and Technology Best Paper Award for "Boundary-Work in the Health Research Field: Biomedical and Clinician Scientists' Perceptions of Social Science Research" (Minerva, 2009), and the 2001 Sheffield Prize awarded by the Canadian Society for the Study of Higher Education for his paper entitled "Funding Agencies' Coping Strategies to the Canadian and Quebec Governments' Budget Cuts" (*Canadian Journal of Higher Education*).

Sulfikar Amir is Assistant Professor in the Division of Sociology, Nanyang Technological University, Singapore. He has conducted research on nuclear politics in Southeast Asia, specifically Indonesia and Thailand. Currently he is working on a research project investigating sociotechnical vulnerability in the Fukushima nuclear disaster. Apart from nuclear politics, his research explores technological nationalism, technopolitics, development, city resilience, and design studies. He is the author of *The Technological State in Indonesia: the Co-constitution of High Technology and Authoritarian Politics*. In 2012, he produced *Nuklir Jawa*, a documentary film that depicts the controversy surrounding Indonesia's desire to build a nuclear power station in Central Java.

Hee-Je Bak is Professor in the Department of Sociology and Director of the Center for Science, Technology and Society at Kyung Hee University. His research centers on the roles of scientists in the transformation of higher education and the politics of scientific and environmental risks. He is the co-author of *The Scientific Community in Korea* (2010). Currently he is leading a research group for a study of scientific governance in Korea. His email address is hbak@khu.ac.kr.

Susan E. Bell is A. Myrick Freeman Professor of Social Sciences/Professor of Sociology, Bowdoin College, 1983–present. She has published widely about medicalization, women's health, experiences of illness, visual and performative representations of cancer, medicine and women's bodies, and the circulation and transformation of biomedical knowledges. She is an editorial advisor for *Sociology of Health & Illness*, on the editorial boards of *Qualitative Sociology* and the *Journal of Health and Social Behaviour*. She is the author of *DES Daughters: Embodied Knowledge and the Transformation of Women's Health Politics* (Temple, 2009) and the guest editor with Alan Radley of a special issue of *Health*, "Another Way of Knowing: Art, Disease, and Illness Experience" (2011). Recent publications include her work with Anne E. Figert, "Disrupting Scholarship", an invited chapter for *Open to Disruption: Practicing Slow Sociology* R. Hertz, A. Garey and M. Nelson (Eds.) (Vanderbilt, forthcoming 2014); "Claiming justice: Knowing mental illness in the public art of Anna Schuleit's 'Habeas Corpus' and 'Bloom'" (*Health*, 2011); and "Artworks, collective experience, and claims for social justice: the case of women living with breast cancer" (with Alan Radley), *Sociology of Health & Illness* (2007). She is the Chair of the Department of Sociology and Anthropology at Bowdoin College and the incoming Chair of the Medical Sociology section of the American Sociological Association.

Kelly Bergstrand is a Ph.D. candidate at the University of Arizona and specializes in social movements, environmental sociology, and social psychology. Her current research projects include examining the role of grievances in mobilization, investigating activism on the U.S.-Mexico border, and mapping the community impact of environmental disasters.

Whitney Erin Boesel is a Ph.D. student in sociology at the University of California, Santa Cruz. She is a weekly contributor for the blog *Cyborgology*, and serves on the planning committee for the Theorizing the Web conferences. Her dissertation work focuses on self-quantification and mood tracking as part of what she terms "biomedicalization 2.0"; she also writes regularly about social media.

Michael Borowy is a researcher at the Centre for Policy Research on Science and Technology (CPROST). He studied human geography and English at the University of British Columbia and holds an M.A. in Communication from Simon Fraser University. His published work includes an article exploring the history of marketing and conceptual thought related to "eSport" in the *International Journal of Communication*. His current research interests include the video and computer games industry, digital policy, and the development of new sports.

Phil Brown is University Distinguished Professor of Sociology and Health Sciences at Northeastern University, where he directs the Social Science Environmental Health Research Institute, which extends the work of the Contested Illnesses Research Group, which started in 1999 at Brown University. He is the author of *No Safe Place: Toxic Waste, Leukemia, and Community Action*, and *Toxic Exposures: Contested Illnesses and the Environmental Health Movement*, and co-editor of *Illness and the Environment: A Reader in Contested Medicine, Social Movements in*

Health, and *Contested Illnesses: Citizens, Science and Health Social Movements.* His current research includes biomonitoring and household exposure, social policy concerning flame retardants, the ethics of reporting back research data to participants, data privacy, and health social movements.

Patrick Carroll is Associate Professor of Sociology and a member of the Science and Technology Studies Program at the University of California, Davis. He is the author of *Science, Culture and Modern State Formation* (University of California Press, 2006). He is currently working on a book entitled *California Delta: The Engineered Heart of a Modern State Formation.*

Hsin-Hsing Chen is Associate Professor in the Graduate Institute for Social Transformation Studies at Shih-Hsin University, Taiwan. His research interests include agriculture and rural development, labor studies, anthropology of work, and more recently, scientific causation in law. He is also a labor activist with special attention on labor conditions in Taiwanese-owned factories abroad. He received his Ph.D. in Science and Technology Studies from Rensselaer Polytechnic Institute, and he teaches courses ranging from contemporary East Asian political thoughts to organizing methods in social movements.

Joseph Dumit is Director of Science and Technology Studies and Professor of Anthropology at the University of California, Davis. He is the author of *Drugs for Life: How Pharmaceutical Companies Define Our Health* (Duke, 2012) and *Picturing Personhood: Brain Scans and Biomedical Identity* (Princeton University Press, 2004). Dumit has also co-edited *Cyborgs & Citadels: Anthropological Interventions in Emerging Sciences and Technologies*; *Cyborg Babies: From Techno-Sex to Techno-Tots*; and *Biomedicine as Culture.* He was associate editor of *Culture, Medicine & Psychiatry* for ten years. He is a founding member of the Humanities Innovation Lab (*http://modlab.ucdavis.edu*) and is currently studying how immersive 3D visualization platforms are transforming science (*http://keckcaves.org*). He has begun work on a new project on the history of flow charts, cognitive science, and paranoid computers. His website is *http://dumit.net.*

Kristin R. Eschenfelder is Professor and Director at the School of Library and Information Studies at the University of Wisconsin-Madison. She is also an affiliate of the Holtz Center for Science and Technology Studies, the School of Journalism and Mass Communications and a founding board member of the Wisconsin Digital Studies program. Her research interests focus on access and use regimes – or the complex, multi-level networks of laws, customs, technologies and expectations that shape what information we can access in our daily lives and how we can make use of it. Her recent work examines the development of and changes to access and use regimes for digital scholarly works including electronic publications (journals, books, citation databases), digital cultural materials, (such as museum, archival or anthropological works) and data sets. Her past work explored web-based government information and policy and management issues inherent in digital production of government information and records. She has also published in the areas of public libraries and financial literacy.

Anne E. Figert is Associate Professor of Sociology at Loyola University Chicago (1991–present). Her research interests are in the sociology of diagnosis, medicalization, and the construction of expertise and authority. She is the author of *Women and the Ownership of PMS: The Structuring of a Psychiatric Disorder* (Aldine de Gruyter 1996) and the co-editor of two volumes: *Building Community: Social Science in Action* (Pine Forge Press, 1997) and *Current Research on Occupations and Professions, volume 9* (JAI Press, 1996). Recent publications include her work with Susan E. Bell, "Doing God's Work and Doing Good Work(s): Unique Challenges to Evaluation Research in Ministry Settings" in *Public Sociology*, P. Nyden, G. Nyden and G. Hossfield (Eds.) (Pine Forge Press, 2011), "The Consumer Turn in Medicalization: Future Directions with Historical Foundations" in *The Handbook of the Sociology of Health, Illness and Healing*, B, Pescosolido, J.Martin, J. McLeod and A. Rogers (Eds.) (Springer Publishing, 2011) and "White Coats: The Shape of Scientific Authority and its Relationship to Religion and Religious Authority" in *Christianity, Gender, and Human Sexuality*, P. Jung and A. Vigen (Eds.) (University of Illinois Press, 2011). She is the Chair of the Medical Sociology section of the American Sociological Association.

Jane E. Fountain is Distinguished Professor of Political Science and Public Policy, and Adjunct Professor of Computer Science, at the University of Massachusetts, Amherst. She directs the National Center for Digital Government and the Science, Technology and Society Initiative, both based at the University of Massachusetts, Amherst. Previously, she served on the faculty of the John F. Kennedy School of Government at Harvard University. Fountain is the author of *Building the Virtual State: Information Technology and Institutional Change* (Brookings Institution Press, 2001), among many other publications. She holds a Ph.D. from Yale University and has been a Radcliffe Institute Fellow, a Mellon Foundation Fellow, and a Yale Fellow. Fountain is an elected fellow of the National Academy of Public Administration.

Nathaniel Freiburger is a doctoral candidate in the Department of Sociology and the Science and Technology Studies Program at the University of California, Davis. His research interests include the political dimensions of artifacts, state formation and theories of the state, materiality, resource topographies, and object-centered ethnography. His dissertation titled "Lithium: Object, Concept, Event" explores the intersection of technoscience and state formation in Bolivia around lithium resources in the Salar de Uyuni.

Scott Frickel is Associate Professor of Sociology at Brown University. He predominantly works on environment, science, and the politics of knowledge. In addition to studying the non-production of knowledge, his current research projects include a relational sociology of interdisciplinarity, a comparative environmental sociology of cities, and the political sociology of science, with a specific focus on the impacts of disasters on scientific networks and knowledge practices.

Anthony Ryan Hatch is Assistant Professor in the Department of Sociology at Georgia State University. His teaching and research interests are in critical social theory, cultural studies, and social studies of science, medicine, and technology. His research investigates how biomedical researchers, government agencies, and pharmaceutical corporations use social categories of race – producing new forms of both race and racism – in the scientific management of disease and illness. His work has received support from the National Institute of Mental Health.

Gabrielle Hecht is Professor of History at the University of Michigan. She is the author of *Being Nuclear: Africans and the Global Uranium Trade* (MIT Press and Wits University Press, 2012) and *The Radiance of France: Nuclear Power and National Identity after World War II* (MIT Press, 1998; new edition, 2009), both of which received awards from the American Historical Association and elsewhere. She has held visiting positions in France, the Netherlands, Norway, and South Africa. Her new research focuses on transnational toxic trash.

David J. Hess is Professor in the Sociology Department at Vanderbilt University, where he is the Associate Director of the Vanderbilt Institute for Energy and Environment and Director of the Program in Environmental and Sustainability Studies. He publishes widely on science, technology, and publics and on the politics of the green economy, and his most recent book is *Good Green Jobs in a Global Economy* (MIT Press). His website is *www.davidjhess.org*.

Dal Yong Jin received his Ph.D. from the Institute of Communications Research at the University of Illinois at Urbana Champaign. He has taught in several institutions, including the University of Illinois in Chicago, Korea Advanced Institute of Science and Technology (KAIST), and Simon Fraser University. His major research and teaching interests are on globalization and media, social media and game studies, transnational cultural studies, and the political economy of media and culture. He is the author of three books: *Korea's Online Gaming Empire* (MIT Press, 2010), *De-Convergence of Global Media Industries* (Routledge, 2013), and *Hands On/Hands Off: The Korean State and the Market Liberalization of the Communication Industry* (Hampton Press, 2011). Jin also edited two books, entitled *The Political Economies of Media* (with Dwayne Winseck, Bloomsbury, 2011) and *Global Media Convergence and Cultural Transformation* (IGI Global, 2011).

Abby J. Kinchy is Associate Professor in the Department of Science and Technology Studies at Rensselaer Polytechnic Institute. She specializes in the study of political controversies surrounding changes in the systems that produce food and energy. Her research examines the unequal distribution of the negative consequences of agricultural and energy systems, as well as the varying capacity of communities and social movements to participate in making decisions about technological change.

Daniel Lee Kleinman is Associate Dean for Social Studies in the Graduate School at the University of Wisconsin-Madison, where he is also Professor in the Department of Community and Environmental Sociology. Kleinman spent five years as Director of the Robert F. and Jean E. Holtz Center for Science and Technology Studies at the University of Wisconsin and has served as chair of the Science, Knowledge and Technology section of the American Sociological Association and as a member of the council of the Society for the Social Studies of Science. Kleinman is the author of three books, including *Impure Cultures: University Biology and the World of Commerce*. His work has also appeared in a wide array of journals ranging from *Issues in Science and Technology* to *Theory and Society*. Among his current projects are investigations of emerging knowledge about and government regulation related to Colony Collapse Disorder, the malady associated with massive die-offs of honey bees, the commercialization of higher education, and the organizational dynamics of interdisciplinary scientific research.

Marianne de Laet is Associate Professor of Anthropology and Science, Technology, and Society at Harvey Mudd College in Claremont, California. She has observed the emerging collaboration of the California Extremely Large Telescope (CELT) (now TMT) at the California Institute of Technology, and has published on astronomy and physics, appropriate technologies, patenting practices, music, and dogs. Her current research concerns the relationship between tasting and knowing. Her email address is delaet@g.hmc.edu.

Brian Mayer is Associate Professor in the School of Sociology at the University of Arizona. His research interests focus on the contestations that emerge around environmental problems in the areas of science, policy, and medicine. Recent research projects include a National Institute of Environmental Health Sciences-funded project to examine the long-term psychosocial and community health impacts of the BP Oil Spill in the Gulf of Mexico, an investigation of the use of community-based science in social movement organizations, and a project funded by the National Science Foundation to explore the interactions of labor and environmental social movement organizations in the United States.

Kelly Moore is Associate Professor of Sociology at Loyola University, Chicago. She studies the relationship between science, morality, and politics. She is the author of *Disrupting Science: Scientists, Social Movements and the Politics of the Military, 1945–1975* (Princeton University Press, 2008), winner of the 2009 Charles Tilly Prize from the American Sociological Association section on Collective Behavior and Social Movements and the 2011 Robert K. Merton Prize from the American Sociological Association Section on Science, Knowledge and Technology, and co-editor with Scott Frickel of *The New Political Sociology of Science* (2006). Her work has appeared in sociological and cross-disciplinary journals, including *Theory and Society, American Journal of Sociology, Geoforum, The Scholar and the Feminist*, and *Research in the Sociology of Organizations*. She has served as Co-director of the National Science Foundation Science, Technology and Society Program, Director of the National Science Foundation Ethics Education in Science and Engineering Program, and Chair of the American Sociological Association Section on Science, Knowledge and Technology. She serves on the Council of the Society for Social Studies of Science and on the editorial board of *Science, Technology and Human Values*. Her current research project, "*Pleasuring Science*", is on gendered and raced forms of scientific eating and exercise under neoliberalism.

Rachel Morello-Frosch is Professor in the Department of Environmental Science, Policy and Management and the School of Public Health at the University of California, Berkeley. Her scientific work examines the combined, synergistic effects of social and environmental factors in environmental health disparities. She also studies the ways in which health social movements (re)shape scientific thinking about environmental health issues. She is co-author of the book *Contested Illnesses: Citizens, Science and Health Social Movements*.

Alondra Nelson is Professor of Sociology and Director of the Institute for Research on Women and Gender at Columbia University. Professor Nelson is the author of *Body and Soul: The Black Panther Party and the Fight against Medical Discrimination*. She is co-editor of *Genetics and the Unsettled Past: The Collision of DNA, Race, and History* and of *Technicolor: Race, Technology, and Everyday Life*.

Elise Paradis is Assistant Professor in the Department of Anesthesia and Scientist at the Wilson Center at the University of Chicago. She obtained her Ph.D. from Stanford in 2011. Her research focuses on two things: the evolution of medical ideas over time, using a sociology of knowledge approach and the factors that impact group dynamics in closed settings. She has published on body fat, the "obesity epidemic", the social context of medical education, the rise of interprofessional research in medicine, interprofessional relations in intensive care units and women in boxing.

Shobita Parthasarathy is Associate Professor of Public Policy at the University of Michigan. She is the author of *Building Genetic Medicine: Breast Cancer, Technology, and the Comparative Politics of Health Care* (MIT Press, 2007). Findings from this book helped to inform the 2013 Supreme Court case over gene patents in the United States. Her next book explores the controversies over patents on life forms in the United States and Europe, focusing on 1980 to the present.

Roopali Phadke is an Associate Professor in Environmental Politics and Policy at Macalester College. Her research and teaching interests focus on domestic and international environmental politics and policies. Her work over the last decade has focused on how energy technologies, such as dams and wind turbines, can be more sustainably and democratically designed and managed. Her recent work appears in the journals *Science as Culture*, *Antipode*, and *Society & Natural Resources*. She received her Ph.D. in Environmental Studies at University of California, Santa Cruz. She also served as a National Science Foundation Postdoctoral Fellow in the Science, Technology and Society Program at the Kennedy School of Government at Harvard University.

PJ Rey is a sociology Ph.D. student at the University of Maryland. He is a founding editor of the *Cyborgology* blog and co-founder of the Theorizing the Web conferences. His work examines the social construction of online/offline and the affordances of digital media.

Joan H. Robinson is a doctoral candidate in the Department of Sociology at Columbia University. Her interests include interactions among women's studies, law, political economy, economic sociology, organizations, and science, technology, and medical studies. She received a BA (*magna cum laude*) from Syracuse University, where she garnered awards for her research in both political science and women's studies, and a JD from Brooklyn Law School, where she was recognized for her extensive work for women and families and in international law.

Katrina Running is Assistant Professor in the Department of Sociology, Social Work and Criminal Justice at Idaho State University. Her primary research interests include the relationship between economic development and environmental concern, the effects of climate change on rural communities in the American West, and the role of science in informing public policy.

J.P. Singh is Professor of Global Affairs and Cultural Studies at George Mason University. Singh has authored four monographs, edited two books, and published dozens of scholarly articles. He has advised international organizations such as UNESCO, the World Bank, and the World Trade Organization, played a leadership role in several professional organizations, and served as Editor from 2006 to 2009 and dramatically increased the impact of *Review of Policy Research*, the journal specializing in the politics and policy of science and technology. He holds a Ph.D. in Political Economy and Public Policy from the University of Southern California.

Laurel Smith-Doerr is the inaugural Director of the Institute for Social Science Research, and Professor of Sociology at University of Massachusetts at Amherst. She held faculty appointments in the Boston University Department of Sociology from 1999 to 2013. Her research and teaching interests focus on science and technology communities, issues in gender and work, organizations, and social networks. She investigates collaboration, implications of different organizational forms for women's equity in science, gendering of scientific networks and approaches to responsibilities, and tensions in the institutionalization of science policy. Results of this research have been published in her book, *Women's Work: Gender Equity v. Hierarchy in the Life Sciences* and scholarly journals including *Nature, Biotechnology, Administrative Science Quarterly, Minerva, Regional Studies, American Behavioral Scientist, Sociological Forum, Sociological Perspectives* and *Gender & Society*. In 2007–2009 she was appointed as a Visiting Scientist and Program Officer in Science, Technology and Society at the National Science Foundation. She received the NSF Director's Award for Collaborative Integration for her work in leading the Ethics Education in Science and Engineering program and on the committee implementing the ethics education policies of the America COMPETES Act of 2007. She has been elected as a Council Member at-large to both the Society for Social Studies of Science (4S) (2011–2013) and the American Sociological Association (2012–2014). She received her BA from Pomona College and her MA/Ph.D. from the University of Arizona.

Laura Stark is Assistant Professor at Vanderbilt University's Center for Medicine, Health, and Society, and Associate Editor of the journal *History & Theory*. She is author of *Behind Closed Doors: IRBs and the Making of Ethical Research*, which was published in 2012 by University of Chicago Press. She is also author of several articles and chapters on the science, the state and social theory. Her current research explores research settings through the lives of "normal control" research subjects enrolled in the first clinical trials at the U.S. National Institutes of Health after World War II. Laura received her Ph.D. in Sociology from Princeton University in 2006, was a Postdoctoral Fellow in Science in Human Culture at Northwestern University from 2006 to 2008, and held a Stetten Fellowship at the Office of NIH History at the National Institutes of Health from 2008 to 2009.

Kim TallBear is Associate Professor of Science, Technology, and Environmental Policy at the University of California, Berkeley. She studies how genomics is co-constituted with ideas of race and indigeneity. More recently, she is engaged in an ethnography of indigenous bio-scientists, examining how they navigate different cultures of expertise and tradition, including both scientific communities and tribal communities. She is also beginning a new ethnographic project looking at tribal, federal, and scientific cultural and policy practices related to the Pipestone Quarries in southeastern Minnesota, a national monument and chief source of stone used to carve ceremonial pipes.

Jaita Talukdar is Associate Professor of Sociology at Loyola University, New Orleans. Her research explores the effects in urban India of social forces of gender, social class, culture, and globalization on the body, bodily processes, and health. At present, she is investigating how new upcoming gyms in the city of Kolkata are shaping perceptions and ideas of a fit and healthy body among urban Indians.

Itai Vardi is a sociologist whose research focuses on the intersections of power, culture, and science and technology. As a Postdoctoral Research Associate at Boston University working on an NSF-funded project with Laurel Smith-Doerr and Jennifer Croissant, he developed an interest in understanding the social dynamics and contexts that enable and inhibit collaborative productions of scientific knowledge. He is currently also working on a book project, which analyzes the public discourse on traffic accidents in the United States. His previous co-authored book, *Driving Forces: The Trans-Israel Highway and the Privatization of Civil Infrastructure in Israel* (Jerusalem, Van Leer, 2010), explores the relation between sociotechnical systems and resource inequality in Israel. By examining the Trans-Israel Highway, the largest infrastructure enterprise in the country's history and its first toll road, the book illuminates the troubling social consequences of Israel's current transportation and environmental policies. He received his BA and MA from Tel Aviv University and his Ph.D. from Boston University. He has published his work in scholarly journals such as *Journal of Social History* and *Food, Culture, and Society*.

Tom Waidzunas is Assistant Professor in the Department of Sociology at Temple University. His areas of interest are at the intersections of sociology of sexuality and gender, science studies, and the sociology of social movements. In addition to his research on knowledge controversies and the ex-gay movement, he has also written about the construction and circulation of gay teen suicide statistics, and the experiences of lesbian, gay, and bisexual engineering students. He is currently working on a book on the ex-gay movement and the relegation of reorientation to the scientific fringe in the United States, also examining transnational dimensions of controversies over sexual reorientation therapies and religious ministries.

Acknowledgements

We thank Soulit Chacko for her expert technosocial coordination of the manuscript preparation. Her organizational skills, ability to anticipate and solve problems, and attention to detail have been critical to the project. We also thank Beth Dougherty for her coordination of the early stages of the project and for her contribution to our discussions about the content of the manuscript.

Introduction

Science, technology and society

Daniel Lee Kleinman and Kelly Moore

In the last decade or so, the field of science and technology studies (STS) has become an intellectually dynamic interdisciplinary arena. Concepts, methods, and theoretical perspectives are being drawn from long-established disciplines, such as history, geography, sociology, communications, anthropology, and political science, and from relatively young fields such as critical race studies and women's and gender studies. From its origins in philosophical and political debates about the creation and use of scientific knowledge, STS has become a wide and deep space for consideration of the place of science and technology in the world, past and present. STS is a field characterized by lively debates centered on foundational ideas and by work shaped by engagement with contemporary and past techno-scientific issues. If scholars energetically pursued questions of what it means to talk about the construction of knowledge and science in the early days of the field, today STS covers a wider range of topics, from the experiences of nuclear power workers to the hydrological construction of nation-state borders and the politics of scientific dieting among India's economic elites. Beyond those who identify primarily as STS scholars are researchers whose work speaks forcefully and compellingly to the technoscience-society/culture nexus. Thus, for example, there are scholars who associate primarily with the field of communications, but do engaging work on new social media, and law professors who write provocatively on the politics of the knowledge commons.

We take a big tent approach to STS, and in this handbook we seek to capture the dynamism and breadth of the field by presenting work that pushes us to think about science and technology and their intersections with social life in new ways. The book is not comprehensive in the sense of covering every corner of the field, but rather, highlights a range of theoretical and empirical approaches to some of the persistent – and new – questions in the field. Rather than soliciting traditional handbook review essays concluding with research questions pointing the field forward, we asked prospective authors, including senior and early career scholars doing some of the most vital work in the broad science, technology and society arena, to contribute pieces that are explanatory and weave clear and cogent arguments through important empirical cases.

The chapters are organized around six topic areas: embodiment, consuming technoscience, digitization, environments, science as work, and rules and standards. Over the past half-decade, these areas have yielded especially important empirical insights and/or generated new theoretical ferment. New developments, such as digitization, the role of markets and corporations in the sociotechnical shaping of human bodies (especially gendered, sexualized, ethno-racialized, and classed bodies), and increasing citizen participation in local, national and international science and technology governance, have been important impetuses for extending and challenging STS theory, including by using insights from other disciplines. Other topics, such as science as work, embodiment, and environments, are of long-standing concern in STS, but are the location where some of the fields' freshest ideas are emerging.

We also aimed to create a collection that, despite the diversity of the chapters, has an element of unity across essays. To generate thematic coherence throughout the book, authors, some more explicitly than others, address one or more of three key processes: how and why ideas, artifacts, and practices come to be **institutionalized or disrupted**; what explains the **scale** at which technoscience comes to have meaning, is struggled over and travels; and by what means **materiality and cultural value(s)** shape science and technology. These themes are not exhaustive but they represent some of the enduring and newer questions that STS addresses. Contributors approach questions of institutionalization and disruption by attending to canonical explananda, and to newer (for STS) foci, such as advertising and market mechanisms, and that most distinctive technology, the internet, that have been ever more important over the past decade. Some of our writers, consistent with recent developments in STS, bring renewed awareness to contests over the meanings of "global" and "local," exploring how at least some aspects of technoscience-infused social life are especially compressed in time and space, and others vastly expanded. Finally, materiality and culture have long been at the heart of STS. Authors in this book attend closely to how the material is realized, and when and how social meanings and symbols matter in questions of technoscience. Materiality, once dismissed by the field, is now a topic of central concern, as is culture, in shaping individual and group identities, careers, nation-states, and scientific objects and networks. These themes are not exhaustive of the strands of work currently being done in STS; they do, however, offer ways of capturing how contemporary STS scholars are grappling with new questions and topics. Our big tent approach to STS thus seeks to capture concerns that are at the center of the field, and some that are at its (current) edges, but which address the crucial issues surrounding the relationship among science, technology, and society.

Situating the work within the book in the context of social studies of science and technology over the past three decades illuminates some of the specific and enduring insights from the field, on which some of our contributors build, some challenge, and others use only indirectly. The canon in STS as a disciplinary arena (as opposed to a wider network of scholarship) was built on a specific set of debates that emerged in the 1970s and 1980s; yet some of the insights and concerns from this early period, such as the relationship between ethnoracialization and science, sexuality, and inequality were never built into the canon. Excellent overviews of the field of STS exist (Hess 1995, Biagioli 1999, Bauschpies et al. 2006, Sismondo 2004, Lynch 2012); our goal in this introduction is more limited: to provide context for the chapters that follow, by tracing out some of the key developments in STS, and some of the current directions of the more expansive contemporary field. Rather than rehearse a canonical story, we also include some debates and contributions that were dormant but have now come to fore, and some of the new areas in which we see considerable dynamism in the field.

From constructivism to politics

Long before there was a field of science and technology studies or even science, technology, and society, there was a small group of American researchers in the area of the sociology of science.

Writing from the middle of the twentieth century forward, these scholars, most especially Robert K. Merton, reflected a dominant strain of sociology at the time, focusing on science as a distinctive institution with unique dynamics (Merton 1957, 1968, Cole and Cole 1973, Zuckerman 1967). Starting from a functionalist perspective, scholars in this tradition were concerned with understanding what about the institutional character of science permitted it to produce a distinctive kind of knowledge, what factors could interfere with the inherently progressive dynamic of science, and how rewards were distributed among scientists. By the late 1960s and early 1970s, a much smaller group of Marxist-influenced social scientists examined how class relations were reproduced through scientific work (Whalley 1978), via the suppression of particular research topics (Martin 1981), and via technology more generally (Noble 1977). Marxists, however, never had the same influence on studies of science as the functionalists did, and indeed, were not associated with the sociology of science *per se*. Their concerns with the social inequalities associated with technology continue to be reflected in other ways in the field, as we discuss below. At the same time that Marxist-influenced analysts were asking questions about technologies and work, race, gender and sexuality analysts, some of them scientists themselves, began to raise questions about how social identities and power relations were produced through technopolitical means (Haraway 1982, 1991, Fausto Sterling 1978, Duster 1981, Cowan 1983, Harding 1986).

It was Merton, and his students and followers, who became the touchstone against which a new wave of European science scholars defined themselves. The Mertonian institutionalist approach, according to its critics, treated the core of science – the production of knowledge – as a black box, and presumed, rather than demonstrated, that science operated according to a unique set of rules (Merton 1957, 1971). Influenced in part by the postmodern philosophical turn in which discursive assumptions and foundations of knowledge claims were scrutinized, STS scholars theoretically examined scientific knowledge-making processes, and engaged in a series of laboratory-based ethnographic studies. Sometimes seeking to show that scientific knowledge is symmetrical with other beliefs (Barnes 1977, Bloor 1976), that technology, not nature, played a key role in vetting what counted as scientific fact, and at other times seeking to understand how qualifications (modalities) surrounding scientific claims are jettisoned in the establishment of "facts" (Latour and Woolgar 1979), much of this work was unconcerned with the role of institutional and organizational characteristics, such as government established laws, bureaucratic rules, and personnel pipelines, in shaping the content of knowledge (but see, for example, Barnes and MacKenzie 1979). The broad rubric under which this work falls is "constructivism," and it re-centered the social analysis of science at a particular scale and around a specific set of questions and possible answers. One of its major contributions to STS is to have demonstrated that knowledge is not simply "discovered," but *made* through the actions and work of scientists.

In the U.S., working in the broad constructivist tradition, Thomas F. Gieryn (1982) introduced the concept of "boundary work," giving STS scholars a way to analyze scientific knowledge production as a form of discursive contestation, in which "people contend for legitimacy, or challenge the cognitive authority of science" (Gieryn 1995: 405). He asked how discursive boundaries are drawn between what is science and what is not. Gieryn's analysis drew attention to contestations not only among scientists, but between scientists and non-scientists over what should count as credible scientific knowledge. The social practice of maintaining science as a rarefied work, then, was one of the many activities that scientists undertook to produce certified scientific claims.

Some post-Mertonian work trained its attention on ontological matters, like whether the boundaries between context and content, inside and outside, science and society, humans and non-humans, and actors and environments are anything more than analysts' constructions. Bruno Latour, most famously associated with "actor network theory," stressed these matters, and rejected the possibility of understanding discrete factors and causal relations in patterns of technoscience (Latour and Woolgar 1979, Latour 1987, Latour 1993). Along with many other analysts, Latour insisted on the mutuality of interactions and the co-construction of such entities as nature and society, human and non-human. Forget cause and effect, independent and dependent variables, social structure, and power relations in any traditional sense, said Latour. Study the networks and the humans and non-humans who constitute them. Latour argued that analysts should capture how actors enroll other actors in a process of interest translation. Methodologically, Latour's approach owes much to Michel Foucault's (1977) genealogical technique. Following Foucault, Latour introduced the notion that just as science was an outcome, so, too, were the "social" and the "political." That is, they could not, in his view, be used to explain anything, since they were made at the same time that science was being made. Latour's work contributed key ideas to the field, including the notion of non-humans as actants, the claim that scientists seek credit rather than truth itself, the contention that enrollment is a key process in scientific credibility contests, and the position that a modernist perspective of unidirectional "causality" is an inappropriate underpinning for analysis.

While Latour's work became a virtually obligatory citation (or should we say obligatory passage point?) in science and technology studies in the 1980s and 1990s, and Gieryn's signature concept has generated thousands of citations and research studies, theirs was not the only work in the spotlight. Among other prominent scholarship was the *social worlds* approach brought to prominence most especially by Adele Clark and Joan Fujimura's *The Right Tools for the Job* (1992). Reflecting the influence of symbolic interactionism (see Blumer 1969), the social worlds approach, like actor network theory, emphasizes the blurring of boundaries between such common dichotomies as context and content, structure and agency, and inside and outside. Much emphasis in this research tradition is placed on the mechanisms – like boundary objects and standardized packages – that enable *cooperation among diverse actors* (Star and Griesemer 1989, Fujimura 1987, 1988, Figert 1996). Social worlds analyses tend to describe how things happen, not explain why *per se*, and to draw attention to the multiple ways that a particular sociotechnical object can be seen by different actors. Like Latour, social worlds analysts are interested in construction processes and the work that it takes to create the common meanings that functionalist analysts took for granted as something that scientific training took care of, and that Marxist analysts assumed were given by capitalism. They were not focused on "culture," *per se* – for symbolic interactionists there is no separate world of "culture," since symbols are constitutive of social life. In this context, they have helped us to see the instability of the meanings of objects of scientific study, the problems of cooperating and coordination, and the ways that taken for granted ideas about the scientific object itself shape scientists' questions, answers and methods.

At the same time, a group of scholars less concerned with the *discursive* construction of science began to investigate large technological systems. Well before the development of STS as a field, social scientists and historians largely saw large-scale technological systems as projects whose accomplishment was the result of the straightforward application of engineering skills and available resources. It was not until several decades later that scholars of technology began to systematically investigate how the technological and social were knitted together across expanses of time, geography, and scale. Historian Thomas P. Hughes' foundational 1983 study of the simultaneous creation of large-scale technical and social systems of power – in his case, electrification – called into question the long standing distinction between the technical and the social (Hughes 1983). Elaborated by Trevor Pinch and Wiebe Bijker (1984), this tradition emphasizes how discursive and material processes produce particular cultural understandings of the material and social worlds.

Diane Vaughan (1996) brought these kinds of cultural concerns to the analysis of technical systems, but through a different lens. Her study of the NASA Challenger disaster sought to understand the place of *organizational* culture in explaining *why* failures occur in large-scale technosystems. What was especially innovative about Vaughan's work was that she showed that cultures that develop in complex technosocial systems can build in high potential for failure. Like the work of Ruth Schwartz Cowan (1983), whose study of how household innovations that were supposed to lighten women's work loads but simply transformed the type, but not the amount, of work that women did, Vaughan's investigation explicitly called into question the long-standing association between technology and progress. Methodologically, STS analysts of large-scale technical systems introduced something new to the field: that these systems must be studied at the very local and everyday scale, and as large, trans-geographic and trans-temporal systems.

Taken together, this foundational scholarship has played an important role in teaching us that we cannot understand knowledge as the product of a simple reading of nature or the development of technological systems as a reflection of an immanent logic in the technology itself. It has also trained us to see that we cannot ignore the local and the contingent, or the role of language, or assume the stability of objects. Methodologically, through careful attention to the everyday practices of scientists, early STS scholars highlighted that the spread of knowledge as an accomplishment that requires work, that technologies and language shape scientific knowledge, and that scientific objects must be created and may have multiple meanings and disparate interpretations. Many of the scholars in this book draw on tools, methods and frameworks from these traditions, often creatively expanding them and joining them together with other ways of analyzing science.

Since the 1990s, STS has expanded and become more intellectually and demographically diverse. One of the key developments was that scholars began to pay much greater attention to the role of other institutions, but especially the state, and to other logics, such as profit, that shaped what was studied, how it was studied, and with what consequences for scientists and other relevant groups. Work by Kleinman (1995) and Guston and Keniston (1994) showed the power of the state can profoundly shape the direction of scientific research via funding patterns and policy decisions that have direct effects on which groups would benefit from science, and which would not. Mukerji (1989) and Moore (1996) showed that scientists' close relationships with the state, as recipients of funding and as advisors in a variety of capacities, including on military matters, simultaneously served to give scientists authority and approbation and to challenge the notion that science was an intellectually and morally autonomous community. In moving outside the laboratory and the study of technological systems, state-centered work moved the social study of science away from investigation of the everyday lives and work of scientists, and toward other social systems and relations of power that shaped scientific knowledge formation. The move from lab to other institutions was not the only move made possible by this body of work: it also opened up new possibilities for understanding why some groups benefited more than others from the fruits of scientific research. While early STS placed scientists at the center of analysis, this work placed them in the context of other social systems that shaped content, methods, and outcomes. More recent work in this vein explores how and why some scientific endeavors result in harm to people and environments, and how government and corporate political-scientific categorizations, rules and guidelines can reproduce, create, or undermine extant inequalities (Frickel and Edwards 2011, Hess 2007, Suryanarayan and Kleinman 2013).

Epstein's now-classic book on AIDS activism and its politics (1996) set a new baseline for studying intersections among states, scientists and scientific research subjects. Epstein showed how scientific credibility could be gained by non-scientists, and how they could transform scientific practices and rules. His book played an important role in prompting discussions about the role of non-experts in highly technical realms (see also Wynne 1992). In addition to opening up new veins of discussion in STS, Epstein's work combined cultural and institutional analysis in a way rare in STS in the mid-1990s. Earlier STS scholars had asserted, through careful attention to the language and practices of scientists, that science is not as distinctive a practice as is often believed. But these earlier scholars made these assertions through their own analytic categories. Epstein used ethnographic evidence from the study of AIDS treatment activism, scientists' practices, and the mixes between the two to demonstrate how science might be revised to incorporate the needs and interests of its subjects, including through changes in formal rules that shape who gets to have a say in what is studied, how it is studied, and for what purposes.

Questions of power and challenges by or on behalf of people who are not themselves scientists, by scientists with concerns about social justice, or more generally, attention to power from below, is now a thriving component of STS (Frickel 2004, Ottinger 2013, Benjamin 2013, Nelson 2011, Allen 2003, Mamo and Fishman 2013, Corburn 2005). These analysts take conflict and inequality as enduring features of social life. While attentive to the agency involved in creating nature/society and technical/social divides, these scholars also see extant arrangements as constraining social action. Questions of inequality are, for these researchers, crucially important concerns. In this context, conceptualizing power becomes a matter of explanatory capacity (Albert et al. 2009, Albert and Kleinman 2011, Fisher 2009, Frickel and Moore 2005). Much of our own scholarship fits into this tradition (Moore 2013, Kleinman 2003).

One of the most striking aspects of recent work in STS is its attention to social aspects of science in places other than the European Union and the U.S. (Varma 2013, Moore et al. 2011, Philip et al. 2012, *East Asian Science, Technology and Society*, Kuo 2012, Petryna 2009). This scholarship has taught us, first, that the standard stories of innovation in which it flows from the technologically and scientifically rich U.S. and the EU to poorer countries fails to take into account the reverse of those flows from less wealthy nations to the EU and the U.S. It also reveals how and why science and technology are integrated into projects of colonial, postcolonial and neoliberal systems of culture and governance in ways that shape the qualities and types of relations we have with each other. Adriana Petryna, for example, has demonstrated how the global spread of clinical trials, particularly from wealthier countries to those less wealthy, has had complex effects that are not neatly captured by the dichotomies of harm and benefit (2009). This set of investigations has illuminated how and why governments include science and technology into systems of governance in very different ways from what the first generation of STS scholars recognized.

It is impossible in a few short pages to capture the movement of STS from a relatively narrowly focused research arena to a broad, diverse and dynamic scholarly field. Indeed, we have not come close to describing all of the prominent work on this intellectual terrain, let alone that vast array of scholarship that has led us to think about technoscience differently. We have, however, sought to provide some sense of the movement in STS over the past several decades, highlighting important conceptual and empirical areas that have been pursued.

In this handbook, we seek to build on the great array of methods, concepts, and empirical foci that that now define STS. At the outset, our aim was to cast a broad net. Among other things, our authors are not only trained as sociologists, as we are and many early STS scholars were, but also as anthropologists, historians, and political scientists. Some of the authors in this book received their degrees in science and technology studies programs, suggesting how far STS has come from its early days. While all of the contributors to this book seek explanations and all, in one way or another, are concerned with politics, power, and inequality, some approach their empirical focus using the tools of organizational and political sociology, others are inspired by action research, and still others take their lead from poststructuralist theory.

Key topics and themes

Too often edited collections are connected by little beyond the glue or thread that holds their pages together. While we sought diversity in bringing the authors in this handbook together, we also aimed to achieve coherence and hoped that directly and indirectly the chapters in this Handbook would speak to one another. As we noted earlier, each of the chapters in this book addresses one or more central sociocultural processes: institutionalization and institutional disruption; scale; and the role of culture and the material world in shaping science and vice-versa. Authors were also asked to read and comment on others' chapters, in order to bring further unity to the book. Finally, we have organized the book around several key topic areas. We briefly return to our key themes and then move on to an overview of the major topics that animate the book and the ways that the chapters engage them.

As we noted earlier, one of the key developments in STS over the past two decades has been a turn toward social systems that constrain and enable action. For social scientists, institutionalization is the process by which norms, worldviews, or social or cultural structures come to be taken for granted and be powerful forces in shaping social life. While institutions are by definition typically stable, an array of factors can undermine or disrupt this stability, creating the space for social transformation of various sorts. In the section entitled "Technoscience and/as Work," Vardi and Smith-Doerr, for example, contribute a chapter that explores the ways in which gender works in the knowledge economy. They show that in some contexts collaboration in research may disrupt gender inequality, while in others it may institutionalize it.

Analysts of scientific ideas, technological innovation, and the spread of technoscientific "goods" and "bads" have drawn attention to problems of aggregation, or scale. Geographers have drawn our attention to this concept, and although analysts do not agree on the meaning of the term, broadly speaking we might think of scale as specifying degrees of aggregation of social action that often point to territorial variation. Capturing the contested meaning of the term, our authors deploy the concept of aggregation in a variety of ways. Among those who make scale a central piece of their analysis in this collection is Abby J. Kinchy. In her chapter, Kinchy uses scale as a way to understand the politics of science-related regulation. She considers the relationship between the scale of environmental governance and the production of knowledge of shale gas development in the United States.

Finally, many of our chapters explore how established cultural values and dimensions of the material world shape science. A long-standing concern in STS – if not *the* preoccupation of the field for its first few decades – our authors come at this issue in innovative and provocative ways. Some of this newer work is more attentive to questions of social identities other than those of scientists, and to relations of power than the first generations of STS researchers were. PJ Rey and Whitney Erin Boesel, for example, examine how deeply established perspectives on subjectivity and organic flesh as the basis for understanding ourselves as humans prompt us to maintain a rigid boundary between "the online" and "the offline," a boundary that devalues digitally mediated experiences and their role in forming our subjectivity. Rey and Boesel argue against any notion of bifurcated subjectivity, suggesting that in the twenty-first century we need to understand ourselves as enhanced subjects.

The chapters to come

Beyond considering one or more of the broad processes we outlined above, the research collected in this book fits empirically into one of six topic areas: Embodiment, Consuming Technoscience, Digitization, Environment, Technoscience as Work, and Rules and Standards. While we might have selected many possible empirical foci to ask prospective authors to address, we chose these areas for two reasons. First, we see the issues raised by each of them as crucial matters of broad public debate. Surely, a vital STS should be adding to central societal discussions about issues such as cyber-bullying and the role of hydraulic fracturing in contributing to environmental degradation. Second, much (although certainly not all) of the most intriguing work in STS today covers the broad empirical terrain mapped out in this book.

Embodiment

As AIDS spreads in the global south, constraints on drug access inhibit its containment. In the U.S., low income citizens are exposed to toxins in their communities and confront limits to medical care. At the same time, some dream of winning the fight against cancer with targeted nanotechological treatments, and implants of various types vow to sharpen our sight, brainpower and endurance. Promises to "sort out" and group people by sexuality, ethno-"racial"-geography, and gender via technologies and scientific definitions abound, yet the proffered simple scientific solutions are increasingly contested by the very groups that are the subjects of intervention. Some of the most innovative and important scholarship in STS-related fields today involves investigations of health, medicine and the body. In this section, authors explore some of the most topical issues at the intersection of technoscience, medicine and health, but reach beyond typical subject matter by examining an array of issues centering on the politics of the body.

In the opening chapter of this section, Kimberly Tallbear uses feminist-indigenous standpoint theory to question whether the DNA ancestry tests offered to Native American groups to help them "solve" questions of ancestry actually serve Native American interests. The tests are often marketed as means of clearing up messy social questions of membership via ancestry, but as she demonstrates, not only are the tests offered with little information about what the tests cannot do, they contribute to undermining Native American cultural and political sovereignty.

Technoscientific means of dividing bodies into biologically-based "racial" groups argues Anthony Hatch, is a precondition for explaining racial inequality in biological terms. In his chapter, Hatch demonstrates that institutionalized racism is filtered through biomedical practices and assumptions. One result, he suggests, is that African Americans are now far more likely than other groups to be diagnosed with "metabolic syndrome." Metabolic syndrome, he demonstrates, compresses group-based differences in the economic and political resources that are necessary for good health into a term that only references the biological processes within the body, effectively turning the interior of bodies into sites of racial categorization.

Tom Waidzunas' chapter takes on just the opposite process: the undoing of a diagnostic standard that divided people into homosexuals and heterosexuals. Debates between scientists and their (often conservative Christian) supporters who see sexual orientation as a set of social behaviors and choices, and thus potentially changeable via "conversion therapy," and scientists who think of orientation as fundamentally biological, raged for several decades. But in a surprising turn of events, the two camps have converged on a new definition of sexual orientation as a biological process, and created a new concept – sexual orientation identity – to capture one's recognition of one's orientation. As Waidzunas shows, however, the new standard is based on a thin understanding of (social) sexuality and is incommensurate with what is known about female sexual desire.

Like the other chapters in this section, Marianne de Laet and Joe Dumit's chapter addresses the relationship between symbolic representations of bodies, and bodies themselves. Using the cases of baby growth charts and calorie counters, they demonstrate how graphs – since Latour, a longstanding topic in STS – tell us not only what to know, but how to know, and materially shape interventions on human bodies by setting biostandards that users apply to the monitoring and production of the bioself.

Consuming technoscience

We think of consumers as those who purchase products in retail environments, but consumption has a vastly broader set of meanings. Governments are consumers of new technologies and their habits of consumption shape production. The U.S. defense establishment's demand for information technology explains the birth of the internet. We consume when we ingest foods and medicines. Different actors consume technoscience and do so in different ways, and new developments in technoscience shape consumption itself. Local stores are challenged by website offerings. Our consumption practices are tracked by internet retailers whose promotions are then geared specifically to us. In this section, authors explore a number of the topics at the intersection of technoscience and consumption, looking at consumption as "ingestion," as the taking in of information, and as the creation and purchase of particular kinds of technoscientific products.

Shobita Parthasarathy's chapter begins this section, by asking to what extent direct-to-consumer medical genetics testing empowers "somatic" individuals, as not only its marketers but other STS analysts have claimed. She challenges four assumptions: that such tests are easily understood and accurate, that all somatic knowledge is valuable, that consumers are sufficiently homogenous that the interpretations of the tests are unproblematic, and that the tests are unproblematically useful for making health decisions. In a surprising move, given STS' new emphases on non-scientists' roles in the development and use of technoscience, Parthasarathy raises doubts about whether consumers are well-positioned to make independent decisions based on these tests and calls for a greater role for biomedical experts in interpreting test results.

It is hard to underestimate the global enthusiasm for personal communication technologies such as computers, but especially, for mobile phones. The excitement over the consumption of what Hsing-Hsin Chen calls "gadgets," however, has not been matched by attention to the human costs of the *manufacturing* of these gadgets in places like Taiwan. As was the case in earlier industrial expansions, factory workers who make gadgets face socially and biologically dangerous work conditions. Only some of the labor controversies that emerge in high-tech manufacturing become visible to publics. Using the case of the 2010 suicides at FoxConn, the maker of the iPhone, and the less well-known legal fight over cancer rates among RCA workers, Chen argues that political institutions shape the way that technopolitical controversies are made visible, or invisible, via the *theatrical* and the *mechanical*. The visibility of the FoxConn case, he shows, had much to do with the ability of activists to harness the profit-driven framing of the mobile phone as a consumer fashion item to bring labor issues to light through theatricality

Jaita Talukdar's chapter concludes this section by investigating the consumption not of material products, but of technoscientific ideas about health and eating. Earlier, crude analyses of the flow of food ideals and products from the U.S. to other parts of globe decried the "destruction" of local ways of eating. Through an analysis of women in different parts of the Indian economy,

Talukdar shows that "scientific eating" ideals from the west are hardly taken up wholesale. Urban women in new economy jobs are most likely to encounter the ideals and their purported benefits, and when they consume the ideas, cognitively and through eating, they are filtered through gendered Indian ideals about fasting and the social value of foods such as rice and oils.

Digitization

From social media to web 2.0, commerce, politics, and social life are being transformed by the revolution in digital technology. Shopping online is commonplace, retailers, banks and governments track consumers'/citizens' habits. U.S. candidates for public office announce their bids on the web, fundraise online, and circulate trial balloons via Twitter. Sexting is new to our lexicon, and Facebook is a means of party planning. Of course, these practices are not distributed equally within countries and across the globe, and the steamroller-like spread of some of these technologies raises questions about their virtues and drawbacks and their broad meaning. STS scholars and researchers in allied fields are contributing to our knowledge of these developments and can help us understand them.

STS scholars have assumed that low costs help account for the spread of new information and communication technologies such as mobile phones and software, yet they rarely investigate how and why price matters. Kristen Eschenfelder shows that pricing effects are far from straightforward in the case of digital databases that contain published works, because vendors' and diverse users' expectations about database use, such as the seemingly simple idea of "downloading," are not always aligned. She uses the concepts *inscription* and *anti-program* to explain users' backlash against pricing schemes, and *constant renegotiation* – rather than static terms like disruption – to characterize the interactions between digital publication database users and vendors.

Just as digitization complicates the meanings of the appropriate use of information, it complicates our ideas about subjectivity. Debates about how digital communications have affected selfhood usually center around the benefits and drawbacks of being able to represent ourselves "online," but have less to say about the idea of personhood and subjectivity that underlie these normative questions. PJ Rey and Whitney Boesel argue against the idea that digitization splits us into distinct "offline" and "online" subjects, wrenching a whole subject apart and making her into two uneasily coexisting subjects. They argue, instead, online and offline subjectivity are not ontologically distinct. Quite the opposite: our subjectivity is now "augmented," such that online and offline are extensions of each other.

Digital gaming is both an increasingly common subjective experience, and very big business. South Korea exemplifies both trends: gaming is extremely popular among young men and is a source of national identity, and the industry has been so successful that other countries are looking to it for clues to global market success. Dal-Yong Jin and Michael Borowy argue that Korean gaming industry's market history results in part from neoliberal government interventions. The industry's trajectory cannot be understood, they argue, without reference to neoliberal interventions to promote the industry *and* social protectionist measures to limit game use because of social concerns over the effects of internet addiction on family life. Political culture, not purely economic logics, must be taken into account to explain the ways that digital industries are created and how they are institutionalized.

The governance and spread of digital communications, argues J.P. Singh, is profoundly shaped by political cultures at the global level, which are formalized and expressed in treaties, agreements, and rules. Some cultural ideas are shaped by the interests of actors involved in governance, such as firms and states, others by the cultural ideas about how information should flow that were inherited from the governance of early communication technologies, such as telephones. Attentive to the multidirectionality of governance, Singh demonstrates that popular participation in digital governance and use, such as indymedias, are increasing, and that wealthy countries' ability to dominate global governance is being challenged by regional governance arrangements and voting power blocs among and by lower income countries.

Environments

We often speak of one environment – nature – which humans have callously destroyed. But there is not one environment. Environments vary across countries, regions, and economies. Environments, but especially those composed of biological entities and physical forces such as wind, fire, and water, are macro, micro, inside and outside and everything in between. What is the relationship between their fates and technoscience? In this section, our authors explore the relationship between an environment and science and technology, addressing some of the crucial questions that face citizens and policymakers across the globe.

Roopali Phadke's chapter introduces what she calls *"place-based technology assessment"* (PBTA), an approach which allows researchers to develop strategies that connect local policy actors with social movement demands through multisite, multiscale research collaborations. Phadke describes a three-stage process working with a local government in Michigan in which her team sought to capture public concerns and opinions about the emergence of wind energy on their landscape. Phadke's chapter clearly shows the viability of a place-based approach to the anticipatory governance of technology.

Abby Kinchy's chapter explores the politics of the regulation of fracking, or hydraulic fracturing natural-gas extraction. Like Phadke, scale figures centrally in Kinchy's analysis. Kinchy shows that scientific knowledge claims may be used to justify and institutionalize the scale at which regulatory decisions are made and at which the concerns of residents and others are attended. Looking at the fracking debate in Pennsylvania, Kinchy also provides evidence that the scale of political engagement can shape the way in which scientific understanding takes shape by affecting the framing of discussion and *"orienting perspectives."*

While questions of the production of non-knowledge (or ignorance) figure in Kinchy's chapter, it is Scott Frickel's central concern. In his contribution, Frickel explores the relational dynamics of knowledge production and non-production. His chapter is a call for sustained and careful study of the problem of ignorance, or the absence of knowledge. Most centrally, Frickel is interested in why ignorance, once produced, gets institutionalized within and beyond science. Ignorance is a matter of absence, making it difficult to study. Accordingly, Frickel spends a good deal of his chapter on methodological matters, and illustrates his methodological and conceptual insights with data from his own work on the relationship of soil quality to environmental risk following hurricane Katrina and the resulting floods.

David Hess' contribution returns us to the subject of green energy technology. Drawing on and extending the work of Pierre Bourdieu and Fligstein and McAdam on field theory, and empirically examining green energy transitions, Hess takes up several of the themes that bind this collection. He considers the relationship between institutionalization and disruption of technology transitions, political values and material culture, and scalar and spatial dynamics of the politics governing the energy transitions. He considers the ideological tensions at stake in U.S. national- and state-level policy conflicts related to green energy and explores the positions and capacities of the relevant stakeholders.

Prior to the Fukushima disaster, nuclear power was making something of an international comeback, and nuclear power is the empirical focus of Sulifkar Amir's chapter. Amir's central contribution is to our understanding of the development of complex technological systems in what he calls "*less advanced modernity*." Amir focuses on the development of nuclear power in Indonesia and argues the Indonesian state has contributed to the escalation of risk in the development of nuclear power as a result of its institutional incapacity. Relatedly, he suggests that in the face of the contribution nuclear power can make to national economic development, the state makes ignorance a "norm of governance that clouds good and clear judgment." Here, national culture and institutionalization and disruption intersect.

While Amir's work provides insights into the limitations and value of the "risk society" literature in the analysis of countries in the global south, Samer Alatout's chapter seeks to bring together concepts from postcolonial studies and STS. Alatout suggests that understanding the politics of empire can help us grasp technoscience, and he suggests that empire is itself a product of scientific knowledge. Alatout takes the Jordan River as his empirical focus. He explores how that waterway was transformed as a result, first of the politics of empire, and later of nation-states, and their intersection with hydrogeological and archeological knowledge. Attentive to the problem of scale, Alatout suggests that it was the encounter with imperialism that reframed knowledge of the archaeology and hydrology of Palestine in such a way that produced the Jordan River as a border and transformed the quality and quantity of the water in it, along the way.

Carroll and Freiburger conclude the Environments section. These authors reject the distinction between the state and the environment altogether and suggest instead we should think of the state-environment as a gathering that is simultaneously human and non-human, material and discursive, practice and process. This reconceptualization, according to Carroll and Freiburger, necessitates that we understand the practices involved in what they term "environing." They illustrate the environing process and the idea of the state-environment through a discussion of a set of boundary objects – land in Ireland from about 1600 to 1900, water in California from about 1850 to 1980, and lithium in Bolivia from about 1970 to 2012.

Technoscience as work

The labor force required for the production of our technoscientific world is diverse, stratified and global. While PhD scientists run their labs in boutique biotechnology companies living in a world that may seem more comfortable and stable than a life in U.S. higher education, in 2010, immigrant workers in the Silicon Valley are exposed to highly toxic chemicals as they process computer chips, and workers in the global south disassemble tons of computer hardware and are poisoned by mercury. Who does technoscientific labor? What does it mean to do scientific or technological work? How are different actors affected by the social organization of the new knowledge economy? In this section, authors explore the many meanings of scientific work.

Questions of the environmental damage and the risk to nearby residents resulting from the nuclear power disaster in Fukushima, Japan, in 2011 has received a great deal of attention in the popular press as well as in an array of scholarly venues. Much less attention has been paid to the nuclear plant workers that Gabrielle Hecht describes in her chapter as marginal to the technoscientific enterprise. Hecht takes us from Fukushima to Gabon, characterizing the experiences of those who maintain nuclear power plants and the workers who mine the uranium used to power them. The implications of Hecht's study extend well beyond nuclear power, providing insight into contract labor, the transnational distribution of danger, and the regimes by which industrial hazards become (im)perceptible.

From marginal workers who are neither scientists nor engineers, Mathieu Albert and Elise Paradis' chapter turns our attention to scholars who experience marginalization in their work environment. Albert and Paradis consider the experience of social scientists and humanists working in medical schools with an explicit commitment to interdisciplinary research. They find that the commitment to interdisciplinarity and collaboration notwithstanding, social scientists and humanities do not garner the same legitimacy or authority accorded to natural scientists. Albert and Paradis' analysis, guided by a framework influenced by Pierre Bourdieu in which "actors are embedded in a social universe where social and symbolic structures, which predate their own entry into this universe, influence actions and social relations," provides a valuable way in which to understand how deeply institutionalized values shape scientific practice.

Itai Vardi and Laurel Smith-Doerr extend the theme of researchers' marginalization. In their chapter, they explore both the benefits and disadvantages of research collaboration for women in the new knowledge economy. The authors find that collaboration, so lauded in discussions about the changing face of science, is a complex phenomenon. Drawing on secondary sources as well as their own preliminary data on collaboration in chemistry, Vardi and Smith-Doerr contend that "The gendered organization of science still favors dominant masculinity and hierarchies led by men in ways that usually go unnoticed." That said, these authors suggest that research collaboration, which demands a different mode of work than individually-oriented research, may undercut cultural and structural patterns of discrimination, and new forms of research organization – like networks – may "disrupt traditional masculine power structures."

Like Albert and Paradis, and Vardi and Smith-Doerr, Hee-Je Bak is interested in the role of institutionalized values in shaping scientific practice. While Albert and Paradis train our attention on the varied work experiences of social scientists, humanists, and natural scientists and Vardi and Smith-Doerr are interested in the politics of gender, Bak is concerned with the ways science policies and commitments vary as we traverse national boundaries. Focusing on South Korea, Bak shows how that nation's government has mobilized science and technology for the purposes of economic development, and how the government's utilitarian focus has affected the norms and practices of Korean scientists. While much early STS scholarship challenged the applicability of Merton's framing of scientific norms to scientists' everyday work, Bak shows how national culture and the state can affect scientists' commitments to those norms.

We move, finally, from researchers central to scientific practice to workers at the margins of science – in this case, seafood workers and their role in testing seafood in the wake of the Deepwater Horizon oil spill in the Gulf of Mexico in 2011. Truth be told, the contribution of Brian Mayer, Kelly Bergstrand, and Kartina Running might have been placed in one of several sections in the handbook. Their investigation of seafood testing in the Gulf shows that, in this case, science was used as a tool to reduce uncertainty for both consumers and seafood workers. Thus, their contribution would fit nicely in the "Consuming Technoscience" section of the handbook. Mayer, Bergstrand, and Running's chapter is also about government responses to environmental disaster and the threat posed to the institutional credibility of science in the face of disaster. Thus, their contribution is relevant to matters of regulation and standards. These authors examine so-called sniff test science used to evaluate the contamination of seafood and the ways in which this easy to learn procedure, to a certain extent, reassured consumers and facilitated seafood worker processing and thus constituted "comfort science."

Rules and standards

What's in? What's out? What's good? What's bad? A mainstay of STS scholarship for the past several decades has been boundary construction and the creation of systems of classification in science and technology. Scholars have taken up these topics for good reason. What science gets done, how it is undertaken and who has access to it are all affected by the ways boundaries are drawn and standards are set. Rules and standards can sharply affect not only what can be undertaken, but who can participate, and in what ways. In this section, authors explore processes through which drug standards are established and institutionalized, how environmental regulations protect and harm different communities, and what it means to be a human subject.

STS scholars have often relied on a key assumption about the intersection of scientistic thought and practice and the running of bureaucratic operations: the main role of scientists is to make impersonal judgments based on technical rules that enable organizations to carry out their goals. Laura Stark shows that this is neither the only role that scientists play in these systems, nor does it capture the basis of their decision making. Working in temporary *deliberative bodies* such as review panels and Institutional Review Boards, scientists, Stark shows, use discretion, rely on multiple kinds of knowledge, and act on the basis of group agreements to create rules and make judgments. In doing so, Stark challenges the conception of scientist-as-machine, and the narrow set of roles assigned to scientists as rule-makers.

Like Stark, Figert and Bell are interested in the practices and outcomes of scientists as judges. Institutional Review Boards in the U.S. and parallel organizations in other countries must sort out ethical questions about who should be exposed to risks and acquire benefits in the rapidly expanding array of pharmaceutical trials around the world. Controversy over reliance on international standards, and its opposite, ethical variability, has caused scientific, philosophical, political, and legal controversy. Figert and Bell show that ethical variability ought to be welcomed, for it upholds *and* disrupts the primacy of western ethical ideals, offering more possibilities for pharmaceuticalization to redress rather than reproduce medical global medical inequalities.

Stark and Figert and Bell ask about how human interpretations shape the formation and enactment of sociotechnical rules, but Jane Fountain trains her attention on digitally mediated systems governing science and technology. In the U.S. and the EU, digitization is increasingly used as a means to coordinate science-related activities, including patenting and grants, across agencies and nations and among constituents, with highly variable success. Governance outcomes are powerfully shaped by institutional processes that include path dependence, the cultural values of software engineers and legislators, and extant structures of coordination. The uneven results of digital coordination of science governance, Fountain concludes, must be understood through close attention to these and other mechanisms of institutionalization.

In the U.S., standards meant to protect human subjects are formalized in federal laws that are based on principles laid out in the 1977 Belmont Report. Meant to ensure that all research subjects were treated justly, the report set a baseline in which human subjects are treated as individuals rather than as members of groups, that potential participants are equals at the point of entry into the research study, and that ethics boards and researchers, understand the scientific problems and probably ethical issues better than participants. Rachel Morello-Frosch and Phil Brown show that in socially disadvantaged communities, these assumptions powerfully limit the kinds of benefits that communities might gain from research, even when scientists are purportedly working on behalf of a community. In this final chapter, they show that *post-Belmont research ethics* can address these problems. Characterized by community-scientist research partnerships that include joint decisions on questions and methods, the interpretation of evidence, and agency reporting, as well as reflexive ethics that include individual report-backs, post-Belmont research ethics can contribute to health improvements for communities and more knowledge of the limits and possibilities of medical knowledge for addressing health issues.

References

Albert, M. and Kleinman, D.L. 2011. "Bringing Bourdieu to Science and Technology Studies." *Minerva* 49(3): 263–273.

Albert, M., Laberge, S., and Hodges, B.D. 2009. "Boundary Work in the Health Research Field: Biomedical and Clinician Scientists' Perceptions of Social Science Research." *Minerva* 47(2): 171–194.

Allen, B. 2003. *Uneasy Alchemy: Citizens and Experts in Louisiana's Chemical Corridor Disputes*. Cambridge, MA: MIT Press.

Barnes, B. 1977. *Interests and the Growth of Knowledge*. London: Routledge.

Barnes, B. and MacKenzie, D. 1979. "On the Role of Interests in Scientific Change." In R. Wallis, ed., *On the Margins of Science: The Social Construction of Rejected Scientific Knowledge. Sociological Review Monograph*, Keele, UK: Keele University Press.

Bauschpies, W., Croissant, J., and Restivo, S. 2006. *Science, Technology and Society: A Sociological Approach*. Boston: Blackwell Publishing.

Benjamin, R. 2013. *People's Science: Bodies and Rights on the Stem Cell Frontier*. Stanford, CA: Stanford University Press.

Biagioli, M., ed. 1999. *The Science Studies Reader*. London and New York: Routledge.

Bloor, D. 1976. *Knowledge and Social Imagery*. London: Routledge.

Blumer, H. 1969. *Symbolic Interactionism: Perspective and Method*. New York: Prentice-Hall.

Clark, A. and Fujimura, J. 1992. *The Right Tools for the Job: At Work in the Twentieth Century Life Sciences*. Princeton, NJ and Oxford, UK: Princeton University Press.

Cole, J. and Cole, S. 1973. *Stratification in Science*. Chicago: University of Chicago Press.

Corburn, J. 2005. *Street Science: Community Knowledge and Environmental Health Justice*. Cambridge, MA: MIT Press.

Cowan, R.S. 1983. *More Work for Mother: The Ironies of Household Technology from the Hearth to the Microwave*. New York: Basic Books.

Duster, T. 1981. "Intermediate Steps Between Micro- and Macro-Integration: The Case of Screening for Inherited Disorders." In Karin Knorr-Cetina and Aaron Cicourel, eds., *Advances in Theory and Methodology: Toward an Integration of Micro- and Macrosociologies*. London: Routledge and Kegan Paul.

Epstein, S. 1996. *Impure Science: AIDS Activism and the Politics of Knowledge*. Berkeley, CA: University of California Press.

Fausto-Sterling, Anne. 1978. *Sexing the Body: Gender Politics and the Construction of Sexuality*. New York: Vintage Books.

Figert, A. 1996. *Women and the Ownership of PMS*. Piscataway, NJ: Transaction.

Fisher, J. 2009. *Medical Research for Hire: The Political Economy of Pharmaceutical Clinical Trials*. New Brunswick, NJ: Rutgers University Press.

Foucault, M. 1977. *Discipline & Punish: The Birth of the Prison*. New York: Pantheon.

Frickel, S. 2004. *Chemical Consequences: Environmental Mutagens, Scientist Activism, and the Rise of Genetic Toxicology*. New Brunswick, NJ: Rutgers University Press.

Frickel, S. and Edwards, M.L. 2011. "Untangling Ignorance in Environmental Risk Assessment." Pp. 111–129 in Soraya Boudia and Nathalie Jas, eds., *Powerless Science? The Making of the Toxic World in the Twentieth Century*. Oxford and New York: Berghahn Books.

Frickel, S. and Kelly Moore, eds. 2006. *The New Political Sociology of Science: Networks, Institutions, and Power*. Madison, WI.: University of Wisconsin Press.

Fujimura, J. 1987. "Constructing 'Do-Able' Problems in Cancer Research: Articulating Alignment." *Social Studies of Science* 17(2): 257–293.

Fujimura, J. 1988. "The Molecular Bandwagon in Cancer Research: Where Social Worlds Meet." *Social Problems* 35(3): 261–283.

Gieryn, T. 1982. "Boundary Work and the Demarcation of Science from Non-Science: Interests in Professional Ideologies of Science." *American Sociological Review* 48(6): 781–795.

Gieryn, T. 1995. *Cultural Boundaries of Science: Credibility on the Line*. Chicago: University of Chicago Press.

Guston, D. and Keniston, K. 1994. *The Fragile Contract: University Science and the Federal Government.* Cambridge, MA: MIT Press.

Haraway, D. 1982. "Sex, Race, Class, Scientific Objects of Knowledge: A Marxist-Feminist Perspective on the Scientific Generation of Productive Nature and Some Political Consequences." *Socialism in the World* 29: 113–123.

Haraway, D. 1991 [1988]. "Situated Knowledges: The Science Question in Feminism and the Privilege of Partial Perspective." Pp. 183–202 in *Simians, Cyborgs, and Women: The Reinvention of Nature.* New York and London: Routledge.

Harding, S. 1986. *The Science Question in Feminism.* Ithaca, NY: Cornell University Press.

Hess, D.J. 1995. *Science and Technology in A Multicultural World.* New York: Columbia University Press.

Hess, D.J. 2007. *Alternative Pathways in Science and Industry: Activism, Innovation and Environment in an Age of Globalization.* Cambridge, MA: MIT Press.

Hughes, T.P. 1983. *Networks of Power: Electrification in Western Society, 1880–1930.* Baltimore: Johns Hopkins University Press.

Kleinman, D.L. 1995. *Politics on the Endless Frontier: Postwar Research Policy in the United States.* Durham, NC: Duke University Press.

Kleinman, D.L. 2003. *Impure Cultures: University Biology and the World of Commerce.* Madison, WI: University of Wisconsin Press.

Kuo, W.-H. 2012. "Transforming States in the Era of Global Pharmaceuticals: Visioning Clinical Research in Japan, Taiwan, and Singapore." Pp. 279–305 in Kaushik Sunder Rajan, ed., *Lively Capital: Biotechnologies, Ethics, and Governance in Global Markets.* Durham, NC: Duke University Press.

Latour, B. 1987. *Science in Action: How to Follow Scientists and Engineers through Society.* Cambridge, MA: Harvard University Press.

Latour, B., with C. Porter, translator. 1993. *We Have Never Been Modern.* Cambridge, MA: Harvard University Press.

Latour, B. and Woolgar, S. 1979. *Laboratory Life: The Social Construction of Scientific Facts.* Princeton, NJ: Princeton University Press.

Lynch, M., ed. 2012. *Science and Technology Studies, Critical Concepts in the Social Sciences* (4 Volume Set). London and New York: Routledge.

Mamo, L. and Fishman, J. 2013. "Why Justice? Introduction to the Special Issue on Entanglements of Science, Ethics and Justice." *Science, Technology and Human Values* 38(2): 159–175.

Martin, B. 1981. "The Scientific Straightjacket: the Power Structure of Science and the Suppression of Environmental Scholarship." *Ecologist* 11(1): 33–43.

Merton, R.K. 1957. "Priorities in Scientific Discovery." *American Sociological Review* 22(6): 635–659.

Merton, R. K. 1968. "The Matthew Effect in Science." *Science* 159(3810): 56–63.

Moore, K. 1996. "Doing Good While Doing Science." *American Journal of Sociology* 101: 1121–1149.

Moore, K. 2013. "Fear and Fun: Science, Gender and Embodiment Under Neoliberalism." The Scholar and the Feminist 11.1–11.2 (Spring 2013) *http://sfonline.barnard.edu/gender-justice-and-neoliberal-transformations.*

Moore, K., Kleinman, D.L., and Frickel, S. 2011. "Science and Neoliberal Globalization: A Political Sociological Approach." *Theory and Society* 40(5): 505–532.

Mukerji, C. 1989. *A Fragile Power: Scientists and the State.* Princeton, NJ: Princeton University Press.

Nelson, A. 2011. *Body and Soul: The Black Panther Party and the Fight Against Medical Discrimination.* NYU Press, NY.

Noble, D.F. 1977. *America By Design: Science, Technology and the Rise of Corporate Capitalism.* New York: Alfred A. Knopf.

Ottinger, G. 2013. *Refining Expertise: How Responsible Engineers Subvert Environmental Justice Challenges.* New York, NY: NYU Press.

Petryna, A. 2009. *When Experiments Travel: Clinical Trials And The Global Search For Human Subject.* Princeton, NJ: Princeton University Press.

Philip, K., Irani, L., Dourish, P. 2012. "Postcolonial Computing: A Tactical Survey." *Science, Technology & Human Values* 37(1): 3–29.

Pinch, T.J. and Bijker, W.E. 1984. "The Social Construction of Facts and Artefacts: Or How the Sociology of Science and the Sociology of Technology Might Benefit Each Other." *Social Studies of Science* 14: 399–411.

Sismondo, S. 2004. *An Introduction to Science and Technology Studies,* 2nd ed. New York: Wiley-Blackwell, 2010; 1st ed. New York: Blackwell.

Star, S.L. and Griesemer, J. 1989. "Institutional Ecology, 'Translation,' and Boundary Objects." *Social Studies of Science* 19(3): 387–420.

Suryanarayanan, S. and Kleinman, D.L. 2013. "Be(e)coming Experts: The Controversy Over Insecticides in the Honey Bee Colony Collapse Disorder." *Social Studies of Science* 43(2): 215–240.

Varma, R. 2013. *Harbingers of Global Change: India's Techno-Immigrants*. Lanham, MD: Lexington Books.

Vaughan, D. 1996. *The Challenger Launch Decision: Risky Technology, Culture and Deviance at NASA*. Chicago: University of Chicago Press.

Whalley, P. 1978. "Deskilling Engineers?" *Social Problems* 32(2): 17–32.

Wynne, B. 1992. "Misunderstood Misunderstandings: Social Identities and Public Uptake of Science." *Public Understanding of Science* 1(3): 281–304.

Zuckerman, H. 1967. "Nobel Laureates in Science: Patterns of Production, Collaboration and Authorship." *American Sociological Review* 32(3): 391–403.

Part I
Embodiment

1

The Emergence, Politics, and Marketplace of Native American DNA[1]

Kim TallBear

UNIVERSITY OF TEXAS, AUSTIN

Scientists and the public alike are on the hunt for "Native American DNA."[2] Hi-tech genomics labs at universities around the world search for answers to questions about human origins and ancient global migrations. In the glossy world of made-for-television science, celebrity geneticist Spencer Wells travels in jet planes and Land Rovers to far flung deserts and ice fields. Clad in North Face® gear, he goes in search of indigenous DNA that will provide a window into our collective human past.

Others – housewives, retirees, professionals in their spare time – search for faded faces and long-ago names, proof that their grandmothers' stories are true, that there are Indians obscured in the dense foliage of the family tree. Some are meticulous researchers, genealogists who want to fill in the blanks in their ancestral histories. They combine DNA testing with online networking to find their "DNA cousins." Some have romantic visions of documenting that "spiritual connection" they've always felt to Native Americans. A few imagine casino pay outs, free housing, education, and healthcare if they can get enrolled in a Native American tribe. Applicants to top-ranked schools have had their genomes surveyed for Native American DNA and other non-European ancestries with the hope of gaining racial favor in competitive admissions processes. Former citizens of Native American tribes ejected for reasons having to do with the financial stakes of membership have sought proof of Native American DNA to help them get back onto tribal rolls (Bolnick et al. 2007, Harmon 2006, Harris 2007, Koerner 2005, Simons 2007, Takeaway Media Productions 2003, Thirteen/WNET 2007, Wolinsky 2006, Marks 2002).

Genetic scientists, family tree researchers, and would-be tribal members – often with little or no lived connection to tribal communities – have needs and perspectives that condition the production and use of Native American DNA knowledge in ways that rarely serve Native American communities themselves. How did it come to be that Native American bodies are expected to serve as sources of biological raw materials extracted to produce knowledge that not only does not benefit them, but may actually harm them by challenging their sovereignty, historical narratives, and identities?

Science and Technology Studies (STS) scholars and those in related fields have addressed the co-constitution of racial and ethnic identities with genetic markers – both ancestry markers and those that are biomedically relevant (for example, Fullwiley 2008, Kahn 2006, Montoya 2011,

Nash 2003, 2004, Nelson 2008a, 2008b). Such analyses often focus on cultural, social, and economic differences historically of such communities in relation to dominant populations. STS scholars have dealt less so with issues of Native American identity as implicated by genetics, with several notable exceptions (for example, Reardon 2005, Reardon and TallBear 2012, TallBear 2008, 2013). This is no doubt because the Native American racial/ethnic category has the additional aspect of being characterized by the unique government-to-government relationships of tribes with the U.S. state. Like most academic fields STS does not seem particularly cognizant of this aspect of indigenous life in the U.S. On the other hand, Native American and Indigenous Studies (NAIS) is the field in which Native American identity and governance issues are expertly treated. In its U.S. formation, NAIS engages little with the biological and physical sciences. Where it focuses on governance, it emphasizes the law over governance through science. Approaches from both fields are needed to understand the topic at hand. I inhabit both fields.

What is Native American DNA?

To understand Native American DNA we must understand not only contemporary genome science practices but also how Native American bodies have been treated historically. Native American bodies, both dead and living, have been sources of bone, and more recently, of blood, saliva, and hair, used to constitute knowledge of human biological and cultural history (Benzvi 2007, Bieder 1986, Reardon and TallBear 2012, Thomas 2000). In the nineteenth and early twentieth centuries, the American School of Anthropology rose to worldwide prominence through the physical inspection of Native American bones and skulls plucked from battlefields or from recent gravesites by grave-robbers-cum-contract-workers, or scientists. It was certainly distasteful work to scavenge decomposing bodies, and boil them down so bones could be sent more easily to laboratories clean and ready for examination.

But two justifications emerged for the work, justifications that will ring familiar in my analysis of contemporary genetic scientists' treatment of Native Americans' DNA. First, the Indians were seen as doomed to vanish before the steam engine of westward expansion. The idea was then and is now that they should be studied before their kind is no more. Second, this sort of research was and is purported to be for the good of knowledge, and knowledge was and is supposed to be for the good of all. Indeed, the notion that "knowledge is power" continues to be used to justify practices that benefit technoscientists and their institutions but with questionable returns to the individuals and communities who actually pay for such knowledge with their bodies or their pocketbooks. Extractive genetics research and profit-making is not of course limited to marginalized Native Americans. Shobita Parthasarathy in Chapter 5 of this volume, "Producing the Consumer of Genetic Testing," questions the knowledge-is-power narrative that informs Myriad Genetics' and 23andMe's marketing of genetic tests to the public as they seek to turn patients into "healthcare consumers." But the knowledge that companies sell is of questionable use to individuals who do not have relevant scientific expertise, yet who pay hundreds or thousands of dollars for test results that are difficult to interpret, and which are sometimes based on unreplicated scientific studies. Thus we must ask for whom are particular forms of genetic knowledge power (or profit), and at whose expense?

Genetic ancestry

What in technical terms is Native American DNA? In the early 1960s, new biochemical techniques began to be applied to traditional anthropological questions, including the study of ancient human migrations and the biological and cultural relationships between populations. The new subfield of "molecular anthropology" was born, sometimes also called "anthropological genetics" (Marks 2002). Sets of markers or nucleotides in both the mitochondrial DNA (mtDNA) and in chromosomal DNA were observed to appear at different frequencies among different populations. The highest frequencies of so-called Native American markers are observed by scientists in "unadmixed" native populations in North and South America. These markers are the genetic inheritance of "founder populations," allegedly the first humans to walk in the lands we now call the "Americas."

The so-called Native American mtDNA lineages A, B, C, D, and X, or Y chromosome lineages M, Q3, or M3 are not simply objective molecular objects. These molecular sequences or "markers" – their patterns, mutations, deletions, and transcriptions that indicate genetic relationships and histories – have not been simply uncovered in human genomes. Native American DNA could not have emerged as an object of scientific research and genealogical desire until individuals and groups emerged as "Native American" in the course of colonial history. Without "settlers" we could not have "Indians" or Native Americans. Instead, we would have many thousands of smaller groups or peoples defined within and according to their own languages, by their own names.

It is the arrival of the settler a half millennia ago, and many subsequent settlements, that frame the search for Native American DNA before it is "too late," before the genetic signatures of the "founding populations" in the Americas are lost forever in a sea of genetic admixture. "Mixing" is predicated on "purity," a presumed state of affairs in the colonial pre-contact Americas, which informs the historical constitution of continental spaces and concomitant grouping of humans into "races" – undifferentiated masses of "Native Americans," "Africans," "Asians," and "Indo-Europeans." This view privileges Europeans' encounter of humans in the Americas. Standing where they do – almost never identifying as indigenous people themselves – scientists who study Native American migrations desire to know the "origins" of those who were first encountered when European settlers landed on the shores of these continents.

On the order of millennia, anthropological geneticists want to understand which human groups or "populations" are related to which others, and who descended from whom. Where geographically did the ancestors of different human groups migrate from? What were their patterns of geographic migration? When did such migrations occur? In the genomes of the living and the dead, scientists look for molecular sequences – the "genetic signatures" of ancient peoples whom they perceive as original continental populations: Indo-Europeans, Africans, Asians, and Native Americans. Native American DNA, as a (threatened and vanishing) scientific object of study, can help answer what are for these scientists, pressing questions.

In human genome diversity research in anthropology and other fields, origins get operationalized as "molecular origins," ancestral populations inferred for an individual based on a specific set of genetic markers, a specific set of algorithms for assessing genetic similarity, and a specific set of reference populations (Lee et al. 2009). But each of those constitutive elements operates within a loop of circular reasoning. Pure bio-geographic origins must be assumed in order to constitute the data that supposedly reveals those same origins. Notions of ancestral populations, the ordering and calculating of genetic markers and their associations, and the representation of living groups of individuals as reference populations each require the assumption that there was a moment, a human body, a marker, a population that was a bio-geographical pinpoint of originality. This faith in originality would seem to be at odds with the doctrine of evolution, of change over time, of becoming.

The populations and population-specified markers that are identified and studied are informed by the particular cultural understandings of the humans who study them. Native

American DNA, sub-Saharan African, European, or East Asian DNAs are constituted as scientific objects by laboratory methods and devices, and also by discourses of race, ethnicity, nation, family and tribe. For and by whom are such categories defined? How have continental-level race categories come to matter? Why do they matter more than the concept of "peoples" that condition indigenous narratives, knowledges, and claims?

An Anishinaabee with too many non-Anishinaabee ancestors won't count as part of an Anishinaabee "population," thus bringing a tribal/First Nation category of belonging into conflict with a geneticist's category. To make things even more complicated, genetic population categories themselves are not even consistently defined. For example, a scientist may draw blood from enrolled members at the Turtle Mountain Band of Chippewa Indians reservation in North Dakota and call her sample a "Turtle Mountain Chippewa" sample. At the same time, she may have obtained "Sioux" samples from other scientists and physicians who took them at multiple sites (on multiple reservations or in urban Indian Health Service facilities) over the course of many years. In the Turtle Mountain Chippewa instance, we have a "population" circumscribed by a federally-recognized tribal boundary. In the "Sioux" instance, we have a population circumscribed by a broader ethnic designation spanning multiple tribes. Histories of politics *inhere in* the samples. Politics are also *imposed onto* the samples by researchers who enforce subsequent requirements for the data – for example, that usable samples come only from subjects who possess a certain number of grandparents from within said population. Thus, the categories favored by scientists are not "objectively true."

The DNA profile

I stretch the definition of Native American DNA beyond its usual reference to "New World" genetic ancestry, which is traceable through female mtDNA and male Y chromosome lines and through more complex tests that combine multiple markers across the genome to trace ancestry. I include the "DNA profile" or "DNA fingerprint." This form of DNA analysis is often treated as a standard parentage test and thus used by Native American tribes to verify that applicants are indeed the biological offspring of already enrolled tribal members. As a parentage test, the same form of analysis shows genetic relatedness between parent and child, or other close biological kin. When used to confer citizenship in U.S. tribes and Canadian First Nations, the DNA fingerprint becomes a marker of Native American affiliation.

In order to follow the complex weave created when threads of genome knowledge loop into already densely knit tribal histories and practices of identity-making, one must grasp complex knowledges simultaneously: molecular knowledges and their social histories and practices of tribal citizenship. DNA testing company scientists and marketers do not have a deep historical or practical understanding of the intricacies of tribal enrollment. Nor do they tend to understand the broader historical and political frame circumscribing their work, and precisely how their disciplines have fed from marginalized bodies. Tribal folks know these politics and histories well – we live day in and out with enrollment rules, and we all know about the Native American Graves Protection and Repatriation Act (NAGPRA). But we do not tend to know the molecular intricacies of DNA tests.

To date, DNA tests used by tribes are simply statements of *genetic* parentage. Tribal governments make regulatory decisions privileging these tests, instead of or along with other forms of parent–child relationship documentation, such as birth or adoption certificates. Tribes increasingly combine DNA tests with longer-standing citizenship rules that focus largely on tracing one's genealogy to ancestors named on "base rolls" constructed in previous centuries. Until now, tribal enrollment rules have been articulated largely through the symbolic language of blood.

Like many other Americans, we are transitioning in Indian Country away from blood-talk to speaking in terms of what "is coded in our DNA" or our "genetic memory." But we do it in a very particular social and historical context, one that entangles genetic information in a web of known family relations, reservation histories, and tribal and federal government regulations.

A feminist-indigenous standpoint analysis of Native American DNA

The entangled technical imprecisions and troublesome politics of genetic relatedness techniques are not apparent to most people, thus do little to undermine the authority of genomics on questions of human "origins" and identity, including for Native Americans. In our world of power and resource imbalances, in which the knowledge of some people is made more relevant than others', genetic markers and populations named and ordered by scientists play key roles in the history that has come to matter. Native American DNA is both supported by and threads back into our social-historical fabric, (re)scripting history and (re)constituting the categories by which we order life, with real material consequences.

This is where feminist epistemology enters the picture. Feminist science studies scholars have called out especially physical and biological science disciplines for representing their views as universal and objective – for misrepresenting what tend to be masculinist, Western standpoints and values as being value neutral (Whitt, in Green 1999). At the same time, as Sandra Harding explains, the modern/traditional binary that continues to shape both social and natural scientific research as well as philosophy and public policy "typically treats the needs and desires of women and traditional cultures as irrational, incomprehensible, and irrelevant – or even a powerful obstacle to ideas and strategies for social progress" (Harding 2008: 3). But for Donna Haraway the strength of all knowledges lies precisely in their not being universal, but rather "situated" – produced within historical, social, value-laden, and technological contexts (Haraway 1991: 188). In this context, Harding argues for "stronger objectivity" produced by beginning inquiry from the lives of the marginal, for example, from the lives of "women and traditional cultures" (Harding 2008: 3). This is not just a multicultural gesture to pay greater attention from without, but it is a call to inquire from within the needs and priorities articulated in marginal spaces.

The conversation between feminists and indigenous critics of technoscience should be obvious. Harding argues that both are "valuable 'strangers' to the social order" who bring a "combination of nearness and remoteness, concern and indifference that are central to maximizing objectivity." The outsider sees patterns of belief or behavior that are hard for the "natives" (in this case, scientists), those whose ways of living and thinking fit "too closely the dominant institutions and conceptual schemes," to see (Harding 1991: 124). Standpoint theorists also emphasize that subjectivities and lives lived at the intersections of multiple systems of domination become complex. Thus they recognize that individuals can be oppressed in some instances, yet privileged in others. This is another reason why feminist theorizing is beneficial for doing indigenous standpoint analyses in the twenty-first century.

Feminist objectivity helps shape a critical account of Native American DNA that does not simply invert and prioritize one side of nature/culture, science/culture, modernity/tradition binaries as we try to see things from an indigenous standpoint. I do not argue that only indigenous people can speak while scientists have no legitimate ground to speak. I do not contend that only "indigenous cultures" or "traditions" matter in circumscribing what it is to be Native American. Nor does federal government policy alone matter. But when we look from feminist and indigenous standpoints we become more attuned to the particular histories of privilege and denial out of which the concept of Native American DNA emerged. We might then convincingly argue against the misrepresentation of this molecular object as an apolitical fact, which ignores what indigenous peoples have suffered in its constitution. We might thus find ways to take more responsibility for the everyday effects, both material and psychic, of this powerful object, and its sister objects – African, Asian, and European DNAs.

Native American intellectuals, policymakers, and concerned tribal or First Nation citizens should find much of interest in this topic. Genetically-determined identity can operate without reference to the federal-tribal legal regime that is critical for contemporary indigenous governance, including rights to determine citizenship. In some instances, DNA testing can support indigenous governance, for example, using a parentage test as part of a suite of enrollment or citizenship criteria. But *genetic ancestry* tests are irrelevant to existing indigenous citizenship criteria, while across-the-membership application of *parentage tests* can contradict hard-won legal battles for indigenous self-determination. Such tests foreground genetic identity as if it were always already a fact central to Native American tribal belonging, even while tribes struggle to emphasize tribal membership as a political designation, with familial relations – documentation of which changes over time – used to support it. As genetic identities and historical narratives command increasing attention in society, they may come to rival existing historical-legal foundations of indigenous governance authority in determining who is and isn't a citizen.

The Native American DNA marketplace

To assess this growing genetic fetishism, I examine the work of three DNA testing companies – DNAPrint Genomics, DNA Today, and Orchid Cellmark – and consider how their work has been co-constituted with two categories of conceptual and social organization, "race" and "tribe." I examine how these historically and culturally contingent categories have informed the marketing and interpretation of company technologies, and to what effect, and then contemplate how the technologies loop back to reconfigure and solidify "race" and "tribe" as genetic categories.

DNAPrint genomics and race

DNAPrint declared bankruptcy in March 2009. But its patented and popular AncestrybyDNA™ test was quickly licensed to DNA Diagnostics Center (DDC).[3] AncestrybyDNA™ provides a detailed and highly visible example of how the concept of race gets understood and deployed in the direct-to-consumer (DTC) genetic testing industry. From a technical perspective, AncestrybyDNA™ is unique because it is an autosomal ancestry DNA test that surveys for "an especially selected panel of Ancestry Informative Markers" (AIMs) across all 23 chromosome pairs, not on just the Y chromosome or mtDNA. Mark Shriver (former technical advisor to DNAPrint) defines AIMs as "genetic loci showing alleles with large frequency differences between populations." Shriver and his colleagues propose that "AIMs can be used to estimate biogeographical ancestry (BGA) at the level of the population, subgroup (for example, cases and controls) and individual" (Shriver et al. 2003). Because AIMs are found at high frequencies within certain groups and at lower frequencies in others, DNAPrint surveyed markers and used a complicated algorithm to estimate an individual's BGA percentages.

DNAPrint's final version of its test, AncestrybyDNA™ 2.5, surveyed approximately 175 such AIMs (175 markers is a minute sample of the total number of base pairs in the human genome, between 3 and 4 billion). The full range of AIMs used to calculate "ancestral proportions" is proprietary information, which makes peer and public review of the science difficult (Lee et al. 2009). Only a handful of scholarly articles were ever cited in reference to AncestrybyDNA™ on the company website, which collectively disclosed just over one dozen such markers (Parra et al. 1998, Parra et al. 2001, Frudakis et al. 2003, Pfaff et al. 2001). Do the 175 markers fairly evenly cover different ancestry categories? Does DNAPrint sample randomly across the genome in choosing markers to survey? Despite the limited number of known AIMS (Shriver et al. 2003: 397), we don't have any evidence other than the fact that they say they do.

Deborah Bolnick has drawn attention to additional technical problems with AncestrybyDNA™. In following the company website for one year, she noted problems with individual test results posted to the site. First, some individuals claiming to have Native American ancestry were shown to have East Asian ancestry, which undermines the idea that such categories are genetically distinct. Bolnick also noticed drastic changes in particular indi- viduals' "precise ancestral proportions" over the course of the year despite no official changes being made to the test during that time (Bolnick 2003).

Critiques of DNAPrint can, in theory, be addressed by more rigorous scientific practices and with larger data sets gathered more evenly across the genome and across genetic populations. However, these problems too have social and cultural aspects to them. It is both logistically and ethically difficult to sample evenly across populations. Some populations require more work to sample. Some populations don't want to be sampled. In addition, samples have been gathered with longstanding racial categories in mind, and not at random or evenly across even the groups that researchers do manage to access.

While many scientists and social scientists today disclaim the possibility of race purity or biological discreteness among populations, DNAPrint has openly promoted popular racial cate- gories as at least partly genetically determined. The DNAPrint website described individuals of "relatively pure BioGeographical Ancestry" versus "recently admixed peoples." As of June 2005, DNAPrint explicitly claimed that their test "measures 'the biological or genetic component of race'" as it detects four "BioGeographical ancestries" or "four lineages or major population groups of the human population." As of July 2009, the DNAPrint website offered the follow- ing definition of BGA:

> BioGeographical Ancestry (BGA) is the term given to the biological or genetic compo-
> nent of race. BGA is a simple and objective description of the Ancestral origins of a person,
> in terms of the major population groups. (e.g. Native American, East Asian, European, sub-
> Saharan African, etc.) BGA estimates are able to represent the mixed nature of many
> people and populations today.

The company cannot help but reinforce the possibility of racial genetic purity when it repeat- edly refers to categories commonly understood to be races – European (EU); Sub-Saharan African (AF), Native American (NA), and East Asian (EA) – as "major [genetic] populations." DNAPrint repeatedly referred, for example, on its Products and Services Page, to "mixed heritage" or the "mixed nature of many people and populations today." Importantly, the notion mixture is predicated on the idea of purity.

DNAPrint included, as late as July 2009, a visual representation of the notion of population purity. Four prototypical portraits were arranged across the top of the page. They were unla- beled. First was European ancestry: a fair and slender young woman smiles into the camera, wearing a hat to cover her hair. To her right was Native American ancestry: a photo of a youngish male with brown skin and long black hair falling down his back. Third was African ancestry: a portrait of a dark-skinned (by U.S. standards), handsome, middle-aged black male.

Fourth was East Asian ancestry: a middle-aged woman about the same color as Native American ancestry, with well-cut chin length black hair. When I first I laid eyes on this page, I took her to represent Native American ancestry. Then I saw the prototypical Native American-looking male. My initial confusion was fitting. AncestrybyDNA™ sometimes has a hard time telling Native American from East Asian ancestry.

On the FAQ page, "What is race?" DNAPrint described recent disciplinary discussions that race is "merely a social construct" as oversimplified: "While, indeed this may often be true, depending on what aspect of variation between people one is considering, it is also true that there are biological differences between the populations of the world. One clear example of a biological difference is skin color."[4] DNAPrint takes this biological difference as part of the genetic component of race. If society then builds important and sometimes discriminatory values and practices around such differences, that is *social* race, and supposedly external to the science of genetic race.

I agree that race as "a social construct" has been oversimplified, but for reasons different from those suggested in the DNAPrint website (Lee et al. 2008, Reardon 2005). Race-associated medical problems illustrate the point. Take elevated levels of hypertension, diabetes, heart disease, or prostate cancer that are found in higher and lower rates in different racial groups or populations. There are scientists who search for genes that occur in elevated frequencies in order to help explain disease risk, or differential responses to medical treatment (for example, Adeyemo et al. 2009, Mathias et al. 2010). Eventually, some researchers hope that such knowledge will lead to cures or treatments (for example, Burchard et al. 2003, Satel 2002). But others emphasize that elevated disease is also, sometimes more strongly, tracked to particular histories of deprivation and stress. Genetic factors are not *the* key. Rather, poverty and discrimination lead to health disparities, and society should not neglect such factors in favor of funding sexier genetic research (for example, Sankar et al. 2004: 612). I have heard participants at Native American health research conferences charge funding agencies and scientific communities with focusing too much on genetics. Genetic research may even compound social problems and historical injustice by constituting the oppressed or deprived body or population as genetically deviant.

Let's return to DNAPrint's claim that the biological difference of skin color is dramatically different "across *populations*" (my emphasis). (Note the conflation of race and population here.) Leading research in the field indicates that the relationship between genes and skin color is not so deterministic. As Nina Jablonski puts it, "pigmentation is a trait determined by the synchronized interaction of various genes with the environment." Thus, near the equator darker skin is selected in order to protect against cancer-causing UV rays. At higher latitudes, lighter skin color is advantageous. Therefore, populations living in similar environments around the world have independently evolved similar skin pigmentation (Jablonski 2004: 612–613).

Simply put, recent science on skin color does not support but confounds DNAPrint's overly simplistic representation of race as a continental-level category with a "genetic component," demonstrable in part by calculating distributions of skin pigmentation markers among different populations. DNAPrint assumes that genetic differences have produced visible differences among people that we can objectively label as "race." They continue to over-simplify when they intimate that genetics can be severed from social aspects of race. Skin color was used to break the biological continuum of humanity into continental races long before we knew anything about underlying genetic differences. Skin color shapes scientists' choices today about which other biological characteristics to look for. DNAPrint's 175 ancestry informative markers do not present us with objective scientific evidence that our race boundaries are genetically valid. Markers were chosen with those racial boundaries in mind and not at random (Bolnick 2008, Fullwiley 2008, Weiss and Long 2009). For DNAPrint, continental ancestry equals population equals race.

DNAToday, Orchid Cellmark, and "Tribe"

DNAToday and Orchid Cellmark have marketed the DNA profile or parentage test to U.S. tribes and Canadian First Nations for purposes of enrollment. The parentage test is both more and less informative than the genetic ancestry tests discussed so far because it gauges related-ness at a different biological level. DNA profiling can be used to confirm close relations (for example, mother, father, siblings) with very high degrees of probability. It does not analyze markers judged to inform one's "ethnic" ancestry.

This technology is commonly used in paternity testing, although it can also be used to examine maternal lineages. The technology examines repeated sequences of nucleotides such as "GAGAGA," called "short tandem repeats" (STRs). We inherit STRs from both parents, but our total individual STR pattern is, in practical terms, unique. DNA fingerprinting examines very specific patterns of multiple STRs. A single such sequence is not unique. But when viewed in combination with other STRs, the pattern becomes increasingly distinctive.

Genetic ancestry test findings do not generally illuminate the biological relationships that tribes care about, i.e., an applicant's close biological relationship to an individual who is/was already on the tribal rolls, or to individuals who were on the tribe's "base rolls." Genetic ances-try tests document an individual's descent from unnamed "founding ancestors" who first settled the Americas. Such genetic information is interesting, but generally irrelevant from a tribal enrollment point of view. On the other hand, parentage tests are useful in certain individual enrollment cases. They are already commonly used when an applicant's biological parentage is in question and the tribal status of that parent is essential for the applicant to be enrolled. However, in the majority of tribal enrollment cases, biological parentage is not in question. Theoretically, parentage testing would be used only occasionally. But due to onslaughts of enrollment applications especially in wealthier tribes, we see increasing numbers of tribes moving to a system in which a DNA parentage test is applied across the board, especially for new applicants. This is done in part to reduce the work burden of tribal enrollment offices related to managing large volumes of identity documents. There are risks to this approach, however, including finding "false" biological parentage where it was not originally in question. When gaps in biological descent are uncovered, sometimes decades after the fact, individuals or entire branches of families can be ejected from the tribal citizenry. When genetic facts get misconstrued as objective truths about kinship they cast a shadow over generations of lived rela-tions, profoundly disrupting families and a people.

OriginsToday™ and the Tribal Identity Enrollment System (TIES)™

DNAToday, LLC, until it declared bankruptcy in 2006, marketed the parentage test in combi-nation with its TIES System that includes an ID card, a "smart card" with an individual's photograph and embedded computer chip that can hold a "DNA profile and an individual's other tribal government records" (i.e., enrollment, voter registration, and health and social serv-ices data). Such records would be managed with the OriginsToday™ Sovereignty and Sovereignty Plus Edition software.[5] Tribal government customers of OriginsToday™ span the country and include tribes in Oregon, Washington State, Arizona, Wyoming, Nebraska, Oklahoma, Minnesota, Mississippi, and New York State.[6] This technologically savvy suite of products aimed to "factor out political issues" and establish "clear answers" for "all future gener-ations."[7] Thus tribal identity can be technologized, made more precise by the merging of two types of information – computer code and genetic data.

I first encountered DNAToday at a 2003 national conference for U.S. tribal enrollment office staff attended by approximately 200 individuals. I was informed about the conference from the enrollment director of the tribe in which I am a member. Coincidentally, DNAToday had already approached our tribe to pitch their services and were declined. The conference hosts, DCI America, allowed me to enroll – at full fee ($459) – as a graduate student who was studying the role of DNA testing in Native American identity. As I entered the brightly-lit ballroom on that October morning with its tapestries and sconces on the walls and its massive chandeliers, the large circular tables were already full with conference goers – many middle-aged and elderly women – eight to a table. I later learned that tables were largely populated by groups of staff from individual tribal enrollment departments. Arriving too late to grab a seat at a table, I sat down next to several individuals dressed in business attire whom I later learned were DCI America staff.

Geared toward the directors and staff of federally-recognized tribal enrollment departments, the conference is held annually at different locations around the U.S. In 2003 it played out like a three-day infomercial for DNA testing services. DNAToday was front and center, leaving the impression that genetic testing is key to solving existing inefficiencies in enrollment. DNAToday company president Steven Whitehead (a self-declared former insurance salesman) sat on the keynote panel in the conference's opening session, and took all questions related to the use of DNA in enrollment, giving the impression that the genetic science at play somehow only involves questions of genetics and not the politics and histories of enrollment. He was accompanied by a tribal chairman, a tribal court judge, and a BIA field office superintendent who addressed questions of law and policy, but with no reference to genetics.

Company scientists and software technicians also staffed a trade-show-like table in the registration lobby where they offered a product demonstration. Company executives and scientists were given the title of "Instructor." DNAToday hosted "workshops" throughout the three-day event, some of which played out more like glitzy product demonstrations. At one point a conference attendee was chosen through a raffle to receive a free DNA test. As her name was called out, the room erupted in applause. The conference goer walked to the front of the ballroom and sat down at a table. Whitehead explained that she and a witness were signing an informed consent form. Whitehead then narrated the taking of her DNA deliberately and in a hushed voice into his microphone, like a sportscaster narrating a quiet, tense moment in a sports match: "He is swabbing one side, now he's swabbing the other."

Workshop overheads included "Reasons to Select the T.I.E.S.™ System," "Your Investment Options," "3 Ways to Get Started," and the name of a contact for further information.[8] Despite the focus on DNA testing services, conference organizers did not offer basic genetics education, especially as it might be read against current tribal blood rules. It was clearly necessary. One attendee asked if DNA testing could be used to scientifically determine the blood quantum of tribal members. DNAToday President Whitehead responded that DNA testing says nothing about blood quantum, "but you can prove a child is related to a mother."[9] Despite this clarification, during the keynote panel, Whitehead claimed that DNAToday's technology is "100% reliable in terms of creating accurate answers" to questions of tribal enrollment.[10] Whitehead did not respond to the loaded and complicated symbolic meanings ascribed to blood that go beyond lineal biological relations that DNAToday and other companies focus on. He took the blood quantum question as purely a molecular knowledge question.

In another session Whitehead advocated tribal-wide paternity testing so that "only Native Americans who deserve to be members of your tribe will be."[11] Another company's spokesperson has claimed that a DNA parentage test "validates tribal identity" or "can preserve a tribe's heritage."[12] These statements gloss over a much more complex process in which tribal communities entertain deep philosophical and political-economic disagreements about who should rightly be a citizen and who should not, who belongs and who does not. Kirsty Gover (2008) has shown that those relations in the twentieth century became increasingly biological or genealogical and less based on residence, marriage, and adoption. The boundaries and definitions of the "tribe" will continue to change. By glossing over the initial decision that tribes make about which relationship(s) to count as pivotal, the DNA spokesperson fetishizes molecules, making certain shared nucleotide sequences stand for a much more complex decision-making process. They make the DNA test appear to be scientifically precise and universal instead of a technique chosen as part of a broader set of political maneuvers in which there are deep philosophical, cultural, and economic disagreements within tribes about who should count, for example, as Dakota or Pequot or Cree.

In 2003, DNAToday offered its "legal" paternity test for a group of 2–3 individuals (an individual plus one or both biological parents) for $495.[13] The price of such services remained the same through 2010.[14] Take a 10,000-member tribe. The number of tests required to retroactively test every member of that tribe might be 4,000 (an average of 2.5 people included per test). At $495 per test the cost to test all members would be nearly $2 million. Add to that the costly ID card sold to accompany tribal-wide DNA testing. DNAToday generated each individual card in their facilities at a cost of $320 per tribal citizen. Thus it would cost $3.2 million for a 10,000 member tribe to equip each tribal member with a programmed card.

Such products and services also present concerns for tribes related to privacy and legal self-determination. For example, while the tribe purchasing the DNAToday/DCI America products would determine the specific data to be programmed to cards and would maintain the data base, DNAToday would retain control of issuing original and replacement cards. DNAToday did note that they purged data from their system after generating cards, and a tribe's confidentiality agreement would no doubt require such safeguards. But DNAToday also proposed to store tribal DNA samples, which raises privacy concerns. Whitehead noted that after identity markers were analyzed individuals could request that their samples be destroyed and receive affidavits certifying as much. Enrollment applicants could also sign as "registered agents" on their samples, which would then require that notarized permission be obtained from agents for others to obtain access to their DNA. But it is difficult to anticipate, plan for, and regulate the array of potential privacy risks that emerge related to the storage and protection of genetic data. Federal and state governments find this work politically thorny, expensive, and onerous. Tribal governments have barely begun to tackle issues of property and control that run deep in this domain.

Given privacy concerns and that DNA storage is fairly low-tech and not too expensive, tribes would probably want tribally-controlled management of samples. I raised this possibility during the question and answer period of a DNAToday workshop. The company representative's response was again a non-sequitur – simply that the industry standard was to include twenty-five years storage at no extra cost. So why not just let the company handle it? For the DNAToday executive, technology and cost were the golden keys to solving the most pressing issues of sovereignty in Indian Country. A few individual tribal enrollment officers and one tribal community panelist expressed hesitation about DNA testing and the smart card system, either because they seemed to contradict notions of tribal citizenship based on indigenous cultural concepts or because the technologies seemed invasive. One elderly woman in the audience was clearly irritated when Whitehead explained how far the smart card and its attendant software could be made to reach across tribal programs and facilities to monitor individual tribal member engagements with different programs, their records, histories, and identities. She asked Whitehead, "Does anyone else carry around this type of info? Do you?" I took her to mean, "Do we expect non-Natives to be so monitored in their day-to-day lives?" DNAToday noted that they provide the same technologies to the U.S. State Department and Homeland Security. Question answered. But the vast majority of attendees seemed either unworried and unimpressed by the technology, or hopeful that it would provide a scientific solution to their considerable and messy political problems in the enrollment office.

In October 2010, I returned to the DCI America National Tribal Enrollment Conference,[15] this time hosted at the tribally-owned Hard Rock Casino & Resort Hotel in Albuquerque. In 2010 I was invited to speak on a panel presentation on DNA and tribal enrollment. I was grateful for the opportunity to speak to a tribal audience about genetic ancestry testing and the DNA profile as they relate to tribal enrollment and more broadly to tribal sovereignty. This perspective was absent from the 2003 tribal enrollment conference. Like in 2003, the audience largely lacked a basic understanding of genetics. Before my presentation, I heard a tribal enrollment office staff member ask of another enrollment officer from a tribe that does DNA parentage testing, "How does DNA work? How do they determine if your DNA is from a particular tribe?" The representative from the DNA testing tribe responded, "No. It just says Indian." Ouch.

During this conference, the DDC spokesperson did a quick, efficient overview of the basic science. I also covered some basic science in my talk. The DDC representative contradicted his careful elucidation of the science during the formal presentation with genetically fetishistic statements that seemed to condition tribal identity on a DNA test. Indeed, there was confusion displayed in conversations between different enrollment staff during question and answer sessions about exactly what the DNA tests reveal about genetic relationships and, by extension, about Native American or tribal-specific identity.

Unlike in 2003, multiple tribal enrollment officers at the 2010 conference noted their tribes' use of DNA parentage testing for enrollment. Listening to participant comments over the two days, I heard confirmed Gover's (2008) insights about the turn to genealogical tribe-making. Any talk I heard of genetics at this conference showed that DNA profiles are clearly being used to support the move to constitute the genealogical tribe. Interestingly, symbolic blood loomed much larger than gene talk in discussions of tribal enrollment processes and debates. Unlike with genetics-based knowledge, the officers who spoke out loud were expert in the nuances of blood rules in enrollment, and potential or hoped-for links to cultural affiliation. They recounted tribal community debates about whether to retain or lower blood quantum requirements. Some talked about the turn to lineal descent criteria instead, but they worried then about the enrollment of too many individuals who really had no cultural connection to their tribes. On the other hand, several participants expressed concern about increasing numbers of reservation residents, including children, who cannot be enrolled due to having too many tribal lineages in their backgrounds. They cannot meet the blood quantum requirements of any tribe from which they are descended. As such, they are disenfranchised when they actually do live within, and are connected socially to, the tribal community.

The exact same predicaments were recounted by tribal enrollment staff during the 2003 enrollment conference. Nothing had changed on that front. DNA testing was clearly of greater interest by 2010. Participants could embrace the use of the DNA profile because it continues to support the broader understanding of relatedness that is already figured through symbolic blood. The question now seems to be less about how "useful" DNA will become in enrollment. But rather, the question seems to be "For how long will blood continue to dominate in the tribal imaginary and condition the take up of DNA tests?" Also, "How will blood and genetic concepts increasingly work together to (re)constitute the notion of the tribe?"

Orchid Cellmark: Which truth is in the DNA?

In Orchid Cellmark advertisements that ran from May to November 2005 in a national news magazine targeted to tribal governments, *American Indian Report*, an Indian in silhouette faced the setting sun. The image in purples, pinks, and blues recalled the classic "End of the Trail" depiction, the broken nineteenth-century Indian on horseback. But Orchid's Indian sat upright. Did he envision a more hopeful future for his progeny than did his nineteenth-century counterpart? After centuries of predicting their demise, could scientists now testify to the American Indians' survival? Orchid's ad announced that genomic technologies reveal "the truth." Was the audience to assume the ad referred simply to DNA markers or were they meant to think that Orchid's science can reveal – via those markers – an ethnic essence, the stereotypical nobility portrayed in the image?

In a unique move, Orchid Cellmark entered the Native American identity market offering a more complete array of DNA testing services than did other companies.[16] Like DNAToday, Orchid promotes the standard parentage test for U.S. tribal and Canadian First Nations enrollment.[17] But it also markets tests for Y chromosome and mtDNA Native American markers and – when it was available – the DNAPrint test that supposedly determined one's "percentage of Native American-associated DNA."[18] By offering both the DNA fingerprint and the genetic ancestry tests to the Native American identity market, Orchid Cellmark presents a more complex set of implications for Native American identity than the products offered by other companies. Like DNAToday, Orchid stresses that its DNA parentage test can "confirm the familial relationship of specific individuals to existing tribal members."[19] Again, the parentage test is technically helpful for tribal enrollment. Native American markers tests, in contrast, are bad technical matches for tribal enrollment. They do not confirm enrollment applicants' relationships with named ancestors who are also documented tribal members, the central requirement in all tribes' enrollment processes. A company spokesperson confirmed in a 2005 interview that tribes and First Nations had purchased only paternity tests for enrollment purposes.[20] However, genetic ancestry tests are important symbolically. They promote the notion of shared ethnic/racial genetic material by constituting objects such as "Native American," "European," "African," and "East Asian DNAs." The AncestrybyDNA™ test also provided a percentage calculation of "biogeographical ancestry" allocated between such categories.

Conclusion

The terminology of "tribe" (or First Nation) and "race" contradict and overlap each other as they help construct the different Native American identities to which DNA testing companies respond. Advocates for tribal government sovereignty typically talk about the fundamental difference between the concept of race and that of the tribe/First Nation in justifying self-governance. They commonly argue that tribes are nations (also implying culture), and not races, and nations decide who gets citizenship and who does not. Tribe and race share common ground. Part of that common ground is symbolic blood, increasingly complemented by DNA, although blood and gene talk take different forms when used to support different positions (i.e., reckoning race versus reckoning tribal membership). Orchid Cellmark simultaneously appeals to DNA as an indicator of race and as representative of one's relatedness to tribe. Because tribes operate in a world that is both tribalized and racialized, they struggle to mediate a Native American identity according to those sometimes contradictory, sometimes overlapping categories.

More than any other company discussed, the work of Orchid Cellmark reflects the breadth of ways in which DNA testing and gene talk can expand into territory previously claimed by blood-quantum regulations and symbolic blood. Unlike blood, DNA testing has the advantage of claims to scientific precision and objectivity. Orchid Cellmark's Director of North American Marketing noted that in using DNA fingerprint analysis, there is "no possibility of incorporating a subjective decision into whether someone becomes a member or not."[21] Of course, whether or not someone is verifiable biological kin of the type indicated by a parentage test is not an "objective" enrollment criterion. Allowing a DNA profile to trump other ways of reckoning kin (for example, blood quantum as a proxy for cultural affiliation by counting relatives, or a signed affidavit of family relatedness) for purposes of enrollment prioritizes technoscientific knowledge of certain relations over other types of knowledge of the same and other relationships. Nonetheless, the idea of scientific definitiveness attached to genetic testing is influential, even if it is not realized. The DNA profile may increasingly look like a good complement to traditional blood (quantum) and other non-genetic documentation – especially if traditional documentation of named relations is difficult to obtain or if enrollment applications are politically contentious.

The increasing use of the DNA profile may condition tribes' eventual acceptance of DNA knowledge as a substitute for tracking blood relations. Some will see such a move as advantageous, as scientifically objective and less open to political maneuvering. DNA testing will not solve what is the most divisive problem in contemporary enrollment debates: in the majority of cases parentage is not in question, but due to out-marriage, increasing numbers of tribal members' offspring cannot meet blood requirements. They simply do not have enough sufficiently "blooded" parents and grandparents to meet standards set by tribes. Using DNA parentage tests can be seen as supporting Gover's (2008) genealogical tribe in which blood rules move away from the broader racial category of "Indian blood" to constitute blood as tribal-specific. In that, she argues, tribes are relying less on race than have federal agents in the past. On the other hand, the increasing tribal practice of DNA testing across-the-membership can also pave the way for a re-racialization of Native Americans by promoting the idea that the tribe is a genetic population. The incommensurable nature of the DNA profile with the genetic concept of "population" and its re-articulations of older notions of race will be lost on many observers. In addition, if genetic ancestry tests come to be coupled with the DNA profile – the DNA Diagnostics Center spoke-scientist at the 2010 tribal enrollment conference noted the use sometimes of mtDNA lineage tests to ascertain maternal lineages in tribal enrollment cases – "race" is certain to loom larger in our conception of Native American tribal and First Nations identity in the U.S. and Canada.

If race in the form of genetic ancestry comes to have greater influence in how we understand Native American and Aboriginal identity, federal government obligations to tribes and First Nations established historically as treaty obligations between nations may fade further from view. In turn, claims to land and self-governance that refer to such nation to nation agreements may be denied and justified by the absence or presence of Native American DNA in individual claimants. This would be the result of privileging the priorities and situated knowledge of dominant society – its misguided application of race to a situation that (also) requires knowledge of tribal governance rights. This could have two effects: first, anti-indigenous interests will have ammunition to use against tribes whom they already view not as beneficiaries of treaty obligations, but of special race-based rights; second, people heretofore racially identified as other than Native American, may increasingly claim indigenous nation authority and land based on DNA. Clearly, the marketing to tribes and First Nations of DNA tests originally developed by extracting indigenous biological resources for the benefit of researchers, and later for consumers mostly without lived connections to tribes, has much broader implications than revisions in tribal enrollment policies. Given the commercialization of genetic-ethnic identities and capitalism's ubiquitous reach, we must consider that the scientific object of Native American DNA (or the definitive absence of such markers) will be important in re-making Native American identity in the twenty-first century.

In October 2010, I left the tribal enrollment conference in Albuquerque thinking that Native Americans selling cappuccinos, serving slightly upscale steak dinners, and renting hotel rooms decorated in a tranquil and vaguely "Asian" theme seemed a preferable kind of cultural hybridity to reconfiguring the tribe as a genetic entity. I can't comment on the gaming facilities. I prefer café Wifi to slot machines. Of course, gaming is central to the constitution of the twenty-first-century capitalist Indian, and that wealth challenges well-worn ideas of authentic indigeneity (Deloria 2002, Cattelino 2010). It felt particularly troublesome to see wealthy gaming tribes with little knowledge of the social and technological contexts that give rise to DNA testing technologies and likewise DNA testing company executives with little knowledge of the social, technological, and historical contexts that frame dynamic blood rules together reconfiguring the tribe into a genetic entity. Will the advent of tribal DNA testing serve to unsettle claims to indigenous sovereignty and implicate current tribal members in that politically complicated perturbation?

Notes

1 Acknowledgements: Portions of this chapter were previously published in *Native American DNA: Tribal Belonging and the False Promise of Genetic Science* (2013) Minneapolis and London: University of Minnesta Press. Thanks to David S. Edmunds and to volume co-editors Daniel Kleinman and Kelly Moore for feedback and suggested edits on this chapter.

2 I often refrain from using scare quotes with "Native American DNA," which can be tedious for the reader. But in every instance I mean the constituted and not simply found nature of that object.

3. AncestrybyDNA, accessed June 13, 2012, *www.ancestrybydna.com*.

4 AncestrybyDNA FAQs, "What is Race?" DNAPrint Genomics, accessed November 5, 2012, *http://web.archive.org/web/20060709021118/http://www.ancestrybydna.com/welcome/faq/#q1*.

5 Steven Whitehead, "The Future of Tribal Enrollment Software," *DCI America Tribal Enrollment Conference* (New Orleans, LA2003). See also "DNAToday Genetic Identification Systems, Solutions, Smartcards," Internet Archive: Way Back Machine, accessed June 13, 2012, *http://web.archive.org/web/20041225035717/www.dnatoday.com/smart_cards.html*.

6 Tribal Net Online. Industry News. February 20, 2006. "Many Tribes Make the Switch to Origins Enrollment Software,"accessed June 13, 2012, *www.tribalnetonline.com/displaynews.php?newsid=33*

7 "DNAToday Genetic Identification Systems, Solutions, Articles: How Would DNA Work for Our Tribe?" Internet Archive: Way Back Machine, accessed June 13, 2012, *http://web.archive.org/web/20041227092646/www.dnatoday.com/article_01.html*.

8 DNAToday overhead presentation, October 2003.

9 Ibid.

10 Ibid.

11 Steven Whitehead, DNA Today, "The Future of Tribal Enrollment Software," *DCI American National Tribal Enrollment Conference* (New Orleans, Louisiana 2003).

12 DNA Diagnostics Center, "Powerpoint Presentation," *DCI America National Tribal Enrollment Conference* (Albuquerque, New Mexico, 2010).

13 TallBear fieldnotes, October 28, 2003. DNAToday sold two types of paternity tests. One was performed by a "disinterested third party" and was to be used to "ascertain legal relationship status." That was the "legal" paternity test promoted at the DCIAmerica conference. They also sold at a lower cost an at-home test to be used for "information only situations."

14 DNA Diagnostics Center, Powerpoint Presentation.

15 "The Sixteenth Annual Tribal Enrollment Conference," DCIAmerica, accessed June 13, 2012, *www.dciamerica.com/pdf/16THATEC.pdf.*

16 In a July 7, 2005 interview with the author, an Orchid spokesperson clarified that the company was unique in selling all four types of tests for Native American identity.

17 "Tribal Testing," Orchid Cellmark, accessed June 13, 2012, *www.orchidcellmark.com/tribal.html* and

Orchid Cellmark (Canada); "Aboriginal Testing," Orchid Cellmark, accessed June 13, 2012, *www.orchidcellmark.ca/site/aboriginal-testing.*

18 TallBear interview with Orchid company spokesperson, July 7, 2005.

19 "Press Releases: Orchid Cellmark Launches New DNA Testing Service to Confirm Native American Tribal Membership, 6/17/2005" Bionity.com, accessed June 13, 2012, *www.bionity.com/en/news/46940/orchid-cellmark-launches-new-dna-testing-service-to-confirm-native-american-tribal-membership.html.*

20 TallBear interview with Orchid spokesperson, July 7, 2005.

21 Author interview with Jennifer Clay, via telephone July 7, 2005. On July 8, 2005 Clay also made a similar comment on "Native America Calling," a national daily radio show on which we were both guests. Native America Calling, Archives, "DNA Testing for Tribal Enrollment," *www.nativeamericacalling.com*, July 8, 2005.

References

Adeyemo, A., Gerry, N., Chen, G., Herbert, A., Doumatey, A. et al., 2009. "A Genome-Wide Association Study of Hypertension and Blood Pressure in African Americans." *PLoS Genetics* 5(7). doi:10.1371/journal.pgen.1000564.

Ben-zvi, Y. 2007. "Where Did Red Go? Lewis Henry Morgan's Evolutionary Inheritance and U.S. Racial Imagination." *CR: The New Centennial Review* 7(2): 201–229.

Bieder, R.E. 1986. *Science Encounters the Indian, 1820–1880: The Early Years of American Ethnology.* Norman and London: University of Oklahoma Press.

Bolnick, D.A. 2003. "Showing Who They Really Are: Commercial Ventures in Genetic Genealogy." American Anthropological Association Annual Meeting (Chicago, IL).

Bolnick, D.A. 2008. "Individual Ancestry Inference and the Reification of Race as a Biological Phenomenon." In *Revisiting Race in a Genomic Age*, edited by B. Koenig, S.S.-J. Lee, and S. Richardson, 70–88. New Brunswick: Rutgers University Press.

Bolnick, D.A., Fullwiley, D., Duster, T., Cooper, R.S., Fujimura, J.H., Kahn, J., Kaufman, J., Marks, J., Morning, A., Nelson, A., Ossorio, P., Reardon, J., Reverby, S.M., and TallBear, K. 2007. "The Science and Business of Genetic Ancestry." *Science* 318(5849), (October 19): 399–400.

Burchard, E.G., Ziv, E., Coyle, N., Gomez, S.L., Tang, H., Karter, A.J., Mountain, J.L., Perez-Stable, E.J., Sheppard, D., and Risch, N. 2003. "The Importance of Race and Ethnic Background in Biomedical Research and Clinical Practice." *New England Journal of Medicine* 348(12): 1170–1175.

Cattelino, J.R. 2010. "The Double Bind of American Indian Need-Based Sovereignty." *Cultural Anthropology* 25(2): 235–262.

Deloria, S. 2002. "Commentary on Nation-Building: The Future of Indian Nations." *Arizona State Law Journal* 34: 55–62.

Frudakis, T., Venkateswarlu, K., Thomas, M., Gaskin, Z., Ginjupalli, S., Gunturi, S., Ponnuswamy, V. et al. 2003. "A Classifier for the SNP-Based Inference of Ancestry." *Journal of Forensic Science* 48(4): 771–782.

Fullwiley, D. 2008. "The Biologistical Construction of Race: 'Admixture' Technology and the New Genetic Medicine." *Social Studies of Science* 38: 695–735.

Gover, K. 2008. "Genealogy as Continuity: Explaining the Growing Tribal Preference for Descent Rules in Membership Governance in the United States." *American Indian Law Review* 33(1): 243–310.

Haraway, D.J. 1991. "Situated Knowledges: The Science Question in Feminism and the Privolege of Partial Perspective." In *Simians, Cyborgs, and Women: the Reinvention of Nature*, 183–201. New York and London: Routledge.

Harding, S. 1991. *Whose Science? Whose Knowledge?: Thinking from Women's Lives.* Ithaca, NY: Cornell University Press.

Harding, S. 2008. *Sciences from Below: Feminisms, Postcolonialities, and Modernities.* Durham, NC, and London: Duke University Press.

Harmon, A. 2006. "DNA Gatherers Hit Snag: Tribes Don't Trust Them." *The New York Times* (December 10): U.S. section.

Harris, P. 2007. "The Genes That Build America." *Observer.* London (July 15): 22–27.

Jablonski, N.G. 2004. "The Evolution of Human Skin and Skin Color." *Annual Review of Anthropology* 33: 611–613.

Kahn, J. 2006. "Race, Pharmacogenomics, and Marketing: Putting BiDil in Context." *American Journal of Bioethics* 6(5): W 1–5.

Koerner, B. 2005. "Blood Feud." *Wired* 13(9).

Lee, S.S.-J., Koenig, B., and Richardson, S. 2008. *Revisiting Race in a Genomic Age*. Piscataway, NJ: Rutgers University Press.

Lee, S.S.-J., Bolnick, D., Duster, T., Ossorio, P., and TallBear, K. 2009. "The Illusive Gold Standard in Genetic Ancestry Testing." *Science* 325(5936): 38–39.

Marks, J. 2002. "What Is Molecular Anthropology? What Can It Be?" *Evolutionary Anthropology* 11(4): 131–135.

Mathias, R.A., Kim, Y., Sung, H., Yanek, L.R., Mantese, V.J., Hererra-Galeano, J.E., Ruczinski, I., Wilson, A.F., Faraday, N., Becker, L.C., and Becker, D.M. 2010. "A Combined Genome-Wide Linkage and Association Approach to Find Susceptibility Loci for Platelet Function Phenotypes in European American and African American Families with Coronary Artery Disease." *BMC Medical Genomics* 3(22). doi:10.1186/1755-8794-3-22.

Montoya, M. 2011. *Making the Mexican Diabetic: Race, Science, and the Genetics of Inequality*. Berkeley: University of California Press.

Nash, C. 2003. "Setting Roots in Motion: Genealogy, Geography and Identity." In *Disputed Territories: Land, Culture, Identity in Settler Societies*, edited by Trigger and Gareth Griffiths, 29–52. Hong Kong: Hong Kong University Press.

Nash, C. 2004. "Genetic Kinship." *Cultural Studies* 18(1): 1–33.

Nelson, A. 2008a. "Bio Science: Genetic Genealogy Testing and the Pursuit of African Ancestry." *Social Studies of Science* 38(5): 759–783.

Nelson, A. 2008b. "The Factness of Diaspora." In *Revisiting Race in a Genomic Age*, edited by Barbara Koenig, Sandra Soo-Jin Lee, and Sarah Richardson, 253–268. Piscataway, NJ: Rutgers University Press.

Parra, E.J., Marcini, A., Akey, J., Martinson, J., Batzer, M.A., Cooper, R., Forrester, T., Allison, D.B., Deka, R., Ferrell, R.E., and Shriver, M.D. 1998. "Estimating African American Admixture Proportions by Use of Population-Specific Alleles." *American Journal of Human Genetics* 63: 1839–1851.

Parra, E.J., Kittles, R.A., Argyropoulos, G., Pfaff, C.L., Hiester, K., Bonilla, C., Sylvester, N., Parrish-Gause, D., Garvey, W.T., Jin, L., McKeigue, P.M., Kamboh, M.I., Ferrell, R.E., Politzer, W.S., and Shriver, M.D. 2001. "Ancestral Proportions and Admixture Dynamics in Geographically Defined African Americans Living in South Carolina." *American Journal of Physical Anthropology* 114(1): 18–29.

Pfaff, C.L., Parra, E.J., Bonilla, C., Hiester, K., McKeigue, P.M., Kamboh, M.I., Hutchinson, R.G., Ferrell, R.E., Boerwinkle, E., and Shriver, M.D. 2001. "Population Structure in Admixed Populations: Effect of Admixture Dynamics on the Pattern of Linkage Disequilibrium." *American Journal of Human Genetics* 68(1): 198–207.

Reardon, J. 2005. *Race to the Finish: Identity and Governance in an Age of Genomics*. Princeton: Princeton University Press.

Reardon, J. and TallBear, K. 2012. "'Your DNA is *Our* History.' Genomics, Anthropology, and the Construction of Whiteness as Property." *Current Anthropology* 53(S12): S233–S245.

Sankar, P., Cho, M.K., Condit, C.M., Hunt, L.M., Koenig, B., Marshall, P., Lee, S.S., and Spicer, P. 2004. "Genetic Research and Health Disparities." *Journal of the American Medical Association* 291(24): 2985–2989.

Satel, S.L. 2002. "I Am A Racially Profiling Doctor." *New York Times Magazine* (May 5).

Shriver, M.D., Parra, E.J., Dios, S., Bonilla, C., Norton, H., Jovel, C., et al. 2003. "Skin Pigmentation, Biogeographical Ancestry, and Admixture Mapping." *Human Genetics* 112: 387–399.

Simons, J. 2007. "Out of Africa." *Fortune* 155(3) (February 19): 39–43.

Takeaway Media Productions. 2003. "A Genetic Journey." *Motherland*, London.

TallBear, K. 2008. "Native-American-DNA.coms: In Search of Native American Race and Tribe." In *Revisiting Race in a Genomic Age,* edited by Barbara Koenig, Sandra Soo-Jin Lee, and Sarah Richardson. Rutgers University Press: 235–252.

TallBear, K. 2013. "Genomic Articulations of Indigeneity." *Social Studies of Science*. Published online before print May 30, 2013, doi: 10.1177/0306312713483893: 1–25.

Thirteen/WNET New York. 2007. *African American Lives,* "Episode 2: The Promise of Freedom," Press Release (July 27).

Thomas, D.H. 2000. *Skull Wars: Kennewick Man, Archaeology, and the Battle for Native American Identity*. New York: Basic Books.

Weiss, K.M. and Long, J.C. 2009. "Non-Darwinian Estimation: My Ancestors, My Genes' Ancestors." *Genome Research* 19: 703–710.

Whitt, L.A. 1999. "Indigenous Peoples and the Cultural Politics of Knowledge." In *Issues in Native American Cultural Identity*, edited by M.K. Green, 223–271. New York: Peter Lang.

Wolinsky, H. 2006. "Genetic Genealogy Goes Global." *European Molecular Biology Organization EMBO Reports* (7)11: 1072–1074.

2

Technoscience, Racism, and the Metabolic Syndrome

Anthony Ryan Hatch

GEORGIA STATE UNIVERSITY

Introduction

Heart disease, diabetes, obesity, and stroke have become epidemic health problems and constitute major causes of death and disability in the United States and globally. Each of these chronic metabolic health problems impacts millions of lives, costs healthcare consumers billions of dollars each year, and has provoked large-scale societal responses of disease prevention, biological testing, and medical treatment. Taken together, these conditions represent and embody something altogether different. As the scope and impact of these epidemics has grown, and the science of human metabolism has become an important site of technoscience, scientists and clinicians have developed new ways of thinking about and acting on the interrelationships among the biological processes that encompass human metabolism. Biomedical researchers, physicians, government agencies, and pharmaceutical companies are increasingly using the term *metabolic syndrome* to describe the combination of biological risk factors that are statistically correlated with the incidence of heart disease, diabetes, and stroke: high blood pressure, blood sugar, body fat/weight, and cholesterol.[1] Supported by robust epidemiological findings that the group of people with three or more abnormally high biological risk factors are more likely to experience heart attacks, strokes, and the onset of diabetes compared to those with two or fewer factors, biomedical researchers are in the process of institutionalizing metabolic syndrome as a legitimate, diagnosable, and treatable disease.

Metabolic syndrome is *metabolic* because it concerns the biological processes by which bodies metabolize nutrients derived from food and describes these processes in terms of physiological or biochemical indicators of disease processes that are measured at the level of an individual body. Specifically, metabolic syndrome is comprised of so-called abnormal levels of several clinical and laboratory measurements that, if present in one body, represent a substantially increased risk of serious metabolic health problems: elevated blood pressure, elevated cholesterol, elevated blood sugar, and elevated weight.[2]

Metabolic syndrome is a *syndrome* precisely because it is an aggregation of clinical and laboratory measurements that has not yet reached designation as a disease (Hall et al. 2003: 414). Metabolic syndrome represents the co-occurrence of hypertension, dyslipidemia, hyperglycemia, and obesity, each of which has been identified as so-called risk factors for heart disease, diabetes, and stroke. Thus, the International Diabetes Federation (IDF) authored its own *iteration* of metabolic syndrome and drew upon a definition of *syndrome* from a 1995 dictionary of epidemiology (Last 1995), which states that what distinguishes syndromes from diseases is their lack of a clearly defined cause. The Federation notes:

> A syndrome is defined as a recognizable complex of symptoms and physical or biochemical findings for which a direct cause is not understood. With a syndrome, the components coexist more frequently than would be expected by chance alone. When causal mechanisms are identified, the syndrome becomes a disease.
>
> *Alberti et al. 2006: 473*

Currently, the National Library of Medicine's online medical dictionary defines a *syndrome* as "a group of signs and symptoms that occur together and characterize a particular abnormality."[3]

Despite its ontological status, the fact that so many people can be classified with metabolic syndrome has helped to establish a context where a range of biomedical, government, and corporate social actors have taken up metabolic syndrome in their research. According to an analysis of the 2003–2006 National Health and Nutrition Examination Survey (NHANES), a nationally representative study of adult populations in the United States, 34 percent of American adults could be classified as having metabolic syndrome (Ervin 2009). Global prevalence of metabolic syndrome ranges between 20 percent and 30 percent of adults (Grundy 2008). In 2000, the World Health Organization's *International Classification of Disease* (ICD-9) included an iteration of metabolic syndrome named "dysmetabolic syndrome X." In 2002, a group of biomedical researchers started The Metabolic Syndrome Institute, an independent and not-for-profit organization that is the first organization dedicated to the dissemination of knowledge of metabolic syndrome.[4] In 2003, a new academic journal, *Metabolic Syndrome and Related Disorders*, was established to publish research articles specifically on metabolic syndrome. Metabolic syndrome is also the subject of numerous medical books and monographs intended for the lay public (Reaven et al. 2000), physicians (Grundy 2005), mental health professionals (Mendelson 2008), and animal and biomedical researchers (Hansen and Bray 2008). In the years since, metabolic syndrome research has continued to accelerate in the United States and globally.

In this chapter, I argue that metabolic syndrome not only constitutes a new way of understanding the metabolism of bodies in technoscientific terms, it has also become a new way for technoscientific practices to be placed in service of racism – *technoscientific racism*. This term connotes the sense that the contemporary human sciences of molecular biology, genomics, and biomedicine reproduce race as a materially real object to be discovered through the investigation of bodily materials quite literally beneath the skin. It operates through the use of various biotechnologies on the body itself (diagnostic, experimental, medicinal, genetic, molecular, endocrine, reproductive, and so on) that are deployed to produce new knowledges of race and the body. Grounding race in the biology and/or genetics of human beings is an epistemological and technical precondition to claiming that racial inequalities can be explained by group-based differences in biology and/or genetics, or, what has been alarmingly called the return of scientific racism (Carter 2007, Duster 2003 [1990]).

Race and ethnicity are both socially constructed systems of categorization that are used to identify, group, and rank human beings, albeit based on different criteria. Race is a social category that emerged in the 1700s to classify individuals into so-called races based on presumed biological differences between population groups. Ethnicity is a social category that emerged in the 1920s to classify individuals into so-called ethnic groups based on presumed differences in culture, geographic origin, and ancestry (Weber 1978). Race and ethnicity are related in that ethnicity emerged in large part in response to early critiques of race concepts that were grounded in biology. Given this historical relationship, race and ethnicity are not interchangeable systems; however, there is meaningful overlap between what constitutes particular racial and ethnic groups. For example, African Americans are widely considered to be both a racial and an ethnic group. Race and ethnicity are both controversial systems of categorization, especially in the context of biomedical research, because individual biological and genetic differences do not fall neatly along the boundary lines circumscribed by racial and ethnic classifications. In other words, despite their shared origins in response to biological interpretations of human differences, race and ethnicity are social constructions whose meanings are shaped by historical and social circumstances (Roberts 2011).

In 1997, the United States Office of Management and Budget (OMB) provided the definitions of race and ethnicity that must be used in all biomedical and health policy research funded by the federal government (Office of Management and Budget 1997).[5] Consistent with the OMB guidelines, the sampling frame, analytic strategy, and research findings of metabolic syndrome research studies are often contextualized using these racial and ethnic categories. In this regulated scientific environment, it is also common to see published research focused exclusively on particular racial and ethnic groups. In this context, many researchers also reframe their research on particular racial groups using culturally-based categories of ethnicity or geography seemingly to avoid talking explicitly about race in ways that could be interpreted as racial bias, or worse, scientific racism (Bliss 2011, Fujimura and Rajagopalan 2011).

As a form of technoscientific racism, metabolic syndrome draws on and circulates racial meanings that construct race as a biological and genetic feature of bodies that can now be understood through technoscience. I situate the relationship between technoscience, racism, and metabolic syndrome in the context of the resurgence of what many commentators term the "reification," "biologization," or "geneticization" of race (Azoulay 2006, Duster 2005, Gannett 2004). This resurgence is fueled by the new institutional knowledge-making practices that signal how technoscience shapes societies. Because of long-standing historical conventions and current federal regulations on the use of race in biomedical research, statistical information about a body's race is also routinely collected along with anthropomorphic, biological, and genetic information about a body's metabolism. Then, by using this new concept of metabolic syndrome, researchers are positioned to argue that racial groups have unequal rates of metabolic health problems because metabolic processes are fundamentally different across racially categorized bodies. Metabolic syndrome folds group-based differences in the economic and political resources that are necessary for good metabolic health into a term that only references the biological processes within the body that create and use energy, a move that obscures the operation of institutionalized racism.

As a result of the implicit and explicit institutionalization of race, all racial and ethnic minority groups now seem to comprise high-risk populations that require permanent forms of metabolic examination, surveillance, and regulation. How, then, do race concepts that serve technoscientific racism become institutionalized in the production of knowledge of metabolic syndrome? One way to think about the puzzle that race presents for metabolic syndrome researchers is in terms of how to measure and interpret metabolic differences across individual bodies and population groups in ways that are consistent with prevailing cultural ideas about racial differences between bodies and populations. To examine this puzzle, I trace the production of racial meanings across three specific domains of metabolic syndrome science: (1) the use of race concepts to organize the measurement of metabolic syndrome; (2) the use of race concepts to analyze racially categorized bodies' "susceptibilities" to metabolic syndrome; and (3) the use of race concepts to construct explanatory theories of "genetic admixture" in racial disparities research on metabolic syndrome.[6] These mappings each contribute in turn to one of the substantive aims of this volume, namely, to understand how technoscience is institutionalized, embodied, and shaped by cultural values.

Theoretical frameworks

In order to understand how metabolic syndrome now serves as a site for technoscientific racism, it is important to situate the emergence of metabolic syndrome in the context of biomedicalization and the processes through which human metabolism came to be treated as a biomedical problem. Two interrelated theoretical approaches frame my analysis of the racial meanings that accompany metabolic syndrome: biomedicalization and technoscience. *Biomedicalization* is a historical and analytic framework for understanding the series of institutional, scientific, and technological processes that have transformed American biomedicine on multiple levels of social organization, especially since the mid-1980s (Clarke et al. 2003). Whereas medicalization refers to a process whereby social practices, bodily processes, and bodily materials were subsumed under the jurisdiction of clinical medicine (Starr 1982, Zola 1972), biomedicalization refers to the ways that medicalization itself is being transformed by an increasingly biological and technological approach to medicine and health.

As a social process, biomedicalization cannot be understood outside the relations of technoscience that constitute it. Technoscience represents the theoretical and material implosion of two terms, technology and science, and reflects the recursive ways through which technologies shape the practice of science and the meanings of scientific knowledge (Haraway 1997, Latour 1987). In terms of biomedicine, this framing of the relationship between science and technology requires analyzing the social practices by which biomedical discourses about the body are culturally and collectively produced by scientists and their use of biotechnologies (Haraway 1997, Jasanoff 2004, Oudshoorn 2002). Stated differently, bodies have to be manipulated to make them produce biomedical knowledge, and this embodied knowledge must be interpreted in the context of the technologies applied in that manipulation. This bodily manipulation occurs through the use of biotechnologies such as diagnostic tools, screening tests, drugs, and other regulatory devices that function as ways of reading the body as a kind of text. Thus, an important understanding of technoscience, and one shared by many scholars in science and technology studies (STS) (Orr 2006, Oudshoorn 2002) is that biomedical scientists gain cultural authority and manufacture scientific objectivity by concealing the institutionalized practices that construct and constitute their knowledge and the unequal power relationships in which those practices are embedded.

An important idea for analyzing how metabolic syndrome operates as a site of technoscientific racism within the contexts of biomedicalization and technoscience is the increasing emphasis on risk in biomedicine (Skolbekken 1995). The so-called risk factor paradigm has been the dominant theoretical framework for chronic disease epidemiology in the second half of the twentieth century (Susser 1998, Susser and Susser 1996). In the risk factor paradigm, researchers produce risk statistics from population-level surveillance data that show that particular variables, often conceptualized at the molecular level, are statistically associated with an undesirable health outcome. Analysts then interpret these population-level risk statistics as individual-level risk factors that, by virtue of their measurement as molecular processes, become understood as biologically meaningful causes of poor health at the individual level (Shim 2000, 2002). As researchers analyze surveillance data that have been collected using the population categories of race and ethnicity, these practices contribute to the construction of race as an individual-level cause of disease. Because variables for race and ethnicity are often statistically associated with undesirable health outcomes, race and ethnicity are often interpreted as individual-level risk factors for those health outcomes. Bodies marked with risk *as* race suggest that race itself becomes an indicator of risk.

In the sections that follow, I interpret metabolic syndrome as a biomedical project, an unfolding representation of bodily and population difference that continually draws upon racial meanings to make sense of human metabolic differences. Historically, the ways in which race informed metabolic syndrome was explicit, and, in other moments, it seems to be implicitly woven into the everyday practice of doing biomedical research in the United States. In the broadest terms, the specific approaches to race within metabolic syndrome research were consistent with the broader cultural understanding of race as a marker of difference in American society. Metabolic syndrome also became a new technoscientific object about which researchers could author genetic conceptions of race that serve to explain racial inequalities in metabolic health.

In the first section, I describe the emergence of metabolic syndrome and then show how early metabolic syndrome researchers' approaches to the study of human metabolism both explicitly and implicitly targeted bodies and populations based upon prevailing racial categorizations that marked the metabolic processes of whites as the standard against which other groups would be compared. Next, I examine how metabolic syndrome researchers aimed to establish the causation of metabolic syndrome in terms of race as a genetic concept. In the final section of the chapter, I use the notion of "special populations" to show how race concepts are used to construct explanatory theories of "genetic admixture" in racial disparities research on metabolic syndrome.

The racial measurement of metabolic syndrome

In the early 1920s, several European physicians were the first to document and publish research about the clustering of metabolic problems they observed in their patients and the potential risks such clustering could pose to metabolic health (Hitzenberger 1922, Kylin 1923, Maranon 1922). While none of these physicians explicitly codified a *metabolic syndrome*, they had similar ideas about how different metabolic processes worked together in the body. In his 1936 study of insulin action, endocrinologist H.P. Himsworth created the distinction between insulin sensitivity and insensitivity, the latter being most likely to precede and then accompany the development of type II diabetes (Himsworth 1936). Throughout the 1960s and into the 1970s, given the increasing proliferation of laboratories across the U.S., and the increasing availability of epidemiological data, more researchers gained access to the technologies and interpretive frameworks required to produce knowledge of risk-based syndromes. In the 1960s, there were several noteworthy contributions to the emergence of what came to be termed "the metabolic syndrome." First, in 1966, French researcher J. Camus suggested that gout, diabetes, and hyperlipidemia comprised "a metabolic trisyndrome" (Camus 1966). The following year in 1967, two Italian researchers advanced the notion of a "plurimetabolic syndrome" that included diabetes, obesity, and hyperlipidemia (Avogaro et al. 1967).[7] In 1968, Dutch researchers Mehnert and Kulmann published an article in a prominent Dutch medical journal about the relationships between hypertension and diabetes (Mehnert and Kuhlmann 1968).

In 1976, Gerald Phillips, drawing heavily on Vague's earlier work (which I discuss later), theorized that the "constellation of abnormalities" that comprised increased heart disease risk could be explained by sex hormones (Phillips 1978, Phillips et al. 2003). In 1977, three studies were published that each codified specific formations of "the metabolic syndrome" into the biomedical literature for the first time (Haller 1977, Singer 1977, Ziegler and Briggs 1977). A few years later, in 1981, two German researchers were also among the first to publish research on "the metabolic syndrome" (Hanefeld and Leonhardt 1981). The increasing technological focus on the clustering of a set of metabolic conditions in bodies formed the basis for the emergence of a litany of biomedical concepts that conceptualized metabolism as a systemic bodily process that could not be understood in terms of individual independent factors.

These early studies are silent on issues of race. None of the samples that formed the evidentiary basis for early studies of metabolic syndrome contained any visible racial minorities, and there was no reference to whether the observed clustering of metabolic abnormalities varied across population groups classified according to race (Hitzenberger 1922, Maranon 1922, Vague 1956). Thus, the white European body comprised the empirical data for the construction of early ideas about metabolic syndrome. In other words, the metabolism of the European body became the norm against which other bodies would be compared.

University of Marseilles physician Jean Vague is routinely cited as one of the primary originators of metabolic syndrome concept because of his investigation of the potential causal relationships between obesity, heart disease, and diabetes (Vague 1947, 1956).[8] Vague's core concept, the index of masculine differentiation, purportedly measures the likelihood of developing metabolic problems by interpreting anthropometric measurements of "the thickness of the fatty tissue on the surface of the body" through the prism of sex (Vague 1956: 20). Vague defines his construct through a binary logic based on sex, and he believed that the more "masculine" the bodily form, the higher the likelihood of heart disease. However, although the naked white skin of Vague's research subjects was published on the pages of the French medical journal *Presse Medicine*, race is not mentioned by Vague or any of his followers.

In contrast to this implicit racial image in Vague's research, race became an organizing principle of the earliest American population-based research on metabolic health. Beginning in the 1940s, the federal government took on a new role in monitoring the metabolic health of the population of the United States. While epidemiological research, like the Framingham Study, provided the empirical basis for biomedical information about risk factors for heart disease in white populations, it was not until the 1980s that the U.S. government began to fund studies on non-European population groups specifically in terms of metabolic health problems (Pollock 2012). This more explicit focus on race, and use of race to study large populations' health, was in part a response to smaller community studies of diabetes, heart disease, and stroke that showed rates of disease on the rise in communities of color beginning in the 1960s and 1970s (Williams and Collins 1995). This new focus on race, and new use of race to structure population research, was also a consequence of broader efforts to include racial and ethnic minorities in clinical and biomedical research (Epstein 2004, 2007).

Population studies were instrumental to the emergence of metabolic syndrome because they provided institutional mechanisms *by* which and a discursive framework *through* which conceptions of race and ethnicity could become attached to metabolic syndrome. Here, I highlight technical and racial frameworks of four of the earliest and most prominent of these federally funded studies, which were all modeled on the 1948 Framingham Study.[9] These four studies were significant theoretically and practically for three reasons: (1) they each included the specific examinations and laboratory tests necessary to classify metabolic syndrome; (2) they sampled and collected data from populations they conceptualized as racial; and (3) their data has been widely used to analyze metabolic syndrome and its connections to race.

The first study, the San Antonio Heart Study, 1979–1988, was a longitudinal cohort study that sampled 5,000 residents of three areas of San Antonio, Texas that were further sorted into low socioeconomic status (SES) "Mexican American", middle SES "Mexican and White," and high SES "White" research subjects (Gardner et al. 1982, Hazuda et al. 1981). The study was designed to identify factors beyond obesity that contribute to diabetes and cardiovascular risk in Mexican immigrants and Mexican Americans as compared to whites. The physical examination data collected in this study included "blood pressure, obesity, body fat distribution, [and] skin color, the latter to estimate percent Native American genetic admixture."[10] Measurements of insulin resistance were compared to skin color "to test the hypothesis that at any given level of adiposity Mexican Americans will be more insulin resistant than Anglos and that the insulin resistance in Mexican Americans is proportional to the degree of Native American ancestry."[11] The San Antonio Heart Study is important because it was the first major study after Framingham to measure all of the components of metabolic syndrome *and* to focus on a particular ethno-national group: Mexicans.

A second study, the Coronary Artery Risk Development in Young Adults Study (CARDIA), 1985–2006, was a prospective, longitudinal, multi-site, cohort study that sampled 5,115 black and white men and women aged 18–30 in Birmingham, Alabama, Chicago, Illinois, and Minneapolis, Minnesota (Hughes et al. 1987). The CARDIA Study has been used to evaluate the relationship between racial discrimination and blood pressure (Krieger and Sidney 1998), as well as the relationships between dairy consumption and the insulin resistance syndrome (Pereira et al. 2002). This study is significant because an explicit effort was made in the sampling strategy for CARDIA to achieve approximately balanced subgroups of race, gender, and education across age and geographic groups.[12] An analysis of the list of publications on the study website shows that as of February 2009 at least sixteen studies have used CARDIA data to analyze the relationship between metabolic syndrome and race.

The Atherosclerosis Risk in Communities Study (ARIC), 1987–1998, constituted a third study that was significant because it was designed to "investigate the etiology and natural history of atherosclerosis, the etiology of clinical atherosclerotic diseases, and variation in cardiovascular risk factors, medical care and disease by race, gender, and location."[13] ARIC was a prospective longitudinal study that sampled 15,792 individuals (aged 45–62) across Minneapolis, Minnesota, Washington County, Maryland, Forsyth County, North Carolina, and Jackson, Mississippi (Williams 1989) (Schmidt et al. 1996). According to the study's website, the ARIC data have been used to publish at least eighteen articles on metabolic syndrome, metabolic syndrome X, and multiple metabolic syndrome since the publication of its first wave of data in 1989.[14]

The fourth study is the Jackson Heart Study (JHS), 1987–2003, the largest prospective study ever of the "inherited (genetic) factors that affect high blood pressure, heart disease, strokes, diabetes and other important diseases in African Americans."[15] JHS initially began as one site of the aforementioned ARIC study. It sampled 6,500 African Americans, aged 35–84, living in Jackson, Mississippi (Taylor et al. 2008). According to the study description at the National Heart, Lung, and Blood Institute, the Jackson Heart Study included an extensive examination, including a questionnaire, physical assessments, and laboratory measurements of conventional and emerging risk factors that may be related to cardiovascular disease. The physical assessment of subjects in JHS includes height, weight, body size, blood pressure, electrocardiogram, ultrasound measurements of the heart and arteries in the neck, and lung function. The laboratory measurements collected from subjects in JHS includes cholesterol and other lipids, glucose, indicators related to clotting of the blood, among others. With these techniques, the Jackson investigators have been able to examine the "physiological relations between common disorders such as high blood pressure, obesity, and diabetes, and their influence on CVD."[16]

Waidzunas (see Chapter 3 of this book), analyzes how population studies serve as an important technoscientific site for working out competing meanings of sexual orientation and identity. The population studies I highlighted in this section provided a technoscientific means for inscribing the metabolic syndrome with racial meanings. By linking race, population, and metabolism in specific ways, these studies constituted an important body of evidence that was analyzed to document and explain racial group disparities in heart disease, diabetes, and obesity with reference to differential access to medical care and inequitably distributed exposure to stressful life circumstances like interpersonal racism. However, as in the case of the San Antonio Heart Study, these population studies also served as a platform for the theory that racial group differences in metabolic risk could be explained by genetic admixture, as understood through the prism of race as biologic difference. Thus, racial disparities in metabolic risks are simultaneously interpreted as one outcome of living in racially stratified societies and the effect of genetically differentiated populations sorted into and known through race.

Race as a cause of metabolic syndrome

While these and other studies produced the surveillance data used to study metabolic syndrome at the level of populations, clinical researchers continued to use racial categorization in their research on metabolic syndrome at the level of individual bodies. In fact, the data that emerged out of race-based population studies provided a basis upon which practicing physicians might understand their patients' metabolic health status differently depending on their specific racial classification. From the epidemiological perspective that shaped government funded race-based population studies, there was a need to understand whether and to what extent risks for metabolic health problems might differ across the major population groups of the nation. As will become apparent in this section, these questions about the *distribution* of metabolic health problems across racially categorized groups began to intersect with new theoretical questions about the *causes* of metabolic syndrome.

In addition to the four studies I have discussed, Stanford University endocrinologist Gerald Reaven's research is also useful for examining how scientists conceptualized race and ethnicity in relation to the population dynamics and individual-level causes of metabolic syndrome. Along with Jean Vague, Reaven is revered as a second so-called founder of metabolic syndrome. In his 1988 Banting lecture,[17] Reaven defined "syndrome X" as a series of six related variables that tend to occur in the same individual – resistance to insulin-stimulated glucose uptake, hyperglycemia, hyperinsulinemia, an increased plasma concentration of VLDL triglyceride, a decreased plasma concentration of HDL-cholesterol, and high blood pressure (Reaven 1988). Reaven did not note any explicit racial or ethnic distinctions in the syndrome X construct nor in the etiological theories that he proposed explained the relationships between insulin resistance, cholesterol, blood pressure, and heart disease risk (Reaven 1988).

However, a close examination of the clinical studies that formed the evidentiary basis for Reaven's theories about syndrome X reveal a pattern similar to Vague's ideas with respect to race. In his earlier research on insulin resistance during the 1970s, Reaven seems to have drawn upon mostly white European research subjects when he was part of a group of medical researchers at Stanford. Different members of the group (both including Reaven) published two studies in the *Journal of Clinical Investigation*, one in 1970 that tested a new technique for measuring insulin-mediated uptake (Shen et al. 1970), and another in 1975 that demonstrated that this new method of measuring insulin resistance tends to identify subjects with diabetes (Ginsberg et al. 1975). The descriptions of the sample, which contains people with diagnosed diabetes and those without diabetes, are different in each paper in one exceptional way. In the 1970 paper, the authors describe how the diabetics in the sample were selected from their patient referral group, matched by weight, age, and percent adiposity with the normal control group. In this brief passage, they describe the sampling procedure for the normal population paper:

> Normal individuals were selected after interviews with a group of volunteers who had recently been discharged from a local minimum-security prison. Volunteers responded to a notice asking for assistance in a research project which would furnish their living expenses during a 2 week hospital stay.
>
> *Shen, et al. 1970: 2151*

In the 1975 paper, the recently released inmates who likely participated in the study in order to get shelter are described simply and neatly as "healthy adult male volunteers." While neither study reveals nor refers to the race or ethnicity of its subjects, both the age and sex of each subject is noted in the printed tables. Without any evidence to the contrary, it is safe to assume that Reaven's subjects were predominantly white. This is so given two facts: the overwhelming number of white Americans in the U.S. in 1970, and the fact that the mass incarceration of black men had yet to begin, so the prison population was composed mainly of whites. By failing to consider how race mattered in the formulation of his signature concept, Reaven seemingly excludes the possibility that race is in any way related to the development of insulin resistance.

By 2000, Reaven explicitly links race and the syndrome, in a book titled *Syndrome X: overcoming the silent killer that can give you a heart attack*. He states that people with genetic abnormalities, people of non-European origin, people with a family history of diabetes, heart attack, and hypertension, and people who eat poorly and exercise little are at a much greater risk for developing syndrome X (Reaven et al. 2000: 20). To support this claim, Reaven cites three lines of evidence, two of which are drawn from research in which he participated, that treat race and ethnicity as constructs that identify disease-relevant genetic differences between groups.

For the first line of evidence, Reaven cites a 1985 study he co-authored that compared fifty-five Pima Indian men living near Phoenix with thirty-five Caucasian men living in California (Bogardus et al. 1985).[18] The investigators measured the levels of obesity, physical fitness, and insulin resistance in the two groups (who are not explicitly labeled as racial groups in any way) and used statistical techniques to determine the degree to which differences in their levels of obesity and physical fitness contributed to the variability of their insulin action. Reaven, writing in 2000, claims that this 1985 study showed that "half of the variability of insulin action was due to lifestyle, *the other half presumably to our genes*. Of the 50 percent attributed to lifestyle, half was due to fitness, half to obesity" (Reaven et al. 2000: 57). Here, Reaven and his co-authors claim that the other half was due to population differences in genetics. This contention is based on the assumption that comparing Pima Indians and Europeans is equivalent to comparing underlying genetic differences between these population groups.

The second line of genetic evidence upon which Reaven draws also comes from research conducted on the same sample of Pima Indians (Lillioja et al. 1987). This study compared levels of insulin resistance within Pima families to levels of insulin resistance across families and demonstrated, again according to Reaven in 2000, that the clustering of insulin action is greater within families than it is across families. In effect, this claim constructs familial heritability and genetic susceptibility as the same biomedical phenomenon when it plays out within a tribal group known to have high rates of intermarriage.

The third line of evidence that Reaven cites to substantiate the role he sees for genetics in causing syndrome X purportedly shows that American Indians, South Asian Indians, Japanese Americans, African Americans, Mexican Americans, Australian Aboriginals, and various Pacific

Islander populations are more insulin resistant than populations of people of European ancestry (Reaven et al. 2000: 57). Reaven does not cite any studies after making these claims, but instead inserts a parenthetical statement that crystallizes his ideas about the causes of racial difference: the observed differences in insulin resistance reflect genetic differences between racial groups. Reaven and his collaborators argue that while it is *possible* that some racial groups might be more insulin resistant because of lifestyle habits and other factors, several studies *did take* group differences in all known factors into account and conclude that differences in insulin resistance result from heritable genetic differences between groups (Reaven et al. 2000: 58). Thus, the evidence that Reaven cites to support the claim that non-European bodies are more likely to develop syndrome X than European bodies *assumes* that bodily differences in insulin resistance result from heritable genetic differences between racial groups that cannot be explained by other ostensibly non-racial factors. This form of essentialism positions race as a heritable genetic structure that governs the functioning of the endocrine system. The work that essentialism accomplishes with respect to race in this case is analogous to Waidzunas' account in this book. He shows that scientists working to oppose the sexual reorientation movement locate sexual orientation within the body, thereby standardizing sexuality as a fixed biological feature of bodies. This essentialist move precludes ways of understanding sexual fluidity that cannot be known through the rigid examination of the biological body.

Special metabolic populations

In the contemporary moment, the uses and conceptions of race in biomedical research on metabolic syndrome seemingly have expanded in new ways. These expansions have taken place through the increasing interaction between new forms of clinical biomedicine and government public health research, both of which are focused on documenting and explaining racial health disparities. Due to these converging forces, there is no lack of biomedical data about race and health in American biomedicine. The term *special populations* is used specifically within government biomedicine to refer to pregnant women, children, racial and ethnic minorities, the elderly, and any other population group that is not white/European and male. The examination of racial health disparities constitutes one rationale for studying special populations that are sampled and targeted using constructions of race and ethnicity. People who are classified with metabolic syndrome or who think they have it comprise a new special population that is constructed out of and reproduces genetic meanings of race. In this final section, I use the notion of "special populations" to show how race concepts are used to construct explanatory theories of "genetic admixture" in racial disparities research on metabolic syndrome.

Since 2001, scientists have increasingly raised questions about the use, measurement, and interpretation of metabolic syndrome when comparing different racial and ethnic populations. These new questions about the relationship between race and metabolic syndrome emerge from the enactment of several important technoscientific practices on these populations. First, since the World Health Organization recommended standardizing obesity measurements in different racial and ethnic groups first in 1997 (WHO 1997) and again in 2004 (WHO 2004), race and ethnicity have explicitly been used in the practice of validating group-specific empirical cutoff points (endpoints) for one of the main physical examinations that comprise metabolic syndrome. Because of the ways that some definitions of the syndrome use race to determine the statistical cut-points of obesity, these definitions classify different proportions of racial and ethnic minority groups with the syndrome. For example, the prevalence of metabolic syndrome in Mexican Americans varied up to 24% between the WHO and NCEP definitions of the syndrome (Kahn et al. 2005: 2291). The argument for using race-based endpoints is that they improve the generalizability and validity of comparisons of disease risk across individuals and populations. Statistical validity is determined with respect to the outcome, metabolic syndrome, by evaluating whether the syndrome successfully identifies all of the individuals who are at increased risk within specific populations groups. Thus, for example, the body mass index for an individual who is classified "African American" would be statistically adjusted to account for the differential relationship between obesity and CVD risk in African Americans as compared to other groups. These standardizations construct valid statistical norms against which racial and ethnic populations can be compared to one another. As Waidzunas argues convincingly in this volume, processes of standardization are often organized via logics of exclusion that privilege some bodies over others and work to reinforce the hierarchical social order that justifies such exclusions.

A second technoscientific practice concerns how different research institutions use ethnicity to construct standardized cut-points for obesity in their definitions of metabolic syndrome across racially conceptualized subpopulations. These cut-points represent statistical thresholds against which bodies' sizes are measured in metabolic syndrome research. In 2003, in their joint definition of the insulin resistance syndrome, the American Association of Clinical Endocrinologists and the American College of Endocrinology provided optional standardizations of obesity for different ethnic groups (Einhorn 2003). Three years later, in 2006, the International Diabetes Federation incorporated racial and ethnic measurements of waist circumference because "…there are clear differences across ethnic populations in the relationship between overall adiposity, abdominal obesity, and visceral fat accumulation" (Alberti et al. 2006: 473). The IDF elaborates a list of country of origin and ethnicity-specific values for waist circumference for "Europids," "South Asians," "Chinese," and "Japanese" populations. Several other groups do not yet have their own standardized values: "Ethnic South and Central Americans," "Sub-Saharan Africans," and "Eastern Mediterranean and Middle East." In the meantime, the authors advocate that the South and Central American ethnic groups should use "South Asian" values, the Africans and the "Arab populations" should use "European" values until "more specific data are available." The authors of the IDF study provide special instructions for applying these country and ethnic specific values in clinical and epidemiological research. They write,

> It should be noted that the ethnic group-specific cut-points should be used for people of the same ethnic group, wherever they are found. Thus, the criteria recommended for Japan would also be used in expatriate Japanese communities, as would those for South Asian males and females regardless of place and country of residence.
>
> *Alberti et al. 2006: 476*

These recommendations imply that these standardizations should be used to compare "ethnic" populations that transcend "place and country of residence." This stated emphasis on ethnic populations obscures questions of race and assumptions about bodily differences that accompany race. To name these populations "ethnic groups" but then to sidestep the country-specific cultural dynamics that shape bodies' response to metabolic environmental exposures reveals that this use of ethnicity functions to deflate the potential criticism that these standardized cut-points are unabashedly racial and potentially racist.

Since 2005, these institutional practices have resulted in a new body of biomedical research that investigates the implications of using metabolic syndrome to compare heart disease risk across different racially categorized groups (Banerjee and Misra 2007, Unwin et al. 2007). Scholars in this emerging field of research have investigated racial and ethnic differences in the relationships between obesity and heart disease risk (Zhu et al. 2005), body composition and metabolic risk factors (Desilets et al. 2006), the power of triglycerides to predict insulin resistance (Bovet et al. 2006, Sumner and Cowie 2008, Sumner et al. 2005), and the relationship between HDL cholesterol levels and CVD risk (Amarenco et al. 2008) in explicitly racial terms.

African Americans, and theories of African American health, occupy a prominent place in special populations research that links race and metabolic syndrome. A review article on metabolic syndrome in African Americans was published in the journal *Ethnicity & Disease* in 2003. All of the authors of this review article are members of the African-American Lipid and Cardiovascular Council (AALCC), a non-profit health professional advisory group that is supported through an unrestricted educational grant from Bristol-Myers Squibb Company, and many of them have published widely on metabolic syndrome and African Americans (Clark and El-Atat 2007, Ferdinand et al. 2006, Grundy et al. 2004, Smith et al. 2005).

In the leading article, Hall et al. (2003) situate their review of metabolic syndrome and African Americans in the context of the epidemiological fact that African Americans have the highest overall CHD mortality and out-of-hospital coronary death rates of any racial group in the United States. However, to explain the racial disparities in metabolic health between "Native Americans," "Mexican Americans," and "African Americans" *as compared to* "European Americans," the group advances a "genetic admixture theory" (Hall et al. 2003: 415).[19] Theories of genetic admixture assume that individual level risk for disease is related to their shared genetic admixture with populations known to be susceptible to the disease. According to this theory, pre-1960s European Americans historically had higher rates of diabetes than African Americans, Hispanics, and Native Americans, but increasing racial miscegenation that has occurred since colonialism explains the increasing rates of diabetes in these racial and ethnic groups (Tull and Roseman 1995: 614). The central assumption of this theory is that racial groups at an earlier moment were pure and segregated, and it is their intermingling since the "discovery" of race that explains racial disparities in modern times. They argue that the degree of genetic admixture is related to the susceptibility of different racial groups to the risk factors that constitute metabolic syndrome. They write,

> Whites of European origin appear to have greater predisposition to atherogenic dyslipidemia [high levels of LDL or bad cholesterol], whereas Blacks of African origin are more prone to HBP [high blood pressure], type 2 diabetes and obesity. Native Americans and Hispanics are less likely to develop HBP than Blacks, but appear particular susceptible to type 2 diabetes. Of particular note is the considerable genetic admixture among Native Americans and Mexican Americans.
>
> *Hall et al. 2003: 415*

As Duana Fullwiley makes clear in her analysis of how scientists use this logic of genomic admixture to make sense of Latinos' development of asthma (Fullwiley 2008), the new biotechnological tools used by these scientists mobilize old logics of racial difference in ways that affirm racial categorization and obscure the cultural imprint of the scientists on the knowledge-making process. The focus on race within special metabolic populations research on metabolic syndrome has served as a location for the application of new technologies that map old racial categories onto groups defined by "genetic ancestry" (Fujimura and Rajagopalan 2011). Genetic admixture analysis becomes a way of knowing racial identity as well as a strategy for determining metabolic risks. The focus on documenting racial differences in metabolic syndrome within biomedicine has accompanied the survival of theories of disease that attempt to link the biologies of racial groups to metabolic risk.

Conclusion

The frameworks of biomedicalization and technoscience provide a set of powerful analytic tools through which to analyze the relationships of metabolic syndrome and race. Technoscience provides a way of understanding how the increasingly technological and scientific aspects of metabolic syndrome come together to shape its racial meanings. In the political context of biomedicalization, molecular processes and genetic differences provide the authoritative scientific explanations for racial differences in health; racism has nothing to do with it. By grounding race in the body along with a predominantly biomedical understanding of metabolism (as opposed to a sociological or political one), these technoscientific practices have the unintended effects of obscuring the ways in which racialized social structures produce poor health for all people.

These technoscientific practices remain linked to historical formations of scientific racism that explained racial inequalities as biological, natural, and immutable. Racial essentialism is central to the operation of scientific racism, a set of discourses and practices that served to explain and justify racial inequalities using the tools and authority of science. Racial health disparities have long constituted a major site of struggle over the meaning of race and explanations of racial inequality. When analyzed in the specific context of racial health disparities, the use of essentialist notions of race to explain away racial inequality takes on special significance in the history of comparative racial biology and eugenics (Zuberi 2001).

While metabolic syndrome can be understood as yet another iteration of racialized practices that have drawn substantial scrutiny like racialized forms of pharmacology and drug marketing (Kahn 2008, Lee 2005), metabolic syndrome has thus far avoided serious discussion among STS scholars, especially in terms of the ways in which it explicitly and implicitly incorporates race into its very definition. One of the unique features of metabolic syndrome, as compared to these other objects of biomedical research, is that its practitioners do not address the ways in which it is both racialized and racializing, a remarkable omission given how central racial disparities in metabolic health have become to national debates on health injustice and the warnings about the return of scientific racisms through technoscience.

Notes

1 I use the term the metabolic syndrome as an umbrella term to encompass many different yet closely related concepts advanced by biomedical researchers to describe these relations including the metabolic syndrome, metabolic syndrome X, dysmetabolic syndrome X, insulin resistance syndrome, multiple metabolic syndrome, and syndrome X.

2 This particular definition of metabolic syndrome represents the 2001 National Cholesterol Education Program (NCEP) definition and includes elevated blood pressure, or hypertension, is defined as having systolic pressure of at least 140 mmHg and diastolic pressure of at least 90 mmHg, elevated cholesterol, or dyslipidemia, is defined as having total serum cholesterol higher than 240, elevated blood sugar, or hyperglycemia, is defined as having fasting blood glucose of at least 126 mg/dL and elevated weight, or obesity, is defined as having a body mass index (BMI) greater than 30.

3 Interestingly, the National Library of Medicine online definition of "syndrome" is drawn from the Merriam Webster online dictionary (*www2.merriam-webster.com/cgi-bin/mwmednlm*) accessed on March 5, 2009 at 4:25pm.

4 (*www.metabolicsyndromeinstitute.com/about/mission*) accessed on March 5, 2009 at 4:15pm.

5 These categories were further institutionalized through the 1993 Revitalization Act of NIH that issued strict guidelines for the inclusion of women and racial and ethnic minorities in NIH funded clinical research and trials (see also Bliss, C. 2011, "Racial taxonomy in genomics." *Social Science & Medicine* 73: 1019–1027, Epstein, S. 2007, *Inclusion: The Politics of Difference in Medical Research*. Chicago: University of Chicago Press, Shields, A.E., Fortun, M., Hammonds, E.M., King, P.A., Lerman, C.,

Rapp, R., and Sullivan, P.F. 2005, "The use of race variables in genetic studies of complex traits and the goal of reducing health disparities." *American Psychologist* 60: 1 77–103.).

6　Theories of genetic admixture assume that an individual's susceptibility for disease is proportionally related to their inherited genetic admixture from ancestral populations known to be susceptible to the disease.

7　In 1993, this construct gets revived by Descovich and colleagues in a book edited by Crepaldi himself (Descovich, G.C., Benassi, B., Canelli, V., D'Addato, S., De Simone, G., and Dormi, A. 1993, "An epidemic view of the plurimetabolic syndrome." In *Diabetes, Obesity, and Hyperlipidemia: The Plurimetabolic Syndrome*, Crepaldi, G., Tiengo, A., and Manzato, E. Amsterdam: Elsever Science.).

8　Researchers at the Metabolic Syndrome Institute, a web-based organization of biomedical researchers whose primary goal is to promulgate the idea of the metabolic syndrome, attribute the concept to Dr. Vague. Several prominent metabolic syndrome researchers belong to this group, including Dr. Scott Grundy (*www.metabolic-syndrome-institute.org/medical_information/history/#lien_a*) accessed December 20, 2006.

9　Several other studies illustrate the general argument presented here. See, for example: *MESA* (Multiethnic Study of Atherosclerosis) – Bild, D.E., Bluemke, D.A., Burke, G.L., Detrano, R., Diez Roux, A.V., Folsom, A.R., Greenland, P., Jacob, D.R.J., Kronmal, R., Liu, K., Nelson, J.C., O'Leary, D., Saad, M.F., Shea, S., Szklo, M., and R.P.T. 2002, "Multi-ethnic Study of Atherosclerosis: objectives and design." *American Journal of Epidemiology* 156: 9 871–81. and *IRAS* (Insulin Resistance and Atherosclerosis Study) Festa, A., D'Agostino, R., Jr., Howard, G., Mykkanen, L., Tracy, R.P., and Haffner, S.M. 2000. "Chronic subclinical inflammation as part of the insulin resistance syndrome: The Insulin Resistance Atherosclerosis Study (IRAS)." *Circulation* 102: 1 42–47.

10　I will discuss the theory of genetic admixture in greater detail in the following section of the chapter, however, it is important to mention that the focus on Native American admixture is commonplace given the exceptionally high rates of insulin-resistant diabetes in some Native American population like the Pima.

11　(*www.clinicaltrials.gov/ct/show/NCT00005146*) retrieved on February 13, 2009.

12　(*www.cardia.dopm.uab.edu/lad_info.htm*) accessed February 16, 2009 at 2:08pm.

13　(*www.cscc.unc.edu/aric*) accessed February 16, 2009 at 2:51 pm.

14　See fn1 on terminology for "metabolic syndrome."

15　(*http://jhs.jsums.edu/jhsinfo*) accessed February 16, 2009 at 3:23 pm.

16　(*www.nhlbi.nih.gov/about/jackson/2ndpg.htm*) accessed February 16, 2009 at 2:59 pm.

17　The Banting Lecture is published annually in the journal *Diabetes*, which is the flagship journal of the American Diabetes Association. As of March 4, 2013, Reaven's published lecture had been cited 8,241 times (search conducted by author).

18　According to an NIDDK website on the special role the Pima have played in government biomedical research on diabetes, "This cooperative search between the Pima Indians and the NIH began in 1963 when the NIDDK (then called the National Institute of Arthritis, Diabetes and Digestive and Kidney Diseases), made a survey of rheumatoid arthritis among the Pimas and the Blackfeet of Montana. They discovered an extremely high rate of diabetes among the Pima Indians. Two years later, the Institute, the Indian Health Service, and the Pima community set out to find some answers to this mystery." (*http://diabetes.niddk.nih.gov/dm/pubs/pima/pathfind/pathfind.htm*) accessed February 16, 2009.

19　Recall that the San Antonio Heart Study was also designed to assess the degree of genetic admixture.

References

Alberti, K., Zimmet P., and Shaw J. 2006. "Metabolic syndrome – a new worldwide definition: A Consensus Statement from the International Diabetes Federation." *Diabetes Medicine* 23: 5 269–480.

Amarenco, P., Labreuche P., and Touboul P.-J. 2008. "High-density lipoprotein-cholesterol and risk of stroke and carotid atherosclerosis: A systematic review." *Atherosclerosis* 196: 489–496.

Avogaro, P., Crepaldi G., Enzi G., and Tiengo, A. 1967. "Associazione di iperlipidemia, diabete mellitoe obesita di medio grado." *Acta Diabetol Lat* 4: 36–41.

Azoulay, K.A. 2006. "Reflections on race and the biologization of difference." *Patterns of Prejudice* 40: 4 353–379.

Banerjee, D. and Misra A. 2007. "Does using ethnic specific criteria improve the usefulness of the term metabolic syndrome? Controversies and suggestions." *International Journal of Obesity* 31: 1340–1349.

Bild, D.E., Bluemke, D.A., Burke, G.L., Detrano, R., Diez Roux, A.V., Folsom, A.R., Greenland, P., Jacob, D.R. J., Kronmal, R., Liu, K., Nelson, J.C., O'Leary, D., Saad, M.F., Shea, S., Szklo, M., and Tracy, R.P. 2002. "Multi-ethnic study of atherosclerosis: objectives and design." *American Journal of Epidemiology* 156: 9 871–881.

Bliss, C. 2011. "Racial taxonomy in genomics." *Social Science & Medicine* 73: 1019–1027.

Bogardus, C., Lillioja, S., Mott, D.M., Hollenbeck, C., and Reaven, G. 1985. "Relationship between Degree of obesity and invivo insulin action in man." *American Journal of Physiology* 248: 3 E286–E91.

Bovet, P., Faeh, D., Gabriel, A., and Tappy L. 2006. "The prediction of insulin resistance with serum triglyceride and high-density lipoprotein cholesterol levels in an East African population." *Archives of Internal Medicine* 166: 11 1236–1237.

Camus, J. 1966. "Gout, diabetes, and hyperlipidemia: a metabolic trisyndrome." *Rev Rhum Mal Osteoartic* 33: 1 10–14.

Carter, R. 2007. "Genes, genomes and genealogies: The return of scientific racism." *Ethnic and Racial Studies* 30: 4 546–556.

Clark, L.T. and El-Atat, F. 2007. "Metabolic syndrome in African Americans: Implications for preventing coronary heart disease." *Clin Cardiol* 30: 4 161–164.

Clarke, A.E., Mamo, L., Fishman, J.R., Shim, J.K., and Fosket, J.R. 2003. "Biomedicalization: Technoscientific transformations of health, illness, and U.S. biomedicine." *American Sociological Review* 68: 2 161.

Descovich, G.C., Benassi, B., Canelli, V., D'Addato S., De Simone, G., and Dormi, A. 1993. "An epidemic view of the plurimetabolic syndrome." In *Diabetes, Obesity, and Hyperlipidemia: The Plurimetabolic Syndrome*, edited by G. Crepaldi, A. Tiengo and E. Manzato. Amsterdam: Elsever Science.

Desilets, M.C., Garrel, D., Couillard, C., Tremblay, A., Despres, J.P., Bouchard, C., and Delisle, H. 2006. "Ethnic differences in body composition and other markers of cardiovascular disease risk: Study in matched Haitian and white subjects from Quebec." *Obesity* 14: 6 1019–1027.

Duster, T. 2003 [1990]. *Backdoor to Eugenics*. New York: Routledge.

Duster, T. 2005. "Race and reification in science." *Science* 307: 5712 1050–1051.

Einhorn, D. 2003. "American College of Endocrinology Position Statement on the Insulin Resistance Syndrome." *Endocrine Practice* 9: 3 236–239.

Epstein, S. 2004. "Bodily differences and collective identities: The politics of gender and race in biomedical research in the United States." *Body and Society* 10: 2/3 183–203.

Epstein, S. 2007. *Inclusion: the politics of difference in medical research*. Chicago: University of Chicago Press.

Ervin, R.B. 2009. "Prevalence of metabolic syndrome among adults 20 years of age and over, by sex, age, race, and ethnicity, and body mass index: United States, 2003–2006", in National Health Statistics Reports, No.13. Hyattsville, MD: National Center for Health Statistics.

Ferdinand, K., Clark, L., Watson, K., Neal, R., Brown, C., Kong, B., Barnes, B., Cox, W., Zieve, F., Ycas, J., Sager, P., and Gold, A. 2006. "Comparison of efficacy and safety of rosuvastatin versus atorvastatin in African-American patients in a six-week randomized trial." *American Journal of Cardiology* 97: 2 229–235.

Festa, A., D'Agostino, R., Jr, Howard, G., Mykkanen, L., Tracy, R.P., and Haffner, S.M. 2000. "Chronic subclinical inflammation as part of the insulin resistance syndrome: The Insulin Resistance Atherosclerosis Study (IRAS)." *Circulation* 102: 1 42–47.

Fujimura, J.H. and Rajagopalan, R. 2011. "Different differences: The use of 'genetic ancestry' versus race in biomedical human genetic research." *Social Studies of Science* 41: 1 5–30.

Fullwiley, D. 2008. "The biologistical construction of race: 'Admixture' technology and the new genetic medicine." *Social Studies of Science* 38: 695–735.

Gannett, L. 2004. "The biological reification of race." *British Journal for the Philosophy of Science* 55: 2 323–345.

Gardner, L., Stern, M., Haffner, S., Relethford, J., and Hazuda, H. 1982. "Diabetes, obesity and genetic admixture in Mexican-Americans: The San-Antonio Heart-Study." *American Journal of Epidemiology* 116: 3 559–559.

Ginsberg, H., Kimmerling, G., Olefsky, J.M., and Reaven, G.M. 1975. "Demonstration of insulin resistance in untreated adult onset diabetic subjects with fasting hyperglycemia." *Journal of Clinical Investigation* 55: 3 454–461.

Grundy, S.M. 2005. *Contemporary Diagnosis and Management of the Metabolic Syndrome*. Newton, PA: Handbooks in Health Care.

Grundy, S. M. 2008. "Metabolic sydnrome pandemic." *Arteriosclerosis, Thrombosis, and Vascular Biology* 28: 629–636.

Grundy, S.M., Cleeman, J.I., Merz, C.N.B., Brewer, H.B., Jr., Clark, L.T., Hunninghake, D.B., Pasternak, R.C., Smith, S.C., Jr., and Stone, N.J. 2004. "Implications of recent clinical trials for the National Cholesterol Education Program Adult Treatment Panel III Guidelines." *Circulation* 110: 2 227–239.

Hall, W.D., Wright J.T., Wright E., Horton W., Kumanyika, S.K., Clark, L.T., Wenger, N.K., Ferdinand, K.C., Watson K., Flack J.M. 2003. "Metabolic Syndrome in African Americans: A review." *Ethnicity & Disease* 13(4): 414–428.

Haller, H. 1977. "Epidemiology and associated risk factors of hyperlipoproteinemia." *Z Gesamte Inn Med* 32: 8 124–128.

Hanefeld, M. and Leonhardt, W. 1981. "Das metabolische syndrom (the metabolic syndrome)." *Dt Gesundh-Wesen* 36: 545–551.

Hansen, B.C. and Bray, G.A. (eds) 2008. *The Metabolic Syndrome: Epidemiology, Clinical Treatment, and Underlying Mechanisms.* Totowa, NJ: Humana Press.

Haraway, D.J. 1997. *Modest_Witness@Second_Millennium. FemaleMan_Meets_OncoMouse.* New York, NY: Routledge.

Hazuda, H.P., Stern, M.P., Gaskill, S.P., Hoppe, S.K., Markides, K.S., and Martin, H.W. 1981. "Ethnic and social-class differences relating to prevention of heart-disease. The San-Antonio Heart-Study." *American Journal of Epidemiology* 114: 3 418–418.

Himsworth, H.P. 1936. "Diabetes mellitus: A differentiation into insulin-sensitive and insulin-insensitive types." *Lancet* 1: 127–130.

Hitzenberger, K. 1922. "Uber den Blutruck bei Diabetes Mellitus." *Weiner Arch Innere Med* 2: 461–466.

Hughes, G.H., Cutter, G., Donahue, R., Friedman, G.D., Hulley, S., Hunkeler, E., Jacobs, D.R., Liu, K., Orden, S., Pirie, P., Tucker, B., and Wagenknecht, L. 1987. "Recruitment in the Coronary-Artery Disease Risk Development in Young-Adults (Cardia) Study." *Controlled Clinical Trials* 8: 4 S68–S73.

Jasanoff, S. 2004. *States of Knowledge: The Co-production of Science and the Social Order.* London: Routledge.

Kahn, J. 2008. "Exploiting race in drug development: BiDil's interim model of pharmacogenomics." *Social Studies of Science* 38: 737–758.

Kahn, R., Buse, F.E., and Stern M. 2005. "The metabolic syndrome: Time for a critical appraisal: Joint statement from the American Diabetes Association and the European Association for the Study of Diabetes." *Diabetes Care* 28(9): 2289–2304.

Krieger, N. and Sidney, S. 1998. "Racial discrimination and blood pressure: The CARDIA Study of Young Black and White Adults." *American Journal of Public Health* 86: 10 1370–1378.

Kylin, E. 1923. "Studien uber das Hypertonie-Hyperglykemie-Hypoerurikemie syndrome." *Zentrablatt fur Innere Medizin* 7: 44 105–127.

Last, J.M. (ed.) 1995. *A Dictionary of Epidemiology.* Oxford: Oxford University Press.

Latour, B. 1987. *Science in Action: How to Follow Scientists and Engineers through Society.* Cambridge, MA: Harvard University Press.

Lee, S.S.-J. 2005. "Racializing drug design: Implications of pharmacogenomics for health disparities." *American Journal of Public Health* 95: 12 2133–2138.

Lillioja, S., Mott, D.M., Zawadzki, J.K., Young, A.A., Abbott, W.G.H., Knowler, W.C., Bennett, P.H., Moll, P., and Bogardus, C. 1987. "Invivo insulin action is familial characteristic in nondiabetic Pima-Indians." *Diabetes* 36: 11 1329–1335.

Maranon, G. 1922. "Uber Hyperonie and Zuckerkrankheit." *Zentrablatt fur Innere Medizin* 43: 169–176.

Mehnert, H. and Kuhlmann, H. 1968. "Hypertonie und Diabetes Mellitus." *Deutsches Medizinisches Journal* 19: 16 567–571.

Mendelson, S.D. 2008. *Metabolic Syndrome and Psychiatric Illness: Interaction, Pathophysiology, Assessment, and Treatment.* San Diego: Academic Press.

Office of Management and Budget 1997 "Standards for maintaining, collecting, and presenting federal data on race and ethnicity." *Federal Register* 62: 58781–58790.

Orr, J. 2006. *Panic Diaries: A Genealogy of Panic Disorder.* Durham: Duke University Press.

Oudshoorn, N. 2002. *Beyond the Natural Body: an archeology of sex hormones.* London: Routledge.

Pereira, M.A., Jacobs, D.R., Jr, Van Horn, L., Slattery, M.L., Kartashov, A.I., and Ludwig, D.S. 2002. "Dairy consumption, obesity, and the insulin resistance syndrome in young adults: The CARDIA Study." *Journal of the American Medical Association* 287: 16 2081–2089.

Phillips, G.B. 1978. "Sex hormones, risk factors and cardiovascular disease." *American Journal of Medicine* 65: 7–11.

Phillips, G.B., Jing, T.J., and Heymsfield, S.B. 2003. "Relationships in men of sex hormones, insulin, adiposity, and risk factors for myocardial infarction." *Metabolism* 52: 784–790.

Pollock, A. 2012. *Medicating Race: Heart Disease and Durable Preoccupations with Difference.* Durham and London: Duke University Press.

Reaven, G.M. 1988. "Banting lecture 1988: Role of insulin resistance in human disease." *Diabetes* 37: 1595–1607.

Reaven, G.M., Strom, T.K., and Fox, B. 2000. *Syndrome X: The Silent Killer: The New Heart Disease Risk.* New York: Simon & Schuster.

Roberts, D. 2011. *Fatal Invention: How Science, Politics, and Big Business Re-create Race in the 21st Century.* New York: The New Press.

Schmidt, M.I., Duncan, B.B., Watson, R.L., Sharrett, A.R., Brancati, F.L., and Heiss, G. 1996. "A metabolic syndrome in whites and African-Americans: The Atherosclerosis Risk in Communities baseline study." *Diabetes Care* 19: 5 414–418.

Shen, S.W., Reaven, G.M., and Farquhar, J.W. 1970. "Comparison of impedance to insulin-mediated glucose uptake in normal subjects and in subjects with latent diabetes." *Journal of Clinical Investigation* 49: 12 2151–2160.

Shields, A.E., Fortun, M., Hammonds, E.M., King, P.A., Lerman, C., Rapp, R., and Sullivan, P.F. 2005. "The use of race variables in genetic studies of complex traits and the goal of reducing health disparities." *American Psychologist* 60: 1 77–103.

Shim, J.K. 2000. "Bio-power and racial, class, and gender formation in biomedical knowledge production." *Research in the Sociology of Health Care* 17: 175–95.

Shim, J.K. 2002. "Understanding the routinised inclusion of race, socioeconomic status and sex in epidemiology: The utility of concepts from technoscience studies." *Sociology of Health & Illness* 24: 2 129.

Singer, P. 1977. "Diagnosis of primary hyperlipoproteinemias." *Z Gesamte Inn Med* 32: 9 129–133.

Skolbekken, J. 1995. "The risk epidemic in medical journals." *Social Science & Medicine* 40: 3 291–305.

Smith, S.C., Daniels, S.R., Quinones, M.A., Kumanyika, S.K., Clark, L.T., Cooper, R.S., Saunders, E., Ofili, E., and Sanchez, E.J. 2005. "Discovering the full spectrum of cardiovascular disease: Minority Health Summit 2003: Report of the Obesity, Metabolic Syndrome, and Hypertension Writing Group." *Circulation* 111: 10 e134–139.

Starr, P. 1982. *The Social Transformation of American Medicine: The Rise of a Sovereign Profession and the Making of a Vast Industry.* New York: Basic Books.

Sumner, A.E. and Cowie, C.C. 2008. "Ethnic differences in the ability of triglyceride levels to identify insulin resistance." *Atherosclerosis* 196: 696–703.

Sumner, A.E., Finley, K.B., Genovese, D.J., Criqui, M.H., and Boston, R.C. 2005. "Fasting triglyceride and the triglyceride-HDL cholesterol ratio are not markers of insulin resistance in African Americans." *Archives of Internal Medicine* 165: 12 1395–1400.

Susser, M. 1998. "Does risk factor epidemiology put epidemiology at risk? Peering into the future." *Journal of Epidemiology & Community Health* 63: 608–611.

Susser, M. and Susser, E. 1996. "Chosing a future for epidemiology: II. From Black Box to Chinese Boxes to Eco-Epidemiology." *American Journal of Public Health* 86: 5 674–677.

Susser, M. and Susser, E. 1996. "Chosing a future for epidemiology: I. Eras and Paradigms." *American Journal of Public Health* 86: 5 668–673.

Taylor, H., Liu, J., Wilson, G.T., Golden, S.H., Crook, E., Brunson, C., Steffes, M., Johnson, W., and Sung, J. 2008. "Distinct component profiles and high risk among African Americans with metabolic syndrome: the Jackson Heart Study." *Diabetes Care* 31: 6 1248–1253.

Tull, E.S. and Roseman, J.M. 1995. "Diabetes in African Americans." In *Diabetes in America*, 2nd Edition, NIDDK National Diabetes Data Group, Bethesda, Maryland: National Institutes of Health: 613–630.

Unwin, N., Bhopal, R., Hayes, L., White, M., Patel, S., Ragoobirsingh, D., and Alberti, G. 2007. "A comparison of the new international Diabetes Federation definition of metabolic syndrome to WHO and NCEP definitions in Chinese, European and South Asian Origin Adults." *Ethnicity & Disease* 17: Summer 522–528.

Vague, J. 1947. "La différenciation sexuelle, facteur déterminant des formes de l'obésité." *Presse Medicine* 30: 3 39.

Vague, J. 1956. "The degree of masculine differentation of obesities: A factor determining predisposition to diabetes, atherosclerosis, gout and uric calclous disease." *American Journal of Clinical Nutrition* 4: 20–34.

Weber, M. 1978. *Economy and Society: An Outline of Interpretive Sociology.* Berkeley and Los Angeles: University of California Press.

WHO. 1997. *Obesity: Preventing and Managing the Global Epidemic.* Geneva: World Health Organization.

WHO. 2004. "Appropriate body-mass index for Asian populations and its implications for policy and intervention strategies." *Lancet* 363: 157–163.

Williams, O.D. 1989. "The Atherosclerosis Risk in Communities (Aric) Study: Design and Objectives." *American Journal of Epidemiology* 129: 4 687–702.

Williams, D.R. and Collins, C. 1995. "U.S. socioeconomic and racial differences in health: Patterns and explanations." *Annual Review of Sociology* 21: 349–386.

Zhu, S., Heymsfield, S.B., Toyoshima, H., Wang, Z., Pietrobelli, A., and Heshka, S. 2005. "Race-ethnicity-specific waist circumference cutoffs for identifying cardiovascular risk factors." *American Journal of Clinical Nutrition* 81: 409–415.

Ziegler, V. and Briggs, J. 1977. "Tricyclic plasma levels: Effect of age, race, sex, and smoking." *Journal of the American Medical Association* 14: 238 2167–2169.

Zola, I.K. 1972. "Medicine as an institution of social control." *Sociological Review* 20: 487–504.

Zuberi, T. 2001. *Thicker Than Blood: An Essay on How Racial Statistics Lie.* Minneapolis: University of Minnesota.

3

Standards as "Weapons of Exclusion"

Ex-gays and the materialization of the male body

Tom Waidzunas

TEMPLE UNIVERSITY

Since the 1970s, the ex-gay movement has brought together professional therapists, ministry leaders, and people struggling with "unwanted same-sex attractions." Ex-gays in the United States are predominantly male, and the movement tends to attract white Evangelical Christians. In many ex-gay movement worldviews, "gender shame," an underlying fear of being one's assigned gender, is often thought to cause homosexuality, and confronting it is supposed to allow the expression of universally natural heterosexual feeling and behavior. These ideas are generally rejected in mainstream science as being based on stereotypes. Nonetheless, practitioners guiding clients through this process often blend theological and psychological concepts, theorizing that becoming heterosexual means aligning oneself with a vision of God's design for men and women (Moberly 1983). Although the movement in the United States grew in recent decades, it has recently undergone serious setbacks as Exodus International, the leading ex-gay ministry in the U.S., disbanded in June 2013.

The movement began to fracture in 2012, the result of a division between ministry leaders, who claimed that leaving homosexuality meant lifelong struggle with same-sex attractions, and professional therapists, who claimed that full reorientation was possible. These divisions can be traced, in part, to a standard established by the American Psychological Association (APA) pertaining to scientific measurement. In a task force report published in 2009, the APA joined other professional mental health associations declaring that there is no evidence for the efficacy of sexual orientation change efforts, and these efforts are potentially harmful (APA Task Force 2009). To make these claims, the APA established a terminological standard that undermined the validity of the self-reports of ex-gays claiming to have become heterosexual as a basis of reorientation research. In the past, "sexual orientation" had been measured as a composite of identity, behavior, and attraction. The APA made a new distinction between "sexual orientation" as a set of physiological attractions and "sexual orientation identity" as the willingness or ability to recognize one's sexual orientation. In effect, the APA's new standards meant that self-reports of sexual orientation became nothing more than an expression of sexual orientation identity, unacceptable as evidence for the efficacy of sexual reorientation therapies. Moreover, in order to demonstrate reorientation, a physiological measure was now deemed necessary.

Through rhetorical calls for physiological tests, the APA enacted what Michelle Murphy has described as a process of "materialization," making matter – in this case the matter of male bodies – relevant and perceptible in the world (Murphy 2006: 7). But at the same time, establishing notions of fixed sexual orientations based on the male body has a number of problematic consequences, including the erasure of female desire and the preclusion of conceptualizing multiple possible ways of being sexual.

Many recent difficulties faced by the ex-gay movement are linked to the APA standards through a complex and layered temporal process, since one standard may be based upon another (Lampland and Star 2009). For example, the APA's claims of "no evidence for efficacy" and "potential for harm," substantiated by the new standards, form an important basis for a new California law that bans reorientation therapy for minors, setting a new legal standard for therapeutic practice. Because the terminological standards became a basis for raising standards of methodology in reorientation research, they effectively debunked a sexual reorientation efficacy study conducted by influential psychiatrist Robert Spitzer (2003), leading him to write an apology to the gay community and to claim that he had misinterpreted his findings (Spitzer 2012). Thus, although remnants of the ex-gay movement persist, the APA standard has acted as a means for excluding ex-gay claims from scientific, public policy, and cultural arenas. In the past, ex-gay claims had been excluded from science primarily on the ethical basis that they were anti-gay, but terminological standards have provided a new reasoning for exclusion that seems to have even undermined the concept of reorientation in many contexts beyond science alone.

The terminological shift described in this chapter must be seen as a *standardization process* that makes concrete particular ways of seeing, manipulating, and accounting for the world (Timmermans and Epstein 2010). The terminological standardization of "sexual orientation" and "sexual orientation identity" has emerged through a dialectic encounter between the ex-gay movement, including experts predominantly relegated to the scientific fringe, and gay affirming mental health professionals in the scientific mainstream. Thus, these debates must be understood as part of a broader set of religious and political contestations, not as walled-off separate processes that took place "inside" science. One important result of standardization is that it has enabled mental health professional associations to engage in a type of boundary work (Gieryn 1999): by maintaining the view that sexual orientation cannot be therapeutically changed, they have effectively marginalized those reorientation practitioners and researchers who believe that attractions can be altered. To trace these terminological standardization processes, I draw on interviews with major players in sexual reorientation therapy controversies in the United States, including members of the APA Task Force, participant observation at relevant conferences, and content analysis of scientific and activist literature.

More broadly, this chapter extends Timmermans and Epstein's framework for the sociological study of standardization by further elaborating on ways that standards can serve to exclude. They assert that "every standard necessarily elevates some values, things, or people at the expense of others, and this boundary-setting can be used as a weapon of exclusion" (Timmermans and Epstein 2010: 83). Being "sensitive to exclusions" was also a major concern of Bowker and Star (1999), who laid a foundation for the study of standards as a family of processes including classification and infrastructure development. Bowker and Star argue that standardized classification systems "always have other categories, to which actants (entities or people) who remain effectively invisible to the scheme are assigned" (1999: 325). Such exclusions become part of the social order, yet in a case such as struggles over sexual reorientation, the domains from which ex-gays and associated experts have been excluded are complicated by the religious culture of the United States and the role of the media in promoting controversy. That is, while reorientation opponents within science have been able to keep reorientation claims out of science, they have had a more difficult time extending that closure to society more broadly.

By focusing on processes of exclusion, I am by no means advocating opposition to standardization, as though it is always oppressive and dehumanizing. Rather, this chapter elaborates on ways that exclusion has operated in cases of standardization found in the science and technology studies (STS) literature, further supplementing understandings of standardization as means of coordination or promoting efficiency for those included. As standards embody values, exclusion through standardization may be necessary to promote justice or some other cherished goal, although it may have ironic effects, too. In the sections that follow, I first discuss the relationship between standards and social life, summarize Timmermans and Epstein's framework for the sociological study of standardization, and review some ways in which standards can exclude. Then, I tell the story of how the standards of "sexual orientation" and "sexual orientation identity" have come about using that framework. Next, I discuss the case in terms of exclusion as the discrediting of people and practices, and then analyze how standards can exclude in the sense that they can preclude possibilities for being and knowing. Finally, I conclude by discussing the limits of considering standards as "weapons of exclusion," as they rely on implementation and institutional authority.

Standards, social life, and exclusions

The power of standards, whether they be classification systems, rules for the production of objects, or guidelines for professional practice, lie in the ways they render the world uniform across time and space. Much of modern life would not be possible without standards, as they coordinate systems and infrastructure, and render people intelligible and subject to regulation by the modern state. Standards are not only just about the mundane and substantive things under consideration; rather, they entail ways of organizing people and things that are extremely consequential for personal biographies. Quite often, standards remain invisible, as a backdrop within infrastructures that we take for granted. Bowker and Star assert that the design of standards is always rooted in ethical judgments, and rendering them visible is a project with an "inherently moral and political agenda" (Bowker and Star 1999). This not only brings them to life from their fossilized state where they might be considered negotiable again, but also brings attention to those entities or people who do not fit into standardized systems, allowing us to reconsider the broader consequences of exclusions.

In some cases, the study of standards does not require rendering them visible, as they have already become targets of wide public contestation. The removal of "Homosexuality" from the *DSM* in 1973 is a case in which social movements successfully drew attention to the values embedded in a nosological standard, resulting in its demise (Bayer 1987). In another case, the U.S. Bureau of Reclamation planned to construct the Orme Dam near Phoenix, Arizona, flooding land of the Yavapai tribe. The Yavapai contested a commensuration process which subjected their land and relocation to a cost-benefit analysis within a standardized system of dollar value, claiming instead that their land was part of their identity. Their advocacy made the values inherent in a standardization process visible, and effectively halted the dam project through protest (Espeland 1998). These examples show not only how visibility is a key aspect of standard contestation, but also illustrate how standards are part of the organization of social life.

Science and technology studies have disagreed about the relationship between standardized classification systems and the social order. Bloor (1982) and Douglas (1986) both assert that standardized systems of classification are really projections of political values onto the natural world, such that the natural world can be read back in a way that makes social values seem inevitable and embedded in nature. On the other hand, Bowker and Star (1999) have argued, and Timmermans and Epstein (2010) concur, that standardization and the social order are simultaneously emergent, as standards themselves have agency in the social domain. This perspective aligns with Jasanoff's idiom of co-production (2004), as the creation and implementation of standards and the development of social order are mutually constitutive phenomena. For example, Epstein (2007) observes that legal requirements for the standardized inclusion of racial minorities in clinical trials that embed race within medical knowledge simultaneously shape the lived order of racial classification.

Like Bowker and Star, Timmermans and Epstein agree that standardization is "a process of constructing uniformities across time and space through the generation of agreed upon rules" (Timmermans and Epstein 2010: 71). As such, a "standard" might be a rule, a definition, or a measurement, developed in order to make the unruly world more predictable and manageable. Timmermans and Epstein provide a typology of standards including: (1) design standards, which define features of tools and products in infrastructure; (2) terminological standards, which "ensure stability of meaning over different sites and times and are essential to the aggregation of individual elements into larger wholes"; (3) performance standards, which set acceptable levels of complications or problems with systems; and (4) procedural standards, which express how procedures are to be performed (2010: 72). Rather than characterizing standardization processes as part of a grand trend of rationalization, these authors note that each standard has its own history, and must be evaluated in terms of its benefits and disadvantages. To analyze the life cycle of a standard, they propose three archetypal phases, which may overlap: creation, implementation, and outcome. Standard creation is a social act requiring the buy-in of multiple parties, and may proceed in a top-down fashion or be based on the formation of consensus. Once created, a standard must be implemented to survive, and is often reliant on some institutional authority for that to occur. The outcomes of standards may promote a range of interests from the most democratic to the most authoritarian, which must be empirically investigated. In sum, the Timmermans and Epstein approach involves considering a broad range of actors and outcomes, including considering those actors excluded, and acknowledging the full range of complexities and contingencies in any particular standardization process.

Timmermans and Epstein assert that standards can be "weapons of exclusion," yet this idea raises questions about what exactly is being excluded, by what mechanisms, and from what domains. In this chapter I consider two families of exclusions. The first involves the discrediting of people, practices, or claims, effectively excluding them from some socially legitimated arena. The second involves precluding possibilities for being or knowing altogether, as ontological and epistemological exclusions that may undermine what Stephen Collier and Andrew Lakoff (2008) call "regimes of living." Such regimes are defined as "tentative and situated configuration[s] of normative, technical, and political elements that are brought into alignment in situations that present ethical problems" (2008: 23). Thus in one family of exclusions, people or things might be kept out of some arena, but in the other family, the very existence of kinds of people may be overlooked, ways of knowing may be made impossible, and regimes of living may be disassembled or reconfigured.

When considering the first family of exclusions in relation to the domain of science, the attribution of credibility to people, claims, and methods is largely shaped by culture (Shapin 1995). In this context, Gieryn (1999) explores how the cultural boundaries between science and non-science are maintained through the setting of standards for scientific expertise, a major component for attributing scientific credibility to people. This involves delineating appropriate professional qualifications and the standards of practice such as participation in peer review to draw the lines between scientists and non-scientists. Research practices themselves may also be discredited, resulting in the crystallization of a "hierarchy of evidence" in which one method may be considered a gold standard for the production of knowledge, to the exclusion of others.

The second family of exclusionary forms involves the preclusion of possibilities for being and knowing. When taking a standardization path, whether it entails standardizing measurements, technologies, or categories, there are always myriad roads not taken (including having no standard at all). Each road leads to a possible regime of living, including an assemblage of standards and ways of being human. In the case of standardized classification systems, Bowker and Star (1999) argue that once a path has been chosen, people who do not fit the standard will often experience distress. They use the term "torque" to describe biographies of people cast into residual categories, as tension builds between the unruliness of lived experience and the rigidness of categories.

In addition to precluding ways of being, when particular category systems become standardized in infrastructures for producing knowledge, they can lead to the preclusion of different ways of knowing. Bowker and Star (1999) argue that terminological standards can establish "causal zones" that allow for construction of a constrained range of kinds of facts. For example, defining an illness in the *International Classification of Diseases (ICD)* in terms of biological causes only allows for the collection of some kinds of data instead of others. As a result, biological agents will come to be understood as the only possible causes of illness disparities, as opposed to social disenfranchisement or marginalization. The problem of the reduction of the range of causal zones to biological properties is also present in research on biomedicine and race. In this book, Hatch shows how the standardized diagnosis of "metabolic syndrome" was consolidated through the practice of population research studies predicated on socially defined racial categories. This research has the effect of reducing causal understanding of these conditions to racial biology, not taking into account the role of social determinants of health. Hatch's case also illustrates how "niche standardization" (Epstein 2007) has taken hold in research on metabolic syndrome, as different bodily thresholds have been assigned to different racial groups for diagnosis and risk assessment, further entrenching a reductionist biological understanding of race.

Finally, while standards often involve the creation of ethical norms in a regime of living, especially for following the standards themselves, they can also relegate previously normal behavior to the domain of abnormality. For example, the profit-based pharmaceutical industry has increasingly promoted the ethic to assess health risk, subjecting oneself to numerous tests and taking lifelong drug regimens even when feeling well. The threshold of normalcy is now based on test scores and risk assessments developed through clinical trials, and to be normal in developed countries is often to be on multiple preventive prescription drugs. Gone are the days when patients went to the doctor simply when they were not feeling well (Dumit 2012). Thus, as the pharmaceutical industry has gained increasing influence over the medical profession, and as medicine is increasingly based on evidence-based standards of care (Timmermans and Berg 2003), older styles of doctor-patient relationships have been replaced with ones technologically equipped to enforce higher thresholds of normalcy. This example shows how standards can reassemble a regime of living by constricting the range of what is considered "normal," supplanting it with a new regime based on a different rationality.

While I have described two families of ways that standards can exclude, it should be noted that this distinction frequently applies to *components* of an exclusionary process that cannot be disaggregated. For example, Fausto-Sterling (2000) describes how the International Olympic Committee, concerned that men might compete in women's sporting events, shifted from a system of determining sex by visual inspection to one based on genetic testing. Within the first family, this standardization process excludes the technical practice of visual inspection, and also keeps people who test positive for Y-chromosomes from participating in women's events. In other words, certain people and practices are kept out of the domain of women's Olympic events due to the new standard. Within the second family of exclusions, this standard also precludes the possibility that a person with Y-chromosomes could officially be considered a "woman" competing in the Olympics, and also precludes the possibility of knowing how such people might fare in women's Olympic events.

With this conceptual terrain of exclusions established, I now apply the Timmermans and Epstein framework to the terminological standardization of "sexual orientation" and "sexual orientation identity," tracing the formation of the standard and then further discussing forms of exclusion that ensued.

The ex-gay threat and the terminological standardization response

The process of terminological standardization began when the ex-gay movement developed a public presence in the late 1990s. This was propelled by large Religious Right organizations that funded a mass advertising campaign and by media coverage of newly conducted self-report research studies. Ex-gay groups challenged mainstream science claims that there is "no evidence" for reorientation efficacy by evoking representations of the self that Tanya Erzen (2007) has described as "testimonial politics." Testimonies took different forms in each of these genres. Advertisements in newspapers and on billboards featured ex-gays telling personal stories of transformation, while researchers utilized self-report data in which ex-gays testified about the extent to which they believed they were now heterosexual within the confines of carefully phrased research questions and response scales. Three particularly prominent large-scale ex-gay self-report studies emerged from the movement at this time.

The studies are important to consider in some detail because of the different ways that they advance notions of sexual orientation change. First, Nicolosi et al. (2000) conducted a self-report "consumer satisfaction" survey published in 2000 with 882 respondents. This has come to be known as the "NARTH study" because of its authors' affiliation with that organization, National Association for Research and Therapy of Homosexuality. Using retrospective self-report ratings of "sexual orientation" before and after reorientation treatment but without defining that term, the authors asserted that respondents experienced statistically significant shifts from homosexuality to heterosexuality, in addition to improvements in overall well-being. In contrast to allowing respondents to define "sexual orientation" themselves, Spitzer (2003) defined this variable using a composite of measures of identity, behavior, and attraction, also with retrospective measures. While his sample was not representative, he claimed his study showed that reorientation was possible for some highly motivated individuals. Many people in his sample reported significant shifts toward feeling heterosexual attractions in addition to identity and behavior. Finally, a study by evangelical researchers Stanton Jones and Mark Yarhouse (2007) used a longitudinal self-report design which they argued improved on previous studies' usage of retrospective measures. This came to be known as the "Exodus study," as Exodus International funded it and was the primary source of research subjects. Jones and Yarhouse also measured identity, behavior, and attraction upon entering an ex-gay ministry and then one year later. However, in this study, ex-gays tended to report little shift in sexual attraction, but large shifts in identity and behavior. While one third of the cases were considered to have been "successfully" reoriented by the authors, half of these subjects were celibate, and most of the "success" cases reported lingering same-sex attractions.

In sum, while the meaning of sexual orientation change remained ambiguous in the NARTH study, Spitzer claimed that the significant attraction changes of his subjects demonstrated reorientation was possible, while Jones and Yarhouse claimed successful reorientation with much less change in attractions in their data. Thus, it was Spitzer's study that made the strongest claim about the efficacy of reorientation therapies. His study garnered immense publicity because he had been on the American Psychiatric Association Nomenclature Committee in 1973, played a central role in the decision to remove "Homosexuality" from the *DSM* (Bayer 1987), and had also been central in the development of later versions of the *DSM*. Thus, his credibility as a supporter of gay rights and famous leader in psychiatry made him a powerful spokesperson for the ex-gay movement.

According to Steven Hilgartner, the popularization of a fringe view can have an effect of "feeding back" on mainstream science, moving researchers forward by pressuring them to clarify concepts and assertions (Hilgartner 1990). While such a feedback dynamic occurred in the case of the APA Task Force responding to popularized ex-gay research and publicity campaigns in 2007, events leading up to the Task Force did not play out in exactly the way that Hilgartner might anticipate. Instead, the development of terminological standards involved a complex convergence process that drew in participants from both sides of the debate. In Timmermans and Epstein's terms, this convergence involved buy-in from multiple parties. Although some activists and professionals maintained stalwart positions on the strict efficacy or inefficacy of therapies, some began converging on the idea that a mismatch between same-sex attractions and heterosexual identity might be typical and even desirable for some interested in living in accordance with religious values. Ex-gay ministry leaders, less invested in the complete elimination of same-sex attractions than more secular professional therapists, began claiming that such attractions are likely to linger for people who go through ex-gay programs.

In part, these confessions were a response to the increasing public visibility of "ex-ex-gays," people who had gone through ex-gay programs, claimed that they failed, and often claimed they experienced some kind of harm. In the early 2000s, the visibility of ex-ex-gays was furthered by research conducted by Shidlo and Schroeder (2002), concluding that many ex-gays experience significant harms when their treatments do not work. In 2007 the "ex-gay survivor" movement held its first national meeting in the United States. Rather than challenging the testimonies of these activists, ex-gay ministries were now forthcoming about the existence of lingering same-sex attractions. Ex-gay ministries lowered expectations of change for new members in hopes of preventing people from leaving ministries and becoming outspoken "ex-ex-gays" or "ex-gay survivors."

However, reorientation ministries were also confronted with critical calls for phallometric testing, both from anti-reorientation activists (for example, Besen 2003) and from scientists responding to the Spitzer study (for example, Beckstead 2003). This method involves measuring the erection level of the penis with a "penile plethysmograph" device while a male research subject views erotic imagery designed to represent homosexuality and heterosexuality. The fact that ex-gays became more forthcoming about lingering same-sex attractions in the wake of these calls suggest that this technique, like the polygraph (Alder 2007), may act as a "truthing technology," calling forth confessions as long as people believe in the methodology. Phallometric testing has become part of a family of technologies that might be considered "truth machines," techniques for extracting truth from the body, such as DNA testing, which may trump testimonial evidence in various contexts (Lynch et al 2008). While a vaginal photoplethysmograph device exists for women, it is considered a less reliable indicator of sexual orientation, as researchers argue that women's subjective and physiological arousal do not correspond (Fishman 2004). Moreover, because men are the predominant clients of reorientation, discussions in this domain frequently use male sexuality as representative of all human sexuality with no qualification.

At the same time as ministry leaders became more forthcoming about lingering same-sex attractions, gay affirmative therapists began to promote "middle path" compromises for clients experiencing conflict between same-sex attraction and religious values. A special issue of the journal *The Counseling Psychologist* published in 2004 captured many of these efforts. Psychologists were particularly moved by working with clients or research subjects that experienced profound conflict, and in some cases, pursuing a heterosexual identity despite their attractions became a client's chosen path. For example, Haldeman (2004) reported on a case in which an African-American male client married to a woman chose a heterosexual identity because he wished to keep his heterosexual family together and maintain his connections to his conservative religious community despite his same-sex attractions. While Haldeman suggests alternative possibilities for such a client, including the possibility that he might come back to therapy in the future, or seek help from evangelical groups that work to reconcile religion and same-sex sexual orientation, he acknowledges that the polarized options of gay affirmative therapy that might devalue religious experience, on the one hand, and sexual reorientation, on the other, are insufficient alternatives. In a commentary in this special issue, Worthington (2004), who would later serve on the APA task force, also proposed that sexual orientation identity be severed from sexual orientation as a means to clarify the language used in discussions over sexual reorientation therapy research.

Furthermore, research on ex-gays emanating from both sides of the debate suggested that reorientation changed attractions very little while behavior and identity did change. This was particularly the case with the Exodus study conducted by evangelical researchers Jones and Yarhouse. Psychologist A. Lee Beckstead, who would later become a member of the APA task force, had conducted an in-depth interview study of Mormons who attempted reorientation and argued that ex-gays experienced little change in attractions, even though they did experience some benefits from reorientation such as reduced isolation. Moreover, he argued that the heterosexual attractions that they felt were particularly weak. One respondent claimed that his response to men was like a "forest fire," while the response to his wife is more of a "campfire" involving a sense of emotional closeness (Beckstead and Morrow 2004). Beckstead, who had worked in a phallometric laboratory as part of his training, argued that these kinds of characterizations of heterosexual attraction do not constitute sexual orientation change, but rather, reveal that living as an ex-gay involves living with incongruity. Beckstead had himself gone through a failed reorientation attempt at Evergreen International, a Mormon reorientation ministry. Being part of this world gave him particular insight into the ways that religious values can be more important to a person than their sexual attractions, and his role on the task force involved foregrounding issues in the psychology of religion (interview with Beckstead, Salt Lake City, UT, 2010).

With this convergence of some parties on both sides, the publication of the American Psychological Association's *Task Force Report on Appropriate Therapeutic Response to Sexual Orientation* in 2009 marks the established terminological standardization of "sexual orientation" and "sexual orientation identity." While other mental health organizations had created position statements opposing reorientation, the APA had become a gatekeeper on the construction of knowledge of reorientation efficacy, especially because the American Psychiatric Association had largely ceded jurisdiction over talk therapy to psychology following the pharmaceutical revolution in that field. The task force was formed in 2007 after gay rights activists raised concerns with the APA about the growing public presence of the ex-gay movement. In making their determination that "no evidence" exists for reorientation, the task force reviewed relevant literature over six decades, and establishing terminological standards effectively made self-report the "wrong tool for the job" (Clarke and Fujimura 1992) of measuring sexual orientation.

Building on these standards, they also established a "middle path" compromise therapeutic approach they called "sexual orientation identity exploration." This therapeutic guideline advises mental health practitioners working with clients experiencing conflicts over same-sex attractions to let those clients determine the path of their sexual orientation identity development, as long as it is not based on anti-gay stereotypes, and as long as clients acknowledge that sexual orientation is unlikely to change (APA Task Force 2009).

Exclusion as discrediting claimants and practices

Within the first family of exclusions that involve undermining credibility, the APA standardization process has led to several outcomes. These entail the discrediting of reorientation practitioners who claim they can change sexual attractions, ex-gays who claim that they now have a heterosexual sexual orientation, and self-report as a means to demonstrate reorientation change. As Timmermans and Epstein note, the different phases of standard creation, implementation, and outcomes may overlap, and in this case, overlaps can be seen in the way NARTH nominees were excluded from the task force itself.

When the APA announced that a task force would be formed in 2007, NARTH put forth four nominees. These included evangelical researchers Stanton Jones and Mark Yarhouse, authors of the Exodus study, and Joseph Nicolosi and A. Dean Byrd, both former NARTH presidents and authors of the NARTH study. Despite their experience conducting ex-gay research, all four were rejected. Task force chair Judith Glassgold explained that Jones and Yarhouse could not be on the task force primarily because their flawed research undermined their scientific credibility. In addition to problems with their statistics and sampling methods, she claimed that Jones and Yarhouse also had errors in how they define sexual orientation in their work. While the problem of definitions was also a basis for excluding Nicolosi and Byrd, Glassgold claimed that those nominees were rejected primarily because they could not likely adhere to the APA Code of Ethics, which requires affirming gay people as equal to heterosexuals (interview with Glassgold, Washington D.C., 2010). Furthermore, the nominees' reorientation research had not been published in adequately peer-reviewed journals. Jones and Yarhouse's study had been published in the religious Intervarsity Press, while Nicolosi et al.'s study was published in *Psychological Reports,* considered by many psychologists to be a "pay to publish" journal (telephone interview with psychologist Gregory Herek 2009).

Nicolosi, charismatic former president of NARTH, protested this exclusion, claiming that because the task force was composed of "activists in gay causes, most of whom are publicly self-identified as gay," they could not be objective; they had already made a personal commitment to the idea that sexual orientation is fixed (Nicolosi 2009). This claim was matched by Glassgold arguing that Nicolosi could not be objective due to his anti-gay views. Thus, the terminological standardization of "sexual orientation" and "sexual orientation identity" and the outcome of undermining the credibility of reorientation researchers using self-report methodology occurred in a simultaneous process. Furthermore, these negotiations reveal that embedded within this terminological standard is an ethical commitment to affirmative equality on the basis of sexual orientation.

In its review of the literature, the APA task force also contributed to the formation of a hierarchy of evidence that favored physiological testing over self-report. Phallometry had been used in the 1960s and 1970s in attempts to condition arousal through aversion therapy, including experiments in which electric shocks and noxious chemical odors were applied to subjects as they viewed same-sex erotic imagery (for example, McConaghy 1969). At that time, behavior therapists found they could diminish same-sex arousal but could not induce heterosexual arousal. Because it had been a means of demonstrating therapeutic failure back then, phallometric testing had gained some credibility among contemporary reorientation opponents. The task force pointed to this genre of research as the only reorientation research with any validity, effectively elevating phallometry over self-report as a measurement tool in this context. In interviews three members of the APA Task Force, including the chair, psychologist Judith Glassgold, psychologist A. Lee Beckstead, and psychiatrist Jack Drescher, endorsed of the need for phallometric testing for male ex-gays. Glassgold stated, "I think if you wanted to have a real empirical study, you hook people up with a plethysmograph… That would be the only way to study sexual orientation change. Everything else is just sexual orientation identity" (interview with Glassgold 2010). By citing phallometric studies as the best evidence that reorientation does not work, the APA rendered male bodies perceptible in a process of materialization.

While these factors point toward a standardization of phallometry in sexual reorientation measurement, this standardization has not coalesced. Even the task force described the limitations of the technique, including high numbers of men who cannot become aroused in a laboratory setting, discrepancies between types of measurement devices, and the possibility of faking arousal (APA Task Force 2009: 31). Whether to use a Kinsey-type continuum scale or some kind of orthogonal scale measuring different dimensions of attraction, such as emotional attachment and sexual desire, also remains in question. Complicating matters further, Beckstead (2012) has recently argued that measures of aversion to heterosexuality or homosexuality should be included in addition to attraction. Some psychologists, including prominent sexuality researcher Gregory Herek, remain skeptical of conducting any reorientation efficacy research at all (personal interview with Herek 2008). Nonetheless, the terminological severing of "sexual orientation" and "sexual orientation identity" undermined self-report methods in the context of reorientation research even if a gold standard did not fully come into being.

Exclusion as precluding possibilities of being and knowing

A second family of exclusions involves ontological and epistemological foreclosures, as various ways of being and knowing can become unworkable due to standards. Creating a terminological standard which locates sexual orientation as a fixed entity within the body is useful for opposing anti-gay politics (Waites 2005). This move bears many similarities to the severing of "gender" and "sex," which has been subject to feminist critique for foreclosing a number of ways of being (for example, Butler 1990). As a form of essentialism, fixing sexual orientation within the body also designates biology as a "causal zone" in determining sexualities, precluding some ways of being sexually fluid as well as ways of knowing human malleability.

Second wave feminists utilized the sex/gender dichotomy to argue that the malleability of the cultural construct of gender proved that inequality between the sexes was contingent and could be changed. However, Butler (1990) has argued that gender does not merely map onto a fixed sex binary, but rather, the cultural construction of sex, seen in the surgeries on intersex infants, is based on gender first and foremost. That is, gender is not mapped onto sex, but rather, binary sex is a product of gender, resulting in the exclusion of intersex categories. This critique of the sex/gender dichotomy applies to sexual orientation identity/sexual orientation as well, as the contemporary notion of "sexual orientation" relies on an essentialist binary of sex for its very existence. Like "gender" of second wave feminism, identities of "gay," "lesbian," "bisexual," and "heterosexual," predicated on binary sex, are themselves performative enactments constituting divisions among human categories and erasing other experiences and possibilities such as intersexuality, or being attracted to a person who does not fit into the categories of "male" or "female." In other words, notions of fixed sexual orientation based on attractions to sexed object choices reinscribe the delineation of binary biological sexes. Further extending Butler's logic to the newly standardized dichotomy, "sexual orientation" categories delimited within scientific experiments, even those based on physiology, cannot exist without a cultural system of "sexual orientation identity" – a set of cultural understandings of what constitutes "gay," "straight," etc. – that precedes them. In phallometric testing, for example, sexual orientation identity categories are built into the test through the selection of erotic imagery. As such, cultural constructs of sexual orientation identities, and the "sexual orientations" produced by them, are predicated on cultural practices of gender division (Butler 1990, see also Epstein 1991).

By invoking behaviorist studies using phallometric testing as the best evidence that reorientation does not work, and using this as a basis for the terminological standards of "sexual orientation" and "sexual orientation identity," a notion of male sexuality has effectively been used as a stand-in for the definition of all human sexuality. Female sexuality is often theorized in science as being primarily emotion based, with emotional connection leading to sexual desire, and women's sexuality is often understood to be more fluid because of this. While the APA does include both dimensions of "sexual and emotional" attraction in its terminological standard of sexual orientation, it is evidence of male sexual attraction that has become the primary basis for ascertaining the properties of "sexual orientation." Emphasis on the phallometric technique, based on a narrow understanding of male sexuality (as defined by erection prompted by visual erotic imagery), potentially devalues a range of physical and emotional alternate bases for defining how human sexuality operates, and contributes to the erasure of female sexual subjectivity.

Even when the fluidity of female sexuality has been theorized in science, a male-centered notion of fixity has still been used as a foundational basis, especially given the need to maintain the idea that sexual orientation is "fixed" in public policy contexts. Psychologist Lisa Diamond (2008) has developed a model of women's sexuality in which women are understood to have an underlying fixed sexual orientation, but a layer of fluidity rides on top of this fixed foundation. However, Diamond's model, like all models of human sexuality that theorize fixed sexual orientation, effectively excludes sexual subjectivities characterized by spontaneous fluidity without any fixed sexual orientation foundation. Critical of how essentialism can preclude our understanding of human plasticity, Fausto-Sterling claims that she lived "part of her life as an unabashed lesbian, part as an unabashed heterosexual, and part in transition" (2000: ix). This is not a possible narrative within the current regime of fixed sexual orientation, even with considering a "layer" of fluidity as Diamond has theorized. Thus, while notions of fixed sexual orientation undermine the therapeutic reorientation of people into heterosexuality, they come with a cost of undermining possibilities of conceptualizing spontaneous fluidity of sexual orientation that may occur in either direction. Indeed, the idea that sexual fluidity is a layer on top of a fixed sexual orientation inverts and even precludes Freudian theory, in which a person's inherent "polymorphous perversity" is later repressed into a fixed sexual orientation. For Freud, this fixity exists as a layer on top of an unconscious potential for sexual fluidity (Freud 1962).

Furthermore, establishing sexual orientation as a causal zone, fixed within the body, undermines research practices that might explore sexual plasticity more deeply. Even given Diamond's model which allows for acknowledging change in attractions over the life course, any evidence that might suggest a change in a person's sexual orientation would be ascribed to a secondary level of fluidity, rather than to sexual orientation change. Providing an alternative model of human development, Fausto-Sterling draws on Grosz' metaphor of the Möbius strip, a ribbon with a single twist that is joined in a loop, to represent the relationship between biological and social forces throughout life. If we consider the outside of the strip to be the environment, and the inside to be biological factors, then infinitely walking along this looped ribbon would represent the relationship between these entities in human development. While Fausto-Sterling does acknowledge that there are windows in a person's development when plasticity is more or less possible, this model would allow for conceptualizing and researching a broader range of human sexualities. Such a conceptual shift in science might then be accompanied by alternate arguments for gay rights. Rather than being based on immutability, such claims could be pitched as being more analogous to rights of freedom of religion, where the state has no authority to force a person to change religions in order to obtain rights. Such conceptual possibilities are erased by establishing sexual orientation, fixed in a body, as a causal zone.

Beyond the exclusion of forms of fluidity and in-between categories, the collapse of Exodus International invites considering how the new APA standard may have contributed to this development. The technique of sexual reorientation therapy and the experience of being ex-gay have co-existed within a regime of living in the ex-gay world. In the wake of such a powerful statement about the fixity of same-sex attractions by the APA, Chambers disbanded Exodus claiming the group had been "imprisoned in a worldview" (Exodus International 2013). Even with the APA's acceptance of incongruity between same-sex attractions and sexual orientation identity in certain circumstances, it appears that such a regime of living might involve too much "torque" to be sustained. That is, with so much public knowledge that ex-gays maintain same-sex attractions and develop little heterosexual attraction, the idea of leaving homosexuality with little sexual satisfaction in one's life may be too difficult to promote, especially in an era when sexual satisfaction is increasingly considered a criterion for fulfillment (Giddens 1992).

Further illustrating how the new standard reconfigures a regime of living, those who continue to challenge the APA position within the remnants of the ex-gay movement have interpreted the standard as an affront to a system of value and way of life. Former NARTH President Joseph Nicolosi claims that the APA position "implies that persons striving to live a life consistent with their religious values must deny their true sexual selves," and ex-gays "are assumed to experience instead a constriction of their true selves through a religiously imposed behavioral control." However, he claims that this position "misunderstands and offends persons belonging to traditional faiths." Utilizing biblical values as guides and inspiration on a "journey toward wholeness," such values lead ex-gays "toward a *rightly gendered wholeness*" that is "congruent with the creator's design" (Nicolosi 2009). However, the materialization of male bodies, establishing the notion of fixed "sexual orientation" embedded within them, precludes thinking that living in a way that completely denies the experience of embodied sexual attraction could be a true form of "wholeness" in the way that Nicolosi describes.

Standards as limited "weapons of exclusion"

Using the case of the ex-gay movement and the terminological standardization of "sexual orientation" and "sexual orientation identity" by the APA, this chapter has explored a range of ways in which we might understand how standards can be "weapons of exclusion" (Timmermans and Epstein 2010). These standards have produced exclusions from both families in an interwoven process, as the exclusion of NARTH nominees and self-report methodology has happened alongside the reinscription of essentialist binaries. However, including these nominees and methods would not necessarily have brought about the inclusion of a broader range of possibilities for genders and sexualities, as these nominees were apparently involved in a project of promoting the idea of an underlying heterosexual essence for all.

Instead, this chapter invites the consideration of alternate possibilities for conceptualizing the categorization of sexualities.

As part of APA policy, these terminological standards might be thought of as part of a broader "intellectual opportunity structure," or those features of knowledge-producing institutions which enable or constrain social movements in the construction of facts (Waidzunas 2013; see also Moore 1996). However, even though the organization Exodus International has disbanded, remnants of the ex-gay movement persist. NARTH continues in its advocacy, and many of the smaller religious ministries and religious reorientation camps continue to operate. Jones and Yarhouse (2011) also published an updated version of their longitudinal study in the *Journal of Marriage and Family Therapy,* making similar assertions but qualifying their conclusions as speculative in light of the APA's position that undermines self-report as evidence of sexual orientation change. Thus, just because the APA has created this standard does not necessarily dictate how researchers should conduct popular studies or how people should fashion their own sexuality. The therapeutic guidelines in the APA report are not in any way legally binding, but rely on an APA Code of Ethics for their implementation, as practitioners are advised to keep up with these statements. While the state of California has passed a law against reorientation for minors, enforcement is particularly difficult as therapeutic offices are private spaces, and the law only applies to minors in one state. Although the APA is influential in other national contexts, it remains to be seen how far the standard might be considered relevant outside the United States. Psychological science may be authoritative in the U.S., but Evangelical Christianity remains a strong cultural challenge to its jurisdiction. Lack of closure also has been bolstered by the media, which often represents this issue as a controversy despite a strong scientific consensus on the matter (for example, Spiegel 2011). Thus, the implementation of this standard remains in question. Beyond its instability as a standard, the APA's terminological standardization of "sexual orientation" and "sexual orientation identity" resonates with and reinforces cultural strains in the United States, including a strong heterosexual/homosexual dichotomy, especially for men (Sedgwick 1990), and a tendency toward essentializing gender and sexual orientation differences (Fausto-Sterling 2000).

Thus, looking at the particularity of the history of these standards exemplifies how standardization is often an ongoing process lacking absolute closure. The recent publication of the Jones and Yarhouse study in a scientific journal reveals how self-report can linger in this domain, despite the APA's definitions, although the authors describe their findings as suggestive rather than definitive. While the terminological standardization that severs "sexual orientation" and "sexual orientation identity" has been achieved to a large extent, further questions remain about how these entities are to be more specifically defined and measured. For "sexual orientation," physical and emotional attraction may mean many things to many people, and measures of attraction, fantasy, or laboratory-induced desire can take many forms. Whether sexual orientation should be measured on a continuum scale or some sort of orthogonal grid remains unclear, and this problem also exists for "sexual orientation identity." The standardization of identity measures also faces the dilemma of whether the most important identity is the one held by an individual or one attributed by others.

Overall, standards may act as 'weapons of exclusion' in many different ways, and just because there will be exclusions does not necessarily mean that these exclusions are problems. Indeed, they may be warranted given the need to assert particular values, but their long term effects must be fully considered. While this chapter has shown some ways that standards can exclude, it is by no means exhaustive, and as it has emphasized, terminological standards, especially those related to biomedicine, raise questions about what kinds of exclusions may emerge with the development of other types of standards.

References

Alder, K. 2007. *The Lie Detectors: A History of an American Obsession*. New York, NY: Free Press.

APA Task Force on Appropriate Therapeutic Response to Sexual Orientation 2009. *Report of the Task Force on Appropriate Therapeutic Response to Sexual Orientation*. Washington, DC: American Psychological Association.

Bayer, R. 1987. *Homosexuality and American Psychiatry: The Politics of Diagnosis*, 2nd Edition. Princeton, NJ: Princeton University Press.

Beckstead, A.L. 2003. "Understanding the Self-Reports of Reparative Therapy 'Successes.'" *Archives of Sexual Behavior* 32 (5): 421–423.

Beckstead, A.L. 2012. "Can We Change Sexual Orientation?" *Archives of Sexual Behavior* 41: 121–134.

Beckstead, A.L. and Morrow, S. 2004. "Mormon Clients' Experiences of Conversion Therapy: The Need for a New Treatment Approach." *The Counseling Psychologist* 32 (5): 651–690.

Besen, W. 2003. *Anything But Straight: Unmasking the Scandals and Lies Behind the Ex-Gay Myth*. Binghamton, NY: Haworth Press.

Bloor , D. 1982. "Durkheim and Mauss Revisited: Classification and the Sociology of Knowledge." *Studies in the History and Philosophy of Science* 13 (4): 267–292.

Bowker, G. and Star, S.L. 1999. *Sorting Things Out: Classification and its Consequences*. Cambridge, MA: MIT Press.

Butler, J. 1990. *Gender Trouble: Feminism and the Subversion of Identity*. London: Routledge.

Clarke, A. and Fujimura, J. 1992. *The Right Tools for the Job: A Work in Twentieth-Century Life Sciences*. Princeton, NJ: Princeton University Press.

Collier, S.J. and Lakoff, A. 2008. "On Regimes of Living." In *Global Assemblages: Technology, Politics, and Ethics as Anthropological Problems*, edited by Aihwa Ong and Stephen J. Collier, 22–39. Malden, MA: Blackwell Publishing.

Diamond, L.M. 2008. *Sexual Fluidity: Understanding Women's Love and Desire*. Cambridge, MA: Harvard University Press.

Douglas, M. 1986. *How Institutions Think*. Syracuse, NY: Syracuse University Press.

Dumit, J. 2012. *Drugs for Life: How Pharmaceutical Companies Define our Health*. Durham, NC: Duke University Press.

Erzen, T. 2007. "Testimonial Politics: The Christian Right's Faith-Based Approach to Marriage and Imprisonment." *American Quarterly* 59 (3): 991–1015, 1044.

Epstein, S. 1991. "Sexuality and Identity: The Contribution of Object Relations Theory to a Constructionist Sociology," *Theory and Society* 20 (6): 825–873.

Epstein, S. 2007. *Inclusion: The Politics of Difference in Medical Research*. Chicago, IL: University of Chicago Press.

Espeland, W.N. 1998. *The Struggle for Water: Politics, Rationality, and Identity in the American Southwest*. Chicago, IL: University of Chicago Press.

Exodus International 2013. "Exodus International to Shut Down." *http://exodusinternational.org/2013/06/exodus-international-to-shut-down* (accessed 15 July 2013).

Fausto-Sterling, A. 2000. *Sexing the Body: Gender Politics and the Construction of Sexuality*. New York, NY: Basic Books.

Fishman, J.R. 2004. "Manufacturing Desire: The Commodification of Female Sexual Dysfunction." *Social Studies of Science* 34 (2): 187–218.

Freud, S. 1962. *Three Essays on the Theory of Sexuality*. New York, NY: Basic Books.

Giddens, A. 1992. *The Transformation of Intimacy: Sexuality, Love, and Intimacy in Modern Societies*. Oxford: Polity Press.

Gieryn, T. 1999. *Cultural Boundaries of Science: Credibility on the Line*. Chicago, IL: University of Chicago Press.

Haldeman, D.C. 2004. "When Sexual and Religious Orientation Collide: Considerations in Working with Conflicted Same-Sex Attracted Male Clients." *The Counseling Psychologist*. 32(5): 691–715.

Hilgartner, S. 1990. "The Dominant View of Popularization: Conceptual Problems, Political Uses." *Social Studies of Science* 20: 519–539.

Jasanoff, S. 2004. "The Idiom of Co-production." In *States of knowledge: The Co-production of Science and Social Order*, edited by Sheila Jasanoff, 1–12. New York: Routledge.

Jones, S. and Yarhouse, M. 2007. *Ex-Gays? A Longitudinal Study of Religiously Mediated Change in Sexual Orientation*. Downers Grove, IL: InterVarsity Press.

Jones, S. and Yarhouse, M. 2011. "A Longitudinal Study of Attempted Religiously Mediated Sexual Orientation Change." *Journal of Sex & Marital Therapy* 37 (5): 404–427.

Lampland, M. and Star, S.L. 2009. "Reckoning with Standards." In *Standards and Their Stories,* edited by Martha Lampland and Susan Leigh Star, 3–34. Ithaca, NY: Cornell University Press.

Lynch, M., Cole, S.A., McNally, R., and Jordan, K. 2008. *Truth Machine: The Contentious History of DNA Fingerprinting.* Chicago, IL: University of Chicago Press.

McConaghy, N. 1969. "Subjective and Penile Plethysmograph Responses Following Aversion-relief and Apomorphine Aversion Therapy for Homosexual Impulses." *British Journal of Psychiatry* 115: 723–730.

Moberly, E. 1983. *Homosexuality: A New Christian Ethic.* Cambridge: James Clarke & Co., Ltd.

Moore, K. 1996. "Organizing Integrity: American Science and the Creation of Public Interest Science Organizations, 1955–1975." *American Journal of Sociology.* 101: 1592–1627.

Murphy, M. 2006. *Sick Building Syndrome and the Problem of Uncertainty: Environmental Politics, Technoscience, and Women Workers.* Durham, NC: Duke University Press.

Nicolosi, J. 2009. "The APA Task Force Report: Science or Politics?" *www.narth.com/docs/scienceorpol.html* (accessed 1 July 2013).

Nicolosi, J., Byrd, A.D., and Potts, R.W. 2000. "Retrospective Self-reports of Changes in Homosexual Orientation: A Consumer Survey of Conversion Therapy Clients." *Psychological Reports* 86 (3, pt2): 1071–1088.

Sedgwick, E.K. 1990. *Epistemology of the Closet.* Berkeley, CA: University of California Press.

Shapin, S. 1995. "Cordelia's Love: Credibility and the Social Studies of Science." *Perspectives on Science* 3 (3): 76–96.

Shidlo, A. and Schroeder, M. 2002. "Changing Sexual Orientation: A Consumer's Report." *Professional Psychology: Research and Practice* 33 (3): 249–259.

Spiegel, A. 2011. "Can Therapy Help Change Sexual Orientation?" *www.npr.org/blogs/health/2011/08/01/138820526/can-therapy-help-change-sexual-orientationthe-jurys-still-out* (accessed 1 August 2012).

Spitzer, R. 2003. "Can Some Gay Men and Lesbians Change Their Sexual Orientation? 200 Participants Reporting a Change from Homosexual to Heterosexual Orientation." *Archives of Sexual Behavior* 32 (5): 403–417.

Spitzer, R. 2012. "Spitzer Reassesses His 2003 Study of Reparative Therapy of Homosexuality." *Archives of Sexual Behavior* 41: 757.

Timmermans, S. and Berg, M. 2003. *The Gold Standard: The Challenge of Evidence-Based Medicine and Standardization in Health Care.* Philadelphia, PA: Temple University Press.

Timmermans, S. and Epstein, S. 2010. "A World of Standards But Not a Standard World: Toward a Sociology of Standards and Standardization." *Annual Review of Sociology* 36: 69–89.

Waidzunas, T.J. 2013. "Intellectual Opportunity Structures and Science-Targeted Activism: Influence of the Ex-Gay Movement on the Science of Sexual Orientation." *Mobilization: An International Journal* 18 (1): 1–18.

Waites, M. 2005. "The Fixity of Sexual Identities in the Public Sphere: Biomedical Knowledge, Liberalism, and the Heterosexual/Homosexual Binary in Late Modernity." *Sexualities* 8 (5): 539–569.

Worthington, R.L. 2004. "Sexual Identity, Sexual Orientation, Religious Identity, and Change: Is it Possible to Depolarize the Debate?" *The Counseling Psychologist* 32 (5): 741–749.

4

Curves to Bodies
The material life of graphs

Joseph Dumit and Marianne de Laet

UNIVERSITY OF CALIFORNIA DAVIS AND HARVEY MUDD COLLEGE

Introduction: image and graph

Figure 4.1a Aicher's lavatory sign[1]
Source: www.nps.gov/hfc/carto/map-symbols.cfm.

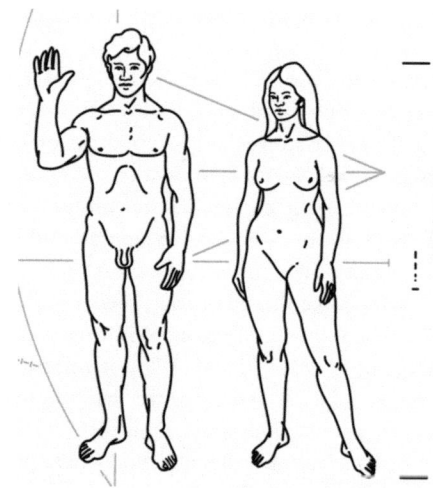

Figure 4.1b Image sent on the starships Pioneer 10 and 11 to inform aliens about diversity on Earth (1972/3)[2]
Source: http://en.wikipedia.org/wiki/File:Pioneer-plague.svg.

Two bifurcated images: a bathroom sign with universal appeal, and a depiction of human life on Earth with cosmic aspirations. You have probably seen them before, and you know instantly what to make of them. In an effort to pare us, humans, down to our essence, these images represent the sexed state of our affairs: there are women, and there are men. This difference not only shows up on bathroom signs and information bulletins for aliens in outer space; it organizes the sociological imagination, defining the basic dimensions in which human interactions and behaviors can and must be understood. There are men, and then there are women, and for many intents and purposes they are different from one another.

Or are they? Perhaps yes. Importantly, yes. In the social and medical sciences, sex is the differentiator by default; the first box to fill out on a survey and a standard dimension in almost any statistical procedure in human research. The recognition of the importance of sex differ- ence has had an enormous, overwhelmingly positive impact on the medical and sociological practices that affect women's (and men's) lives. But the predilection to rely on sex difference in research procedures and treatment protocols hides the normative and performative effects of its use (Fausto-Sterling 1992, 2000, Epstein 2007, Hamilton 1995, Jordan-Young 2010). A force- ful categorizer, sex difference not only eclipses other dimensions, it also, in certain circumstances, produces sexed bodies where in the absence of sex-based statistical operations they would not exist – which you may find a counter-intuitive contention precisely because sex difference is such a pervasive statistical "fact." But, as thirty years of research and debate in Science and Technology Studies have shown over and again, facts – even the most thoroughly naturalized ones – are made; invariably, they have a history and conditions of production. So let's examine the practices, statistical and everyday, through which sex differences are made to be meaningful, and normative, facts of life.

Created by German graphic designer Otl Aicher for the Olympic Games in Munich in 1972, the first sign directs everyone, everywhere, to the lavatory. While it does not reveal specifics about restrooms in the various locales in which the sign is used – one might find them uni-sexed or segregated, depending on the moral and geographical economies of place and space; plumbed and connected to a sewage system, or covering a hole in the ground, depend- ing on other social materialities – this sign *does* lend an odd particularity to the human bodies that these lavatories serve. Triangular or rectangular, humans appear to be of two – and only two – types. So, the sign divides and conquers: sorting humans – but not necessarily lavatories – into sexes, it commands its constituents with unassailable authority where to go. Equally forcibly, it orders how to categorize, and how to think about bodies. Triangular humans, after all, are not the same as rectangular ones.

Much like the first image, the second insists on sorting humans according to sex, and, like the first, it insists on universality.[3] Sent into the voids of the universe on unmanned spaceships Pioneer 1 and 2 in the 1970s, it informs target aliens about life on Earth. A representative image, indeed, standing in for lively human beings, it tells, somehow or other, of what such beings are like. This universal is particular, too. Without visiting Earth, aliens might imagine us all naked, buff, fair-skinned, and operating in oddly coupled pairs. Of all diverse life on the planet, this image draws a telling set of types, informing "us" about ourselves as much, perhaps, as alerting "them" to what and who "we" are. For we know how to read this man and woman.

Or do we? As we see them, do we take note of what they are not? Do we notice that they are adults – and not children or seniors? That they are not injured or disabled in obvious ways? That they are white – not of other races, and athletic instead of obese or anorexic? That they have shaved bodies, and uncovered, groomed, flowing hair – so are not Muslim or Orthodox Jewish? Do we register their unmarked-ness; their nakedness? Do we question whether, in their unmarked nakedness, they can be representative of anything at all?

There is much more to be said about these images, but for now let's note that each has designs, accomplishments, agency, intended and unintended consequences, and that both fore- ground sexual characteristics as a central categorizing factor, upon which a moral economy rests. While you may agree with us, authors, that the first image, if it were sent into space, might somewhat mislead our extra-terrestrial interlocutors about what and who humans are, some among us – NASA engineers in the 1970s – were hopeful that the second one would not. Aicher's icon categorizes; its sexed representation of a generic human necessity helps us find our way. The Pioneer image illuminates this human proclivity for categorization. Rather than providing our alien friends a handle on how to act once they find themselves in our midst, it teaches them about this particular proclivity, giving them precisely the handle *on* us that they need.

Images do things. They move us and, as they embody how we think, they move others to know about us. Perhaps it *is* of significance to let "them" know that "we" are divided into two distinct shapes, with triangular folks very different from rectangular ones. For while Star Trek and Star Wars may suggest aliens to be ordered in sexes as much as we "are"; to groom in the same way, to similarly privilege fit adults low in body fat, and to love classification even more than we do, it is precisely this latter characteristic that is salient in humans: the tendency to value ideal-types over variation; to figure three-dimensional things in two dimensions; to rigorously categorize. To classify is human, to quote Bowker and Star's argument in *Sorting Things Out* (2000). This is why we are paying attention to the categorization practices that inform statistical operations – perhaps what matters is not so much the *knowing* that is represented in our categorical imagery, but the *practices of knowing* that shape these images as the telling icons we think they are.

Statistical operations, then. Each of our two signs tames variety; both trade variation for straightforward types or trends. They are thus the exact opposite of diversity: the artifact, and the iconic representation, of a statistical practice that orders variation to the extent that only ideal types remain. We argue in this chapter that images such as the ones above *epitomize the work of statistics*: they are the end result, the product, and the goal of the statistical graph. At the same time, the power of statistics to sort unruly humans and produce tables, charts, and graphs that map for example caloric needs and baby growth norms – our examples, below – take for granted sex differences, and depend on the pervasiveness of sexed imagery. There is nothing sinister here, and this is not an accusation, allegation, or dismantling of a plot; the problem of how to streamline and order variety is precisely what demographic and other graphs are made to solve. We take this argument one step further, however, to suggest that these bifurcated images translate into sexed and, as such, idealized practices of objective self-fashioning: biometric and demographic statistical operations are agents, in that they *perform* idealized, typed bodies and selves (Dumit 2012). Again, we are pointing to effects, rather than schemes or plots. But that is not to say that the statistical operations we have in mind are not, at times, pernicious in their consequences.

So, we must ask what the path is from population variety via graph to bifurcated image to sexed practices of the self. It is not a hard sell or a counter-intuitive argument that the graphs that illustrate demographic distributions in biometric and social scientific accounts are, themselves, the products of distinct and intricate laboratory operations. It is perhaps less evident, however, that such graphs, as they articulate trends and extrapolate tendencies into solid and stable truths, *produce* the very bodies that populate the accounts which are illustrated by the graphs. No ordinary self-fulfilling prophecy, but rather a body-shaping practice, these graphs shape the physical substance that we live with, in, and through.

That is precisely what this chapter aims to demonstrate: how, and to what effect, graphs do their body-sculpting work. We discuss two cases in which normative distributions produce very specific, normed, and sexed, bodies: calorie counters and baby growth charts; we explain how sex, sexed bodies, and normative comparative logics are the *product* of the graphs that document them, as much as they form their *frame*. So this is our point: that while graphs are products with histories and specific, local, and material conditions of production – the outcomes of certain, by now very well documented, practices – they are also, at the same time, performative and productive; material prompts for how to act. As artifacts to live by, they lay people out along a line, so dividing them into categories; showing a range, while making the middle of the distribution typical. They are normative, in the sense that they *make* the norm and *make* us take it into account. This is precisely what "norming" means: the graph defines, enacts, and so at once produces the abnormal. Both the normal and the abnormal body are a result of statistical operations; and (ab)normality is an artifact – the material and semiotic product – of the norm, and of the graph, itself.

As much as charts and graphs make the body, in as much as they make us forget that things might be imagined differently, and in as much they suggest that variability can be tamed, it is by looking at their material effects and the conditions of their production that we learn to disrupt and destabilize them. As we shall see, it is precisely the normative work of statistical operations that points up how variability remains at large.

Case 1: sexing calories

Calorie counting takes up a prominent place in the moral and digestive economies of today's middle-class northern-hemispheric self. The calorie – like, as we shall see, the baby growth chart – is an agent in public heath discourse; it makes subjects of this discourse do things, and it does so in sex-related ways.

You have seen the tables: plotting weight, age, and activity level, sexed matrices prescribe the daily number of calories one must ingest in order to maintain a healthy weight. Calorie counting for weight control is a matter of comparing two readily available numbers: the number of calories that a body needs, and the calorie content of the food that body ingests. The challenge to consumers is to connect the two – to add and subtract and, on the basis of the metrics that are so readily at hand, to make responsible individual choices of what to ingest. The body's sex is a mediator: it decides which numbers to list in which table or graph, which numbers to identify with, which to plug into one's personal calculations. If calorie information is taken in and taken seriously, if sex is attributed correctly, if successful calculations are made, and if the tallies of calories needed and calories consumed are not too far apart, then – or so one might imagine – a healthy, trim, and responsible citizen body must ensue.

There are always two graphs or tables: a "male" and a "female" chart (see Figures 4.2a and 4.2b). They reiterate that, when it comes to calories required to subsist, there are two kinds of bodies: men and women. One, the male body, needs more calories than the other, the female – or so the charts say. That is what those who are familiar with such charts have learned to assume. With the help of the charts, men and women, separately, can find out *precisely how many* calories they need per day. Looking at the table that is appropriate to one's sexed body, one's number, which secondarily depends on age, weight, and activity level, appears. Bodies and tables are distinct: if we belong in the pink table, we do not need to know about the blue. In this sexed world that we unreflexively inhabit, the tables bear no relation to each other. The bodies that they represent are separate and separated; grouped in two strictly differentiated kinds. This is problematic – but how and why?

At first sight, these tables may rather seem quite un-problematically useful. For once, we have a statistical tool that is not only easy to use, but that, in addition, does male and female bodies justice; that does not measure women with the generic metrics based on the male half of the human species. It seems as if these charts have absorbed the feminist critique that the male in medicine has counted as the standard for all of us, for much too long.

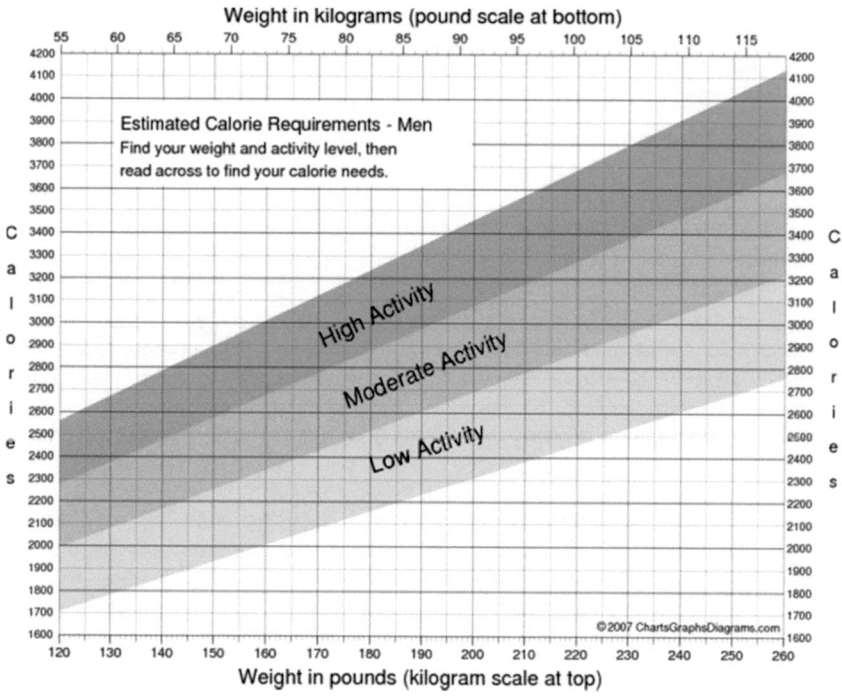

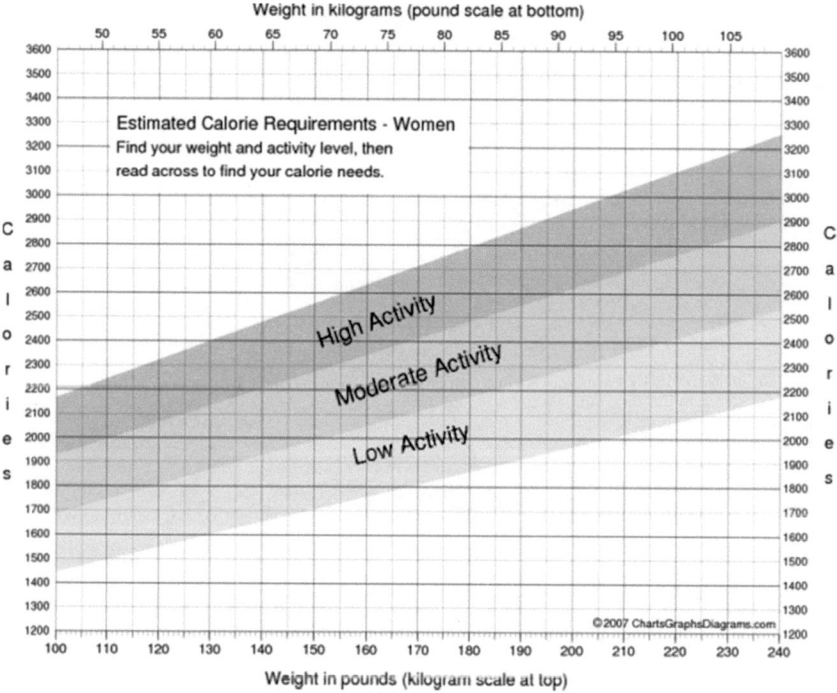

Figure 4.2a and b Calorie requirements by weight and activity level[4]

Source: *www.chartsgraphsdiagrams.com/HealthCharts/calorie-requirement.html*, based on the Harris-Benedict equation (Harris and Benedict 1918).

Better yet, within this general framework is opportunity for individualization: I, Joe, plug in my numbers, in matrices based on constants and formulas provided by nutritional science, and my individual caloric need shows up; a number that tailors the general to the individual's specificity, and that allows for the reassuring certainty that an outside, objective standard will tell me how much – or, at least, how many calories – to eat.

Table 4.1 A typical matrix of men's and women's individual needs, based on an age-specific algorithm

Age	Men	Women
10–18	(Bodyweight in kilograms × 17.5) + 651	(Bodyweight in kilograms × 12.2) + 746
19–30	(Bodyweight in kilograms × 15.3) + 679	(Bodyweight in kilograms × 14.7) + 496
31–60	(Bodyweight in kilograms × 11.6) + 879	(Bodyweight in kilograms × 8.7) + 829

We could also notice that by taking sex categories as our starting point, we reify statistical correlations about calories as primarily about sex characteristics. We put the cart before the horse. Only because we care about sex (and gender) differences socially, does it make sense to say that "men need more calories than women," or, to take another example, the icon reproduced in the Pioneer image, that "men are taller than women" (see Figure 4.3). Staying with height for a moment, even the more precise claim that "men are on average taller than women" still participates in this sexing, since other sentences, such as this one: "almost all men fall within the same height ranges as all women," are more accurate (if accuracy is appropriate to invoke here.)

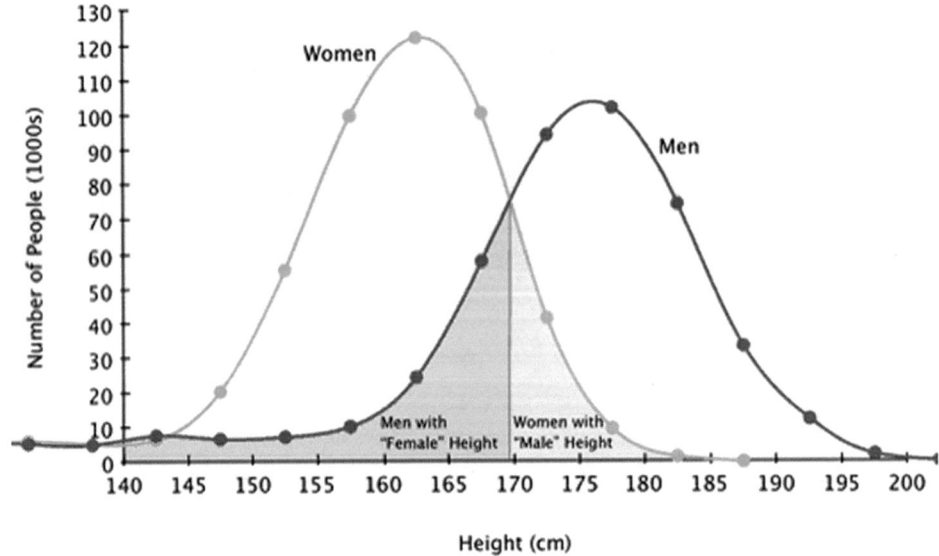

Figure 4.3 Why sex differences don't always measure up[5]

Source: *http://sugarandslugs.wordpress.com/2011/02.*

In addition, the very existence of charts like the one in Figure 4.3 implies that there is something universal about sex difference; that it is more important than all other characteristics. To the extent that tallness is attributed to men and shortness associated with women, taller women become "man-like" and shorter men "women-like." The side of each curve away from the other one generates a typicality, which comes to have socially normative effects, even though the chart itself shows that "typical" men and women come in a great range of heights.

The graph shows two distributions, one for men and one for women, each roughly following the classic bell-curve shape of a statistical normal distribution. At a little before 5'7" (169.5 cm), the curves cross – anyone taller than that is statistically more likely to be a man, and anyone shorter is more likely to be a woman. This suggests that we might use this value as a threshold between "female" heights and "male" heights. But quite a few people are "on the wrong side" of the cut-off point. While more than 1 in 9 women (11.6%) are taller than 169.5 cm (the shaded section of the graph), putting them on the side we might have described as "more likely to be male," even more men have "girly" heights: almost exactly one third (33.4%) are shorter than 169.5 cm.

Similar graphs exist that outline men's and women's caloric needs. The practice of clustering bodies into those two calorie-consuming kinds suggests great clarity; we know what men and women are, and as each kind has its own height curve so, in calorie graphs, it is allotted its own caloric needs. From scattered distributions (as, again, in Figure 4.3), two distinct (but overlapping!) curves are plotted, that appear to justify the two distinct categories that were taken as starting points. But this very practice hides the fact that some people do not fall neatly within one category or the other – the presumption of the chart is that there is nothing to be counted except bodies that are obviously either women or men (Krieger 2003, Fausto-Sterling 2012). As we shall see, the distribution also hides that people's caloric needs are more variable within the two default categories than across them.

The calorie, as we know it, is a product of its material relationships.[6] For anyone who has read some STS literature, this will not come as a surprise: a product, early on, of the intricate laboratory practices that sustained thermodynamics and, today, of those that define nutritional science, the calorie itself, and the matrices that mobilize it, are products of scientific practices – instruments, infrastructures, and negotiations – they form the tangible results of the relational materialities that define the sites of their production. Here we argue that a reverse of sorts is true, too: the calorie produces the sexed materiality of the bodies in which we live. Nutritional accounts, of how many calories a body needs, help define what "women" and "men" are. The calorie is one of many co-producers of what we consider female and what we think is male.

Feminist research has demonstrated that appropriate weight, body image, and other physical ideals are not just physical; they are the result of social negotiations as well.[7] So it is with knowledge – in our case, with knowledge of the differing calorie needs of women and men. The knowledge that calorie needs differ according to sex is not "given" in those differently sexed bodies, but depends on a statistical practice that begins by dividing those bodies into two sexes. While population graphs suggest that individual bodies "have" marked sex differences, which in turn "have" different calorie needs, the actual process by which these graphs are produced follows a reverse path: onto a scattered spectrum of data points, the categories man/woman are superimposed, so identifying two distinct sets of bodies, *now primarily defined by sex*. It is only later that group characteristics come to define and explain the caloric requirements of the individuals who happen to "belong" in either of the different groups.

Consider once again the sex-specific tables that connect weight, activity levels, and calorie need. But now, imagine them arranged differently, with the graphs placed on top of each other (Figure 4.4).

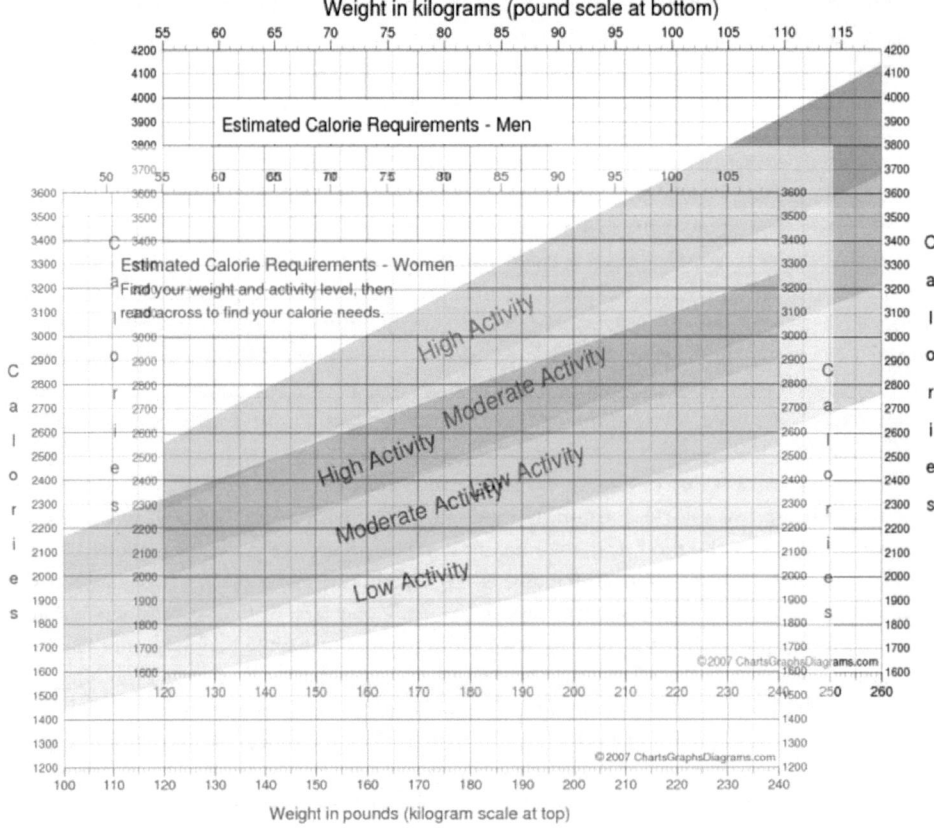

Figure 4.4 Sexed calorie graphs, superimposed

Source: *www.chartsgraphsdiagrams.com/HealthCharts/calorie-requirement.html*, based on the Harris-Benedict equation
(Harris and Benedict 1918).

If we compare the graphs, and so, by this act of comparing, relate them to one another, it turns out that both men's and women's caloric requirements are varied and that perhaps that variation is more telling than clustering in a "male" and a "female" grouping would suggest. Such variation depends on a host of complicating factors, which can have no place in the one-size fits all health advisory: body shape, height, hormone levels, fat percentage versus muscle mass, environment, and type of activity. These particular graphs also leave out age. The matrices hide the aspects of nutritional need that are tied to these other factors, and as long as these aspects remain hidden they do not receive the public, academic, or policy attention they deserve.[8]

So it is that, while serving the healthy bodies discourse – but not necessarily serving all bodies in their individual quests to get healthy – bifurcated sexed statistics also serve a sexist imagination, fleshing out men and women in bodily characteristics that are, purportedly, typically "female" or "male." Our reading, then, of sexed graphs in the context of public health practices denaturalizes sex categories, showing how these categories are the *product* of – among other practices – calorie counting, rather than its *source*. The calorie contributes to the making of sex; sex is its *product*, as much as it is its *frame*. Deployed in health policy, this product may not serve all bodies well.

Case 2: better babies 2.0

Marianne's sister-in-law takes her one-year old son Figo to the Dutch consultation bureau for infants. From before birth to around age 4, children in The Netherlands are regularly and carefully monitored for height, weight, motor development, speech pattern, and such – as part of the country's system of "social medicine"[9] and in the service of the early detection of possible problems with development and growth. Marianne's nephew is a healthy, beautiful, happy baby. Note that healthy, happy, and beautiful are not independent markers of this child; they are wrapped into each other, and one signifies the next. But it is not only Figo's demeanor and looks that make him a healthy, happy, beautiful baby; it is the charts against which he is measured every month or so that do so, as well.

Perpetually controversial, baby growth charts show the percentile distribution of height and weight by age for boys and girls. Quite explicitly, they turn average into norm, implying – for example – that, somehow, weights in the 90th and 10th percentiles fall outside of the spectrum of the normal; that they require, if not intervention, in any case scrutiny and extra care. By definition, the 10th percentile, which includes the 10% of babies with the lowest weights, and the 90th percentile, comprising the 10% with highest weights, are outliers. This is not to be taken for granted as a "natural" state of things; it is the statistical procedure which is at the center of the baby growth charts that decides, *a priori*, that 10% of babies will be classed as underweight and another 10% as too heavy for their sex and age.

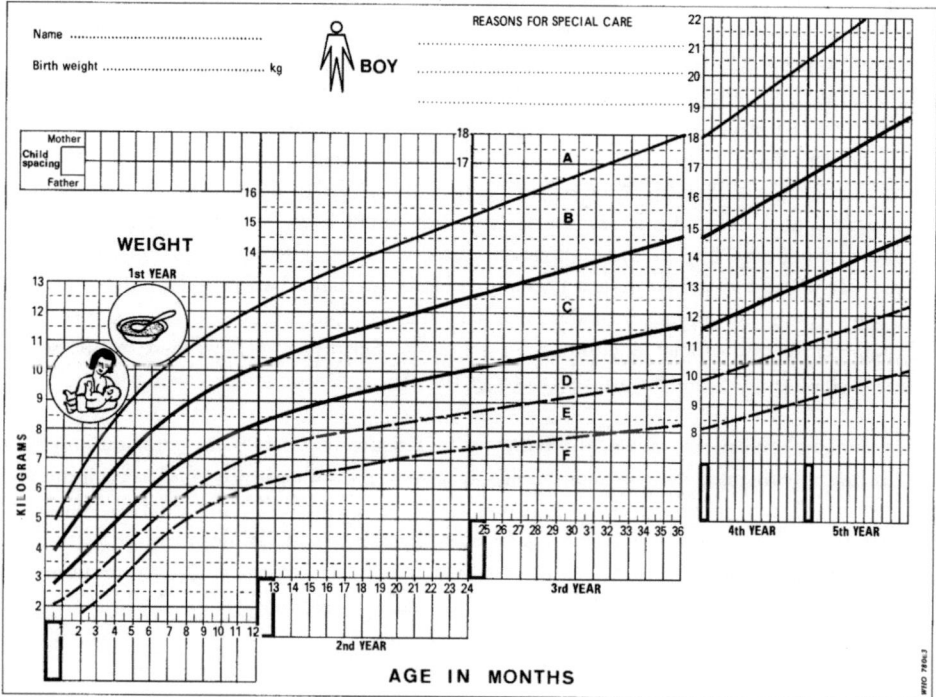

Figure 4.5 1977 growth chart for boys
Source: WHO 1978.

These charts have agency: they *make* Figo a happy, healthy, and beautiful baby – and other children, less so. Take his little, somewhat younger, somewhat smaller cousin, for instance. Melle was born prematurely just a few months after Figo came into the world. Melle, according to the reports, struggled; it is, after all, difficult to eat if you are born prematurely, because the muscles you need to do so are not quite there yet in all their devouring strength. So Melle has perhaps "lagged behind" a bit – literally when he is plotted on the baby growth chart, he is smaller than he "should" be – and people have been worried about him as he continued to fall below on the charts. But when Marianne met him for the first time, six months old and beaming, she saw nothing but a happy, healthy, beautiful little baby. The comparison is there, though; already, family members and day-care teachers have adopted the language of the nurses and charts and talk about Melle as being "small" and "struggling to catch up" to his "normal height and weight." Such discourse is part of the caring practice to help babies like Melle become healthier, and the charts are agents that quite literally "make" them healthy or not.

In the U.S. too, where Joe's child grew up, baby growth charts form a constant threat, and they act as a potent marker of parenting. For the apparent transparency of the three characteristics that matter when one faces the charts – age, weight, and sex – almost guarantees the charting of babies in doctor's offices, at home, and elsewhere. Not surprisingly, while these graphs help diagnose growth problems early on, they are also – both – a source of satisfaction for parents whose children's physiology behaves well, and a tremendous cause of anxiety for those whose babies do not perfectly fit the curves. Intended or not, with their ease of use, the charts launch new parents into the spiraling, self-fulfilling wormhole of the comparative moral imperative.

We argue that the charts have agency, as commanding parts of these babies' lifeworlds; they speak and order, imposing a will to act on those involved in the infants' care. Surprisingly, as we show below, this performative agency was not, initially, the charts' intention. Nevertheless, here it is: the sexed chart itself, with its crystal clear percentile lines separating normal from above average and from below average infants, shapes the kinds of worry and love, care and comparison, nutrition and attention that shapes these babies' lives.

The text that accompanies Figures 4.6 reads: 'The "Best Parent Ever" is better than you because their child is in the 95th percentile. "The 95th percentile of what?" you ask. Whatever the pediatrician says! Height? Weight? Intelligence? Body hair? It doesn't matter. Just as long as their child is at the top. This is because, for the Best Parent Ever, Pediatric Growth Charts are a competitive sport. They are the highly-anticipated weekly box scores or quarterly results that let the "Best Parent Ever" know just how much better they and their brood are than everyone else. Why all the fuss? Because, for the "Best Parent Ever", the Pediatric Growth Chart is one of their child's first official documents in a lifelong Scripture of Betterness.'[10] But wait a moment. Ease of use? It is not so easy to enact the charts. As one father explains:

> I am an engineer, statistician, and father of 4. … squirming babies are notoriously hard to measure with any degree of repeatability whatsoever. I noticed early on that each individual health care provider's own measurement technique radically distorted our children's percentile placement. I have no doubt that the analysis of development patterns is an important diagnostic tool. I am however very much inclined to weight the result against my personal experience as to the health and vitality of each individual child.
>
> *Green 2012*

Measuring a baby, and calibrating it against the metric of the graph is no picnic. It is a practice, and one has to be practiced in order to do this practice well. Children in The Netherlands, in the United States, and in other places where growth charts anchor the methods of child

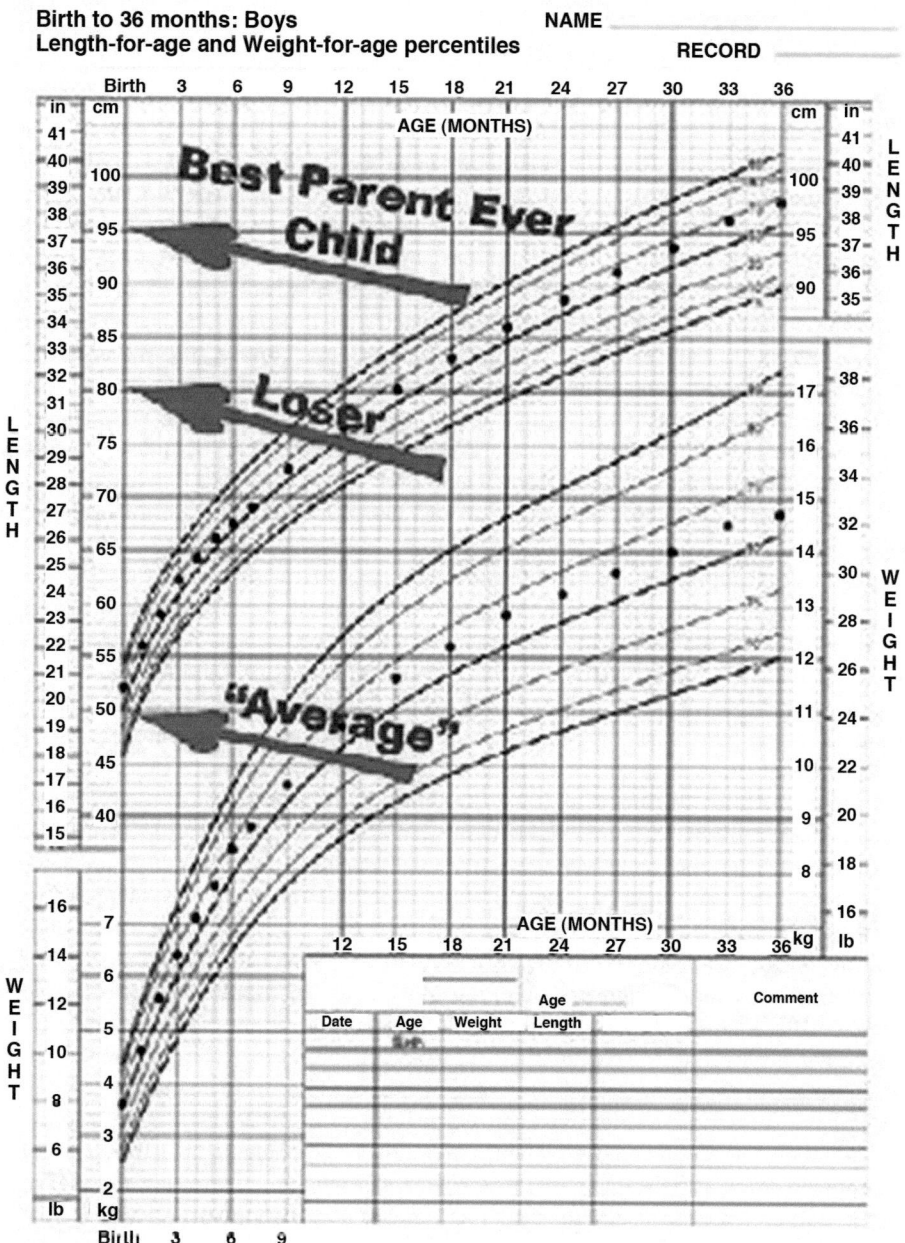

Figure 4.6 Illustration from Best Parent Ever blog post: Growth Charts
Source: http://bestparentever.com/2008/06/02/37-growth-charts/

development monitoring, are weighed against the charts as a matter of fact. So a self-fulfilling prophecy is conjured into being: the more children are measured against the charts, the more stable the charts become, the more they can be used with confidence, and the more children are weighed against the charts. After all, statistics become more solid as the numbers on which they are based get larger; an interesting example of how a practice makes itself increasingly "true," scale enhancing its stability.[11]

Let's look for a moment at the mechanics that, in this self-fulfilling prophecy, are at work. More than likely, the father-engineer above used as his baseline a particular set of charts, produced in Ohio in 1977, that by the early 2000s had become the standard all over the United States. Based on a dataset that originated in a "single longitudinal study of *mainly formulafed*, white middle-class infants in a limited geographic area of southwestern Ohio, collected 1929–1975" (CDC 2002: 2–3, emphasis ours, Hamill et al. 1979) – in a singular culture, defined by particular material attributes – these charts were devised as a *referent* of average infant growth. But once circulated, as they traveled to pediatrician offices and baby-raising handbooks, they were readily at hand as precisely the kind of material attribute that is taken to *transcend* culture.[12] They were put in place as if they already were the standard. And they became in everyday practice a norm – the ideal against which to judge new babies.

This alternate approach to the charts forms a type of "antiprogram," a term coined by Akrich and Latour (1992) and used by Eschenfelder (Chapter 9, this book); users mobilize the charts in ways that conflict with what the designers of the charts intended. This had unintended and insidious consequences: if a breast-fed baby fell into a low percentile on the chart, monitoring health care personnel would interpret the results as proof that the baby was under-nourished, and recommend that breast milk be replaced with formula. As it turns out, breast-fed babies tend to have a slower early growth curve than those nurtured on formula, only to, on average, catch up later in their development. Health researchers confirmed that rather than acting as a reference, the charts were, in effect, transforming childcare – when nurses and pediatricians told the parents of "underweight" breastfed babies to switch to formula. According to later 1996 WHO analysis, as a result of a choice made at the outset in the Ohio study, namely to use only formula-fed babies, the weight distribution in the study's data is skewed toward obesity (de Onis et al. 1997). An "unhealthy" characteristic of the NCHS reference produced a metric that misclassified overweight children as "normal," so setting the baseline too high and rendering breast-fed babies malnourished.

In retrospect, we can say that the chart literally produces – and stabilizes – the practices that it assumes; by moving more children to formula, increasing the scale of its use, it reproduces in infants the obesity that it assumes as the norm. This, then, is how baby growth charts perform baby bodies. Delegating the adjudication of normality to the charts, parents unwittingly invite comparison and its attendant anxieties into their lives; anxieties about the baby's qualities and, by extension, quality – and, by yet another extrapolation, about the quality of parenting itself.

Substituting an experience-based sense of baby health with measurement, the chart enacts a medicalized parent-child relationship as the normal state of affairs. What we mean by the term "medicalization," here, is the process by which "a healthy baby" comes to be defined exclusively by measurement and its relation to statistical norms; and while such norming and measuring serves its purposes, the term suggests that in this process norms and measures assume a life of their own. They come to be institutionalized as virtuous norms. With consequences: learning to delegate the adjudication of what constitutes health in a baby to charts, we learn to distrust our sense of what a healthy baby looks and feels like; trusting the charts implicitly, parents talk about feeling better once their babies are on track – which is to say, once they score in the proper percentiles.

The realization that the "Ohio average" had in everyday use turned into a type of "world baby norm" eventually led to questions about whether there should be a new set of charts, based on population research including both formula-fed and breast-fed babies. New charts raise new questions: given that average formula-fed babies and average breastfed babies are known to develop differently, should they have been averaged together at all?[13] Don't the new charts, by way of a statistical technique, force two distinct classes into one – so producing outliers who, in their own categories, would register as "normal" (De Onis et al. 2007)? Aren't the charts then introducing new parental anxieties – inducing yet another set of perhaps unwarranted changes in infant nutrition practices? Indeed, calls ensued to create special "breastfed-only charts." Should these be made the standard, then, reversing the previous practice – perhaps to be complemented by special "formula charts"?

Responding to these questions in the late 1990s, the World Health Organization (WHO) commissioned a series of studies to "generate new growth curves for assessing the growth and developments of infants and young children from around the world." Thus began a process, which is continuing today, to implement new charts with new effects (see Figure 4.7). The challenge for the Multicentre Growth Reference Group Study (MGRS) is especially alert to the difficulties of measuring with reliability. How to enact standard measurements in a study that, by design, has many centers – where, undoubtedly, many measuring practices exist? (de Onis et al. 2004a, S27). How to stabilize a field whose practices are, in their nature, disrupted? Given how difficult it is to tame variability of growth factors within an – admittedly already quite heterogeneous – "place" like the United States, how can a sensible metric for comparing children around the world come about?

The MGRS solution was not to try to average children at all. Rather than trying to find standard babies, they would make them! The new chart would then act as a norm to prescribe care-taker behavior *and* baby growth; as an objective guide for intervention, so to speak. In other words, the WHO protocol explicitly aims to enact changes in the population on which its standard is based, so turning the unintentional effect of the Ohio study into intentional achievements of the very implementation of the protocol, itself. Prior anti-programs become here part of a new, anticipatory program, performing a socio-technical fix where, as we shall see later, perhaps a cultural one might have been more appropriate (see Layne 2000, for a discussion of a case with a similar trade-off). The new standard, then, is a metric not only for

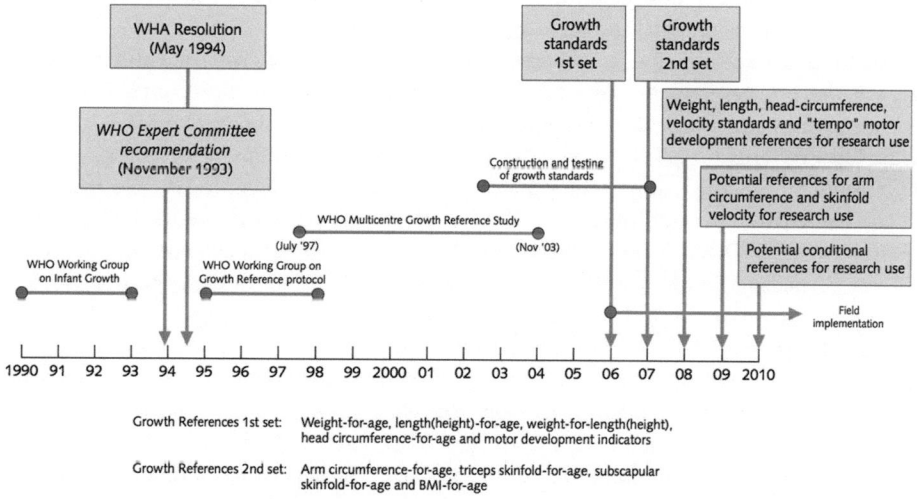

Figure 4.7 Timeline of the new international growth references[14]
Source: De Onis et al. 2004b.

infant growth conditions, or even for care and intervention; it sets a new standard for what a protocol is and should achieve. Interestingly, this new WHO procedure for making charts demonstrates how standards make worlds – precisely as, again, Bowker and Star point out in *Sorting Things Out*. But it goes a step further, designing policy that incorporates, explicitly, the recursive, reflexive, and self-fulfilling nature of standards and norms; their stabilizing effect. The new WHO approach no longer considers growth standards as referring to existing babies, nor to group differences, nor to any sort of average; instead it aims to produce what it considers to be the ideal babies reared in the most ideal ways in the most ideal places so that the new standard growth chart is an *explicit* device for making value judgments (Garza and de Onis 2004: S9–10).

The MGRS thus set out to create "the most standardly reared babies ever." One of its sites was Davis, California, where a rather Borgesian list of interestingly specific selection criteria was used to determine which babies to include in the research population: the mothers of infants who can participate in the study must intend to breastfeed; they must be willing to do so exclusively for first four months and continue for the first twelve; parents must be non-smokers and on average highly educated; parents must be prepared to talk to a lactation consultant and submit to a detailed protocol regarding the introduction of foods and limits on juice; they commit to making "mealtime a happy, pleasant experience [and not to force the child] to eat certain foods or finish everything on the plate." The study mentioned, moreover, that the children in its population share a low altitude living environment in a town with a middle-class median income (Onyango et al. 2004: 587). By attending to these specifics, the study standardized in advance many of the factors that may – or may not be – influential in child development.

But with all this taming of variability, complication does not disappear. The only difference left in the charts is sex – baby boy and baby girl – yet as the charts were being made, evidence suggested that "differences in growth are primarily due to environmental and socioeconomic constraints" (Kuczmarski et al. 2002: 13); precisely the factors for which the study aims to control. With the problem of infant growth and child development a major concern of the World Heath Organization, the challenge to growth charts – specifically, that of taming diversity – has become much more complicated, indeed. Not only because the factors that influence development – environmental and socioeconomic constraints – vary so widely that they escape the ambitions at standardizing that are embedded in efforts to make charts, but also because the very practice of measurement may elude stabilization, and escape all efforts to tame it.

New metrics and protocols respond to this recognition of variability. The MGRS takes on the problem of variation in baby growth by national and ethnic groups in part through a rigorous screening technique that uses the same criteria as the one in Davis. Its effort to standardize the absolute environment of babies around the world, by choosing five sites as the baseline for the study – Pelotas Brazil, Muscat Oman, Oslo Norway, and "selected affluent neighborhoods" of Accra Ghana and South Delhi India – appears to pay off; babies in these locations seem to conform to a single standard that could become the world standard. "[B]reastfed infants from economically privileged families were very similar despite ethnic differences and geographic characteristics," except for China and India: "compared with the arbitrarily selected reference group (Australia), Chinese infants were approximately 3% shorter and Indian infants were approximately 15% lighter at 12 months of age" (Garza and de Onis 2004: 59). In India, the mothers' education seems to do away with the variance. The report makes no further mention of China.

Making global growth charts may be an attempt to confine a ghost in a wire-mesh cage. Given the fact that most babies will not be reared at sea level, in middle class homes, amidst minimal pollution, in smoke-free environs, with fresh food and lactation consultants, what will the prescription of health care workers be – move, find new parents, eat more or less? With emerging concerns as much about overweight babies as about underweight ones, the ideal height-weight ratio may become even more tightly regulated. As in Mol and Law's study of how people "do" hypoglycemia, particularity and specificity trump generalization. Where they found that tight regulation, trying to stay on target too much, too often, "is not good or bad for the body as a whole [but rather] good for some parts of the body and bad for others" (2004: 55),[15] we suggest that babies' material circumstances may be more telling about their prospective health and growth than their adherence to any generalized chart.

The problem of specificity extends to actual practice. Recall our engineer-statistician father who, earlier on, noted that squirmy babies are difficult to measure. In the MGRS, those who conduct the measurements have to be rigorously trained to match the measures obtained by expert "anthropometrists" (De Onis et al. 2004a). This is not an easy procedure; "novices" make many "mistakes." While training may reduce such mistakes in creating the charts, to make training protocols uniform and regiment measurement practices into global uniformity may be too much to ask. But variability in measuring practices obviously compromises the use of the charts in unintended ways. As the problem that the MGRS hopes to solve, namely to create general standards for nutrition-related child development and growth, travels beyond the confines of the places where the initial metrics originated – as the problem becomes "globalized," so to speak – we find that its solution – the growth chart – points up untamable variability, rather than affording the harmonization – and stabilization – that the term globalization suggests.

These new charts are new creatures: they are prescriptive, rather than descriptive. No longer meant to represent baby populations at a specific time and place, they are instead explicitly *designed* to intervene in baby rearing practices, and will do so effectively precisely because they have reduced the variability of babies to two sexes and clean, easy-to-read lines of normality. Depending on how they are implemented, they have the potential to push, shame, or force parents to conform to the kinds of practices that produced the chart. But the new charts are similar to the previous ones, in that the initial decisions about the population from which the chart emerges come to define the conditions of the production of what the charts normalize as healthy babies. These charts embody particular cultural and material circumstances; breast-feeding, being well-off, having educated mothers, living in urban areas, at low altitude, and having access to lactation consultants. In addition, their use requires regimented personnel who execute standard procedure. The norms are thus inscribed in the charts, as they are in the practices that the charts are expected to engender.

There is, then, something about norming charts that – precisely because of their reflexivity – is inherently unstable, or so it would seem. But while the questions we raised before may seem to point up a statistical, norming problem, they reveal that we are dealing here with moral, ethical, political, and practical, material problems, too: how to make charts that are somehow representative of the diversity of baby growth under different dietary, measuring, and other conditions – in the knowledge that they will produce new norms, with attendant new practices and policies, and therefore new babies.

Conclusion: bio-graphs need biographies

Graphs and categories are ubiquitous, especially in health care practices. Even when we are not confronted with them directly, they are behind the scenes – shaping the way in which decisions are made, protocols developed, and facts made to circulate. In our two cases, these graphs may be understood as subject to what Eschenfelder (Chapter 9, this book) calls "continuous renegotiations": as much as we can point to times and places when specific new graphs are introduced, it is also the case that their design, implementation and dissemination is ongoing; so, too, is their meaning and use. In the case of baby growth charts, what is represented as the "average baby" becomes, in practice, the "normal baby," while non-average babies become abnormal ones that have to be changed. As baby-rearing practices develop, the very notion of what is a "healthy baby" alters, and when new standards are implemented to create better normal babies, they create the new type of the abnormal baby, too.

In our two cases, the relation between "taking an average" and imposing it as a norm is clear – at least in hindsight; in a constant cycle of stabilization and disruption, ideals materialize. The making of norms follows the path of the making of facts (Latour and Woolgar 1986): once they circulate, statistics about health will become norms whether their makers want them to or not. Reporting or creating, then, may be the question. Do our graphs and images of body types show factual biological differences or do they create different bodies? Are these signs us, or do we become them? What, precisely, is problematic about the confluence of self and sign? While we cannot point to when and where any of the practices we described become pernicious, we can offer a heuristic, a set of rules of method, for noting, listing, pointing up, making strange, and articulating where in the process from graphing to reifying we are, and in what kind of performance we are taking part. So here goes: how to "make strange."

First move. When you come across a graph, or stumble upon a differential claim about bodies, consider both the conditions of its production, and its use and actual effects in the world. If the data are already separated into categories – such as male and female – find out where, and for what reasons, in the process of designing the study those categories were imposed. Consider whether the effects of the categories is to reinforce the apparent value of the average of each (since the averages differ), and ask what would happen if you looked at the range of individuals who are being lumped together in just one of the categories. Might other differences between those individuals (such as age, wealth, disability) be more salient than the category used? What does the chart fail to take into account? What would the graph look like if a different category were used? How might different categories be more telling? Of what would you want them to tell?

Second move. Look again at the construction of the graphs at hand, and consider who were studied in order to create them. Like iconic images, the final graphs are meant to be generic markers, understandable to, and representative of, all. As we saw in both cases – in the manufacture of baby growth charts and the framing of calorie metrics – it is not an easy task to choose which people might stand in for all. For reasons of expense, labor, and accessibility, research studies have to use, and generalize from, a relatively small sample group. Research papers usually provide a clear justification for sample choice; they qualify the results, reporting on the conditions of their production. But their cautions about how to interpret results, especially graphs, typically get lost as such graphs travel into the world. Especially when picked up by mass media news accounts and general practices, qualifiers disappear. While in general use, graphs and images may gain a measure of facticity (Latour 1988), they do not gain in accuracy. Generalizability, then, does not have anything to do with representative force. The general use of a graph or iconic image does not mean it is representative; it just means that it is – without wanting to seem circular – generally used.

Third move. Then watch how the graphs and results are mobilized. Who uses them, where and when? Notice where an average turns into a recommendation, and when it becomes norm based on how the user reads the graph. An anxious parent worried about his or her child may want to be told what to do. Step back again and ask how the graphs traveled to get into the hands of these users (on this method, see Dumit 2004, 2012). In the case of baby growth charts and calorie charts, we saw that reports based on research articles were then turned into simplified and "handy" charts, the kind that can appear on posters and websites and be turned into apps. Often it is in these moments of making charts "useful" that qualifications disappear, while scientific medical authority – which rests precisely on the making of such qualifications – remains (cf. Dumit and Sensiper 1998). In this way graphs and images become "black boxed" as self-contained facts and norms about categories of people (Latour 1988).

Finally, note where and when you take graphs, tables, images, or apps, at face value and do the work of problematizing them, so that you can begin to recognize something new about yourself, in addition to something interesting about the graphs.

Without this work, without reflection on the kinds of reflexivities we have described, without an account of the performative effects of using, for representative purposes, images and graphs that result from select samples, such representations seem natural, necessary, and innocuous. We suggest that the image sent into space on the Pioneer spaceships, sexed calorie metrics, common-use baby charts, and even Aichler's globally understood bathroom sign, are cases in point. When, in the name of convenience, idealized snapshots prevent us from imagining what else might matter, we all lose out. What we enjoy about our STS training is that it presses us to make strange the production and circulation of artefacts such as graphs, charts, and images, and to account for their agency and social life.

Notes

1   Rathgeb 2007.
2   Turnbull 1994.
3   Of course, this reduction-to-universality is precisely what graphic design *does*. According to another German designer, Christoph Niemann, design *is* the graphic language of data, of charts, of icons, of the visual reduction. "What does a graphic designer do? They create bathroom signs so the women know which room to go to, and the men know which room to go to. For me, that's like the ultimate reason for being a graphic designer, and everything is a more complex derivative of that."
4   Charts from *www.chartsgraphsdiagrams.com/HealthCharts/calorie-requirement.html*, based on the Harris-Benedict equation (Harris and Benedict 1918). The original equation was based primarily on young, healthy adults. Subsequent reviews of the equations have suggested various corrections based on age, health, malnutrition, and metabolism (Roza and Shizgal 1984).
5   Image retrieved July 1, 2013 from *http://sugarandslugs.wordpress.com/2011/02*, "Why Sex Differences Don't Always Measure Up," Posted February 13, 2011, by "Nebulous Persona." NP continues, "It's tempting to imagine we might be able to salvage the simple threshold idea by saying that people taller than 176.5 cm are usually male and a those shorter than 159.5 cm are usually female, and abandoning everyone the middle (more than half of our sample!) as living in a gray area. But even that doesn't work. Once we get to people shorter than 4'9" (145 cm), there is no reliable sex difference. There are 60,447 women less than 145 cm tall, and 63,690 men, making it pretty much a wash. But because there are more men than women, there are proportionately more particularly short men (13% of all men) than short women (11.8% of all women)."
6   See Mol and Law 2002.
7   Hesse-Biber and Leavy (eds.) 2007.
8   A related analysis of the problem of reducing variability to one or two categories can be found in Cohen 2001.
9   What in the United States is labeled as "social" medicine – in the current political climate somewhat derogatively – is simply called "healthcare" in The Netherlands. In other words, Dutch political discourse acknowledges no choice of healthcare models, enabling only one system, in which generic screening practices are built in. There is nothing particularly "social" about it, except perhaps in the sense that *all* healthcare systems are social artefacts – which makes the term obsolete. We refer here to Bruno Latour and Steve Woolgar's objection to the word social in "(social) construction" in the 2nd edition of *Laboratory Life* (Latour and Woolgar 1986).
10   *http://bestparentever.com/2008/06/02/37-growth-charts*.

11 As immutable mobiles, in actor-network terms – see Bruno Latour's *Science in Action* (1988).
12 Ibid.
13 Yet according to the CDC FAQ, "The 2000 CDC Growth Chart reference population includes data for both formula-fed and breast-fed infants, proportional to the distribution of breast- and formula-fed infants in the population… Healthy breastfed infants typically put on weight more slowly than formula fed infants in the first year of life. Formula fed infants gain weight more rapidly after about 3 months of age. Differences in weight patterns continue even after complementary foods are introduced (Dewey, 1998)." (CDC 2009).
14 De Onis et al. 2004b.
15 For a similar analysis of the obfuscation of specificity in the implementation of generic health standards, see de Laet and Mol 2000.

References

Akrich, M. and Latour, B. 1992. "The De-scription of Technical Objects." In *Shaping Technology, Building Society*, edited by Wiebe E. Bijker and John Law, 205–224. Cambridge, MA: MIT Press.

Bowker, G.C. and Star, S.L. 2000. *Sorting Things Out: Classification and Its Consequences*. Cambridge, MA: MIT Press.

Callon, M. 1986. "Some Elements of a Sociology of Translation: Scallops and Fishermen in St. Brieuc Bay." In *Power, Action and Belief: A New Sociology of Knowledge*, edited by J. Law. London: Routledge and Kegan Paul.

CDC. 2002. Kuczmarski, R.J., Ogden, C.L., Guo, S.S., Grummer-Strawn, L.M., Flegal, K.M., Mei, Z., Wei, R., Curtin, L.R., Roche, A.F., and Johnson, C.L. "2000 CDC Growth Charts for the United States: Methods and Development." *Vital and health statistics*. Series 11, Data from the national health survey 246: 1–190.

CDC. 2009. "Frequently Asked Questions about the 2000 CDC Growth Charts," retrieved on July 10, 2013 from *www.cdc.gov/growthcharts/growthchart_faq.htm*.

Cohen, J.S. 2001. *Overdose: The Case against the Drug Companies*. New York, NY: Tarcher-Putnum.

De Laet, M. and Mol, A. 2000. "The Zimbabwe Bush Pump Mechanics of a Fluid Technology." *Social Studies of Science*, 30.2: 225–263.

De Onis, M., Garza, C., and Habicht, J.-P. 1997. "Time for a New Growth Reference." *Pediatrics* 100.5: e8–e8.

De Onis, M., Garza, C., Onyango, A.W., and Borghi. E. 2007. "Comparison of the WHO Child Growth Standards and the CDC 2000 Growth Charts." *The Journal of Nutrition* 137.1: 144–148.

De Onis, M., Onyango, A., Van den Broeck, Wm., J., Chumlea, C., and Martorell, R. for the WHO Multicentre Growth Reference Study Group. 2004a. "Measurement and Standardization Protocols for Anthropolometry Used in the Construction of a New International Growth Reference." *Food and Nutrition Bulletin*, 25.2 (supplement 1): The United Nations University.

De Onis, M., Onyango, A., Van den Broeck, Wm., J., Chumlea, C., and Martorell, R. 2004b. "The WHO Multicentre Growth Reference Study: planning, study design, and methodology." *Food & Nutrition Bulletin* 25 (supplement 1): 15S–26S.

Dumit, J. 2004. *Picturing Personhood: Brain Scans and Biomedical Identity*. Princeton: Princeton University Press.

Dumit, J. 2012. *Drugs for Life: How Pharmaceutical Companies Define Our Health*. Durham: Duke University Press.

Dumit, J. and Sensiper, S. 1998. "Living with the 'Truths' of DES: Toward an Anthropology of Facts." In *Cyborg babies: From Techno-Sex to Techno-Tots*, edited by Joseph Dumit, and Robbie Davis-Floyd. Routledge, New York: 212–239.

Epstein, S. 2007. *Inclusion: The Politics of difference in Medical Research*. Chicago: University of Chicago Press.

Fausto-Sterling, A. 1992. *Myths of Gender: Biological Theories about Women and Men*. (Revised edition). New York: Basic Books.

Fausto-Sterling, A. 2000. *Sexing the Body: Gender Politics and the Construction of Sexuality*. New York: Basic Books.

Fausto-Sterling, A. 2012. *Sex/Gender: Biology in a Social World*. New York: Routledge.

Garza, C. and de Onis, M. 2004. "Rationale for Developing a New International Growth Reference." *Food & Nutrition Bulletin* 25 (supplement 1): 5S–14S.

Green, K. 2012. Comment on Beck, M. "Is Baby Too Small? Charts Make It Hard to Tell." *The Wall Street*

Journal. July 24. Retrieved from *http://online.wsj.com/article/SB10000872396390443437504577 544861908329668.html#articleTabs%3Dcomments*

Hamill, P.P.V.V., Drizd, T.A., Johnson, C.L., Reed, R.B., Roche, A.F., and Moore, W.M. 1979. "Physical growth: National Center for Health Statistics percentiles." *American Journal of Clinical Nutrition* 32.3: 607–629.

Hamilton, J.A. 1995. "Sex and Gender as Critical Variables in Psychotropic Drug Research." In *Mental Health, Racism and Sexism*. Pittsburgh (PA): Taylor & Francis, Inc, 297–350.

Harris, J.A. and Benedict, F.G. 1918. "A Biometric Study of Human Basal Metabolism." *Proceedings of the National Academy of Sciences of the United States of America* 4.12: 370.

Hesse-Biber, S.N. and Leavy, P.L. 2007. *Feminist Research Practice: A Primer*. London: Sage.

Jordan-Young, R.M.2010. *Brain storm: The flaws in the Science of Sex Differences*. Cambridge, MA: Harvard University Press.

Krieger, N. 2003. "Genders, Sexes, and Health: What Are the Connections – and Why Does It Matter?" *International Journal of Epidemiology*, 32.4: 652–657.

Kuczmarski, R.J., et al. 2000. "2000 CDC Growth Charts for the United States: Methods and Development." *Vital and Health Statistics. Series 11, Data from the National Health Survey* 246: 1.

Latour, B. 1988. *Science in Action, How to Follow Scientits and Engineers through Society*. Cambridge, MA: Harvard University Press.

Latour, B. and Woolgar, S. 1986 (2nd edn). *Laboratory Life: The Construction of Scientific Facts*. Princeton: Princeton University Press.

Layne, L.L. 2000. "The Cultural Fix: An Anthropological Contribution to Science and Technology Studies." *Science, Technology & Human Values*, 25.4: 492–519.

Mol, A. and Law, J. 2004. "Embodied Action, Enacted Bodies: the Example of Hypoglaecemia." *Body & Society* 10: 43–62.

Onyango, A.A.W., Pinol, A.J., and de Onis, M. 2004. "Managing Data for a Multicountry Longitudinal Study: Experience from the WHO Multicentre Growth Reference Study." *Food Nutrition Bulletin* 25.1.

Rathgeb, M. 2007. *Otl Aicher*. London: Phaidon.

Roza, AA.M. and Shizgal, H.M. 1984. "The Harris Benedict Equation Reevaluated: Resting Energy Requirements and the Body Cell Mass." *American Journal of Clinical Nutrition* 40.1: 168–182.

Turnbull, D. 1994. *Maps are Territories, Science is an Atlas*. Chicago: University of Chicago Press.

World Health Organization 1978. *A Growth Chart for International Use in Maternal and Child Health Care: Guidelines for Primary Health Care Personnel*. Geneva: World Health Organization.

Part II
Consuming Technoscience

5

Producing the Consumer of Genetic Testing

The double-edged sword of empowerment

Shobita Parthasarathy

UNIVERSITY OF MICHIGAN

In America's healthcare marketplace, the user has long been treated as both a patient and a consumer. She has had some choice over her physician and could demand access to some services, and was often left paying exorbitant prices. But her power has always been somewhat limited, constrained by government regulations over drugs and medical devices, the rules of insurance companies, and the expertise and authority of science policy advisors, physicians and other healthcare professionals. The technologies of the twenty-first century, facilitated by particular social, political, and economic systems, have pushed the identity of the healthcare user further toward that of a consumer who is presumed to be empowered by increased access to information and can, therefore, make independent decisions about herself, her healthcare, and her life. Scholars have argued, for example, that the internet allows users to come to medical appointments informed with information that diagnoses maladies and suggests treatments (Broom 2005, Fox et al. 2004). These technologies have shifted the power dynamics of the doctor-patient relationship, destabilizing the traditional role of the physician whose expertise was based on access to specialized knowledge.

In this chapter, I argue that genetic testing is playing an important role in producing the twenty-first century healthcare consumer. Once a service available only through specialized clinics inside research-based university hospitals and used to predict or diagnose a handful of severe diseases, an array of genetic tests are now offered by companies through the internet and used to identify a wide variety of characteristics and conditions. Some tests predict an individual's risk of developing breast or ovarian cancer, while others, as Kimberly Tallbear discusses (Chapter 1 in this book), claim to map an individual's genetic ancestry. Genetic testing companies, health policymakers, and some scholars herald the potential of these tests to empower users. Nikolas Rose and Carlos Novas, for example, have suggested that with the rise of genetic medicine has come the birth of the "somatic individual," who, rather than being passive and disadvantaged by biological destiny, is able to use the information generated through genetic testing to "increase the quality of their lives, self-actualize, and to act prudently in relation to themselves and to others" (Novas and Rose 2000: 487). Responding to scholars who argue that the rise of genetic technologies will lead to the stigmatization of genetically disadvantaged individuals, these authors and their followers remind us that these users should not be understood simply as helpless victims. By producing knowledge about their bodies, these technologies can also help users better govern themselves and their futures. Novas and Rose's "somatic individual" can also be understood as an autonomous consumer, who makes rational choices in the marketplace about whether to purchase a genetic test and which test to take. Once she is equipped with the additional information generated by the test, she can become a better and more rational actor in the marketplace.

But those who emphasize the empowerment potential of genetic technologies, including these scholars who have developed the idea of the "somatic individual," assume that the experiences of users will be the same regardless of the technology's provenance and design, and the position of the user. Indeed, the tantalizing prospect of genetic testing as a tool of empowerment rests on four key assumptions: that the information generated by this technology is accurate, easily understood, and benign; that the information produced is similarly empowering regardless of its utility; that it will have similar meaning and consequences regardless of the demographics of the user; and that offering this information directly to the end user will allow her to become sufficiently expert to make independent decisions about her life. If we look deeply at these genetic testing systems, however, we see that these assumptions are problematic. Rather, we see that the design of a genetic testing system shapes the utility of the information provided and its capacity for empowerment. In addition, treating the users of genetic tests as typical consumers may have significant drawbacks for their health and welfare, which public health professionals and policymakers would do well to address.

This chapter brings together two strands of scholarship from the field of science and technology studies (STS). The first demonstrates that a technology's historical, social, and political environment, including its users, shapes how it is built and marketed (Bijker et al. 1989, Wetmore 2004, Winner 1980). My own research has demonstrated how national context, including its political culture, institutional arrangements, and social norms and values shape the development of science and technology (Parthasarathy 2007). I have also suggested that this context influences both the design of technologies and their implications. To facilitate this analysis, I introduced (Parthasarathy 2007) the concept of "sociotechnical architectures" – the human and technical components of innovations and the way developers fit them together to perform specific functions. Identifying sociotechnical architectures and tracing their development illuminates how choice over the form of each component and its assembly into a functioning whole influences a technology's social consequences.

Just as a building's architecture is the orderly organization of materials and components to achieve a functional, economical, and sometimes environmentally sustainable entity, a technology's "sociotechnical architecture" incorporates functional, social, and ethical considerations to serve human needs. Thus, to conduct an architectural assessment, we must first identify the functions of the technological system being assessed. We then identify the human and technical components that fulfill those functions – how the technology is packaged, how it is offered, inherent safeguards and limitations in the offering, and its cost. Finally, we study how the choice, design, and assembly of these components "structure" the technology's implications, by influencing the way system participants interact with, and think and make decisions about, the technology. A building's architecture shapes who can enter, how they move inside, and what and where activities can be performed. The placement of windows or doors, for example, influence the types of discussions that occur in a room, and often have symbolic meaning regarding an occupant's power and authority. Similarly, the architecture of a genetic testing system enables and constrains user empowerment, familial relationships, and the healthcare professional's level of authority, among other things. In other words, we cannot understand the empowerment potential of a technology without investigating, in detail, its sociotechnical architecture.

Second, a growing body of science and technology studies literature has made the concept of expertise an object of analysis. This work has caused us to question how we identify an expert in a particular subject or policy domain, arguing that academic credentials and professional qualifications should not necessarily be seen as more relevant to decisionmaking than the knowledge that comes from experience. Many patient advocates, for example, have argued that patients possess experiential expertise that is important for policymakers to consider (Callon and Rabeharisoa 2008, Epstein 1996, McCormick et al. 2004). This work suggests that the expertise proffered by highly specialized scientists, for example, is only partial and is sometimes subjective, offering an overview based on research among large samples rather than a deep understanding of an individual case. This shapes sociotechnical arrangements in particular ways, toward more scientifically elegant or medically efficient solutions rather than those that consider the needs and experiences of users (Epstein 1996, Lerner 2001). More and more, scholars observe, citizens are challenging the expertise of the scientific and technological establishment in favor of their own experiential understandings (Parthasarathy 2011). This strand of research generally applauds efforts to consider expertise in a more multi-faceted way and to democratize decision making about science and technology. The analysis provided in this chapter, however, cautions us against accepting the expert capacity of healthcare users. First, it suggests that because these genetic tests generate complex and often minimally digested information, their utility is not obvious. Therefore, there is still a role for the traditional genetics expert who can help the user understand her genetic risk in the context of both her individual circumstances and aggregated scientific data. Second, it shows us that although the users of genetic testing seem like somatic individuals who now have the power to use risk information to make better decisions about their lives, genomics companies, who have an interest in increasing access to and demonstrating the utility of their technologies, are actually quietly replacing healthcare professionals and science policy advisors in the expert role. In other words, the visible experts are being replaced by the often invisible producers of innovation.

In what follows, I explore the definition – and re-definition – of the user of genetic testing as a healthcare consumer. I argue that the sociotechnical architectures of specific genetic testing systems shape their somatic individualism. I begin with an overview of the history of genetic testing and a picture of the early genetic testing user in the United States. I then compare two currently available genetic testing systems – offered by Myriad Genetics and 23andMe – and demonstrate that each offers a different kind of empowerment potential, which is produced by its sociotechnical architecture. I also explore the benefits and risks of the empowered consumer that each envisions. I conclude with a brief discussion of how the ongoing controversy over human gene patents affects the production of the healthcare consumer, and I offer suggestions for how we might deal with the complex implications of user empowerment in the new era of genetic testing.

History of genetic testing in the United States

The first genetics clinics emerged in the first half of the twentieth century in hospitals connected to universities. At this time, no laboratory tests were available to analyze the genetic makeup of individuals and their families. Rather, the genetics analysis was a consultation between a genetics expert (either MD or PhD trained) and the user (and, perhaps her spouse or family members). It usually occurred when an individual or her healthcare professional suspected a hereditary component to a particular condition or disease present in a family member. While some of these users would consult with pediatricians or obstetricians, most of them, seeking information about whether to marry or reproduce, would eventually meet with specialists in medical genetics, discuss their family history, and learn about the hereditary dimensions of their condition or disease. These genetics specialists, however, operated quite differently than most physicians of the day; not only did they not have quick cures for genetic diseases, but they also opted to take a non-directive approach in their consultations. Trying to distinguish themselves from an earlier eugenics movement in which governments, healthcare professionals, and scientists sought to take control over reproductive decision-making, geneticists did not prescribe a specific course of action, but rather tried to simply provide information and discuss options about the meaning of an individual's family history of diseases and the consequences of particular marital and reproductive decisions (Paul 1995, Kevles 1988).

These services changed significantly in the 1960s and 1970s, as laboratory tests to detect chromosomal and DNA anomalies became available and the U.S. Supreme Court legalized abortion (Lindee 2000). More and more diseases could be diagnosed (and some even treated) biochemically, leading to the addition of laboratory testing to the clinical consultation. For example, biochemists developed karyotyping techniques that they could use to identify conditions that resulted from extra or missing chromosomes such as Down's Syndrome and various sexual anomalies. Clinicians could now use these laboratory services to offer a more refined diagnosis of genetic disease as well as more "treatment" options, including pregnancy termination and dietary changes. Sometimes they even held out the possibility of future cures. Geneticists remained the primary advisors, gathering family history information and counseling individuals about particular genetic conditions, but laboratories at academic medical centers now played an important role as well, confirming or rejecting diagnoses made using family history information.

As genetic medicine began to expand and gain force, the U.S. government took notice. Many states developed newborn screening programs for diseases such as phenylketonuria and sickle cell anemia, and, in the early 1980s, the President's Commission for the Study of Ethical Problems in Medicine and Biomedical and Behavioral Research (1983) suggested that genetic medicine be developed with explicit attention to five ethical principles: confidentiality, autonomy, increasing knowledge, respect for well-being, and equity. As it defined these principles, the President's Commission acknowledged that this new area of medicine would raise regulatory questions; in a country with a private health insurance system, it was likely to be rapidly commercialized, for example, and would thus require the development of professional and quality standards.

As the President's Commission predicted, genetics did grow rapidly in the 1980s, as researchers began to find genes and build DNA-based genetic tests for diseases such as sickle-cell anemia, cystic fibrosis, and Huntington's disease. Most of these tests, like the karyotyping and biochemical techniques that had emerged earlier, were built in and provided by research laboratories at academic medical centers. As tests became available for more and more common diseases, however, larger diagnostic laboratories across the country began to recognize that demand for genetic testing services was likely to increase and considerable revenues might be available for those who provided genetic testing services on a wide scale. Large laboratories at academic medical centers that already offered a variety of diagnostic tests, including the Mayo Clinic in Minnesota and Baylor College of Medicine in Houston, began to develop infrastructures to offer DNA analysis services to clinics beyond their institutional walls. Private companies and stand-alone laboratories began to develop such services as well. Economies of scale worked to the advantage of these larger laboratories. While most academic medical centers might be reluctant to build up services to test for each rare genetic condition, such testing might be lucrative for large clinical laboratories that were able to test samples from all over the country. These companies also applied for and received patents on these tests and on the genes themselves, which allowed them to develop potentially lucrative monopolies. Despite these changes, users interested in these services still had to access them through genetics specialists based at academic medical centers.

While the growing availability of these services led to extensive public discussions about the impact of genetic medicine, very few new policies were devised to deal with this set of technologies. Rather, genetic testing was fitted into existing regulatory frameworks. In 1988, the Clinical Laboratories Improvement Act (CLIA) was amended to cover the analytic validity (accuracy and reliability) of genetic testing. But this policy did not address genetic testing's clinical complexities. How would doctors and patients interpret the complicated risk information generated by these tests? Were all of these tests useful, and how should we judge utility? If they weren't useful in improving health outcomes, then should they still be available?

Over the next decades, multiple advisory committees recommended a comprehensive approach to govern genetic testing (Committee on Assessing Genetic Risks 1994, Task Force on Genetic Testing 1997, McCabe 2000). They developed various standards to assess the clinical utility of the information generated by a genetic test, and suggested that these be used to develop a regulatory framework. These recommendations, however, have had virtually no effect. At the dawn of the twenty-first century, as the number of genetic testing services exploded, only CLIA governed this area of technological development. CLIA's advisory committee developed regulations covering analyte-specific reagents (ASRs), the active ingredients of diagnostic tests, which were increasingly available for sale to laboratories setting up genetic testing services. Of course, these proposed regulations still only focused attention on the laboratory aspects of the test. Meanwhile, physicians (sometimes through professional organizations) continued to oversee themselves and make their own determinations about the clinical utility of genetic tests and the interpretation of genetic risk information.

By the beginning of the twenty-first century, a large number of genetic tests were available. Although a mixture of academic and private laboratories offered them, they were almost always accessed through genetics specialists at academic medical centers. Thus, there seemed to be considerable informal control over access to these services and the types of services provided. In particular, with the increasing availability of tests that generated complex risk information, specialists provided targeted counseling to users and their families. Although the government had thus far refused to step in, genetics specialists played pivotal roles as gatekeepers and experts who could shape the meaning and utility of the technology for users in the context of their family history and circumstances. In this era, the user looked more like a traditional patient, reliant on help and expertise from the referring physician and the genetics specialist, than a well-informed and prepared consumer.

Myriad Genetics, Inc.

The picture described above changed dramatically when Myriad Genetics, a publicly-held biotechnology company based in Salt Lake City, Utah, began to offer genetic testing for breast and ovarian cancer in the United States. Although it was not the first commercial provider of genetic testing, it was the first to offer commercial services for such a widespread disease, to provide such services through any physician, and to market its technology directly to users. Myriad built its testing system by constituting the users of genetic testing as virtually autonomous consumers who could be empowered by simply purchasing DNA analysis, even though the risk information it generated was rather complex and its clinical utility unclear and controversial.

Formed to capitalize on the genealogical data that had been collected over centuries from large Mormon families who lived in the state, Myriad invested initially in a large-scale research effort to find the BRCA genes, which were thought to be linked to inherited susceptibility to breast and ovarian cancer. It looked for, and eventually found and developed tests for, other cancer genes as well, but the BRCA gene investigations were clearly its most important investments; breast cancer was a common disease in much of the developed world, quite visible in the public consciousness, and caused millions of women considerable anxiety. Thus, a test that looked for inherited mutations that might cause the disease would have a very large market.

The company's efforts were successful. In 1994, Myriad announced it had mapped and sequenced the BRCA1 gene and by 1995, the company announced that it had mapped and sequenced the BRCA2 gene as well.[1] Despite considerable excitement over these breakthroughs, however, the picture that emerged from the gene discoveries was quite unclear. Mutations in BRCA 1 and 2 were only relevant to a small fraction of breast cancers (about 5%), and an individual with a BRCA mutation seemed only to have an increased risk (rather than a certainty) of future disease incidence (a risk that ranged widely from 30–85% over a lifetime) (Easton et al. 1994, 1995, Hartge et al. 1997). This contrasted with most of the genes that had been discovered at the time, including for cystic fibrosis and Huntington's disease, which when mutated demonstrated that an individual had or would get a particular disease. Furthermore, except for surgery to remove breasts and/or ovaries (which have serious side effects), no preventive options seemed clearly effective. Despite the complexity of such "susceptibility" genes, Myriad immediately set about developing a test to search for mutations in the BRCA genes (BRCA testing). At the same time, it applied for patents on both genes, adopting the same strategy as many other U.S. gene discovery companies before it. In fact, these patents proved to be particularly important for Myriad. Once they were granted, the company used them to shut down all other providers of BRCA testing and become the sole provider of the technology in the United States (Parthasarathy 2007).

By late 1996, the company had set up BRACAnalysis™, the first commercial service to test for inherited mutations in both BRCA genes. It decided to offer prospective users with a choice of four laboratory tests. Those curious about their BRCA mutation status but with no known BRCA mutation in the family may have been interested in the company's "Comprehensive Analysis" which offered DNA sequencing of both BRCA genes – a technique, it noted, which was considered the "'gold standard' of genetic testing because it identifies mutations that cannot be found by any other method" (Myriad Genetics, Inc. 1996). This test cost approximately $3000.

For a user who already knew that a particular disease-causing mutation was present in her family, a "Single Site," mutation-specific, analysis was available. This test was considerably cheaper than the comprehensive analysis, costing about $250. For approximately $450, users of Ashkenazi Jewish descent could take advantage of the company's "Multisite" analysis, which searched for three mutations known to be common among the Ashkenazim. Finally, the company offered Rapid BRACAnalysis™, a full-sequence analysis of both BRCA genes that would be done quickly, in approximately two weeks. The company targeted this analysis to users who had recently learned of a breast or ovarian cancer diagnosis, suggesting that additional information about their BRCA gene status could help them make decisions about what kind of treatment avenue to take (for example, whether to remove a malignant tumor through lumpectomy or to have a complete mastectomy of both breasts).

Although it expended considerable effort in developing its extensive menu of laboratory tests, the company did not pay the same kind of attention to the clinical dimensions of its testing system. Many scientific and medical organizations, patient advocacy groups, and government advisory committees had recommended that genetic tests in general, and in particular complicated genetic analyses such as those for the BRCA genes, should only be available to a limited group of people with significant family histories of breast and/ovarian cancer and also integrated with specialized counseling by trained professionals such as medical geneticists or genetic counselors. In other words, they advocated embedding BRCA testing into the existing system that provided access through genetics clinics at academic medical centers. But the company argued that it, like many of the other private diagnostic laboratories that offered tests for conditions such as high cholesterol and pregnancy, was simply responsible for the conduct of its laboratory (for example, whether or not the chemicals and machines used to conduct the analysis were working properly). It refused to involve itself either in who used its laboratory analyses or the type of counseling that was provided. A company official noted, for example, how difficult it would be to assure the quality of genetic counseling:

> You know, are we going to have to have people pass some sort of exam? How are we going to ensure the quality of genetic counseling? ... And I can tell you, there were many, many discussions about, let's just hire a bunch of genetic counselors and provide genetic counseling. Sort of the way, sort of the way that Genzyme [another genomics company] does it. And we did think about that very closely as well, and didn't feel that that was meeting the goals of where we wanted to go in the laboratory.
>
> *Myriad Genetics Counselor, 2000*

In sum, the central component of Myriad's testing system architecture was the DNA analysis. Any physician could provide access to the test, and the company encouraged users to visit another physician if the first would not provide access. The average primary care physician, of course, would not be able to provide specialized counseling that might help users understand how to interpret gene mutation status such as insight into how a specific mutation caused disease in a particular family and how to use the gene mutation information to influence healthcare or lifestyle choices. Thus, Myriad built a system that ensured broad access but did not include specialized genetics counseling as a necessary part of the system's architecture. Requiring the involvement of genetics specialists would have made them gatekeepers and limited access, since they are traditionally linked to academic medical centers and there are far fewer of them than other specialists such as oncologists or surgeons.

In developing this architecture, the company thus determined that the *expertise* of the genetics specialist was less important than *access* to the DNA analysis. It characterized its DNA analysis as just like any other consumer product that did not require users to access it through a credentialed expert. In fact, the company argued, limiting access would restrict the test's "empowerment" potential. Myriad also privileged access, and the empowerment potential of the test, over the expertise of the science policy advisor: as discussed earlier, Myriad's DNA analysis had received only limited government oversight. Thus, it might be more fruitfully compared to a toothbrush or toothpaste than a home pregnancy test kit that has undergone extensive Food and Drug Administration (FDA) review.

Meanwhile, although the company was reluctant to influence *how* users gained access to and were counseled about the test, it encouraged demand by marketing the test directly to the public. The company produced brochures and videos that served dual purposes of promotion and education, while using traditional advertising routes as well. In the late 1990s, it began with advertisements in locations as diverse as *The New York Times* magazine, *USAirways Magazine*, and a Broadway playbill (Myriad Genetics Laboratories 1999, Myriad Genetics Laboratories 2000). By September 2002, it had launched an ambitious radio, television, and print marketing campaign in popular magazines such as Better Homes & Gardens and People and during television shows such as "Oprah," the "Today Show," "ER," and the series premiere of "CSI: Miami" (Myriad Genetics, Inc. 2002b, Myriad Genetics, Inc. 2002c).

The company's advertisements and educational materials focused on how its DNA analysis technology could empower its users. Echoing the language of the women's health movement of the 1970s, an educational video produced by Myriad showcased a woman who stated "Knowledge is Power," when describing how the company's BRCA testing service had influenced her life (Myriad Genetics, Inc. 1999). The company also expressed similar sentiments in educational brochures that were available to potential users through physicians or directly from the company upon request. One such brochure stated, "Given a choice, would you rather deal with the known or the unknown?" while the back of the brochure offered "Answers" (Myriad Genetics Inc. 2000). Advertisements reiterated this connection between genetic mutation information and empowerment. The New York Times magazine advertisement, for example, showed a woman boldly staring straight at the camera with her arms crossed, declaring, "I did something today to guard against cancer" (Myriad Genetics Inc. 1999). The 2002 media campaign emboldened women to both "choose to do something now" and "be ready against cancer now" (Myriad Genetics Inc. 2002a, Myriad Genetics Inc. 2002b, Myriad Genetics Inc. 2002c). In a sustained and comprehensive strategy, and in multiple venues, the company relayed the same clear message: by taking the accurate and informative genetic test women would be empowered to take charge in the delivery of their own healthcare.

At first glance, the user envisioned by Myriad through both the architecture of its testing system and marketing strategy seems to fit the description of the "somatic individual" articulated by Rose and his colleagues, an engaged and active consumer. She can demand access to the test, and use the risk information generated to make decisions about her life, her health care, and her future. But upon further consideration, the picture becomes far more complicated. The user is a consumer whose power to take advantage of her BRCA risk information is constrained by both the company and the design of its testing system. First, although Myriad offered users the opportunity to "choose to do something now," the choice is by no means unlimited. When it used its patent position to shut down all other testing providers, a totally unsurprising move for a for-profit provider, Myriad limited the prospective user's choices. She would only be able to choose among the options offered by Myriad. In this way, the genomics company took on an authoritative, expert role. In addition, access to Myriad's laboratory analyses was constrained by cost; not only were they expensive, but most users paid for them themselves because they were reluctant to tell their insurers and employers about their genetic status.[2] Second, Myriad's stand-alone laboratory technology offered a specific kind of empowerment. Its technology, DNA analysis of the BRCA genes, focused attention on the empowering potential of genetic mutation information alone. Rather than suggesting that users would be empowered through a combination of DNA analysis and specialized counseling, which would provide users not only with mutation information but also insight into the meaning of her family history for her disease risk, the benefits and risks of testing, and the prevention options available, it suggested that simply knowing one's mutations status would help users make better decisions. Finally, while Myriad suggested that information about BRCA mutation status would allow the user to make better choices about her life, it also placed the burden of these decisions squarely upon her shoulders. Users were expected to navigate through the murky world of risk statistics and treatment options without advice from a specialist in genetics or cancer. Consequently, she would ultimately be accountable for the decisions that she made. Healthcare professionals were facilitators, providing access to the technology, while the user was envisioned as a rational actor who was fully prepared and capable of assessing the options at her disposal.

23andMe

Ten years after Myriad began offering BRACAnalysis™, biotechnology investment analyst Anne Wojcicki founded 23andMe in Mountain View, California. Located in Silicon Valley, the company seeks to bring computing power together with the promise of genomic science and is guided explicitly by values of empowerment and democratic engagement. It is no surprise that Wojcicki has married to one of Google's founders, as 23andMe's philosophy – not unlike the ideas that drove the creation of the internet (Turner 2006) – suggests that increased access to knowledge provides the path to democratic liberation. Of course, 23andMe also adopts a similar approach to Myriad, announcing "Knowledge is Power" when the potential user arrives at the webpage through which she can order its test. "23andMe empowers you to better manage your health and wellness," it notes further, through access to more than 200 personalized test reports, "easy access" to your genetic information, and "the largest genealogical DNA database in the world" (23andMe 2013a). The company underscores this approach when describing its "core values," which include: "that having the means to access one's genetic information is good"; "that people's similarities are just as important as their differences"; and that everyone should have "the opportunity to contribute to improving human understanding" (23andMe, 2013b).

While Myriad's test finds known deleterious mutations in two genes, 23andMe's service provides users with risk information based on tests for genetic variants linked to "more than 200 diseases, traits, and health conditions and we regularly add new information" (23andMe 2013c). This "new information" is based on research conducted by genetics researchers around the world, but also by 23andMe. When users sign up for its service, they are asked to answer surveys about their personal and family histories and circumstances. This data, combined with the information gathered from DNA analysis, allows the company to conduct its own research. Although Myriad has also built a proprietary database with which it can conduct its own investigations (and presumably offer unique insights, thereby increasing the value of its services, without patent protection), 23andMe's databases are more comprehensive and are central to its business model. The user is thus, simultaneously, a consumer, a research subject, and a patient, and her opportunity for self-actualization depends both on the services that 23andMe currently offers and on the findings that the company generates in the future. While 23andMe markets its research as benefitting its customers, it does not, understandably, emphasize how the user's multiple identities might conflict or shape her empowerment (for example, how the data it gathers might be used in ways that the consumer/research subjects may not like).

This approach goes far beyond rhetoric. Like Myriad, 23andMe embeds its values in the sociotechnical architecture of its testing system in a variety of ways. As it does this, empowerment takes on a particular meaning. It is quite simple and cheap to access 23andMe's test, with a relatively low cost ($99), and can be ordered with a few clicks on the company's website and the input of one's basic information. After the user orders the test, a kit arrives in the mail. It contains instructions, a vial to store a saliva sample, consent forms that cover both the testing process and the use of the genetic information for research purposes, and pre-paid shipping labels. Once the user sends the consent forms and saliva samples back to 23andMe, she must wait anywhere from four to eight weeks for her results. In making the test directly available to the user, the company foregoes both a gatekeeping and an expert role for the healthcare professional. While the company does have genetics experts on its staff, it does not offer their services to users as they decide whether to take the test or after they receive the results. Like Myriad's system, this architecture envisions the test as similar to a regular consumer product. But 23andMe's system is even cheaper and easier to access than Myriad's test. It does not require that the user visit a physician, and instead facilitates direct access to the DNA analysis. The company envisions the user as autonomous, capable of deciding whether and how to use it on her own. Meanwhile, as described earlier, unlike over-the-counter drugs or home pregnancy tests, 23andMe's test is subject to very little government oversight.

Once the user's saliva sample arrives at 23andMe's laboratory, the company conducts its DNA analysis. While it tests for some variants that are linked to significantly increased risk of disease, many of its tests are based on single nucleotide polymorphisms (SNPs) associated with only mildly increased risk of getting the disease (3–5%; some SNPs seem to have no effect at all). This contrasts with BRCA and other susceptibility genes, which put those who have mutations at high risk of contracting the disease. However, using genome-wide association studies, scientists have connected particular SNPs with the increased likelihood that a person has a particular trait or is at mildly increased risk for a particular condition. As mentioned above, 23andMe is also conducting this kind of research. In other words, the company provides users with risk information based on multiple genetic variants, and each confers a different level of risk. Although 23andMe does provide some information about the nature of each of these risks, this information is quite limited and likely difficult for the average user to understand – particularly because it provides aggregate, rather than individual-level, risk.

23andMe adds a SNP to its testing platform as soon as researchers have established its association with a particular trait or health condition. This allows the company to offer users results from a large number of tests, but the utility of these tests may be limited. For example, 23andMe tests for two genetic variants related to schizophrenia, each based on a single study. Meanwhile, competing SNP testing company Navigenics does not offer these tests because, "recent research into the genetic risk markers for this mental illness has yielded inconsistent results." 23andMe deals with the variable quality of the research that supports its analyses and the lack of replicated results by generating two kinds of research reports for users: "Established Research Reports" and a "Preliminary Research Result."

"Established Research Reports" are based on "genetic associations supported by multiple, large, peer-reviewed studies" and "widely regarded as reliable" (23andMe, Inc 2013d). Therefore, the company can provide some quantitative estimates and explanations of their meaning. For example, in addition to describing the disease itself, 23andMe's Established Research Result for Celiac Disease calculates the probability that someone with the user's genotype will get the disease and compares the user's probability to the average probability across all men with the same genotype and within the same age range (23andMe, Inc. 2013f). This result also calculates the importance of genetic susceptibility for Celiac disease (in comparison to environmental factors).

For most of the variants it analyzes, 23andMe generates only a "Preliminary Research Result." In these cases, scientists have conducted "peer-reviewed, published research," but they have not replicated the results in other studies (23andMe, Inc 2013d). In many cases (like schizophrenia) the test is based on a single study, and occasionally, there is "contradictory evidence" of the importance of a particular genetic variant. The Preliminary Research Result for Schizophrenia provides a very short description of the disease and the two studies on which 23andMe's test is based (23andMe, Inc 2013e). It describes the user's two relevant genetic variants and then compares them to the findings of the two studies. So, in the sample report available online, the user has a CT variant at one location on the genome, which one study has found is associated with "slightly higher odds" for schizophrenia. The report also says that the study was conducted among – and is therefore likely to be most useful to – "Asians," but it does not define who might be included this category.[3] It does not, furthermore, provide information about the importance of genetic versus environmental risk (presumably because researchers do not yet know). 23andMe has tested for these particular variants since at least 2009 even though no scientists have replicated either of these studies. In sum, the company values making the results available to users over ensuring that the results are reliable or useful. The presumption is that the user can assess reliability and utility in the context of her own life better than the company or a healthcare professional can. In other words, 23andMe views empowerment in terms of potential, based on a user's ability to interpret its results. In addition, of course, by continuing to analyze these variants and collecting survey data from users, 23andMe may be able to replicate these results itself sometime in the future.

After 23andMe completes its analyses, it invites users (via email) to review the results online. When users log in, they see an overview of their genetic risk and carrier status for all of the conditions analyzed. The risk results compare the user's risk to "average risk," in percentage terms. Information on carrier status simply states whether the relevant variant is present or absent. Users can click through to generate either the "Established" or "Preliminary" Research Report. The overview page does not address the relevance of genetic makeup for a particular disease; for example, while it may report that a user has a 30% risk of contracting coronary heart disease, the user must access the Established Research Report to discover that only 39–56% of all coronary heart disease is attributable to genetics. As stated above, if only a Preliminary Research Report is available then the company does not provide any information about how to evaluate the genetic information it has generated. This approach assumes that the user will interpret both the genetic risk probabilities and the relative importance of genetic versus environmental information in ways that are the most useful to her, and that she does not need assistance from a healthcare professional.

Even after test results have been conveyed, 23andMe reinforces its approach to user expertise and empowerment through its social networking site, 23andWe. Rather than providing opportunities to chat with genetics specialists, the site facilitates connections and conversation among users. Users share their knowledge and experiences with 23andMe's system, and with healthcare more broadly, and teach one another how to interpret risk information and respond to test results. In 2012, the company emphasized this approach by acquiring CureTogether, a crowd-sourcing platform that encourages users to communicate with one another to compare symptoms and treatments to increase knowledge of the disease. In sum, 23andWe has replaced the healthcare professional and science policy advisor. Through the site, the user's anonymous peers provide expertise on how to understand genetic risk information and what to do about it. Together, with 23andMe's assistance, they are taking back the power to interpret information about their bodies. But each peer's individual circumstances and experience shapes her expertise, and 23andMe – and its users – must assume that in the aggregate, these experiences provide better information than a traditional expert who has the benefit of standardized and systematically analyzed data but lacks the insights that come from experiential understanding. The problem, of course, is that users must rely on other users' understanding of complicated risk information and knowledge of genetic science and terminology. They must also weigh the relative merits and drawbacks of other users' experiences for their own lives. On the one hand, this kind of crowd-sourcing can provide users with information that traditional experts are unlikely to provide about how to incorporate genetic risk information into everyday life. But on the other, it exposes users to the self-reported experiences of a self-selected and interested group of strangers without much information from an outside, disinterested perspective.

It is important to note that this will help 23andMe as well, by providing the company with additional information to refine the meaning of its test results; in 2012 the company also announced that it had received its first patent on variants associated with Parkinson's disease, based on its DNA analysis and survey data information. Although the company has yet to exercise its patent rights, it is possible that it will adopt a similar strategy to Myriad and prevent other providers from offering genetic tests. If it does this, it will reinforce its own expert and gatekeeping role and continue to diminish the role of the healthcare professional.

Conclusion

Genetic testing is often seen, understandably, as the quintessential technology of empowerment. It generates information that patients can use to make decisions about their healthcare and lives. The empowerment promised by these technologies is particularly popular in an era when citizens are eager to participate more directly in decisions about science, technology, and medicine, and when scholars from a variety of disciplines are encouraging the democratization of healthcare decision-making.

The initial provision of DNA analysis in conjunction with counseling in the United States, made available through specialists at academic medical centers, limited access to genetic testing and restricted the power of the individual to interpret the test results in the context of her own life. Medical professionals played an active role in shaping access to the testing system and the interpretation of results. The user resembled a traditional patient. But as the genomics industry grew, companies expanded their markets by transforming the user from a patient into a consumer and, today, into a research subject as well. Taking advantage of a regulatory vacuum and offering DNA analysis outside the context of specialized counseling, both Myriad and 23andMe have managed to increase *access* to their tests and offer users more interpretive opportunity while diminishing the traditional role and expertise of the healthcare professional and the science policy advisor. But this opportunity is shaped and constrained by the testing providers and the system's sociotechnical architectures. The generalist physician who helps a Myriad user access the test may have some insights about the user's family background and current circumstances, but she probably does not have the conceptual tools to interpret complex genetic risk information. Access to BRCA risk information without the expertise of a genetics specialist could actually undermine the power of the patient because it could lead to less informed medical and life decisions. Meanwhile, the 23andWe community offers users the benefits of both varied and collective experience (in a format that is likely more accessible than dry scientific explanation), but these experiences are self-selected and self-reported and lack the benefits of systematic and comprehensive analysis. In this system, the user's potential for empowerment depends on her own ability to understand test results and the quality of the information provided by 23andWe. In this system, there is even more potential for disempowerment because users could make health and life decisions based on test results that have not been validated by replicated scientific study. Finally, efforts by both systems to collect user data, while sold as improving future services and therefore increasing empowerment potential, may not be in users' best interests.

Novas and Rose's analysis of genetic technology and the production of the "somatic individual" assumes substantial similarities across testing systems. But my analysis suggests that there are important differences in sociotechnical architectures, even between two companies that have dissolved the traditional link between DNA analysis and specialized counseling. A physician must still authorize access to Myriad's BRACAnalysis™ (although the company's marketing materials encourage her to demand access and switch physicians if necessary), while 23andMe's user can simply order its testing portfolio online, and the cost is much, much lower than Myriad's, which is likely to produce more consumer demand. The apparatus that shapes the DNA analyses themselves also differs. Although each company authorizes how the DNA is analyzed, the importance of the BRCA genes in causing breast and ovarian cancer has been validated by multiple independent scientific studies (users suffered, however, when Myriad's method of DNA analysis initially missed some disease-causing mutations and the company refused to change its test accordingly). 23andMe, by contrast, authorizes DNA analysis based on a single, un-replicated study, arguing that this increases interpretive power to consumers. It also encourages users to share their experiences and expertise in order to crowd-source the interpretation of genetic knowledge and prevention and treatment options.

These differences have important implications for users, including their empowerment potential. Myriad maintains a minor gatekeeping role for a physician, whereas 23andMe almost completely eliminates this role except for vague suggestions to discuss results with a doctor. Because Myriad's test is far more expensive unless an insurance company (which usually imposes some access restrictions based on family history) reimburses the user, the rights and responsibilities of the consumer are still somewhat restricted. In many respects, 23andMe's system produces a more traditional consumer, with unfettered rights to access the test at a relatively low cost and to interpret the results. But she is also specially burdened, as she must deal with the test results not only without the involvement of a healthcare professional but also without the often invisible infrastructure of scientific advice and regulation. 23andMe trusts this burdened consumer to understand – with the help of like-minded users – the results of individual scientific studies in the context of her own life. As we look into the design of each of these testing systems, the costs of *somatic individualism* come into focus. Although users of both Myriad and 23andMe's testing systems now have access to genetic risk information that may help them make better decisions about their lives, they must also bear different levels of burden in interpreting this information. Since this burden is embedded in the system's design, they may not be aware that they are now expected to do the advisory and interpretive work left previously to both independent and government specialists. If the user does not understand the knowledge to which she now has access, is it really empowering? In-depth analysis of a testing system's architecture can also reveal the meaning and consequences of the DNA ancestry testing systems that Kim Tallbear analyzes in Chapter 1. As Tallbear notes, scientific studies that guide these kinds of DNA analysis are often quite limited and based on erroneous understandings of race, ethnicity, and population. But as these tests become more popular, they are likely to re-inscribe particular definitions of those categories to negative effect.

The design and growing popularity of this next generation of genetic testing systems also raise important questions for scholars and advocates who encourage greater lay involvement in healthcare decision making. As the systems discussed here provide users with undigested risk information and validate their experiential expertise in the interpretation of test results, they also diminish the roles of both scientists and healthcare professionals who can understand and interpret scientific studies. In other words, the move toward the supposedly empowered healthcare consumer seems to be disempowering not only individual healthcare professionals but also the advisory and policy apparatus associated with them. Is there no place for aggregate understanding, and interpretive insights, that specialists provide? If there is a place, how should it be balanced with the experiential expertise of average users? To what extent should the user be burdened with the responsibility of understanding and interpreting a genetic test's results? Will the move toward somatic individualism have ripple effects for policymakers, making it increasingly difficult to make policy decisions regarding research priorities or unsafe food and drugs? Will it delegitimize regulatory institutions like the FDA?

While it is easy to assume that genetic tests do not need to be tightly regulated because they are relatively straightforward, non-invasive technologies that simply produce information to facilitate decision making, we have seen that the reality is much more complex. The way a technology is built shapes the type of information that is produced, how this information is accessed, and thus, how an individual can use genetic information to enhance her healthcare or life. The lack of regulation in this arena has led not only to technological variation, but also to variation in the interpretive burden that users must bear. This can have serious negative consequences for individual users in terms of the utility of genetic risk information, and for society, as DNA seems to validate particular disease definitions and social categories. There is an opportunity for policymakers to intervene, however; the FDA routinely makes decisions about how drugs should be made available, to whom, and under what conditions, and has already acknowledged its regulatory authority over genetic testing. The knowledge that genetic testing produces is not, as I have shown, automatically empowering. If we want it to maximize this potential, the government must step in.

Postscript: In June 2013, the US Supreme Court invalidated patents on isolated human, including the BRCA, genes. But Myriad is likely to maintain its monopoly on BRCA gene testing, at least in the short term, because of pending litigation, and its mutation database. In late 2013, 23andMe suspended its health-related tests temporarily, in response to an FDA letter questioning whether they were analytically valid and based on compelling scientific evidence. If the company is forced to alter the suite of tests it offers, this could redefine its approach to empowerment. But, if it simply tones down its rhetoric, the effects are likely to be negligible because the user's power is embedded in the service's sociotechnical architecture.

Notes

1   There is dispute over priority of the BRCA2 gene discovery. Robert Dalpé, Louise Bouchard, Anne-Julie Houle, and Louis Bédard. 2003. "Watching the Race to Find the Breast Cancer Genes." *Science. Technology, and Human Values*. 28: 187–216.
2   This situation has changed somewhat, since the passage of the Genetic Information Non-Discrimination Act in 2008. GINA, as it is called, prohibits employers and health and life insurers from discriminating on the basis of genetic information (broadly defined). Today, more insurers pay for BRACAnalysis™, but they often have rules about the family history an individual must have to take the test (so individuals will pay for the technology themselves in order to gain access).
3   As both STS scholars and geneticists have argued, and Kimberly TallBear discusses in detail in this volume, race cannot and should not be understood as a biological category. Therefore, its utility in understanding genetic test results is unclear. See, for example, Wailoo, Keith, Alondra Nelson, and Catherine Lee, eds. 2012. *Genetics and the Unsettled Past: The Collision of DNA, Race, and History*. Rutgers, NJ: Rutgers University Press.

References

23andMe, Inc. 2013a. "Store – Cart – 23andMe." *www.23andme.com/store/cart* (downloaded May 12, 2013).
23andMe, Inc. 2013b. "About Us: Core Values – 23andMe." *www.23andme.com/about/values* (downloaded May 12, 2013).
23andMe, Inc. 2013c. "What Health Conditions Are Not Coveredby 23andMe's Personal Genome Service?" *https://customercare.23andme.com/entries/21979637-What-health-conditions-are-NOT-covered-by-23andMe-s-Personal-Genome-Service* (downloaded May 12, 2013).
23andMe, Inc. 2013d. "Health and Traits – List of Conditions – 23andMe." *www.23andme.com/health/all* (downloaded May 12, 2013).
23andMe, Inc. 2013e. "Schizophrenia Genetic Risk – 23andMe." *www.23andme.com/health/Schizophrenia* (downloaded May 12, 2013).
23andMe, Inc. 2013f. "Celiac Disease Genetic Risk – 23andMe." *www.23andme.com/health/Celiac-Disease* (downloaded May 12, 2013).
Bijker, W., Hughes, T., Pinch, T., eds. 1989. *The Social Construction of Technological Systems: New Directions in the Sociology and History of Technology*. Cambridge, MA: MIT Press.
Broom, A. 2005. "Virtually He@lthy: The Impact of Internet Use on Disease Experience and the Doctor-Patient Relationship." *Qualitative Health Research*. 15(3): 325–345.
Callon, M. and Rabeharisoa, V. 2008. "The Growing Engagement of Emergent Concerned Groups in Political and Economic Life: Lessons from the French Association of Neuromuscular Disease Patients." *Science, Technology, and Human Values*. 33: 230–261.
Committee on Assessing Genetic Risks, Division of Health Sciences Policy, Institute of Medicine. 1994. *Assessing Genetic Risks: Implications for Health and Social Policy*. Washington, DC: National Academies Press.
Dalpé, R., Bouchard, L., Houle, A.-J., and Bédard, L. 2003. "Watching the Race to Find the Breast Cancer Genes." *Science, Technology, and Human Values*. 28: 187–216.

Easton, D.F. et al. 1994. "Cancer Risks in A-T Heterozygotes." *International Journal of Radiation Biology.* 66: 177–182.

Easton, D.F., Ford, D., Bishop, D.T., et al. 1995. "Breast Cancer Linkage Consortium: Breast and Ovarian Cancer Incidence in BRCA1-Mutation Carriers." *American Journal of Human Genetics.* 56.

Epstein, S. 1996. *Impure Science: AIDS, Activism, and the Politics of Knowledge.* Berkeley, CA: University of California Press.

Fox, N.J., Ward, K.J., and O'Rourke, A.J. 2004. "The 'Expert Patient': Empowerment or Medical Dominance? The Case of Weight Loss, Pharmaceutical Drugs and the Internet." *Social Science & Medicine.* 60(6): 1299–1309.

Hartge, P., Struewing, J.P., Wacholder, S., et al. 1997. "The Risk of Cancer Associated with Specific Mutations of BRCA1 and BRCA2 among Ashkanazi Jews." *New England Journal of Medicine.* 336.

Kevles, D. 1988. *In the Name of Eugenics: Genetics and the Uses of Human Heredity.* Cambridge, MA: Harvard University Press.

Lerner, B.H. 2001. *The Breast Cancer Wars: Hope, Fear, and the Pursuit of a Cure in Twentieth-Century America.* New York: Oxford University Press.

Lindee, M.S. 2000. "Genetic Disease Since 1945," *Nature Reviews Genetics* 1: 236–241.

McCabe, E.R., Chair, Secretary's Advisory Committee on Genetic Testing to Donna E. Shalala, Secretary of Health and Human Services. 2000. "On Behalf of the Secretary's Advisory Committee on Genetic Testing (SACGT), I Am Writing to Express Our Support…" Letter.

McCormick, S., Brody, J., Brown, P., and Polk, R. 2004. "Public Involvement in Breast Cancer Research: An Analysis and Model for Future Research." *International Journal of Health Services.* 34: 625–646.

Myriad Genetics Counselor. 2000. Personal Interview, April 3, 2000.

Myriad Genetics, Inc. 1996. "Myriad Genetics Introduces the First Comprehensive Breast/Ovarian Cancer Susceptibility Test." Press Release.

Myriad Genetics, Inc. 1999. "…I Did Something New Today to Guard against Cancer." Advertisement. *The New York Times Magazine.* September 1999.

Myriad Genetics, Inc. 1999. *Testing for Hereditary Risk of Breast & Ovarian Cancer: Is It Right for You? An Informational Program for People Considering Genetic Susceptibility Testing.* VHS. Salt Lake City, UT: Myriad Genetics Laboratories.

Myriad Genetics Inc. 2000. *Breast and Ovarian Cancer: Given the Choice. Wouldn't You Rather Deal with the Known Than the Unknown?* Salt Lake City, UT: Myriad Genetics, Inc., 2000.

Myriad Genetics, Inc. 2002a. "BRACAnalysis™: Be Ready Against Cancer," Television Advertisement.

Myriad Genetics, Inc. 2002b. *Media Guide Atlanta: BRACAnalysis™.*

Myriad Genetics, Inc. 2002c. *Media Guide Denver: BRACAnalysis™.*

Myriad Genetics Laboratories. 1999. Advertisement for BRACAnalysis™, *The New York Times* magazine. September.

Myriad Genetic Laboratories. 2000. Advertisement for BRACAnalysis™. *US Airways Attaché.* January.

Navigenics. 2009. "Conditions We Cover." *www.navigenics.com/visitor/what_we_offer/conditions_we_cover* (downloaded November 3).

Novas, C. and Rose, N. 2000. "Genetic Risk and the Birth of the Somatic Individual." *Economy and Society.* 29.4: 485–513.

Parthasarathy, S. 2007. *Building Genetic Medicine: Breast Cancer, Technology, and the Comparative Politics of Health Care.* Cambridge, MA: MIT Press.

Parthasarathy, S. 2011. "Breaking the Expertise Barrier: Understanding Activist Strategies in Science and Technology Policy Domains." *Science & Public Policy.* 37: 355–367.

Paul, D. 1995. *Controlling Human Heredity: 1865 to the Present (Control of Nature).* Amherst, NY: Humanity Books.

President's Commission for the Study of Ethical Problems in Medicine and Biomedical and Behavioral Research. 1983. *Screening and Counseling for Genetic Conditions: The Ethical, Social, and Legal Implications of Genetic Screening, Counseling, and Education Programs.* Washington, DC: Government Printing Office.

Task Force on Genetic Testing. 1997. *Promoting Safe and Effective Genetic Testing in the United States.* Bethesda, MD: National Human Genome Research Institute.

Turner, F. 2006. *From Counterculture to Cyberculture: Stewart Brand, the Whole Earth Network, and the Rise of Digital Utopianism.* Chicago: University of Chicago Press.

Wetmore, J. 2004. "Redefining Risks and Redistributing Responsibilities: Building Networks to Increase Automobile Safety." *Science, Technology, and Human Values.* 29: 377–405.

Winner, L. 1980. "Do Artifacts Have Politics?" *Daedalus.* 109: 121–136.

6

The Social Life of DTC Genetics
The case of 23andMe

Alondra Nelson and Joan H. Robinson

COLUMBIA UNIVERSITY

The decoding of the human genome in the summer of 2002 was accompanied by the swift commodification of direct-to-consumer (or DTC) genetic tests – that is, DNA data analyses for sale to the lay public. DNA-based paternity testing had been publicly available since the early 1990s first, at select laboratories, and later through online commerce. In the late 1990s, medical genetic tests such as Myriad's BRACAnalysis (for hereditary predisposition to breast and ovarian cancers) were introduced in clinical settings. But the subsequent decade saw a watershed of DTC genetic testing services aimed at a far broader market than potential parents and possible cancer sufferers. Readily available for purchase on the internet, these new commercial technologies targeted luxury consumers, genealogy buffs, and DIY-science geeks, among many others, and promised to tell us who we are, where we come from, and how we can live optimally.

The trajectory of DTC genetic testing over the last dozen years offers science and technology studies (STS) scholars a rich site at which to examine institutionalization – the process by which objects or practices circulate in regulatory and other types of organizations and through this process come to be understood as normative or "regular" facets of the social world. The DTC genetic testing case is informative as well because it takes place in the context of today's robust neoliberalism and thus sheds light on the effects of the twinned-forces of deregulation of various institutional domains concurrent with the diminution of social welfare programs for healthcare and other services (Moore, Kleinman, Hess and Frickel 2011).

Additionally, the DTC genetic testing case provides a productive contrast with one of the more well-studied trajectories of institutionalization: the pharmaceutical market. With pharma, a significant aspect of the institutionalization process *precedes* the introduction of a product into the marketplace. When a drug enters the market, the product reflects the outcome of months or years of institutionalization, including in the form of laboratory science, clinical trials, the scrutiny of regulatory agencies (e.g., the United States Food and Drug Administration or FDA), and the framing of an illness and its treatment on the part of varied stakeholders (for example, social movements, patient advocates, and professional associations) (Epstein 1996, Dumit 2012).

Both over-the-counter and prescription pharmaceuticals have been directly advertised to consumers for decades and, in some ways, this practice both anticipated and precipitated the rise of DTC genetics. The origins of DTC DNA services lie at the juncture of two sociotechnical processes – molecular biology and supercomputing; these institutional predecessors of today's commercial genetic testing are analogous to some initial aspects of pharmaceutical development. Yet there are important differences in the institutionalization processes of these products that are worth noting: In contrast to pharma, the DTC genetic services industry was introduced by a set of actors – including businesspersons, investors, and scientists – whose scientific claims and products went mostly uninterrogated by outside reviewers or other types of checks and balances and received scant governmental regulation and ethical oversight. Consequently, efforts to institutionalize DTC DNA testing have mostly come *after* products and services have entered the marketplace. And, in contrast with pharma, customers and industry leaders have been able to play sizeable – if uneven – roles in this process.

Whether institutionalization occurs before or after the introduction of a commercial product, it enables classification – the sorting of new or contested objects and entities into classes, categorical boundaries, or architectures of social meaning. Because the introduction of commercial DNA testing proceeded with little external oversight, the classification of this new commercial entity – by state and federal agencies, consumers, industry professionals, and others – remains in formation. Boundaries are actively under negotiation. With DTC genetic testing, institutionalization is evolving as the tests do, making the dynamics of this process readily observable by scholars.

In the face of this regulatory lag, purveyors and consumers of DTC genetic testing may seek to shape the course of institutionalization. When enterprises and organizations are established with a low regulatory threshold, it may be the industry insiders themselves who initiate the institutionalization process. Anticipating regulation, they may seize the opportunity to set the terms of their own surveillance, as did some entrepreneurs who pioneered some of the first commercial genetic ancestry testing services in the United States (e.g., Kittles and Shriver 2004, also see Wagner 2012). On the other hand, genetic testing companies may capitalize on the lack of clarity about the classification of their products to resist institutionalization and create their own boundaries and norms: Some purveyors of DTC genetics have claimed that tests should be understood as personal, leisure pursuits that are non-medical or recreational and, therefore should not fall under the stringent regulatory schemes of agencies like the FDA (e.g., Lee 2013). Similarly, as we describe, consumers may want to keep regulatory institutionalization at bay for fear that it will restrict their access to genetic data, as was the case when federal agencies held hearings on DTC DNA testing several years ago. Consumers testified powerfully about their "right" to their own genetic information, free from government oversight (see our discussion below and also FDA 2010, Vorhaus 2010, Lee 2013).

DTC genetic testing's categorical dynamism presents STS researchers with a challenge. How can scholars study a social phenomenon that is in formation, that may defy classification, or that vacillates between numerous institutions and organizations? One tried-and-true strategy for dealing with this challenge is to fix an object and study it within a single institutional location. For example, medical sociologists are most likely to study diagnostic genetic technologies and may do so at a physician's office or among one group of patients (Atkinson, Parsons and Featherstone 2001). But genetic data is never simply one kind of information. Even if the outcome of genetic testing is supposed to be solely for medical use, the inherent nature of DNA means that it also always contains information about one's health and may also be deemed to be informative for ancestry inference or in a criminal justice setting, even if these uses are not intended. The growing, problematic use of "familial searching" in criminal investigations, such as that leading to the apprehension of the BTK and Grim Reaper serial murderers – that brings the relatives of crime suspects who are disproportionately members of poor communities of color, under unwarranted police surveillance – is a case in point.

A flexible analytical approach is needed to account for the inherent characteristics of DNA that make it informative in numerous contexts and for the emergent, liminal nature of forms of DTC genetic testing. Much like the shift from sociology *in* medicine (e.g., sociologists serving an uncritical supporting role to physicians and medical education) to the sociology *of* medicine (a perspective that brings sociological approaches to bear on medical professions, claims, expertise, authority, etc.) (Chaska 1977), STS scholars should not take all of our analytic cues from the genetic testing industry and the categorical claims it makes about its DTC services.

"The social life of DNA" (Nelson 2010, Wailoo, Nelson and Lee 2012) perspective is a more apt way of describing and analyzing the relatively recent phenomenon of DTC genetics. In keeping with anthropologist Arjun Appadurai's methodological mandate that it is by attending to "the social life of things" – "things in motion" – that we can bring "human and social contexts" into view (Appadurai 1986), in this chapter, we track one DTC genetic test product in order to understand how meaning and norms accrue to it through this flow. Here we also follow Sarah Franklin and Celia Roberts' elaboration of "the social life of PGD" (preimplantation genetic diagnosis) in their book *Born and Made: An Ethnography of Preimplantation Genetic Diagnosis*. Here the social life approach involves "researchers immers[ing] themselves in a range of different contexts to collect data about a particular object of inquiry, 'following it around' to build up a kind of hyperstack of definitions, images, representations, testimonies, description, and conversations…" (Franklin and Roberts 2006: xix). Franklin and Roberts offer a model of "how to account for the social dimensions of new biomedical technologies" (2006: xv) by thickly describing and analyzing these entities and their social circulation.

Our understanding of the institutionalization of an emergent technology and social practice such as DTC genetic testing can be enhanced by the "social life" approach. Because DTC genetic testing is both emergent and transverses categories and boundaries, the descriptive and analytic moves proposed by Appadurai, Franklin and Roberts are apt. Moreover, for genetic testing in particular, a social life of DNA perspective can also help to highlight the symbolic qualities with which we imbue genes and which partly derive from its use as a social explanation in many fora simultaneously (Nelkin and Lindee, 1995).

And, most importantly, this perspective attends to the particular physical properties of DNA that help to constitute how we make meaning of and with it. For, genes are omnibus; they contain multitudes. A social life of DNA perspective offers a way to conceive how the techniques and logics of genetics (especially, the centrality of ideas of kinship; bio-banks and the database; statistics and probability; and molecular scale) are engaged in myriad social projects that may both abide and confound institutionalization. A second property of genes is that they are transitional. Genetic tests and the data they yield move between institutions and organizations, being engaged in various uses ranging from "optimization" to health to "security."

A social life of DNA perspective also helps to account for the boundary blurring that attends the low institutionalization of DTC genetic tests and, by following the circulation of them at varied sites, helps to bring into relief how institutional boundaries take shape, recombine, and collapse. Additionally, following these tests and the contexts in which they draw meaning is precisely what allows us to see how DTC genetic testing does not abide the domains and boundaries that both entrepreneurs and social scientists – for very different reasons – endeavor to put around them.

Focusing on the well-known genetic testing company 23andMe, this chapter charts one course of the institutionalization of DTC genetics. We first briefly describe the technical facets of the spectrum of DTC genetic tests and the socio-cultural meanings that they engender. Next, we describe the boundary crises produced around so-called "recreational" genetic tests that are not merely an idle pursuit and that, furthermore, do not abide the categorical distinctions. In the second half of the chapter, we explain the current framework for regulation of DTC genetic testing in the U.S., returning to the specific case of 23andMe and describing struggles over the regulation of this (and similar) company's services. In closing we discuss what the regulatory struggles over 23andMe suggest about the institutionalization of new technologies.

Direct-to-consumer genetic testing

The human genome is a composite. A central conceit of the Human Genome Project was that we could derive a great deal of information – indeed, life's ultimate data – by deciphering the genetic signatures of a select, unidentified and multicultural group of five persons: three women and two men. (This group was rumored to have included Craig Venter, the scientist who was a driving force behind the completion of this ambitious research endeavor). On the summer day in 2000 when the successful drafting of the human genome was announced, President Bill Clinton proclaimed that this multicultural, multiracial sample of DNA signatures highlighted "our common humanity" (White House 2000). The Human Genome Project confirmed that humans are 99.9% alike. But in the arenas of biomedical and scientific research, it was our supposedly *uncommon* genetic traits that were said to matter most of all, because even a 0.1% difference is meaningful in the context of the more than three billion base pairs that are the building blocks of human DNA. Concomitant with this development in genetic science was growing concern in the 1990s about disparities in health outcomes and research inclusion by race and gender and, soon after, rising support for research into the causes of these inequalities. To some minds, genetics research was poised to offer a biological explanation for the persistence of this form of inequality.

DTC Genetics: not just fun and games

The emergence of direct-to-consumer genetic testing in recent years has been considered of limited use to researchers across the disciplines. Natural and social scientists note that DTC genetic testing is not *real* science compared to survey-based genetic testing. This position has some merit, as these tests do not meet clinical research validity standards, and the company databases to which DTC genetics companies compare customer DNA are proprietary and therefore not subject to verification or refutation from other researchers or genetic testing companies who use different statistical assumptions, algorthims, or reference databases. Bioethicist and legal scholar Hank Greely, for example, contends that DTC DNA testing companies too often "invoke science's power while skip[ping] the caveats" and "without accepting its limits."[1] The purveyors of these commercial tests support this position as well, but for entirely different reasons: they market testing as recreational or personal – and *not* as medical testing or clinical research – in part to avoid regulation of their practices. As a result, DTC genetics testing occupies a complex social space: it uses the language of science for marketing, yet the testing systems do not generate data that other scientists can verify.

It is perhaps unsurprising then that some of the earliest work in the social studies of genetics has explored its medical implications. STS scholars have focused intently on medical genetics, including some of the most influential work in the social sciences of genetics. For example, Abby Lippman coined the term "geneticization" and defined it as "an ongoing process by which differences between individuals are reduced to their DNA codes, with most disorders, behaviors and physiological variations defined, at least in part, as genetic in origin" (Lippman 1991).[2] Notably, she elaborated this concept through feminist, sociological analyses of new reproductive technologies. Sociologist Troy Duster's (1990) classic book, *Backdoor to Eugenics*, forewarned of the detrimental implications of a turn to DNA as an explanatory catchall, and of the consequences of an emergent public health genetic screening apparatus. The work of Lippman, Duster and others is obviously concerned with the big questions about the social implications of geneticization, and these works set a critically important research agenda. Yet, these ideas emerged from scholars' engagement with a *medical* genetics perspective that failed to anticipate the wider uses of genetic analysis, especially the ways in which they would operate without the expertise of medical professionals and without regulatory oversight.

The medical focus of STS scholars of genetics has meant that our tools for analyzing genetic ancestry testing and other forms of commercial genetic testing are underdeveloped. Writing in *Genetics in Medicine*, Jennifer Wagner and collaborators underscore this point, noting that "[e]xpert discussions and formal reviews of the DTC genetic testing industry have generally omitted an entire sector of the industry: companies that offer DNA ancestry tests" (2012: 586). Although DTC tests may be a leisure pursuit to some, they are not simply fun and games. The introduction of direct-to-consumer genetic testing over the past 15 years has spurred an evolution (if not, a revolution) in how we think about our selves and our communities. Like DTC pharmaceutical advertising introduced in the U.S. in the 1990s, some genetic testing can be requested by a health consumer but must be provided by a physician, such as Myriad Genetics' BRACAnalysis test which was first introduced in 1993. Yet DTC genetic testing has produced new relationships within consumers and between consumers and experts. Here we focus on DTC tests that do not require mediation by a third expert party in either the testing process or in the disclosure of results.

The DTC genetic testing field

DTC genetic testing companies give their services colorful brand names such as Ancestry Painting and AncestrybyDNA. For the purposes of this essay, these tests may be sorted into two broad, overlapping clusters: health or "optimization" of one's biology (Rose 2007: 6), and genealogy, ancestry and identity. In the U.S., commercial DTC genetics ventures began to emerge in 2000 with the founding of the genealogical testing company Family Tree DNA. In 2003, one study reported that there were seven DTC testing companies that broadly provided health information and close to sixty that offered some form of "identity" testing.[3] By 2008, another report documented that "more than two dozen websites (including three of the original seven) offer more than 50 health-related tests to consumers" (Hogarth, Javitt and Melzer 2008: 165).[4] There has been comparable growth in DTC genetic ancestry testing companies. In 2004, there were eleven companies offering this service; three more companies had entered the field by 2008. By 2010 "there were 38 companies selling a wide variety of DNA ancestry products, packages, and services."[5] In addition to Family Tree DNA, other early players in the genetic ancestry testing field include (or included) African Ancestry, deCODEme (deCode Genetics), the Genographic Project (a Family Tree DNA partner) and Gene Tree (which is no longer in operation). 23andMe is now one of the leaders in health-related testing, but like deCode it provides genealogical and health analysis under the same umbrella.

Rather than taking up the DTC genetic testing companies' technical or brand descriptions, for STS scholars the tests are perhaps better classified according to the type of information each imparts and thus, the social meaning or action it enables on the part of the consumer. Nelson's multisited ethnographic fieldwork in the U.S., that took place between 2003 and 2009 with African Americans consumers of DTC genetic testing, revealed that consumers purchase specific genetic tests in order to fulfill specific "genealogical aspirations" that may include affirmation of a multiethnic ancestry or evidence of "membership" in a certain racial or ethnic community (Nelson 2008a, Nelson 2008b). With these desires in mind, companies that sell

DNA analysis for genealogical purposes can be said to offer three principal tests: ethnic lineage, racio-ethnic composite, and spatio-temporal. These three titles are a better fit than the brand names for what the tests actually offer by way of information to consumers. And consumers purchase the tests that best fit with the information they seek. All DTC ancestry tests do not provide the same information. Some allow affiliation with a particular nation-state, for example, while others offer inferred membership in a racial group or ethnic community.

Ethnic lineage testing draws on the distinctive features of *Y-chromosome* DNA (Y-DNA) and *mitochondrial* DNA (mtDNA) to infer consumers' ancestral links to contemporary nation-states or ethnic groups. Y-DNA is transmitted virtually unchanged from fathers to sons and can be used to trace a direct line of male ancestors. mtDNA, which is understood to be the energy catalyst of cells, is passed to both male and female children exclusively from mothers and is useful for tracking matrilineage. With both types of ethnic lineage testing, a consumer's DNA is searched against a testing company's reference database of genetic samples. If the sample and the reference DNA match an established number of genetic markers (typically eight or more), an individual can be said to have shared a distant maternal or paternal ancestor with the person who was the source of the matching sample in the reference group. Most DTC genetic testing companies offer ethnic lineage testing, including African Ancestry, Family Tree DNA, and 23andMe. A typical ethnic lineage result may inform a test-taker that her mtDNA traced to the Mende people of contemporary southern Sierra Leone or, more generally, to a region in Western Europe.

With spatio-temporal testing, a consumer's DNA sample is classified into a haplogroup (sets of single nucleotide polymorphisms [SNPs] or gene sequence variants that are inherited together) from which ancestral and geographical origins at some point in the distant past can be inferred. This form of analysis was made possible by the ambitious Y-DNA and mtDNA mapping research that resulted in theories about the times and places at which various human populations arose (Cann et al. 1987, Cavalli-Sforza et al. 1994). Several DTC companies offer this haplogroup information, including National Geographic's Genographic Project. Based on a match with the mtDNA-derived H haplogroup, for example, a customer employing this test can receive a result indicating that her ancestors lived in Southwest Asia or the Middle East 20,000 years ago or more. Because these tests provide very broad results that apply to large portions of human communities, consumers using them are less interested in being matched to a specific ethnic group or nation-state than to their regional origins or their place in the larger history of human migration. This is one example of how technology and genealogical aspirations are co-constituted.

Racio-ethnic composite testing involves the study of nuclear DNA – which is unique to each person (identical twins excepted, although this is now being debated) and consists of the full complement of genetic information inherited from parents – for the purpose of making claims about one's ancestry. A DNA sample is compared with panels of proprietary SNPs that are deemed to be 'informative' of ancestry. Algorithms and computational mathematics are used to analyze the samples and infer the individual's admixture of three of four statistically constituted categories – sub-Saharan African, Native American, East Asian, and European – according to the presence and frequency of specific genetic markers said to be predominate among, but importantly, not distinctive of, each of the original populations. This form of analysis was developed and is principally offered by the AncestrybyDNA division of DNAPrint Genomics as well as by other companies that use its techniques, such as the Genetic Testing Laboratories in New Mexico and UK-based International Biosciences, and more recently, it has been offered by 23andMe. A hypothetical customer might learn that his or her composite is 80% European, 12% Native America and 8% East Asian, for instance. Each of these tests thus offers a different window into the past, and roots-seekers demonstrate different interests and preferences based on their genealogical aspirations.[6]

Similar techniques can be used to elicit health-related information for an individual consumers' DNA samples. This testing tends to analyze sets of SNPs that are deemed to provide information about propensity to some diseases. Although the tests are similar, regulatory controversy has mostly surrounded health-related testing. Critics have noted that these tests lack ethical and regulatory oversight and do not provide consumers with adequate information about the limitations of the tests. As Shobita Parthasarathy's chapter describes, 23andMe has started to use its proprietary customer information for research studies, a development that raises concerns about informed consent and bioethical oversight.

Regulation of DTC genetic tests

The current framework

DTC genetic testing remains a largely unregulated field, either by federal or state authorities. What consumers are purchasing can vary widely from company to company, and the information that they are obtaining can range from a precise analysis of a SNP linked by extensive research with the development of a disease, to data about a number of other characteristics unrelated to health or illness that are, at best, not well-supported by research and generally falls under the umbrella of "recreational genomics."

Legal and regulatory experts have been calling for the development of a federal regulatory schema for DTC genetic tests, and genetic tests generally, for nearly a decade, but despite the widespread concern, little action has been taken to address this growing field until quite recently[7] (e.g., Javitt, Stanley and Hudson 2004, Solberg 2008, Conley, Doerr and Vorhaus 2010, Schlanger 2012).

In the absence of a federal regulatory framework, states have written their own laws, and these vary widely. The state legal landscape looks like a patchwork of mismatched pieces. For instance, some states only permit physicians and medical professionals to order DTC tests, and this effectively prohibits this testing by negating any benefits to consumers by directly ordering them. Some states, on the other hand, are actively regulating DTC testing, while other states remain silent on the issue (effectively permitting them) (Novy 2010, Drabiak-Syed 2010). Further, how each state defines the tests – "medical," "clinical," or "laboratory," for instance – impacts who is permitted to order DTC tests and how. This legal patchwork leaves both consumers and companies wary about what laws apply to them and how. If a multi-state family, for instance, wanted each person to order a DTC genetic test, it may not be technically permissible for each of them to do so from their home states. The enforcement of these varied laws is questionable, which leaves consumers with little to rely on for assurance about the quality of the tests and possible social ramifications at the family level.

The federal government has the authority to regulate DTC genetic tests through various agencies, but has not yet done so systematically. Currently, there are at least three possible avenues for regulation that are not being fully utilized. First, the Clinical Laboratories Improvement Act of 1988 (CLIA) requires the federal government to certify laboratories that perform testing regarding the diagnosis, prevention, or treatment of any disease; however, the scope of CLIA and its requirements do not match the needs of DTC genetic tests. Next, the Federal Trade Commission (FTC) has the ability to regulate false and misleading claims, but it has not taken action to regulate the field of DTC genetic testing. Finally, the Food and Drug Administration (FDA) has the authority to regulate medical devices and laboratory developed tests (see, e.g., Javitt 2007). Though the FDA has done little to date to regulate the vast majority of DTC genetic tests, the agency has recently taken steps in this direction, which will be discussed below.

One possible, but unlikely avenue of further regulation is additional governance of the laboratories. All laboratories performing "clinical genetic testing" must be certified under the Clinical Laboratories Improvement Act of 1988 (CLIA), which is implemented by the Centers for Medicare and Medicaid Services (CMS). CLIA imposes "quality standards for all laboratory testing to ensure the accuracy, reliability and timeliness of patient test results regardless of where the test was performed" (42 U.S.C. § 263a). Unfortunately, however, "accuracy" and "reliability" as they have been applied to other laboratory tests are not defined in such a way as to be useful for genetic tests. In particular, CLIA only assures analytic validity and "does not address clinical validity or claims made by the laboratory regarding the tests." In other words, though a lab may decode a particular gene sequence accurately, CLIA does not certify that the string of As, Cs, Ts, and Gs are actually linked to the disease or trait that the company claims that they are, let alone whether this information is usable by a patient or a health professional. Ensuring the analytic validity, the clinical validity, and the utility of these tests is of central concern to those calling for regulation (Solberg 2008).

As we noted, the FTC has the capacity to prohibit false and misleading claims made by DTC genetic testing companies. Though the FTC has the jurisdiction to administer consumer protection laws, and it exercises that jurisdiction in a variety of arenas, it has neglected to take meaningful action against claims of clinical validity made by DTC genetics companies. Despite consumer complaints, the only action taken by the FTC to date is issuing a warning in 2006 to consumers to be skeptical of claims made by genetics testing companies and to speak with a physician before and after taking such a test (Drabiak-Syed 2010, Novy 2010).

The most likely avenue for regulation of DTC genetic tests is by the Food and Drug Administration (FDA). Understanding several regulatory schemas of the FDA is critical to understanding why DTC genetics tests have heretofore not been regulated, and how they are likely to be regulated in the future. Due to questions surrounding genetic information itself – what does it measure and how? is it risky? can it harm you? – as well as loopholes in the regulation, shown below, genetics companies have not yet been governed by the same provisions required of other laboratory tests transmitting biological information.

The Medical Device Amendments of 1976 gave the FDA broad authority to regulate the safety and effectiveness of medical devices, which it defines as any "instrument, apparatus, implement, machine, contrivance, implant, in vitro reagent or other similar or related article, including any component, part or accessory" that is "intended for use in the diagnosis of disease or other conditions, or in the cure, mitigation, treatment or prevention of disease" (The Medical Device Amendments, 21 U.S.C. § 301 et seq.). How much regulation is imposed upon a manufacturer of a device depends on how risky a device is deemed. Medical devices are grouped into three categories of risk, Class I devices, which have the least oversight, Class II devices, which have moderate oversight, and Class III devices, which have the most oversight (The Medical Device Amendments). Class I devices are subject only to "general controls," which include good manufacturing practices, record keeping, and filing specified reports with the agency. Class II devices are subject to "special controls," such as performance standards, ongoing surveillance, and specific guidance and interventions by the agency. Class III devices are subject to pre-market approval (PMA), which means that the agency must approve the device before it is distributed to the public.

Devices that entered the market after 1976 were presumptively Class III, and thus required PMAs, unless they could show that they were "substantially equivalent" to a device that came on the market before 1976 – a "predicate device." To circumvent the lengthy and expensive PMAs and show "substantial equivalence," a manufacturer would submit what is known as a "510(k)." As it became more and more difficult for manufacturers to show substantial equivalence with pre-1976 predicate devices, it became clear that new, less-risky devices should be able to apply for 510(k) without a predicate device. In 1997, Congress amended the law to permit "de novo" 510(k) classification for low- and moderate-risk devices (The FDA Modernization Act of 1997, 21 U.S.C. § 513 (f)(2)). As we will show below, this amendment became critical for DTC genetic tests nearly 15 years later.

Any in vitro diagnostic (IVD) can be regulated as a medical device. An IVD is defined as:

> The reagents, instruments and systems intended for use in the diagnosis of disease or other conditions, including a determination of the state of health, in order to cure, mitigate, treat or prevent disease or its sequelae. Such products are intended for use in the collection, preparation and examination of specimens taken from the human body.
>
> *21 C.F.R. § 809.3(a), In Vitro Diagnostic Products*

The FDA regulates IVDs in three categories: general purpose reagents, analyte specific reagents (ASRs), and test kits. General purpose reagents are more typical chemical reagents that are available in a laboratory, and they are categorized as Class I devices. ASRs are more complex reagents, such as antibodies, specific receptor proteins, and nucleic acid sequences, which function through a specific binding mechanism to a biological sample. Though most ASRs are Class I, some ASRs are Class II and III. Regardless of classification, however, the FDA restricts ASRs sale, distribution, and use, to specifically permitted manufacturers and laboratories. A third IVD regulated by the FDA are "test kits," which can include bundled reagents, a microassay, or another testing platform (Javitt 2007).

From these regulations, it might appear logical that the FDA could regulate DTC genetic tests as test kits, but historically, the FDA has not regulated what are considered laboratory developed tests (LDTs), beyond regulating their ASRs. LDTs are those test kits that are created and used completely in-house, and as such are sometimes called "home brews" (Patsner 2008). The FDA has made statements that it, sometime in the future, may take steps to regulate LDTs more broadly, but it has not yet done so (Schlanger 2012).

Though many options may be available to the FDA to regulate DTC genetic tests, no widespread action has yet been taken. Proposed regulatory revisions to the current FDA schema include modifications to streamline DTC and CLIA oversight and to create an entirely new division in the FDA, separate from the medical device regulations, that would oversee and evaluate all "advanced personalized diagnostics" (Schlanger 2012: 404). In sum, though many experts have called for more regulation of DTC genetic tests and it appears that there is the legal authority to do so, little action has been taken at the state or federal levels to broadly regulate this growing field. This leaves consumers with, at best, an ambiguous patchwork of regulation covering a wide variety of products, and at worst, unscientific, misleading, or incorrect information about themselves and their families.

FDA and Congressional Action, 2006 to present

In 2006, the FDA sent out a series of "untitled" letters (a low-level warning) to several genetic testing companies requesting a variety of actions ranging from meetings and consultation with the FDA to submission of a de novo 510(k). The companies responded by altering their tests to make them completely in-house laboratory developed tests (LDTs), questioning the jurisdiction of the FDA, or ignoring its requests altogether.

In September 2006, in response to concern about genetic LDTs, the FDA issued draft guidance on its role regulating a subset of LDTs, in vitro diagnostic multivariate index assays

(IVDMIAs). These assays are characterized by their use of proprietary algorithms, often run through software, that generate patient-specific results based on multiple pieces of data derived from one or more in vitro assays (Javitt 2007). The FDA approved its first IVDMIA in February 2007, the Mammaprint test, which uses mRNA to determine the likelihood of breast cancer returning within five to ten years after an initial diagnosis. The company submitted data about a clinical trial, and the FDA approved Mammaprint with narrow intended use that reflected the limitations of that trial (Javitt 2007).

In 2006, the General Accountability Office (GAO), the investigative arm of Congress, made an unexpected intervention by doing a "sting" operation into four unidentified DTC genetic testing companies. GAO posed as fourteen individual consumers, although twelve of the DNA samples came from a nine-month-old female baby and two came from a forty-eight-year-old male. The results and recommendations of the companies investigated varied widely – in one case, the company recommended the same expensive "personalized" nutritional supplements for both the baby and the man (Piehl 2011). The sting revealed that there was little consistency between the testing companies. But even these alarming results of GAO's sting operation did not result in the regulation of the DTC genetic testing industry.

Although critics of the unregulated growth of the industry continued to raise concerns from 2006 on, significant FDA and Congressional activity did not resume until 2010. In May of 2013, Pathway Genomics (Pathway) announced that it planned to sell mail-in saliva sampling kits directly to consumers at Walgreens stores nationwide (Schlanger 2012, Mullard 2012). The FDA responded by sending a letter to Pathway, warning it that its product was a medical device, and that it must comply with the standard regulatory obligations placed upon medical device manufacturers. Pathway defended its proposed action, but pharmacies refused to carry the test, temporarily resolving the FDAs role. Pathway's action also prompted the House Committee on Energy and Commerce (House Committee) to open an investigation into 23andMe, Navigenetics, and Pathway. Among the reasons for the investigation were to find out what exact tests companies were offering, how accurate the tests were, how consumers were being protected, and whether the companies were FDA compliant (Schlanger 2012).

The FDA sent letters to five more genetic testing companies in June 2010, 23andMe, Navigenetics, deCODE Genetics (deCODE), Knowme (pronounced "know me"), and Illumina, which were similar to the letter sent to Pathway. The letters stated that the companies' tests were medical devices and thus subject to regulation by the FDA, but that the tests were not considered LDTs, because the tests were not made and used in-house. One month later, in July 2010, the FDA sent similar letters to fourteen additional genetics companies (totaling 20 altogether) (Schlanger 2012).

During the same summer, the FDA called a public meeting regarding the oversight of LDTs, during which it acknowledged its intent to regulate them but without stating how it would do so. At the public meeting in July, the FDA suggested it would take a "risk-based application of oversight" (Schlanger 2012: 394).

The same week as the FDA meeting, the House Committee held a hearing about the regulation of genetic tests and their effects on public health. At the hearing, GAO announced it had conducted a follow-up to its 2006 sting operation (Schlanger 2012, GAO Highlights 2010). This investigation involved four named companies – 23andMe, Navigenetics, Pathway, and deCODE Genetics. GAO used five people and submitted two samples for each person to each company (40 total tests). For each individual, one DNA sample was submitted with the individual's age, race, and medical history, and one DNA sample was submitted with a fictitious age, race, and medical history. One participant's DNA, for instance, was submitted both as a thirty-seven-year-old Caucasian woman with colon cancer and as a sixty-eight-year-old African

American woman with hypertension and diabetes. As with the 2006 study, the disease risk predictions received from the companies were "misleading and of little or no practical use" (GAO Highlights 2010). The disease risk predictions differed greatly both within the same company for the same DNA and disease and between the companies for the same DNA and disease. The companies justified the former by arguing that different algorithms are available to determine disease susceptibility for different racial and ethnic populations (23andMe 2010). This argument, however, contradicts other claims made by the companies, which purport to be able to determine racial and ethnic heritage as part of their genetics package. The results also differed greatly between companies. For instance, one donor, being tested for risk of leukemia, was found to have "below average," "average," and "above average" risk of developing the disease by the various companies (GAO Highlights 2010). Through its blog *The Spitoon*, 23andMe did not challenge any of the specific findings of the GAO report, but defended its results as sound – not only were the "As, Cs, Ts, and Gs" accurate and reliable, but disease risk predictions should be expected to vary from company to company – and called the GAO investigation "unscientific" (23andMe 2010).

Thus, while federal agencies tried repeatedly over the last several years to regulate the DTC genetics industry, confusion over the categorical boundaries of the tests – along with resistance on the part of both purveyors and consumers – held institutionalization at bay. Consumers were vocal during the FDA hearings, declaring their right to have access to genetic information about themselves and also their right to be free of state and federal constraint[8] (FDA 2010, Vorhaus 2010). Industry leaders, for their part, countered claims that their tests were medical (thereby falling into a well-established regulatory apparatus) by resisting any label but the amorphous "non-medical" or "recreational." Moreover, at least one company responded to the federal investigators claims that the tests were not scientifically valid, by calling into question the science of the government's own researchers.

23andMe's De Novo 510(k) Application

Several years later, after it had collected thousands of consumer samples and was poised to enter the market for medical and drug patents, 23andMe unfolded a new strategy. Once critics of the institutionalization of DTC genetics, this company became the state's partner in this process. Following several years of behind-the-scenes conversations with the FDA, 23andMe applied in June 2012 for de novo 510(k) review of seven of its approximately 240 offered SNP-based genetic reports. The company stated that it intended to submit up to 100 more reports by the end of 2012 (The Burrill Report 2012, Mullard 2012). Such reports would be evaluated for both analytic and clinical validity, meaning that 23andMe can consistently and accurately identify gene sequences, and more challengingly, that it can validate its claims about correlations between specific genes and associated risks for developing diseases. These submissions were the first of their kind in the DTC genetics market. Moreover, according to Daniel Vorhaus of The Genomics Law Report, it could "represent the way forward for certain components of the DTC industry" (The Burrill Report 2012). Such a route to approval would have several advantages for 23andMe, primarily its ability to claim that they are the first, and for a time, the only FDA-approved DTC genetics company, and it would allow the company to set the terms of its own surveillance (and of competitors that will follow it). In the industry broadly, establishing that de novo 510(k) is a possible avenue for DTC genetic tests could encourage other companies that may have been wary of the more onerous pre-market approval process *and* it may make it a requirement for 23andMe's competitors that may be less well-prepared to clear this stringent hurdle. In other words, it would set a course for institutionalization for the industry.

On November 22, 2013, however, the FDA issued a warning letter to 23andMe to immediately stop marketing its DTC genetic test or risk regulatory action such as seizure, injunction, and civil money penalities," with which 23andMe stated it would comply (FDA 2013, 23andMe 2013). Under the regulatory framework, 23andMe failed to meet analytical and clinical validity requirements for de novo 510(k) approval, that is, it failed to validate the claims about correlations between specific genes and associated risks for developing diseases. This failure rendered the test a Class III device, which requires pre-market approval.

The warning letter details the lengths to which 23andMe went to receive a de novo 510(k) approval from the FDA — there were over fourteen meetings, hundreds of emails, and dozens of written communications (FDA 2013). The letter states that the FDA provided "ample detailed feedback" regarding analytical and clinical validity requirements (as well as suggesting modifications to the device's label to meet requirements for certain uses), but that 23andMe failed to provide any of the studies necessary for de novo 510(k) approval (FDA 2013). Instead, 23andMe simultaneously ceased communications with the FDA and expanded the marketing claims and consumer base for its test. This extensive relationship between 23andMe and the FDA may reveal a belief, on the part of both organizations, that such tests could ultimately garner approval within the current regulatory framework. Indeed, this belief was common in the regulatory community (The Burrill Report 2012). Alternatively and perhaps cynically, this unusual alliance could belie 23andMe's success in keeping regulatory boundaries blurred, allowing the company to proceed unregulated for as long as possible while it gained popular acceptance.

Can recreational genetic testing become institutionalized?

Several years of behind-the-scenes conversations and lobbying provided reason for those in the industry to believe that regulatory oversight could provide more benefits than costs. Overcoming the state-by-state patchwork of regulations, mollifying the uncertainty for some investors, and gaining the claims of "first" and "only" had the potential to outweigh the financial and ultimately regulatory costs of applying for de novo 510(k) approval. As one expert in the field stated, "Getting it through an FDA review process … can have a validating effect for 23andMe in the eyes of thought leaders, opinion leaders, policy makers in the field that have been, many of whom have been, critical of the DTC model" (The Burrill Report 2012).

To pass muster under 510(k), the reports submitted by 23andMe would need to show both analytical and clinical validity. For prospective prediction of diseases, proving clinical validity could be possible – studies can be done to examine who, with that SNP, develops a particular disease. Moreover, the companies' claims can be compared and contrasted with existing clinical research. But, for retrospective determination of racial and ethnic ancestry, it would be impossible to ever show a link between genes and the past, given that our ancient ancestors are not readily available for examination and the companies create proprietary algorithms that make idiosyncratic historical claims and categorical claims (e.g., the presence of a higher percentage of this SNP means one is Asian, with no accounting for how "Asian" is socially constituted) that cannot be validated without undermining the secrecy that contributes to brand identity and market share. It seems from this model that "clinical validity" as applied to prospective disease markers is irrelevant to retrospective ancestry testing.

This distinction between health-related disease prediction tests and recreational ancestry tests is highlighted by 23andMe's response to the FDA letter. 23andMe stated that it will comply with the FDA by providing "raw genetic data without interpretation" for health-related tests, but that it would continue providing "ancestry-related information" (23andMe 2013).

119

Because the 510(k) process exists under the framework of medical device regulation, it is unclear if ancestry tests would ever be approved under the same process. Nevertheless, the legitimacy that would be afforded to 23andMe by an approval of its disease markers would likely have an effect, one way or the other, on the perceived legitimacy of its ancestry tests. On one hand, the imprimatur of the FDA on 23andMe's health-related tests could, in the minds of consumers, trickle down to the company's recreational genetic tests. It is also possible that some consumers would be turned off by ancestry tests that are unable to gain the validating stamp of the federal government. It seems unlikely, however, that all consumers would stop purchasing the ancestry tests, given their willingness to use the technology to date and eagerness to find "workaround[s]" to the FDA's 2013 action (Hensley 2013).

The DTC genetics testing companies have often claimed that their products are "informational" and "educational," and not intended to provide a medical or health assessment (Popovsky 2010). It is likely that those products, such as ancestry testing, would continue to be defensively labeled "informational" by the companies.

Following its 510(k) submission, 23andMe wrote in their blog *The Spitoon* that they continue to believe that consumers "have a fundamental right to their personal genetic data" and that the data "will power a revolution in healthcare. But we also recognize that appropriate oversight of this industry can be a stepping stone on the path to realizing that revolution" (23andMe 2012). In the past, 23andMe has argued, "genetic information is a fundamental element of a person's body, identity and individuality. As such, the rights that people enjoy with regard to financial, medical and other forms of personal information should apply to genetic information as well" (see Popovsky 2010: 76). In this way, the company articulates arguments from both the industry side and the consumer perspective to forestall institutionalization, even as it has sought a regulatory partnership with the FDA.

Conclusion

23andMe's de novo 510(k) application for their DTC genetic tests highlights an era of rapid growth in the history of DTC genetic testing. By exploiting the boundary crisis between health and recreational genomics during a period of regulatory lag, the genetic testing company sought to negotiate the terms of its own surveillance and in the process attempted to define and expand the meanings and understandings of all of its products. By specifically applying for a 510(k) rather than PMA, 23andMe not only indicated that there is no substantially equivalent device – that it provides a sui generis service that has no regulatory analogue – but also that they believe that the product they offer is not risky and does not warrant stringent oversight. From a regulatory standpoint, an approval under these conditions would open the DTC genetic testing field quite a bit – by avoiding the higher categories of risk, many more applications would be sent to the FDA, with a regulatory imprimatur serving to vouchsafe the market. Further, being the first company in such a position, 23andMe would gain more credibility as being a pioneer in the field. As such, smaller companies having neither the FDA's seal of approval nor the means to gain it may find it hard to survive in such a marketplace. Moreover, by gaining some federal regulatory stability that supersedes the state-by-state patchwork, 23andMe could attract more investors.

Through these various mechanisms of institutionalization, 23andMe sought to gain the legitimacy of the bureaucracy while they actually built it themselves. While consumers eager to gain insight into their health and ancestry, and believing it is their "right" to the tests, may, against the backdrop of a neoliberal assault on the welfare state, appear to be the primary beneficiaries of the growth of the DTC genetic test industry, the fact that 23andMe seeks to set the terms of its own governance calls into question who or what is in the position of power. The tests that could to gain approval under the current regulatory framework are medical, but, as mentioned previously, such approvals are likely to have a "validating effect" on all products, that is, consumers would likely deem the entire company regulated, reliable, and trustworthy (The Burrill Report 2012). Thus, consumers are more likely to trust a company that has the FDA's imprimatur on any of its products, including that company's products for ancestry testing that are not currently regulated. Given what we know about why consumers purchase DTC genetic ancestry tests – in order to fulfill genealogical aspiration – and what we know about geneti-cization – how consumers can substitute DNA meanings for other explanations and understandings of society – the institutionalization of DTC genetic testing should proceed with, if not caution, at least a broad awareness of the biosocial implications.

Notes

1 Greely, H.T. 2008. "Genetic Genealogy: Genetics meets the Marketplace." In *Revisiting Race in a Genomic Age*, edited by Koenig, B.A., Lee, S.S.-J., and Richardson, S.S. New Brunswick: Rutgers University Press: 215–34.
2 Lippman, A. 1991. "Prenatal Genetic Testing and Screening: Constructing Needs and Reinforcing Inequalities," *American Journal of Law and Medicine* 17: 15–50.
3 Gollust, S.E., Wilfond, B.S., Hull, S.C. 2003. "Direct-to-consumer Sales of Genetic Services on the Internet." *Genet Medicine* 5(4): 332–37.
4 Hogarth, S., Javitt, G. and Melzer, D. 2008. "The Current Landscape for Direct-to-Consumer Genetic Testing: Legal, Ethical, and Policy Issues," *Annual Review of Genomics and Human Genetics* 9: 161–82.
5 Wagner, J.K., Cooper, J.D., Sterling, R. and Royal, C.D. 2012. "Tilting at Windmills No Longer: A Data-Driven Discussion of DTC DNA Ancestry Tests." *Genetics in Medicine* 14: 586.
6 Obviously, these categories are constructed by the DTC genetics companies in a fashion that seldom takes into account the fact that these categories are socially constructed or that the nations and borders to which they refer have changed repeatedly over human history. A thorough discussion of the limitations of genetic ancestry testing is provided by Nelson and collaborators; see Bolnick et al. 2007.
7 As this essay was going to press, the FDA ordered 23andMe to immediately stop marketing its DTC tests (FDA 2013). This development is discussed below.
8 23andMe's Anne Wojcicki also testified against regulation on this day.

References

21 C.F.R. § 809.3(a). In Vitro Diagnostic Products For Human Use.
21 U.S.C. § 301 et seq. The Medical Device Amendments of 1976, as amended at The Federal Food, Drug and Cosmetic Act of 1938. Classes of devices can be found at 21 U.S.C. § 360(a)(1).
21 U.S.C. § 513 (f)(2), The FDA Modernization Act of 1997.
42 U.S.C. § 263a. The Clinical Laboratory Improvement Act of 1988 (CLIA). See also U.S. Food and Drug Administration, CLIA. Retrieved October 18, 2012 (*www.cms.hhs.gov/clia*).
23andMe. 2010. "GAO Studies Science Non-Scientifically." The 23andMe Blog. Retrieved October 18, 2012 (*http://blog.23andme.com/23andme-research/gao-studies-science-non-scientifically/*).
23andMe. 2012. "23andMe Takes First Step Toward FDA Clearance." The 23andMe Blog. Retrieved October 18, 2012 (*http://blog.23andme.com/news/23andme-takes-first-step-toward-fda-clearance/*).
23andMe. 2013. "23andMe, Inc. Provides Update on FDA Regulatory Review." The 23andMe Media Center. Retrieved January 15, 2014 (*http://mediacenter.23andme.com/press-releases/23andme-inc-provides-update-on-fda-regulatory-review/*).
Appadurai, A. 1986. *The Social Life of Things: Commodities in Cultural Perspective*. Cambridge, UK: Cambridge University Press.
Atkinson, P., Parsons, E., and Featherstone, K. 2001. "Professional Constructions of Family and Kinship in Medical Genetics." *New Genetics and Society* 20 (1): 5–24.
Bolnick et al. 2007. "The Science and Business of Ancestry Testing." *Science* 318 (5849): 399–400.
Burrill Report 2012. "Regulate Me." Audio podcast interview with Daniel Vorhaus. Retrieved October

18, 2012 (*www.burrillreport.com/article-4808.html*).

Cann, R.L., Stoneking, M., and Wilson, A.C. 1987. "Mitochondrial DNA and Human Evolution." *Nature* 325: 31–36.

Cavalli-Sforza, L. Luca, P.M., and Piazza, A. 1994. *The History and Geography of Human Genes*. Princeton: Princeton University Press.

Chaska, N.L. 1977. "Medical Sociology for Whom?" *Mayo Clinic Proceedings* 52 (12): 813–18.

Conley, J.H., Doerr, A.K., and Vorhous, D.B. 2010. "Enabling Responsible Public Genomics." *Health Matrix* 20: 325–85.

Drabiak-Syed, K. 2010. "Baby Gender Mentor: Class Action Litigation Calls Attention to a Deficient Federal Regulatory Framework for DTC Genetic Tests, Politicized State Statutory Construction, and a Lack of Informed Consent." *Michigan State University Journal of Medicine and the Law* 14: 71–92.

Dumit, J. 2012. *Drugs for Life: How Pharmaceutical Companies Define Our Health*. Durham: Duke University Press.

Duster, T. 1990. *Backdoor to Eugenics*. New York: Routledge.

Epstein, S. 1996. *Impure Science: AIDS, Activism and the Politics of Knowledge*. Berkeley: University of California Press.

FDA 2010. "Public Meeting on Oversight of Laboratory Developed Tests." Retrieved July 20, 2010 (*www.fda.gov/downloads/MedicalDevices/NewsEvents/WorkshopsConferences/UCM226204.pdf*).

FDA 2013. Warning Letter to 23andMe, Inc. Document Number: GEN1300666, dated November 22, 2013. Retrieved January 15, 2014 (*www.fda.gov/iceci/enforcementactions/warningletters/2013/ucm376296.htm*).

Franklin, S. and Roberts, C. 2006. *Born and Made: An Ethnography of Preimplantation Genetic Diagnosis*. Princeton: Princeton University Press.

GAO Highlights 2010. "Direct-To-Consumer Genetic Tests: Misleading Test Results Are Further Complicated by Deceptive Marketing and Other Questionable Practices." Highlights of GAO-10-847T. Retrieved Janury 15, 2014 (*www.gao.gov/new.items/d10847t.pdf*).

Hensley, S. 2013. "23andMe Bows To FDA's Demands, Drops Health Claims." NPR. December 6. Retrieved January 15, 2014 (*www.npr.org/blogs/health/2013/12/06/249231236/23andme-bows-to-fdas-demands-drops-health-claims*).

Hogath, S., Javitt, G., and Melzer, D. 2008. "The Current Landscape for Direct-to-Consumer Genetic Testing: Legal, Ethical, and Policy Issues." *Annual Review of Genomics and Human Genetics* 9: 161–82.

Javitt, G.H. 2007. "In Search of a Coherent Framework: Options for FDA Oversight of Genetic Tests." *Food and Drug Law Journal* 62: 617–52.

Javitt, G.H., Stanley, E., and Hudson, K. 2004. "Direct-to-Consumer Genetic Tests, Government Oversight, and the First Amendment What the Government Can (and Can't) Do to Protect the Public's Health." *Oklahoma Law Review* 57(2): 251–302.

Lee, S.S.-J. 2013. "Race, Risk and Recreation in Personal Genomics: The Limit of Play." *Medical Anthropology Quarterly* 27(4):1–20, DOI: 10.1111/maq.12059.

Lippman, A. 1991a. "Prenatal Genetic Testing and Screening: Constructing Needs and Reinforcing Inequalities," *American Journal of Law and Medicine* 17(1/2): 15–50.

Lippman, A. 1991b. "The Geneticization of Health and Illness: Implications for Social Practice. *Endocrinologie* 29: 85–90.

Moore, K., Kleinman, D.L., Hess, D., and Frickel, S. 2011. "Science and Neoliberal Globalization: A Political Sociological Approach." *Theory and Society* 40: 505–32.

Mullard, A. 2012. "Consumer Gene Tests Poised for Regulatory Green Light." Council for Responsible Genetics. Retrieved October 18, 2012 (*www.councilforresponsiblegenetics.org/blog/post/Consumer-gene-tests-poised-for-regulatory-green-light.aspx*).

Nelkin, D. and Lindee, M.S. 1995. *The DNA Mystique: The Gene as a Cultural Icon*. Ann Arbor: University of Michigan Press.

Nelson, A. 2008a. "Bio Science: Genetic Genealogy Testing and the Pursuit of African Ancestry." *Social Studies of Science* 38(5): 759–83.

Nelson, A. 2008b. "The Factness of Diaspora: The Social Sources of Genetic Genealogy," in *Revisiting Race in a Genomic Age*, edited by B. Koenig, S.S.-J. Lee, and S. Richardson. New Brunswick: Rutgers University Press: 253–70.

Nelson, A. 2010. "The Social Life of DNA." *The Chronicle of Higher Education*. August 20, 2010. Retrieved August 20, 2010 (*http://chronicle.com/article/The-Social-Life-of-DNA/124138*).

Novy, M.C. 2010. "Privacy at a Price: Direct-to-Consumer Genetic Testing & the Need for Regulation."

University of Illinois Journal of Law, Technology & Policy 2010: 157–80.

Patsner, B. 2008. "New 'Home Brew' Predictive Genetic Tests Present Significant Regulatory Problems." *Houston Journal of Health Law and Policy* 9: 237–77.

Piehl, M. 2011. "Regulating Hype and Hope: A Business Ethics Model Approach to Potential Oversight of Direct-to-Consumer Genetic Tests." *Michigan State University Journal of Medicine & Law* 16: 59–94.

Popovsky, M. 2010. "Exaggerated Benefits and Underestimated Harms: The Direct-to-Consumer Genetic Test Market and How to Manage It Going Forward." *Dartmouth Law Journal* 8: 65–87.

Rose, N. 2006. *The Politics of Life Itself: Biomedicine, Power, and Subjectivity in the Twenty-First Century.* Princeton: Princeton University Press.

Schlanger, S.J. 2012. "Putting Together the Pieces: Recent Proposals to Fill in the Genetic Testing Regulatory Puzzle." *Annals of Health Law* 21: 382–405.

Shriver, M.D. and Kittles, R.A. 2004. "Genetic Ancestry and the Search For Personalized Genetic Histories." *Nature Reviews Genetics* 5: 611–18.

Solberg, L.B. 2008. "Over-the-Counter but Under the Radar: Direct-to-Consumer Genetic Tests and FDA Regulation of Medical Devices." *Vanderbilt Journal of Entertainment and Technology Law* 11: 711–42.

Sunder Rajan, K. 2006. *Biocapital: The Construction of Postgenomic Life.* Durham: Duke University Press.

Vorhaus, D. 2010. "The Conversation Continues: Recap from Day Two of FDA's Regulatory Meeting." *Genomics Law Report.* Retrieved July 21, 2010. (*www.genomicslawreport.com/index.php/2010/07/21/fda-ldt-day-2-recap*).

Wagner, J.K., Cooper, J.D., Sterling, R., and Royal, C.D. 2012. "Tilting at Windmills No Longer: A Data-driven Discussion of DTC DNA Ancestry Tests." *Genetics in Medicine* 14: 586–93.

Wailoo, K. Nelson, A., and C. Lee. 2012. "Introduction: Genetic Claims and the Unsettled Past," in *Genetics and the Unsettled Past: The Collision of DNA, Race and History*, edited by K. Wailoo, A. Nelson and C. Lee. Rutgers: Rutgers University Press: 1–12.

White House. 2000. "White House Remarks on Decoding of Genome." *New York Times* (27 June): F8.

7

Cultures of Visibility and the Shape of Social Controversies in the Global High-Tech Electronics Industry

Hsin-Hsing Chen

SHIH-HSIN UNIVERSITY, TAIWAN

How can we explain the different degrees of visibility among sociotechnical controversies? In today's world, the social "visibility" of an issue is almost always figurative and mediated, especially through technology. A "visible" event might be reported on television, quickly and widely shared through social media, and commented upon in blogs, although the number of direct eye witnesses might be few. What makes controversies visible is not only technology, the so-called "newsworthiness" of events, or even the high visibility of certain actors involved in them. Using two ongoing cases of controversy about manufacturing in the global electronics industry, I demonstrate that historically constituted institutional infrastructures have a profound influence on the trajectories and outcomes of controversies, chief among them national legal institutions and international political–economic systems. These systems, I argue, shape controversy trajectories via *cultures of visibility*.

The first case is the controversy over cancers suffered by the former employees of Radio Corporation of America (RCA) in Taoyuan, Taiwan. Many believe the cancers to be a result of exposure to toxic chemicals at the workplace while the company operated between 1969 and 1992. Former workers are now fighting Taiwan's first collective lawsuit against RCA. Court hearings started in late 2009, five years after the suit was initially filed, and are still ongoing in the district court. A coalition of activists, lawyers, scientists, and other academics was built in support of the RCA workers' struggle, and continue to participate in this unprecedented science-intensive litigation in Taiwan.

The second case is the Apple/Foxconn controversy. It started in 2010 with a series of suicides of young Chinese workers at a factory complex owned by the largest electronics manufacturing firm in the world today, the Taiwanese-owned Foxconn Technology, chief supplier of Apple Inc. Media reports of seventeen successive suicides in the first seven months of that year aroused public outcry and intensive debates. Initially, prevailing public opinion blamed the young workers themselves for their inability to endure hardships. It shifted to condemning the excessive stress Foxconn, in the service of Apple, imposed on its employees as the root cause of workers' suicides. To cope with the negative publicity, Apple undertook investigations on plants in its supply chain, following practices developed by the global apparel industry in the name of Corporate Social Responsibility (CSR), a set of private labor and environmental codes enforced by the brand name on its suppliers. So far, actual changes in labor practices at Foxconn's plants are still limited, and anti-sweatshop NGOs are closely monitoring the situation.

Comparing these two cases, we can see similar underlying issues playing out in sharply contrasting ways, despite some commonalities between them. Both cases involve mass-produced high-tech electronics in Asia mainly for U.S. consumers, workforces chiefly composed of young women workers with rural backgrounds, global relocation of the assembly line in order to tap sources of cheap and docile labor, and fragmentation of the labor process into simple repetitive tasks enabling management to hire fresh hands off the street while burdening workers with monotonous and stressful workdays. Furthermore, many of the international activist networks and organizations involved in the RCA and Apple/Foxconn campaigns overlap.

However, the ways these two controversies play out with respect to public visibility are drastically different. The RCA controversy is now fought out mainly in the arena of scientific discourse and in arduously long court sessions. The social worlds of science and law often appear opaque and the lay public finds them hard to understand, despite each world's explicit doctrines of openness and transparency. In the Apple/Foxconn controversy the actual tragedies and everyday sufferings of the workers take place deep inside the gigantic walled factory that are physically impossible for the public to literally see with their own eyes. Making this world visible is precisely the central area of contestation. Labor advocates and the companies have fought with media exposés, news releases and other publicity tools, each trying to move the focus of public attention to where they hope it to be. By contrast, in the RCA controversy, the action has played out in courtrooms and through legal procedures that are supposed to be transparent, but which in practice make it difficult for outsiders to "see" the debate.

The ways that social visibility and invisibility are produced and reproduced in sociotechnical controversies, I argue, can be most clearly understood using conceptual tools from STS for analyzing institutional infrastructures. Chief among them are the "visual culture" metaphor developed by Yaron Ezrahi (1990) through his elaboration of the work of Steve Shapin and Simon Schaffer (1985), and the social-worlds frameworks developed by analysts such as Susan Leigh Star and Adele Clarke (Clarke and Star 2007, Star 2010). Visual culture, in Ezrahi's view, refers to "cultural codes that in any given political world fix the political functions of visual experience and govern the meanings attributed to the relations between the visible and the invisible" (1990: 70). Visual cultures are historically variable, and are shaped by legal, technological, and economic systems that prevent or enable particular ways of seeing, whether by design or not. The visibility of the Apple/Foxconn controversy was shaped, I argue, by a "theatrical" culture, and the RCA controversy by a "mechanical" culture that must be understood in terms of their historical and geographic contexts. Looking at specific contents of each controversy and its intersection with visual culture allows us a better view of its dynamics as well as its underlying historical contexts.

The invisible machine and the theatrical spectacle

The "*theatrical*" culture in the Apple/Foxconn case and the "*mechanical*" in the RCA case are both constitutive parts of what Ezrahi calls "attestive visual culture." It characterizes, he argues, both natural science and the liberal-democratic state (including modern legal systems). Attestive visual cultures value "public witness," as opposed to private, unaccountable settings of mysticism and speculative philosophy, as a more reliable context for producing and confirming "fact" (Ezrahi, 1990). Following Shapin and Shaffer (1985), Ezrahi builds his case via the struggle between Robert Boyle and Thomas Hobbes in Restoration-era England over issues of ontology, epistemology, ethics, religion, politics, and science. Debates in these now-segregated fields were actually the same fight for the actors in that particular historical context. In the Boyle-Hobbes controversy, the ideal polity for the resolution of disputed political and scientific issues was defined by each side's view of how reliable knowledge can be obtained. For Boyle, the preferred way was direct observation in a controlled environment by a group of reliable witnesses who shared certain necessary ways of seeing, and through open and collegial debates among them. The experimental community of educated gentlemen in the Royal Society is Boyle's model polity.

Hobbes rejected Boyle's experimental philosophy and preferred mathematical-geometric reasoning as the way to certainty. He argued that when the rational method of geometry is followed, it yields irrefutable and incontestable knowledge, because "[a]ll men by nature reason alike, and well, when they have good principles" (quoted in Shapin and Schaffer 1985: 100). Self-evident mathematical certainty, however, requires prior agreement upon a set of definitions and axioms. This corresponds to Hobbes's ideal polity in Restoration-era England: that everyone concedes part of his rights to one sovereign in exchange for peace and order and preservation of her or his own private space.

Translated into politics, Boyle's model is parliamentary democracy and Hobbes' the authority of the state. In law, Boyle's legacy has been inherited by the common-law practice of adversarial procedures in jury trial, in which a "public eye," represented by the jury, is called upon to decide the "matter of fact" in the courtroom based on what they see in the proceeding; both the space and the proceedings in the court are kept orderly by a presiding judge. Hobbes' ideal, on the other hand, is inherited, via Spinoza's *Ethica Ordine Geometrico Demonstrata (Ethics, Demonstrated in Geometrical Order*, published posthumously in 1677), by the civil law (also known as Continental Law) traditions in Continental Europe in the jurists' pursuit of rigorous, rational codification of the law (Wieacker 1996: 302–304), and in the procedural norms that seek to limit the court's discretion vis-à-vis the legislature.

Thus, as Ezrahi argues, in spite of the ostensible antagonism between Boyle and Hobbes, both of their legacies have become constitutive parts of political orders (including the law) in the modern states of our time. He uses these legacies to analyze contemporary scientific and political cultures. Political culture, in this context, encompasses not only explicit institutions and ideas, but also implicit patterns of behavior and meaning. He calls the scientific metaphor of politics articulated by Hobbes and practiced by Boyle the *"theatrical"* metaphor. Theatrical metaphor distinguishes the enacted persona and the actor's actual person, thus enabling modern social sciences to study dynamics of society, culture, and polity without invoking moral judgments on people who "play the roles."[1] Theatrical acts appeal to common perception; a drama is believable as long as the audience finds it convincing, and part of the convincing power lies in the audience's tacit agreement not to peek behind the stage curtain.

Dissatisfied with the implied centralization of power in the theatrical metaphor used by Hobbes to justify monarchy and by Boyle to justify elitism, liberal-democratic thinkers in the Enlightenment-era West, argues Ezrahi, upheld the *"mechanical"* metaphor of politics, which is aptly expressed in Thomas Jefferson's idea of a constitution as "a machine that would go by itself" without hierarchical and centralized notions of power and authority. Checks and balances created by the separation of government powers were most convincingly expressed with the metaphor of the clockwork. Clockwork is transparent; it needs no curtain and can be examined by anyone keen on it from every angle.

These naive Jeffersonian notions about the "transparency" of the machine became less and less convincing in the late nineteenth century industrial West, as the specialization of scientific disciplines rendered material machines increasingly opaque to outsiders,[2] and the actual machinery in social lives grew from the discernible assembly of gears and fixtures found on shop floor machines and farm implements into big, complex, and largely invisible technological systems, such as long waterways, railroad and telegraph systems, and later the electric power grid. Users and even most of the workers operating those systems saw only parts of them, but knew the systems worked regardless of how well they understood the way they functioned. Meanwhile, with the expansion of franchise and establishment of permanent specialized government agencies, the state was becoming increasingly complex, and indeed, the technical, material and political came to operate as one system (Carroll 2006).

In the milieu in which using complex technological systems with only partial views of them became people's daily experience, science and other expertise tended to be called upon in cases in which the contested facts are not believed to be immediately perceivable by ordinary eyes. Causation, especially in the intersections between social worlds of science and law, is the central "boundary object," or focus on contestation and attention. The 1782 English legal case *Folks vs. Chadd* is often referred to as a landmark case in which "men of science" were allowed to testify about "chain of causes imperceptible to lay observations, a chain perceptible only to those intimated with the hidden operation of nature" (Golan 2008: 890). The RCA controversy has also become one such case. With most of the relevant events in the past, both parties invoke science to piece together their own versions of the facts. In addition, the style of work in the civil-law system in contemporary Taiwan, with its traditional aversion to theatrical spectacles, renders the science-intensive RCA litigation even more unseen by the public eyes.

By contrast, the Apple/Foxconn controversy takes place mostly in the arena of theatrical spectacles, in the virtual but highly visible courtroom of public opinion. The theatrical character of the Foxconn/Apple case is greatly shaped by the theatrical style of the Apple brand name itself, which successfully brings electronic gadgets into the fashion business but also renders it especially vulnerable to negative public perception.[3] This character is generated by corporate strategies designed to cope with the global industrial overproduction looming over contemporary capitalist economy. At least for the time being, Apple successfully masters market volatility by "manipulation of taste and opinion" (Harvey 1989: 287). This allows anti-sweatshop activists to apply practices and toolkits developed in campaigns against global garment and shoe brand name companies (a.k.a. the fashion business), to the Apple/Foxconn campaign. In response, Apple Inc. deployed its own CSR system, a set of private labor and environmental codes enforced by the brand name on its suppliers. The CSR system was developed largely by the garment and shoe industries in the 1990s, to shift public attention from criticisms of sweatshop exploitation and environmental degradation to a set of more manageable boundary objects: standards (Corporate Codes of Conduct) and methods (CSR auditing) (Klein 2000, AMRC 2012).

RCA and the search for a pliable workforce

Initially established as a government-supported consortium in 1919, pooling together patents and capital from American corporate giants such as Westinghouse, AT&T, and the General Electric, RCA later became an independent corporation and bought its own production facility at Camden, New Jersey in the mid-1930s. Applying the then-pervasive technique of "Scientific Management," industrial engineers at RCA broke down its radio production process into fragments, in which each worker repeats a small set of tightly defined tasks hundreds of times in a shift, and the pace of all operators' work on the assembly line is regulated by the speed of the moving conveyor belt carrying work pieces through hundreds or even thousands of work stations on the line. Young women, most of them new immigrants from rural Europe, comprised 75% of its 9,800 workers at RCA's Camden plant. Men mostly filled the jobs of inspectors and technicians (Cowie 1999: 17).

Soon after workers at RCA's Camden plant organized their trade union and began to struggle for higher wages and better working conditions, the company started what Cowie (1999) calls a "seventy-year quest of cheap labor." Similar to its contemporaries, RCA kept relocating its production to what are often called "greenfield sites" in labor relations: places where there are abundant supplies of workers who are freshly coming out of non-industrial communities, and are thus less likely to be influenced by trade unions (for example, Kochan et al. 1984). The 22-year sojourn of RCA's production lines in Taiwan is but one link of a long chain of similar moves, initially inside the United States and then throughout the world. Similar to RCA's Camden plant in its early decades and the Foxconn plants in China today, the RCA plant in Taiwan employed mostly young women workers with rural backgrounds on its production lines. Most of them lived in company dormitories for years, moving out only after marriage.

RCA arrived in Taiwan as a result of a strategy of economic development that was still new in 1970. The Export Processing Zone (EPZ) program was implemented in earnest in Taiwan, South Korea, and the Philippines in the late 1960s. These authoritarian, pro-business governments instituted a series of tax incentives and other forms of subsidies for foreign investors, mainly Japanese and American, who were willing to import parts and components, utilize cheap and obedient local workforces to do the labor-intensive segments of manufacturing, and export finished or semi-finished products back to where the consumer market was. Similarly, Mexico started its "maquiladora" program in 1968. Such "export-oriented industrialization" programs were seen as a radical break away from the dominant development paths of import substitution and national industrialization pursued by most developing countries after World War II (Petras and Hui 1991).

As the EPZ model was widely replicated throughout underdeveloped countries, labor-intensive swaths of the global assembly line yielded less and less under intensifying competition. In search of ways out of such a conundrum, since the late 1970s, governments in countries like Taiwan almost invariably turned to "industrial upgrade," that is, shifting manufacturing to capital-intensive "high-tech" segments of the global assembly line. RCA cooperated with the Taiwanese government in a famous technological transfer project in the 1970s and trained selected young engineers from Taiwan at its semiconductor and integrated circuit (IC) factories in the United States. These members of the "RCA Project" played vital roles in establishing Taiwan's highly-subsidized IC and semiconductor companies in the early 1980s. Many of them are now business magnates in Taiwan's electronics-related industries, whose revenue accounts for half of the country's exports. With a handful exceptions, most of Taiwan's electronics-related firms play similar roles today as those in the early EPZ-style industries – subcontractors and suppliers for foreign brand names. While successful Taiwanese entrepreneurs feature the bright side of RCA's legacy, cancer among RCA's former workers is a much more obscure and deadly side. It took workers and their supporters decades to render the issue visible.

Cancer: the invisible legacy of the global assembly line

Causation is at the center of the RCA controversy. To manage the vast variety of workplace health hazards, national occupational health and safety authorities usually maintain a list of known hazardous factors in the workplace and the diseases each factor is known to cause. The list grows with the accumulation of knowledge particularly in the discipline of occupational and environmental medicine. Once a worker's exposure to a listed factor is measured to exceed legal standard, and the worker is diagnosed with the listed disease, the regulatory agency will deem the causation established. However, many issues fall outside the list, and become controversies. The RCA case is one of them.

RCA does not dispute that more than twenty-eight toxic chemicals have been identified by the Taiwan Environmental Protection Agency in the groundwater and soil at the plant site, and that many former RCA employees are suffering from or have already died of an array of ailments, including cancer. These are "established facts" with authoritative documentation in the form of chemical analysis and medical diagnosis. Both parties also agree that some of those chemicals, by themselves or in combination, may cause harm on human bodies through contact, inhalation, digestion or other possible channels of exposure. What is up for debate, is with what certainty we know about the types of exposures and their health effects in this specific situation.

In the late 1990s, when workers started to pressure the government to investigate whether their exposures and illnesses constituted occupational hazards, RCA claimed that all of its operational records were destroyed in a warehouse fire sometime after the plant closed. Soon afterward, all factory buildings were demolished, along with the last material traces of RCA's actual production processes. This is a typical asymmetric dispute that appears in many techno-scientific controversies; the side possessing vital information can get an upper hand by simply denying its opponent such information (Markowitz and Rosner 2002).[4] Without access to factory records to establish exposure history, scientists investigating the RCA case have no alternative but to rely on some criteria existing in the sketchy Labor-Insurance records of RCA workers as proxies to reconstruct their pictures.[5] Epidemiologists evaluate causation by calculating statistical correlation, with various indices, between factors from this reconstructed history of exposure with those from workers' health records.

Two scientific teams carried out studies of cancer among the RCA workers after workers started protesting. Both studies were led by authoritative experts in occupational and environmental medicine in Taiwan. After three years of study, one team reported few significant findings, while the other suggested some adverse health effects among RCA workers. Research findings by the two teams were later published in English-language peer-reviewed international journals. Since the first team was funded by the Council of Labor Affairs (Taiwan's ministry of labor), their report of low statistical correlation was used by RCA's lawyers to argue that the Taiwanese government has "proven" that workers' health problems are not caused by the chemicals used at RCA.

The first team, led by Saou-Hsing Liu at the National Defense Health Center, found some statistically significant correlations between women's history of work at RCA and cancer, but they qualified this finding by reporting that the Standardized Incidence Ratio of female breast cancer exhibits "no significant dose-responsive relations on duration of employment" (Chang et al. 2003a; Chang et al. 2003b; Chang et al. 2005). Essentially, this team uses "duration of employment" as a proxy for exposure dosage, assuming those who worked longer for RCA would receive a larger dosage of exposure.

The second team, led by Jung-Der Wang of the National Taiwan University, used "wage level" as proxy for exposure dosage, assuming low-wage employees tended to be production workers and thus receive higher dosage on the job, while high-wage employees tended to be office staff who receive much less exposure. They found increased incidences of cancer among offspring of female RCA workers, as well as increased rate of infant mortality and birth defects among the offspring of male RCA workers (Sung et al. 2008, 2009). Having differentiated former RCA workers in the labor-insurance record into cohorts and wage level, the team also found a significant increase in the rate of breast cancer among female workers who had been hired by RCA prior to 1974, and among those with lower wages (Sung et al. 2007). They also assumed that, because the use of TCE in electronics production was banned in Taiwan in 1974, the cohort who entered RCA before 1974 were more likely to be exposed to large dosages in their youth.[6] Taking these social-historical factors into consideration, the NTU team was able to render the pattern more visible than the previous team, who largely abided by general conventions in previous epidemiological studies of workplace (C.L. Chen 2011). However, none of those studies produced significant correlations comparable to the Relative Risk (RR) > 2 threshold, which RCA's parent companies held up as the scientific standard of proof, following many U.S. corporate defendants under toxic tort litigations[7]. Thus the battlefield shifted from scientific journal to the court of law.

After five years of argument on procedural matters, a Supreme Court decision in 2009 finally allowed the RCA case to be heard in the district court. Since the court hearing started in late 2009, the RCA trial has proceeded very slowly. Court sessions are typically scheduled with one-month intervals. By July 2013, the court, consisting of a panel of three judges was still hearing from the eighth witness, and the second scientific expert witness for the plaintiffs. The beginning of the RCA trial was reported by all major media outlets in 2009. However, ironically, as the open trial sessions proceeded, media attention quickly faded away; it became very difficult for reporters to write a news story about the RCA case.

Producing invisibility: mechanical cultures of visibility and trial procedure in the Civil-Law Court

One reason for the lack of news-worthiness in the RCA controversy lies with the strongly mechanical character of the legal system of Taiwan. Modeled after Germany and Japan, legal procedure in Taiwan's civil-law system is markedly different from its common-law counterpart. Merryman and Pérez-Perdomo (2007) summarize three characteristics of common-law procedures in comparison with those in the civil-law system: concentration (in time), immediacy (in space) and orality (in preferred way of expression). By contrast, despite many reforms inspired by common-law practices, civil procedures in judicial systems with a civil-law tradition still exhibit an anti-theatrical tendency: lack of temporal concentration, mediacy, and value placed on written documentation over oral speech.

These characteristics developed out of a particular way of enhancing fairness in court – what Merryman and Pérez-Perdomo call a "documentary curtain" – blocking the view between the judge and the parties and witnesses, so that the judge can be free from biases arising from seeing and hearing the demeanor of those people, and, more practically, be free from persuasion, bribery and threats by the rich and powerful. The "documentary curtain" has roots in procedural practices in Roman law as well as medieval canon law, and is still common in many civil-law systems, especially those in which evidence-taking and rendering of judgment are done by two different judges or groups of judges. In addition, by making the judicial process as machine-like as possible, the civil-law systems seek to narrow the space for judicial discretion and prevent the judges from usurping the power of the legislature – a legacy of the French Revolution's distrust of the "aristocracy of the robe" in the *ancien régime* (Merryman and Pérez-Perdomo 2007).

These document-centered characteristics make the courtroom scene in the civil-law system radically different from that of the common-law court. In the latter, if the plaintiffs' counsels make it through the usually arduous pre-trial procedures, the dispute between the parties will play out as a public spectacle in an intense series of trial sessions, often involving a jury. In today's common-law adversarial system, expert witnesses are expected to be partisan, and their testimonies will be openly scrutinized through cross-examination. So dramatic is such events that "courtroom drama" has now become a thriving genre of movie and TV drama, providing everyday entertainment to a global audience.

Such court dramas, however, are rarely found in a typical civil-law court. Instead, trial procedures in the civil-law courts are emphatically anti-dramatic. As legal scholars Glendon et al. summarize, in a typical civil action in such a court:

> ...[O]n the one hand, there is no real counterpart to [American] pre-trial discovery and motion practice, while on the other hand there is no genuine "trial" in the [American] sense of a single culminating event. Rather, a civil law action is a continuous process of meetings, hearings, and written communications during which evidence is introduced, testimony is taken, and motions are made and decided.
>
> *2007: 185*

Lack of concentration in the procedure is but one factor. Also contributing to the anti-dramatic and mechanical character of the situation are the statutory requirements and norms that expect, in such an "inquisitory system" (as opposed to the common-law "adversarial system"), all professionals – judges, lawyers, and expert witnesses – should act collegially and rationally in a collective pursuit for the truth (Bal 2005). Animosity, passion, and other emotions are often considered unhelpful, and the judges are expected to cleanse them from the courtroom. Lawyers' questions for the witness, for instance, are required by the court to be submitted in writing weeks before the hearing, so that the witness can prepare in advance, and elements of surprise, along with theatricality resulting from surprise, can be minimized.

As in the German judicial systems, the court record in Taiwan is never verbatim. Instead, the presiding judge usually dictates a summary of testimony to the clerk, after deliberating with the lawyers and the witness about the accuracy of the summary. Text, in sharp contrast with the common-law preference for "authenticity" of orality, is obviously privileged; carefully worded articulations of one's "genuine meaning," approved with the authority of the judges, are considered closer to truth than momentary outbursts.

During the testimony-taking sessions of ailing former workers in the RCA trial, the presiding judge periodically interrupted the witness and asked the witness, other judges in the panel, and the lawyers on both sides to look at the computer monitors in front of them, and clarify whether the clerk's typed record was indeed what the witness had just said. Especially when the witness was recounting her hardship and suffering, and apparently on the brink of bursting into tears, the judge invariably interrupted her, and all parties spent several minutes looking at the monitors and discussing the wording of the record of her testimony. After this, the witness was allowed to continue, and the emotion effectively disappeared.

With practices such as this, the courtroom oftentimes resembles a machine, whose sole purpose seems to be translating testimonies from speech to text, and to produce proper textual records. The presiding judge resembles a chief technician tending this translation machine, leading other actors in the courtroom in a collaborative effort to make it operate smoothly. This character would most likely please Enlightenment-era thinkers, with their machine metaphor of the state.

In spite of its definitely un-exciting appearance, a civil-law system is not necessarily archaic. The Japanese judicial system, for example, is often credited with the creation of "one of the world's most advanced environmental liability regimes" (Rolle 2003: 151). A series of judicial innovations were produced as a result of a series of four seminal environmental litigations, starting with the famous Kumamoto Minamata Disease Lawsuits in 1956 (A.C. Lin 2005, Osaka 2009). The Japanese precedents are especially important for the RCA workers and their supporters to establish the claim that a mixture of multiple chemical compounds, through multiple channels of exposure, has resulted in multiple diseases, including cancers in various sites of the human body. This may fall outside the range of what Joan Fujimura (1987) calls "doable problems" in the convention of today's medical science; nonetheless, it is closer to our daily reality in an industrial society.[8]

Co-existing with mechanical culture is theatrical culture. While the RCA trial lingers on outside the public view, a series of events culminated into a highly visible social controversy around another electronic manufacturer – Foxconn.

From personal problem to social controversy: Apple/Foxconn suicides in theatrical visual culture

In the social worlds of industrial hygiene and public health in general, suicide, along with psychological sufferings that may be pretexts for suicide, has long been a residual category – something "not otherwise categorized." In the context of liberal, individualistic legal systems, the necessarily intentionality in self-inflicted harm often leads to the conclusion that responsibility lies mainly with the victim her/himself, although most people would instinctively accept that circumstances do lead to suicide (and social scientists and psychologists have demonstrated that this is the case). Unlike the RCA case, mired in decades of disputes about causation and sequestered in mechanical invisibility, public perception of the series of suicides at Foxconn that began in 2010 quickly shifted the blame from the victims themselves and their individual psyches to the terrible conditions the employer and its main corporate customer had created for the workers. Theatricality abounds in this controversy.

The first widely-known suicide case at Foxconn was a 25-year old engineer, Sun Danyong, who jumped to his death at around 3:30 am, July 16, 2009. He left an SMS text message to his friend, saying he was beaten and searched by men from the company's security services. Pictures of Sun's suicide note can be easily found by any major search engine. Several days earlier, his Foxconn work team lost an iPhone 4 prototype in their product-development lab. If a reporter or even a common blogger had gotten hold of this phone, Steve Job's official product launch would be spoiled. For years, Apple's newest product has always been kept top secret, revealed only on the launch day by Jobs himself, on the stage, clad in his hallmark black turtleneck and blue jeans. This dramatic introduction of Apple products has long been among the most eagerly awaited media and marketing events. A spoiler, Foxconn's management might have been thinking when they interrogated Sun Danyong, would not be tolerated.

Initially, when the chain of suicides at Foxconn started to be reported by the Chinese media, one overwhelming question among media pundits was: What is wrong with those young people? If their parents' generation could cope with the hardship before, why couldn't they? It was an open letter by nine sociologists released on May 18, after the ninth suicide in 2010 that started to turn the direction of public opinion. Led by Shen Yuan of Tsinghua University, the sociologists argued:

> The use of cheap labor to develop an export-oriented economy may have been a strategic choice for China in the first period of its reforms, given restrictions and capital deficiencies due to historical conditions. But this kind of development strategy has shown many shortcomings. Low wage growth of workers has depressed internal consumer demand and weakened the sustainable growth of China's economy. The tragedies at

Foxconn have further illustrated the difficulty, as far as labour is concerned, of continuing this kind of development model. Many second generation migrant workers, unlike their parents' generation, have no thought of returning home to become peasants again. In this respect, they have started out on a road to the city from which they won't return. When there is no possibility of finding work by which they can settle in the city, the meaning comes crashing down: the road ahead is blocked, the road back is already closed. The second generation of migrant workers are trapped. As far as dignity and identity are concerned, there is a grave crisis, from which has come a series of psychological and emotional problems. These are the deeper social and structural reasons we see behind the Foxconn workers who walk on the "path of no return".

Shen et al. 2010

In effect, the sociologists were opposing blaming the victims by invoking the classic line of inquiry in their trade: seeking explanation of the event not in the individual psyche, but in the social context, as Émile Durkheim did in his 1897 work *Suicide*. This is also a narrative of causation, a "sociological causation," so to speak. Following the social-scientific framework made possible by Hobbes's separation of the person of the actor from the persona of the character, sociological analyses are able to frame events in a larger structural context, instead of individuals' intentions, personality, and morality. Such interpretation necessarily points to social reform, instead of personal improvement, as the domain where appropriate remedies can be found.

Their interpretation was quickly echoed by journalists and academics in Mainland China, Hong Kong, and Taiwan. Chinese newspapers ran undercover investigative reports on Foxconn. Trade unionists and labor-right advocates organized rallies in front of Foxconn's Hong Kong headquarters and the corporate headquarters of Foxconn's parent company Hong Hai Precision in Taiwan, mourning the deaths of Foxconn workers. Chinese-American activists mounted simultaneous protests in front of an Apple flagship store in San Francisco. An open letter signed by more than 150 academics in Taiwan demanded that government-operated funds, such as the Labor Pension Fund and Labor Insurance Fund, sell all Hong Hai stocks lest the savings of all Taiwanese workers be tainted by Foxconn's blood money. Coordinated chiefly by the Hong Kong-based anti-sweatshop NGO Students and Academics against Corporate Misbehavior (SACOM), existing networks of anti-sweatshop, occupational health and safety advocacy, and environmental organizations in North America and Western Europe quickly responded with their protest actions and statements (see, for example, Good Electronics 2010).

Since the summer of 2010, academics and university students from Mainland China, Hong Kong and Taiwan have organized joint research teams to investigate working and living conditions of Foxconn workers throughout China every year. So far, in spite of Foxconn's constant reiteration of their commitment to improve those conditions, the joint investigative team has observed little change on the ground. With Taiwan's RCA case in mind, the team is especially alarmed by occupational safety and health issues constantly mentioned by interviewed employees. Because of the company policy to punish first-line plant managers for workplace accidents and violation of health and safety codes, those managers have a strong incentive to under-report or even cover up such incidences. However, with only anecdotal testimonies from interviewees, with the high employee turnover rate at Foxconn, and with the sheer size of the workforce, it is difficult to determine in any systematic way exactly how bad the situation is (Pun 2011),[9]

Foxconn president Terry Guo tried to meet this PR disaster, first, by counter attack, claiming that those workers who committed suicide did so to get the hefty compensation money the company gave to their families, and that the suicide rate among Foxconn workers was actually lower than the national average in China. Supporters of Foxconn in the media and governments in both China and Taiwan decried singling out Foxconn, asserting that many other factories have worse working conditions, and that the attacks on Foxconn represented the West's anxiety about the rise of China as an economic superpower, or were possibly the result of a conspiracy by competitors to defame this successful enterprise. However, as one after another Foxconn employees jumped to their death in June and July 2010, the pro-Foxconn sentiment quickly lost its strength. Foxconn then scrambled to put up some rudimentary material measures to prevent further suicides – for example, setting up nets all around its dormitory buildings.

In addition to launching criticisms of the Chinese government (by Chinese media and scholars) and the Taiwanese company (by Taiwanese protesters), almost all sides demanded that Apple Inc. take its share of responsibility, as a slight increase in the profit margins Apple allotted to Foxconn would be more than enough to pay living wages to Foxconn workers. As the world's biggest company today in stock market net worth, and with its highly publicized huge profit margin, Apple could easily afford such an increase.

In accord with its secretive style, Apple never officially issued any public statement specifically on the Foxconn controversy. Instead, it sent auditors to its suppliers' factories to inspect whether they complied with Apple's Corporate Code of Conduct, and published their reports, admitting in general terms that they found a series of violations in some of those factories and that they were working on the issues. As of spring 2013, Apple and Foxconn jointly declared their intention to establish "freely-elected democratic trade unions" in Foxconn's plants to promote workers' wellbeing. Students and Academics against Corporate Misbehavior published a report made by the joint investigative teams of university students from Mainland China, Hong Kong, and Taiwan, denouncing the companies' "democratic union" proposal as a mere cosmetic measure, as most of the workers surveyed had no knowledge of union elections, and the existing state-controlled unions at Foxconn plants did not show much effort in defending workers' interests (SACOM 2013).

In sum, the controversy over the Foxconn suicides played out as theater, with multiple audiences throughout the world, situated in different social settings, but their attentions are all drawn into this human drama.

The making of theatrical culture in the Apple/Foxconn controversy

One of the conditions that made this culture of theatricality possible was that computer manufacturers had themselves drawn intense scrutiny to the computer by treating it as a fashion item. When the personal computer entered mass production as a high-tech, high-earning alternative for the electronics industry that was mired in chronic overproduction, it was not meant to be a household product with a memorable trademark. Indeed, even after the advent of the internet, computing devices were widely perceived as productivity-enhancing equipment, more similar to a photocopy machine – most of them were painted in generic grey, signifying technical reliability and, by implication, lack of personality. Although, through planned obsolescence, the PC industry has managed to drive affluent consumers to replace their computers with greater and greater frequency since the 1980s, such speed never seems to keep up with the global expansion of PC production. Since the mid-1990s low profitability has forced one after another once-leading PC brand name companies, Packard Bell, Compaq, and so on, to be sold to its competitors, culminating in the 2004 sales of IBM's PC department to the much smaller Chinese company Lenovo. Such circumstances make promoting fashion, long enjoyed by the apparel companies, highly desirable for the computer makers.

Versatile as it is, fashion, from a business point of view, is simply "[product] turnover not dictated by the durability of garments" (Ross 2004: 25); that is, you'd buy a new one before the old one is broken. To put it in cultural terms, fashion is "cultural production and the formation of aesthetic judgments through an organized system of production and consumption mediated by sophisticated division of labor, promotional exercises, and marketing arrangements" (Harvey 1989: 346). As Naomi Klein shows, in the late twentieth century, apparel brand names successfully introduced rapid, periodic change of taste, and thus the constant urge to buy, to all walks of life. Offshore outsourcing to EPZs in countries like Taiwan and South Korea enabled the brand-name garment and shoe companies to drastically cut costs, extend their reach to the low-end markets, and to shed themselves of the operational overhead of actual production. Nike is one of the most visible examples of such companies; it has never owned any production facility since its founding in the 1960s (Klein 2000), long before computer companies like Apple closed their U.S. plants and outsourced manufacturing to companies like Foxconn.

Apple Inc. successfully appropriated the operational model of the apparel brand names and emerged from the conundrum of saturated PC markets with a series of relatively inexpensive hand-held devices. Its stock market net worth skyrocketed from a modest US$ 10 billion around the turn of this century to a record high of US$ 614 billion in late 2012, surpassing all other corporations in the entire history of capitalism. Aside from the actual profitability of its product sales, Apple's attractiveness for investors is magnified many times over against a background of depressed capital markets after the 2008 financial crisis. Stock market net worth of Apple Inc. reached record high at US$ 620 billion in August 2012, surpassing all companies in the history of capitalism.[10] There were simply no good ways to park one's money elsewhere in the midst of the economic depression.

Public perception is central to profitability for fashion businesses, which now include gadget makers like Apple. Brand name images meticulously constructed by daily marketing campaigns make customers pay extra for their products. As anti-sweatshop activists have demonstrated, tainted brand names diminish desirability, and translate into lost sales. This makes branded fashion products especially vulnerable to exposés of the dreadful working conditions faced by workers making them. By discursively connecting the sufferings of these workers in the global South with sufferings of people on the depressed job markets in the global North, and contrasting them with the euphoric images projected by brand name advertisements, the anti-sweatshop movement has built a workable trans-border alliance between consumers and production workers that pressures brand name companies to address the labor and environmental issues on their production line. The complex quasi-legal Corporate Social Responsibility systems that are corporate responses to these challenges, have arisen from numerous confrontations and tentative agreements between the anti-sweatshop alliances and the brand names since the early 1990s (Klein 2000, Ross 2004).[11] Thus, when the Foxconn chain suicides make it to the public, a transnational instituional infrastructure is in place to contest over public perception of the issue, and to make this contestation vital for companies like Apple.

Conclusion

After examining the centrality of the theatrical and mechanical visual culture in the development of modern liberal democracy for the past three centuries, Ezrahi expresses his wariness of the decline of the attestive visual culture "in which [public] actions are treated as observable factual events, as causes which have discernible consequences in a public space of perception" (1990: 286). He links this with what he calls the "privatization of science" – the loss of science as a model for rational public actions. He cautions against the "aesthetization" of politics in which the theatrical metaphor looms large and the boundary between reality and fiction becomes increasingly blurred. While his concern is important especially in our age of identity politics and mass media, what we have seen in this chapter is that both mechanical and theatrical visual cultures still have their place in today's public life; each dominates different social domains and frames different forms and degrees of visibility for the public. Through these frames of visual culture, specific contexts and contents of social controversies become visible, invisible, or with some combination of both. Different visibility delineates different courses of action for people in these controversies.

Causation and perception mark two ends of a spectrum in which the RCA and Apple/Foxconn controversies appear to be on the opposite sides. The RCA case has come to be virtually invisible for the public eye. Although chemical substances in the groundwater and soil and physical ailment on workers' bodies are both material, linking these phenomena, determining causation, and assigning responsibility have to be mediated by institutionalized epistemic authorities of science and law, who are believed to possess the ability to see the chain of causation beyond common perception. Mechanical characteristics of the civil-law system, including practices of its procedural norms, render the issue even more opaque for outsiders. Ironically, both modern legal system and scientific establishment take transparency of some sort as one of their principal virtues.

By contrast, the Apple/Foxconn controversy takes place in the highly theatrical "tribunal of public opinion" over public perception. Theatricality of the event brings home a sense of immediacy for the audience; exposés appeal to the public's common sense about work-related stress and suffering, which has also been constructed by activists like anti-sweatshop groups. Highly visible episodes of cumulative and dramatic events, including suicide, mark this continuous controversy. This sense of immediacy is created in large part by Apple's successful marketing that transformed its electronic gadgets into fashion items; this success makes the company especially vulnerable to negative public perception. This enables anti-sweatshop activists to fight their case in an arena already built by their opponents, and shift the blame from suicide victims to those whom they worked for – Foxconn and Apple.

These two cases also exhibit different social configurations shaped by two consecutive periods of recent global expansion of industrial production, which are often termed "Fordist" and "Post-Fordist." RCA was a vertically-integrated corporation, directly owned and operated in virtually every step of the production – from research and development, parts and components, all the way to final assembly and packaging. Its overseas plants assumed the legal form of foreign direct investment (FDI). Controversy about its legacy can therefore take place in the legal system of a nation-state, whose jurisdiction covers a specific physical space. Apple, by contrast, outsources most of its production to foreign suppliers like Foxconn. Legally, shipments from Foxconn's plants to Apple's retail stores are market transactions, and oftentimes international trade; a legal firewall as well as national boundaries and physical distance separate the savvy Apple Inc. from the relentless daily operations of Foxconn, as in the older separation of the fashion brand names from the global apparel sweatshops.

However, as the Apple/Foxconn controversy shows, a legal firewall does not work well in the "tribunal of public opinion," especially when the way people see becomes so central to the company's business. Although mediated by certain visual cultures and through historically constituted infrastructure, the "public eye" is still a central contested ground in the early twenty-first century.

Notes

1   Goffman (1959) is perhaps the most widely cited sociological work explicitly based on the theatricality of social life. Its influence became more pertinent in cultural studies and related fields in recent decades.

2　Ezrahi (1990) calls this process "privatization of science" and contends that this is central to the prevailing disillusion with technocracy.

3　Unlike their twentieth-century predecessors such as RCA, multinational electronics companies today, such as Apple, increasingly contract out product manufacturing to myriads of overseas suppliers, instead of owning and operating their own factories – either at home or abroad. This has long been the operating model of global garment and shoes industries since the 1980s. See, for example, Duhigg and Bradsher (2012) for discussions of such a large-scale change.

4　See Tu (2007) for another example of an electronic manufacturer meeting pressure from environmentalists by refusing to disclose content of chemical mixtures used in the plant to regulatory authorities on the ground of "trade secret." Another similarly controversy is the dispute over undisclosed or under-disclosed chemicals used in the U.S. "fracking" oil drilling since 2010. See, for example, Jaffe (2013).

5　Starting in 1960, large private employers in Taiwan are required to insure their employees with the government-operated Labor Insurance, which covers medical care (later transferred to National Health Insurance), occupational injury compensation, and other benefits for workers. Insurance records, if filed and kept properly, should include information about each worker's duration of employment, wage level, insurance claims, and so on. Researchers often find errors, omissions, and other problems in this database, especially for years before computerization.

6　The 1973 ban on TCE in Taiwan was in response to the death of several young women workers at the U.S.-owned Philco Electronics after prolonged exposure to large dosage of TCE. Before the Taiwan RCA case, there was no toxicological research on TCE's health effect on women. RCA workers thus became the first female cohort to be studied for effects of occupational TCE exposure. See Y.-P. Lin (2011) for detailed gender analysis of TCE research. Also see Jobin and Tseng (2014) which locates the RCA case in a wider recurrent pattern of legal-scientific disputes.

7　Prominent epidemiologists, such as Sander Greenland of UCLA, have repeatedly argued that equating RR > 2 to the legal notion of "more likely than not" is a logical error. However, it is still commonly used in expert testimonies at court and accepted by some judicial authorities (Greenland 1999).

8　Despite the slow progress, the litigation serves as a rallying point and helps various support groups for RCA workers to become more and more organized. Regular meetings have been held since 2007 among an academic support group, the RCA workers' organization, supporting activist groups, and the lawyers, who, among other things, seek out scholars and scholarly work that might become useful in the trial. In the summer of 2011, the support groups successfully mobilized hundreds of pro bono lawyers and volunteers for a massive undertaking demanded by the court: a comprehensive questionnaire-based interview of every plaintiff about her/his detailed work history and medical history (Chen 2011). Many volunteers for the RCA litigation are also involved in a wider rising wave of labor and environmental protest movements. By allowing members of various backgrounds and concerns to cooperate, the RCA litigation strings together a series of social worlds with different commitments, allowing members of these worlds to understand the issue in different ways.

9　See the websites of SACOM (*sacom.org*) and the Maquiladora Health and Safety Support Network (mhssn.igc.org) for continuous updates.

10　Benzinga Editorial, *Forbes.com* (2012) "Apple Now Most Valuable Company in History." August 21, 2012, *www.forbes.com/sites/benzingainsights/2012/08/21/apple-now-most-valuable-company-in-history.*

11　See AMRC (2012) for criticisms of the CSR system from the anti-sweatshop movement.

References

AMRC (Ed.) (2012). *The Reality of Corporate Social Responsibility: Case Studies on the Impact of CSR on Workers in China, South Korea, India and Indonesia*. Hong Kong: Asia Monitor Resource Centre.

Bal, R. (2005). How to Kill with a Ballpoint: Credibility in Dutch Forensic Science. *Science, Technology and Human Values*, 30(1): 52–75.

Benzinga Editorial (2012). Apple Now Most Valuable Company in History. *Forbes.com*, 4/08/2012, Retrieved August 13, 2013, from *www.forbes.com/sites/benzingainsights/2013/04/08/google-might-be-planning-free-nationwide-wi-fi-since-the-fcc-isnt.*

Carroll, P. (2006). *Science, Culture, and Modern State Formation*. Berkeley: University of California Press.

Chang, Y.-M., Tai, C.-F., Lin, R.-S., Yang, S.-C., Chen, C.-J., Shih, T.-S., and Liou, S.-H. (2003a). A Proportionate Cancer Morbidity Ratio Study of Workers Exposed to Chlorinated Organic Solvents in Taiwan. *Industrial Health*, 41: 77–87.

Chang, Y.-M., Tai, C.-F., Yang, S.-C., Chen, C.-J., Shih, T.-S., Lin, R.S., and Liou, S.-H. (2003b). A Cohort Mortality Study of Workers Exposed to Chlorinated Organic Solvents in Taiwan. *Annals of Epidemiology,* 13: 652–660.

Chang, Y.-M., Tai, C.-F., Yang, S.-C., Lin, R.S., Sung, F.-C., Shih, T.-S., and Liou, S.-H. (2005). Cancer Incidence among Workers Potentially Exposed to Chlorinated Solvents in an Electronics Factory. *Journal of Occupational Health,* 47: 171–180.

Chen, C.-L. (2011). 〈流行病學的政治：RCA流行病學研究的後設分析〉 (The Politics of Epidemiology: A "Meta-Analysis" on RCA's Epidemiological Research), 《科技、醫療與社會》 *Keji, Yiliao yu Shehui,* 12: 113–157.

Chen, H.-H. (2011). Field Report: Professionals, Students, and Activists in Taiwan Mobilize for Unprecedented Collective-Action Lawsuit against Former Top American Electronics Company. *East Asian Science, Technology and Society,* 5(4): 555–565.

Clarke, A.E. and Star, S.L. (2007). The Social Worlds Framework: A Theory/Methods Package. In E.J. Hackett, O. Amsterdamska, M. Lynch and J. Wajcman (Eds.), *The Handbook of Science and Technology Studies* (3rd ed., pp. 113–137). Cambridge, MA: MIT Press.

Cowie, J. (1999). *Capital Moves: RCA's Seventy-Year Quest for Cheap Labor.* New York: The New Press.

Duhigg, C. and Bradsher, K. (2012, January 12). How the U.S. Lost Out on iPhone Work, *New York Times.* Retrieved from *www.nytimes.com/2012/01/22/business/apple-america-and-a-squeezed-middle-class.html*

Ezrahi, Y. (1990). *The Descent of Icarus: Science and the Transformation of Contemporary Democracy.* Cambridge, Mass.: Harvard University Press.

Fujimura, J.H. (1987). Constructing "Do-able" Problems in Cancer Research: Articulating Alignment,. *Social Studies of Science,* 17: 257–293.

Glendon, M.A., Carozza, P.G., and Picker, C.B. (2007). *Comparative Legal Traditions: Texts, Materials and Cases on Western Law* (3rd ed.). St. Paul, MN: West Publishing.

Goffman, E. (1959). *The Presentation of Self in Everyday Life.* Harmondsworth: Penguin.

Golan, T. (2008). A Cross-Disciplinary Look at Scientific Truth: What's the Law to do?: Revisiting the History of Scientific Expert Testimony. *Brooklyn Law Review,* 73: 879–942.

GoodElectronics (2010). Foxconn: Prolonged Protests and Clear Proposals. *http://goodelectronics.org/news-en/foxconn-prolonged-protests-coupled-to-ambitious-solutions.*

Greenland, S. (1999). Relation of Probability of Causation to Relative Risk and Doubling Dose: A Methodologic Error That Has Become a Social Problem. *American Journal of Public Health,* 89(8): 1166–1169.

Harvey, D. (1989). *The Condition of Postmodernity: An Enquiry into the Origins of Cultural Change.* Oxford and Cambridge, MA: Basil Blackwell.

Jaffe, M. (2013). Colorado Fracking Database Questioned by Harvard Study, *The Denver Post,* April 23, *www.denverpost.com/breakingnews/ci_23091371/colorado-fracking-database-questioned.*

Jobin, J. and Tseng, Y.H. (2014) "Guinea Pigs Go to Court: The Use of Epidemiology in Class Actions in Taiwan", in S. Boudia, N. Jas (eds.), *Powerless Science? Science and Politics in a Toxic World,* Oxford and New York: Berghahn.

Klein, N. (2000). *No Logo: Taking Aim at the Brand Bullies* (1st ed.). Toronto: Knopf Canada.

Kochan, T.A., McKersie, R.B., and Cappelli, P. (1984). Strategic Choice and Industrial Relations Theory. *Industrial Relations: A Journal of Economy and Society,* 23(1): 16–39.

Lin, A.C. (2005). Beyond Tort: Compensating Victims of Environmental Toxic Injury. *Souther California Law Review,* 78 S., 1439–1528.

Lin, Y.-P. 林宜平 (2011). 〈死了幾位電子廠女工之後：有機溶劑的健康風險爭議〉 (After the Death of Some Electronic Workers: The Health Risk Controversies of Organic Solvents) 《科技、醫療與社會》 *Keji, Yiliao yu Shehui,* 12: 61–112.

Markowitz, G.E. and Rosner, D. (2002). *Deceit and Denial: The Deadly Politics of Industrial Pollution.* Berkeley: University of California Press.

Merryman, J.H. and Pérez-Perdomo, R. (2007). *The Civil Law Tradition: An Introduction to the Legal Systems of Europe and Latin America* (3rd ed.). Stanford: Stanford University Press.

Osaka, E. (2009). Reevaluating the Role of the Tort Liability System in Japan. *Arizona Journal of International and Comparative Law,* 26: 393–426.

Petras, J., and Hui, P.-K. (1991). State and Development in Korea and Taiwan. *Studies in Political Economy,* 34, 179–198.

Pun, N. (2011). Apple's Dream, Foxconn's Nightmare Suicide and the Lives of Chinese Workers. *Public Sociology, Live!,* Spring 2012. *http://burawoy.berkeley.edu/Public%20Sociology,%20Live/Pun%20Ngai/PunNgai.Suicide%20or%20Muder.pdf.*

Rolle, M.E. (2003). Graduate Note: Unraveling Accountability: Contesting Legal and Procedural Barriers in International Toxic Tort Cases. *Georgetown International Environmental Law Review,* 15: 135–201.

Ross, A. (2004). *Low Pay, High Profile: The Global Push for Fair Labor.* New York: New Press: Distributed by W.W. Norton.

SACOM (2013). Promise from Foxconn on Democratic Union is Broken, May 10, *http://sacom.hk/archives/986.*

Shapin, S. and Schaffer, S. (1985). *Leviathan and the Air-pump: Hobbes, Boyle, and the Experimental Life.* Princeton, NJ: Princeton University Press.

Shen, Y., Guo, Y., Lu, H., Pun, N., Dai, J., Tan, S., and Zhang, D. (2010). Appeal by Sociologists: Address to the Problems of New Generations of Chinese Migrant Workers, End to Foxconn Tragedy Now. *http://sacom.hk/archives/644.*

Star, S.L. (2010). This Is Not a Boundary Object: Reflections on the Origin of a Concept. *Science, Technology and Human Values,* 35(5): 601–617.

Sung, T.-I., Chen, P.C., Lee, L.J.-H., Lin, Y.-P., Hsieh, G.-Y., and Wang, J.-D. (2007). Increased Standardized Incidence Ratio of Breast Cancer in Female Electronics Workers. *BMC Public Health* 7: 102.

Sung, T.I., Wang, J.D. and Chen, P.C. (2008). Increased risk of cancer in the offspring of female electronics workers. *Reproductive Toxicology,* 25(1): 115–119.

Sung, T.I., Wang, J.D., and Chen, P.C. (2009). Increased Risks of Infant Mortality and of Deaths Due to Congenital Malformation in the Offspring of Male Electronics Workers. *Birth Defects Research. Part A, Clinical and Molecular Teratology,* 85(2): 119–124.

Tu, W. (2007). IT Industrial Development in Taiwan and the Constraints on Environmental Mobilization. *Development and Change,* 38(3): 505–527.

Wieacker, F. (1996). *Privatrechtsgeschichte der Neuzeit. Unter besonderer Berücksichtigung der deutschen Entwicklung* (2., neubearb. Aufl. ed.). Göttingen: Vandenhoeck u. Ruprecht.

8

The Science of Robust Bodies in Neoliberalizing India

Jaita Talukdar

LOYOLA UNIVERSITY NEW ORLEANS

In general, the body as a machine requires surveillance under appropriate 'Diaetetick Management.'

Turner 1982a: 261

I eat every three or four hours – lots of vegetables and salad. I drink at least ten glasses of water. What I am doing is a kind of balanced and managed eating – you know, a kind of diet management.

20-year-old woman, sales executive in Kolkata, India (Talukdar 2012:113)

These two quotations have an uncanny resemblance. In the first, Turner (1982a) documents the nomenclature "diaetetick management" that was popularized by George Cheyne, a physician in eighteenth-century England, to promote healthy eating habits among affluent British men. Cheyne's contention was that the body was like a hydraulic system and human beings must avoid a heavy diet of meat products to prevent it from clogging. The second quote is from a young Indian woman I interviewed for my study of urban, upper-middle class Indian women's dieting practices. Like many of the women I interviewed, she spoke of dieting and religious fasting as scientific practices that enabled her to become energetic and healthy (Talukdar 2012, Talukdar and Linders 2013). Similar to Cheyne's contention, the women I interviewed believed that their bodies are objects that would yield better results if subjected to the control and modification of a managed diet. In spite of being separated by a large landmass, a span of two centuries, and other obvious differences, the similarity in responses between Cheyne and the women in my study point to critical periods in modern history when cultures discover and embrace "new" sciences of body and food that promise great physical benefits.

The historical moment when the women in my study were discovering new sciences of the body and diet is, however, drastically different from the time when Cheyne was writing his books on dietary restrictions based on food charts. The last several decades – starting perhaps with the discovery of the recombinant DNA (rDNA) technology in the 1970s[1] – has seen the morphing of fields of technology, engineering, medicine, and biology into one large conglomerate field, typically referred to as *biomedical technological sciences* (hereafter *biotechnological sciences*), which produce intricate knowledge of how the human body functions. In this system, biologists, physicians, engineers, and technicians collaborate to develop sophisticated techniques of excavation and analysis that seek solutions and remedies for illnesses and vehicles that promote general wellbeing at molecular and chromosomal levels. The desire to know more about bio-scientific advancements that promise reengineering of the molecular or genetic foundations of life-threatening or debilitating conditions is no longer restricted to a scientific vocation, but is part of a modern cultural imperative. Individuals learn the functioning of the body at its most fundamental biological levels.

Developments in the biotechnological sciences mark a new phase in individualization processes of the modern post-industrial self (Giddens 1991); they have opened up a new frontier in terms of biological solutions to bodily ailments (Rose 2001). These advances also mark a shift in the field of scientific research of the human body, where parameters of investigation are now bound by the use of neoliberal economic principles of "standardized, corporately framed diagnostic and prescribing procedures" (Rose 2007: 11). At first glance, then, it would seem that the women in my study and their desire to manage their food consumption, combine food groups, or learn more about the biological properties of food or their bodies were part of a larger project of self-making (Maguire 2008, Moore 2010) that characterizes modern living across the globe. For the women in my study, weight loss efforts were directed toward feeling "energetic," or productive in their day-to-day life, or making an "investment in their health" (Talukdar 2012) to lead *physically robust lives*. The women were exhibiting what Rose would call a "modern personhood" or a mindset dedicated to being constantly engaged in how to optimize the functions of the physical body and maximize its returns. After all, knowledge of how we function at the very core of our biological being, or our "ability to relate to ourselves as 'somatic' individuals" (Rose 2007: 26) means that we can correct or even enhance our bodies' functioning to determine the kind of individuals we want to be. But it soon became clear to me that the nature of women's engagements with scientific knowledge in my study belied the notion that that there is an identifiable, clear progression in how scientific knowledge spreads; a progression that requires breaking away from old, conventional knowledge systems presumed to be non-scientific, and embarking upon new, radically novel ways of thinking about the inner workings of the human body.

In the accounts of women, scientific engagements in the pursuit of robustness emerged out of a multitude of worldviews and cultural worlds that they inhabited. The women made distinctions between vital and non-vital sciences of the body, which were either anchored in traditional views of bodily wellbeing or in "new" discoveries in laboratories in America they had read about in newspapers. On still other occasions, the women concocted a *hybridized* notion of what they believed was scientific knowledge of healthy and robust bodies. Most of the time, the women were speculatively modern (Talukdar 2012) about how they engaged with new sciences of body and diet. For instance, the women expressed a lot of interest and desire in acquiring knowledge of fitness to improve overall efficiency and productivity of the body to better fulfill responsibilities associated with being good employees, good mothers, and, in general, good and responsible citizens (Talukdar and Linders 2013).

But at the same time, discernable in some of these connections was a certain amount of skepticism toward new discoveries in the field of "aesthetic medicine" (Edmonds 2003) of the body such as cosmetic enhancements and diet supplement usage. Very few women experimented with weight-loss pills or diet supplements, which guaranteed quick fixes and instantaneous results, on grounds that they only served aesthetic functions and hence were not essential or vital to the functioning of the body. This indicates that culturally specific ideologies of wellbeing tend to shape the spread of modern "technologies of optimization" (Rose 2007). Another noticeable feature of the spread of biotechnological sciences in the Indian context was the eagerness of the professional classes or those with aspirations of becoming part of India's urban elite to embrace new sciences of the body, diet, and food items. The purpose of this essay is to offer a thick description of urban Indians' engagements with biotechnological sciences of the body and diet, which alert us to the surprising and diverse ways in which new scientific ideas spread or are embraced by a culture.

To do so, I situate my arguments in a larger epistemological debate over whether there exists a unidirectional causal relationship between "science" and "society," where science is the transformative agent of a culture or if instead both are parts and products of a larger "heterogeneous matrix of culture" (Martin 1998: 30), in which modern science as a social institution does not reside outside the purview of the politics of class distinctions or value judgments (Turner 1982a, 1982b). In other words, the path along which ideas move from the scientific complex into larger society and among lay individuals is crisscrossed with cultural institutions that shape and bend these ideas in varied ways (Rajan 2006). This query also requires that we understand that the "scientific enterprise" as a connotative term is not considered exclusive to western thought systems or that formerly colonized societies lacked educational or scientific institutions prior to their encounters with western knowledge systems (Alter 2004, Harding 2008).

Biotechnological scientific theories of good health and wellbeing that were developed in western contexts are indeed gaining ground in the urban Indian landscape (Moore et al. 2011, Rose 2007). English dailies and a burgeoning fitness enterprise in the country – the main disseminators of new sciences in India – repeatedly remind urban Indians that a "wellness wave" (Sarkar 2009) is about to engulf them, as global companies compete among themselves to introduce new health products in the market. Against this backdrop, the women in my study, and urban Indians in general, were eager to develop a biotechnological approach toward life as a way to express their commitment to strive for physical vitality amidst the risks, anxieties, and biological hazards that characterize contemporary existence (Beck 1992). However, very little research has been done on the actual interface between biotechnological theories of the body and diet and urban Indians, or one that illuminates how adoption of biotechnological sciences unfolds in postcolonial, non-western settings that exhibit some amount of ambivalence, and in some cases resistance, to western knowledge systems. I seek to remedy this.

The science of robust bodies in neoliberalizing India

In this section of the chapter, I lay down my arguments about why a cultural framework is a necessary addition to existing research that has made explicit the links that exist between neoliberalism and biotechnology as knowledge systems of the body (Guthman and DuPuis 2006, Moore et al. 2011). Before I outline my framework, I would like to begin with why biotechnologies of the human body – and what we feed it – flourish in a neoliberal system.

Neoliberalism is a set of economic political practices that proposes that human wellbeing can be best advanced when individuals take responsibility for their economic destinies by perfecting the art of well-calculated risk-taking in their entrepreneurial endeavors (Harvey 2005). Biotechnologists and a sub-group of experts such as nutritionists, personal trainers, science journalists (Saguy and Almeling 2008) who act as intermediary agents between scientific experts and the public, argue that modern individuals must not only become competent in the "molecular vocabularies" of complex biological functions, but must translate this competency to act upon and improve vital, organic functions of the body. Both biotechnological sciences and neoliberalism are, thus, essentially about governing the material body to optimize its functions and physically extend it in ways that increases satisfaction with the self. In neoliberal contexts, the science of optimizing the experiences of life is sought after in the marketplace, and consumption becomes the most powerful embodiment of principles of self-work and growth (Guthman and DuPuis 2006, Maguire 2008). The corollary dictum, though, is that while one should be enthusiastic and "joyful" about the prospect of growth (Moore 2010), self-enhancing projects need to be pursued in a controlled, disciplined manner.

Thus, not surprisingly, the growing popularity of biotechnological sciences as a style of thought among urban Indians has coincided with the emergence of a technically skilled, globally attuned urban population, who have benefited the most from the neoliberal turn that the country took in the early 1900s. Liberalization of the country has seen multinational pharmaceutical and entrepreneurial groups of western industrialized nations setting up shop in the country and entering into partnerships with local entrepreneurs (Philip 2008, Rajan 2006). The major benefactors of the new economic order are (and continue to be) urban elites – or the *new middle class* (Fernandes 2006) – who have channeled their English-speaking ability, and their technical and managerial skills to become sophisticated service-providers of the national and global economy. The far-reaching nature of the global industry has provided both an impetus to improve this group's chances of upward mobility and the backdrop against which personal success is often interpreted purely as an outcome of personal striving, hard work, and individual merit (Radhakrishnan 2011, Upadhya 2011). Furthermore, in stark contrast with the rest of the country's population, easy access to a digital world has not only expanded the worldview of urban Indian elites but has also added to their repertoire of experiences and dispositions considered essential to lead a globally-attuned life.[2] With the ability to traverse multiple spaces – material and digital – at their fingertips, urban Indians have emerged as, what Rey and Boesel in their chapter in this volume have called, "augmented subjects." Deep enmeshments of material and digital worlds have augmented experiences of modern individuals, enabling new means of communication and multiple ways of imagining and reinventing one's sense of self.

The image of the normative Indian in popular media is a globetrotting company executive or software professional who seeks to maximize life's offerings by consuming goods and services in the new economy (Thapan 2004, Chaudhuri 2001). Shopping for self-pleasures or enhancement of the self, however, has had to be legitimized in a society that had kept Gandhian and socialist principles of maintaining a distinctive national identity alive in the face of western capitalist expansion, and a strong culture of empathizing with disenfranchised groups such as peasants, the workers, members of lower castes, and tribal groups. As a result of that, the print and the electronic media embarked on a persuasive campaign to break the mold of the good Indian as modest and thrifty and recast the marketplace as a legitimate social space for urban elites looking for ways to express their neoliberalism (Chaudhuri 2001, Fernandes 2006). These measures have proven to be successful, as consumption has become the dominant mode of expressing progressive, democratic values that urban elites espouse in the country.

Undeniably then both neoliberalism and biotechnologies as "styles of thought" (Fleck 1979) induce an inward gaze, in a Foucaultian sense, that aims to detect and improve upon the functioning of the minutest part of the human body. Cultivating an inward gaze has become indispensable to the project of self-development. However, scholars examining the global movement of neoliberal dispositions from its Euro-American points of origin have questioned the existence of a pure form of neoliberalism. As a set of dispositions, even if it promotes a specific type of "market rationality," neoliberalism is always and necessarily subjected to adaptations, contestations, and modifications in local contexts (Hilgers 2013). If the very process of the spread of neoliberalism is unevenly distributed, and historically and geographically specific (Moore et al. 2011, Philip 2008), then the relationship between neoliberalism, biotechnological sciences, and cultures inherently varies. Fernandes (2000: 622) has shown that the very definition of neoliberalism in the Indian context has been hybridized (Bhabha 1994). Processes of hybridization have not transcended the boundaries of the nation-state, and have grown along the lines of class-based cosmopolitanism of the urban middle-classes.

Cultural displays of consumption follow the trope of a nationalized form of globalism, with an explicit goal of protecting national, regional, and familial solidarity or inter-personal bonds in the face of rapid globalization. For instance, in the late 1990s, new electronic household gadgets such as washing machines and kitchen equipment were presented to women as means of strengthening inter-generational family ties (Munshi 1998) rather than becoming free of family chores or responsibilities. Being "appropriately Indian" means matching core cultural values of simplicity and modesty associated with Indian culture, with practices of conspicuous consumption (Radhakrishnan 2011). Thus, if a scientific approach toward daily living fulfills the "imaginative work" (Baviskar and Ray 2011) associated with a neoliberal ethos toward life, in the Indian context it means being simultaneously nationalist and globalist in nature.

Moreover, engagement with new sciences seems to be inducted into the lifestyles of urban elites as embodiments of superior cultural preferences. The class dynamics that feed the symbiotic growth of neoliberal values and a biotechnological frame of mind are quite apparent in the Indian context. For instance, the sites of consumption are English-language newspapers, weeklies, articles and health magazines that few have access to, both materially and symbolically. This implies that regardless of personal desire or motivation, the prospects of embodying neoliberal values and by implication an inward gaze is not available to everyone. Commodification of a neoliberal lifestyle means removing it from the reaches of people of limited means.

Based on these developments, I ask two broad questions in this essay: (1) what kinds of biotechnological sciences of the body and food enjoy the most patronage in India, and by whom?; (2) how are biotechnological sciences of the body being consumed and used in actual daily practices in a society that has held on to its traditional practices? In order to answer my questions, I rely on two bodies of evidence – my own study of urban Indian women's dieting and fasting practices situated in Kolkata (formerly known as Calcutta) in the year 2005 and English newspaper dailies – in the city, namely the *Telegraph* and *Times of India*, between 2005 and 2010, that have emerged as important public sources of biotechnological knowhow of bodily functions and biological properties of food.

Between these two sources, I cover a time period of approximately five years. The data I present in the rest of this essay are organized around two major themes that aim to uncover some broad trends of lay scientific engagements in the Indian context. In the first few sections, I document the science surrounding what Rose (2007) has identified as *vital functions* of the body, such as fitness, energy and productivity concerns, juxtaposing them to those sciences that serve *non-vital functions* such as cosmetic or plastic enhancements of the body. Plastic enhancers, it can be argued, are not vital to the body's survival; however, in a neoliberalizing context, they are considered to be "good for the self" (Edmonds 2007). In the last section of the chapter, I focus on how scientific knowledge gets hybridized in the Indian context.

My goal, however, is not to determine the accuracy or validity of such scientific claims but draw attention to what is happening "outside the citadel" (Martin 1998) of scientific laboratories and corporate decision-making. That is to say, how are "non-scientists" adapting to, revising, and redefining what it means to be *scientific* toward one's understanding of life.

Scientization of fitness

Neoliberal ideologies of a hard-working, productive individuality often get imprinted on the material body. In western neoliberal systems, it is a toned and fit body that has been assigned "super-value" (Crawford 1980) and is used as a proxy for a productive and healthy body

(Dworkin and Wachs 2009, Kirk and Collquhoun 1989). As Guthman and DuPuis (2006) have argued, the fit body (or the thin body) now enjoys a privileged status as the embodiment of modern individuality, an instant indicator of being in control and charge of one's body, regardless of whether a thin body is being actively or consciously pursued. The women in my study, similarly, invoked the belief that fit bodies were necessarily healthy and energetic bodies. They added another dimension to their beliefs about a fit body – a fit body is a product of a "well-thought-out" and measured process of controlling the body, which the women called "scientific." More than emphasizing the physical aspects of exertion for a toned body or the food that they were eating to become fit, the women were intrigued by and spoke of a scientific frame of mind as indispensable to their quest for fitness.

Predictably then, the fit body, according to the women, was a "muscled" or a "taut" body that could be achieved by dieting or exercising. In order to explain the process of becoming fit, however, the women took refuge in science. For instance, Neera, a forty-six-year-old director of a software company, explained her quest for a fit body by saying, "let me explain it to you *scientifically*. If you can tell a person's weight and the age, then you can tell if the person is in the normal range or if the person is overweight" (quoted in Talukdar 2012: 114). As is evident, Neera was talking about calculating her body mass index (or BMI) to become fit, something she first learned about from her nutritionist. Popular English dailies regularly feature BMI as a scientific approach to assess the normality of the body, and readers are informed of online calculators that help compute BMI and keep track of one's body weight. In this sense, and similar to its popularity in the west, BMI is gaining currency as a viable measure of health and overall wellbeing among urban Indians. What seemed distinctive about the women's interpretation of BMI, however, is that they used the terms "BMI" and "science" interchangeably, as though one uncomplicatedly explained the other.

BMI, calculated by dividing a person's weight in kilograms by the square of her height in meters, serves as a measure of the fat content (or adiposity) in the individual body. Though an individual-level measure, researchers and policy-makers use BMI to determine changes in the average body mass of a population. However, the dust kicked up by debates surrounding the scientific accuracy of BMI as an independent predictor of individual health conditions – such as cardiovascular complications or diabetes – has not yet settled down (Campos et al. 2006); the only consensus on this issue is that individuals with very high body mass index (≥ 35) are indeed at higher risk of contracting certain diseases or health problems than those with lower BMIs. Also, BMI does not account for geographical variations in typical body types, and, in that sense, is a standardized measure. The women, in my study, were unaware of any such debates or concerns about the validity of BMI as a measurement tool of good health or fitness. They were also untouched by the political argument that there is an inherent bias in the ways modern sciences of fitness and dieting use BMI to monitor and control the body and its illnesses in western contexts. Guthman and DuPuis (2006), for instance, argue that the institutionalization of thinness as a measure of good health in the United States has made individuals without access to food and other resources that help prevent weight gain, susceptible to labels such as "lazy" or "incompetent," and open to state regulation and control. Furthermore, the normalcy of a thin body as a fit or healthy body powerfully overshadows the fact that a thin body also succumbs to sickness (Campos et al. 2006) or that a thin body may be symptomatic of chronic hunger issues and its attendant health problems that affect millions of men and women in less affluent parts of the world.

Nevertheless, and as I have shown above, the women in my study were convinced of scientific principles underlying the practice of using BMI to determine good health. The women believed that using charts, numbers, and graphs to calculate BMI made the process "scientific."

Others, like Mallyon and her colleagues (2010), have similarly found that the use of numbers or technology to display and monitor changes in the body are interpreted as being methodical or being in control, which are then tagged as scientific means of becoming fit or healthy. More importantly, the scientific pursuit of a fit body, or the positioning of the body as a site of control and work, was part of a new cultural identity or orientation that the women were expected to acquire in their careers as engineers, managers, or as *techies* (Upadhya 2011). Others have commented on how a rapidly expanding class of IT and management professionals in the country are being encouraged to develop a work ethic that entails better "time management" and adoption of methodical means to push forward their personal careers (Upadhya 2011). Bodies invariably emerge as the template for such self-work. The women in my study, for instance, thought of a fat body in terms of reduced "energy." A body that was not fat was believed to bring more returns associated with them in terms of giving "more mileage," "pushing back one's physical age," or increasing bodily efficiency (Talukdar 2012). The fit body was deliberately linked to a notion of modern womanhood engaged in formal, productive work. But these beliefs were not limited to employed women, even stay-at-home mothers believed that a fit body would help prevent the onset of lethargy associated with the monotony of being a housewife (Talukdar and Linders 2013). A scientific approach to fitness was called upon to simultaneously add method and legitimacy to their efforts of improving the overall productivity or efficiency of the body.

The fact that "science" is used as a catchphrase in order to appeal to urban elites is also readily evident in contemporary cultural messages about fitness that are embedded in the larger discourse of developing a scientific acumen about how the body works. Businesses selling fitness or thinness products to urban Indians rely heavily on the use of the term "science" to explain their methods. New and upcoming gyms in big cities, for instance, advertise the presence of "certified" dieticians and massage therapists trained in "cutting-edge, scientific" techniques to provide hands-on training to their clients. Others advertise the availability of cafés that sell "protein shakes" and "organic drinks" to complete the health experience of gym-goers. English dailies place their articles on weight-loss or fitness tips in the same section as articles on such topics as recent developments in stem cell research and the scientific properties of food that cures physical ailments.

The radical novelty typically associated with the science of fitness, however, needs to be contextualized in terms of the power-knowledge-body nexus that underlies the adoption of new sciences. Historically, privileged groups have used their encounter with new sciences of wellbeing to augment their material interests. As Turner (1982a), and others have shown, the rationalization of diet and the importance of physical exercise were taken most seriously by affluent classes in industrial England and America – the aristocrat, the "learned professions," or those in sedentary occupations looking for ways to strengthen and secure their interests in a rapidly diversifying and prospering economy (Vester 2010, Seid, 1989). Colonized communities greeted modern, western medicine with similar enthusiasm, while still being ruled over by imperialist powers (Abugideiri 2004). The internal development of new sciences was entangled in the politics of distinctions, where privileged groups often imputed outbreak of diseases to careless, unhygienic (and hence unscientific) practices of less privileged groups. Prasad (2006), for instance, showed that urban elites in colonial Bengal were encouraged to become familiar with "germ theories of diseases" because of their distrust of daily practices of rural migrants who had recently moved to the city.

Historical evidence also points to privileged women as ardent supporters of new sciences. Counting calories or thinking about food and body in calculable and measured terms, now associated with women's obsession with thinness, was synonymous with early twentieth century European and American middle class women and their efforts made at "scientifically feeding" (Brumberg 1988, Seid 2010) their families. The fusing of science, bodily wellbeing, and food in my study was most prominent in the narratives of women in the professional classes. This is not to suggest that the women belonging to non-professional classes lacked a scientific acumen, but that their bodies were not implicated in the rhetoric of self-work that pervaded the cultural worlds of India's urban elites.

Scientization of robust bodies

Other than acquisition of a thin-fit body, urban Indians are encouraged to know more about new discoveries in the field of biomedical research that can add to the overall robustness of the body. English daily newspapers across the country and the internet inform urban Indians of recent bio-scientific research on the genetic properties of the human body and food (Heikkero 2004). The goal seems to be to make complex biological mechanisms and intercellular processes comprehensible, palatable, and entertaining – all of which will contribute to individuals intervening on their own behalf to improve the productivity of their bodies. This is in accordance with the Rosian argument that the rhetoric of self-work has now penetrated the surface of the body and that the ability to articulate, judge, and act is happening primarily in the language of biomedicine (Rose 2001, 2007).

If bodies were earlier, as part of a legacy of Descartian rationalism dictating nineteenth-century medicine, compared to machines and envisioned as mechanical systems, now bodies are conceived as organic systems with innate and inter-cellular properties that aid in processes of recovery and healing from life threatening or debilitating conditions. Recognizing that the human body is endowed with properties that in some cases enable it to naturally heal has not, however, minimized the scope of a scientific intervention; rather, it has enhanced bio-scientific investigation to know more about the intricate functioning of body parts to aid the process of healing. The point of bio-scientific knowledge of the body is to push the human body beyond the realm of healing or just being, creating the possibility for the body to be fine-tuned into robustness and perfection (Rose 2007).

Here, as in the previous section, information that has consequences for the vital functioning of the body and improving its overall robustness got the most attention and was considered worthy of consumption among the women I interviewed. Thus, as in the case with fitness concerns, acquiring knowledge of how to attain a robust body was part of the "obligation of self-work" (Maguire 2008) expected from urban elites of India. Not all new sciences were greeted with similar enthusiasm, however. Contrasts were made between knowledge that contributes to vital functions such as finding ways to overcome diseases, conserve energy, or improve memory retention and knowledge that aids non-vital functions such as cosmetic enhancements, which were considered non-essential to personal growth and development.

Robustness of the body entailed being able to prevent the onset of life threatening or debilitating conditions such as cancer, heart diseases, and diabetes, and to increase the body's efficiency and optimize its function. A *Times of India* article, for instance, informed readers of research published in the science journal *Nature,* about an 'assassin' protein called "perforin" that the body was capable of producing, which penetrated and killed cells infected by viruses or had turned malignant (2010b). Other articles linked brain function with intelligence and memory. One claimed that children who exercised regularly had a bigger hippocampus that resulted in improved memory (Pulakkat 2010c), and another article asserted that obese individuals are not able to regulate the intake of high calorie food because of chemical imbalances in their brains (Jain 2010). In both of these pieces, a link was made between body size, physical activity, and the way the brain develops or that fatness and lack of exercise are related to brain impairment.

Typically produced in a newspaper article format, these stories are usually blurbs of information, with just enough space to introduce the findings of research and how it would benefit the body. The goal seems to be to make readers aware of the advancements being made in the field of biotechnological sciences in finding cures for cancer or memory lapses or malfunctions. These news articles need to be viewed as an illustration of the instant value that is placed on scientific knowhow coming out of the field of biotechnological sciences, notwithstanding the fact that research of this kind of sciences are inherently reductionist in nature. Biotechnologically-related sciences, by themselves, commonly produce essentialist notions of the body as they not only isolate a process or mechanism to determine causality, which in reality is typically linked to other biological and chemical processes but rarely take into account environmental factors. Interestingly, consumption of biotechnologies follows a similar path; that is, readers are not provided with much context while being presented with new discoveries in the field. Let me elaborate on this in the case of research presented to urban Indians on obesity.

For the most part, research on obesity presented to urban Indians are about western countries. In case of the obesity debate in western contexts, research about the discovery of a "fat gene" to explain for high obesity rates has been challenged by studies that show socio-structural factors such as lack of access to high-fiber food and low-sugar diet, public parks, health insurance, and other social stressors that combine to create an "obesogenic environment" in which individuals become obese (Saguy and Almeling 2008). Childhood obesity rates among minority groups such as Aboriginal and Pacific Islander children in Australia and African-American and Hispanic children in the United States tend to be disproportionately high, which speaks to the fact that environmental factors have a potent effect in determining obesity rates. In the Indian context, in contrast, it is chronic hunger and not obesity that affects children living in urban or rural poverty.

While urban Indian readers are made to believe that obesity is an emerging problem affecting them, India continues to be a country with very low obesity rates in the world. Wherein then lies the need to consume information about obesity? Obesity, or a fat body, goes against the rhetoric of self-work that characterizes a neoliberal approach toward life. Since the embodiment of a productive and robust individuality is a physically toned, muscular, and lean body (Crawford 1980, Kirk and Collquhoun 1989), expressing a fear of gaining weight is an appropriate and necessary response in such settings. Dworkin and Wachs (2009), for instance, found that in the United States panicking about body fat is acute among those who are very well equipped in their access to resources that prevent accumulation of fat. In this sense, urban Indian elites are similar to their western counterparts in expressing a fear of becoming fat. Acquiring knowledge of fat prevention is deemed essential, even when not faced with the stressors of an obesogenic environment.

News about how scientists have discovered ways to map the gene structure or that gum tissues are an easy and superior source for stem cell research introduce readers to specialized ways of thinking about how the body functions. In the January 2010 issue of the *Knowhow* section of *The Telegraph*, for instance, there was an article encouraging readers to ring in the New Year by making a resolution of walking two thousand and ten kilometers in 365 days to enjoy good health (Mathai 2010). The distance that the readers were being asked to walk was not part of some rigorous mathematical calculation but it was worth paying attention to because research has shown, another article claimed, that walking produced such large amounts of physical energy that scientists were now able to convert it into electricity (Pulakkat 2010a). By implication, then, the very act of walking presented itself with an opportunity to immerse one's body in a scientific experiment, with the prospect of impacting individual health by producing a lot of energy within the body. The appeal of this piece of information does not lie in whether modern individuals act on biotechnological sciences to realize a specific outcome, but that individuals learn to take the science behind such an experiment seriously.

I found more evidence of the body "as privileged sites of experiments with the self" (Rose 2007: 26) when the women in my study spoke of dieting and their diet plans. Dieting, as a practice of administered weight loss, has become hopelessly entangled with a woman's insatiable desire to be thin. The women, in my study purged from their narratives on dieting terms like "starving" or "looks" or being motivated by an appearance as a motive to lose weight (Talukdar 2012). Instead, the women assigned new purposes to the practice of dieting that were by design self-implemented, and an exercise in "mixing and matching" information. For instance, Tina, a twenty-eight year old, market executive very excitedly shared with me her vitamin-C diet. After repeatedly falling sick with the common cold, and getting tired of being treated with antibiotics, Tina incorporated large amounts of food into her daily diet that she believed had high vitamin C content such as *amla* (Indian gooseberry) and cod liver oil.

By sharing information about her vitamin C diet, she was illustrating to me her own breadth of knowledge of the biological properties of food such as the Indian gooseberry that were available locally and those that had been recently introduced in the market such as cod liver oil. This was also an illustration of how Tina was capable of mixing and matching information about these two food groups, which she believed were high in citric content. In general, talking in an ostensibly educated and informed manner about experimenting with diets reflected their sense of understanding of the science of food and bodies and that the women could convincingly and effortlessly lift the information from newspapers, magazine articles, or the internet to recreate it in different settings (Talukdar 2012). Tina, thus, fits the description of an "augmented subject" – she not only engages and moves fluidly between more than one medium, she also "experiences her agency as multifaceted" (Rey and Boesel: Chapter 10).

The women I interviewed considered some practices artificial and not "life-saving" in nature, and hence rendered them non-vital to the functioning of the body. A lot of caution was raised about the use of Botox or anti-obesity pills that are now available in Indian markets. The discomfort with such products reflected a combination of fears typically associated with the dangers of using newly discovered synthetics or chemicals, and the association of these treatments with quick fixes and instant gratification that are part of a consumerist, appearance-obsessed society. In fact, in the news dailies, along with news features that spoke of health benefits of Botox in curing headaches or urinary tract infections, there were also articles that raised concerns about the possibility of indiscreet use of Botox to remove wrinkles (Dey 2005). Elsewhere, I have argued that urban Indians exercise a "speculative modernity" in relying on traditional notions of beauty and wellbeing to filter and selectively adopt new beliefs of food and the body, especially products and procedures that are distinguishably western (Talukdar 2012).

For instance, some of the women in my study on dieting spoke of their skepticism surrounding excessively thin body or an emaciated look, which they associated with the unhealthy, mental condition of anorexia. I found that, in the context of dieting, traditional or familial values and sentiments served as prophylactics in regulating what and how urban Indians consume information about new products and practices. Maaya, a twenty-year-old law student, who wanted to lose some weight, was unsure of the theory that eliminating carbohydrate from meals was a healthy way to lose weight, especially since her mother believed that rice added "glow" and "glamour" to the body, and hence should be part of one's daily diet (Talukdar 2012: 115).

Similar to the speculative sentiments that the women in my study expressed about new technologies of cutting body fat, the use of Botox to look younger falls under the category of new sciences that are blanketed in some amount of ambivalence. One article, for instance, while evaluating the efficacy of Botox as a product weighed the emotional value of a grandmother's wrinkled face over the value of youthfulness that the protein injection promised (Banerjee 2011). This is in accordance with the argument that urban Indians are asked to keep the possibility of infinite material rewards grounded in the values of simplicity, and family values that disavow excessive consumption and profligacy of a hyper-consumerist culture (Radhakrishnan 2011).

Scientization of tradition

A distinguishing feature of the dissemination of biotechnological theories of the body and food in the Indian context is the way they intersect and are made to reconcile with local, traditional practices of life. Peculiar to the Indian context is an overwhelming desire to maintain or arrive at an authenticity along traditional or nationalistic lines, when taking on or adopting new or global practices. Thus, while some of the scientific engagements were directed toward aligning scientific theories with Ayurvedic medicinal beliefs about the body, others required reimagining mundane, colloquial practices as science projects, with concrete health benefits for the body.

Yoga in contemporary India originated in the practices and prescriptions of Hindu ascetics (or the *yogi*), and aptly illustrates how and where tradition meets modernity. In place, even before the neoliberal movement, were yoga teachers who fused spiritual and scientific goals in their yogic, meditative practices. Alter (2007, 2004) documents how a yoga teacher by the name of *Swami Kuvalyanada* in the 1920s took refuge in the trappings of modern science, such as adorning a lab coat and testing yoga poses with machines to determine the physiological benefits of doing yoga on the nervous and endocrine system. Scientizing yoga in the years building up to India's independence from its imperialist ruler was part of a larger culturally based nationalist movement aiming to unveil to the world a modern India (Strauss 2002). In its most recent renditions, though, yoga has been repackaged, yet again, to solve and remedy the rigors of a stress-laden life brought about by the new economy. According to one specialist in a *Times of India* article, yoga "oozes energy and creates aura, removing toxic elements from the body" (2010a). Benefits of yoga are now being repackaged in terms of reduced stress, improved metabolism, and a cleanser of toxic elements in the body. In this sense, to an extent yoga has been stripped of its cultural uniqueness to be aligned to neoliberal values of life.

Similar to the age-old tradition of yoga, readers are asked to rediscover properties hidden in vegetables and victual items that are part of daily Indian diet such as turmeric, yogurt (or curd), and cabbage. Turmeric is a spice that is ubiquitously used in daily meals across the Indian landscape, and in the Ayurvedic culture it is known for its antiseptic properties. Recently, turmeric has been subjected to a clinical gaze. Referred to as the "yellow magic," turmeric is being reimagined as an effective weapon against Alzheimer's disease, as it contains a compound called "curcumin" that supposedly reduces plaque formation on brains (Pulakkat 2010b). Yogurt, like turmeric, is similarly believed to have healing properties since scientists have discovered that yogurt is a natural source of pro-biotics that helps in digestion. Thus, like yoga, food products such as turmeric and yogurt that are otherwise banal and part of daily regimens are now readily available for those interested in personalized, science projects.

Another practice amenable to scientific introspection is that of religious fasting. In traditional Hindu fasting, women practice familial fasting that is inextricably tied to their roles as daughters, wives, and mothers. Common to Hindu women's fasting practices, and a pan-Indian feature, is that fasts embody the vow a Hindu woman takes to be responsible for the material and emotional wellbeing of her family. However, women in the new middle classes have recast religious fasting as a scientific practice endowed with health benefits such as purifying the body of unhealthy toxins (Talukdar 2014). Since religious fasts end with practitioners consuming food such as fruits and milk, some believe that fasting work in ways similar to "detox diets" that recommend somewhat similar restrictions on food. Others like Dora, a thirty-year-old school teacher, reframed religious fasting as a type of "scientific dieting" because it entailed breaking the fast with foods like milk and fruits that she believes helps in digestion. "Scientific" was thus a term Dora was linking to the practice of fasting, based on a loosely formed idea of how the body functions. More importantly, she believed that a scientific reason was "hidden" to women of previous generations, like her mother-in-law, who fasted for religious or familial reasons. Under her clinical or biomedical gaze, however, the science of fasting became clear.

Biotechnological sciences, at least the way they have unfolded in the Indian context, have contributed to a sort of a consciousness-raising among urban Indians, awakening them to the scientific rationales underlying daily practices that preceded western, technological developments. Proponents of new sciences, both experts and lay people, thus, engage in a kind of hybridization work (Bhabha 1994); on one hand, they call upon western scientific institutions to legitimize information being disseminated, and on the other hand, they support adoption of practices where new sciences supplement or add precision to traditional forms of knowledge.

Conclusion

The growth of new sciences in India can be seen as mirroring developments in western contexts such as the symbiotic relationship that exists between biotechnological sciences and neoliberal values, but it is also clear that biotechnological sciences are being subjected to forces of hybridization. That is to say, biotechnological sciences are being contrasted and, in some cases, re-aligned with traditional forms of knowledge, indicating some amount of ambivalence in completely submitting to the authority of western sciences.

Consuming knowledge of biotechnological sciences has indeed become an important part of the lives of members of the professional classes in urban India who want to be seen as modern, engaged participants of a new, global economy. Espousing a scientific frame of mind is part of the "imaginative work" that is expected from individuals living in neoliberal orders where the goal is to maximize the benefits of both one's body and environment. For instance, sciences directed toward improved productivity or enhanced energy, improved memory, or higher intelligence, which have explicit economic benefits tied to them, were most eagerly sought out. Desire for fitness, and acquiring it via scientific means, was another area of self-work popular with neoliberal Indians. A scientific bent of mind has emerged as a commendable social force, rich in cultural capital, capable of diluting and rendering ineffective contradictory evidence. For instance, I have shown in my discussions on the use of BMI, very few women I interviewed were aware of the argument that a standardized measure that does not account for regional and geographical disparities in body types of individuals may not be the best measure of good health or normal bodies among urban Indians. Sciences that contribute to a physically robust and enhanced state of life thrive in neoliberal systems, and urban Indian society is no different from other neoliberal contexts in this regard. Yet, there are some contrasts.

Noteworthy in the Indian context is a certain amount of ambivalence regarding new sciences of the body that promise aesthetic enhancement of the body, which are typically geared toward personal satisfaction. This was most evident in urban Indians engagement with Botox injections or diet pills available in the marketplace. As mentioned earlier, in my study on dieting, I found a great deal of skepticism expressed toward adoption of new technologies of liposuction and stomach stapling. Quick measures of weight-loss were identified as either arti-ficial or as extreme. Some of the women also identified ultra-thin bodies as decisively western and foreign, even calling such bodies anorexic, and a sign of mental depravation. The Indian context, in this aspect, seems to be different from other emerging neoliberal economies such as Brazil, where aesthetic sciences such as cosmetic surgery (or *plastica*) to acquire Euro-American facial and bodily features enjoy immense popularity (Edmonds 2003, 2007). Though there is a need for more research in this area, it appears that nationalistic and cultural ideals of beauty and wellbeing still enjoy a strong patronage among Indians, which may channel and modify how aesthetic sciences make inroads into Indian society.

Scientific discourses of wellbeing, in the Indian context, are being directed toward aligning personal growth with that of the family and nation's economic prosperity. While self-control and regulation is a key feature of maximizing benefits in a neoliberal order, in the Indian context, it seems familial and nationalistic values funnel how these regulations unfold in the new economy. However, a biotechnological orientation toward an enhanced and robust state of life is part of an elite culture in contemporary India, which has not yet penetrated the lives of vast masses of people living in the country.

I would like to end this chapter by revisiting the two quotations that I used to introduce the essay. Both the quotes undoubtedly capture the wonder surrounding scientific knowledge that all of us are predisposed to, but as I hope to have demonstrated, how and through what means scientific knowledge of the body becomes meaningful at a particular historical moment is not only specific to a culture but also indissolubly entangled in it. If, as suggested by Martin (1998: 40), "culture is non-linear, alternately complex and simple, convoluted and contradictory," so is the science that is being produced outside the scientific complex.

Notes

1 Recombitant DNA (rDNA) technology is a technique discovered by Herbert Boyer and Stanley Cohen that enabled the cutting, splicing, and joining of DNA molecules in a laboratory setting. The discovery of cutting and sequencing of genes is widely viewed as a landmark development that resulted in biological sciences becoming technological and paving the way for the emergence of a biotechno-logical industry.
2 In spite of the IT revolution in Indian society, the larger population's access to internet is still very low. According to Thussu (2013), internet penetration remains very low: by the end of 2011, 121 million Indians had internet access, a penetration rate of just over 10 percent of India's population of 1.2 billion.

References

Abugideiri, H. 2004. "The Scientisation of Culture: Colonial Medicine's Construction of Egyptian Womanhood 1893–1929." *Gender & History* 16: 83–98.
Alter, J. 2004. *Yoga in Modern India: The Body between Philosophy and Science.* Princeton: Princeton University Press.
Alter, J. 2007. "Yoga and Physical Education: Swami Kuvalayananda's Nationalist Project." *Asian Medicine* 3: 20–36.
Banerjee, P. 2011 "Club Sandwich: Beauty, Health, and Age-defying Treatments were the Focus at a Ladies' meet at the Bengal Club." *The Telegraph* November 21.
Baviskar, A. and Ray, R. 2011. *Elite and Everyman: The Cultural Politics of the Indian Middle Classes.* London: Routledge.
Beck, U. 1992. *Risk Society: Towards a New Modernity.* London: Sage Publications.
Bhabha, H. 1994. *The Location of Culture.* New York: Routledge.
Brumberg, J.J. 1988. *Fasting Girls: The History of Anorexia Nervosa.* New York: First Vintage Books.

Campos, P., Saguy, A., Ernsberger, P., Oliver, E., and Gasser, G. 2006. "The Epidemiology of Overweight and Obesity: Public Health Crisis or Moral Panic?" *International Journal of Epidemiology* 35: 55–60.

Chaudhuri, M. 2001. "Gender and Advertisements: The Rhetoric of Globalization." *Women's Studies International Forum* 24: 373–385.

Crawford, R. 1980. "Healthism and Medicalization of Everyday Life." *International Journal of Health Services* 10: 365–388.

Dey, A. 2005. "The Great Cosmetic Scam: Unscrupulous Medical practitioners Are Laughing Their Way to the Bank." *The Telegraph Knowhow* October 23.

Dworkin, S.L. and Wachs, F.L. 2009. *Body Panic: Gender, Health and Selling of fitness.* New York: New York University Press.

Edmonds, A. 2003. "New Markets, New Bodies: An Ethnography of Brazil's Beauty Industry." Ph.D. Dissertation, Princeton University.

Edmonds, A. 2007. "The Poor Have the Right to be Beautiful: Cosmetic Surgery in Neoliberal Brazil." *Journal of Royal Anthropological Institute* 13: 363–381.

Fernandes, L. 2000. "Nationalizing 'the Global': Media Images, Cultural politics and the Middle Class in India." *Media, Culture, and Society* 22(5): 611–628.

Fernandes, L. 2006. *India's New Middle Class: Democratic Politics in an Era of Economic Reform.* Minneapolis. University of Minnesota Press.

Fleck, L. 1979. *Genesis and Development of a Scientific Fact.* Chicago: University of Chicago.

Giddens, A. 1991. *Modernity and Self-Identity: Self and Society in the Late Modern Age.* Stanford: Stanford University Press.

Guthman, J. and DuPuis, M. 2006. "Embodying Neoliberalism: Economy, Culture, and the Politics of Fat." *Environment and Planning D: Society and Space* 27: 427–448.

Harding, S. 2008. *Science from Below: Feminisms, Postcolonialities and Modernities.* Durham, NC: Duke University Press.

Harvey, D. 2005. *A Brief History of Neoliberalism.* New York: Oxford University Press.

Heikkero, T. 2004. "The Fate of Western Civilization: G.H. von Wright's Reflections on Science, Technology, and Global Society." *Bulletin of Science, Technology, and Society* 24: 156–162.

Hilgers, M. 2013. "Embodying Neoliberalism: Thoughts and Responses to Critics." *Social Anthropology* 21(1): 75–89.

Jain, T.V. 2010. "High on Calories." *The Telegraph Knowhow* March 29.

Kirk, D. and Collquhoun, D. 1989 "Healthism and Physical Education." *British Journal of Sociology of Education* 10: 417–434.

Maguire, J. 2008. "Leisure and Obligation of Self-Work: An Examination of the Fitness Field." *Leisure Studies* 27: 59–75.

Mallyon, A., Holmes, M., Coveney, J., and Zadoroznyj, M. 2010. "'I'm not Dieting, I'm Doing it for Science': Masculinities and the Experience of Dieting." *Healthy Sociology Review* 19(3): 330–342.

Martin, E. 1998. "Anthropology and the Cultural Study of Science." *Science, Technology, and Human Values* 23: 24–44.

Mathai, G. "2,010 in 2010." *The Telegraph Knowhow* January 11.

Moore, K. 2010. "Pleasuring Science: Nourishment, Habitus, Citizenship in the United States, 1970–2010." Paper presented at the annual meeting of American Sociological Association. Atlanta, Georgia, August 14–17.

Moore, K., Kleinman, D.L., Hess, D.J., and Frickel, S. 2011. "Science and Neoliberal Globalizations: A Political Sociological Approach." *Theory and Society* 11: 505–532.

Munshi, S. 1998. "Wife/mother/daughter-in-law: Multiple Avatars of Homemaker in 1990s Advertising." *Media, Culture and Society* 20: 573–591.

Philip, K. 2008. "Producing Transnational Knowledge, Neo-liberal Identities, and Technoscientific Practice in India." In *Tactical Biopolitics,* edited by B. da Costa and K. Philip, 243–267. Boston: MIT Press.

Prasad, S. 2006. "Crisis, Identity, and Social Distinction: Cultural Politics of Food, Taste, and Consumption in Late Colonial Bengal." *Journal of Historical Sociology* 1: 246–265.

Pulakkat, H. 2010a. "Power Walking." *The Telegraph Knowhow* March 22.

Pulakkat, H. 2010b. "Yellow Magic." *The Telegraph Knowhow* August 2.

Pulakkat, H. 2010c. "Brawns and Brains." *The Telegraph Knowhow* September 27.

Radhakrishnan, S. 2011. *Appropriately Indian: Gender and Class in a New Transnational Class.* Durham: Duke University Press.

Rajan, K.S. 2006. *Biocapital: The Constitution of Post-Genomic Life.* Durham: Duke University Press.

Rose, N. 2001. "The Politics of Life Itself." *Theory, Culture & Society* 18: 1–30.

Rose, N. 2007 *The Politics of Life Itself: Biomedicine, Power, and Subjectivity in the Twenty-First Century.* New Jersey: Princeton University Press.

Saguy, A. and Almeling, R. 2008. "Fat in the Fire? Science, the News Media, and the Obesity Epidemic." *Sociological Forum* 23: 53–83.

Sarkar, S. "Riding the Wellness Wave." *India Today* 2 January 2009.

Seid, R.P. 1989. *Never Too Thin: Why Woman Are at War with Their Bodies?* Englewood Cliffs, NJ: Prentice Hall.

Strauss, S. 2002. "Adapt, Adjust, Accommodate: The Production of Yoga in a Transnational World." *History and Anthropology* 13: 231–251.

Talukdar, J. 2012 "Thin but Not Skinny: Women Negotiating the 'Never Too Thin' Body Ideal in Urban India." *Women's Studies International Forum* 35: 109–118.

Talukdar, J. 2014. "Rituals and Embodiment: Class Differences in the Religious Fasting Practices of Bengali, Hindu Women." *Sociological Focus* 47 (3).

Talukdar, J. and Linders, A. 2013. "Gender, Class Aspirations, and Emerging Fields of Body Work in Urban India." *Qualitative Sociology* 36: 101–123.

Thapan, M. 2004. "Embodiment and Identity in Contemporary Society: Femina and the 'New' Indian Woman." *Contributions to Indian Sociology* 38: 411–444.

Thussu, D.K. 2013. "India in the International Media Sphere." *Media Culture and Society* 35: 156–162.

Times of India 2010a. "Yoga Coordinates Body and Mind" November 28.

Times of India 2010b. "How Protein 'Assassin' Kills Rogue Cells" November 1.

Turner, B. 1982a. "The Government of the Body: Medical Regimens and the Rationalization of Diet." *British Journal of Sociology* 33: 254–269.

Turner, B. 1982b. "The Discourse of Diet." *Theory, Culture and Society* 1: 23–32.

Upadhya, C. 2011. "Software and the 'New' Middle Class in the 'New India'." In *Elite and Everyman: The Cultural Politics of the Indian Middle Classes*, edited by Amita Baviskar and Raka Ray, 167–192. London: Routledge.

Vester, K. 2010. "Regime Change: Gender, Class, and the Invention of Dieting in Post-Bellum America." *Journal of Social History* 44: 39–70.

Part III
Digitization

9

Toward the Inclusion of Pricing Models in Sociotechnical Analyses

The SAE International Technological Protection Measure[1]

Kristin R. Eschenfelder

UNIVERSITY OF WISCONSIN-MADISON

Introduction

Beginning in the 1980s, the development of information and communications technologies (ICT) and computer networking introduced massive change to practices for distributing scholarly and scientific materials like journal articles. ICT and mass digitization led to changes in the nature of the available material, what attributes of content users valued, control over access and use of a work, and pricing and packaging. From a publisher perspective, ICT and networking increased risks of unauthorized copying and redistribution, thereby encouraging development of a controversial set of technologies known as *technological protection measures* (TPM), designed to control how users access or use digital works.

This chapter tells one exemplary story of the effects of these changes using the case of the implementation and subsequent retraction of a TPM and a new "token" pricing model (explained later) in the SAE Digital Library (SAE-DL) in 2006 and 2007. The SAE-DL is published by SAE International (formerly known as the Society for Automotive Engineers) which publishes research and other materials in the transportation engineering fields (Post 2005, Sherman 1980). Implementation of the TPM and a required move to token pricing in 2006 created a backlash among users and led to a widespread cancellation of the SAE-DL by academic engineering libraries. Pressure from academic stakeholders ultimately led SAE to remove the TPM in 2007 and abandon the token pricing requirement a year later.

In telling the story of the SAE-TPM controversy, I have three goals. First, I use the science and technology studies (STS) concepts *inscription* and *anti-programs* to explain stakeholder's negative reaction to the TPM and the token pricing model. I use interview, listserv, and observational evidence to show how stakeholders' work practices and values conflicted with assumptions inscribed in the TPM and the token pricing. In explaining, I demonstrate how analysis of pricing is an integral part of sociotechnical analysis of user interaction with media distribution systems. Second, I trace the *anti-program* actions taken against the TPM and pricing model by stakeholders, highlighting alignments and slippages between the groups. I use the

SAE story to demonstrate a common challenge in scholarly communications reform: the difficulties of maintaining stakeholder unity to press publishers for pricing change. Finally, I complicate this book's theme of "disruption" and suggest an alternative metaphor of *constant renegotiation* to describe how the terms and conditions for acquisition of published works have always been in flux.

To set the background for the story of the SAE-DL, I first describe major changes introduced by mass digitization of scholarly works, including changes to the nature of the available material, how users access materials, valued attributes of content, control over access and use of a work, and pricing. Stakeholders most important to my story include the publishers who distribute the work (SAE), end users (faculty and students) and intermediary users (librarians/information managers) who acquire works from publishers on behalf of a larger end user community.

Changes to the nature of the scholarly materials and what users value

Among publishers and libraries, ICTs and networking produced new products called "digital libraries" (DL) that provide network-based access and search services to databases of content (Borgman 2003). Readers of this chapter likely have extensive experience using digital libraries such as Elsevier's Science Direct, or JSTOR.[2] The subject of this chapter, the SAE-DL, like other DLs, revolutionized how users made use of and valued scholarly materials.

To understand how DLs changed use and valuation of scholarly materials, one must remember the pre-full text DL era. Prior to the 1990s, most DLs held citations rather than full text. Once researchers had obtained useful citations from a DL, they would visit a library that subscribed to the physical copy of the desired journal to make photocopies of the desired content. In this era, users valued the comprehensiveness of a physical collection owned by their library, and convenience meant the user could walk over to the library and get a desired paper from the shelf.

DLs changed what users could use, how they could use it, and what they valued. They radically changed perceptions of convenience. For one, DLs often provided access to a broader range of materials than had previously been purchased on paper including, as time went on, full text. DLs were increasingly network accessible, meaning that you no longer had to visit the library to use them. Finally, DLs incorporated new indexing and search tools and hyperlink-enabled interfaces that allowed speedy browsing of many papers. Over time, *end users came to value speed and network access over physical ownership of material*. Physical ownership decreased in value in the minds of end users, and convenience meant the ability to get the full text of a paper instantly without leaving your office. Yet DLs were often more expensive than earlier print counterparts. This was true of the SAE-DL. Librarians had always considered SAE materials expensive, and the DL was even more expensive than most databases – it "should be gold plated" remarked one librarian on an engineering library listserv.

Another important trend was the development of the Open Access (OA) movement in the 2000s. Academia has had a long informal gift culture (Cragin and Shankar 2006, Hilgartner and Brandt-Rauf 1994). But digitization and networked access led to calls for free access to the output of scholarship, under the assumption that making the output of research widely accessible would democratize science, increase economic development, and improve global health (Willinksy 2006). Within the story of the SAE-DL, arguably the rhetoric of OA influenced some end users' expectations about how publishers ought to treat scholarly material.

The shift from ownership to licensing

The digitization and networking of scholarly output was not without downsides. One major concern has been the shift from the ownership of physical works to the leasing of networked access to digitized works that live on publisher servers. This shift has occurred across content industries, including music and movies, as well as academic publishing. In many ways, a music lover's decision to subscribe to Spotify instead of purchasing CDs has the same affordances and creates the same problems as a librarian's decision to rely on networked access to a DL instead of buying paper or CD-ROMs.[3]

Critics lament that the shift to licensing of networked access allows publishers to create much broader restrictions that those typically protected by U.S. copyright law, including restricting uses that would arguably be protected under the principle of Fair Use. As Lessig explains, because digital use necessarily involves making a copy of a work (i.e., downloading a document from the vendor's server), rights holders argue that they can place greater restrictions on use of that copy than they could place on a paper work (Lessig 1999). Within the context of scholarly publishing, many worry about loss of "perpetual access" to licensed materials. Purchased paper volumes remain available to users if a library cancels its subscription. In contrast, DLs often resemble a subscription-based cable-TV model or a Spotify account: when you cease to pay, you lose access to the content.[4] In the context of the SAE-DL story, this meant that researchers working for institutions that cancelled their DL subscriptions lost access to content which they had not printed or previously saved.[5] No bound volume sat on the library shelves.

Technological protection measures

Due to concerns about unauthorized copying, some publishers began to adopt "technological protection measures" (TPM), or computer hardware- and software-based tools that limit access to a work or use of a work.[6] TPM, in combination with license restrictions and the anti-circumvention provisions of the 1998 Digital Millennium Copyright Act, limit uses of protected digital materials. The arguments in favor of TPM are that publishers may refuse to digitally distribute valuable content unless they can ensure it will not be misused, that TPM facilitate new business models not possible without new forms of control, and that they better ensure a return on investment. Scholarly publishing's expansion of its customer base into inter-national markets with different social copying norms make issues of control more important than in previous years. In this context, the 2011 International Intellectual Property Alliance 301 report on China specifically calls for increased attention to the problem of piracy of "library academic journals" and "usage of books and journals at universities" (IIPA 2011: p 59).

TPM may control access to a work by controlling who can log in to a DL, for example by limiting access to a particular computer or login credential. They also control use of a work, for example, by blocking users' ability to save digital copies or limiting printing. Many different types of TPM exist with varying access and use restriction possibilities. The SAE-DL was what is known as a "container" TPM, similar to those commonly used to protect streaming movies or music (for example, Netflix, iTunes). In a container system, all of the content is kept within a locked (encrypted) container (software reader). A user cannot view the content outside of the container, and the container prevents the user from making copies of the content.

To use content a user must: (1) have access permission via a valid login or an approved IP address, (2) have a copy of the software reader, (3) have an encrypted network connection to the publisher's server, and (4) only seek to enact uses allowed by the software reader. This arrangement is very similar to the Netflix system today where a user can only access and view movies while: (1) she has a Netflix account, (2) she has downloaded the Netflix movie viewer, (3) she has an active network connection to the Netflix server, (4) she is only seeking to use the content in ways allowed by the Netflix movie viewer.

The SAE-DL controversy was shaped by the fact that it was one of the first instances where university campuses faced the prospect of scholarly content embedded in a restrictive TPM. But while the SAE-DL was arguably the first highly restrictive TPM in campus libraries, some less restrictive types of TPM, such as those that require page-by-page printing in order to make copying more onerous, were already widely implemented by scholarly publishers (Eschenfelder 2008).

Pricing models

SAE's introduction of the TPM was coordinated with a required shift from the popular "subscription" pricing model to "token" pricing. Academic library pricing norms for DL products had come to center around the subscription model, where the library pays a set price for all possible uses in a subscription period (Farjoun 2002). Often called the "all you can eat" model, it provides cost predictability for libraries, which is important given campuses' highly variable use. If a professor develops a new assignment that requires her students to find papers from a certain DL, use of that DL may jump without warning. Using the subscription model, variance in use within a given year does not impact the library's bill for that year.[7]

In contrast with token pricing, a library pays in advance for a number of downloads/uses of papers in one year. There is no physical token, rather tokens are virtual permissions, typically managed on the publishers' server.[8] For example, a library may purchase 850 tokens for a year. A publisher's server would keep track of these uses and (ideally) periodically report to the subscribing libraries about how many tokens they had left.[9] The token model works well for low use customers or customers with very predictable/managed numbers of downloads; however, it is generally considered disastrous by academic libraries given the lack of predictability in campus use.[10]

The token pricing and the TPM are strategically linked. In the case of the SAE-DL, the TPM blocked users from saving digital copies of papers (described later). This inability to make digital copies of papers ensured that each digital use of an SAE paper counted against the token count. Downloading a paper once used one token. Re-downloading it the next day (because you can't save an electronic copy) uses another token. Users had to remember to print out paper copies to avoid incurring token charges. In my analysis, I explore and compare the problems created by the token pricing with the problems created by the access and use restrictions built into the TPM itself. Each clashed with stakeholder groups' practices and values in interlinking ways.

Inscription and resistance

The science and technology studies (STS) concepts *de-scription* and *anti-programs* (Akrich and Latour 1992) are valuable in illuminating the SAE-DL TPM story, which includes anti-program actions by researchers, students and librarian/information managers, during a period of description in which ultimately some (but not all) of the assumptions that the publishers embedded into the TPM and the token pricing were challenged. To explain why the de-scription occurred, I demonstrate how the TPM and the token pricing conflicted with the work practices and values of stakeholders.

STS has a long history of showing how system designers' preconceptions influence the design of systems and shape future uses of those systems (Bijker and Law 1992, Winner 1986). Akrich and Latour (Akrich 1992, Akrich and Latour 1992) use the metaphor of a film script to describe how designers' preconceptions delimit what activities users can undertake with a system, such that designers' assumptions become *inscribed* into a system. STS scholars have also explored how users resist this direction and reconfigure systems to suit their own needs (for overviews see Oudshoorn and Pinch 2003, 2007). Akrich and Latour use the term *anti-program* to describe user actions that conflict with designers' scripts about what ought to occur. They also introduce the term *de-scription* to describe the period of adjustment (or failure to adjust) between the system designers' idealized scripted uses and what users are actually doing (or not doing) in the real world. De-scription is a period of renegotiation of the script between some users and systems designers (Akrich and Latour 1992). It may also be a literal de-inscription of the assumptions previously inscribed into a system. STS studies of technology implementations have also shown how different groups of stakeholders may have different conceptions of the degree to which a system is "working" or "not working" based on how the system fits with preexisting work practices and values (Bijker 1995).

Within the world of media and information studies, scholars have described how media industries have inscribed preconceptions about who should use intellectual and cultural property into media distribution systems, and they have demonstrated how those systems script how consumers can use media (Gillespie 2007, Eschenfelder et al. 2010, Eschenfelder and Caswell 2010). Other scholars have outlined the anti-programs undertaken by activists or users to resist industries' inscribed conceptions (Postigo 2012, Eschenfelder et al. 2010).

I present the SAE-DL TPM story as a period in which anti-programs reigned and an ultimate de-scription of some (but not all) of the assumptions that the publishers embedded into the TPM occurred. I illustrate how conflicts between work roles and values and the idealized scripts embedded in the TPM and the token pricing made the SAE-DL "not work" for stakeholders. In doing so I illustrate how pricing is a major design factor of a system that shapes how users interact with the system.

Methods

Data collection occurred in fall 2007 and spring 2008 just after the implementation of the TPM in academic libraries in the United States. Of 31 known U.S. university SAE-DL subscribers, 23 cancelled their subscription immediately after the TPM implementation, while 8 retained their subscription (Thompson 2008) [11] I selected case sites that retained their subscriptions and sites that cancelled their subscriptions.[12] Collecting data from both allowed me to compare the effects of the TPM with the effects of loss of access to the SAE-DL due to cancellation. All my case sites were campuses with undergraduate and graduate engineering programs. I interviewed three sets of stakeholders: engineering faculty, students, and librarians/information managers. I employed face-to-face, phone and email interviews. I also reviewed SAE literature about the TPM and its pricing model, press releases, annual reports, and a copy of the SAE-DL license from the 2007/2008 period. Finally, for 18 months I followed discussion of the controversy on engineering librarian blogs and listservs that discussed the controversy.

For analysis, I used inductive coding to generate greater understanding of the perceived indirect and direct "effects" of the pricing and TPM implementations on student, faculty and librarian stakeholders across and between the case sites.

SAE's technological protection measure and pricing controversy

SAE is an unusual, but important publisher for researchers in transportation engineering. My interviewees perceived SAE's periodicals as having lower prestige than other engineering publishers, but SAE's large membership and niche focus on transportation.[13] made it a valuable venue for attracting grants from manufacturers and gaining a large audience.[13] As one interviewee explained, SAE is "still the best place to publish because of [the] large audience." Another explained, "at an SAE conference, you could have 250 people in the room, at [a competing conference] [there] might be 9." Students I spoke to were aware of the importance of SAE as a publication venue and spoke of their plans to become SAE members after they graduated.

According to librarian/information managers, SAE sales were predominately focused on industry, and this focus limited SAE's interest in understanding academic users. They complained that SAE was less responsive to the needs of the academic community than other engineering publishers; "SAE just doesn't get academic libraries and how student use of its materials differs from corporate use of its materials."

The first mention of the TPM appeared in SAE's 2005 Annual Report which described the development of a system to "ensure protection" for SAE's intellectual property (SAE 2005). SAE's website framed the TPM as a sustainability issue, helping to ensure continuation of SAE's role in developing and distributing information. According to SAE, "Managing access to documents based on copyright ownership and protecting content rights is imperative to the continued development and distribution of technical documents and standards" (SAE 2008). In addition to implementing the TPM, SAE announced it would move from a subscription-based pricing model to a token pricing model for the SAE-DL.

SAE adopted a commercial container TPM system called File Open, which employed an Adobe PDF plug-in viewer. As with all container TPM, SAE documents were in a protected viewer (a secure PDF). In order to use SAE papers, a user needed to meet four conditions:

1   The user had to have permission to use the work/have a login or IP address that was authorized by SAE.
2   The user had to have an authorized copy of the PDF plug-in software reader.
3   The user had to have an active network connection so the software reader could contact the SAE server to request permission to use an SAE document.
4   The user could only seek to enact uses allowable by the software reader.[14]

The TPM scripted a narrow range of uses. Users could not save digital copies of any documents. Each opening of a digital document counted as a use. Users could only access digital documents from an authorized network connected computer. The TPM did not allow documents to be "edited, copied, saved, emailed, or accessed from an unauthorized machine." Linking to the token pricing, if a user wanted to look at the same document on different days, the later view counted against their library's token count. Users could print a document once. SAE instructions advised users to always print documents (given the one allowed printing) to facilitate use over time, and in case they needed to use the document while offline (SAE 2008). In terms of the required token pricing, SAE offered libraries four pre-pay options for tokens: 850, 1800, 2500 and 3300 uses. While prices were not publicly posted, one listserv poster described how SAE charged her library $7,500 for 850 downloads ($8.82 per use) and $15,000 for 1800 downloads ($8.30 per use).

In late 2006 and early 2007 as libraries began to renew their licenses, SAE began to insist on implementation of the TPM and the token pricing. In reaction, engineering libraries and research centers at many universities cancelled their subscription to the SAE-DL. This is when I began data collection.

How did the TPM and pricing impact users?

In implementing the TPM, SAE inscribed values like protection of its intellectual property and created the above described very narrow script for use. SAE's idealized use assumed users would find a relevant paper based on its abstract and then, using a token, print one copy. They would store the printed copy for later use. They would never attempt to save a digital copy. In collaborative situations, each team member would acquire and print their own copy or rely on photocopies made from the authorized print out. In implementing the token pricing, SAE also inscribed assumptions into the SAE-DL. In this script, an information manager would successfully predict use to buy the correct number of tokens and then effectively manage use so as not to run out of tokens. In the sections that follow I detail how these assumptions and scripted uses conflicted with the actual work practices and values of the stakeholders, generating the period of de-scription.

Intermediary users: impacts on librarians/information managers

On professional listervs and blogs, engineering librarians/information managers reacted to the TPM and token pricing implementation by beginning an anti-program encouraging each other to cancel their SAE-DL subscriptions. Intermediary users were concerned about the token pricing and the precedent SAE's TPM would set within the larger scholarly information marketplace.

From an intermediary user perspective, it is almost impossible to separate the TPM from the pricing because the TPM's tracking of all uses and blocking of digital re-use and sharing was the foundation of many pricing problems. The token pricing relied on the TPM to stop users from making digital copies of SAE papers. If users could save electronic copies, or share e-copies among themselves, they could avoid drawing down the library's tokens. But the pricing of tokens also matters for understanding why users reacted negatively to implementation of the token pricing. If the cost of each token/use was low enough, then librarians/information managers may have accepted the new pricing model. But token costs were high, and SAE-DL did not allow rollovers. "Nothing rolls over to the next year, and you get no financial credit" one librarian complained. Alternatively, if librarians/lab mangers could better predict use, they might have been more amenable to the token model.[15] But academic use is irregular, and intermediary users were frustrated by lack of SAE reports on token usage (tracking token usage took place on SAE servers). One complained, "Given how difficult it is to obtain download data from SAE, it is not possible to monitor the usage of downloads through the subscription year." The TPM also caused user support issues. Intermediary users had to explain the new restrictions and network connectivity requirements as well as troubleshoot installation and use problems.

The TPM/pricing combination also conflicted with long-standing business practices between publishers and academic libraries; in particular, it challenged norms about subscription pricing and unlimited use. As described earlier, the subscription pricing model with "all you can eat" use is typically a pricing option that libraries can choose when contracting with journal publishers (Farjoun 2002). Intermediary users feared that SAE's requirement to move to the combination of token pricing and TPM restrictions would set a bad precedent in the market and other publishers would be tempted to follow suit. As one librarian explained, "if these publishers watch our reaction to SAE, and decide that librarians think this publication model is 'OK,' then we've got serious problems down the road." Another concurred, "Wavering within the academic library community at this point and having shops buy the [SAE-DL] with the current or proposed terms is a precedent that we don't want to set. It just opens up the possibility of other publishers following SAE's lead."

There is a parallel here with Dumit and de Laet's work (Chapter 4, this book), which describes the power of norms to create social expectations. Dumit and de Laet critique the normative power of average weight graphs, arguing that the graph creates societal expectations that average weight is "normal" and that other weights above and below the average are not normal. In the SAE case, intermediary users sought to protect the existing norm of subscription pricing and unlimited use from the introduction of an alternative. They feared that academic library acceptance of SAE's actions would weaken the subscription/unlimited use norm by representing the TPM/token pricing combinations as also "normal." Other publishers might then see the TPM/token combination as an acceptable way of arranging their business relationships with academic libraries.

For the intermediary users, the assumptions built into the TPM and the token pricing conflicted with numerous realities of the academic work environment. They also conflicted with long-held professional practices dictating how publishers and libraries normally exchange goods and services. The interlinked nature of the TPM and the token pricing made both an issue for intermediary users; however, the fact that most libraries waited to renew their subscriptions until 2008 when SAE re-offered a subscription "all you can eat" model, suggests that pricing was a significant problem on its own.

I continue by discussing impacts on faculty and students. It is important to remember that I had very limited data from users with actual experience with the TPM. Most campuses cancelled access. For sites that kept their subscription, most end users' experiences were limited to a few months.

End users: impacts on faculty and students

The scripts assuming paper printing and paper based sharing inscribed in the TPM directly conflicted with transportation engineering's team-based work practices and strong norms of sharing important information within teams. Labs maintain shared local collections (LC) of papers that are most relevant to their work and the first stop in any literature review. Faculty referred students to papers in their LC. One student described how students don't "go to the library" instead they go the LC, "which has folders of both old and new key papers." In some cases, these LCs were paper-based, but in many cases LC were shared server files. As one researcher described, "Somebody gets the paper from the library and then they store it in a topically organized set of shared folders." Students working on a project together described how "they have group directories and share both electronic and paper files related to their topics." The LCs were typically only available to team members associated with a particular project or advisor. One faculty described his set-up as an "online paper archives that are IP restricted and password protected."

Most of the faculty I interviewed did not have direct experience with the TPM because their students did most of their literature work, but all the faculty were aware of it. A few faculty reported usability inconvenience problems created by the TPM, but the most dominant problem was that they believed *their* research should not be subject to the TPM restrictions. Faculty used arguments common from the open access movement, pointing out how most of the papers in the DL were created by faculty, and SAE should not place such onerous restrictions on faculty and student use of faculty-produced material. Others noted they were SAE members and that their organization should be more sensitive to their needs. For faculty, the SAE-DL did "not work" because the narrow use scripts allowed by TPM conflicted with faculty expectations about how publishers should treat work produced by faculty and how their students should be able to make use of scholarly materials.

Students had more direct experience with the TPM because they did the bulk of the literature-related work: finding new articles, adding them to the LC and bringing them to their advisor's attention. Students described how the assumptions about paper printing and sharing embedded in TPM made traditional sharing practices more difficult, required extra work to make paper copies, led them to avoid using SAE papers, and even made them question their future membership in SAE. One student described the extra work he had to do to access a paper he was using in his team: "I am doing a project for Engines with my partner and we share a shared drive. I couldn't open a paper so I had to request it." Another student concurred, "The restriction hampers the sharing of papers needed to complete the project or research work." Faculty also lamented the impact of the TPM on sharing between graduate students. Said one professor: "In the past they could swap information/knowledge more readily because they could trade papers more easily. Now they are more compartmentalized." Because of the extra copy labor required by the TPM, some students began to avoid using SAE papers and instead found substitutes. But the ability of students to use alternative sources depended on the exact nature of their research – for some, SAE materials were not fungible. The imposition of the TPM also caused some students to question their possible future membership in SAE. Joining SAE was a professional norm in the field, but some students noted that because of the TPM they no longer wished to join.

The TPM's narrow use scripts constituted a major disconnect with the real values and work practices of faculty and students and users began anti-programs in reaction to these scripts. Scripts that blocked saving digital copies would have precluded inclusion of SAE papers into an electronic LC and precluded peer-to-peer electronic sharing of papers among team members. But for both faculty and students, the TPM was seen as a "nuisance" rather than an absolute impediment because they believed they could use work-arounds.[16] As one student described, "It does not seriously affect the project or research work, but it surely does create nuisance." Similarly, faculty interviewed believed their graduate students could "work around" it to continue their normal practices.[17] These work-arounds constituted anti-programs that conflicted with the SAE idealized use scripts. Avoiding use of SAE materials represents another anti-program.

Imposition of the TPM conflicted with the work practices of student and faculty, requiring them to create anti-program work-arounds. It also conflicted with their expectations about how publishers ought to treat scholarly materials.

The imposition of the TPM and the token pricing had a secondary effect of loss of full text access due to a campus cancelling or reducing access to the SAE-DL. The most dominant theme in the interviews was that reduced full text access slowed down the 24/7 research production now expected in engineering labs research and reduced the quality of literature reviews. On campuses that cancelled the DL, users could only browse abstracts before ordering a paper through document delivery. The time drags in the ordering process interfered with instant access to full text papers now expected in contemporary academic practice. One student noted that ordering a paper, "Pretty much just wasted a day for me." Another described how, "It's a day lag for information that I would like to have at this instant."

The token pricing assumed that users could make good judgments about the value of a paper based on its abstract. But according to participants, reliance on abstracts before ordering or spending a token was problematic because SAE paper abstracts were poor quality. Further, often abstracts were inadequate because users needed to look at images, graphs and equations in an article to judge its utility. The poor quality abstracts, combined with the extra time and mental effort needed to order the paper meant they were less likely to request an SAE paper. One student explained that ordering based on an abstract, "is a too big an investment of my time especially when I don't know if that specific paper may be useful or not." One faculty member explained how he knew he could order papers, but "couldn't be bothered" to order a paper "based on a 'maybe' relevant abstract." Users also testified that because they were less likely to look at SAE papers, they likely had less-thorough literature reviews than they might have had, and this, in theory, affected the potential quality of research. As one student explained: "To be honest, my lit reviews, and papers in general were much better researched before." A faculty member figured his literature reviews were "probably less thorough now than before."

In the second type of reduced access, sites limited access to the SAE-DL at designated computers (sometimes supervised by a librarian) in order to better manage token usage. In this case users had to travel to the designated computer to obtain SAE papers and this required travel also directly conflicted with the instant full text access that academic users have come to take for granted. Users resented having to physically go to the library. One complained how "It would be much more useful if we could look up papers from our desk rather than ... sending someone to the library to get the papers, which is time consuming." One faculty member reported that instead of requesting a paper from the library, he would ask a student in his class for access via his [industry] work accounts because that was more convenient.

The change to the distribution system for SAE materials, which resulted in the loss of convenient full text access and a reversion to pre-digitization work patterns of library-mediated abstract-based access, conflicted with the contemporary academic beliefs and practices about how and when work should occur.

How did the token pricing impact end users?

In this section I consider how the assumptions and values inscribed in the token pricing model interfered with end users work practices and values apart from the TPM. I do so in order to better understand the impact of pricing on users' potential interactions with a system. One can argue that systems designers either assumed that end users would be unconcerned with the cost of papers or that their paper use would be very selective such that individual users would not fear running up high costs. My data shows students were sensitive to the token pricing and became much more selective about using papers. Some faculty used fewer papers, but other faculty resisted librarian/information manager attempts to manage token use.

On campuses that cancelled the SAE-DL, students could order SAE papers through document delivery. In these cases, some student reported limiting their use despite not having been instructed to do so. They explained that the need to order the paper made them more aware of the costs. As one reported, "I try to use it less often due to the cost." Some, but not all, faculty reported using fewer SAE papers in classes because they knew students had to order the papers.

Other campuses kept their subscriptions, but limited access to just one or two computers and sometimes actively sought to manage token use by mediating searches or advising students and faculty to be "judicious" about downloading papers due to their need to manage the token count. Students reported some anxiety about using SAE papers because of these warnings, and a few faculty members reported concerns about assigning SAE papers in classes due to the token concerns. While the system script assumed they would continue to use relevant SAE papers, users instead engaged in the anti-program of avoiding SAE materials.

Not all faculty changed their use of the SAE-DL. Some reported actively resisting calls by librarian/information managers to moderate use due to token concerns. They complained that instructions to "be judicious" were not feasible in the highly distributed and independent academic working environments. Their refusal to self-regulate use illustrates the strength of the current academic expectation for unfettered network access to full text materials. One researcher argued, "How do you control how independent people download the same paper? My student might download it ... then Professor X's student might download it ... then someone else does ... how do you control that?" Librarians/information managers confirmed that it was difficult to manage others' use and to explain to others "how or why people should control their usage."

Interests shift and anti-programs vacillate

In March 2007, in a move that created a huge stir in the engineering library community, MIT engineering faculty, in coordination with the MIT engineering library, very publicly cancelled their SAE-DL subscription citing the implementation of the TPM (Library Journal 2007). Because MIT is seen as a premiere engineering program, its decision to cancel was hugely influential. The MIT cancellation gave libraries/information managers more social capital with which to persuade reluctant faculty that they too should cancel their SAE-DL subscription. Further, the publicity associated with the cancellation likely increased pressure on SAE's publishing unit. My participants reported that around the same time, several prominent faculty who had held leadership positions in SAE complained about the TPM at the spring 2007 SAE Annual Meeting. Shortly after the Annual Meeting, on April 19, 2007, the SAE Publications Board voted to remove the TPM from the academic library product (SAE International, April 2007). Thus, the actual life of the TPM was quite short. An academic user could have experienced the TPM from sometime in early 2007 to April 19 of 2007: approximately 1–4 months. *Importantly – the token pricing was retained for one more year until spring of 2008.*

I argue that the high-profile cancellation by MIT, combined with faculty anti-programs of complaining at SAE Board Meetings and growing awareness of rights issues among faculty stemming the broader open access movement, created a confluence of interests among intermediary and end users to pressure SAE to remove the TPM. But as I later show, it was more difficult to retain that confluence of interest to change the token pricing model. Getting end user buy-in for pricing issues can be difficult due to the current structural arrangement of DL purchasing.

Following up from the April decision, on November 6, 2007 the SAE Publications Board voted to permanently remove the TPM from the academic library DL product (SAE International Nov 2007). But the token pricing model remained in force until spring 2008. Therefore, many of the concerns related to reduced access and token count anxiety were ongoing.

At this stage in the story, we see destabilization of the confluence of interests between intermediary-user and end-user stakeholders. According to librarians/information managers, the suspension of the TPM greatly increased pressure from end-users to re-subscribe to the SAE-DL because faculty and students were focused on the TPM and had less understanding of the pricing issues. Once the TPM was removed, they thought the controversy was over. One librarian complained, faculty "don't understand that the problem is not only the TPM, but also the associated pricing model..." Another complained that faculty "didn't understand ... effects [of the token pricing] on the library budget and how viewing would cost them a lot."

Finally, in spring 2008, SAE sent out an email announcing new "Welcome Back" pricing which included a return to the (still expensive) subscription "all you can eat" pricing model. By March 4, 2008 MIT announced it was resubscribing to the SAE-DL, (Durrance 2008) and many academic engineering libraries followed.

While the TPM was in place, intermediary and end user interests were aligned, but with the removal of the TPM, some faculty and students lost interest. This illustrates how the change in the pricing model did not affect all stakeholders the same way. It directly impacted librarian/information manager budgeting and created concerns about setting a bad precedent in the scholarly information market. The impact of pricing on end users was mixed. Students and some faculty reported reduced use of SAE papers due to concerns about pricing and encouragements to manage use. Further, end users experienced many indirect negative consequences due to cancellation or reduction of access because of price concerns. But, as I explain below, the structure of campus DL acquisition budgets arguably reduces faculty and student sensitivity to broader pricing problems.

Discussion

In telling the SAE-DL story, I have had three goals: to demonstrate how analysis of pricing is an integral part of sociotechnical analysis of media distribution systems, to highlight the slippages in alignment between intermediary and end user stakeholders in scholarly communication reform efforts, and to complicate this book's theme of "disruption" by suggesting an alternative metaphor of constant renegotiation. I draw out evidence from the SAE-DL case to further develop each idea below.

My depiction of the SAE-DL TPM case study demonstrates how analysis of pricing and packaging is an integral part of sociotechnical analysis of media distribution systems – along with consideration of legal, cultural and technical elements (Eschenfelder et al. 2010). The SAE-DL controversy was generated by both the TPM and the token pricing. The scripts created by TPM alone cannot explain why all user groups undertook "anti-programs" to resist or reconfigure the SAE-DL (Akrich 1992, Akrich and Latour 1992, Oudshoorn and Pinch 2003, 2007). The period of de-scription I examined (2006–2008), or the period of renegotiation of the script between users and SAE-DL systems designers, included renegotiation of both the uses controlled by the TPM and ultimately also of the pricing model with the reversion to the "Welcome Back" subscription pricing.

I argue that pricing tends to go underexplored in studies of digital media distribution systems. To understand the importance of pricing in the SAE-DL case, I gave analysis of token pricing equal weight with analysis of the TPM. I analyzed how the idealized use scripts stemming from both the pricing and the TPM interacted and conflicted with work practices and values. Pricing impacted librarians/information managers most directly, complicating their budgeting, conflicting with the established field norm of subscription pricing, and raising concerns about the new model creating a new community norm (see Dummit and de Laet, Chapter 4, this book). Pricing reduced students' and faculty use of SAE materials and interfered with contemporary assumptions about instantaneous access to full text materials held by many faculty at research universities. While continued faculty pressure on SAE no doubt aided in the ultimate return to the subscription pricing model, other faculty resisted calls to moderate usage to control pricing. Further, librarians/information managers perceived that some faculty did not care about pricing. It is important to remember that removal of the token pricing model took almost a year longer than removal of the TPM.

While imposition of restrictive TPM are still relatively rare in the academic library world, difficult pricing models are unfortunately common. Pricing remains problematic in scholarly communications for several reasons. The market-based regime of scholarly communications distribution, where scholarly output is seen as a product that should generate (at least some) revenue, remains dominant. Critics charge that (some) publishers will extract as much revenue from universities as possible instead of merely seeking a reasonable rate of return. While many critique the current model, and alternative models exist (for example, Open Access, government subsidization of publication as a public good), there is still widespread acceptance of the market-based model and institutional acquiescence to the negative aspects of that model.

Faculty could be powerful critics of difficult pricing, but the process by which goods and services are exchanged between libraries and publishers is currently largely hidden from faculty. Budgets to purchase scholarly materials are commonly centralized at libraries rather than being distributed out to academic units. While this arrangement has efficiencies, critics argue that it creates a faculty ignorance of pricing problems that reduces market pressures on publishers (Darnton 2010, Davis 2003). Further, librarians and information managers often do not have the power to cancel an important subscription without agreement of faculty who are unfamiliar with what constitutes a reasonable price (Darnton 2010, Davis 2003).

This brings me to the second goal of this chapter – to highlight the challenge of stakeholder unity facing all scholarly communication reform. The Open Access movement has increased scholars' awareness of the value of their work as input to the scholarly publishing system. My faculty participants were primarily upset with SAE because the use restrictions scripted by the TPM conflicted with their expectations of how their publishers should treat their research work. Faculty believed SAE should make their work available to their institution under reasonable terms and conditions. But, while faculty know an onerous use restriction when they see it, arguably many faculty are ignorant of what constitutes a reasonable price, especially given the sheltered position of faculty in the cost-distribution system (Darnton 2010). The story of the SAE-DL, therefore, illustrates the common structural challenge facing scholarly communication reform. Because faculty and student end users are structurally sheltered from the price of materials, they sometimes are unreliable allies in efforts to force publisher pricing change (Darnton 2010, Davis 2003). Of course, it is dangerous to generalize; many faculty care deeply about scholarly communication reform. In the case of the SAE-DL engineering faculty were highly instrumental in removing the TPM and no doubt aided in the ultimate return to the subscription pricing model. But it is still generally true that faculty members are not directly impacted by pricing due to the structure of budgeting.

My final goal is to complicate this book's theme of "disruption." This book has a theme of *how and why ideas, artifacts and practices come to be institutionalized or disrupted*. I argue the term disruption implies some level of stability that ought to be returned to and that a metaphor of *continuous re-negotiation* better captures the landscape of scholarly communications. I suggest that Akrich and Latour's *de-scription* is a better term than disruption as it could imply an indefinite loop of negotiation and re-negotiation. Publishing (scholarly or otherwise), taken in the long view and from an industry level, has always been changing. Terms and conditions for acquisition of published goods has always been in flux and subject to debate (Eschenfelder et al. 2013, Eschenfelder and Caswell 2010). Restrictions pushed by publishers today may be gone tomorrow and vice versa (Johns 2011). While we are in a particularly visible period of instability and change where Open Access or SPARC advocates are advocating alternative visions of the scholarly communications distribution process, from a long-term perspective, stakeholders have always been in constant negotiation about acceptable terms for the exchange of published works. From this perspective, the story of the SAE-DL TPM is a brief, but educational illustration of one publishers' reconfiguration of distribution systems and end users' anti-programs in reaction to the system.

Conclusion

The SAE-DL TPM case study demonstrates that the TPM alone cannot explain all the antiprograms undertaken to change the SAE-DL. For some stakeholders, pricing was as important as the TPM; moreover the TPM and the pricing were interdependent. More broadly the case study demonstrates that studying changes in pricing is important to understanding users' acceptance or rejection of a system and in explaining variance in stakeholders' relationship to a system – different users may find different types of pricing acceptable based on their unique institutional roles.

The changes seen in the SAE-DL TPM case study are in many respects representative of changes in the larger consumer content publishing industry. The SAE-DL case study demonstrates how DLs like e-journal and e-book distribution platforms allow publishers to exert control over published works in ways not possible in a paper distribution environment. These forms of control include TPM, but also include new pricing models that may tie cost to digitally tracked usage patterns. Further, both TPM restrictions and pricing models can have negative effects on day-to-day research and teaching practices. Everyone should be interested in the potential negative effects of media rights and pricing because both shape how we learn, debate and produce new knowledge. The case study also illustrates challenges to scholarly communication reform efforts. These challenges should be of interest to all readers because scholars are part of the problem. The current budgeting structures for scholarly materials may inure you to the pricing problems in your field.

Notes

1   Data collection for this paper was funded by the Institute of Museum and Library Services Laura Bush 21st Century Research Grant RE-04-06-0029-06. An early version of this paper appeared as Eschenfelder, K.R. (2012) "DRM in the Ivory Tower: The Society for Automotive Engineers Digital Library and Effects of TPM on Research, Learning and Teaching." Proceedings of the I-Schools Conference. Many thanks for the feedback provided by the editors and contributors to this book that improved this paper.

2   Digital libraries have complex variations including publishers who host all their own content, publishers who host their content and other publisher's content, and resellers who host other publishers content.

3   My claims about the shift to licensing in the library field are a generalization. While many electronic products are licensed, some are purchased. For example, purchased e-books may be owned by a library but still hosted from a publisher server. In these cases libraries typically pay an additional yearly maintenance fee. It is still unclear to what degree first sale and other rights apply to e-books acquired by libraries but maintained on a publisher's server. The devil is often in the details of the contract.

4   The question of whether or not subscription to a digital library should include perpetual access to all materials paid for during the period of subscription is still under debate in the scholarly publishing community. Some publishers do now offer perpetual access, but others do not (Zhang and Eschenfelder, working paper).

5   In this case the library would seek to borrow a copy through interlibrary loan or purchase a copy directly from the publisher or another commercial document provider to fulfill the end user's request.

6   Many people use the term "DRM" or digital rights management system to refer to this set of tools; however, the term "digital rights management" refers to the much broader set of concerns and practices associated with managing rights to a work. In order to avoid confusion, this article employs the narrower term TPM.

7   Publishers may choose to increase the subscription fee when the license renewal comes due based on histories of increasing use.

8   Tokens can be implemented as code based permissions keys similar to e-commerce gift certificate codes. In this case, an information manager like a librarian can give a token to a particular user. The SAE-DL tokens tended not to be managed in this way.

9 The token pricing model has numerous variations including instances in which the subscribing library holds token codes that (like gift certificate numbers) it can pass out to end users. This was not so with the cases I examined.

10 Reality is always more complicated. There are examples of very successful uses of token pricing on campuses in instances where intermediary users can successfully manage use, for example, a very limited potential user group that must ask permission to use a token.

11 The survey was undertaken by Thompson (2008). It was conducted through a professional engineering librarian listserv and personal contacts between Thompson and other librarians.

12 It is important for readers to understand that librarian/information managers do not always have the social power to make difficult cancellation decisions. Their jobs exist to manage information resources for end users like faculty and students. They must convince faculty to cancel; and if faculty wish to keep a certain DL product, it can be politically difficult for libraries/information managers to do otherwise.

13 Competitors to SAE for engineering research publication include the American Society of Mechanical Engineers (ASME), and the Institute of Electrical and Electronics Engineers IEEE, the American Institute of Physics (AIP) and the American Physical Society (APS).

14 When an authorized library user requested a document, a client-server session was established between the end user's computer and SAE's "permissions server. " The desired document was encrypted and sent to the user. A key was also delivered through a separately encrypted transmission to the security handler plug in on the user's machine. The SAE-DL plug-in then passed the key to the users Adobe Reader to permit display of the encrypted file. Importantly, users had to have an internet connection in order to view the documents in order to access the key to decrypt the document (SAE 2008).

15 One easy to imagine workaround would be to print the document and rescan it to create a non-TPM protected digital copy.

16 Some may argue that any workaround to a TPM constitutes a violation of federal law. In order to protect research participants, I did not ask any questions about if and how users employed workarounds. U.S. law currently prohibits circumvention of a technological protection measure expect for certain exemptions determined by the Library of Congress (17 USC 1201). The anti-circumvention provisions have been thoroughly critiqued elsewhere (Burk 2003). There is no reason to believe that circumvention of SAE's TPM would have been considered exempt from the anti-circumvention law. In addition, any workaround to the TPM would be a violation of the license agreement signed by universities to gain access to the DL product.

17 It is worth noting here that faculty and students at smaller institutions with more constrained library budgets are more reliant on interlibrary loan and document delivery. They do not enjoy the assumptions of 24/7 instanteous full text access I describe here. Cross-campus inequities are a problem deserving their own paper.

References

Akrich, M. 1992. "The De-Scription of Technical Objects." In *Shaping Technology/Building Society,* edited by Bijker, W. and Law, J. Cambridge, MA: MIT Press.

Akrich, M. and Latour, B. 1992. "A Summary of a Convenient Vocabulary for the Semiotics of Human and Nonhuman Assemblies." In *Shaping Technology/Building Society,* edited by Bijker, W. and Law, J. Cambridge, MA: MIT Press.

Bijker, W. 1995. *Bicycles, Bakelites and Bulbs*. Cambridge, MA: MIT Press.

Bijker, W. and Law, J. 1992. *Shaping Technology/Building Society*. Cambridge, MA: MIT Press.

Borgman, C.L. 2003. *From Gutenberg to the Global Information Infrastructure: Access to Information in the Networked World*. Cambridge, MA: MIT Press.

Burk, D. 2003. "Anti-Circumvention Misuse." *UCLA Law Review* 50: 1095–1140. *http://ssrn.com/abstract=320961*.

Cragin, M.H. and Shankar, K. 2006. "Scientific Data Collections and Distributed Collective Practice." *Computer Supported Cooperative Work* 15,3: 185–204.

Darnton, R. (2010) "The Library: Three Jerimiads." *New York Review of Books*, November 30.

Davis, P.M. 2003. "Tragedy of the Commons Revisited: Librarians, Publishers, Faculty and the Demise of a Public Resource." *Portal: Libraries and the Academy* 3, 4:547–562.

Durrance, E., 2008. "Following Removal of TPM, MIT Resubscribes to SAE Database." *MIT Library News.*

Eschenfelder, K.R. 2008. "Every Library's Worst Nightmare: Digital Rights Management and Licensed Scholarly Digital Resources." *College and Research Libraries* 69, 3: 205–226.

Eschenfelder, K.R. and Caswell, M. 2010. "Digital Cultural Collections in an Age of Reuse and Remixes." *First Monday* 11,1.

Eschenfelder, K.R., Desai, A., and Downey, G. (2010) "The 1980's Downloading Controversy: The Evolution of Use Rights for Licensed Electronic Resources." *The Information Society* 27, 2: 69–91.

Eschenfelder, K.R., Tsai, T., Zhu, X., and Stewart, B. (2013) "How Institutionalized Are Model License Use Terms? An Analysis of E-Journal License Use Rights Clauses from 2000 to 2009." *College and Research Libraries* 74: 326–355.

Farjoun, M. 2002. "The Dialectics of Institutional Development in Emerging and Turbulent Fields: The History of Pricing Conventions in the Online Database Industry." *Academy of Management Journal* 45,5: 848–874.

Gillespie, T. 2007. *Wired Shut: Copyright and the Shape of Digital Culture.* Cambridge, MA: MIT Press.

Hilgartner, S. and Brandt-Rauf, S.I. 1994. "Data Access, Ownership and Control: Toward Empirical Studies of Access Practices." *Knowledge: Creation, Diffusion, Utilization* 15,4: 355–372.

International Intellectual Property Alliance (IIPA). 2011. *Special 301 Report On Copyright Protection And Enforcement: People's Republic Of China. www.iipa.com/rbc/2011/2011SPEC301PRC.pdf.*

Johns, A. 2011. *Piracy: The Intellectual Property Wars from Gutenberg to Gates.* Chicago: University of Chicago Press.

Lessig, L. 1999. *Code and Other Laws of Cyberspace.* New York: Basic Books.

Library Journal. March 28, 2007. "MIT Automotive Engineering Faculty Say No to TPM." *Library Journal.*

Oudshoorn, N. and Pinch, T. 2003. "How Users and Non Users Matter." In *How Users Matter: The Co-Construction of Users and Technologies*, edited by Oudshoorn, N., and Pinch, T. Cambridge, MA: MIT Press.

Oudshoorn, N. and Pinch, T. 2007. "User Technology Relationships: Some Recent Developments." In Hackett, E.J., Amsterdamska, O., Lynch M., and Wajcman, J. (eds) *The Handbook of Science and Technology Studies* (pp. 541–566). Cambridge, MA: MIT Press.

Post, R.C. 2005. *The SAE Story: One Hundred Years of Mobility.* Warrendale, PA: Society of Automotive Engineers.

Postigo, H. (2012) *The Digital Rights Movement: The Role of Technology in Subverting Digital Copyright.* Cambridge, MA: MIT Press.

SAE International. 2005. *Annual Report.* Society of Automotive Engineers. Warrendale, PA: SAE International.

XSAE International. April 19, 2007. "News Release: SAE Publications Board to Review Digital Rights Management Controls for Students, Faculty." Warrendale, PA: SAE International.

SAE International. Nov 6, 2007. "News Release: SAE Removes FileOpen Digital Rights Management for Students, Faculty." Warrendale, PA: SAE International.

SAE International. 2008. "Digital Rights Management (DRM) Overview and FAQs". SAE International: Warrendale PA. *http://store.sae.org/drm-overview.htm.*

Sherman, W.F. 1980. *75 Years of SAE: Springboard to the Future, Freedom Through Mobility.* Warrendale, PA: Society of Automotive Engineers.

Thompson, L. 2008. "SAE and Digital Rights (Mis)management: How to Marginalize Your Product, Alienate Your Customer, and Jeopardize Your Future (in Three Easy Steps)." American Society for Engineering Education Annual Conference, Pittsburgh PA.

Willinsky, J. 2006. *The Access Principle: The Case for Open Access to Research and Scholarship.* Cambridge, MA: MIT Press.

Winner, L. 1986. *The Whale and the Reactor.* Chicago: University of Chicago Press.

10

The Web, Digital Prostheses, and Augmented Subjectivity

PJ Rey and Whitney Erin Boesel

UNIVERSITY OF MARYLAND AND UNIVERSITY OF CALIFORNIA, SANTA CRUZ

The turn of the twenty-first century has been called "the Digital Age," and not without reason. In (post-)industrial nations, most young adults between the ages of 18 and 22 cannot remember a time when computers, cell phones, and the Web were not common features of their cultural landscape. Today we have profoundly intimate relationships not just *through* these newer digital technologies, but *with* them as well. Because we use digital technologies both to communicate and to represent ourselves across time and across space, we express our agency through those technologies; at times, we may even experience our Facebook profiles or our smartphones as parts of ourselves. The way we interpret these subjective experiences has social and political consequences, however, and it is those consequences that we seek to interrogate in this chapter.

We begin with a quick overview of the sociological understanding of subjectivity and two of its key elements: embodiment and the social conditions of subjectification. We argue that contemporary subjects are embodied simultaneously by organic flesh and by digital prostheses, while, at the same time, contemporary society maintains a conceptual boundary between "the online" and "the offline" that artificially separates and devalues digitally mediated experiences. Because we collectively cling to the online/offline binary, the online aspects both of ourselves and of our being in the world are consequently diminished and discounted. The culturally dominant tendency to see "online" and "offline" as categories that are separate, opposed, and even zero-sum is what Nathan Jurgenson (2011, 2012a) terms *digital dualism*, and it leads us to erroneously identify digital technologies themselves as the primary causal agents behind what are, in fact, complex social problems.

In our final section, we use so-called "cyberbullying" as an example of how digital dualist frames fail to capture the ways that subjects experience being in our present socio-technological milieu. We argue that the impact of "cyberbullying" violence stems not from the (purported) malignant exceptionality of the online, but from the very unexceptional continuity of the subject's experience across both online and offline interaction. At best, digital dualist frames obscure the causal mechanisms behind instances of "cyberbullying"; at worst, digital dualist frames may work to potentiate those mechanisms and magnify their harms. For these reasons, we develop the concept of *augmented subjectivity* as an alternative framework for interpreting our subjective experience of being in the world. The augmented subjectivity framework is grounded in two key assumptions: (1) that the categories "online" and "offline" are co-produced, and (2) that contemporary subjects experience both "the online" and "the offline" as one single, unified reality.

Dimensions of subjectivity: social conditions and embodiment

To begin, we need to be clear both about what we mean by "subjectivity" and about our understanding of the relationships between subjectivity, the body, and technology. Speaking broadly, subjectivity describes the experience of a conscious being who is self-aware and who recognizes her[1] ability to act upon objects in the world. The concept has foundations in classical philosophy, but is deeply tied to the mind/body problem in the philosophy of the Enlightenment – i.e., the question of how an immaterial mind or soul could influence a material body. While some Enlightenment philosophers examined the material origins of the subject, idealist philosophers and Western religions alike tended to view subjectivity as a transcendent feature of our being; they believed that the mind or soul is the essence of a person, and could persist even after that person's body died. This quasi-mystical separation of the subject from her own body and experience was typified by René Descartes, who famously observed, "I think, therefore I am" (1641 [1993]). Near the end of the Enlightenment, however – and especially following the work of Immanuel Kant – philosophy began to abandon the idea that it had to privilege either of mind or matter over the other. This paved the way for the modern understanding of subjectivity as a synthesis of both body and mind.

Western understandings of subjectivity continued to evolve over the nineteenth and twentieth centuries, and philosophers came to understand subjectivity as historically situated. They also began to recognize that different bodies experience the world in radically different ways. These points are critical to our argument, so we expand on both.

First, we consider subjectivity to be *historically situated* – or, in other words, that the nature of subjectivity is fundamentally shaped by the subject's particular time and place. For example, the range of objects available to be acted upon – and thus the range of experiences – available to human beings thousands of years ago most certainly differed from the range of objects available to human beings in modern consumer society. We trace this understanding of subjectivity as historically situated back to the work of G.F.W. Hegel, who observes in *Phenomenology of Spirit* that self-awareness – and, therefore, subjectivity – does not emerge in a vacuum; instead, our subjectivity arises from our interactions with both other conscious beings and the objects in our environment (1804 [1977]). Since both the objects and the other individuals in an environment will vary based on historical circumstance, it follows that the nature of subjectivity shaped by those objects and individuals will be particular to that historical moment as well.[2]

Our notions of the subject's historical specificity are further reinforced by the work of Michel Foucault, who focuses on the critical role that power plays in the socially controlled process of "subjection" or "subjectification" that produces subjects (Butler 1997, Davies 2006).[3] Foucault argues that modern institutions arrange bodies in ways that render them "docile" and use "disciplinary power" to produce and shape subjectivity in ways that best promote those institutions' own goals (1975 [1995]). He further argues that, because modern prisons and prison-like institutions act on the body in different ways than did their medieval counterparts, they are, therefore, different in how they control the ways that populations think and act. To illustrate: in medieval Europe, monarchs and religious authorities used public torture and execution of individuals (such as during the Inquisition) as a tool for controlling the behavior of the populace as a whole. In contrast, an example of modern disciplinary power (examined by Jaita Talukdar in this volume) is present-day India, where a liberalizing state with a burgeoning pharmaceutical and entrepreneurial sector has encouraged its urban elite to adopt "biotech-nological sciences as a style of thought," and to discipline their bodies by adopting new styles of eating. Talukdar explains that members of India's new middle classes who focus on consumption and individual self-improvement in pursuit of more efficient bodies are simulta-neously embracing neoliberal ideologies – and in so doing, reshaping their subjectivities in ways that benefit the multinational corporations that are starting new business ventures in India. Needless to say, the experience of adopting an inward, "scientific" gaze in order to cultivate a more "efficient" body is very different from the experience of adopting docile, submissive behavior in order to avoid being accused of witchcraft (and thereafter, being burned at the stake). Different techniques of social control therefore create different kinds of subjects – and as do people and objects, the techniques of social control that a subject experiences will vary according to her historical time and place.

Second, we consider subjectivity to be *embodied* – or, in other words, that the nature of subjectivity is fundamentally shaped by the idiosyncrasies of the subject's body. For example, being a subject with a young, Black, typically-abled, masculine body is not like being a subject with a young, white, disabled, feminine body, and neither is like being a subject with a middle-aged, brown, typically-abled, gender-non-conforming body. This probably seems intuitive. Yet in the mid-twentieth century, the field of cybernetics transformed popular understandings of human subjectivity by suggesting that subjectivity is simply a pattern of information – and that, at least theoretically, subjectivity could therefore be transferred from one (human/robot/animal/computer) body to another without any fundamental transformation (Hayles 1999). The cybernetic take on subjectivity turns up in popular works of fiction such as William Gibson's *Neuromancer* (1984), in which one character is the mind of a deceased hacker who resides in "cyberspace" after his brain is recorded to a hard drive. More recently, in the television show "Dollhouse" (2009–2010), writer and director Joss Whedon imagines a world in which minds are recorded to hard drives and swapped between bodies.

If you had a body that was very different from the one you have now, would you still expe-rience being in the world in the same way? As we write this chapter, the United States is reacting to the acquittal of George Zimmerman in the murder of Trayvon Martin; as article after article discusses the color and respective size of both Zimmerman's and Martin's bodies, as well as the color and sex of the six jurors' bodies, and the role that these bodily attributes played both in Martin's murder and in Zimmerman's acquittal, it seems impossible to believe that bodies do not play a critical role in shaping subjectivity, even in the Digital Age.

We follow N. Katherine Hayles in arguing that bodies still play a crucial role in subjectifica-tion. She observes that the cybernetic framing of subjectivity-as-information leads to a "*systematic devaluation of materiality and embodiment*" (1999: 48, emphasis in original). She asks rhetorically (1999: 1), "How could anyone think that consciousness in an entirely different medium would remain unchanged, as if it had no connection with embodiment?" She then counters by stating that, "for information to exist, it must always be instantiated in a medium," and moreover, that information will always take on characteristics specific to its medium (1999: 13).

Here, Hayles is building on the phenomenological observations of Maurice Merleau-Ponty, who argues that knowledge of the world – and, ultimately, the experience of one's own subjec-tivity – is mediated through the body. As he explains, "in so far as we are in the world through our body ... we perceive the world with our body" (Merleau-Ponty 1945 [2012]: 206). Information is neither formed nor processed the same way in metal bodies as in flesh-and-blood bodies, nor in female bodies as in male bodies, nor in black bodies as in white bodies, nor in typical bodies as in disabled bodies, and so on. Even if we believe that subjectivity is reducible to patterns of information, information itself is medium-specific; subjectivity is therefore embodied, or "sensory-inscribed" (Farman 2011). Because computer sensors or other devices for perceiving the world are vastly different from the human body, it follows that, if a computer were to become conscious, it would have a very different kind of subjectivity than do human beings of *any* body type.

If we pull all these pieces together, we can start to get an idea of what subjectivity is and where it comes from. Put loosely, subjectivity is the experience of being in and being able to act upon the world. This sense is shaped by the people and objects that are available for you to act both with and upon, and by the institutions and agents that act upon you, all of which are specific to your particular time and place. Your body is how you act both in and on the world, as well as the medium through which you experience the world and through which everything in the world acts upon you in return. Because the kind of body you have has such a profound impact in terms of both how you can act and how you are acted upon, it plays a critical role in shaping your subjectivity. Accordingly, we conclude that, for our purposes, subjectivity has two key aspects: historical conditions and embodiment.

If we are to understand contemporary subjectivity (or any particular subjectivity, for that matter), we must therefore examine both the contemporary subject's embodiment and her present-day historical conditions. We begin with her embodiment.

Embodiment: "split" subjectivity and digital prostheses

Your embodiment comprises "you," but what exactly are its boundaries? Where do "things which are part of you" stop, and "things which are not part of you" begin? As a thought experiment, consider the following: Your hand is a part of "you," but what if you had a prosthetic hand? Are your tattoos, piercings, braces, implants, or other modifications part of "you"? What about your Twitter feed, or your Facebook profile? If the words that come from your mouth in face-to-face conversation (or from your hands, if you speak sign language) are "yours," are the words you put on your Facebook profile equally yours? Does holding a smartphone in your hand change the nature of what you understand to be possible, or the nature of "you" yourself? When something happens on your Facebook profile, does it happen to you? If someone sends you a nasty reply on Twitter, or hacks your account, do you feel attacked?

Back when Descartes was thinking and (supposedly) therefore being, what did and did not constitute embodiment seemed pretty clear. The body was a "mortal coil" – that thing made of organic flesh that each of us has, that bleeds when cut and that decomposes after we die; it was merely a vessel for the "mind" or "soul," depending on one's ideological proclivities. Philosophy's move away from giving priority to either the mind or the body, however, necessarily complicated conceptions of embodiment. Over time, our bodies became not just mortal coils, but parts of "ourselves" as well. This conceptualization manifested, for example, in certain feminist (as well as hegemonic masculine) discourses that claim men's and women's desires are irreconcilably different based on how sexual differences between male and female bodies construct experience. Such perspectives might hold, for example, that women are biologically prone to be caregivers, while men are biologically prone to be aggressive and violent. While these naturalistic discourses recognize the significance of the body in shaping subjectivity, they make a grave error in portraying the body as ahistorical (and thereby fall into biological determinism).

Judith Butler contests the idea that the body, and especially the sexual body, is ahistorical (1988). Citing Merleau-Ponty, she argues that the body is "a historical idea" rather than a "natural species." Butler explains that the material body is not an inert thing that simply shapes our consciousness, but rather something that is *performed*. This is because the materiality of the body is both shaped and expressed through the activity of the conscious subject. As Butler explains:

The body is not a self-identical or merely factic materiality; it is a materiality that bears meaning, if nothing else, and the manner of this bearing is fundamentally dramatic. By dramatic I mean only that the body is not merely matter but a continual and incessant *materializing* of possibilities. One is not simply a body, but in some very key sense, one does one's body and, indeed, one does one's body differently from one's contemporaries and one's embodied predecessors and successors as well.

1988: 521

It is the fluidity of embodiment, as Butler describes it here, that we are most interested in. Our perceptions of an essential or natural body (for example, the idea of a "real" man or woman) are social constructs, and as such represent the reification of certain performances of embodiment that serve to reinforce dominant social structures. This is why trans★[4] bodies (for example) are seen as so threatening, and a likely reason why trans★ people are subject to such intense violence: because trans★ bodies disrupt, rather than reinforce, hegemonic ideals of masculinity and femininity.

Following Butler, we understand subjectivity to have a dynamic relationship to embodiment: both the composition of bodies (embodiment) and the way we perform those bodies (part of subjectivity) are subject to change, both over an individual subject's life course and throughout history. Given the fluid relationship between subjectivity and embodiment, it is especially important to avoid succumbing to the fallacy of naturalism (which would lead us to privilege the so-called "natural body" when we think about embodiment).

Recall from the first section, above, that our embodiment is both that through which we experience being in the world and that through which we act upon the world. Next, consider that our agency (our ability to act) is increasingly decoupled from the confines of our organic bodies because, through prosthetics – which may be extended to include certain "ready-at-hand" tools (Heidegger 1927 [2008]) – and also computers, we are able to extend our agency beyond our "natural bodies" both spatially and temporally. Allucquère Rosanne Stone argues that our subjectivity is therefore "disembodied" from our flesh, and that the attendant extension of human subjectivity through multiple media results in a "split subject" (1994). The subject is split in the sense that her subjectivity is no longer confined to a single medium (her organic body), but rather exists across and through the interactions of multiple media.

Stone interrogates our privileging of the "natural body" as a site of agency, and elaborates on the term "split subject," through accounts of what she terms *disembodied agency* (1994). In one such account, a crowd of people gathers to see Nobel laureate physicist Stephen Hawking (who has motor neurone disease) deliver a live presentation by playing a recording from his talking device. Although Hawking does not produce the speech sounds with his mouth, the crowd considers Hawking to be speaking; similarly, they consider the words he speaks to be his own, not the computer's. Implicitly, the crowd understands that Hawking is extending his agency through the device. Of course, Hawking's disembodied agency through the talking device is also, simultaneously, an embodied agency; Hawking has not been "uploaded" into "cyberspace," and his organic body is very much present in the wheelchair on the stage. Hawking's agency and subjecthood, therefore, reside *both* in his body *and* in his machine, while both the "real" Hawking and the "live" experience of Hawking exist in the interactions performed between human and machine. This is why Stone uses the term "split subjects" to describe cases where agency is performed through multiple media: because the subject's agency is split across multiple media, so too is her subjectivity.

Stone refers to Hawking's computerized talking device as a *prosthesis* – "an extension of [his] will, of [his] instrumentality" (1994: 174). Just as Hawking acts upon and experiences the world through both his organic body and his prostheses (his various devices), the contemporary subject acts upon and experiences the world through her organic body, through her prostheses, and through what we will call her *digital prostheses*. Persistent communication via digital technologies is no longer confined to exceptional circumstances such as Hawking's; in fact, it has become the default in many (post-)industrial nations. SMS (text messaging) technologies, smartphones, social media platforms (such as Facebook or Twitter), and now even wearable computing devices (such as Google Glass) extend our agency across both time and space. Digital technologies are pervasive in (post-)industrial societies, and our interactions with and through such technologies are constantly shaping our choices of which actions to take. In addition to conventional prostheses (for example, a walker or an artificial limb), the contemporary subject uses both tangible digital prostheses (such as desktop computers, laptop computers, tablets, smartphones) and intangible digital prostheses (such as social media platforms, blogs, email, text messaging, even the Web most generally) to interact with the people and institutions in her environment. Similarly, she experiences her own beingness through her organic body, her conventional prostheses, and her digital prostheses alike.

Again, to us – and perhaps to you, too – these ideas seem obvious. Yet as we elaborate further in the following section, there is still a good deal of resistance to the idea that interactions through digital media count as part of "real life." Why is this? In part, Stone suggests that "virtual [read: digital] systems are [perceived as] dangerous because the agency/body coupling so diligently fostered by every facet of our society is in danger of becoming irrelevant" (1994: 188). The contemporary subject moves beyond Enlightenment notions of a discrete subject because her embodiment is the combination of her organic body, her conventional prostheses, and her digital prostheses. In other words, your hand is a part of you, your prostheses are parts of you, and your online presences – which are just some of your many digital prostheses – are parts of you, too.

The contemporary subject is therefore an embodied subject, one whose materiality resides not in any one distinct and separate medium (organic body, conventional prostheses, or digital prostheses), but which is performed both within and across multiple media. In order to fully understand what the implications of such an embodiment are for the contemporary subject, however, we must turn to examining the historical conditions that shape her. We pay particular attention to how digital technology, especially the Web, is conceptualized in our present social milieu.

Historical conditions: co-production of "online" and "offline"

In (post-)industrial societies, many of us live in a conceptually divided world – a world split between "the online" and "the offline." We argue, however, that the categories "online" and "offline" are *co-produced*, and so are created simultaneously as the result of one particular attempt to order and understand the world (aka, the drawing of the boundary between them) (Jasanoff 2004, Latour 1993). As such, the categories "online" and "offline" reflect an ongoing process of collective meaning-making that has accompanied the advent of digital technologies, but they do not reflect our world in itself. We further argue that all of our experiences – whether they are "online," "offline," or a combination of both – are equally real, and that they take place not within two separate spheres or worlds, but within one *augmented reality*. In order to understand the significance of digital technologies within our present historical context, we must explore the social construction of "online" and "offline," as well as augmented reality itself.

It has become commonplace for writers (for example: Carr 2013, Sacasas 2013) to describe the world that existed before the advent of the Internet as being "offline." In the version of history that such narratives imply, the world came into being in an "offline" state and then stayed that way until about 1993, when parts of the world began to gain the ability to go "online" and access the Web. But this line of thinking is fundamentally flawed. No one could conceptualize being "offline" before there was an "online"; indeed, the very notion of "offline-ness" necessarily references an "online-ness." Historical figures such as (say) W.E.B. DuBois, Joan of Arc, or Plato could not have been "offline," because that concept did not exist during their lifetimes – and similarly, it is anachronistic to describe previous historical epochs as being "offline," because to do so is to reference and derive meaning from a historical construct that did not yet exist. In short, there can be no offline without online (nor online without offline); these two linked concepts are socially constructed in tandem and depend upon each other for definition.

The issue with configuring previous historical epochs as "offline" is that doing so presents the socially constructed concept of "offline" as a natural, primordial state of being. Accordingly, the concept of "online" is necessarily presented as an unnatural and perverted state of being. This is the fallacy of naturalism, and it poses two significant problems.

The first problem is that, the more we naturalize the conceptual division between online and offline, the more we encourage normative value judgments based upon it. It would be naive to assume that socially constructed categories are arbitrary. As Jacques Derrida suggests, the ultimate purpose of constructing such binary categories is to make value judgments. He explains that, "In a classical philosophical opposition [such as the online/offline pair] we are not dealing with a peaceful coexistence of a vis-à-vis, but rather with a violent hierarchy" (1982: 41). In this case, the value judgments we make about "online" and "offline" often lead us to denigrate or dismiss digitally mediated interaction, such as when we engage in digital dualism by referring to our "Facebook friends" as something separate from our "real friends," or by discounting online political activity as "slacktivism" (Jurgenson 2011, 2012a). Our experience of digitally mediated interaction may certainly be different from our experience of interaction mediated in other ways, but this does not mean that online interaction is somehow separate from or inferior to what we think of as "offline" interaction. Rather, if we think more highly of "offline" interaction, it is because the very advent of digital mediation has led us to value other forms of mediation more highly than we did in the past. For example, in an age of mp3s, e-readers, text messages, and smartphones, we have developed an at-times obsessive concern with vinyl records, paper books, in-person conversations, and escapes into wilderness areas where we are "off the grid." This tendency both to elevate older media as symbols of the primacy of "the offline," and also to obsess over escaping the supposed inferiority of "the online," is what Jurgenson calls "the IRL [in real life] fetish" (2012b).

The second problem that follows from the naturalistic fallacy is the framing of "online" and "offline" as zero-sum, which obscures how deeply interrelated the things we place into each category really are. Most Facebook users, for instance, use Facebook to interact with people they also know in offline contexts (Hampton et al. 2011). Do we believe that we have two distinct, separate relationships with each person whom we are friends with on Facebook, and that only one of those relationships is "real"? Or is it more likely that both our "online" and "offline" interactions with each person are part of the same relationship, and that we have one friendship per friend? Similarly, when people use Twitter to plan a protest and to communicate during a political action, are there really two separate protests going on, one on Twitter and one in Tahrir Square or Zuccotti park? Just as digital information and our physical environment reciprocally influence each other, and therefore cannot be examined in isolation, so too must we look simultaneously at both "the online" and "the offline" if we want to make sense of either.

If the conceptual division between "online" and "offline" does not reflect the world of our lived experiences, then how are we to describe that world? In the previous section, we argued that the contemporary subject is embodied by a fluid assemblage of organic flesh, conventional prostheses, and digital prostheses. It likely will not come as a surprise, then, that the contemporary subject's world is similarly manifest in both analog and digital formats – in both "wetware" and software – and that no one format has a monopoly on being her "reality." Just as the subject exists both within and across the various media that embody her, so too does her world exist both within and across the various media through which her experiences take place. Following Jurgenson, we characterize this state of affairs as *augmented reality* (2011, 2012a).

Why do we use the term "augmented reality" to describe the single reality in which "online" and "offline" are co-produced, and in which information mediated by organic bodies is co-implicated with information mediated by digital technology? After all, engineers and designers of digital technologies have developed a range of models for describing the enmeshment and overlap of "reality as we experience directly through our bodies" and "reality that is mediated or wholly constructed by digital technologies." Some of those conceptual models include: "mediated reality" (Naimark 1991); "mixed reality" (Milgram and Kishino 1994); "augmented reality" (Drascic and Milgram 1996, Azuma 1997); "blended reality" (Ressler et al. 2001, Johnson 2002); and later, "dual reality" (Lifton and Paradiso 2009). "Mediated reality," however, problematically implies that there is an *unmediated* reality – whereas all information is mediated by something, because information cannot exist outside of a medium. "Mixed reality," "blended reality," and "dual reality," on the other hand, all imply two separate realities which now interact, and we reject the digital dualism inherent in such assumptions; recall, too, that "online" and "offline" are co-produced categories, and as such have always-already been inextricably interrelated. "Augmented reality" alone emphasizes a single reality, though one in which information is mediated both by digital technologies and by the fleshy media of our brain and sensory organs. Moreover, "augmented reality" is already in common use, and already invokes a blurring of the distinction between online and offline.

Thus, rather than invent a new term, we suggest furthering the sociological turn that is already taking place in common use of the term "augmented reality" (c.f. Jurgenson 2011, Rey 2011, Jurgenson 2012a, 2012c, Boesel 2012b, Boesel and Rey 2013, Banks 2013a, Boesel 2013). We use the term "augmented reality" to capture not only the ways in which digital information is overlaid onto a person's sensory experience of her physical environment, but also the ways in which those things we think about as being "online" and "offline" reciprocally influence and co-constitute one another. Such total enmeshment of the online and offline not only means that we cannot escape one "world" or "reality" by entering the other; it also means that in our one world, there is simply no way to avoid the influence either of our organic bodies (for example, by transporting oneself to "cyberspace"; see Rey 2012a) or of digitally mediated information and interaction (for example, by "disconnecting" or "unplugging"; see Rey 2012b, Boesel 2012a).

Augmented reality encompasses a unitary world that includes both physical matter (such as organic bodies and tangible objects) and digital information (particularly – though not exclusively – information conveyed via the Web), and it is within this world that the contemporary subject acts, interacts, and comes to understand her own being. In today's techno-social landscape, our experiences are mediated not only by organic bodies, but also by conventional *and* digital prostheses – each of which is a means through which we act upon the world and through which other people and social institutions act upon us in return. In this way, the historical conditions we have described as augmented reality give rise to a new form of subjectivity (as we will elaborate upon in the final section).

Before we explain why a new conceptual frame for subjective experience in the Digital Age is necessary, recall that in this section we have explored how our society conceptually divides the contemporary subject's world into two separate pieces ("online" and "offline"), even as the material embodiment of her subjectivity is expanding (to include her organic body, conventional prostheses, and digital prostheses). This observation invites us to ask: What does it mean for us, as contemporary subjects, that the dominant cultural ideology holds some aspects of our being to be less "real" than others? What happens when we conceptually amputate parts of ourselves and of our experiences, and what follows when some expressions of our agency are denigrated or discounted? We argue that under-recognizing the ways in which augmented reality has affected contemporary subjectivity has grave social and political consequences, as evidenced by the case study in our final section.

"Cyberbullying": a case study in augmented subjectivity

In October 2012, a 15-year-old girl from British Columbia, Canada, named Amanda Todd committed suicide after several years of bullying and harassment. Todd's death gained international attention, both because a video she'd made and posted to YouTube a month earlier went viral after her death, and because subsequent press coverage frequently centered on so-called "cyberbullying" – i.e., the fact that Todd had been persecuted through digitally mediated interaction in addition to other, more "conventional" methods of abuse.

Figure 10.1 A screenshot of Amanda Todd's 2012 YouTube video, "My story: struggling, bullying, suicide, self-harm"

Todd never speaks in her video (which she titles, "My story: struggling, bullying, suicide, self harm"), but nonetheless tells a heartbreaking story across nine minutes of cue cards that she turns in time to music. During her seventh grade year, Todd and her friends would "go on webcam" to meet new people and have conversations with them. Through these interactions, she met a man who flattered and complimented her – and who later coerced her into briefly showing him her breasts via her webcam ("flashing"). A year later, the man had somehow tracked her down on Facebook ("don't know how he knew me"), and – after demonstrating that he now knew her name, her address, the name of her school, and the names of her friends and family members – he threatened to circulate a screen-captured image of Todd showing her breasts through the webcam if she did not "put on a show" for him (a type of blackmail commonly referred to as "sextortion"). Todd refused to be intimidated, however, and told the man no. Not long thereafter, the local police came to her family's home to inform them that "the photo" (as Todd refers to it) had been "sent to everyone."

Traumatized, and ostracized by her classmates at school, Todd developed clinical depression, as well as anxiety and panic disorders; she started using drugs and alcohol in an attempt to self-medicate. She moved, and changed schools. But a year later, the man returned. This time he created a Facebook profile, used the image of Todd's breasts as its profile picture, and got the attention of Todd's new friends and classmates by sending them Facebook "friend requests" from that profile, claiming to be a new student starting at their school. Todd's classmates at her second school were as ruthless as those at her first; exiled and verbally abused, she began cutting herself and changed schools once again. Although she was still isolated and friendless at her third school, Todd says that things were "getting better" – until she had casual sex with "an old guy friend" (another teenager) who had recently gotten back in touch with her. In an eerie echo of the original webcam interaction, Todd's friend seemed to offer the promise of kindness and affection in exchange for sex ("I thought he liked me," she says repeatedly). Instead, the boy subsequently arrived outside Todd's school with a crowd of other teenagers, and merely looked on as his girlfriend physically assaulted Todd. The other teens cheered, encouraged the boy's girlfriend to punch Todd, and recorded video of the assault with their phones. After her father picked her up from school, Todd attempted to commit suicide by drinking bleach; when she came home from the hospital, she found Facebook posts that said, "She deserved it," and "I hope she's dead."

Todd moved out of her father's house and into her mother's, in order to change schools and towns once again. Six months later, she says, people were posting pictures of bleach on Facebook and tagging them as her. As Todd explains across two cards, "I was doing a lot better too... They said... She should try a different bleach. I hope she dies this time and isn't so stupid. They said I hope she sees this and kills herself." "Why do I get this?" she asks. "I messed up but why follow me... I left your guys city... Im constantly crying now... Everyday I think why am I still here?" Todd goes on to explain that her mental health has worsened; that she feels "stuck"; that she is now getting counseling and taking anti-depressants, but that she also tried to commit suicide by "overdosing" and spent two days in the hospital. "Nothing stops," she says. "I have nobody... I need someone," followed by a line drawing of a frowning face. The last card reads, "My name is Amanda Todd..." and Todd reaches forward to turn off the webcam. The last few seconds of the video are an image of a cut forearm bleeding (with a kitchen knife on a carpeted floor in the background), followed by a quick flash of the word "hope" written across a band-aged wrist, and then finally a forearm tattooed with the words "Stay strong" and a small heart. Todd hanged herself at home in a closet one month and three days after she posted the video; her 12-year-old sister found her body.

Todd's suicide sparked a surge of interest in so-called "cyberbullying," and yet the term grossly oversimplifies both what an unknown number of people did to Todd (across multiple media, and in multiple contexts) and why those people did such things in the first place. As danah boyd (2007), Nathan Fisk (2012), David A. Banks (2013b), and others argue, the term "cyberbullying" deflects attention away from harassment and abuse ("-bullying"), and redirects that attention toward digital media ("cyber-"). In so doing, the term "cyberbullying" allows digital media to be framed as *causes* of such bullying, rather than simply the newest type of mediation through which kids (and adults) are able to harass and abuse one another. "The Internet" and "social media" may be convenient scapegoats, but to focus so intensely on one set of media through which bullying sometimes takes place is to obscure the underlying causes of bullying, which are much larger and much more complicated than simply the invention of new technologies like the Web. Such underlying causes include (to name just a few): teens' lack of positive adult involvement and mentorship (boyd 2007); the contemporary conception of childhood as preparation for a competitive adult workforce, and the attendant emphasis on managing, planning, and scheduling children's lives (Fisk 2012); a culture of hyper-individual-ism that rewards mean-spirited attacks, and that values "free speech" more highly than "respect" (boyd 2007); sexism, misogyny, and "rape culture" (Banks 2013b). When we characterize digi-tally mediated harassment and abuse as "cyberbullying," we sidestep confronting (or even acknowledging) any and all of these issues.

While recent conversations about "cyberbullying" have been beneficial insofar as they have brought more attention to the fact that some adolescents (almost always girls, or gender non-conforming boys) are mercilessly tormented by their peers, these conversations have simultaneously revealed just how disconnected popular ideas about the Web, digital media, and human agency are from the contemporary conditions that shape human subjectivity. One study concluded, for example, that cyberbullying is "an overrated phenomenon," because most bullies do not exclusively harass their victims online (Olweus 2012). That study was flawed in its assumption that online harassment ("cyberbullying") and offline harassment ("traditional bully-ing") are mutually exclusive and must occur independently (in other words, that they are zero-sum). Another study (LeBlanc 2012) made headlines by finding that, "cyberbullying is rarely the only reason teens commit suicide," and that, "[m]ost suicide cases also involve real-world bullying as well as depression" (Gowan 2012). Again, it is only the flawed conceptual division between the "online" and the "real world" (or "offline") that allows such conclusions to be "findings" rather than "stating the obvious." Moreover, the term "cyberbullying" has no resonance with the people who supposedly experience it: Marwick and boyd (2011) and Fisk (2012) find that teenagers do not make such sharp divisions between online and offline harass-ment, and that teens actively resist both the term "cyberbullying" and the rigid frameworks that the term implies.

Conclusion

As we have argued, and as recent public discourse about "cyberbullying" illustrates, our pres-ent-day culture separates digitally mediated experiences into a separate, unequal category, and often goes so far as to imply that digitally mediated experiences are somehow "less real." The contemporary subject is a split subject, one whose embodiment and expressions of agency alike span multiple media simultaneously – and some of those media are digital media. What does it mean when some parts of our lived experience are treated as disconnected from, and less real than, other parts of our lived experience? "Online" and "offline" may be co-produced concep-tual categories rather than actual ontological states, but the hierarchy embedded in the online/offline binary wields significant social power. As a result, we accord primacy to some aspects of the contemporary subject's embodiment and experience (generally, those most closely associated with her offline presence), and we denigrate or discount other aspects of her embodiment and experience (generally, those most closely associated with her online presence). As such privileging and devaluing is never neutral, this means that contemporary subjectivity is "split" across media that are neither equally valued nor even given equal ontological priority by contemporary society.

In an attempt to counter the devaluation of those aspects of our subjectivity that are expressed and embodied through digital media, we offer an alternative framework that we call *augmented subjectivity* – so named because it is subjectivity shaped by the historic conditions of augmented reality. Augmented subjectivity recalls Stone's notion of split subjectivity, in that it recognizes human experience and agency to be embodied across multiple media (1994). However, the augmented subjectivity framework emerges from our recognition of the online/offline binary as a historically specific, co-produced social construction, and so emphasizes *the continuity* of the subject's experience (as opposed to its "split"-ness), even as the subject extends her agency and embodiment across multiple media. This is not to suggest, of course, that digital technologies and organic flesh have identical properties or affordances. Rather, we seek to recognize – and to emphasize – that neither the experiences mediated by the subject's organic body nor those mediated by her digital prostheses can ever be isolated from her experience as a whole. All of the subject's experiences, regardless of how they are mediated, are always-already inextricably enmeshed.

Amanda Todd's story perfectly (and tragically) illustrates how both "online" and "offline" experiences are integrated parts of the augmented subject's being, and demonstrates as well the problems that follow from trying to interpret such integrated experiences through a digital-dualist frame. Todd very clearly experienced her world through her digital prostheses as well as through her organic body; the pain she experienced from reading, "I hope she's dead," was not lessened because she read those words on a social media platform rather than heard them spoken face-to-face (in that particular instance). Nor was her pain lessened because the comments were conveyed via pixels on a screen rather than by ink on paper. Neither were Todd's experiences mediated through her Facebook profile somehow separate from her experiences more directly mediated through the sensory organs of her organic body: another girl physically assaulting Todd, mediated through smartphones, became digital video; digitally- and materially-mediated words, remediated by Todd's organic body, became her affective experience of pain, which became cuts in her flesh, and which – mediated again by several digital devices and platforms – became a still image in a digital video. The digital image of Todd's breasts that strangers and classmates alike kept circulating was always-already intimately linked to Todd's organic body. Neither Todd's beingness nor her experience of being can be localized to one medium alone.

Similarly, both Todd and her tormentors are able to express their agency through multiple media. Todd's harassers were able to continue harassing her across both spatial and temporal distance; the man who made the screen-captured image of her breasts, and then repeatedly distributed that image, understood precisely the power he had to affect not just Todd, but also the other teenagers who tormented Todd after seeing the photo.

It is important to note that bullying is not the only type of social interaction that we experience fluidly across online and offline contexts. Support, too, transcends this constructed boundary, as a significant number of scholars have argued (for example: Turner 2006, Chayko 2008, Gray 2009, Baym 2010, Jurgenson and Rey 2010, Davis 2012, Rainie and Wellman 2012, Wanenchak 2012). Todd and her seventh grade friends understood that they could extend themselves through their webcams, and so were able to view webcam sites as ways to connect with other people; of her reasons for making the video she posted to YouTube, Todd writes, "I hope I can show you guys that everyone has a story, and everyones future will be bright one day, you just gotta pull through." Here, Todd clearly understands that she can act upon future viewers of her video, and she hopes that she will do so in a positive, beneficial way.

The characteristic continuity of the augmented subject's online and offline experience is clearly illustrated above, and yet – with respect to teen-on-teen harassment – that continuity becomes markedly less apparent if we fixate on the digitally mediated aspects of such harassment as "cyberbullying." If we look past the reified division between "online" and "offline" however, it becomes clear again that each of us living within augmented reality is a single human subject, and that each of us has one continuous experience of being in the world. Digital prostheses do have different properties and do afford different ranges of possibilities than do their conventional or organic counterparts, but we should not focus on the properties of *any* type of media to the detriment of our focus on the actions and moral responsibilities of human agents. Focusing on one particular means through which people enact violence against each other (for example, through digitally mediated interaction) will never solve the problem of violence; as boyd argues, bullying will continue to occur through whatever the newest medium of communication is, because bullying is not – nor has it ever been – unique to any particular available medium (2007).

Rather than hold digital media as causal, or even as exceptional, we need to recognize that digitally mediated experiences and interactions are simply part of the day-to-day world in which we live. Digital media are a distinctive part of our cultural moment, and they are part of ourselves. When we view contemporary experience through the frame of augmented subjectivity – and, in so doing, recognize the inextricable enmeshment of online and offline experience – it is clear that what happens to our digital prostheses happens to us. For this reason, we state unequivocally that moral regard must be extended to these digital prostheses. Similarly, each action we take through our digital prostheses is an expression of our agency and, as such, is something for which we must take responsibility. That said, your authors are not cyberneticists; we do not believe that subjectivity can or ever will transcend the organic body, or that deleting someone's digital prosthesis is the same thing as killing that person in the flesh. Although contemporary subjectivity is augmented by digital technology, organic bodies are still an essential and inseparable dimension of human experience.

Information depends critically on the medium in which it is instantiated. Accordingly, it is essential that we acknowledge *all* of the media that comprise the augmented subject: her digital prostheses, her conventional prostheses, and her organic flesh. Rather than trade digital denialism for flesh denialism, we argue that subjectivity is irreducible to *any* one medium. Instead, we must recognize that contemporary subjectivity is augmented, and we must grant all aspects of the subject's experience equal countenance.

Acknowledgements

This paper could not have taken the form that it did without ongoing inspiration and support from a vibrant (augmented) community. In particular, the authors would like to acknowledge their colleagues at the *Cyborgology* blog: David A. Banks, Jenny L. Davis, Robin James, Nathan Jurgenson, and Sarah Wanenchak. Conversations with Jeremy Antley, Tyler Bickford, Jason Farman, and Tanya Lokot have also provided the authors with valuable insight.

Notes

1 We refer to the singular subject as "her" (rather than "they" or "him") for purposes of clarity and readability, and have chosen to use the feminine pronoun "her" in order to counter the earlier tradition of using generalized masculine pronouns.
2 This observation was foundational to Karl Marx's understanding of ideology and theory of historical materialism, and to the discipline of sociology more broadly.
3 It is useful to distinguish the socio-historical nature of subjectivity from the concept of the self (even if the two are often used interchangeably in practice). Nick Mansfield (2000: 2–3) cautions: "Although the two are sometimes used interchangeably, the word 'self' does not capture the sense of social and cultural entanglement that is implicit in the word 'subject': the way our immediate daily life is always already caught up in complex political, social and philosophical – that is, shared – concerns."
4 "trans★" is an umbrella term that encompasses a broad range of non-cisgender identities, for example, transgender, trans man, trans woman, agender, genderqueer, etc.

References

Azuma, R. 1997. "A Survey of Augmented Reality." *Presence: Teleoperators and Virtual Environments* 6(4): 355–385.

Banks, D.A. 2013a. "Always Already Augmented." *Cyborgology*, March 1, 2013, *http://thesocietypages.org/cyborgology/2013/03/01/always-already-augmented*.

Banks, D.A. 2013b. "What the Media is Getting Wrong about Steubenville, Social Media, and Rape Culture." *Cyborgology*, March 23, 2013, *http://thesocietypages.org/cyborgology/2013/03/23/what-the-media-is-getting-wrong-about-steubenville-social-media-and-rape-culture*.

Baym, N. 2010. *Personal Connections in the Digital Age*. Cambridge, UK: Polity Press.

Boesel, W.E. 2012a. "A New Privacy (Full Essay)." *Cyborgology*, August 6, 2012, *http://thesocietypages.org/cyborgology/2012/08/06/a-new-privacy-full-essay-parts-i-ii-and-iii-2*.

Boesel, W.E. 2012b. "The Hole in Our Thinking about Augmented Reality." *Cyborgology*, August 30, 2012, *http://thesocietypages.org/cyborgology/2012/08/30/the-hole-in-our-thinking-about-augmented-reality*.

Boesel, W.E. 2013. "Difference Without Dualism: Part III (of 3)." *Cyborgology*, April 3, 2013, *http://thesocietypages.org/cyborgology/2013/04/03/difference-without-dualism-part-iii-of-3*.

Boesel, W.E. and Rey, PJ 2013. "A Genealogy of Augmented Reality: From Design to Social Theory (Part One)." *Cyborgology*, January 28, 2013, *http://thesocietypages.org/cyborgology/2013/01/28/a-genealogy-of-augmented-reality-from-design-to-social-theory-part-one*.

boyd, d. 2007. "Cyberbullying." *Apophenia*, April 7, 2007, *www.zephoria.org/thoughts/archives/2007/04/07/cyberbullying.html*.

Butler, J. 1988. "Performative Acts and Gender Constitution: An Essay in Phenomenology and Feminist Theory." *Theatre Journal* 40(4): 519–531.

Butler, J. 1997. *The Psychic Life of Power: Theories in Subjection*. Palo Alto, CA: Stanford University Press.

Carr, N. 2013. "Digital Dualism Denialism." *Rough Type*, February 27, 2013, *www.roughtype.com/?p=2090*.

Chayko, M. 2008. *Portable Communities: The Social Dynamics of Online and Mobile Connectedness*. New York: State University of New York Press.

Davies, B. 2006. "Subjectification: The Relevance of Butler's Analysis for Education." *British Journal of Sociology of Education* 27 (4): 425–438.

Davis, J.L. 2012. "Prosuming Identity: The Production and Consumption of Transableism on Transabled.org." *American Behavioral Scientist*. Special Issue on Prosumption and Web 2.0 edited by George Ritzer. 56(4): 596–617.

Derrida, J. 1982. *Positions*, trans. Alan Bass. Chicago and London: University of Chicago Press.

Descartes, R. 1641 [1993]. *Meditations on First Philosophy: In Which the Existence of God and the Distinction of the Soul from the Body Are Demonstrated*. Indianapolis, IN: Hackett Publishing Company.

Drascic, D. and Milgram, P. 1996. "Perceptual Issues in Augmented Reality." In *Proc. SPIE Stereoscopic Displays and Virtual Reality Systems III*, 2653: 123–134. San Jose, California.

Farman, J. 2011. *Mobile Interface Theory: Embodied Space and Locative Media*. London and New York: Routledge.

Fisk, N. 2012. "Why Children Laugh at the Word 'Cyberbullying.'" *Cyborgology*, July 14, 2012, *http://thesocietypages.org/cyborgology/2012/07/14/why-children-laugh-at-the-word-cyberbullying*.

Foucault, M. 1975 [1995]. *Discipline & Punish: The Birth of the Prison*. New York: Vintage.

Gibson, W. 1984. *Neuromancer*. New York: Ace Books.

Gowan, M. 2012. "Cyberbullying Rarely Sole Cause Of Teen Suicide, Study Finds." *Huffington Post*, October 25, 2013, *www.huffingtonpost.com/2012/10/25/cyberbullying-teen-suicide_n_2016993.html*.

Gray, M. 2009. *Out in the Country: Youth, Media, and Queer Visibility in Rural America*. New York: New York University Press.

Hampton, K., Goulet, L.S., Rainie, L., and Purcell, K. 2011. "Social networking sites and our lives." *Pew Internet and American Life Project*, June 16, 2011, *www.pewinternet.org/Reports/2011/Technology-and-social-networks.aspx*.

Hayles, N.K. 1999. *How We Became Posthuman: Virtual Bodies in Cybernetics, Literature, and Informatics*. Chicago and London: University of Chicago Press.

Hegel, G.F.W. 1804 [1977]. *Phenomenology of Spirit*, trans. A.V. Miller. Oxford, UK: Oxford University Press.

Heidegger, M. 1927 [2008]. *Being and Time*. New York: Harper Perennial/Modern Thought.

Jasanoff, S. 2004. "Ordering Knowledge, Ordering Society." In *States of Knowledge: The Co-production of Science and Social Order*, edited by S. Jasanoff, 13–45. New York and London: Routledge.

Johnson, B.R. 2002. "Virtuality and Place: The Relationship between Place, Computation, and Experience." Paper presented at *ACADIA 2002: Thresholds Between Physical and Virtual*, Pomona, CA. *http://faculty.washington.edu/brj/presentations/acadia02/0.default.html*.

Jurgenson, N. 2011. "Digital Dualism versus Augmented Reality." *Cyborgology*, February 24, 2011, *http://thesocietypages.org/cyborgology/2011/02/24/digital-dualism-versus-augmented-reality*.

Jurgenson, N. 2012a. "When Atoms Meet Bits: Social Media, the Mobile Web and Augmented Revolution." *Future Internet* 4: 83–91.

Jurgenson, N. 2012b. "The IRL Fetish." *The New Inquiry*, June 28, 2012. *http://thenewinquiry.com/essays/the-irl-fetish*

Jurgenson, N. 2012c. "Strong and Mild Digital Dualism." *Cyborgology*, October 29, 2012, *http://thesocietypages.org/cyborgology/2012/10/29/strong-and-mild-digital-dualism*.

Jurgenson, N. and Rey, PJ 2010. "Bullying Is Never Just Cyber." *Cyborgology*, October 28, 2010, *http://thesocietypages.org/cyborgology/2010/10/28/bullying-is-never-just-cyber*.

Latour, B. 1993. *We Have Never Been Modern*. Cambridge, MA: Harvard University Press.

LeBlanc, J.C. 2012. "Cyberbullying and Suicide: A Retrospective Analysis of 22 Cases." Paper presented at *American Academy of Pediatrics National Conference and Exhibition*, October 2012, New Orleans, LA. Abstract: *https://aap.confex.com/aap/2012/webprogrampress/Paper18782.html*.

Lifton, J. and Paradiso, J.A. 2009. "Dual reality: Merging the Real and Virtual." In *First International Conference, FaVE 2009, Berlin, Germany, July 27–29, 2009, Revised Selected Papers*, edited by F. Lehmann-Grube and J. Sablatnig, 12–28. Berlin, Germany: Springer.

Mansfield, N. 2000. *Subjectivity: Theories of the Self from Freud to Haraway*. New York: New York University Press.

Marwick, A. and boyd, d. 2011. "The Drama! Teen Conflict, Gossip, and Bullying in Networked Publics." Paper presented at A Decade in Internet Time: Symposium on the Dynamics of the Internet and Society, Oxford, UK, September 2011. *http://ssrn.com/abstract=1926349*.

Merleau-Ponty, M. 1945 [2012]. *Phenomenology of Perception*. London and New York: Routledge.

Milgram, P. and Kishino, F. 1994. "A Taxonomy of Mixed Reality Visual Display." *IEICE Transactions on Information Systems* E77–D: 12.

Naimark, M. 1991. "Elements of Realspace Imaging: A Proposed Taxonomy." In *SPIE/SPSE Electronic Imaging Proceedings* 1457. San Jose, CA.

Olweus, D. 2012. *European Journal of Developmental Psychology* 9(5): 520–538.

Rainie, L. and Wellman, B. 2012. *Networked: The New Social Operating System*. Cambridge, MA: MIT Press.

Ressler, S., Antonishek, B., Wang, Q., Godil, A., and Stouffer, K. 2001. "When Worlds Collide: Interactions between the Virtual and the Real." Proceedings on *15th Twente Workshop on Language Technology; Interactions in Virtual Worlds, Enschede, The Netherlands,* May 1999. *http://zing.ncsl.nist.gov/godil/collide.pdf*.

Rey, PJ 2011. "Cyborgs and the Augmented Reality they Inhabit." *Cyborgology*, 9 May 2011, *http://thesocietypages.org/cyborgology/2011/05/09/cyborgs-and-the-augmented-reality-they-inhabit*.

Rey, PJ 2012a. "The Myth of Cyberspace." *The New Inquiry* 1(3):137–151. *http://thenewinquiry.com/essays/the-myth-of-cyberspace*.

Rey, PJ 2012b. "Social Media: You Can Log Off, But You Can't Opt Out." *Cyborgology*, May 10, 2012, *http://thesocietypages.org/cyborgology/2012/05/10/social-media-you-can-log-off-but-you-cant-opt-out*.

Sacasas, L.M. 2013. "In Search of the Real." *The Frailest Thing*, July 4, 2012, *http://thefrailestthing.com/2012/07/04/in-pursuit-of-the-real*.

Stone, A.R. 1994. "Split Subjects, Not Atoms; or, How I Fell in Love with My Prosthesis." *Configurations* 2(1): 173–190.

Todd, A. 2012. "My Story: Struggling, Bullying, Suicide, Self Harm" [video]. Retrieved July 2013 from *www.youtube.com/watch?v=vOHXGNx-E7E*.

Turner, F. 2006. *From Counterculture to Cyberculture: Stewart Brand, the Whole Earth Network, and the Rise of Digital Utopianism*. Chicago and London: University of Chicago Press.

Wanenchak, S. 2012. "Thirteen Ways of Looking at Livejournal." *Cyborgology*, November 29, 2012, *http://thesocietypages.org/cyborgology/2012/11/29/thirteen-ways-of-looking-at-livejournal*.

11

Political Culture of Gaming in Korea amid Neoliberal Globalization

Dal Yong Jin and Michael Borowy

SIMON FRASER UNIVERSITY

Introduction

Since the beginning of the twenty-first century, online games have become a major youth leisure activity, with a market value that now exceeds both the music and film sectors. While western game developers and publishers, such as Electronic Arts (EA) and Blizzard Entertainment, have expanded their global reach, game developers in Korea, including NEXON, NCsoft, and Wemade, have also become important contributors to the global game market. This includes the lucrative massively multiplayer online role-play game (MMORPGs) market. As a result of Korea's success in this industry, many other countries and game corporations have become keenly interested in emulating their development systems.

While there are a number of factors contributing to Korea's success in this industry, as in many other countries that rely on neoliberal logics for economic development, the Korean government has played a key role. In general, the Korean government has provided support for measures intended to harness market forces, and out-sourced more government services, in addition to reducing industrial and government costs by emphasizing, in some cases, citizen self-help instead of reliance on government services (Jeannotte 2010). In terms of the online game industry, neoliberal economic logics have meant extensive institutional and economic support for its development. In a very general sense, the Korean gaming industry case is similar to many other countries in which government support for industry is paralleled by decreases in some social supports for citizens.

But upon closer inspection, the story diverges from many other analyses of neoliberalism and science and technology. Not only has the Korean government advanced new policies as part of its efforts to make information technology (IT) a centerpiece of the national economy, it has also done so because of the importance of gaming for youth culture in Korea. Co-existing with this social value is that many parents and some government officials have become anxious about the content of online games and their appropriateness for youth because of the possibility of gaming-related addictions. Addiction is not simply a metaphor for particular relationships to online games, as it is in some societies; in Korean society, it refers to a specific medical condition and a social relationship. One set of government responses has been to regulate the use of these games.

Thus, in this case, the Korean government operates in ways that are consistent with neoliberalism, but also with the social protectionism characteristic of developmental and redistributional states. It therefore complicates analyses of the role of neoliberalism in the development of the science and technology sector (Moore et al. 2011, Slaughter 2005). Moore et al. (2011) argue that we need to attend to variations in how neoliberal practices and ideologies are carried out in terms of science and technology; this chapter is an effort to do so. As other analysts, such as David Harvey (2005) point out, neoliberalism does not make the state irrelevant but changes its foci; among other new directions, the government plays a key role in supporting industrial growth over and above activities such as taxation, environmental protection and labor regulation that were characteristic of other periods of economic growth in capitalist systems. In the Korean case, neoliberally-inclined Korean officials have used state power in the pursuit of the country's economic goals, in the midst of severe criticism based on social issues related to game overuse, such as school violence and digital bullying. Thus, Harvey and other neoliberal analysts would not have predicted the co-existence of social protectionism and neoliberal economic processes in the very same industry.

Whereas there are several previous works analyzing the role of the government as a primary actor in the growth of the online game industries (Cao and Downing 2008, Dwyer and Stockbridge 1999), there has been relatively little academic discussion about the varied roles of the government and the nature of political culture in the context of the Korean online game industry. This chapter examines the changing cultural politics in the online game sector and its influence on the Korean online game industry. This is accomplished by examining the growth of Korean game development in the midst of a political environment whereby local game developers have enjoyed numerous state-led growth measures, such as subsidies, tax breaks, and education initiatives, but at the same time have experienced the regulatory side of government filtered through cultural concerns. In attending to political culture and the multiple roles of the state, we show that purely economic, and especially, neoliberal logics alone cannot explain the dynamics of the Korean government in the development of IT as a technological and cultural industry. Instead, we argue that attention to national political cultures is of critical importance.

The evolution of the Korean video game industry in the twenty-first century

The video game industry, including online and mobile games, is a burgeoning cultural and technology sector in Korea. In fact, online gaming has become the economic and cultural leader in the Korean video game industry, rivaling films and broadcasting in terms of both its influence on youth culture and its importance in the export of cultural goods (Jin 2011). When the video game industry started to become a major part of the digital economy in 2000, domestic sales were valued at $864.8 million, and online gaming accounted for 22% of the entire game market. During the same year, arcade games were the largest gaming sector (60%), and PC games were the third at 13%, while mobile and console games were marginal (Feller and McNamara 2003). The rapid growth of online gaming, however, has dramatically changed the map of the video game industry. In 2010, the Korean market value of gaming, including console/handheld, online, mobile, arcade, and PC games, was as much as $4.90 billion.[1] Including internet cafés (*bang* in Korean) the total video game market was valued at as much as $6.42 billion in 2010. The online game industry, excluding internet cafés, accounted for as much as 84% of the total, followed by console/handheld (7.5%), mobile (5.6%), arcade, and PC games (Korea Game Development and Promotion Institute 2011).

Meanwhile, the Korean video game industry has continuously expanded its export of games to the global markets. Korean online games are currently well received, not only in Asia, but also in North America and Europe. In the early stage of the growth, Korea exported $140 million worth of games, while importing $160 million worth of games in 2002. However, in 2011, the country exported as much as $2.28 billion worth of games, a 48.1% increase from the previous year, while importing $200 million worth of games. During the period 2002–2011, the export figure increased 17-fold, while game imports increased by only 1.25 (see Figure 11.1) (Ministry of Culture, Sports, and Tourism 2012).

The online game industry consisted of 96.2% of exports in the Korean video game sector in 2011. In 2012, the Korean game industry was expected to export as much as $2.8 billion worth of games, which is more than a 19% increase from the previous year (Ministry of Culture, Sports and Tourism 2012).

More importantly, what makes Korea's online game industry unique is its global penetration. This success is distinctive for at least two major reasons. First, other Korean cultural industries, including the film and television sectors, have not made the transition to the international level to the same degree. While the major target of these Korean cultural genres is East Asia, the online game industry has exported several games to Western countries, including the U.S. and Western European nations. The largest foreign markets for Korean games are still in East Asia, including China, Japan and Taiwan, although in 2010 the U.S. and Europe consisted of 17.2% of Korean online game exports (Korea Game Development and Promotion Institute 2011, 2008: 53). In 2011, the export of online games to the U.S. and Europe decreased to 13%; however, the U.S. became the largest market for domestic mobile games at 57%, followed by Japan (26.2%) and Europe (9.7%) (Ministry of Culture, Sports and Tourism 2012). Second, for much of its history the video game industry (both development and publishing) has been dominated by the U.S., Japan, and more recently some European countries. Korea's continued success in exporting major video game titles internationally has signaled a shifting landscape for

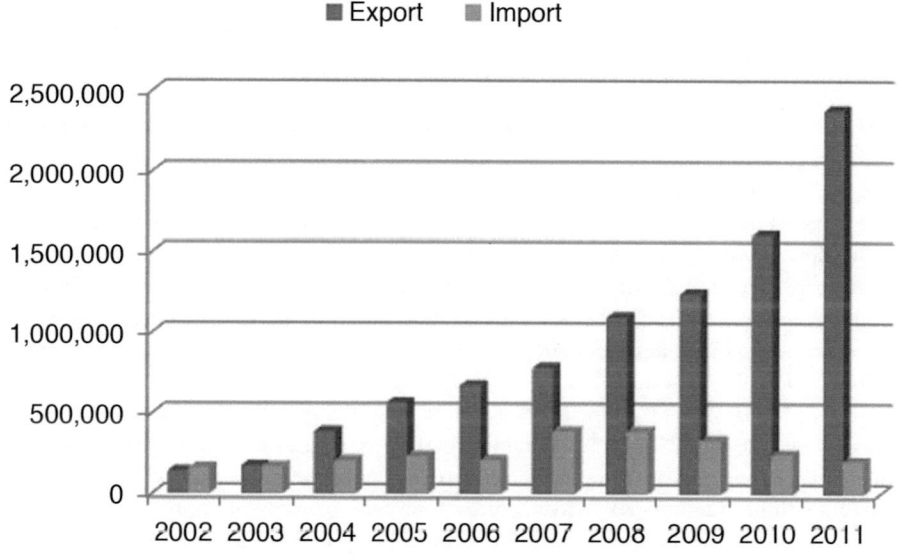

Figure 11.1 Global trade of domestic game products (unit: $1,000)

Source: Ministry of Culture, Sports and Tourism (2012). 2012 Game Industry Whitepaper. Seoul: MCST.

the global video game industry in which new players have emerged during the recent flourishing of both the online and mobile sectors of gaming.

In the Korean case, MMORPGs and casual games are two successful genres. The first to accomplish success in the overseas market was "MIR II," developed by Wemade. Based on a martial arts motif, this MMORPG experienced great popularity in the Chinese market (Seok 2010). Afterwards, stimulated by its success, numerous domestic games, including "Maple Story," "Lineage [I]," "Lineage II," and "AION" began knocking on the doors of foreign markets. Several Korean game developers and investors have thus jumped into the global market, receiving positive responses from consumers in North America and Europe. The NCsoft-developed Lineage games have especially ascertained a unique presence in the global market. NCsoft recorded $338 million in revenues in 2006, and 32% of that income came from the U.S. and Europe (NCsoft 2009). AION, developed by NCsoft, also rapidly penetrated the North American and European markets. AION earned the company $46 million through its U.S. and European sales in the fourth quarter of 2009, which accounted for 24% of its revenue for that quarter (NCsoft 2010). Nexon's Maple Story (2D side-scrolling fantasy MMORPG with cartoony graphics) exceeded three million registered users in North America, making it one of the fastest growing MMO (massively multiplayer online) games, known for its robust community and the ability to personalize game characters in endless combinations (Chung 2010).

Korean online games are now closely entwined with global cultural markets, such that they have begun to achieve stable performances in the U.S. and European markets, as a substantial number of Western publishers are interested in Korean games. Although there are several global game giants, the Korean game industry has gained some credit globally (Jin 2011). Online games produced by Korean developers have kindled a new trend that has rarely been seen in other countries or in other Korean cultural genres. Korea's online game sector has become a unique case of the actualization of contra-flow – cultural flow from developing countries to developed countries – in the twenty-first century. Of course, that is not necessarily to say that Korean popular cultures have become a major part of western culture. Instead, it proves the possibility of the growth of local cultural products in western markets.

Government politics in the video game industries

As mass media and academic research have continuously documented, the Korean economy has become a symbol of successful top-down, state-led economic development (Kim 2013a; Heo and Kim 2000). Just as the Korean government has been a fundamental player in the national economy generally, the role of the government has also been crucial to the video game industry. The Korean government put the video game industry, and in particular online gaming, at the center of its policy for the digital economy and culture. In the twenty-first century, the Korean government has firmly believed that culture is one of the most significant parts of the national economy, and it has simultaneously developed economic and cultural policy.

More specifically, government support was not meant to directly target the online game industry, but broadly the IT sector in the initial stage of the growth of the game industry. The rapid deployment of IT in Korea originated in 1995 when the government enacted the Framework Act on Information, which set up a comprehensive strategy for the Korean Information Infrastructure (KII) (Ministry of Information and Communication 2004). The Korean government has developed its information and communications technologies to establish a high-tech industrial base (Lee 2012). This measure has worked as a fundamental conduit for the growth of the online game industry, because the swift developments of broadband and PC it enabled have facilitated online gaming (Lee 2012, Jin 2010). Unlike other countries in which broadband has not become a primary goal of national infrastructure, it is relatively easy for Korea to develop the online game industry because high-speed internet is key for the growth of online gaming. Korea has become the world's best laboratory for broadband services – and a place to look to for answers on how the internet business, including online gaming, may evolve (Taylor 2006).

Attempting to integrate measures more directly pertinent to the online games industry, over the last several years, the Korean government has provided legal supports and financial subsidies to game developers, and it has increased its investment in a bid to nurture its software industry, with plans to continue. By the end of the 1990s, the government began to actively support the game industry by enacting the Music, Video, and Game Software Act as the legal and policy basis of the game industry, and it founded the Integrated Game Support Center in 1999 in order to develop the domestic game sector into a strategic export-oriented cultural industry. The government also established the Game Industry Promotion Act as a special law in 2006, and it laid down legal grounds for the relevant policies on the promotion and regulation of the game industry. The provisions of the Promotion Act stipulated the creation of a sound game culture, invigoration of e-Sports, and establishment of the Game Rating Board to create a favorable environment for game culture and to secure specialized game rating standards (Ministry of Culture, Sports and Tourism 2008: 6).

The Korean government has also supported the game industry through law. In April 2006, the government enacted the "Relevant Implementation Order and Implementation Rules for the Game Industry Promotion Law" to protect the game industry by differentiating it from speculative games, including poker games. The Promotion Law asserts that the game industry is the core industry of the next generation; therefore, the government wanted to develop an environment of growth for the game industry by providing legal supports, such as tax breaks and copyrights. Later, the government established the Korean Creative Content Agency to effectively support content industries in May 2009. Thus, the Korean government has continuously supported the video game industry with its policy measures, resulting in favorable conditions for online game developers and publishers.

At the same time, the government has also advanced its financial support of the game industry. In 2004, the Korean government announced the so-called "Long-Term Promotion Plan of the Game Industry." According to the Plan, the government was willing to support the game industry to become one of the three so-called gaming empires of the world by increasing its market size to $10 billion by 2007, while increasing the number of employees in the game industry to 100,000 (Ham 2003). Although detailed plans were not articulated, the Plan served as a starting point for direct government support of the online game industry. Based on this Plan, for example, in 2005 the Korean government announced a project to invest $20 million to support the development of graphics and virtual reality technologies in the games sector. As part of this effort, in 2006, the government invested $13.5 million in the growth of the game industry and the creation of a game culture (Ministry of Culture and Tourism 2006). Although the amount of this particular investment is inadequate when compared with actual costs, it illustrates the intention of the Korean government to deliver on its promise (Jin and Chee 2008). In particular, this support was intended to place five local game firms on the U.S. stock market by 2010. The Korean government used Webzen – one of the major online game developers – which raised $97.2 million with a NASDAQ listing in December 2003, as an exemplary case in the global capital market (*Korea Times* 2003). Several online game companies had weighed IPOs (initial public offering) in the U.S. stock market, although these plans were not fulfilled due to the financial crisis that occurred between 2007 and 2009 in the U.S.

These state-led support initiatives are important in countering the notion of unrestrained globalization overtaking geopolitically established national boundaries. Some theoreticians, such as Hjarvard (2003) and Giddens (1999), claim that the idea of a truly sovereign nation-state has been lost in the contemporary political situation amid neoliberal globalization, resulting in a small government, and a weak role for the nation-state and a consequent decrease in the role of domestic culture and cultural identity in non-western countries. However, in the early twenty-first century, several analysts have recognized that nation-states provide high levels of financial support and protection for industry, while lessening their roles in protecting citizens and providing social benefits. This means that nation-states are not shrinking, but shifting their roles from supporters of welfare states to supporters of big corporations. With the case of the Global Governance of Information Technologies and Networks, J.P. Singh (Chapter 12) also points out that

> nation-states are no longer the only actors defining governance and, therefore, the resulting understandings now reveal both top-down and bottom-up prerogatives, although the broadening and deepening of polycentric global governance of information networks does not replace the authoritative prerogatives of powerful actors but it supplants them with increasingly participatory mechanisms.

As such, the Korean government has played a proactive role in the development of its domestic game industry. Online game policies and regulations focusing on culture, technology, and economy have visibly led to the formation of a prosperous new cultural industry on a global scale. Since the cultural sector, and in particular, the game industry has become an emerging area for the national economy and culture, the Korean government has had to develop its cultural policy to initiate and sustain the growth of the game industries. Consequently, the online game sector has become a full-fledged industrial giant in Korea.

Political culture of video gaming: regulation in the game industry

What makes the Korean case different from other cases in which the government directly intervenes to support an industry, is that the swift growth of Korea's online game sector was also shaped by the nation's political culture. Political culture refers to the symbolic environment of political practice, and it has been shaped by historical experiences and philosophical and religious traditions (Kluver 2005). This also includes the assumptions, expectations, mythologies and mechanisms of political practice within a nation, as Kluver and Banerjee (2005) aptly put it.

In many ways, the cultural attention to the content and effects of the gaming industry are no different than they are in other countries. In the Korean gaming industry, the regulatory face of government is evident in such actions as the rating of violent game content and speculative games, which embroils it in public controversy and political conflict, as is the case in many other countries (Dyer-Witheford and Sharman 2005: 187).

Thus, while the Korean government plays a proactive role in facilitating the further development of the domestic game industry, the government has also interestingly had an active part in limiting gaming among young people in the face of societal concerns about the effect of excessive game playing. Video game addiction, which includes symptoms like becoming withdrawn or angry when not allowed to play, has often led to severe cases that have resulted in addicts simply not eating or sleeping until they can satisfy their cravings. Video game addiction plagues Korea – perhaps worse than elsewhere. Occasional reports of compulsive gamers dying or murdering loved ones to satisfy their addictions have given rise to criticism of the industry.

Furthermore, parents worry about their children's academic ability due to game addiction. This trend may be increasingly problematic as Korea continues as one of the most wired countries in the world, but that connectivity comes with a price (Cain 2010).

In fact, many worry about the relationship between game addiction (overuse) and violent behavior among middle and high school students. In many Korean schools, bullying is a serious problem, and some media have reported a connection between online games and violent behavior. Lawmakers and government officials have to respond to increasing concerns from parents and teachers, and the government has developed various methods to prevent game overuse (Korea Creative Contents Agency 2011: 8). About ten years ago, in 2004, educational, religious, and women's groups initiated the first open forum to discuss the negative impacts of video games, such as the lack of sleep due to playing until late night, game addiction, and lack of social life, on young students, including high school students (Kwon 2004). A few earlier attempts to regulate games had failed due to severe opposition by game corporations.

In the contemporary moment, the situation has changed. Effective game regulation has become much more common. Of most importance is that the Ministry of Gender, Equality and Family (MGEF) and the Ministry of Culture, Sports, and Tourism (MCST) together advocated the "Shutdown Law" as a way to prevent online gaming addiction (Caoili 2011a). Since this was the first official attempt by the government to curb online game overuse, the "Shutdown Law," also known as "Cinderella Law," has been the most controversial. The law went into effect on November 20, 2011 and required online games to block children aged under 16 from playing during a late-night six-hour block, between midnight and 6am. The law exempted console and mobile games for the time being, saying it would expand it into those categories, if necessary, following a test period.

However, critics of the law argue that it violates children's civil rights, and that the government has not proven that playing games is more harmful than watching TV or movies, listening to music, or engaging in other indoor activities (Caoili 2011b). A cultural solidarity organization, MoonHwaYunDae, filed an appeal to Korea's Constitutional Court against the law in October 2011, and the effectiveness of the online game curfew itself has also been called into question repeatedly since the law was implemented.

There is no doubt that game corporations vehemently oppose this regulation in the name of the game industry, which has become part of the national economy. One of the major characteristics of neoliberalism at both the global and local levels is the new power of owners of large, multinational corporations that benefit from economic policies associated with innovation, trade liberalization, reduced government spending on entitlements and decreased state restrictions on labor, health, and environmental hazards of production (Moore et al. 2011; Harvey 2005). While civic groups emphasize children's civil rights, game corporations focus on the potential backlash to the knowledge economy that the game sector itself rapidly becomes part of.

Despite this opposition, and in contrast with what many supporters of neoliberal policy argue for, the government has activated gaming regulations. As of January 30, 2012, the average number of people logging on to six different online games by the country's big three online game firms slightly decreased from 43,744 to 41,796 people, which was the average figure recorded ahead of the Shutdown policy implementation. This is only a 4.5% decrease, leading to assumptions that a large portion of teenagers are still logging on to the sites through other methods, which include using their parents' registration numbers (Cho 2012).

However, due to the controversy and opposition by game corporations and civil rights groups, the MCST has shifted its Shutdown Law to "the Selection System of Game Availability Period," which means that the government is mandating that for any large gaming company (with more than 100 employees and $27 million in revenue) their games will have to have a built-in feature that allows parents to set a certain time during which their children are allowed to play. This applies to widely played titles like "League of Legends" and "Starcraft 2," but will not be enforced for games like Blizzard's "Diablo 3," as that game is not allowed to be sold to anyone under 18 in Korea (Tassi 2012). Since about 59% of teens use their parents' registration numbers for restricted games, the new measure shifts the responsibility for restricting game playing by young people to parents. The Education Ministry has also considered the implementation of "the Game Cooling Off system," which forces online game players to take a 5-to-10 minute break following every two hours of play time by disconnecting them from the sites (Cho 2012). These policy measures drew fierce protest from the country's online game industry, which argued that the new regulation will most likely hinder the country's rapidly growing market.

What is most significant is that these changing policy measures are based on Korean culture's heavy emphasis on children's education, in addition to concerns about online games' violent characteristics. Korea is one of the most developed countries in terms of education, and its high rate of literacy and school enrollment are crucial for a country with no particular natural resources. Over-enthusiasm for education in Korea is not new, but for most parents, their children's education is the highest priority, because they firmly believe that going to top-tier universities is the most important asset for their children.

The Korean government could not resist parents' demands once they started to focus on the quality of education, which is hampered by the overuse of online games (Kim 2013b). In Korea, online addiction has long been associated with hardcore gamers who play online games for days on end, isolated from their school and blurring the line between the real and fantasy online worlds. Young children's obsession with being online is a byproduct of being reared in one of the world's most digitally connected societies where 98% of households have broadband internet. Being wired is an icon of South Korea's national pride in its state-directed transformation from economic backwater to one of Asia's most advanced and wealthy nations. But some now fret about the effects that Korea's digital utopia is having on its children, part of the first generation to play online games on smartphones, tablets and other devices even before they can read and write (*Associated Press* 2012). Many teachers and parents charge that online gaming has disrupted Korea's well-established ideas about education; therefore, they vehemently demand that the government must act properly and in a timely fashion.

Against this backdrop, the government has primarily sided with parents and teachers who worry about children's academic ability and their potentially violent behavior in school over economic imperatives. Because the tension among cultural values, in this case, in conjunction with children's education, and economic imperatives has been especially severe in the realm of online gaming, the government has no choice but to emphasize the importance of cultural values surrounding children's education over economic imperatives. As will be discussed in detail later, along with several other regulatory measures, including the game rating system, this contradictory priority has consequently reshaped gaming technologies and industries in the Korean context.

Of course, mechanisms to regulate youth exposure to digital games (both online and offline) have not only been implemented in Korea. Because Korea arguably entered the video game consumption market relatively late, many of its game play-related social problems were congenital and had been persistent since the origins of the gaming industry and its troubles internationally. In other words, Korea's recent regulations targeting online games come in the context of similar governmental interventions within the realm of game play globally. To begin with, in the U.S., there were several attempts to regulate video games. In the early 1980s, many parents, legislators, and teachers joined the police in denouncing gaming arcades as sources of troublemaking and delinquency (Burrill 2008, 7). Several sources cite then-U.S. surgeon general C. Everett Koop in particular, who in 1982 stated that digital games were producing "aberrations in childhood behavior" and causing extensive addiction (Sheff and Eddy 1999; Poole 2000). Williams and Smith (2007) argue that moral panics centered on gaming are part of a long legacy of such alarm in the twentieth century (following, for instance, jazz music or rock 'n' roll), which they largely describe as having been propagated by the media and biased scholarship. They declare that gaming panics took place along two trajectories. The first was through fantasy gaming (such as "Dungeons & Dragons") where fears about the likelihood of occult worship and 'negative psychological conditions' such as attempted suicide were widespread. Second, panics arose because of video game platforms and PCs as new technologies that could bring harmful side effects, including the lack of sleep and school violence (Chalmers 2009, Anderson et al. 2007).

However, even early investigations such as the one by Ellis (1984: 60) found that "a relatively few arcades, by attracting troublemakers and/or facilitating troublesome behavior by young people, seem to be able to establish in the minds of the public a connection between video arcades and drugs, prostitution, theft, vandalism, violence, and truancy." Despite reports suggesting that arcades were not the cause of deviance, these moral panics persisted throughout the 1980s and 1990s. The games featured at arcades caused widespread global panics owing to the combination of overt violence and arcades "as uncontrolled spaces in which young people congregated" (Kline et al. 2003). Sheff and Eddy (1999) also discussed the success of the gaming industry in the face of sustained negative discourses based on video game arcades.[2]

The condemning of public video gaming locations as problem sites was thus international in scope. In Japan, arcades in the mid-1980s began to be described with "the three 'k's" – kurai, kitanai and kowai – meaning dark, dirty and scary (Hamilton 1993). The situation was similar in Hong Kong, despite the fact that popular arcade fighting games like "Street Fighter" and "King of Fighters" resulted in audiences manipulating the Japanese import into a socially and economically significant cultural hybrid.[3] Ng (2006) declared that the "image of game centres perceived by the officials, teachers, parents and the press was relatively negative, associated with smoking, excessive noise, gang activities, and drug dealing." Consequently, gaming at arcades came to be considered entertainment for lower classes and uneducated males.

In Malaysia, the backlash was more severe. Legal action began against video arcades that also promoted gambling in the 1990s, including a "mass demolition of illegal video game slot machines" to squelch public concerns over the government's apparent stagnation on enforcement. One Member of Parliament who led the campaign was quoted as saying "I take this matter very seriously as I do not want schoolchildren and teenagers in my neighbourhood to play truant and indulge in moral misconduct" (Pa'wan 1997). By 2001, the Malaysian government was ordering all video game arcades to be shut down by the end of the year in response to the increasingly negative public opinion of gaming.[4] Malaysia's deputy prime minister asserted that arcades were a "dangerous and large social problem," and the country's most influential newspaper compared the movement against arcades to a jihad, portraying arcade goers as victims in need of saving (Yoong 2001). In Venezuela, the more recent attempts at banning violent video games, which put arcades and internet cafés in danger, were applauded by some, but considered as a "public relations stunt" by critics as a response to the country's extensive violent crime problems. Lawmakers there have argued that the alarmingly high murder rate could be lowered significantly by preventing access to such games (Toothaker 2009).

Of course, arcades and other public gaming sites were not the only targets of opponents to these purportedly harmful games. When digital game technology reached the point where home consoles could rival or even outperform their arcade counterparts, much of the negative attention shifted toward the home as a site of play. In the early 1990s, the gaming industry began to witness rising confrontation in response to the effects of mediated interactive entertainment, with a major goal becoming the censorship of especially violent or offensive games. Among the most prominent cases is the well-documented 1993 campaign led by Senator Joe Lieberman, proposing federal regulations because of the releases of "Mortal Kombat" and "Night Trap" on consoles in 1992, which were deemed overly violent and disturbing by lawmakers. Besides being the catalyst for the creation of the industry's self-regulatory Entertainment Software Rating Board (ESRB), this would set off a series of legislative challenges that would contest the gaming industry and its practices for the coming decades.

Most interestingly, in 2005 the State of California in the U.S. proposed that violent video games should be classed as X-rated entertainment and to make it a felony to sell or to rent such games to anyone aged under 18. The subsequent case, Brown (formerly Schwarzenegger) *vs.* Entertainment Merchants Association, arose out of a challenge to the constitutionality of the 2005 California law. The law contained a detailed definition of such games, applying to those in which the range of options available to a player includes killing, maiming, dismembering, or sexually assaulting an image of a human being and if the games also met other criteria reflecting a lack of positive value to minors. The law also included a labeling requirement on the games, requiring violent video games to be designated as such by including a specific 18+ label on the front cover (Green 2011, 36).

However, the U.S. Supreme Court ruled against California's video game violence law, declaring it unconstitutional in June 2011. The court also dismissed the unconvincing evidence of studies into links between video game violence and violent behavior in players (Maxwell 2011). Supreme Court Justice Antonin Scalia, by way of majority opinion (7–2) in the case of Brown *vs.* EMA/ESA, declared that,

> Psychological studies purporting to show a connection between exposure to violent video games and harmful effects on children do not prove that such exposure causes minors to act aggressively. Any demonstrated effects are both small and indistinguishable from effects produced by other media.
>
> *Supreme Court of the U.S., 2010, 2*

Nevertheless, studies sustaining that the entertainment industries, and gaming in particular, are resulting in increasing levels of youth violence remain influential, especially among policy makers. The 2006 U.S. House subcommittee hearing, "Violent and explicit video games: informing parents and protecting children" similarly criticized retailers and the Entertainment Software Rating Board for not adequately protecting children from games with adult content, while a number of recent books point to the fact that the gaming medium, as inherently interactive, contributes more excessively toward the development of violent tendencies in children than can be seen with other popular media such as TV or film (Chalmers 2009, Anderson et al 2007, Grossman and DeGaetano 1999).

The major difference between the U.S. and Korea in this regard is that the U.S. Supreme Court regards the impact of any regulatory decision on the right to freedom of speech as contained in the First Amendment of the U.S. Constitution, which is not the case in Korea (Supreme Court of the U.S. 2010). In the case of the U.S., this means that many would claim that the government telling parents how to be parents is an overreach (Tassi 2012). In other words, while in the U.S. the game industry has largely been able to follow a path of self-regulation, Korea has developed some measures for the regulation of online games. Most of all, while other countries attempt to regulate video games due in large part to their violent content as the case of the state of California in the U.S. exemplifies, Korea has been more concerned about game addiction because students play too much, and therefore parents who worry about educational goals and their children's health issues make strong demands for the regulations.

As suggested earlier, the Korean government has expanded its regulatory system in the video game sector. The Game Rating Board (GRB) was established as an independent agency in October 2006, under the Game Industry Promotion Act, as a public organization for fostering the growth of game culture and developing the game industry in Korea. The GRB rates games in four major categories (12+, 15+, 18+, and all). For the 12+ category, the guideline explains that the content of the game has no representation of anti-societal ideas, distortion or profanity of religion and public morals that would be harmful to children between ages 12 and 14 emotionally and physically, for instance.

Pursuant to the Game Act, game producers or distributors must report to GRB any and all modifications or updates of the content of online games at least seven days prior to their provision. In 2007, in the first full implementation of the rating system, a total of 2,025 rating applications for online games were filed (Ministry of Culture, Sport and Tourism 2008: 18). In 2010, of the 3,218 ratings assigned by GRB, the vast majority of games received an "all" rating (74.6%), followed by 12+ (7.8%), 15+ (3.2%), and 18+ (14.4%) (Game Rating Board 2012). As announced on its website, "GRB does its utmost to prevent illegal activities by strengthening the observation of ratings. Such activities will lead to a dynamic game environment and the nurturing of the game industry as an engine for the future economic growth of Korea" (Game Rating Board 2012). This mission statement would demonstrate the government's claims to balance the financial viability of this rapidly expanding industry with the public interest. In other words, instead of actually trying to prevent young children from playing games that are not appropriate for them, the rating system is largely symbolic.

In sum, Korea has unexpectedly developed unique financial and institutional supports for the game sector, regardless of the fact that the county has been one of the most active countries actualizing neoliberal globalization. As Peck and Tickell (2002: 381) point out, "the new religion of neoliberalism combines a commitment to the extension of markets and logics of competitiveness with a profound antipathy to all kinds of Keynesian strategies. The constitution and extension of competitive forces is married with aggressive forms of state downsizing, austerity financing, and public service reform." Instead of being anti-statist, however, neoliberally-inclined Korean officials have used state power in the pursuit of the country's own goals in the midst of severe criticism based on social problematics related to game overuse. As David Harvey (2005) aptly put it, we can clearly witness that neoliberalism does not make the state irrelevant, and there are few places where this is more evident than in the Korean game industry. The Korean government, which has been a key player in the growth of video game software, has developed cautious approaches to the online game industry. In order to deal with game addiction problems and increasing concerns from parents, the government has had no choice but to take regulatory action, but at the same time it has weighed this against initiatives to cement the ongoing ascendency of its foremost entertainment industry's operational success by not taking these regulations too far.

Conclusion

The Korean online game industry has been experiencing unprecedented growth with the popularity of several games in the global markets, as well as the domestic market. In the midst of intensifying global competition in the game market, domestic game firms have developed several MMORPGs and casual games, and these games have become some of the most significant products and markers of youth culture and have become successful in the global game markets. While several factors are major contributors to the success of the local online game sector, government policy especially has proven to be a key element for the trajectories of game industries via cultural policies surrounding media economy and youth culture.

The Korean government has applied neoliberal globalization approaches to the national economy; however, the reality has been more complex than a straightforward neoliberal explanation would predict. The government has realized that the local video game industry has substantially grown when the government actively supports it. As a result, the government has to facilitate the marketization of online games, as much through its economic as its cultural policies. While neoliberal norms call for small government in the realm of culture, the Korean government has taken a major role. The Korean government has continued to be involved in the game industry as a major player mainly due to economic imperatives. In other words, the government has initiated the growth of video games for the national economy.

In recent years, the government has also regulated video games. In Korea, online games have been a major entertainment tool and playing online games has become a significant pastime for Korean youth between their mid-teens and late twenties. With school violence and excessive gaming and the decline of academic interest among many young students widely recognized as important social problems, the Korean government's game policies have largely shifted from deregulatory to socially regulatory measures. The government has taken action to restrict youth exposure to violent online games, and doing so has oftentimes hurt the gaming industry. Instead of regulating games, many other countries are trying to resolve some side effects of games through education; however, unlike these cases the Korean government has taken regulatory action in the belief that these rules will improve the situation more quickly.

This means that government policy in the video game sector oftentimes appears contradictory due to the complicated socio-political environment; the government sometimes becomes a major engine for the growth, while at other times, it limits people's liberty by restricting the growth of the video game industry. Regardless of the fact that there is arguably no clear connection between online gaming and violent behavior, the ministries in the Korean government work together to create new rules, because school violence in Korea has become one of the most serious issues in the twenty-first century. The government, instead of more carefully weighing pros and cons, has regulated the game sector regardless of the protests and criticisms. This implies that the political culture of the Korean government as it relates to the video game industries has been constantly amended owing to a number of factors, including but not limited to historical precedents, societal pressures and expectations, and the need to balance economic prosperity with the safeguarding of the country's own populace. The government has had to find a delicate balance between supporting its largest cultural export sector and at the same time not allowing its products to run unchecked, particularly on the country's youth. However, most recently the government seems to be leaning more strongly toward regulation. What the government has to consider is the nature of video gaming as a major part of youth culture. How to develop this unique youth culture is more significant than its temporary reign because youth culture cannot be curbed by these regulatory policy measures.

Notes

1   The actual value will be different due to the currency exchange rate between 2000 and 2010. In order to provide an approximate trend of the growth, the chapter simply calculates the exchange rate the same ($1=1,156.26 won) during the period.

2 For instance, in Paramus, New Jersey, in 1997 a law was proposed to ban free-standing arcades. Furthermore, any game or arcade had to be more than 1,500 feet away from the nearest school, church, synagogue, park, library, hospital, day-care center, nursing home, or from "any other amusement device, even a single video game in the lobby of a diner" (Sforza 1997).

3 Hong Kong youth culture accepted the Japanese games, but they were localized and this hybridization mixed the games and local culture. The character designs in *King of Fighter* particularly influenced youth street fashion styles as players began to dress similarly to the characters in the game. Furthermore, arcade game culture was alleged to have promoted new "combat game-related jargons" especially among the lower classes (Ng 2006).

4 This was mainly under the general impression that arcades were endangering youth. The decision to rid the country of arcades happened "amid claims by many parents and government officials that moral values have plunged among teenagers and young adults. Some academics blamed the increasing popularity of violent video games for a surge in school crimes last year, including stabbings and students setting fire to classrooms. They also said arcades cause teenagers to skip school and waste their allowances. Meanwhile, police argued that many of the dark, smoky game centers function as a front for gambling and money laundering" (Yoong 2001).

References

Anderson, C.A., Gentile, D., and Buckley, K. (2007). *Violent Video Game Effects on Children and Adolescents: Theory, Research, and Public Policy*. Oxford: Oxford University Press.

Associated Press (2012). South Korea sees "Digital Addiction" in 2.5 Million as Young as 3. November, 28.

Burrill, D.A. (2008). *Videogames, Masculinity, Culture*. New York: Peter Lang Publishing.

Cain, G. (2010). South Korea Cracks Down on Gaming Addiction. *Time*, April 20.

Cao, Y. and Downing, J. (2008). The Realities of Virtual Play: Video Games and Their Industry in China. *Media, Culture and Society* 30(4): 515–529.

Caoili, E. (2011a). South Korea's Shutdown Law Goes into Effect. *www.gamasutra.com/view/news/38251/ South_Koreas_Shutdown_Law_Goes_Into_Effect.php*.

Caoili, E. (2011b). New Anti-game Addiction System, This Week in Korean News. *www.gamasutra.com/ view/news/171408/New_antigame_addiction_system_this_week_in_Korean_news.php*.

Chalmers, P. (2009). *Inside the Mind of a Teen Killer*. Nashville: Thomas Nelson, Inc.

Cho, J.H. (2012). Game Use Drops Just 4.5% after Curfew Started. *The Korea Herald*, January 30.

Chung, Y.H. (2010). Korea's Bustling Online Games. *Korean Insight*, April 26.

Dwyer, T. and Stockbridge, S. (1999). Putting Violence to Work in New Media Policies. Trends in Australian Internet, Computer Game and Video Regulation. *New Media and Society* 1(2): 227–249.

Dyer-Witheford, N. and Sharman, Z. (2005). The Political Economy of Canada's Video and Computer Game Industry. *Canadian Journal of Communication* 30: 187–210.

Economist (2011, Apr.14). Game Over: A Liberal, Free-market Democracy Has Some Curious Rules and Regulation.

Ellis, D. (1984). Video Arcades, Youth, and Trouble. *Youth & Society* 16(1): 47–65.

Feller, G. and McNamara, M. (2003). Korea's Broadband Multimedia Marketplace. March 26. Available at *www.birds-eye.net/international/korea_gaming_market.shtml*.

Friedman, M. (1982). *Capitalism and Freedom*. Chicago: University of Chicago Press.

Game Rating Board (2012). Rating Guide. *www.grb.or.kr/english/enforcement/ratingguide.aspx*.

Giddens, A. (1999). Runaway World: 1999 Reith Lecture. Available at *http://news.bbc.co.uk/hi/english/ static/events/reith_99/week1/week1.htm* (accessed 11/03/2012).

Green, D. (2011). Regulation of Violent Video Games Sales to Minors Violates First Amendment. *The News Media and Law*, Summer: 36–38.

Grossman, D. and DeGaetano, G. (1999). *Stop Teaching Our Kids to Kill: A Call to Action Against TV, Movie, & Video Game Violence*. New York: Crown Publishers.

Ham, S.J. (2003). Domestic Game Market to Be $10 Billion: The Government Announced the Long Term Plan. *Hangaeyae Shinmun*, November 13, 31.

Hamilton, D. (1993). Pow Goes Posh as Arcades Zap Old Image. *Wall Street Journal* July 2: B1.

Harvey, D. (2005). *A Brief History of Neoliberalism*. New York: Oxford University Press.

Heo, U.K. and Kim, S.W. (2000). Financial Crisis in South Korea: Failure of the Government-Led Development Paradigm. *Asian Survey* 40(30): 494.

Hjarvard, S. (2003). *Media in a Globalized Society.* Copenhagen: Museum Tusculanum Press.

Jeannotte, S. (2010). Going with the Flow: Neoliberalism and Cultural Policy in Manitoba and Saskatchewan. *Canadian Journal of Communication* 35(2): 303–324.

Jin, D.Y. (2010). *Korea's Online Gaming Empire.* Boston, MA: MIT Press.

Jin, D.Y. (2011). The Digital Korean Wave: Local Online Gaming goes Global. *Media International Australia* 141: 128–136.

Jin, D.Y. and Chee, F. (2008). Age of New Media Empire: A Critical Interpretation of the Korean Online Game Industry. *Games and Culture: A Journal of Interactive Media* 3(1): 38–58.

Kim, H.R. (2013a). *State-centric to Contested Social Governance in Korea: Shifting Power.* New York: Routledge.

Kim, K.T. (2013b). What Is the Hidden Reason for the Consecutive Regulation on the Game Industry? *IT Maeil.* 13 February. *http://itmaeil.com/main/main_news_view.php?seq=16055*

Kline, S., Dyer-Witheford, N., and de Peuter, G. (2003). *Digital Play: The Interaction of Technology, Culture, and Marketing.* Montreal: McGill–Queen's University Press.

Kluver, R. (2005). Political Culture in Internet Politics. In *Internet Research Annual*, vol. 2, M. Consalvo and M. Allen (eds), New York: Peter Lang, pp. 75–84.

Kluver, R. and Banerjee, I. (2005). Political Culture, Regulation, and Democratization: The Internet in nine Asian nations. *Information, Communication and Society* 8 (1): 30–46.

Korea Creative Contents Agency (2011). *2011 White Paper on Korean Games.* Seoul: KCCC.

Korea Creative Contents Agency (2012). *2012 Contents Perspective.* Seoul: KCCA.

Korea Game Development and Promotion Institute (2008). *2008 Korea Game Whitepaper*, Korea Game Development and Promotion Institute, Seoul.

Korea Game Development and Promotion Institute (2011). *2011 Korean Game Whitepaper*, Korea Game Development and Promotion Institute, Seoul.

Korea Times (2003). Asian Online Game Providers Rush to Nasdaq, December 23.

Kwon, H.J. (2004). Online Game Shut Down System is Pursued. October 12. *http://news.inews24.com/php/news_view.php?g_menu=020500&g_serial=126030.*

Lee, K.S. (2012). *IT Development in Korea: A Broadband Nirvana?* London: Routledge.

Maxwell, A. (2011, June 27). Supreme Court Rules against Violent Videogame Regulation. Edge. *www.edge-online.com/news/supreme-court-rules-against-violent-videogame-regulation.*

Ministry of Culture and Tourism (2006, February 10). *Game Industry Policy Exhibition* [press release]. Seoul, Korea: Ministry of Culture and Tourism.

Ministry of Culture, Sports and Tourism (2008). *White Paper on Korean Games.* Seoul: MCST.

Ministry of Culture, Sports and Tourism (2012). *2012 Game Industry Whitepaper.* Seoul: MCST.

Ministry of Information and Communication (2004). *Broadband IT Korea Vision 2007*, Seoul: MIC.

Moore, K., Kleinman, D.L., Hess, D., and Frickel, S. (2011). Science and Neoliberal Globalization: A Political Sociological Approach. *Theory and Society* 40: 505–532.

NCsoft (2009). Investor Relations, 2004–2009. Available at *www.ncsoft.net/global/ir/earnings.aspx?BID=&BC=2009*

NCsoft (2010). *Investor Relations, 4Q 2009*, Seoul: NCsoft.

Ng, B.W.-M. (2006). Street Fighter and The King of Fighters in Hong Kong: A Study of Cultural Consumption and Localization of Japanese Games in an Asian Context. *Game Studies: The International Journal of Computer Game Research* 6(1). *http://gamestudies.org/0601/articles/ng* (accessed May 9, 2012).

Pa'wan, A.A. (1997). Closing in on Illegal Video Arcades. *New Straits Times* June 26: 29.

Peck, J. and Tickell, A. (2002). Neoliberalizing Space. *Antipode* 34 (3): 380–404.

Poole, S. (2000). Trigger Happy: Videogames and the Entertainment Revolution. Available at *http://steven-poole.net/trigger-happy.*

Seok, J.W. (2010). Korean Game Market: The Rising Representative Contents of Korea, Games. *www.hancinema.net/korean-game-market – the-rising-representative-contents-of-korea-games-26901.html.*

Sforza, D. (1997). Paramus Sued; Arcade Wars; RKO Says Curb Would Favor Rival. *The Record* February 11: L01.

Sheff, D. and Eddy, A. (1999). *Game Over: How Nintendo Conquered the World.* Wilton: GamePress.

Slaughter, S. (2005). *Liberty Beyond Neo-liberalism: A Republican Critique of Liberal Governance in a Globalising Age.* New York: Palgrave.

Supreme Court of the U.S. (2010). Brown *vs.* Entertainment Merchants Assn.

Tassi, P. (2012). New Korean Law Lets Parents Decide When Their Kids Can Play Games. Forbes. 2 July. *www.forbes.com/sites/insertcoin/2012/07/02/new-korean-law-lets-parents-decide-when-their-kids-can-play-games.*

Taylor, C. (2006). The Future is in South Korea. *CNN Money*. Available at *http://money.cnn.com/2006/06/08/technology/business2_futureboy0608/index.htm*.

Toothaker, C. (2009). Chavez Allies to Ban Violent Video Games in Bid to Curb Venezuela's Soaring Crime Rate. *The Canadian Press*. 4 October.

Williams, J.P. and Smith, J.H. (2007). *The Players' Realm: Studies on the Culture of Video Games and Gaming*. Jefferson: McFarland & Co.

Yoong, S. (2001). Malaysia to Pull Plug on Video Arcades. *The Globe and Mail*, 11 January, A10.

12

Cultural Understandings and Contestations in the Global Governance of Information Technologies and Networks[1]

J.P. Singh

GEORGE MASON UNIVERSITY

The following pages trace the cultural understandings that informed two sets of important global governance practices at two historical moments: the formation of the International Telegraph Union (ITU) in 1865 among nation-states, and the recent governance controversies about the internet, information technology and development, and intellectual property among states, global firms, and societal actors.[2] Three interrelated arguments inform this essay. First, successive stages of global governance have involved an increasing number of social groups in shaping the ordering principles of global governance. The current principles in no way replace the influence of powerful actors but as a whole they do dilute it, making the global information policy regimes increasingly participatory. Second, in the new systems of governance, increasingly, a greater share involves global mechanisms. Networked technologies proliferate through engineered and human interconnections; this essay goes further in exploring the social construction of international interconnections that have increased the share of the global in their governance. Third, the social construction of these principles moves parallel to the materiality of information technology networks. The calculations of engineers regarding telephone switching protocols in the late nineteenth century reflected the idea of "national" networks, while Facebook's corporate owners in the twenty-first century write technology codes that reflect both their commercial prerogatives and their users' needs and their advocacy.[3]

Global governance has been central in the provision of information technologies and networks, from the rollout of the telegraph in the nineteenth century to the current era in which nearly a billion people worldwide connect through social media platforms such as YouTube and Facebook. Global governance is often understood narrowly as formal arrangements, such as the foundation of international organizations or the inscription of international laws. Importantly, global governance is also a set of informal cultural understandings among participants, which convey the everyday life and the origins of such governance. Collective or cultural understandings among groups – such as nation-states, firms, society, and individuals – represent the rationale for the organization of global governance: who is to be included or excluded in the daily regulative functions; the origins of these principles and their acceptance or contestation among included or excluded groups; and, generalizing in the context of this essay, the collective understandings about technologies among the global governors and the governed. Global governance, therefore, defines the ordering principles of information technology networks and encapsulates both the "textbook" form of these principles, and the political contestations that inform these principles. In the context of this essay, the inclusion of different groups is understood as "participatory," but not necessarily always positive, while their ability to problem-solve and provide public reasons for their solutions, especially when not reflective of what dominant groups' preferences, is deemed deliberative.[4]

The essay will first provide a broad historical overview of how top-down "national" cultural understandings regarding technology were institutionalized through global governance. It will then show how globalizing imperatives, especially those originating in firms, resulted in a market-driven information infrastructural regime epitomized by rules framed at the World Trade Organization and the World Intellectual Property Organization in the period after 1980. The essay then shows how the bottom-up cultural understandings regarding user-driven practices developed slowly, challenging existing governance practices following the proliferation of internet, social media, and multi-use user devices. The final section details how the top-down practices of nation-states or firms are becoming increasingly ineffective in solely defining the shape of global governance. While the top-down practices are becoming ineffective, the bottom-up participatory practices among societal users and global activists remain variable in their incidence and degree of effectiveness. For every successful example of a "crowd-sourced" crisis map, one may count several armchair "slacktivists" clicking 'like' buttons on Facebook, or the latter's owners who shape social behavior through infrastructural codes.

Historical global governance

The market-driven governance of global information networks is relatively new: the World Trade Organization moved the regime close to market principles through a series of multilateral agreements in the 1990s. Historically, information networks were organized and expanded according to state priorities set in place during the nineteenth century when nation-states were the primary and dominant actors in global affairs. The engineering logic of information networks benefitted from and furthered the underlying nationalistic logic, even when there were clear alternatives, such as the proliferation of small-scale "mom and pop" telephony providers all over the United States in the early part of the twentieth century before AT&T consolidated them.

The nineteenth century set in place a cultural understanding regarding the role of the nation-state, which meant that the state set the priorities for telecommunication infrastructures and expanded them according to a logic that regarded such technologies as "progress". From Hobbes to Rousseau, the West European nation-state developed in the context of embodying and reflecting a social contract, which in the shadow of nineteenth century rapid industrialization made technologies almost synonymous with "modernity" even as Luddites and Marxists, among others, questioned this logic (Meltzer et al. 1993). James Scott's (1998) concept of "a high-modernist ideology" that informed the conceptualization and practice of nation-state led social engineering programs in the twentieth century is also apt for describing the large-scale rollout of telecommunications infrastructures that began in the previous century. Large-scale bureaucracies and nationally controlled networks became the mantra of telecommunications: regulatory practices responded to these forces, while a "prostrate civil society", as Scott calls it, could not resist them.

The story of global rules that preserved state-owned telecommunications monopolies predates the invention of the telephone. The International Telegraph Union, founded in 1865, set up rules and protocols for telegraph messages to be exchanged among countries. But, it also accepted and legitimized the position of telecommunications monopolies that, until the early 1980s, dominated communication industries. A set of explicit agreements and implicit understandings at the international level sanctioned the monopoly cartel in telecommunications.

The need for global rules arose shortly after the invention of the telegraph in 1837. By the mid-1860s, telegraph cables spanned the Atlantic and the distance between London and Calcutta. Napoleon III called for a conference in Paris in 1865, leading to the birth of the International Telegraph Union (precursor to the present day International Telecommunication Union or ITU) to ensure that flows of communication would supplement the increasingly freer flows of commerce. It was at this time in Europe that the major powers, including Britain, France, Prussia, and Italy, reduced their tariff barriers toward each other. Telegraph's spread necessitated, and the ITU began to provide, rules for interconnection, equipment standardization, pricing agreements among countries, and a mechanism for decision-making to address all these needs. The latter became even more important with the invention and spread of the telephone after 1876.

The emphasis on national sovereignty in Europe directed the shape of everything that the ITU designed. An early understanding was that each nation would own its own monopoly in telecommunications and, depending on national capacity, its own torchbearer for equipment manufacturing. The monopoly rule would later be buffered by the cost calculation of engineers who argued that network benefits could be optimized only if there was a single "natural monopoly" in every nation (Cowhey 1990).[5] Global governance ensured that these monopolies would be interconnected with each other through subsequent rule formation. The rules of joint provision of services (where two nations were sending messages to each other) and joint ownership (of cables and, later, wireless networks) extended the sovereignty principle to international communications. Bilateral agreements (known as settlements) of division of revenues, usually divided equally between the states involved, were also regularized through the ITU.

Studies of the old telecommunications governance arrangements often overlook the business strategies of national monopolies that sustained the regime. These strategies were also rooted in shared understandings. The basis of every firm's strategic moves can be found in its attempts to maximize profits and minimize risk and uncertainty (Knight, 1921). Telecommunications entities argued for monopoly control emphasizing the economies of scale necessary to provide services, and sought to recover these costs by targeting institutional users such as government administrations and large corporate users. With high (start-up) fixed and high variable (total minus fixed) costs, industries like telecommunications used their economies of scale to make the argument that they were "natural monopolies". However, their focus on large institutional markets such as government and industry who could afford these services, also led to a relative neglect of the needs of the small business sector and the personal "home" market.

Politics informed the engineering logic of monopoly: telecommunication engineers in the monopolies did not contest this logic, but there were alternatives such as the presence of small-scale operators of rural America, or the competitive non-monopoly providers in Finland. Thus the connection between the techno-economics and the business strategies of a monopoly was not a fait accompli. First, businesses quite consciously took advantage of certain features of the infrastructure, such as the necessity of large-scale networks and associated costs, to make the argument about monopoly. Mueller (1998) has brought detailed historical evidence to bear on how AT&T constructed the argument about the so-called natural monopoly to legitimize and protect its market status.[6] Second, there was nothing natural about trying to recover high fixed costs through targeting large institutional users. Most large firms were loathe to recover their costs by focusing on small businesses or large personal markets and had no organizational back-bone to do so. Costly mistakes were made. The Consent Decree AT&T signed in 1956 is an apt illustration. Congruent with its provisions, AT&T agreed not to provide computer based and data processing services. In hindsight, the decision seems myopic given that AT&T was aware of the potential for these new services. However, AT&T was not barred from providing these services to the government (including the Defense Department), which the corporation calculated would be their major customer for these services (Horwitz, 1989: 145). AT&T's organizational culture always prioritized large institutional users.

While the origins of change in monopoly arrangements lie in legal challenges that weak-ened AT&T's technological claims, the main impetus came from the remarkable coalition of powerful states and large users who called for a liberalized competitive marketplace in telecom-munications.[7] Furthermore, human innovation in telecommunications technology by the 1970s, many of them specifically designed to weaken AT&T's claim, had evolved to a point that made the monopoly argument increasingly unsustainable at national or global levels. Judicial and Federal Communication Commission (FCC) rulings in the United States, starting with the late-1950s, had begun to affirm the rights of potential service providers and large users to own and operate their own networks and interconnect with AT&T's monopoly network. For exam-ple, the "Hush-a-phone" decision in 1957 permitted the use of a foreign attachment on the telephone handset to reduce noise and was instrumental in setting the stage for further compe-tition. A broad interpretation of this decision led in turn to the Carterphone decision in 1968, which allowed the interconnection of mobile radiotelephone systems to the public switched telephone network. The latter is widely viewed as the beginning of competition in the telecommunications industry (Schiller 1982: 15). In 1969, Microwave Communications Inc. (MCI) obtained a license to provide service between St. Louis and Chicago. MCI argued that its microwave network was different from AT&T's terrestrial network and, therefore, did not threaten AT&T's position. A 1982 judgment led to the break-up of AT&T in 1984, following an anti-trust (anti-monopoly) lawsuit from the U.S. Department of Justice filed in 1974.

On the whole, the liberalization of telecommunication made possible by these legal-techno-logical changes was played out within individual states. The United Kingdom was the first in 1981 to privatize its monopoly, British Telecom, and introduce duopolistic competition by licensing a second common carrier, Mercury, owned by Cable and Wireless. The 1984 break-up of AT&T introduced competition into long-distance services. Japan began the privatization and competition process in 1985 but the state retained a big oversight role in introducing particular forms of competition in the various service markets. The European Community's technocrats from Brussels began to tout the benefits of liberalization through various reports in order to persuade member states to privatize and move toward market-driven mechanisms in telecom-munications. This came on the heels of several important national and European Commission reports and policy initiatives that touted the benefits of liberalization. A "Green Paper" in 1987 urged telecommunications reform and pushed countries toward adding telecommunications to the creation of the European Union in 1992. While individual countries started moving toward liberalization and privatization in the 1980s, the marketplace did not become competitive until 1998. Nonetheless, as the next section shows, the early European efforts were aided by the transnational user coalition frustrated with monopoly service provision.

The market-driven regime

The different national policies toward competition in the provision for telecommunication services soon had a global governance counterpart. The natural monopoly understandings had been institutionalized internationally through organizations such as the International Telecommunication Union. Beginning in the 1980s, not only were these understandings questioned but global governance took an added "participatory" dimension: instead of merely reflecting national priorities, as this essay will detail, new global understandings shaped by telecommunications providers and users now began to re-order these priorities.

At first glance, the breakdown of the public-monopoly model in telecommunications seems to merely substitute public authority for private control. Nevertheless, this breakdown required amendments to existing cultural meanings among bureaucracies and telecommunication users. First, at the level of global governance, new understandings of liberalizing markets involved, and were reinforced through, lengthy international negotiations. International diplomacy and negotiations are to global governance what conversations are to a household: they shape meanings depending on the language spoken and those participating.[8] Second, there were bottom-up pressures from businesses and consumers who demanded information services that the monopolies were unable to provide. This ushered the awakening of the "prostrate civil society" but their pressures played second fiddle to that of businesses.

A powerful coalition for global regime change arose on behalf of large users (multi-national firms) of telecommunications services who accounted for a majority of the long-distance telecommunication traffic in the world. This is significant: instead of making their case at a national level, these globalized businesses advanced a global argument on the need for change. These users, most of them using data-based networks for their operations, had found themselves increasingly hamstrung by the inefficient way in which most of the telecommunications monopolies operated. They were mostly run as overly bureaucratized government departments not that concerned with either expanding or improving the quality of the infrastructures. The irony was that, given the inelasticity of demand, demand did not fall when prices rose. Thus the government monopolies were often used, especially in the developing world, as "cash cows."

The large users, located in the developed world, put pressure on their home governments for international regulatory reform. Several factors helped their agenda. First, neo-liberal or pro-market ideas were on the rise in policy-making, academia, and international organizations. The large users saw their needs best met through a competitive marketplace rather than through monopolies. Initially, their calls found easy reception in those home governments, which had already begun to liberalize many sectors of the economy. The governments of President Ronald Reagan in the United States and Prime Minister Margaret Thatcher in the United Kingdom are particularly important in this regard. The demand for international reform followed the liberalization of domestic telecommunication in large economies such as the U.S., UK, and Japan, which accounted for nearly two-thirds of the global telecommunications market. Second, as telecommunication markets opened up in these countries, competition began to develop among service providers and the preferred national equipment manufacturers, even in countries which had not yet liberalized, who now wanted to get into international territories in search of new revenue streams. These service and equipment providers joined the large users in an international coalition for reform. Finally, the European Community (now European Union) began to cajole member-states and move them toward liberalizing their telecommunications. This came on the heels of several important national and European Commission (EC) reports and policy initiatives that touted the benefits of liberalization and, as noted earlier, led to competing telecommunication providers in the EU by 1998. Nonetheless, the early European efforts aided the coalition of large users.

Residential users provided another participatory layer. While large business users dominated the global discussions about moving beyond national monopolies, residential users around the world shared their sentiment and also argued for adequate information services. A popular saying in France at the time was that one-half of France was waiting for a phone and the other half for a dial tone. It epitomized the case of societal users in Europe who either lacked these services or received shoddy ones. The old cultural understandings regarding national monopolies stood in the way. Although the nineteenth century "modern" state had responded to new technologies pro-actively by furnishing vast transportation and communication infrastructures and equating communications with breaking societal barriers and hierarchies, the late-twentieth century counterpart of this state presented an irony: the state was comfortable with its centralized bureaucratic apparatus, but often overlooked the societal pressures to which it needed to respond. In 1976, President Giscard d'Estaing of France appointed a commission to advise the state on the computerization of society. The Nora-Minc Report, as it came to be known, pointed to both the centralized French state and the resulting ossified French society as key barriers for information networks, which the report called telematics. In an introduction to this report, sociologist Daniel Bell (1980: xiv) wrote that France was "going through an intensive period of self-scrutiny in which the traditional centralization of administrative power was increasingly under question."

In the developing world, limited access and poor quality telecommunication services generated social pressure to address these issues. Table 12.1 provides an overview of several newly industrializing and developing countries in 1980 and 1995 to show their low teledensity and the long waiting lists for telephones. Although telephones tended to concentrate in large cities, it could take 10–15 years to get a telephone connection in Bombay (now Mumbai). Favelas in Brazil featured illegal telephone connections obtained by tapping into telephone lines; a similar story repeated in other parts of the world, even among affluent customers. South Korea, at that time a newly industrializing country, can now boast of one of the best broadband infrastructures in the world. But in 1980 it had only seven mainlines per 100 people, and the waiting list for telephone lines exceeded five million lines for a country with a total population of 40 million. Nevertheless, in response to middle-class and student protests, President Chun Doo Hwan (1979–1987) undertook telecommunication prioritization and corporatization programs and by 1987 the waiting lists were eliminated.

Societal groups participated in various ways in evolving telecommunication policies. The most immediate demands were economically-based demands for telephone connections, and

Table 12.1 National telecommunications infrastructures

Country	Main lines (Per 100 population)			Waiting list ('000)		Waiting time (years)	Largest city main lines (% of total)
	1980	1990	1995	1980	1995	1995	1995
Singapore	25.98	38.96	48.18	4	0.2	0	100
S. Korea	7.34	30.97	41.47	604	0	0	33.7
Mexico	3.73	6.55	9.58	409	196	0.3	36.1
Malaysia	2.95	8.97	16.56	133	140	0.3	9.5
China	0.2	0.6	3.35	164	1,400	0.2	4.6
India	0.33	0.6	1.29	447	2,277	1.3	12.7
Brazil	3.93	6.5	8.51	3,250	n.a.	0.7	17.4
Myanmar	0.1	0.17	0.35	n.a.	n.a.	n.a	46.3

Note: Adapted from J.P. Singh, *Leapfrogging Development? The Political Economy of Telecommunications Restructuring* (Albany, NY: State University of New York Press, 1999), p. 57.
Source: ITU, *Yearbook of Telecommunication Statistics*, various years.

another was the forceful case for recognizing the communication needs of societies. This case moved into international debates on the New World Information and Communication Order (NWICO) from the late-1970s to the end of the 1980s, which took place in the United Nations Educational and Scientific Organization (Singh 2011a).[9] The first sentence of the MacBride Commission Report (1980: 3) prefaces the intellectual rationale to NWICO: "Communication maintains and animates life. It is also the motor and expression of social activity and civilization…" (MacBride Commission 1980: 3).

At the national level, the economic and social demands for communication were part of larger demands for a better material life, including the case for basic needs. As the South Korea example above shows, these demands held governments accountable directly or indirectly. Riots and social unrest in the developing world in the 1980s reflected these demands for basic economic services. Pressures within countries for telecommunications provision and the inability of the post, telegraph, and telephone departments (PTTs) to meet demands – especially from businesses and urban middle-income consumers – led to a period of telecommunications restructurings and liberalization. However, national public monopolies were administratively ill-prepared to respond to these demands. By the early 1990s, the case for telecommunication restructuring was being made globally.

GATT/WTO Negotiations in Telecommunications[10]

One of the most significant changes in the global governance of information infrastructure is the shifting of authority away from ITU to involve institutions like GATT/WTO. ITU in 1865 reflected national understandings. The GATT/WTO negotiations still feature nation-states but often they reflect the globalizing prerogatives of businesses or, in many cases, states are coordinating their policies in order to join global information networks rather than merely preserving national regulatory control. Furthermore, the push for liberalization in telecommunications globally reflects similar trends and pressures in services including banking, hotels, and airlines – in what is often termed the global services economy that now accounts for a majority of economic output in almost all countries around the world.[11]

The Uruguay Round of the GATT (1986-1994) – undertaken through WTO's predecessor the General Agreement on Tariffs and Trade – was instrumental in establishing a framework for service liberalization through its Group on Negotiation of Services (GNS). This framework served as the backdrop for the WTO telecommunications negotiations from 1994 to 1997. These negotiations unraveled the prior collective understandings on sustaining national monopolies and allowed for international liberalization and competition among telecommunication carriers. The ubiquitous presence of international information carriers around the world at present is chiefly the result of deliberations at the WTO.

The General Agreement on Trade in Services (GATS) that emerged from the WTO is particularly important in the case of telecommunications but also for other information and communication industries such as film and broadcasting. Formally, GATS consists of 29 articles, 8 annexes, and 130 schedules of commitments. Informally understood, the GATS agreement makes services "tradable": no longer the preserve of national monopolies and subject to international regulatory principles. Therefore, the global liberal governance regime, whether shaped by powerful states or international businesses, now supplements and often undercuts national efforts. The WTO dispute settlement mechanism, instituted to arbitrate international disputes has, for example, ruled on Mexican interconnection barriers for competitive carriers (in 2004) and Chinese distribution practices for films (in 2011).

However, these new global arrangements were highly contested at the negotiations in the

1990s. Developing countries, whose cause was spearheaded by India in Geneva, would lose important revenue bases if liberalization schemes were introduced. The European Union eventually wanted to move toward such liberalization but its incumbent telecommunication industry could resist these moves in the early 1990s. Developing countries, therefore, found coalitional partners among the Europeans who were averse to basic services being negotiated just then. Meanwhile, global businesses – such as the Coalition for Service Industries, located in the United States, but representative of global business interests – pushed hard for international liberalization. Countries that had already liberalized – the U.S., U.K., New Zealand and Japan – at the GATT negotiations. In 1994, 67 governments made commitments in the telecommunications annexure of GATS, which covered value-added services.[12] Initially, the annexure was also to cover basic services, and move countries toward cost-based-pricing schemes to hinder national governments from overpricing their telecommunication services or using them as "cash cows."

The WTO telecommunication negotiations, begun in May 1994, took up the unfinished agenda of GATS with respect to the liberalization of basic services. GATT's Uruguay Round of trade negotiations form 1987–1994, which created the WTO and instituted the GATS agreement also called for on-going sectoral negotiations in issues that could not be decided at the Uruguay Round.[13] Three years of complicated negotiations around telecommunications ensued, almost coming undone in April 1996 when the United States responded to weak liberalization offers from others by walking out of the talks. Nonetheless, the February 15, 1997 accord was hailed by the United States and WTO as a major victory. Some 95 percent of world trade in telecommunications, at an estimated $650 billion, would fall under WTO purview beginning January 1, 1998, the date of implementation.

The WTO telecommunications accord signed by 69 countries in 1997 included 40 less developed countries and formalized the new market-led regime in telecommunications. More than 100 governments had joined by 2013. Historically, telecommunications sectors were controlled or operated by national monopolies. The ITU governance mechanisms reflected national priorities. Technically, nation-states also shaped the new cultural understandings of market-liberalization at the WTO. However, these new cultural understandings now reflected not just nation-state priorities but also those of businesses and users. Most importantly, the ITU governance mechanism responded to national priorities; those from the WTO have shaped national priorities, making the evolving arrangement increasingly global in scope, and participatory in including global actors other than the nation-state such as businesses, societal users, and international organizations.

An important feature of the Fourth Protocol of the WTO telecommunications accord was the "Reference Paper" that introduced global "regulatory disciplines" to observe the WTO rules. Until then the inter-governmental "cultural understanding" favored national monopolies and disparate national regulatory structures. The Reference Paper was perhaps the most important outcome of the negotiations in providing regulatory teeth for market liberalization through "market access" and "national treatment" commitments executed by the signatory states.[14] Such commitments would be moot if they were not enforced through national regulatory authorities: the Reference Paper attended to this need and provided the legal and regulatory impetus for the liberalization commitments countries made. While 69 countries signed the accord in 1997, 53 countries signed the Reference Paper appended to the Protocol.[15]

Currently, WTO's Fourth Protocol is understood to be the major global governance regime in telecommunications. International negotiations at the WTO often feature both coalitions of support and also those of protest, both within and outside the conferences of the WTO. Therefore, scholars now argue that international negotiations are far more participatory than international rules made at the behest of a few great powers (Singh 2008, Niemann 2006, Risse 2000). Several well-known disputes adjudicated through the WTO dispute settlement, such as those mentioned above, have further deepened the decision-making procedures governing the regime. WTO is, therefore, the de facto global institution behind this regime. The ITU, which harbored the old monopoly regime, is now involved mostly with technical standards and inter-connection protocols governing the regime. As we will examine later, ITU has recently pushed for involvement in internet governance through encouraging the World Summit on Information Society.

The current regime and bottom-up pressures

The current global information policy regime is polycentric and, depending on the technol-ogy in question, governance takes place through a variety of institutions and processes. It is participatory, because as noted above, it now includes multiple players other than nation-states, such as businesses and global civil society, and international organizations such the ITU, WTO, UNESCO, and the World Intellectual Property Organization. More importantly, these partici-patory mechanisms are deliberative, in the sense of providing spaces for giving public reasons and sometimes for problem-solving in favor of weak players. An example of the former is that international diplomacy is no longer the kind of closed-door secretive exercise that led to the creation of the ITU. An example of the latter is that the increasingly assertive Global South now shapes global governance rather than being a passive recipient of global rules.

Great powers and global businesses still dominate many of the global institutions involved, but several bottom-up and civil society processes are important. The developments that have brought about these changes include liberalization of the telecommunication regime, diffusion of mobile telephony, and proliferation of the internet and social media. Importantly, the infor-mation policy regimes now encompass issue areas that at one time were not included in this regime. Intellectual property is a well-known example. In general, the rise of digitized content has affected everything from archival practices in museums to international treaties governing creative industries signed in various agencies of the United Nations. Creative industries – such as broadcasting and film – are now often understood as the important "content" industries that flow over the "conduits" of information networks. The distinction is hard to make in practice: telephone calls, Facebook posts, and cultural content on YouTube are all digital forms.

Information technologies also continue to facilitate participation from grassroots actors. The concept of technological "affordances" describes "actions and uses that technology makes qual-itatively easier or possible when compared to prior like technologies" (Earl and Kimport 2011: 32). One of the affordances from growing digitization and information technologies is the way it has allowed individuals to network with each other for advocacy and governance purposes. Earl and Kimport (2011:71) call it "supersizing participation," while in the words of Clay Shirky (2008) organizing can now take place without organizations. Such affordances pose a challenge to existing governance processes, although they may not replace them.

Before turning to how these affordances for networked individuals play out in specific instances of global governance, it is worth considering some evidence from the "grassroots". There are more than five billion mobile phone users worldwide, nearly one billion people on Facebook, and one billion people who watch four billion videos on YouTube each month.[16] Nearly half the people in U.S. and Western European countries are believed to own smart phones, which act as multi-media devices, and the fastest growth rates for smart phones are in the developing world (*Financial Times* April 8, 2012). This exponential growth in mobile teleph-ony and social media use have led to affordances such as online advocacy, organizing and protests, content creation and sharing, and participation in global governance processes. This essay now turns to three illustrations of these affordances at the global level: debates on intellectual property, internet governance, and information and communication technologies for development (ICT4D).

Intellectual property

A set of intellectual property agreements emerged at the same time as the Uruguay Round of trade negotiated new rules for services such as telecommunications. These agreements are mostly characterized by giving global firms the ability to get knowledge and information termed and understood as "property". Over the last decade, grassroots opposition has questioned such notions of intellectual property, and the "supersized civil-society participation" against restrictive intellectual property provisions, as detailed below, recently led to embarrassing defeats for the proponents of intellectual property both in the U.S. and the EU.

Brand name clothiers and accessories, pharmaceuticals, and software firms initially pushed for intellectual property protection as part of U.S. trade interests in the 1970s.[17] By the mid-1980s, cultural industries led by the film, television and music industries began to join in. Collective efforts, by global businesses' Intellectual Property Coalition (IPC) resulted in a far-reaching agreement at the Uruguay Round: TRIPS, or the Trade Related Aspects of Intellectual Property. TRIPS remained controversial. The beginning of the next trade round, the Doha Round, was held up in November 2001 until developed countries made concessions on an array of intellectual property matters. For example, paragraph 6 of the so-called Doha Health Declaration instructed the WTO's Council for TRIPS to find make certain that countries with pubic health emergencies that lacked manufacturing capacities could gain access to patented drugs and treatments. This could involve compulsory licensing of patented drugs or breaking of international patents to meet national needs. This task was completed in August 2003 after some contentious negotiations where the U.S. sought to reduce the list of emergencies that would be covered and assurances that the compulsory licenses would not lead to the export of drugs from the developing to the developed world.

As firms have sought to obtain international property rights for knowledge embedded in patented, copyrighted or trademarked products, various transnational actors have contested the claim that the reward for inventions and innovations should be permanent ownership of the resulting "property," rather than a temporary grant of ownership for a limited number of years. Second, they have contested the claim that, even if inventions and innovations are viewed as "property," they should belong to firms as opposed to being freely circulated or assigned commons rights, especially in internet spaces (such as Creative Commons).

Activist opposition to expansive notions of intellectual property for inventions and innovations led to the defeat of two bills in U.S. Congress in January 2012. These were the House of Representatives Stop Online Piracy Act (SOPA) and Senate Protection of Intellectual Property Act (PIPA). The rise of the Pirate Party in Sweden, Germany, and the European Parliament – to advocate against copyright and for internet freedoms – also reflects opposition to industry efforts to expand notions of intellectual property. The multilateral agreement known as the Anti-Counterfeiting Trade Agreement (ACTA), negotiated between 2008–2010, similarly produced heated debates. ACTA is often called "TRIPS-plus" in putting even more restrictive provisions on intellectual property than those negotiated at the WTO. Kader Arif, the European rapporteur for ACTA, resigned his post in protest after the EU officials signed the agreement in January 2012 before planned European Parliament and civil-society debates had taken place. The European Parliament is unlikely to ratify the agreement, and similar developments are noticeable in other states that signed the agreement, including the Mexican Parliament. More than perhaps other arena of global governance discussed in this essay, activism against intellectual property has showcased the clout of global civil society activists and their coalitions against global corporations.

Internet corporation for assigned names and numbers (ICANN)

ICANN, located in Marina Del Rey, California, was founded in November 1998 to regulate internet domain names and the associated Internet Protocol addresses used for transferring messages. It is a private organization, which is overseen by a 19-member Board of Directors. Hailed as a model of self-regulation, it is sometimes seen as a major shift away from intergovernmental organizations serving as governance institutions. The corporation is housed in the U.S. and provides its government with considerable oversight, because it is chartered through its Department of Commerce. Critics have charged that its regular conferences with the U.S. government make ICANN unlikely to take a position that would harm U.S. national interests. Pressures from the EU led to the creation of a global Governmental Advisory Committee (GAC) that diluted somewhat the U.S. government's insistence that the corporation remain totally private. Nevertheless, critics of the U.S. domination of ICANN continue to propose alternatives, especially through the United Nations auspices.

A closer look at this emergent form of global internet governance reveals the influence of powerful political and economic interests (Mueller 2002). ICANN would not have been possible had the U.S. government, through its Department of Commerce, not intervened to arbitrate the claims of rival coalitions seeking to assert their dominance over the internet. Such struggles can be traced back to 1992 but they became especially severe after the introduction of the world wide web in 1994 led to a proliferation of domain names. Network Solutions Inc. (NSI), a private firm, received a five-year contract from the U.S. National Science Foundation to provide these domain name addresses. A rival coalition, centered around Internet Assigned Names Authority (IANA), started with academics and engineers but was able to get, after several failed attempts at establishing its cause, a number of important European governments on board, after which it was called the International Ad-Hoc Committee (IAHC). NSI represented a private alternative, while the IAHC was a public alternative. The U.S. Department of Commerce White Paper in 1998 on domain names and IP addresses largely legitimized the IAHC-led coalition's position on making the internet domain authority participatory in an international sense, rather than dictating the governance from Washington, DC. Primarily at the behest of the European Commission, the U.S. Department of Commerce also requested that the World Intellectual Property Organization (WIPO) set-up a service to resolve domain name disputes (known as the Uniform Domain-Name Dispute-Resolution Policy or UDRP).

The early history of ICANN's "participatory" global functioning has been messy politically, even if the actual task of assigning domain names is somewhat easier. The direct elections for the five directors for At-Large Membership were seen by some as international democracy, and by others as messy populism, and were soon discontinued. Critics also do not think that ICANN reflects bottom-up practices in decision-making as it claims to do. The U.S. government has considerable de facto power in the decision-making.

The World Summit for International Society (WSIS) is particularly important in questioning U.S. domination. WSIS begun in 1998 as an International Telecommunication Union initiative to examine the slow diffusion of internet in the developing world. WSIS was a multistakeholder forum bringing together governments, considerable number of civil society activist groups, businesses, and international organizations. Quite soon, it became the forum for addressing the grievances of developing countries for being left out of domain name governance and a host of other issues, many of which – spam, child pornography, data privacy, freedom of speech – went far beyond the ICANN mandate. The main demand of the international coalition, to which the EU lent support in mid-2005, was to bring ICANN under the United Nations. The U.S. government and ICANN, supported by business groups worldwide, resisted these moves and without the support of the incumbents, the moves eventually failed.

Internet governance began to emerge as an important issue for developing countries in 2003, and WSIS has become a symbolic arena for broadening the field of stakeholders who participate in internet governance. The move was led by influential developing countries such as China, Brazil, India and South Africa, but successive global meetings have brought in additional governments and, more importantly, civil society and international organizations. At the second WSIS summit, held in Tunis mid-November 2005, the United Nations Secretary General created an Internet Governance Forum led by Special Advisor Nitin Desai to convene "a new forum for multi-stakeholder policy dialogue" reflecting the mandate from WSIS processes (Internet Governance Forum 2009). IGF features a unique gathering of multiple-stakeholder diplomacy, with the forum convening annual meetings and consultations among states, business, and civil society organizations. However, critics dismiss IGF as a talking shop and WSIS as ineffective in challenging ICANN dominance. On the issue of internet governance itself, the United States government remains opposed to considering any alternatives to ICANN, while the European Union tries to balance its business groups' support for ICANN and member-states' varying levels of support for WSIS and IGF.

The most recent challenge to ICANN came from ITU's World Conference on International Communications, concluded in Dubai in December 2012. The conference tried to push through an international treaty – at the behest of Russia, China, and many Middle-Eastern and African states – that would have replaced private and U.S. control of internet governance with the UN and ITU supervising a set of agreements enhancing state surveillance of the internet for cybersecurity and controlling spam. The internet, and social media firms such as Google and Facebook that profit from it, would then be brought under a set of technical arrangements at the ITU similar to those governing traditional telecommunications firms. This includes a pricing plan for internet-based firms for their use of the telecommunications networks. Without the support of the U.S., the EU, Japan, and many developing countries, the Dubai summit for changing the parameters of internet governance was unsuccessful. The lead-up to the technical arrangements proposed in Dubai were kept secret from all but a few nation-states opposed to the U.S. position, (although many details leaked out); civil society and business users were thus kept out of the deliberations. Interestingly, many nation-states oppose the United States and its businesses in the name of civil-society, yet these states seldom consult with their own civil societies, and unlike the United States, do not bring them along to international forums. China, an avid critic of the United States at many international forums, opposes WSIS and IGF precisely because they include civil society representatives in their deliberations.

Information and communication technology for development (ICT4D)

The role of information technologies in international development project implementation offers another instance of the shift from top-down global interventions to bottom-up advocacy and ingenuity as detailed below. The top-down interventions were enshrined in expertise-driven agencies such as the World Bank. These global agencies tended to reflect a rather mechanistic understanding of Europe's industrialization, drawing from both West European and the Eastern Communist bloc economies. These interventions became known as the "modernization approach": development economists isolated magic bullets for development, such as a high savings rate or Soviet-style central planning, which would induce a process of economic growth. Grassroots practices the world over now challenge these "external" interventions from global or national planning agencies. In terms of communication infrastructures, these practices question communication messages delivered from expertise-led institutions such as the ITU or the World Bank, the role of the nation-state in infrastructural provision, and the exclusion of societal actors in decision-making. Furthermore, ICT4D practices expand the scope of information technologies to include many forms of mediated communications such as through broadcasting and film.[18]

In traditional development communication models, radio was supposed to bring modernity to the developing world by awakening "traditional" societies to "modern" forms of communication (Lerner 1958). We have come a long way in thinking of traditional and new media in terms of voice, aspiration and social change. Radio's monologic and, at times, propagandistic nature notwithstanding, community radio often provides for interactivity that was missing earlier. The proliferation of mobile telephony may also account for the popularity of call-in shows. Scatamburlo-D'Annibale et al. (2006) note the important role of "alternative media" in fostering dialogic communication. As opposed to corporate media, which they posit as legitimizing capitalist oppression, people-owned radio and television can lead to an "active, engaged, informed political participation." In particular, they note the importance of the Independent Media Center movement, a global progressive network "dedicated to 'horizontal' and 'non-hierarchical' forms of communication and organization… As in many other parts of Latin America, Indymedia Argentina represents media for the oppressed" (p. 7). Indymedia offers broadcasting possibilities through multimedia uses such as the internet, radio and pirated TV signals, but most importantly its proponents celebrate its ability to provide alternative forms of story-telling about social and economic life.

Radical analysts, such as the one describing the Indymedia movement, regularly scoff at representational politics offered through "corporate media," but these alternative story-telling possibilities, nonetheless, are commonplace, and offer alternative understandings of social life and its possibilities. Take the example of Latin American telenovelas, or soap operas. Two billion people all over the world, and not just in the Spanish-speaking world, now watch telenovelas, and the cultural establishments backing them challenge Hollywood's dominance, itself the sine qua non of media imperialism for radical writers. More importantly, while telenovelas are often critiqued for offering escapist and commodified fantasies revolving around romance and beauty, the emphasis in the plotlines on structural obstacles such as poverty, class and bureaucracy distinguishes them from U.S. soap operas (Ibsen 2005). Even these supposedly monologic representations open up dialogic possibilities in allowing the audiences to imagine a different world. The popularity of media-savvy Arab female singers from Lebanon and Egypt offers another example. The lyrics and the sex appeal of these music videos are now seen as challenging patriarchal practices in the Arab world. These scattered examples do not amount to an unquestioned acceptance of all media messages but an exploration of these messages in dialogic terms, cautioning against throwing out the proverbial baby with the bathwater.

The governance processes underlying these creative or cultural industries provide alternative avenues for societal actors to express themselves in narrating their stories (Singh 2011B). Whether through the ITU or the WTO, global governance has been narrated from above. Creative industries, whether in the form of a YouTube video or a Facebook protest, thus provide another participatory dimension and an alternative to "hegemonic" global governance practices of international organizations and nation-states.[19] Admittedly, there are still only a few instances of ICT4D practices involving rigorous citizen engagement and deliberation in a rigorous problem-solving sense. However, in indirect and more general ways, ICTs are often deployed to mobilize populations and provide feedback to policymakers prior to program implementation. SMS [Short Message Service] may be used to provide citizen feedback or report cards to government. Social media in general are well suited for crowdsourcing. In such cases, ICT4D might be seen as approximating ideal forms of deliberation. Several factors need emphasis. First, a few features of ICTs make them well-suited for development purposes, either in the form of enhancing participation and deliberation, or through cutting transaction costs for fulfilling particular objectives. Second, like any other human endeavor, technology invites creativity and human ingenuity and there are many innovative applications in ICT4D. Taking these factors into account points to the human potential for utilizing technology for development.

The United Nations Development Program's (UNDP) report, *Making Technologies Work for Human Development*, is instructive in this regard:

> Development and technology enjoy an uneasy relationship: within development circles there is a suspicion of technology-boosters as too often people promoting expensive, inappropriate fixes that take no account of development realities. Indeed, the belief that there is a technological silver bullet that can "solve" illiteracy, ill health or economic failure reflects scant understanding of real poverty.
>
> Yet if the development community turns its back on the explosion of technological innovation in food, medicine and information, it risks marginalizing itself and denying developing countries opportunities that, if harnessed effectively, could transform the lives of poor people and offer breakthrough development opportunities to poor countries.
>
> *United Nations Development Programme 2001: iii*

ICT4D also showcases human ingenuity in the ways that communities have redefined artifacts for use and used them for problem solving.[20] There are many examples. The open source movement, led by software engineers in academic institutions and private firms, has led to a variety of applications. In a prominent move in 2004, the Brazilian government moved toward adopting open source software and applications in order to cut technology expenditures. Taking another example, the software platform Ushahidi, named after the Swahili word for testimony or witness, was first used in the December 2007 Kenyan elections. It allowed citizens to use a variety of media such as mobile, landlines, radio or the internet to monitor elections and report cases of violence that were then centrally collected and reported on Google maps. Since then, the Ushahidi platform has had a variety of applications including reporting from conflict and disaster zones and was even used in the 2010 winter snowstorm in Washington, DC. Entertainment industries have also been widely utilized for applications. The practice of crisis mapping, spurred by initiatives such as Ushahidi, now employs crowd-sourcing and multimedia devices to provide humanitarian intervention, disaster updates, and violence reports. A final example of problem solving comes from "old-fashioned" broadcasting technologies employed in new political contexts. The BBC TV program Sanglap in Bangladesh introduced politicians and public officials to questions and discussions among live audiences and citizens, and was one of the most popular programs on Bangladeshi TV. The history of entertainment education – to convey socially progressive messages, through song and telenovelas, for example – is an old one. More recently, changes in technology have allowed for these messages to be accessed over a variety of devices and networks and allow interaction with others who are doing so. Livingston and Walter-Drop (2014) note that information technologies now increasingly allow for accountability and transparency in the developing world, and perform functions of information gathering, which used to be limited to large bureaucracies earlier.

It must be remembered that just because applications are widely available does not mean they meet societal aspirations. The *Bhoomi* system that digitized land records in the South Indian state of Karnataka (*Bhoomi* means land) is criticized in policy and scholarship for cutting out the functions of village assemblies or local governance for land registration. Farmers now must spend money and time to travel to sub-district headquarters, known as taluks, to access the service (Prakash and De 2007). In other applications, Indian matrimonial and astrology markets have experienced a boom in applications, where the social productivity of these endeavors may be questionable. In this sense technology is neutral: it can have effects that are socially useful and economically productive, but the opposite can also be true. The Ushahidi platform did not prevent all of the violence among Hutus and Tutsis following the election; in fact, other rumors circulating over mobile phones and social media may have also contributed to violence.

In the global governance of ICT4D, the collective understandings of expertise-led development is not supplemented with increasing examination of participatory practices and the ways in which societal users redefine technologies for use. The UNDP report mentioned above is, therefore, an exceptional instance of such "human development" practices. Many of the technology applications are donor-driven and the official development agencies or private firms with technology foundations have built-in biases toward narrating successful stories and overlooking the failures. While human ingenuity is an important factor for judging the usefulness of ICT4D, we may also need to account for the many failed projects in ICT4D to provide a balanced story.[21] We also need to account for the participatory story-telling practices through various media to account for alternative forms of governance practices that fly below the radar of global governance.

Conclusion

A polycentric regime with different but intersecting institutional networks informing different aspects of information governance has now replaced the old forms of "telecommunications" global governance. The collective understanding around the old regime sanctioned national monopolies and a web of engineering protocols that connected them. The International Telecommunications Union was the de jure global actor, though the de facto preeminent actors were nation-states.

Beginning with the move to market liberalization and developments in digital technologies, global governance has become variegated. On one hand, it has endowed enormous power to global firms who may be viewed as replacing the nation-states as the primary actors in the regime. On the other hand, civil society actors and networked individuals now regularly contest the top-down understandings among firms and nation-states. Important global rules on intellectual property, framed at the behest of global firms, have recently been "defeated" in important forums such as the European Union and the U.S. Congress. Private governance through ICANN is similarly debated in important alternative forums. In the debates on international development, participatory and community-driven practices are fast becoming the norm.

The cultural understandings of the global governance of information networks have shifted in three significant ways. First, nation-states are no longer the only actors defining governance and, therefore, the resulting understandings now reveal both top-down and bottom-up prerogatives. Second, global governance of information issues has broadened to include a variety of issues no longer limited to telecommunications, and central to the way societies communicate and live with each other or order their economic lives. These communication nodes now include broadcasting, social media, creative industries, contested notions of intellectual property, and internet governance. Third, the idea of global governance has become the de facto ordering principle for information networks: instead of merely reflecting national priorities, as the ITU did in the nineteenth century, global governance now increasingly shapes national or domestic priorities. The broadening and deepening of polycentric global governance of information networks, therefore, does not replace the authoritative prerogatives of powerful actors but it supplants them with increasingly participatory mechanisms.

Notes

1   A previous version of this essay was presented at the International Studies Association Convention in San Francisco, April 2013. I am grateful to the audience, the discussant, David Blagden, for detailed comments, and the editors of this volume for extensive feedback.
2   Information networks are understood in this essay to include "pipelines" such as telecommunications and the internet, but also the cultural information "content" that flows over them. Considering the former without the latter would be akin to speaking of society without people. See Jin and Borowy, Chapter 11 in this book for an example of the online video-gaming entertainment industry that is as much about pipelines as content.
3   Global governance understood in this essay builds upon sociological traditions. Ruggie (1993) writes of nominal versus qualitative forms of global multilateral processes: the former accounts for the practices of coordination among groups, while qualitative processes account for the *origins* and *forms* of multilateralism. Similarly, Ernst Haas (1982) distinguishes between mechanical and organic worldviews, and Andrew Hurrell (2007) between international orders analyzed as facts versus socially constructed values.
4   These understandings reflect the literatures on participation as involvement or advocacy from representative groups, and deliberation as the giving of public reasons (Baiocchi 2011, Mansbridge et al. 2010, Mutz 2006), or problem-solving in the public sphere (Friere 2000, Habermas 1989).
5   For Cowhey (1990), the epistemic community of engineers and bureaucrats of the ITU, and national authorities in telecommunications held the system in place.
6   AT&T was a private monopoly, and unique among publicly-owned telecommunication carriers, but as the following analysis will show, its regulatory protections from U.S. government made it a quasi-public monopoly.
7   Thus, while AT&T targeted large users, it was not that successful even in meeting their needs.
8   Bull 1995, Singh 2008, Sharp 2009.
9   NWICO is often remembered as a cold-war debate between the East and the West, and the one that provided the impetus for the U.S. to leave UNESCO in 1986. Although, NWICO indeed pitted the West against the East and the South, at its heart was also the vision of a contemporary society in which communication infrastructures played a central role. See also essays in Frau-Meigs et al. (2012).
10 This section is adapted from Singh (2008).
11 In the old economic jargon, the services sector – as opposed to agriculture or manufacturing – was known as the "tertiary" sector. The term "tertiary" has fallen out of use as the services sector has become the most cutting-edge and dynamic for the economy and for the ways in which societies view themselves in the information age.
12 Value-added services alter or enhance the content of the message as it leaves the sender, while basic services leave it unchanged. An example of value-added services is stored voicemail, and that of basic is a telephone conversation.
13 Typically, GATT/WTO must formally launch "rounds" of negotiations to proceed in all economic sectors simultaneously. The sanctioning of on-going sectoral negotiations was different but also spoke to the importance of creating global governance arrangements in these issues.
14 Market access commitments dealt with the number and scope of private and foreign investment providers National treatment clauses in trade mean that foreign investors get the same market privileges as domestic ones.
15 See World Trade Organization, "Telecommunications Services." *http://wto.org/english/tratop_e/serv_e/telecom_e/telecom_e.htm* (accessed August 23, 2013).
16 Statistics from various sources. YouTube statistics from *www.youtube.com/t/press_statistics* (accessed July 14, 2013). Facebook statistics from *www.checkfacebook.com* (accessed July 14, 2013).

17 Intellectual property concerns parallel the rise of commerce in Western Europe. International agreements in intellectual property proliferated since the late-nineteenth century along with the rise of free trade in Europe. The 1980s were unique in putting forth a cohesive international coalition, which made intellectual property a part of U.S. trade agenda, initially through amendments in U.S. trade law through Congressional amendments in the mid-1980s.

18 This case is not limited to the developing world, and has always had global counterparts. ITU governance always affected telecommunications and broadcasting. WTO's communication agreements govern telecommunications and audio-visual industries. Social media now have further blurred the distinction between various forms of communication media.

19 On February 4, 2008, the 33-year old Colombian engineer Hector Morales Guevara, organized a protest against the paramilitary group FARC through Facebook that took place around the world in nearly 200 cities and 40 countries. The estimates range from hundreds of thousands of protesters up to twelve million. It was one of the biggest protests in Colombia against the paramilitary's kidnappings and other forms of violence, and may have brought out nearly 2 million people in Bogota alone (Bordzinski 2008).

20 Pinch and Bijker (2012: 22–23), while not using the language of ingenuity, note a similar relationship between artifacts and problem-solving: "In deciding which problems are relevant, the social groups concerned with the artifact and the meanings that those groups give to the artifact play a crucial role."

21 Between 2000 and 2003, I was involved in implementing an electronic commerce project to assist women's shawl-making cooperatives in a rural area of India. We received a Development Marketplace Award from the World Bank for this project. The e-commerce part of the project was a failure, but the development intervention resulted in other useful effects such as opening up marketing possibilities in domestic markets. See Singh and Alimchandani (2003).

References

Baiocchi, G., Heller, P., and Silva, M.K. 2011. *Bootstrapping Democracy: Transforming Local Governance and Civil Society in Brazil*. Stanford, CA: Stanford University Press.

Bell, D. 1980. "Introduction." In *The Computerization of Society: A Report to the President of France*, Simon Nora and Alain Minc, eds. vii–xvi. Cambridge, MA: The MIT Press.

Bordozinski, S. February 4, 2008. Facebook Used Against Colombia FARC with Global Rally. The Christian Science Monitor. Available at: *www.csmonitor.com/World/Americas/2008/0204/p04s02-woam.html*. Accessed August 23, 2013.

Bull, H. 1977/1995. *The Anarchical Society: A Study of Order in World Politics*. New York: Columbia University Press.

Cowhey, P.F. 1990. "The International Telecommunications Regime: The Political Roots of Regimes of High Technology." *International Organization* 44(2):169–199.

Earl, J. and Kimport, K. 2011. *Digitally Enabled Social Change: Activism in the Internet Age*. Cambridge, MA: The MIT Press.

Financial Times. (April 8, 2012) "BlackBerrys Flourish in the Malls of Lagos." Available at *www.ft.com/intl/cms/s/0/9ed9c5d0-71dd-11e1-90b5-00144feab49a.html#axzz1ref85xYc*. (accessed April 9, 2012).

Friere, P. 2000 [1970]. *A Pedagogy of the Oppressed*. 30th Anniversary Edition. New York: Continuum.

Frau-Meigs, D., Nicey, J., Palmer, M., Pohle, J., and Tupper, P. eds. 2012. *From NWICO to WSIS: 30 Years of Communication Geopolitics – Actors and Flows, Structures and Divides*. European Communication Research and Education Association Series. Bristol, UK: Intellect.

Haas, E.B. 1982. "Words Can Hurt You: Or, Who Said What to Whom about Regimes." *International Organization* 36(2): 207–243.

Habermas, J. 1989. *Structural Transformation of the Public Sphere: An Enquiry into a Category of Bourgeois Society*. Cambridge, MA: The MIT Press.

Horwitz, R.B. 1989. *The Irony of Regulatory Reform: The Deregulation of American Telecommunications*. New York: Oxford University Press.

Hurrell, A. 2007. *On Global Order: Power, Values, and the Constitution of International Society*. Oxford: Oxford University Press.

Internet Governance Forum (2009). *About the Internet Governance Forum*. *www.intgovforum.org/cms/index.php/aboutigf* (accessed January 5, 2012).

Knight, F. 1921. *Risk, Uncertainty, and Profit*. Boston: Houghton Mifflin.

Lerner, D. 1958. *The Passing of Traditional Society*. Glencoe, Il: Free Press.

Livingston S. and Walter-Drop, G. eds. 2014. *Bits and Atoms: Information and Communication Technology in Areas of Limited Statehood*. Oxford: Oxford University Press.

MacBride Commission. 1980. *Many Voices, One World: Towards a New, More Just, and More Efficient World Information and Communication Order*. Lanham, MD: Rowman & Littlefield.

Mansbridge, J., et al. 2010. "The Place of Self-Interest and the Role of Power in Deliberative Democracy." *Journal of Political Philosophy* 18(1): 64–100.

Martinez, I. "Romancing the Globe." *Foreign Policy* (November 1, 2005).

Meltzer, A.M., Weinberger, J., and Zinman, M.R. 1993. *Technology in the Western Political Tradition*. Ithaca, NY: Cornell University Press.

Mueller, M.L. 1998. *Universal Service: Competition, Interconnection, and Monopoly in the Making of the American Telephone System*. Washington, DC: American Enterprise Institute.

Mueller, M.L. 2002. *Rooting the Root: Internet Governance and the Taming of Cyberspace*. Cambridge, MA: The MIT Press.

Mutz, D.C. 2006. *Hearing the Other Side: Deliberative versus Participatory Democracy*. Cambridge, UK: Cambridge University Press.

Niemann, A. "Beyond Problem-Solving and Bargaining: Communicative Action in International Negotiations." *International Negotiation* (December 2006).

Prakash A. and De, R. 2007. "Importance of Context in ICT4D Projects: A Study of Computerization of Land Records in India." *Information, Technology and People* 23(3): 262–281.

Pinch, T.J. and Bijker, W.E. 2012. "The Social Construction of Facts and Artifacts: Or How the Sociology of Science and Sociology of Technology Might Benefit Each Other." In W.E. Bijker, T.P. Hughes, and T. Pinch, eds. *The Social Construction of Technological Systems" New Directions in the Sociology and History of Technology*. Cambridge, MA: The MIT Press.

Risse, T. Winter 2000. "'Let's Argue!': Communicative Action in World Politics." *International Organization* 54(1): 1–39.

Ruggie, J.G., ed. 1993. *Multilateralism Matters: The Theory and Praxis of an Institutional Form*. New York: Columbia University Press.

Scatamburlo, D,V., Suoranta, J., Jaramillio, N., and McLaren, P. (November 2006) "Farewell to the 'Bewildered Herd': Paulo Freire's Revolutionary Dialogical Communication in the Age of Corporate Globalization." *Journal of Critical Education Policy Studies* 4(2) 1–14. Available from *www.jceps.com/index.php?pageID=article&articleID=65*.

Schiller, D. 1982. *Telematics and Government*. New York: Ablex Publishing.

Scott, J.C. 1998. *Seeing Like a State: How Certain Schemes to Improve the Human Condition Have Failed*. New Haven, CT: Yale University Press.

Sharp, P. 2009. *Diplomatic Theory of International Relations*. Cambridge, UK: Cambridge University Press.

Shirky, C. 2008. *Here Comes Everybody: The Power of Organizing Without Organizations*. New York: The Penguin Press.

Singh, J.P. 2008. *Negotiation and the Global Information Economy*. Cambridge, UK: Cambridge University Press.

Singh, J.P. 2011a. *United Nations Educational Scientific and Cultural Organization: Creating Norms for a Complex World*. London: Routledge.

Singh, J.P. 2011b. *Globalized Arts: The Entertainment Economy and Cultural Identity*. New York: Columbia University Press.

Singh, J.P. and Alimchandani, S. 2003. "Development as Cross-Cultural Communication: Anatomy of a Development Project in North India." *Journal of International Communication* 9(2): 50–75.

United Nations Development Program. 2001. *Human Development Report 2001: Making Technologies Work for Human Development*. New York: Oxford University Press.

Part IV
Environments

13

Green Energy, Public Engagement, and the Politics of Scale

Roopali Phadke

MACALESTER COLLEGE

Introduction

Concerns about climate change, energy security, and green jobs have produced a resounding public call for renewable energy development in the U.S. Consequently, federal investments in renewable energy research and development have multiplied at an unprecedented rate. These investments presume broad public support for the new green energy economy. Yet national opinion polls often mask strong local resistance to the installation of new energy projects, including hydro, solar, biofuels and wind. Communities across rural America are challenging the transparency and accountability of the government decision-making that is vastly resculpting energy geographies. The news media has termed this conflict the "green civil war" between climate change mitigation and landscape preservation goals.

Green energy programs have focused mainly on innovation and investment pipelines with little attention to how the risks and benefits of a new clean energy economy are being redistributed across society. In particular, rural communities at the forefront of new energy development are asking why they are disproportionately expected to carry the burden of a low-carbon future while urban residents continue their conspicuous use of energy. Wind energy opposition groups are concerned about economic, wildlife, health and visual impacts. This opposition is in part the consequence of outdated modes of public participation that fail to fully capture the socio-cultural dislocations and vulnerabilities that occur when vast new energy geographies emerge. Although national and local governments are rapidly investing in scientific and technological innovation, little attention has been given to constructing community-based research models that address the challenges of the new energy economy. Such models are necessary if residents and policy makers are to recognize, understand, and weigh carefully the opportunities and risks that accompany these landscape transformations.

These political responses and debates come as no surprise to those who study technology decision-making. Despite decades of analysis in Science and Technology Studies (STS) of the need for participatory technology assessment and community-based research (see section II), scholars are still at a loss for how to methodologically offer interventions. How do we actually *do* community-based research that matters to local communities and policy makers? Answering this question requires serious reflection on the social, emotional and temporal contours of action-research engagements.

This chapter describes how local communities are struggling over the meaning, scale and impact of wind energy transitions.[1] Wind energy thus provides the backdrop for my main goal: to illustrate a new model for STS action-research that I call "place-based technology assessment" (PBTA). PBTA has the potential to enhance the scope of public debate and public participation. In particular, PBTA enables researchers to develop placemaking strategies that connect local policy actors with social movement demands through multisite, multiscale research collaborations.

The case study I report on here is based on a collaborative field research project conducted by my research team in Michigan.[2] I describe three distinct but connected phases of empirical research that were aimed at registering public concerns and opinions about the emergence of wind energy on the landscape. These included a community survey, a landscape symposium and a set of Town Hall meetings. The research aimed at providing local government officials with recommendations for developing wind energy regulations based on community preferences. The STS literature presents many models for engaged scholarship (see later in the chapter). In the following section I connect our research with previous analyses of "citizen science" and "anticipatory governance." These bodies of work have much in common, but as I demonstrate, are rarely linked in thinking about research methodologies for developing action research.

STS public engagement scholarship

Decades of STS scholarship has examined the history and politics of energy and environment planning, including debates over the impacts of visible and invisible hazards like power plants and nuclear waste. This literature has focused on why and how infrastructures emanate from state organizations and take root in communities (Hughes 1983, Nelkin 1992, Nye 1992, Hecht 1998). I am interested in connecting research on technoscience controversies with emergent STS scholarship on public engagement, specifically citizen science and anticipatory governance initiatives. While these bodies of work share similar normative goals for extending public participation, this scholarship has been only loosely connected. In this literature review, and the empirical research description that follows, I bridge these bodies of research toward developing a new model for grassroots anticipatory governance.

The "participatory turn" in STS scholarship has meant two important shifts in practice. First, as scholars embrace a more normative approach to science and technology policy research, they have been increasingly working in partnership with their subjects. Whether examining the work of breast cancer activists or "fenceline" communities facing toxics contamination, STS scholars reflexively cite the political commitments that guide their practice (Hess 2001, Forsythe 2001, Brown 2007). Second, STS scholars are also increasingly being called upon to act as "practitioners, organizers and evaluators" of engagement exercises (Delgado et al. 2011: 827). As Davies and Selin describe, playing the double role of "practitioners and researchers" raises a set of uncharted frustrations, challenges and tensions for STS scholars (2012: 121).

The scholarship on citizen science, including citizen-scientist alliances, has developed a rights-based, social justice inspired discourse (Coburn 2005, Mamo and Fishman 2013). A generation of research has focused on thick descriptions of the work of environmental health social movements. These scholars practice a mode of multisited ethnography based on participant observation, interviews and ethnographies of institutions, events and communities (Brown 2007, Hess 2001). Grassroots case studies of "bucket brigades," street science and popular epidemiology have documented the coming together of local knowledge and expert science in the interest of environmental and procedural justice concerns (Ottinger 2012, Ottinger and Cohen 2011). Researchers have drawn attention to how community activists set research agendas and advocate for the knowledges that should and should not be produced (Frickel et al. 2010).

While citizen science scholars examine partnerships, and often work in collaboration with their informants, they have rarely adopted participatory or action research modes where they are orchestrating engagement exercises. Citizen science scholars also seldom intervene in direct policy making. Instead, this scholarship has focused on recording and analyzing the obstacles, as well the methodological solutions, social movements and scientists face when challenging the status quo. I do not intend to minimize the inspirational work of these scholars. For example, Phil Brown's studies on toxic exposures and "popular epidemiology" for/in/with environmental health justice movements has been ground-breaking (2007). Morello-Frosch and Brown's chapter (Chapter 28 in this book) on community-based participatory research continues this trajectory. Similarly, Sabrina McCormick's studies on breast cancer activism and citizen science after the BP oil spill, have focused on developing a community-based research model for examining environmental health justice movements (McCormick 2009, 2012). These scholars describe how participatory projects can serve the interests of science, advocates and policy makers. Their work parallels, but never directly intersects with, STS scholarship on the orchestration of public deliberation through techniques like consensus conferences, citizen juries, and deliberative polls. I describe this literature below.

In contrast to citizen science scholarship, anticipatory governance research has emerged in the last decade as a model for considering the role of public engagement in technological futures. As Barben et al. describe, anticipatory governance seeks to enhance "the ability of a variety of lay and expert stakeholders, both individually and through an array of feedback mechanisms, to collectively imagine, critique, and therefore shape the issues presented by emerging technologies before they become reified in particular ways" (2008: 993). As it has matured, anticipatory governance has involved a range of techniques for engaging lay and expert publics in deliberating concerns, challenges and opportunities for policy interventions. These include public events, scenario planning with local leaders, museum outreach and consensus conferences (Davies and Selin 2012).

The anticipatory governance scholarship is part of the broader STS interest in experimenting with the "real-life examples of participatory science" at work in consensus conferences, citizen juries, participatory budgeting and science shops (Irwin 1995, Sclove 1995, Guston 1999, Smith and Wales 2000, Brown 2006, Kleinman 2007 et al.). STS analysts have been actively constructing and evaluating exercises, including national scale state-sponsored deliberations, such as the recent GM Nation and Stem Cell Dialogues in the UK (Reynolds and Szerszynski 2006, Mohr and Raman 2009). This research has also documented the spectrum of meanings that are attributed to public participation activities, from constrained public consultation to engaged "upstream" decision-making about problem framing and technological design (Brown 2006, Wilsdon and Willis 2004). It also draws attention to how shifts toward greater public deliberation in science and technology policy can come at a cost to broader notions of citizenship and the public good (Irwin 2006, Jasanoff 2005, Stirling 2008).

Yet this form of interaction is not fully institutionalized, nor are there a wide range of models. Three American think-tanks and research centers are committed to advancing anticipatory governance; the Loka Institute, the Stanford Center for Deliberative Democracy, and the Arizona State University Consortium for Science, Policy and Outcomes. While these organizations have sponsored a range of deliberative dialogues, particularly recently on the societal dimensions of nanotechnology and energy ethics, these efforts often focus on prioritizing public values on national scale policy issues, as well as technologies that remain largely invisible and immaterial to publics. Davies and Selin describe the challenges around engaging publics in nanotech futures, considering the uncertain effects of "world-shaping technologies and their as-yet unknown impacts" (2012: 125). They also contrast public comfort with discourses and discussions of global energy disasters and risks, with the public's lack of familiarity with the mundane, obdurate realms of local energy use. This may suggest why anticipatory governance approaches have had less impact on how government agencies approach the everyday, local politics of energy planning. Routine decisions, like environment impact assessment and land use planning, are rarely thought of as transformative sites of technoscience development. It is these issues that spark environmental justice movements into formation and where new forms of STS informed grassroots participatory research interventions hold significant promise.

The gap between citizen science and anticipatory governance scholarship can be bridged through a demand-side, grassroots model of engaged STS research that responds to the concerns of local policy actors as well as environmental justice movements. As Delgado et al. have argued, we need not focus on either the top-down organized state-sponsored processes or bottom-up grassroots phenomena (2011). We can attend to both invited and uninvited models of public engagement. By combining multisited ethnography with public engagement exercises, our wind energy research is situated at this crossroads in theory and practice. This model of community-based anticipatory governance may be especially well suited for asking questions about the scales at which technoscience has meaning for local communities. I call this model place-based technology assessment (PBTA) because it privileges the ways that local policy actors and social movement organizations work at "place making." Place making is defined by practitioners in human geography, urban planning and other allied disciplines as the process of asking questions of people who live, work and play in particular spaces to discover their values, needs and hopes for the future.[3]

Energy and society research

Shifting the American economy toward renewable and responsible energy policy requires a massive effort to gain a level of social and political assent that is unparalleled in the postwar era. To date, most political efforts have focused on innovation pipelines and economic incentives. For example, the Apollo Alliance, a labor and environment coalition, concentrated its efforts on raising the $500 billion investment it believes is necessary to build a "green" energy economy over the next ten years. In another example, the American Wind Energy Association and the Solar Energy Industries Association published their "Building a Green Power Superhighway" report to guide federal policy making on the green energy transition. These industries employ the highway analogy to signal the need for a massive federal mobilization akin to that which built the American interstate highway system in the mid-twentieth century, a sociotechnical system that was created by eminent domain rather than through deliberative democracy. Industry representatives rarely consider the disturbing resonances of this analogy, particularly the severe social dislocations the interstate highway system occasioned for racial and ethnic minorities, as well as the major economic declines that towns on parallel but smaller routes experienced (Lewis 1997, Rose 1990).

Unlike the development of transportation and energy infrastructure in the postwar period, the new energy economy presents administrative agencies with unprecedented challenges for engaging publics. First, the implementation of these clean energy technologies is bearing little resemblance to the decentralized and distributed generation models that were popularized in the alternative energy, "small is beautiful" counterculture. The large-scale development of renewable energy anticipated by the Department of Energy will require massive reshaping of rural landscapes through new transmission corridors and new electricity generating projects.

Second, new policy regimes, including those being currently debated in the U.S. Congress around climate change, call for a far greater level of transparency and accountability to the American public than was ever required of comparable industries in their path-breaking stages. Third, social movement organizations, operating within a highly networked information economy, are challenging and blocking projects at the earliest stages of planning, even before regulatory frameworks get triggered into action. These movements are not waiting to voice their concerns through conventional public hearings and post-project law suits. Instead, they use social networks to broadcast their message via alternative and mainstream media to create a political storm policy actors cannot ignore.

These shifts in public meanings and engagements with energy infrastructure decision-making have raised expectations of better public engagement mechanisms. American public agencies remain loyal to administrative tools that came into being in a post-World War II political culture. The legal regimes that direct public participation, such as outlined in the 1969 National Environmental Policy Act's requirement for impact reviews, have created a "decide, announce and defend" approach to participation that excludes the substantive and creative engagement of citizen perspectives (Hendry 2004). These mechanisms also fail to legitimate the personal, emotive and culturally-situated forms of political rationality that get activated when local landscapes are at stake.

Similarly, the extensive political science and communications scholarship on community participation in American environmental decision-making has identified many of the shortcomings of traditional state-driven "notice and comment" engagement processes, such as open houses, public hearings and comment periods (Depoe et al. 2004). It also describes a desperate need for new modalities of public engagement that breathe life, and bring greater transparency, accountability and humility, into the staid politics of project planning (Poisner 1996, Carr and Halvorsen 2001).

The shortcomings of extant models of participation are evident across a wide range of renewable energy projects, including solar, geothermal and biomass where local concerns about environmental and aesthetic issues are evident. Wind energy presents even more challenges to local communities and to extant forms of governance. First, many U.S. states are relying almost solely on wind power to make their renewable energy targets because it is cost competitive, readily deployable and scalable. At that same time, wind energy facilities are extremely difficult to disguise and mitigate; their giant towers and expansive footprints are highly visible. While hydro and nuclear power plants are spatially concentrated, a utility-scale wind project can involve hundreds of individual turbines, each impacting a unique viewshed. Third, the anti-wind opposition movement is not only blocking individual projects but also moving upstream to effect broad legislative work through the enactment of restrictive local zoning codes and development moratoria. For these reasons, wind energy is proving to be one of the most contentious new energy sectors.

Given the shortcomings of extant governance models and STS analysts' emphases on the need for approaches to public engagement that bring greater transparency, accountability and humility to the staid politics of project planning, STS scholars have an opportunity to extend their insights and interests in participation toward community-based efforts that interface with pivotal issues in American energy administrative decision-making. In this context, I develop the model of place-based technology assessment by drawing on STS insights to examine the "social gap" between what public opinion polls tell us about national support for wind energy and the realities of local opposition (Bell et al. 2005). While the popular press often attributes this social gap to NIMBY politics (not-in-my-backyard), wind energy scholars have aimed to tell a more complex story. These analyses suggest that community concerns for the siting of wind energy are often expressions of concern about the changing character of place, as well as a sense of disenfranchisement over planning decisions, and can be a response to misconduct by developers who site projects from a great distance with little concern for how these new technological systems interact with livelihoods and landscapes (Bell et al. 2005, Barry et al. 2008, Phadke 2011).

Likewise, despite these realizations, the wind policy scholarship has been less attentive to ways to enfranchise local communities in energy siting and development practices. The place-based technology assessment model I describe works in close collaboration with local government clients. The research I have undertaken with my collaborators has combined qualitative and quantitative methodologies, such as using opinion surveys alongside consensus conference-like citizen symposia, to reveal spatialized understandings of the new vulnerabilities and opportunities that are being coproduced with the renewable energy economy. This research asks local residents: What are your main concerns about wind energy's landscape and livelihood impacts? Are these concerns based on experience "living with" these technologies? Are certain technological configurations, locations and scales of development more favorable than others? Are mitigation options available and acceptable? The research also aims to contribute to a set of broader STS concerns that relate to balancing engagement processes and policy outcomes. The research asks: What are the appropriate "upstream" moments to engage communities in articulating their hopes and fears with emerging technologies? What kinds of research methods and facilitation efforts nurture broad, inclusive and thoughtful public engagement?

The following section presents a case study from the state of Michigan. In this instance, our research group was hired by the elected leaders of six townships to provide research on public concerns about wind energy development. I present background on Michigan's energy economy, describe our research method, and discuss the intellectual and personal challenges we experienced when simultaneously playing the roles of conveners, translators and evaluators. In these ways, the research has brought into vivid relief public values and expectations of neutral expertise.

Wind energy in western Michigan

The promise of a green energy based economic recovery has led many American states headlong into the pursuit of renewable energy. The state of Michigan, fraught with economic stagnation after a loss of industrial manufacturing, has invested heavily in a renewable energy future. These investments have been part of a four-billion-dollar-program to create green public and private sector investments for energy efficiency, mass transit and renewable energy (Pollin et al. 2008).

In 2008, Michigan passed the Clean, Renewable, and Efficient Energy Act that mandated that utilities in the state generate 10% of electricity from renewable sources by 2015. To meet this target, Michigan will need to produce 5,274 MW from renewable energy resources by 2015. Although Michigan has created a legally binding renewable portfolio standard (RPS) and the Department of Labor and Economic Growth has developed zoning guidelines for wind energy, the state has delegated zoning, permitting, and environmental review of wind energy projects to local governments and townships.

Wind energy has been the poster project for economic recovery in Michigan because it allows retooling of advanced manufacturing and engineering expertise linked to the auto industry. Federal studies have ranked Michigan among the top four states in industrial capacity to develop and manufacture wind energy systems. Michigan has also been identified as having significant wind resources – enough to generate more than 70% of Michigan's electricity.

Despite this potential, there are less than a dozen installed wind energy facilities in Michigan. In total, wind energy currently supplies just 0.3% of electricity in the state. Michigan continues to rely heavily on other sources of energy, most notably coal, nuclear, and natural gas. Because Michigan produces only about 30% of its electricity in state, it spends, on average, $26 billion annually importing energy from other states and countries.

The hype around wind energy has created high tensions among rural communities. In particular, Michigan's best wind resources are located either on or near Lake Michigan (see Figure 13.1). While jobs may be created for a Detroit-based new energy industry, those who

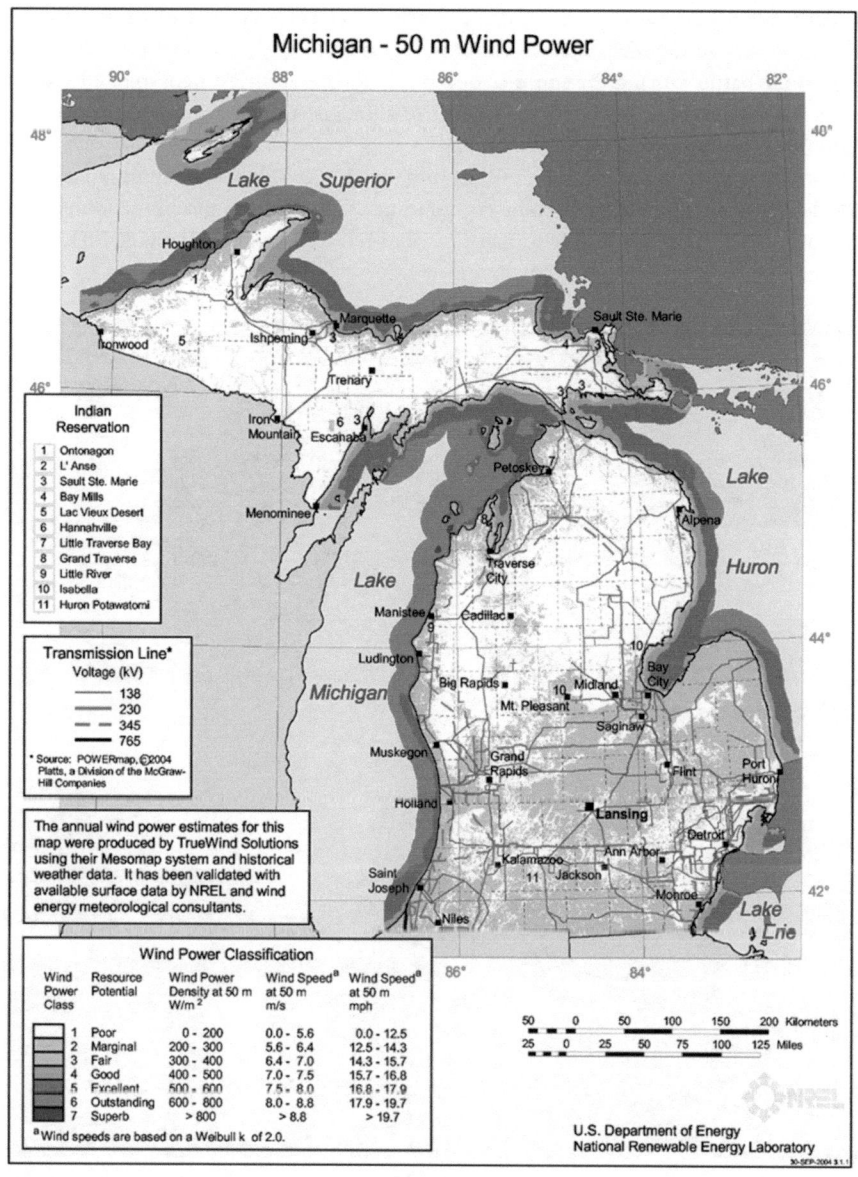

Figure 13.1   Wind resources in Michigan

depend on the tourist economy along Lake Michigan have feared the landscape and viewshed impacts associated with new energy infrastructures. The following case study describes our role in providing support to local governments in Michigan as they struggled to understand public opinion about wind energy.

This Michigan research was done in collaboration with an initiative called "Understanding Wind Energy." This initiative, funded by the C.S. Mott Foundation, established a committee of elected leaders from six townships across two counties in western Michigan. The initiative was a direct response to a proposed project by Duke Energy to place more than 160 turbines on ridgelines within a mile of Lake Michigan. As a local institution, the C.S. Mott Foundation has been expressly concerned with Great Lakes landscape preservation. This scale of development was particularly unsettling because not a single turbine had been erected in these counties before the project was announced. Debates about the merits of this project sent these communities into vitriolic battle. During the summer of 2011, billboard and yard signs greeted visitors to the region. The signs either heralded the loss of "pristine" beauty or promoted the promise of new economic development (see Figure 13.2).

I was approached by the Understanding Wind Initiative to provide background research on public attitudes toward wind energy in these communities. Our research group recommended, and was hired to implement, three overlapping phases of research. First, we provided responses

Figure 13.2   Billboards and yard signs present in summer 2011

to questions that were generated in a series of Town Hall meetings across these six communities. Second, we were contracted to create, implement, and analyze a community mail survey of residents' public opinions of wind energy. Finally, we held a citizen symposium with a small but representative sample of local residents. For each phase, we provided a draft report back to sponsors that included time for public comment.[4] This project was attractive to our research group because each phase allowed us to use a different research method for getting at the same question: how can wind energy policy respond to citizen concerns? I describe each of these phases and analyze our findings. While each phase presented us with a unique data set, by bringing the three phases together we were also able to provide our local clients with a more nuanced understanding of public concerns and ways forward.

Alongside these three phases of research, we were engaged in a broader ethnographic project to understand the contours of the wind energy debate in Michigan. This included interviews with local environment, planning and landscape preservation NGOs, participant observation at community events and fairs, phone and email interviews with members of local anti-wind organizations and driving tours with elected officials. We were particularly interested in understanding the land use history of the region and the recent demographic changes that have produced the more recently established tourist economy. In addition to deepening our understanding of the political ecology of the region, these methods allowed us to build trusting personal relationships with those involved in all sides of the controversy. While our clients expected us to provide "objectivity" – which was consistently a challenge in our conversations given our STS constructivist orientation – the leaders of the Understanding Wind Initiative said that our passion, knowledge and credibility were the reason they contracted with us.

Town Hall Forums Report

The townships involved in the "Understanding Wind Initiative" were tasked with writing new planning ordinances to guide wind energy siting and development. Elected officials were struggling to understand the range of topics that these ordinances needed to cover. They wanted to know what kinds of questions residents had about wind energy and how those concerns could be addressed through new legal ordinances.

Local residents were invited to attend any of the four wind energy Town Hall forums held in spring 2011 or to submit questions through email, fax, or mail. The notice for the forums and the questions was sent through email blasts, and was also picked up by local newspapers. At these forums, local conveners collected questions about wind energy development. All questions were taken straight from the source, word-for-word. In total, close to 100 different participants attended the forums, of which about a third submitted some form of a question. Additionally, 60 individuals submitted questions by email. The questions were compiled by local staff from the Manistee-based organized Alliance for Economic Success (AES). The completed notes and questions were sent to all of the forum participants through email to make sure the questions and ideas were being conveyed correctly. Only a handful of questions were adjusted.

Our research team received thematically categorized questions in summer 2011. We examined the published and grey literature to provide basic, and regionally appropriate, responses to a subset of the questions.[5] The team then identified and recommended to the Wind Energy Initiative Leadership Team topical experts who could provide additional expertise, particularly on acoustic ecology and township zoning laws.

More than 450 questions were received and organized into 38 thematic categories. These questions included issues from the very general, "How does the environmental impact of having wind energy compare with those of solar, coal, natural gas, oil, and nuclear?", to the very specific, "What percentage of people living within 1500 foot of a large industrial turbine reported sleep loss?" As would be expected many questions included passionate statements such as:

> Those who are grieving the loss of pristine panoramas, unspoiled natural acreage and loss of a peaceful, undisturbed environment will need to listen to and accept the mandate to support reasonable local development of clean energy. There will be losses for everyone, but greater gains for all are a hope worth striving for.

The questions served as a dataset for helping us understand how concerns were being articulated and knowledge gaps self-identified. In our analysis, we observed four important categories of questions (although there were 38 topical categories). First, many questions addressed the basics of energy infrastructure (such as "How much electricity is produced by foreign oil in the United States?"). Often, these were not about wind energy but attempts to compare and contrast a range of electricity alternatives. The questions also demonstrated that many residents simply wanted general information about the technical aspects of energy infrastructure. Hundreds of questions asked about electrical transmission, federal and state energy policy, and different types of electricity generation and consumption.

Second, residents displayed curiosity and confusion about the role of township governments in making legal decisions regarding wind energy. This occurred in the context of a "home rule" state, where permitting authority rests with local officials and not a state utility commission. For example, one resident asked "What type of authority do townships have to regulate and control wind farm developments?". Many people asked whether townships "are allowed" to create ordinances in their community regarding the height of towers, zoning of wind farms, and other public policy related to wind energy projects. Others inquired about how state or agency guidelines (such as the U.S. Fish and Wildlife Service recommendations) could be transformed into local law. Generally, the questions indicated a desire to understand township-level governance and a need for education about township planning processes, the extent of a township's legal authority, and the role of a master plan in guiding wind energy development.

Third, residents were concerned about preserving existing rural character. Many questions suggested that residents anticipated the worst possible outcome from wind energy development in the area. One question asked "What forms of renewable energy might respect the independence and pride of small rural, agricultural communities such as Frankfort?" In some cases, residents asked how warranted their concerns were. They asked whether or not wind energy "belonged" in places like Manistee or Benzie counties and worried that it wouldn't fit "the ways" of the community. Many residents had a very clear idea of how they wanted their communities to look and were concerned that wind energy development threatened that ideal.

Finally, we found questions about social and procedural justice dominated these submissions. These included "Are the landowners hosting the windmills being fairly compensated?" and "How can a township protect itself from a large corporation suing and bankrupting the township in order to get what it wants?" In these justice-oriented questions, citizens expressed several common emotions; one of which was a feeling of unfairness in the lack of transparency and integrity in the wind energy development process. In this context, several questions indicated a desire for more input from citizens, or asked about how they could get involved.

Our research team prepared a 110-page report over a six-month period that responded to these questions. This was a herculean task for a small research team that included just two faculty members and four undergraduate research assistants. Because there was a public comment period for the report, it took an additional three months to accept, respond and incorporate information provided by members of the public. We were routinely challenged, both by our clients and local citizens, to describe how we were presenting "unbiased, neutral" information. There was a great deal of concern among residents over our level of expertise, and one entire category of questions asked about the depth of our credentials and whether we could be trusted to be independent from the wind industry. Residents were also frustrated that scientific knowledge of many topics, including the impact of wind energy on wildlife or the public health impacts of exposure to low frequency vibrations, was contingent and continually evolving. They asked how township ordinances could be written in this context.

Community survey

In addition to the Town Hall forums, a second phase of the project involved writing and analyzing a community survey, with the goal of understanding how much general support and opposition there was for wind energy development in this region. We also aimed to break down support/opposition into a series of specific issues. Finally, we were interested in gauging how engaged local citizens were on the topic already. The Understanding Wind Initiative leaders who hired us were most interested in this phase of research because they simply had no sense of how widespread support or opposition was in their towns. This was the part of the project that involved the largest participant sample size and provided us the widest snapshot of the region. Compared with writing a question on a piece of paper in a Town Hall forum, it takes more time to complete a three-page survey, yet respondents had a limited range of responses to choose from in reply.

The survey and a pre-stamped return envelope were mailed to all property owners in the five township area (total of 7015). Of that, 1268 residents responded. While the response rate for the overall sample was 18%, it was as high as 28% in one township (Arcadia) and as low as 9% in another (Onekama). Sixty percent of respondents were male, and the mean age was 62. Of the respondents, 28% were seasonal residents. The residency status is important to emphasize given the role of tourism and a "second home" economy in the area. Seasonal residents come from the Chicago, Detroit, and Lansing area. Figure 13.3 displays survey results.

The survey also used a Likert scale to ask about the effects of wind energy on 23 different wind energy impact categories. Sixty-five percent of respondents told us their top three

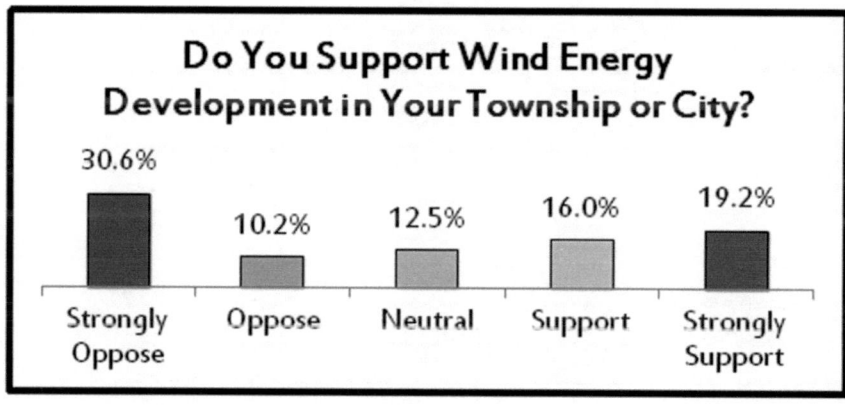

Figure 13.3 Data from the community mail survey

concerns were about impacts of wind energy production on local scenic beauty, property values and wildlife. On the other hand, nearly 70% reported that they perceived the main benefit as "producing clean energy." We were interested in whether respondents chose to qualify their support or opposition as "strong." More respondents labeled themselves as "strongly" opposing than "strongly" supporting. Figure 13.4 includes a sample survey from a respondent who self-identified as "strongly opposed." Across the 23 categories of potential impacts, those who "strongly" opposed tended to circle "very negative" across most of the categories. In addition to the surveys, our interviews and participant observation suggested that those who said they oppose wind energy seemed to be more deeply invested in their position, compared with those who told us they supported wind energy. This included a greater sense of frustration and anger about the nature of the political process around project development.

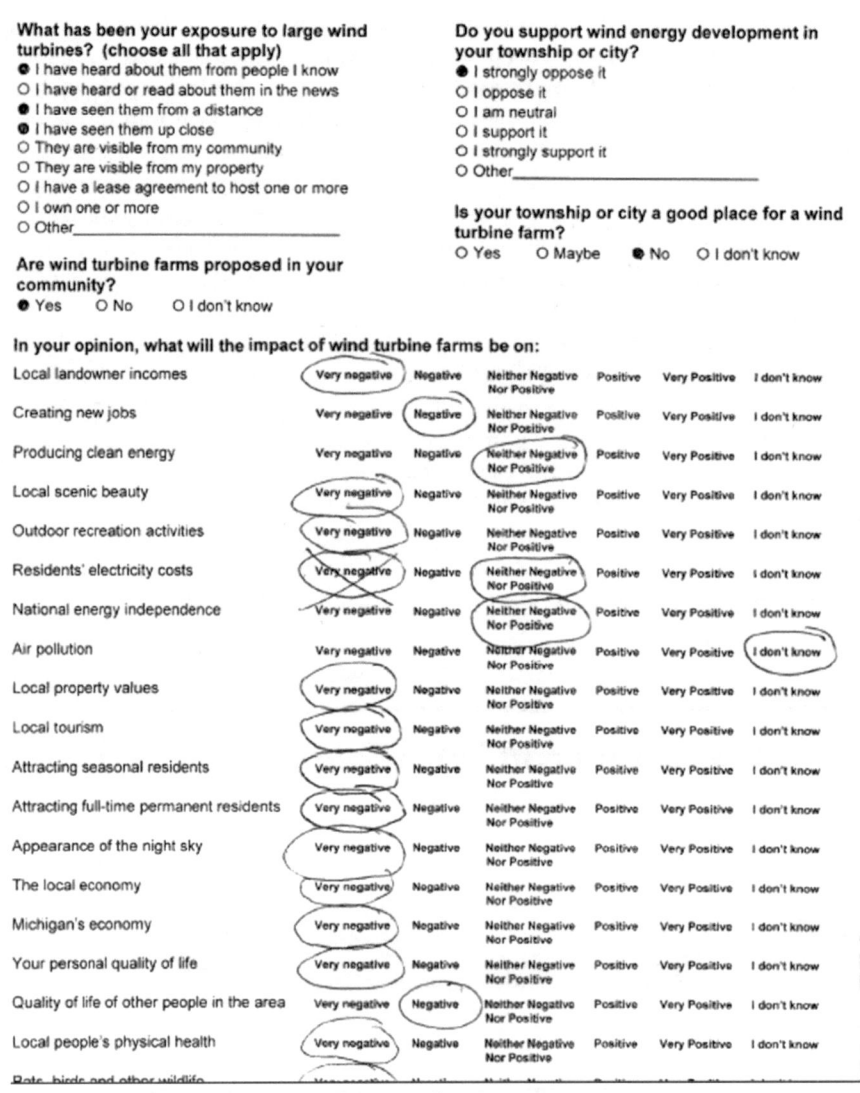

Figure 13.4 Sample survey from someone "strongly opposed"

It is important to note that wind energy opponents were far more organized and better resourced than wind energy supporters. Two vocal organizations, the Arcadia Wind Study Group and Citizens for Responsible Wind Development, led campaigns to write op-ed letters and blogs, and to host film screenings and house meetings. They continue to maintain an active website even though the main project they were fighting against has been canceled. In comparison, while there were citizen advocates for wind energy, they never developed a countervailing political machine.

We were also interested in the issues that respondents felt most confident and least certain about. As Figure 13.5 suggests, residents seemed certain they understood how wind turbines would negatively affect some aspects of their community quality of life, such as recreational activities and scenic beauty. They were more uncertain about personal economic impacts, such as potential increases in incomes or electricity costs. Our impression was that residents thought they understood the potential global benefits of wind energy and its local costs, while remaining skeptical about potential positive local impacts.

As with the questions report, the surveys themselves became material artifacts for examination. Many respondents wrote extensive comments in the margins about scale, subsidies, local character, siting, and clean energy. They also included photographs and drawings with their surveys. For example, one respondent wrote "I am in favor of small individually owned systems on your house or yard.", another told us "The country is far behind in wind energy and we need to get with it!", and another stated "If you build it we will sell the property or abandon it."

Landscape symposium

The last phase of research involved conducting a wind energy consensus conference that we titled a "wind energy landscape symposium." Participants were charged with recommending a set of regional best practice siting principles to township officials. Participants were recruited for the event through news releases, media, announcements at community meetings and advertisements on listservs. Interested participants were asked to submit an application form that asked basic questions about demographics and one question about wind energy attitude/perception. Using census data, we chose applicants aiming for balance across income, age, gender, and educational status. The wind opinion information was used to avoid a polarization of participants.

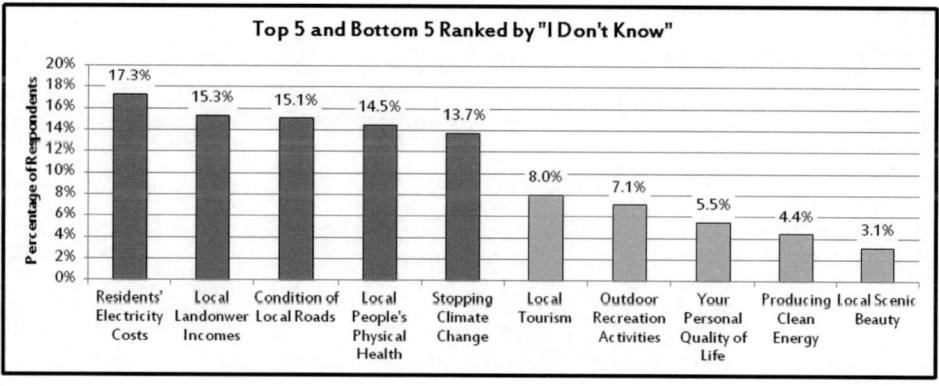

Figure 13.5   Confidence and uncertainty

Of the 21 participants, 66% were aged 50–64, 14% were aged 30–49, and 19% were aged 29 or under. Equal numbers of men and women were at the symposium. In addition, 76% of participants lived in the area full-time and 76% had lived in Michigan for 20 years or more.[6]

While this phase involved the smallest sample size, we were able to engage participants in a daylong event. This part of the research was funded through a separate NSF grant (SES #1027294). Participants received $100 stipends for their participation. The event was held at a regional medical center, which was selected to be equidistant from all participants homes and because it could serve as a neutral venue.

The symposium engaged participants in a range of deliberative exercises including real-time interactive keypad polling, photographic analysis, and computer visualizations of hypothetical projects. Through facilitated small group discussions and brief open-ended writing exercises, participants reflected on questions like: What are the most valued local landscapes? Are there places that should remain protected from new energy development? Are there landscapes/sites appropriate for wind development? How can new energy developments be designed to avoid negative impacts? How might those impacts be mitigated (visually, socially, economically)?

The symposium began with a photo essay depicting local landscapes and various forms of industrial and economic development infrastructure in Manistee and Benzie counties. This was meant to put the discussion of wind energy in the context of other types of historical and contemporary land use. It was followed by a morning pre-symposium keypad poll, which collected data about the participants' perceived concerns about wind energy development before the discussions began. This survey mirrored the mail survey we used. Following the keypad poll, participants shared their initial perspectives about the opportunities and challenges they thought wind energy development would bring.

Participants then broke into three smaller groups based on which part of the region they resided in. The small group activity had participants identify places in Manistee and Benzie counties on an aerial map that were most significant to them and record what qualities made those places meaningful and valuable. They then discussed the relevance of those places with each other. Figure 13.6 includes photographs from the event. A graphic artist was on hand to respond to the small groups and sketch different configurations of wind energy on local land-scapes. Figure 13.7 represents one such drawing requested by a participant who wanted to imagine many small-scale turbines in his town, rather than the 2 MW scale ones that had been proposed. This led to an interesting discussion of the visual, economic and energy impacts of many small towers, as opposed to a handful of very tall ones.

Figure 13.6 Images from the symposium

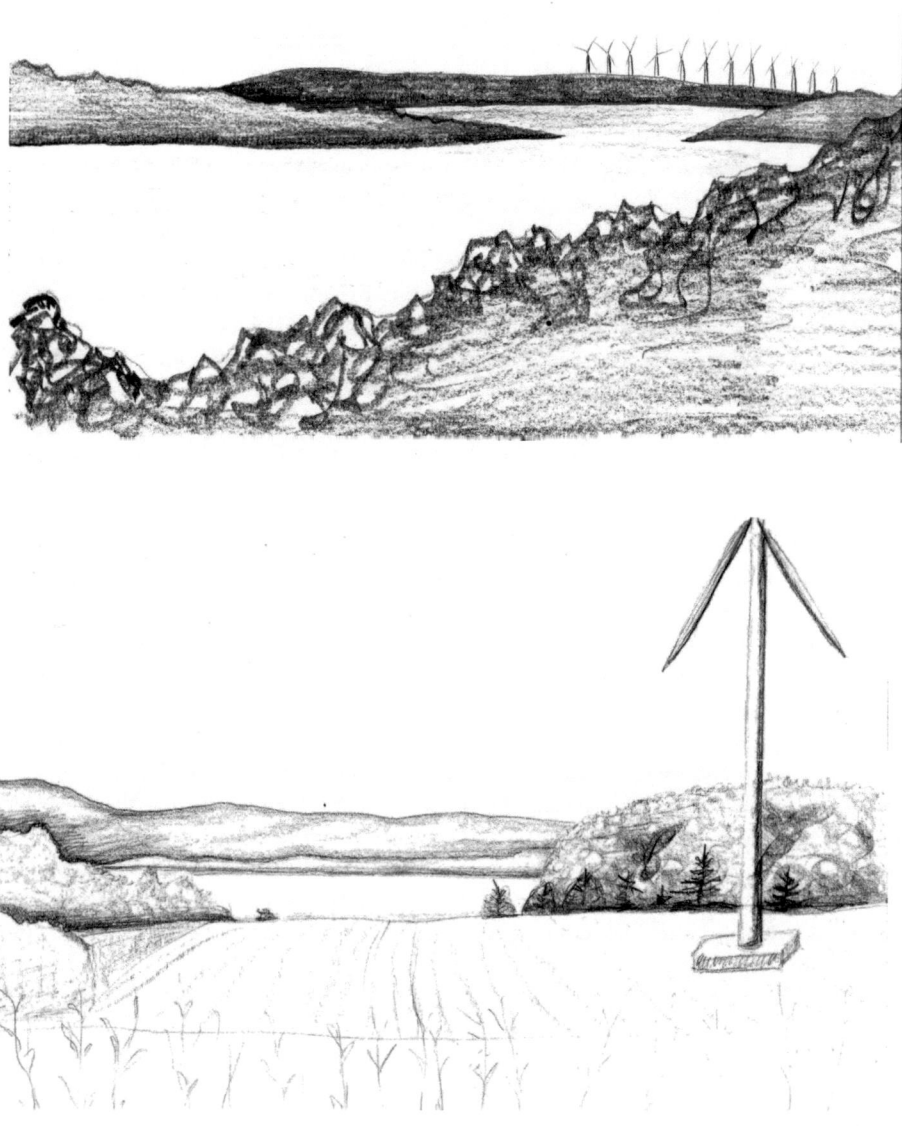

Figure 13.7 Sketches by graphic artist Emma Anderson made during the symposium

After a lunch break, participants heard from a stakeholder panel. Panelists were chosen in concert with the local leadership of the Understanding Wind Initiative to address the issues we thought were of greatest concern to participants. The panelists included representatives from Detroit Edison (a Michigan utility), a four-season resort owner with prior experience, a biology professor from Grand Valley State University, a member of a local concerned citizens group and a Native American tribal planner. Participants asked panelists about the development process, wildlife impacts, sound and health impacts, and appropriate setbacks.

After the panel, participants again broke into small groups for the afternoon discussion session. This part of the symposium sought to extend the conversation beyond landscape concerns and generate discussion of the wider socio-economic impacts of wind energy. An exercise involving three hypothetical wind projects was designed to get participants thinking about what aspects of wind development they liked and disliked for their communities. From there, each group created a list of several best design principles they would like developers and townships to consider. Finally, the full group reconvened and voted on a compiled list of best practice principles.

Our summary symposium report described our observations about public preferences. From the mapping exercise, we learned about which places in the two counties held the most meaning for residents, and therefore which should be protected from development. The visualizations created mixed reactions including firm opinions that no wind turbines should be built regardless of size and scale and others arguing that if they were to be built, power output should be maximized. The best practice principles were largely procedural, describing the need for greater transparency in decision-making and the need for independent studies about impacts.

Conclusion

While each of the three research phases (questions report, community survey and symposium) had its methodological strengths and weaknesses, when examined together this combination of qualitative and quantitative data provided a good snapshot of public concerns. The overall results found divergent concerns across and within townships. The citizen responses were also consciously or unconsciously based on the configurations of the pending project (in terms of its scale, location and process). Survey data suggested that yard signs and billboards may have exaggerated opposition. The survey showed there was still considerable support for some form of wind energy development. By analyzing the questions that were submitted at the Town Hall forums, we got the sense that there were also a substantial number of citizens in the middle who wanted to understand the contours of specific proposals and options other than wind energy.

Our client, the elected township officials who sponsored the research, reported to us that the data and analysis we provided was extremely useful in getting a better sense of public opinion but also in examining how to focus their limited resources in investigating town ordinances. One local elected leader wrote,

> As a member of the Understanding Wind Energy Initiative Leadership Team and Chairman of the Arcadia Township Planning Commission I want to thank and commend you on a very successful Michigan Wind Energy Symposium. I am confident that the knowledge gained will serve the participating townships, and all of Michigan, well as they proceed with master plan and ordinance development. While I learned a great deal observing the Symposium, I am looking forward to the final report and better understanding acceptance for wind energy systems in Northwest Lower Michigan. I also got some great ideas observing the Symposium exercises that I plan to implement for community roundtables in upcoming master plan workshops.

Township representatives and officials also learned about the topics around which there is still great uncertainty and need for additional information, research and/or education projects (such as basic energy science).

Our research team conducted two public presentations after the release of our study that reported back the results to community members. These were well attended and provided us a chance to answer questions face-to-face rather than over email. It also provided us a sense of closure. We encountered many hostile voices over our research period. These included those who questioned our role and expertise. Duke Energy officials openly criticized our research methodology, including directly undermining our research protocol on their website before any of findings were made public. Questions and threatening comments were often written to us and to the sponsors of our research, including officials at the National Science Foundation. One symposium participant wrote an op-ed letter in the local press chastising us as "pro-wind" posers.

Our public meetings at the end of the project incited very different reactions. We were literally given hugs and thanks for doing this work. Our clients told us that our presence in their communities created a "calming effect." More than one participant commented that our work had helped heal their communities by providing a third-party research team that was looking carefully at public opinion. One participant wrote to me,

> I am glad to have the opportunity to tell you how valuable I thought your process was for our community. I felt much lighter and very hopeful when I left. I have confidence that we can reach agreement about the wind turbines with most people.

This sentiment was also evident in the letter that a symposium participant presented to me at one of our public meetings. The letter read,

> Dear Dr. Phadke,
> At the end of the symposium I had written some comments – at that time I didn't have good feelings as a pro-wind supporter. In retrospect, I realize my discomfort with the process was probably spot on in gathering your data, and my comments were likely unfair to your process. I appreciate all the work it has taken to gather info. I had a bad attitude going into the symposium and felt outnumbered. I apologize... Had I a chance to do it again, I might try to be more open-minded going into the process.

We also tracked the policy impacts of this research after our research was complete. Individual townships did produce zoning ordinances to set height and density restrictions for development. The most meaningful result of our work extended beyond wind energy. The leaders of the UWI went on to create the Lakes to Land Initiative to address the need for a regional land use master planning process that erased municipal boundaries so that they may view the region as a whole. Fifteen townships joined together to develop a shared vision for rural development. Given how polarized townships were on the wind energy issue just two years ago, it is remarkable to see how the Lakes to Land Initiative has created food innovation districts and fisheries improvements in the region. The leadership has held dozens of visioning sessions that bring residents together to prioritize the future of their communities. Our research helped build the foundation for this initiative.

Contributions to STS action research scholarship

Decades of STS scholarship has produced many models for understanding and guiding public engagement. When it comes to recording the impact of these efforts, STS scholars have been concerned with better understanding both the instrumental policy outcomes and the substantive impacts of citizen learning. Our research collaboration demonstrates how engaged STS research can directly impact local policy, and illustrates the possibilities for broad social and institutional learning through such projects. How issues of scale impact both understandings of technological options, and opportunities for public engagement is of importance.

STS scholars have discussed how public engagement presents opportunities for anticipatory governance. For example, Barben et al. discussed the importance of lay citizen deliberation about the unarticulated assumptions of nanotechnology development and its governance. Barben et al. also describe how an anticipatory approach may "rethink the scope, scale and organization of STS research" (Barben et al. 2008: 980). They argue that "a multi-method, mission-driven, action-oriented research characterizes a potentially new form of STS research" (2008: 990). Described as upstream engagement, this form of STS research is seen as increasing the potential for citizens to shape decision-making, as well as increase transparency in the policy process.

Most attempts at anticipatory governance have focused on national scale policy discussions, about technologies that remain invisible and immaterial. What are the benefits of our place-based approach to anticipatory governance? First, we are focused on the everyday politics of land use planning. Our multi-method approach provided many different data points for thinking about how wind energy came to have meaning for citizens at different scales. This included struggles over the technical attributes of wind technology and the socio-political contours of governance. We also learned about the limited capacity of local decision-makers to sift through competing claims, particularly about what scale means for wind energy development. Zoning officials told us they struggled to understand the impacts of 12 2 MW turbines, compared with 25 500 KW turbines. They wanted to know which was better for the local tourist economy, for generating clean energy and for attracting new jobs? Our research provided resources to think through these questions, as well as pushed the issue out to local residents for their consideration.

Barben et al. have argued that the anticipatory governance approach implies an awareness of the co-production of sociotechnical knowledge and "richly imagining sociotechnical alternatives that inspire its use" (2008: 992). Our research reports provided this kind of rich detail. By combining data we collected from comments, surveys and deep dialogues, we reported back to policy makers about the kinds of wind energy their communities were likely to support and the deeply seated concerns they had about landscape transformations. This extended beyond just hard numbers indicating how many supported or opposed wind energy. Our data set included photographs, hand-drawn sketches, and the transcripts of heated personal discussions of the merits of a wind energy future. This pointed at new ways of thinking about how wind energy can be deployed so that it fits with the cultural character of these communities.

Scale is important not just for thinking about how large or how many wind turbines are on the landscape; it is also useful for thinking about methods for inviting public participation. We attempted different scales of citizen engagement to expand the reach and depth of our inquiry. Our survey reached all local property owners. In comparison, our symposium brought together a carefully selected group of 21 citizens. By scaling up and down research methods, we were able to respond to the interests of local policy makers for an aggregate understanding of concerns, as well as a microview into the kinds of conversations neighbors were having with each other.

Not every technoscience decision requires this scale of citizen engagement. However, Richard Sclove has argued that this level of engagement is important when a technoscience development promises to fundamentally or enduringly affect social life (1995). Installing a wind turbine on the landscape is a transformative political act. For some residents, the coming of wind energy signified the ability to save the family farm. Each additional turbine significantly added to their annual income from potential lease payments. For others, even a single turbine was deemed an industrial intrusion on the landscape that could cause economic harm to those who were dependent on a tourist economy. Our research needed to go beyond simply reporting public opinion. We needed to forge a way forward for thinking about how this technological intervention might be co-producing society itself. The co-production of a wind-energized landscape is fundamentally different from deliberations over nanotechnology or synthetic biology. Wind energy is about place politics, and thus demands a place-based approach toward technology assessment. While the benefits of wind energy may be global, the impacts to livelihoods are geographically concentrated. Governance issues are thus more localized as well.

Barben et al. also ask how STS scholarship can respond to the generous invitations of policy makers to partake in, and describe with integrity and credibility, the work of action research. Our answer to this question draws inspiration from the citizen science scholarship about how to go about working in partnership with communities on issues local residents deem worthy of examination. For us, this meant engaging in multi-sited ethnographic groundwork that supports public engagement. It was vital that our research method involved extensive interviews with local stakeholders, observation and collaboration with social movement organizations (including the anti-wind groups), and an investment in understanding the environmental history of the region. We could then tailor our engagement exercises to reflect community concerns. This also allowed us to understand what was at stake, essentially the meaning-making work that was happening, as residents came to their positions around wind energy.

We took away one other important lesson from this collaborative research project. Given the great mistrust residents have of the energy industry, it was vital that the funding for the research came from public and private foundations. For other communities facing similar challenges over energy development, the "Understanding Wind" approach may be a model for public engagement that gets upstream of conflict. Foundation support provided us a source of legitimacy when we were accused of being shills of either the industry or anti-wind organizations. More importantly, the residents, government clients and our foundation supporters were all interested in helping support a research project whose goals extended beyond just their immediate communities. They have been as focused as we have on learning about public engagement techniques that help heal communities in conflict over technological futures.

After our Michigan case study, our research group went on to conduct similar studies in three other states working in collaboration with county level governments. Our comparative results demonstrate even further how important scale is to the politics of new energy development. Each community we worked with had very different desires and concerns about the scale of wind energy that was "right" for their community. As a result of our research, we recommend that developers and planning officials carefully assess public land use preferences, rather than rely on a blueprint approach for energy development that fails to recognize the spatial politics of participation, local knowledge and perceptions of risks.

My enthusiasm for STS action research is not naïve to the challenges cited by STS scholars (Delgado et al. 2011). This literature discusses the competing normativities at work in public engagement practice, including tense characterizations of the public as "a feckless mass whose actions must be changed or responsible stakeholders who should be consulted" (Davies and Selin 2012: 130). To their list, I would add that working with political leaders has a disciplining effect on the goals of research. We may be limited to asking those questions deemed useful and actionable. We also have to live with the practically strategic, but ethically questionable, portrayal of ourselves as "neutral" interveners in controversy studies. Finally, in the Michigan case, we learned that engaged research fundamentally changes our relationship to communities. Although our research contract is over, we have become a part of the community we study and implicated in the choices they make for their futures. We do not take this long-term responsibility lightly. The

243

opportunities to revisit these communities in the future rests on our ability to stay relevant to, and invested in, these partnerships.

Notes

1   For another chapter that takes up the matter of scale in STS scholarship, see Kinchy (Chapter 14 in this book) on natural-gas fracking.
2   My debts to the members of my Macalester-based research team for their extensive work on this project: Dr. Christie Manning, Brianna Besch, Ava Buchanan, Natalie Camplair, and Erica De Jong. Another six students contributed to the data entry tasks. I also thank Daniel Kleinman, Kelly Moore and Abby Kinchy for their constructive comments on drafts of this chapter. I also acknowledge support for this research from the STS Program NSF Grant (SES #1027294).
3   This is the definition offered by the nonprofit Project for Public Spaces. See their explanation at *www.pps.org/reference/what_is_placemaking.*
4   All of these individual reports are available at *www.macalester.edu/understandingwind.*
5   The term "grey" literature is used by library and information experts to refer to informally published material, such as academic, NGO, government committee or industry reports, that are an important and timely source of information for researchers yet difficult to trace through conventional publication channels.
6   Additional demographic information about the participant pool is available in our symposium report. Accessible at *www.macalester.edu/understandingwind.*

References

Bell, D., Gray, T., and Haggett, C. 2005. "The Social Gap in Wind Farm Siting Decisions." *Environmental Politics* 14(4): 460–477.

Barben, D., Fisher, E., Selin, C., and Guston. D.H. 2008. "Anticipatory Governance of Nanotechnology: Foresight, Engagement and Integration," in E. Hackett et al. (eds), *The Handbook of Science and Technology Studies.* Cambridge: MIT Press. Pp. 979–995.

Barry J., Ellis, G., and Robinson, C. 2008. "Cool rationalities and hot air." *Global Environmental Politics* 8(2): 67–98.

Brown, M. 2006. "Survey Article: Citizen Panels and the Concept of Representation." *Journal of Political Philosophy* 14(2): 203–225.

Brown, P. 2007. *Toxic Exposures: Contested Illnesses and the Environmental Health Movement.* New York: Columbia University Press.

Carr, D.S. and Halvorsen, K.E. 2001. "An Evaluation of Three Democratic, Community-based Approaches to Citizen Participation: Surveys, Conversations with Community Groups, and Community Dinners." *Society and Natural Resources* 14: 107–126.

Corburn, J. 2005. *Street Science: Community Knowledge and Environmental Health Justice.* Cambridge: MIT Press.

Davies, S.R., Selin, C., Gano, G., and Pereira, Â.G. 2011. "Citizen Engagement and Urban Change: Three Case Studies of Material Deliberation," *Cities* 29(6): 351–430.

Davies, S.R. and Selin, C. (2012). "Energy Futures: Five Dilemmas of the Practice of Anticipatory Governance." *Environmental Communication: A Journal of Nature and Culture* 6(1): 119–136.

Delgado, A., Kjølberg, K.L., and Wickson, F. 2011. "Public Engagement Coming of Age: From Theory to Practice in STS Encounters with Nanotechnology." *Public Understanding of Science* 20(6): 826–845.

Depoe, S.T., Delicath, J.W., and Elsenbeer, M.A. (eds). 2004. *Communication and Public Participation in Environmental Decision-making.* Albany: SUNY Press.

Forsythe, D. 2001. *Studying Those Who Study Us.* Stanford: Stanford University Press.

Frickel, S., Gibbon, S., Howard, J., Kempner, J., Ottinger, G., and Hess, D.J. 2010. "Undone Science: Charting Social Movement and Civil Society Challenges to Research Agenda Setting." *Science, Technology & Human Values* 35(4): 444–473.

Guston, D.H. 1999. "Evaluating the First U.S. Consensus Conference." *Science, Technology & Human Values* 24(4): 451–482.

Hecht, G. 1998. *The Radiance of France: Nuclear Power and National Identity After World War II.* Cambridge: MIT Press.

Hendry, J. 2004. "Decide, Announce, Defend," in S.P. Depoe, J.W. Delicath and M.A. Elsenbeer (eds),

Communication and Public Participation in Environmental Decision Making. Albany: SUNY Press, Pp. 99–111.

Hess, D. 2001. "Ethnography and the Development of Science and Technology Studies," in P. Atkinson et al. (eds), *Sage Handbook of Ethnography.* Thousand Oaks, CA: Sage Publications.

Hughes, T. 1983. *Networks of Power: Electrification in Western Societies, 1880–1930.* Baltimore: Johns Hopkins University Press.

Irwin, A. 1995. *Citizen Science.* London: Routledge Press.

Irwin, A. 2006. "The Politics of Talk: Coming to Terms with New Scientific Governance." *Social Studies of Science* 36(2): 299–320.

Irwin, A. and B. Wynne. 1996. *Misunderstanding Science.* Cambridge: Cambridge University Press.

Jasanoff, S. 2005. *Designs on Nature: Science and Democracy in Europe and the United States.* Princeton: Princeton University Press.

Kleinman, D.L., Powell, M., Grice, J., Adrian, J., and Lobes, C. 2007. "A Toolkit for Democratizing Science and Technology Policy: The Practical Mechanics of Organizing a Consensus Conference." *Bulletin of Science, Technology & Society* 27(2): 154–169.

Laird, F. 1993. "Participatory Analysis, Democracy, and Technological Decision Making." *Science, Technology & Human Values* 18: 341–361.

Lewis, T. 1997. *Divided Highways.* New York: Penguin Books.

Mamo, L. and Fishman, J. 2013. "Why Justice? Introduction to the Special Issue on Entanglements of Science, Ethics, and Justice." *Science, Technology & Human Values* 38(2): 159–175.

McCormick, S. 2009. *No Family History: The Environmental Links to Breast Cancer.* New York: Rowman & Littlefield.

McCormick, S. 2012. "After the Cap: Risk Assessment, Citizen Science, and Disaster Recovery." *Ecology and Society* 17(4): 31.

Mohr, A. and Raman, S. 2009. "Capturing the Public or Evoking the Moral Codes of Science? Reflections on the Politics of Public Engagement". Working Paper for Science and Democracy Network (SDN) Annual Conference, June 29–July 1, 2009. Harvard University, Cambridge, MA.

Nelkin, D. 1992. *Controversy: Politics of Technical Decisions.* Newbury Park, CA: Sage. (3rd edition).

Nye, D. 1992. *Electrifying America: Social Meanings of a New Technology, 1880–1940.* Cambridge: MIT Press.

Ottinger, G. 2012. "Changing Knowledge, Local Knowledge, and Knowledge Gaps: STS Insights into Procedural Justice." *Science, Technology & Human Values* 38(2): 250–270.

Ottinger, G. and Cohen, B. (eds). 2011. *Technoscience and Environmental Justice.* Cambridge: MIT Press.

Phadke, R. 2011. "Resisting and Reconciling Big Wind: Middle Landscape Politics in the New American West." *Antipode* 43(3): 754–776.

Poisner, J. 1996. "A Civic Republican Perspective on the National Environmental Policy Act's Process of Citizen Participation." *Environmental Law* 26: 53–94.

Pollin, R., Garrett-Peltier, H., Heintz, J., and Scharber, H. 2008. "Green Recovery: A Program to Create Good Jobs and Start Building a Low-Carbon Economy." Washington, DC: Center for American Progress.

Reynolds, L. and Szerszynski, B. 2006. "Representing GM Nation." Working paper presented at the Participatory Approaches in Science and Technology conference, The Macaulay Institute, June 4–7. Available at: *www.macaulay.ac.uk/pathconference/outputs/PATH_abstract_1.2.3.pdf.*

Rose, M.H. 1990. *Interstate: Express Highway Politics 1939–1989.* Knoxville: University of Tennessee Press.

Sclove, R. 1995. *Democracy and Technology.* New York: Guilford Press.

Smith, G. and Wales, C. 2000. "Citizens' Juries and Deliberative Democracy." *Political Studies* 48: 51–65.

Stirling, A. 2008. "'Opening Up' and 'Closing Down': Power, Participation, and Pluralism in the Social Appraisal of Technology." *Science, Technology & Human Values* 33: 262–294.

Wilsdon, J. and Willis, R. 2004. *See-through Science: Why Public Engagement Needs to Move Upstream.* London: Demos. Available at: *www.demos.co.uk/publications/paddlingupstream.*

14

Political Scale and Conflicts over Knowledge Production

The case of unconventional natural-gas development

Abby J. Kinchy

RENSSELAER POLYTECHNIC INSTITUTE

Geographical concepts – both metaphorical and analytical – have a prominent role in Science and Technology Studies (STS). Geographic metaphors are pervasive in STS (frontiers, situated knowledge, standpoints, boundary-work, immutable mobiles, the view from nowhere). Analytically, "places" and "sites" of knowledge production are key categories for many STS scholars. The "rise of a geographical perspective" (Shapin 1998) in STS began in the early 1970s with studies of scientific cultures in particular countries or regions. Over the subsequent decades, several waves of research on the significance of place for the practices and products of science have emerged (Henke and Gieryn 2008), and scholars are increasingly working at the intersection of geography and science studies (Goldman et al. 2011). For example, a recent volume on interdisciplinary geographies of science (Meusburger et al. 2010) demonstrates the ongoing value of studying science in relation to geographic space, place, and mobility.

This chapter considers the geographical concept of *scale* and its relevance in understanding the politics of science – particularly in the domain of environmental regulation. Since at least the early 1990s, scale has become a central and much-debated concept in the field of human geography. Scale is one of several key socio-spatial concepts, used in relation to place, positionality, mobility, and networks. Scale is often thought of as a "nested hierarchy of differentially sized and bounded spaces" (Marston et al. 2005: 416–17). Another way to think of scale is as the "level" at which particular processes operate, in contrast to other levels with different territorial scopes (for example, the neighborhood versus the city). Among human geographers, there appears to be a consensus that scale is always socially constructed and contested; therefore, the research focus is often on how a particular scale came into being or how political actors seek to shift decision-making processes from one scale to another. In STS, Fortun (2009: 75–76) has argued that notions of scale "require ethnographic scrutiny," particularly as they operate in the natural, social, and computer sciences, because they provide an "inevitably limited" "way of seeing that frames and orients perspective."

In this chapter, I will treat scales as practical categories that are more or less taken for granted in social life. Scales are "variably powerful and institutionalized sets of practices and discourses"

(Moore 2008: 213) that have observable effects on social relations. That is, they are socially constructed categories that vary in the degree to which they are taken to be naturally "given" (as opposed to fluid, artificial, or transient). I focus on *political scale*, the nested levels of government that serve as venues or entry points for people taking political action. I make two interrelated arguments about why this understanding of political scale is important for STS. First, scientific knowledge claims may be used to justify and institutionalize the scale at which decisions are made and problems are addressed. For example, science-based claims that the health or ecological impacts of an industry are felt in highly variable, locally-specific ways may bolster arguments that county, municipal, or individual levels of oversight are more appropriate than regulation under state or national law. Second, the scale of political engagement is likely to affect the construction of scientific knowledge, by framing and orienting perspectives. The kinds of questions asked, and knowledge produced, about the natural environment can be radically different, depending on whether they are viewed from a global, national, or local point of view. Therefore, conflicts over political scale may also, implicitly or explicitly, be conflicts over scientific research agendas and the recognition of particular sources of knowledge.

I examine the relationship between political scale – in this case, the scale of environmental governance – and the production of knowledge (and ignorance) about the environment, using the case of shale gas development in the United States. A new method of natural-gas extraction, hydraulic fracturing (or "fracking") is facilitating an energy boom across the United States, and there have been major social conflicts over its ecological, health, and socio-economic impacts. These clashes involve a debate about the appropriate scale at which political decisions about gas development and regulation should be made – the individual property, the municipality, the state, the nation, or even the watershed. These and other disputes over the scale of governance are important not only because of their regulatory outcomes, but also because they have important consequences for what is knowable, or even "askable" about the ecological, social, and economic implications of gas development.

First I discuss the relevance of the human geography literature on the "politics of scale" to STS. I then provide some background on the conflict over fracking, focusing on development of the Marcellus Shale in the northeastern U.S. Finally, I discuss three cases from the fracking debate that illustrate different ways in which political scale is linked to environmental knowledge production. In the conclusion, I reflect on the lessons these cases offer on how to create "less partial and distorted" (Harding 1991: 186) understandings of the consequences of extractive industries like shale gas.

Why study scale?

The concept of "scale" is pervasive in social and political life. We speak of the "global scale" of economic trade and environmental problems and the desirability of "small-scale" farming. In the United States, debates about schooling, social welfare, and even marriage laws hinge on whether decisions should be made at the state or national "level" – that is, whether or not deliberation and control should be "devolved" to the states and municipalities. In contemporary politics, efforts to shift between scales are common. Social movements try to "scale up" their efforts by gaining national and international allies, in order to increase their political power, find receptive venues, build coalitions, and put pressure on unresponsive governments (Tarrow 2005, Keck and Sikkink 1998). Advocates of neoliberal economic policies argue for devolution of state responsibilities to localized governments and individuals, as well as scaling up to politically-weak international governance bodies, in order to weaken state power over capitalist activity (Peck and Tickell 2002, McCarthy and Prudham 2004).

The institutions that dominate political life give the impression that scales are distinct levels, corresponding to territorially-defined areas that are nested within increasingly expansive and encompassing spaces (the town, the county, the state, the nation, the planet). An individual may inhabit all of these spaces simultaneously, but orient his or her actions toward particular scales at different times. For example, a worker may view herself as competing for a job on a global scale, advocate for educational reforms at the scale of the city school district, and perceive threats to security on a national scale.

Despite the apparent naturalness of these scales of social engagement, none of them are fixed or given. This is most obvious in cases such as the creation of new nation-states, in which territorial boundaries are drawn, a state is established, and national identity is forged (successfully or not). The construction of a new political scale is also visible in geopolitical processes like the creation of the European Union, meant to create a new scale of economic governance and to forge a "European" political identity. The social construction of scale is evident in more subtle examples, as well, such as efforts to orient consumers toward the "local" or "regional," as in Buy Local campaigns and the 100-Mile Diet, or in efforts by environmental groups to raise awareness of the watersheds in which people live, thus reorienting perceptions toward the ecological scale at which polluting activities have an effect.

Human geographers have intensely debated the concept of scale over the last few decades (for a detailed discussion of these debates, see Herod 2009). While it is not possible to review the debate fully here, a few key points are important to highlight. Generally speaking, this research has "questioned concepts of geographical, political and analytical scale to ask where notions of scale came from, how particular scales became entrenched in social science research, and how processes operating at overlapping scales shape social life" (Kurtz 2003: 893).[1] Among geographers, there is considerable disagreement about whether scales are constructed in a material sense or as a representational trope. Neil Smith, for example, describes scales as the "materialization of contested social forces" (Smith 1993: 101). Scales, in this view, are physically produced through human activities, particularly through industrial processes and trade. For example, we can observe processes of "glocalization" (Swyngedouw 2004), in which companies that operate across multiple countries enact "localized" versions of "global" practices – such as McDonald's meals that are strategically flavored to suit the palettes of consumers in particular locales. McDonald's thus simultaneously constructs a global scale (of capitalist expansion) and a local scale (of culinary preferences) through its pursuit of profits.

In contrast, some geographers characterize scale as a discursive frame, a trope for representing the socio-spatial order (2003: 894). As Jones (1998: 27) writes, "participants in political disputes deploy arguments about scale discursively, alternately representing their position as global or local to enhance their standing." Kurtz (2003: 894) further advances this notion with the concept of "scale frames," discursive practices in social movement struggles "that construct meaningful (and actionable) linkages between the scale at which a social problem is experienced and the scale(s) at which it could be politically addressed or solved." For example, Mexican opponents of genetically-engineered crops have appealed to state and federal agencies as well as intergovernmental agencies in order to find authorities who are receptive to their grievances (Kinchy 2012). To do this, they have framed the scale of the problem as a threat to "local" agricultural traditions, "national" foodways, and "global" biodiversity.

Research on scale frames points to a frequently overlooked role of science in political processes. Scientific claims (for example, about the causes and scope of environmental degradation) are often used to defend a particular scale of government oversight. Harrison (2006) demonstrates this in a study of the politics of pesticide drift in California. In that case, regulatory officials and scientists justified the devolution of regulatory oversight to county government bodies by making arguments that ill effects of pesticide drift are infrequent and localized "accidents" (Harrison 2006: 519). Residents of areas affected by pesticide drift, primarily farm workers, contested the experts' claims, but had little political power, largely because of "the historical invisibility of farm labor in California" (Harrison 2006: 525). As Harrison concludes, the longstanding disenfranchisement of California farm workers meant that their knowledge claims about the scale of the problem were ignored, resulting in a decision to treat the county as the appropriate scale of governance.

Another perspective on the relationship between scale and scientific knowledge comes from a line of analysis in human geography that examines "the ways in which scalar narratives, classifications and cognitive schemas constrain or enable certain ways of seeing, thinking and acting" (Moore 2008: 214). In a widely-cited essay, Jones (1998: 27) argues that "scale is an epistemological category." Epistemology, used in this sense, refers to "ways of knowing," which are shaped by cultural categories. A shift in scale does "not merely shift politics from one level to another. Rather, it recast[s] what [is] true or knowable... Certain questions ... simply become unaskable" (Jones 1998: 28). As an example, Jones cites research on the history of urban planning:

> Urban planners introduced practices such as the geometrical plan, zoning, and social cartography, and these practices were instrumental in changing the way the city was known and represented. The more that urban information was presented through maps and zones, the more the city was understood only by way of these sorts of spatialized and geometrical systems, until what was considered 'true' about the city was altered in practice. Aggregate maps of poverty, delinquency, and housing, to name a few, came to be accepted as *the* most accurate understanding of what the city truly *was*. ... The truth of an 'ordinary gaze' became less 'true', while other questions about zones, for example, became more readily askable.
>
> *Jones 1998: 27–28*

Scale categories not only produce certain kinds of knowledge, but also generate ignorance. In Jones's example, as cities came to be understood as aggregations of observations about zones, the observations of ordinary people about their neighborhoods were excluded from official knowledge; what might be known with an ordinary gaze was ignored and forgotten. This observation resonates with a growing body of research in STS that analyzes the social dimensions of "ignorance" and "unknowns" (Proctor and Schiebinger 2008, Gross 2007, Frickel 2008). Scale concepts provide ways of seeing the world, and the perspectives they provide are "partial" (Haraway 1988). For STS scholars, this calls for analysis of the perspectives that are ignored or rendered impossible when a particular scale is the dominant frame in the analysis of environmental problems.

Might scientific research be steered by assumptions about the scale at which problems occur, the scale at which decision-making is appropriate, and the scale at which affected people have a political identity? To take one example, critics of the dominant approach to research on breast cancer causation have argued that progress toward cancer prevention has been stymied by an overemphasis on processes that occur at the scale of the body, at the expense of studies at the various environmental scales at which cancer-causing pollution is produced (McCormick et al. 2003). In this volume, Phadke demonstrates the importance of scale in deliberations about technological change, since local experiences of energy projects are often ignored in or distanced from the global debates about fossil fuels and renewable energy.

The case of shale gas development in the northeastern United States provides a rich set of examples in which the relationship between scale concepts and the production of environmental knowledge (and ignorance) takes different shapes. In these examples, scientific credibility contests have implications not only for what is known about gas drilling, but also which level of government has authority to govern the industry. Supporters and critics of gas drilling strategically work to shift the scale of political oversight in order to bring knowledge about the impacts of the industry to light, or ensure certain things remain unknown.

Marcellus Shale development

U.S. production of natural gas from "unconventional" sources, like low permeability shale, has burgeoned in the past decade. In part, this is due to recent technological advances that have made the extraction of gas from shale technically and economically feasible. Drilled in a traditional fashion (a hole straight down in the ground), gas releases from shale very slowly. Therefore, drilling companies are now using a combination of techniques, including drilling horizontally through the shale layer and using high-volume hydraulic fracturing, or "fracking." Fracking involves injecting millions of gallons of "frack fluid" into a gas well, under very high pressure, causing the shale to fracture and release its gas. Frack fluid is a mixture of water, multiple chemicals, and a special kind of sand that works as a "proppant" to hold open the fractures.

Unconventional gas development has also been facilitated by a favorable regulatory environment, particularly a set of exemptions from federal oversight and substantial government subsidies that were included in the 2005 Energy Policy Act. For example, oil and gas wells developed using hydraulic fracturing were excluded from the category of injection wells that are regulated under the Safe Drinking Water Act. This and other exemptions to federal environmental laws have had the effect of devolving regulatory authority over many aspects of unconventional gas drilling to the states, which have approached shale gas development in diverse ways (Wiseman 2012). The Energy Policy Act also created massive new subsidies and tax breaks, worth billions of dollars to the oil and gas industry, to incentivize domestic fossil fuel exploration (Lazzari 2007).

There are numerous shale formations across the U.S. The focus of this chapter is shale gas development in the state of Pennsylvania, where intensive development of the Marcellus Shale is now underway. The Marcellus Shale extends throughout much of Pennsylvania, West Virginia, New York and Ohio. Exploration of the Marcellus began around 2006, and by 2012, Pennsylvania alone had more than 6,000 Marcellus Shale gas wells, with permits for nearly twice as many.

There is considerable controversy over the environmental impacts of extracting gas from the Marcellus Shale. The extraction of shale gas is a complex, multi-stage process with environmental impacts that differ from conventional gas drilling, particularly because of the large quantity of contaminated wastewater that is produced in the drilling process (Lutz et al. 2013), and because of the possibility of polluting underground aquifers (Osborn et al. 2011).

A growing anti-fracking movement, driven primarily by grassroots, community-based citizens associations, has posed a formidable challenge to the gas industry, particularly in New York, where well permitting has been repeatedly delayed. Like many conflicts surrounding natural resource extraction and environmental pollution, the debate about fracking involves complex scientific questions and a large number of unknowns. For example, the fate of frack fluid that does not return to the surface of a well, remaining underground, is poorly understood. Also, large gaps exist in data about surface water pollution, as monitoring for pollution is carried out in only limited and piecemeal ways. Furthermore, the health effects of air, water, and soil pollution resulting from shale gas development are not understood, although there are many cases in which individuals have reported illnesses that they believe are linked to natural-gas production.

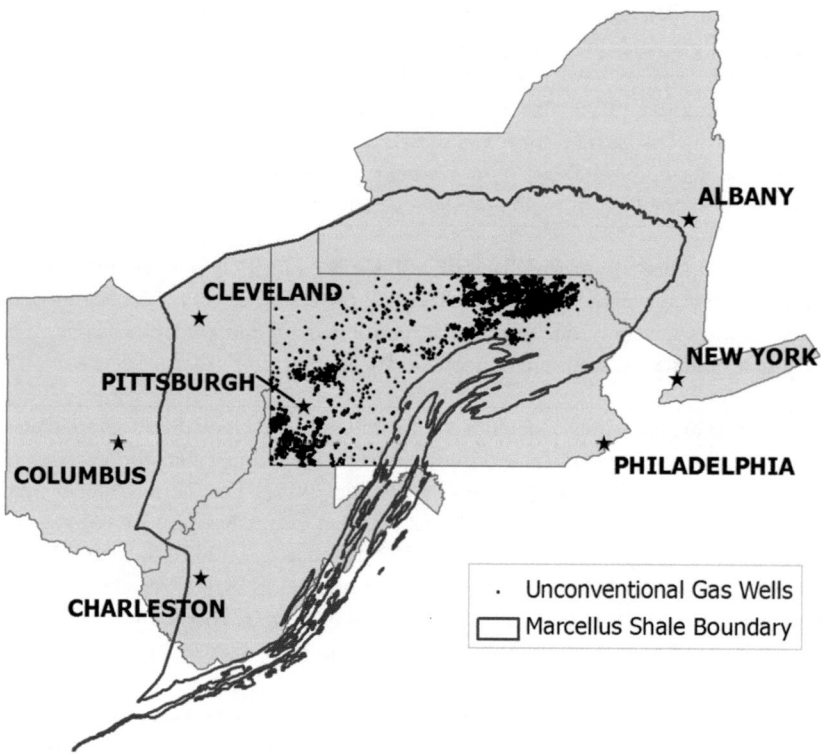

Figure 14.1 Shale gas development in Pennsylvania

Note: In Pennsylvania, Marcellus Shale gas development activity is concentrated in the northeastern and southwestern areas of the state. The Marcellus Shale is also being drilled in Ohio and West Virginia (gas wells not represented here). New York has not granted permits for high-volume hydraulic fracturing.

Source: Pennsylvania Department of Environmental Protection (PADEP), "January–June 2012 (Unconventional wells)," www.paoilandgasreporting.state.pa.us/publicreports/Modules/DataExports/DataExports.aspx, last accessed November 25, 2012.

An important factor affecting what is known and not known about the impacts of the shale gas industry is the scale at which gas development is scrutinized, as illustrated in the three following examples.

National climate policy and ignorance of the impacts of fracking

A prevailing view of natural-gas development is that it supports a transition away from the "dirtier" fossil fuels – coal and petroleum. Recently, some scientists have offered dissenting views of the carbon impacts of natural gas, based on calculations of the amount of methane leaked into the atmosphere during the different stages of gas development (Howarth et al. 2011). However, it has been assumed, for many years, that natural gas is "better" than other fossil fuels, from a climate perspective. Gas industry advocates, such as the American Gas Association and the National Gas Council, have promoted natural gas as a lower-carbon energy alternative to coal and petroleum. A number of academic reports lend support to this view of natural gas as a "less harmful" fossil fuel. A multidisciplinary report published by MIT in 2011, for example, indicated that natural gas could be a carbon-reducing substitute for coal in many applications and that it would help support wind power generation (Moniz et al. 2010).

Since the 1990s, industry players, environmentalists and policymakers have viewed natural gas as a "bridge fuel" – a less harmful fossil fuel source that can be used until alternative energy sources like wind, solar, and biofuels are more readily available. In 2006, the Sierra Club articulated its position on natural gas in an Energy Policy Statement:

> Despite supply and price concerns, [natural gas] is still a much cleaner fuel than coal and emits less CO_2 per unit of energy produced. During the transition to a clean energy future, the Sierra Club is generally not opposed to continued production from existing fields following best practices to limit environmental damage.[2]

Major environmental organizations, including the Sierra Club, the Natural Resources Defense Council, and the Environmental Defense Fund had, for some time, taken the position that switching from coal to natural gas was a necessary transitional step. The Sierra Club, in particular, advanced this position, accepting more than $26 million from Chesapeake Energy, a major natural gas company, to support its "Beyond Coal" campaign from 2007 to 2010. When it became apparent that the U.S. had untapped natural-gas reserves in shale formations across the country, climate activists like Bill McKibben viewed it favorably. In *Eaarth*, McKibben (2010: 59) offered a list of good news about the climate, including "new discoveries of natural gas in the United States that could help wean us off dirtier coal."

When viewed at the scale of the global climate, new sources of natural gas appeared to be unquestionably beneficial to the environment. However, this position betrayed the national environmental organizations' ignorance about the consequences of gas development at the local scale, in the places where shale gas is produced. By considering shale gas development with a global climate frame, or a national energy policy frame, the Sierra Club and other groups failed to ask questions about environmental change at other scales. In reaction, communities threatened by shale gas development began to question environmentalists' favorable stance on natural gas, arguing that, at the local scale, it was polluting water, soil, and air, and posing unacceptable hazards to human life and community well-being. In one notable example, regional chapters of the Sierra Club came into conflict with the national organization. In this case, regional chapters were not questioning the national organization's assessment of the climate impacts of natural gas (although some would do so, over the ensuing months and years); instead, they were drawing attention to environmental impacts at a different scale.

The Atlantic Chapter of the Sierra Club, which draws its membership from New York State, formed a Gas Drilling Task Force in 2008, in response to concerns raised by its grassroots members.[3] According to a member who recounted the debate at that time,

> By October, 2009, the Chapter was convinced that fracking represented a major threat to our environment and public health, and that no amount of regulation could change that. The Chapter further determined that even if hydrofracking could be made safe, it represented a big distraction from what should be our major goals: moving away from fossil fuels and toward energy efficiency, conservation and renewables. The Chapter [executive committee] passed a resolution calling for a ban on hydrofracking and became the first major environmental group in Albany to do so.[4]

Subsequently, however, "The Atlantic Chapter was told that its resolution was 'not consistent with existing Club policy.' Thus, we began the Atlantic Chapter's long, arduous task of educating National."

Ultimately, the national Club came to see the chapter's point of view, first by advocating greater regulation of unconventional natural-gas development, and eventually dropping altogether its advocacy of gas as a transition fuel. Today, the Sierra Club has a national campaign against fracking. While the details about how and why this transition occurred require further study, the case suggests that *bringing two scaled perspectives into conversation within an organization can produce more complete and nuanced environmental knowledge,* leading to revised policy positions. The organizational form of the Sierra Club – an umbrella body attuned to environmental problems at global and national scales, with chapters attuned to local and regional environments – may be a useful model for future efforts to apprehend and respond to complex, multi-scalar problems.

Private landownership and knowledge of groundwater

Numerous reported cases of well water pollution – particularly methane contamination – have been associated with natural-gas drilling. Research investigating the link between gas drilling and groundwater pollution has been highly controversial. The Environmental Protection Agency (EPA) has undertaken a national study, and other scientists have attempted to understand the general processes through which such pollution might occur (Osborn et al. 2011, Vidic et al. 2013). However, it is not fully understood how methane and other contaminants from a gas well could pass into groundwater, and both industry lobbyists and academic researchers have pointed to methodological flaws in studies that show a link between gas drilling and well water pollution (Roach 2013).

One of the reasons for why it has been so difficult to establish the facts about the impacts of shale gas development on groundwater is that private drinking water wells are not regulated by government entities; they are not required to meet water quality standards under the Safe Drinking Water Act (SDWA) or other environmental laws. In Pennsylvania, oversight of well water quality has long been devolved to the private landowner. Unlike routine monitoring of municipal water supplies, there is no public system for analysis and treatment of groundwater that feeds private water wells, so knowledge of the quality of groundwater is typically inaccessible to anyone but the landowner who has voluntarily chosen to have his or her water analyzed. Some groundwater quality data is gathered by the U.S. Geological Survey, but the agency has very few monitoring locations, so it is difficult to use this data to establish whether polluted groundwater is a pervasive consequence of shale gas development (Kappel and Nystrom 2012).

When drinking water wells become polluted in the vicinity of gas drilling operations, these are usually treated as isolated incidents. Gas drilling companies operating in Pennsylvania have responded to water wells that have been contaminated by providing payments, drinking water deliveries, or a water filtration system directly to the landowner, typically through legal settlements that come with gag orders that silence the victim. As a *Bloomburg* report summarizes, "The strategy keeps data from regulators, policymakers, the news media and health researchers, and makes it difficult to challenge the industry's claim that fracking has never tainted anyone's water" (Efstathiou and Drajem 2013). However, anti-fracking activists have challenged the industry's claims. Critics of gas drilling have called on government agencies to provide more detailed investigations of pollution incidents and to aggregate that data to better understand how unconventional gas drilling affects groundwater. In addition, there are several efforts by civil society groups to amass groundwater quality data themselves, with the aim of scaling up individual well water pollution incidents to broader geographical scales. For example, one grass-roots effort attempts to compile individual reports of harm into a national database.[5] Through this aggregation of reports, experiences of harm at the individual scale are connected to the rapid expansion of shale gas production that is occurring nationally. Another organization is testing private wells and creating a map of "baseline" groundwater quality throughout a region of New York where drilling has not yet occurred.[6] This not only makes regional groundwater quality data available to the public in a way that has otherwise been lacking; it also reframes the scale of groundwater quality as regional, rather than individual.

One case provides a complex picture of how the scale of water governance affects the production of knowledge of gas drilling. The rural village of Dimock, located in Susquehanna County in northeastern Pennsylvania, has risen to national prominence because of well water problems that occurred there. In 2009, an elderly woman's drinking water well filled with methane gas, causing an explosion. Several other Dimock residents complained that their well water was fizzy, discolored, and had a bad smell. Pennsylvania's Department of Environmental Protection (DEP) determined that Cabot Oil and Gas, a company that had drilled several natural-gas wells in Dimock, had constructed its wells poorly, allowing methane to migrate into groundwater. While Cabot acquiesced to the DEP's requirements, the company denied that it was responsible for the problems with the Dimock residents' drinking water, claiming that the pollution could have had other sources. Cabot maintained that the groundwater already had fluctuating levels of naturally-occurring methane.[7]

Most Dimock residents did not have baseline, pre-drilling water quality data that they could use to prove a change in water quality. This made it easy for Cabot to speculate that methane and other contaminants were in the wells water prior to gas drilling, a claim that was widely repeated by supporters of the industry. The company attempted to discredit the scientific evidence presented by the EPA and other environmental agencies. For example, an investigation by the EPA reviewed water quality data gathered by Cabot and concluded that several wells contained not only methane but also several other contaminants. Cabot denied that these pollutants had any link to its drilling operations. When the EPA indicated, in January 2012, that some Dimock residents should have water delivered to their homes, Cabot replied with a statement criticizing the agency's analysis:

> It appears that EPA selectively chose data on substances it was concerned about in order to reach a result it had predetermined. EPA chose to include specific data points without adequate knowledge or consideration of where or why the samples were collected, when they were taken, or the naturally occurring background levels for those substances throughout the Susquehanna County area. The end result is an unwarranted investigation and unnecessary delivery of water.[8]

This kind of argument is common in conflicts over environmental pollution, in which those accused of doing harm challenge the credibility of the science used against them, in order to avoid legal or regulatory consequences. Notable about this case, however, is the relationship between the scale of governance of groundwater and the production of knowledge of ground-water pollution. As in the pesticide drift case discussed earlier (Harrison 2006), the conflict over groundwater pollution was not only about establishing truthful claims about the impacts of an industry, but also about justifying a particular scale of governance. In Harrison's study, claims that harmful pesticide drift occurred only infrequently and as isolated accidents justified local oversight rather than statewide regulation of the industry. Similarly, in this case, Cabot sought to discredit the scientific evidence against it in order to ward off a federal agency, the EPA, which was taking a growing interest in the impacts of fracking. Cabot used scientific arguments to frame the scale of the problem as localized – *not* a federal environmental concern. Cabot suggested that the Dimock area has a local, longstanding problem of methane pollution, unrelated to gas drilling. The company was able to make this claim because of the absence of baseline data about private well water quality, the result of much earlier decisions to scale well water oversight to the individual landowner.

Cabot succeeded at creating uncertainty about whether its gas wells really did cause groundwater pollution, which has made it more difficult for critics of the industry to advocate for increased state and federal interventions. For example, in an early reaction to the incident, DEP called for all affected households to be moved on to a municipal water system, which would involve constructing a twelve-mile waterline, to be paid for by the state, which would then sue Cabot for the cost. But there were protests from many members of the community whose well water quality was unaffected by gas drilling and who supported Cabot's drilling activities and did not believe the company caused the pollution of their neighbors' wells. As summarized in a local newspaper article, a group calling itself Enough Already took the position that "The state has 'gone amok' by siding with families who are suing Cabot for contamination the driller says it did not cause and using the opportunity to demand a 'handout' from a wealthy company in order to expand a public utility."[9] The DEP subsequently dropped its plans to build the waterline, in exchange for Cabot agreeing to pay $4.1 million directly to residents affected by methane contamination. Many of the affected residents refused the payment, pursuing a lawsuit against the company instead. Ultimately, in August 2012, after three years of strain, the weary families agreed to accept a settlement with the company.[10]

Despite the ongoing denials by Cabot and its supporters, the Dimock case brought public attention to problems in industry practices, prompting changes that may increase knowledge of how gas drilling affects groundwater. After the Dimock controversy, it has become standard practice for gas drilling companies to take pre-drilling well water samples for laboratory analysis. In terms of policy, a 2012 change in the Pennsylvania Oil and Gas Act makes it such that gas companies are presumed responsible for any well water pollution occurring within 2,500 yards of a gas well.[11] Under the law, if a landowner reports well water contamination, the DEP takes samples for analysis. Furthermore, the law requires the state to maintain a registry of incidents of groundwater pollution definitively caused by hydraulic fracturing. These policy changes, if fully implemented, could lead to more comprehensive knowledge of how gas drilling affects groundwater.

Despite these changes, there are still major obstacles to using groundwater data to yield useful knowledge of the effects of the gas industry. First, water quality data gathered by gas companies is not made public; thus, baseline levels of methane and other contaminants are secret. Second, the DEP reportedly has not fully disclosed the contaminants found in well water samples, misleading private landowners about the safety of their drinking water.[12] Finally, the database of groundwater pollution incidents has not yet materialized. Recently, a journalist had to use the Freedom of Information Act to obtain DEP's determinations on hundreds of complaints about polluted drinking water wells.[13]

In sum, despite the fact that groundwater quality is now being gathered extensively by centralized institutions (gas companies and the state), it is not being compiled in ways that produce a scaled-up picture of regional groundwater quality; it remains individualized, private information. This case suggests that *political scale does not dictate the scale of environmental knowledge that is produced; scale frames do.* Even though a state agency, DEP, is tasked with statewide governance of the gas industry, its questions about groundwater pollution have been framed at the scale of the individual property-owner's water well, which is treated as personal information. Thus, the answers it has generated have failed to generate any region- or state-wide conclusions about how gas drilling affects groundwater.

State regulation and local knowledge

One debate in the conflict over shale gas development centers on the rights of municipalities to make decisions restricting gas industry activity, through zoning or local drilling bans. Again, this conflict involves questions of scale, in this case, the right to regulate the gas industry at the municipal (city or town) scale. Advocates of local governance of gas development argue that if decisions are made at the local level, questions will be asked about quality of life, community character, and local visions of the future, which are not well addressed at the state regulatory level. Thus, the call for local decision-making is not just about shifting power to the municipalities; it also means producing situated knowledge of the social and place-based consequences of gas development.

In many towns and cities, local decision-making is being embraced, and may pose a significant obstacle to the gas industry. More than 200 municipalities in fifteen states have banned unconventional natural-gas drilling and associated activities, such as wastewater storage.[14] However, while community organizers have sought to frame gas drilling as a problem rightfully addressed by local communities, advocates of gas drilling have, in many cases, succeeded at stripping municipalities of power. In some cases, gas companies and pro-development landowners have challenged local bans by suing the municipal governments, with mixed outcomes. A New York trial judge ruled against Anschutz Exploration Corporation in a case involving a local fracking ban in Dryden, NY, thus bolstering the right of municipalities to pass such bans.[15] However, several states, including Ohio, Idaho, Colorado, Texas, and Pennsylvania have proposed, and in some case enacted, legislation to preempt local land-use planning and zoning and to prevent municipalities from enacting local gas drilling bans.[16] The influence of the gas industry is evident in these bills, working in part through the American Legislative Exchange Council, a corporate-funded organization that writes "model" legislation on a wide range of issues.[17]

In February 2012, Pennsylvania's Governor, Tom Corbett, signed into law a piece of legislation known as Act 13, which amends the state's Oil and Gas Act. This law brought about the changes discussed in the previous section pertaining to the protection of well water quality. However, other portions of the legislation were struck down in December 2013 by the Pennsylvania Supreme Court. Those parts of the Act required that local ordinances "allow for the reasonable development" of the Marcellus Shale, disallowing attempts to ban gas drilling activity. Furthermore, the legislation replaced local planning and zoning decisions with standard, statewide (and generally weak) rules. For example, under the new legislation, gas wells, fluid impoundments, and pipelines had to be allowed in all zoning districts, including residential areas. As legal scholar Nancy Perkins observed, the law:

> tells municipalities what type of gas operations must be authorized in various zoning districts, and forbids them from imposing numerous other conditions on gas operations. By dictating with remarkable precision the contours of allowable land use ordinances, the amendments remove any rationale for constructing an inclusive process to achieve sustainable land use controls. It would be illogical for individuals to invest time and effort in a local legislative process when nearly the entire outcome has already been pre-determined.
>
> *Perkins 2012: 75*

In sum, the law aimed to shift the scale of deliberation and decision-making away from local municipalities to the state level, where regulators had already demonstrated their eagerness to allow gas development to proceed with minimal obstacles.

For advocates of local decision-making, at stake in this scale-shift was not just the character of the democratic process, but also the kinds of questions that became askable and the forms of expertise that were deemed relevant. Perkins (2012: 76) criticizes, from a feminist sustainability perspective, the knowledge consequences of scaling regulation at the state level:

> There is no opportunity for local residents to question the data and research that led state leaders to allow drilling operations in every zoning district in the state. Neither is there an opportunity for local residents to bring to bear the interests of future generations and intersection of race, class, gender, age, education, and place of their experience.

Perkins was not alone in observing that this move to strip municipalities of their right to govern gas development was also a move to delegitimize local knowledge and perspectives. This idea has been consistently expressed by lawyers and community activists from the Community Environmental Legal Defense Fund (CELDF), an organization based in Mercersburg, PA. CELDF aids communities in the pursuit of local self-governance, through "Democracy Schools" and the writing of local rights-based ordinances that prohibit activities such as gas drilling, factory farms, or sewage sludge dumping. CELDF argues that state and federal regulatory processes exclude ordinary citizens by relying only on expert assessments of environmental risk. Therefore, the organization maintains, there is a need for new democratic processes in which the knowledge and expertise of ordinary citizens is a legitimate basis for decisions about industrial developments. CELDF contends that the scale at which ordinary citizens have the most pertinent expertise is in their own towns and cities, particularly on matters related to the harmful effects of industrial developments on "workers, the local economy, livability, property values, or the environment."[18] As stated in a CELDF publication entitled "Common Sense: Banning Fracking at the Local Level," "It makes sense for us to make decisions about the communities in which we live. In our communities, it is we who are the experts. Is there anyone more qualified to make these decisions?"[19]

CELDF and other advocates of local bans on gas development deploy localist scale frames, not only to make their case to state authorities, but also to convince residents of areas where gas drilling is occurring to demand municipal-level governance. A key justification for the localist scale frame is that locals have social and environmental knowledge that is essential to good decision-making. Clearly, this argument has been compelling to many communities, as the numerous municipal moratoria on gas drilling suggest. However, shifting decision-making authority to the municipalities does not necessarily mean that opponents of gas drilling will have an advantage. Geographers have questioned the notion that "local" politics necessarily yields better outcomes for disadvantaged groups, given that municipalities may lack resources or capacity to adequately respond to powerful industrial actors (DuPuis and Goodman 2005). In the Dimock case discussed above, the people who are fighting for clean drinking water have been accused of dishonesty by their own neighbors, who have defended the gas industry against criticism. Decision-making at the local level would be unlikely to help the victims of pollution in that case. In circumstances where a community is divided on the issue or where local government lacks capacity to challenge a powerful industry, bringing grievances to state- or federal-level regulators may appear to be a more viable strategy for obtaining desired environmental and health protections.

I observed an example of this at a CELDF organizing event near Dimock, Pennsylvania in the summer of 2009. While a community organizer made a well-articulated, compelling case for CELDF's style of activism, many citizens who attended the meeting appeared skeptical. Repeatedly, people attending the meeting claimed that a campaign for a local drilling ban would not work in their communities, because there are too many influential people in their towns who want the drilling to occur. Several people called for alternative, state-level actions such as attending DEP permitting hearings to make their case against gas drilling. The organizer countered that in such hearings, only the official experts' assessments of the risks and benefits have any weight, and that ultimately the permitting process is aimed at allowing, not deterring, gas development. Moving the fight to the state level means grappling with knowledge that pertains to existing permitting requirements, rather than knowledge of social concerns such as "livability" and the character of the local economy.

However, it was clear by the end of the meeting that only a small fraction of the (approximately) fifty people in attendance were convinced of CELDF's approach, which would require persuading the majority of people in their municipalities to ban gas development. The chance of success, in the form of environmental and health protections, seemed to them greater at the state level, where regulators might take their concerns about environmental and health impacts more seriously, because of an official mandate to protect the natural environment of the state – even if this means narrowing the range of criteria on which the impacts of the gas industry are assessed. This case indicates that *when activists and their opponents struggle over the appropriate scale of governance, at stake are the kinds of knowledge and sources of expertise that may be applied to decision-making.* Grassroots activists may not only frame their grievances in ways that emphasize local knowledge; they may also strategically seek to scale up to non-local government agencies in order to achieve their aims. These scale framing efforts contribute to defining the boundaries of relevant expertise in debates about industrial developments.

The need for reflexivity about scale

In this chapter, I showed how scientific knowledge claims have been used to justify or challenge the scale of decision-making about shale gas. For example, while anti-fracking activists used scientific documentation of water pollution to call for greater state and federal scrutiny of the gas industry, Cabot Oil and Gas worked to spread uncertainty about the scientific evidence of groundwater pollution in order to resist the EPA's growing involvement in the issue. I also argued that the scale of political engagement frames and orients perspectives in ways that affect the construction of scientific knowledge. Organizations that approach environmental politics at different scales tend to produce different kinds of knowledge claims about shale gas development, as seen in the example of the national and local chapters of the Sierra Club. As a result, conflicts over political scale may also be disputes over scientific research agendas and the recognition of particular sources of knowledge – as seen in the struggle over the legitimacy of local drilling bans as opposed to expert assessments of risk at the state level.

Bringing two scaled perspectives into conversation within an organization (as in the Sierra Club example) can produce more complete and nuanced environmental knowledge. Perhaps the same thing might occur in government agencies, like the EPA, that have regional branches as well as national offices. Political scale does not dictate the scale of environmental knowledge that is produced; more important are the scale frames that guide the questions that are asked and answered. For example, a state agency like the DEP may still pose questions that frame groundwater pollution as an individualized problem rather than a regional or state-wide one. Similarly, local activist groups may frame their questions in ways that scale up the issue to the state level, even if means giving up their own claims to local expertise.

Attention to political scale and scale framing processes can improve understanding of the social production of ignorance about environmental change. In each case examined in this chapter, political scale affected what was known and unknown about gas development. In the first example, an emphasis on the comparative climate impacts of natural gas and coal for many years – a national energy policy issue – distracted many key environmental advocacy groups from the more localized problems associated with the extraction of natural gas from unconventional sources. In the second example, policies that left private landowners solely responsible for monitoring and maintaining the quality of their well water – water governance scaled to individuals, rather than to the municipality or aquifer – created problematic gaps in knowledge of how gas drilling has changed groundwater quality. In the third example, policies that concentrated decision-making authority at the level of the states, rather than municipalities, made it difficult to bring local knowledge of land use priorities, cultural preferences, and environmental conditions to bear on the regulation of gas industry practices. Generally speaking, the gas industry and its supporters have advocated scales of environmental oversight that result in ignorance about the consequences of shale gas development.

Studying conflicts over scale can also reveal processes through which social movements challenge scientific unknowns. Opponents of shale gas development have sought to shift between political scales in order to bring more diverse knowledge claims and forms of expertise to bear on decisions about gas drilling. In the case of the Sierra Club, the Atlantic Chapter brought knowledge of the impacts of fracking to the national organization, reframing shale gas as a regional pollution concern in addition to a national energy policy issue. Critics of the gas industry have sought to connect and scale up the many "isolated incidents" of well water contamination, thus making the case for greater state- or federal-level scrutiny of the industry. Finally, advocates for local self-government have succeeded, in many cases, at passing bans or moratoria on shale gas development after working locally to assess how the industry might affect quality of life.

What should be clear from these case studies is that there is no single ideal scale for governing unconventional gas drilling. I found that, depending on the context, both opponents and advocates of gas drilling work to shift the scale of governance up and down in pursuit of particular outcomes. However, every scale that is constructed makes some questions more or less possible to ask. Considering that the natural-gas industry operates nationally and transnationally, it may appear desirable to construct a comparably "scaled up" level of oversight. The arguments of CELDF and the Atlantic Chapter of the Sierra Club should make it evident that without questions pitched at a municipal, watershed, or community scale, knowledge of gas development is dangerously incomplete. These observations resonate with the findings of Phadke (Chapter 13 in this book) on community conflicts over wind power. The national conversation about the benefits of wind power has suffered from the absence of attention to how wind farms affect livelihoods in "geographically concentrated" ways. Fully apprehending the impacts of shale gas or wind power – or any industrial development, for that matter – requires reflexivity about the ways that political scales shape what we know, and what remains obscured.

Notes

1   Some geographers have rejected the concept of scale. Marston et al (2005) have proposed replacing the concept of scale in human geography with a "flat ontology" that draws on the insights of actor-network theory. They argue that concepts of scale are too often treated by geographers as "conceptual givens," obscuring other ways that socio-spatial processes might be operating. Critics of this perspective argue that the move to drop the concept of scale would foreclose the possibility of studying how

scale – as a concept mobilized by people in everyday life – is constructed and has effects in politics and culture. One commentator notes:

> ...[T]heir proposal to do away not just with a hierarchical scalar ontology, but to 'eliminate scale as a concept in human geography' (2005: 416) – and thus presumably any reference to scale politics – is a misguided case of throwing the baby out with the bathwater. ...[T]hrough this theoretical manoeuver they are unwittingly reproducing a materialist/idealist binary that: (1) unhelpfully centres debate upon the ontological and theoretical status of scales in human geography at the expense of attention to their existence and use as practical categories, and (2) rests upon the flawed assumption that denying the ontological reality of scales implies that they are merely inconsequential heuristics in the minds of geographers that 'do no work', or have no effect in themselves. (Moore 2008: 213).

2 "Sierra Club's Energy Resources Policy," accessed 8 July 2013 at *http://indiana.sierraclub.org/issues/articles/energy-resources-policy.html*.

3 "Fracking: Is National catching up with the grassroots?" accessed 8 July 2013 at *http://newyork2.sierraclub.org/content/fracking-national-catching-grassroots*.

4 Ibid.

5 The "List of the Harmed" contains many reports of well water pollution, among other harms reportedly caused by gas drilling: *www.fractracker.org/2013/03/pacwas-list-of-the-harmed-now-mapped-by-fractracker*.

6 "Welcome to CSI's Database!" Accessed 8 July 2013 at *www.communityscience.org/database*.

7 Laura Legere, "Cabot argues to resume drilling in Dimock as tests show surges of methane in water wells," *thetimes-tribune.com*. Accessed 8 July 2013 at *http://thetimes-tribune.com/news/gas-drilling/cabot-argues-to-resume-drilling-in-dimock-as-tests-show-surges-of-methane-in-water-wells-1.1220204#axzz1fmqcRIJc*.

8 Cabot Oil & Gas Corporation, "U.S. EPA's January 2012 Position on Water Delivery," accessed 8 July 2013 at *www.cabotog.com/pdfs/Cabot_Statement_EPAWaterDelivery.pdf*.

9 Laura Legere, "Debate over proposed Dimock waterline divides community," *thetimes-tribune.com*. Accessed 8 July 2013 at *http://thetimes-tribune.com/news/debate-over-proposed-dimock-waterline-divides-community-1.1053233*.

10 For details about this case, see the collection of documents at *http://stateimpact.npr.org/pennsylvania/tag/dimock*.

11 For an overview of the law, see "New Pennsylvania Oil and Gas Law Targets Unconventional Gas Operations for Heightened Regulatory Oversight," accessed 8 July 2013 at *www.jdsupra.com/legal-news/new-pennsylvania-oil-and-gas-law-targets-59565*.

12 Don Hopey, "Lawmaker challenges Pennsylvania DEP's reporting of gas well water safety," Pittsburgh Post-Gazette, 2 November 2012, accessed 8 July 2013 at *www.post-gazette.com/stories/local/state/lawmaker-challenges-pa-deps-reporting-of-gas-well-water-safety-660238*.

13 Laura Legere, "Sunday Times review of DEP drilling records reveals water damage, murky testing methods" *thetimes-tribune.com* (19 May 2013), accessed 8 July 2013 at *http://thetimes-tribune.com/news/sunday-times-review-of-dep-drilling-records-reveals-water-damage-murky-testing-methods-1.1491547*. Furthermore, activist groups have compiled reports of groundwater pollution nationwide. See, for example, "Incidents where hydraulic fracturing is a suspected cause of drinking water contamination" on an NRDC blog. Accessed 8 July 2013 at *http://switchboard.nrdc.org/blogs/amall/incidents_where_hydraulic_frac.html*.

14 "Local Officials Standing Up to Protect Their Communities from Fracking," accessed 8 July 2013 at *www.ombwatch.org/node/12150*.

15 Ibid.

16 "Exposed: Pennsylvania Act 13 Overturned by Commonwealth Court, originally an ALEC Model Bill," accessed 8 July 2013 at *www.desmogblog.com/exposed-pennsylvania-act-13-overturned-commonwealth-court-originally-alec-model-bill*.

17 Ibid., and "Fracking Democracy: Why Pennsylvania's Act 13 May Be the Nation's Worst Corporate Giveaway," accessed 8 July 2013 at *www.alternet.org/story/154459/fracking_democracy%3A_why_pennsylvania%27s_act_13_may_be_the_nation%27s_worst_corporate_giveaway*.

18 "Home Rule," accessed 8 July 2013 at *http://celdf.org/-1-87*.

19 "Common Sense – Banning Fracking at the Local Level," accessed 8 July 2013 at *www.celdf.org/-1-80*.

References

DuPuis, E.M. and Goodman, D. 2005. "Should We Go 'Home' to Eat?: Toward a Reflexive Politics of Localism." *Journal of Rural Studies* 21: 359–371.

Efstathiou, J. and Drajem, M. 2013. "Drillers Silence U.S. Water Complaints With Sealed Settlements." *Bloomburg*, June 6. Accessed 8 July 2013 at *www.bloomberg.com/news/2013-06-06/drillers-silence-fracking-claims-with-sealed-settlements.html*.

Fortun, K. 2009. "Scaling and Visualizing Multi-Sited Ethnography." In *Multi-sited Ethnography: Theory, Praxis and Locality in Contemporary Social Research*, 73–86. Farnham: Ashgate Publishing, Ltd.

Frickel, S. 2008. "On Missing New Orleans: Lost Knowledge and Knowledge Gaps in an Urban Hazardscape." *Environmental History* 13: 634–650.

Goldman, M.J., Nadasdy, P., and Turner, M.D. eds. 2011. *Knowing Nature: Conversations at the Intersection of Political Ecology and Science Studies*. Chicago: University of Chicago Press.

Gross, M. 2007. "The Unknown in Process: Dynamic Connections of Ignorance, Non-Knowledge and Related Concepts." *Current Sociology* 55: 742–759.

Haraway, D. 1988. "Situated Knowledges: The Science Question in Feminism and the Privilege of Partial Perspective." *Feminist Studies* 14: 575–599.

Harding, S. 1991. *Whose Science? Whose Knowledge?: Thinking from Women's Lives*. Ithaca, NY: Cornell University Press.

Harrison, J.L. 2006. "'Accidents' and Invisibilities: Scaled Discourse and the Naturalization of Regulatory Neglect in California's Pesticide Drift Conflict." *Political Geography* 25: 506–529.

Henke, C., and Gieryn, T. 2008. "Sites of Scientific Practice: The Enduring Importance of Place." In *The Handbook of Science and Technology Studies*, third edition, edited by E.J. Hackett, O. Amsterdamska, M. Lynch, and J. Wajcman, 353–376. Cambridge, MA: The MIT Press.

Herod, A. 2009. *Scale*. Abingdon: Taylor & Francis.

Howarth, R., Santoro, R., and Ingraffea, A. 2011. "Methane and Greenhouse-gas Footprint of Natural Gas from Shale Formations." *Climatic Change* 106: 679–690.

Jones, K. 1998. "Scale as Epistemology." *Political Geography* 17: 25–28.

Kappel, W. and Nystrom, E. 2012. "Dissolved Methane in New York Groundwater." *U.S. Geological Survey Open-File Report*: 1–6.

Keck, M.E. and Sikkink, K. 1998. *Activists Beyond Borders: Advocacy Networks in International Politics*. Ithaca, NY: Cornell University Press.

Kinchy, A. 2012. *Seeds, Science, and Struggle: The Global Politics of Transgenic Crops*. Cambridge, MA: The MIT Press.

Kurtz, H.E. 2003. "Scale Frames and Counter-scale Frames: Constructing the Problem of Environmental Injustice." *Political Geography* 22: 887–916.

Lazzari, S. 2007. *Oil and Gas Tax Subsidies: Current Status and Analysis*. Congressional Research Service Report for Congress.

Lutz, B.D., Lewis, A.N., and Doyle, M.W. 2013. "Generation, Transport, and Disposal of Wastewater Associated with Marcellus Shale Gas Development." *Water Resources Research* 49: 647–656.

Marston, S.A., Jones, J.P., and Woodward, K. 2005. "Human Geography without Scale." *Transactions of the Institute of British Geographers* 30: 416–432.

McCarthy, J. and Prudham, S. 2004. "Neoliberal Nature and the Nature of Neoliberalism." *Geoforum* 35: 275–283.

McCormick, S., Brown, P., and Zavestoski, S. 2003. "The Personal Is Scientific, the Scientific Is Political: The Public Paradigm of the Environmental Breast Cancer Movement." *Sociological Forum* 18: 545–576.

McKibben, B. 2010. *Eaarth: Making a Life on a Tough New Planet*. London: Macmillan.

Meusburger, P., Livingstone, D., and Jèons, H. eds. 2010. *Geographies of Science*. New York: Springer.

Moniz, E.J., Jacoby, H.D., and Meggs, A.J.M. et al. 2010. *The Future of Natural Gas: An Interdisciplinary MIT Study*. MIT Energy Initiative. Accessed 8 July 2013 at *http://mitei.mit.edu/system/files/NaturalGas_Report.pdf*.

Moore, A. 2008. "Rethinking Scale as a Geographical Category: From Analysis to Practice." *Progress in Human Geography* 32: 203–225.

Osborn, S.G., Vengosh, A., Warner, N.R., and Jackson, R.B. 2011. "Methane Contamination of Drinking Water Accompanying Gas-well Drilling and Hydraulic Fracturing." *Proceedings of the National Academy of Sciences of the United States of America* 108: 8172–8176.

Peck, J. and Tickell, A. 2002. "Neoliberalizing Space." *Antipode* 34: 380–404.

Perkins, N.D. 2012. "The Fracturing of Place: The Regulation of Marcellus Shale Development and the Subordination of Local Experience." *Fordham Environmental Law Review* 23.

Proctor, R. and Schiebinger, L.L. eds. 2008. *Agnotology: The Making and Unmaking of Ignorance*. Stanford: Stanford University Press.

Roach, J. 2013. "Natural Gas Found in Drinking Water Near Fracked Wells." *Science on NBC News*. Accessed 8 July 2013 at *www.nbcnews.com/science/natural-gas-found-drinking-water-near-fracked-wells-6C10433123*.

Shapin, S. 1998. "Placing the View from Nowhere : Historical and Sociological Problems in the Location of Science." *Transactions of the Institute of British Geographers* 23: 5–12.

Smith, N. 1993. "Homeless/Global: Scaling Places." In *Mapping the Futures: Local Cultures, Global Change*, edited by J. Bird, B. Curtis, T. Putnam, G. Robertson, and L. Tickner, 87–119. London: Routledge.

Swyngedouw, E. 2004. "Globalisation or 'Glocalisation'? Networks, Territories and Rescaling." *Cambridge Review of International Affairs* 17: 25–48.

Tarrow, S. 2005. *The New Transnational Activism*. Cambridge: Cambridge University Press.

Vidic, R.D., Brantley, S.L., Vandenbossche, J.M., Yoxtheimer, D., and Abad, J.D. 2013. "Impact of Shale Gas Development on Regional Water Quality." *Science* 340: 1235009.

Wiseman, H. 2012. "Fracturing Regulation Applied." *Duke Environmental Law & Policy Forum* 6: 361–384.

15

Not Here and Everywhere

The non-production of scientific knowledge[1]

Scott Frickel

WASHINGTON STATE UNIVERSITY

In his 1995 *Annual Review of Sociology* essay, Steven Shapin identified the central conceptual problem in the sociology of scientific knowledge as "how to interpret the relationship between the local settings in which scientific knowledge is produced and the unique efficiency with which such knowledge seems to travel" (1995: 290). Coming close on the heels of important edited volumes by Pickering (1992) and Clarke and Fujimura (1992) and marking the cultural turn in science and technology studies (STS), Shapin's essay was widely cited and helped solidify the field's contemporary focus on the cultural practices that first generate and then circulate scientific knowledge. The essay's title "Here and Everywhere" signified the field's focal attention to these dual processes.

But if Shapin's framing accurately portrayed the state of STS research in the mid-1990s, it failed to capture the full range of the conceptual problems facing the field then and today. Specifically, it did not consider how knowledge production is forestalled or how the transmission of knowledge from one place to others is constrained. Lots of knowledge that could be made isn't and when knowledge circulates it does not do so uniformly. In fact, scientific knowledge is not "here and everywhere" and STS should have interesting things to say about how and why that is. I believe a reformulation of the problem is needed, one that better highlights the relational dynamics of knowledge production and non-production: Why are certain kinds of scientific knowledge created, certified and circulated while other kinds are not?

Answers to this seemingly straightforward question are not easy to come by. Half a century has passed since Thomas Kuhn (1962) published his endogenous social theory of knowledge growth, yet there are few empirically-substantiated explanations for the variation we see in rates, processes, and forms of intellectual change (Frickel and Gross 2005). That is, if we think we know quite a lot about why some areas of scientific research and expertise flourish, we still know relatively little about why other areas wither on the vine and still others – no doubt most – simply fail to germinate ... or perhaps not so simply fail.

This chapter considers the hidden half of the knowledge production/non-production equation and calls for greater scholarly attention to the problem of ignorance or the absence of knowledge[2]. It covers a range of issues, but focuses mainly on how, where, and why ignorance, once produced, becomes institutionalized within and beyond science. Like knowledge, the production of ignorance can also be local and situational. However, over time different types of ignorance accumulate and combine to create complex architectures or structures of non-knowledge. These structures reflect and reinforce a range of social values and power relations and are significant for their consequences, both intended and unintended. Studying the processes, structures, and impacts of institutionalized ignorance is not easy, in part because such absences are rarely visible or readily apparent, but also in part because the values and power relations dominating STS in recent decades have tended to channel analytical attention toward the centers of scientific power (see Hess 2011). To render ignorance and its effects more visible and thus more available as a productive element of theory and research design, STS will need to spend more time and energy at the peripheries of scientific action and among the experts and non-experts who inhabit those marginalized spaces and positions (see Hecht 2009 and Chapter 20 of this book).

Why studying ignorance matters to me

My own interest in these issues is personal as well as intellectual, stemming from hurricane Katrina's catastrophic flooding of New Orleans in August 2005. I resided there with my family when the hurricane hit. We evacuated before the storm made landfall and, over the next days, weeks and months we followed the catastrophe unfolding in New Orleans like the rest of the country, glued to televisions, radios, and computers. Nearly four months later we moved back home. Over the next eighteen months we struggled with the stark consequences of our return to a city that seemed at the time to be on the brink of total collapse. We ultimately left a second time, in search of more certain ground on which to build our lives, but paradoxically, it is New Orleans' uncertain ground – and the relationship of soil quality to environmental risk following the flood – that continues to hold my intellectual fascination and helps secure the city's emotional hold on me.

As it turns out, the research question I originally fashioned for what has become an ongoing study of post-Katrina environmental risk – "what is contaminated soil?" – is not a simple one to answer, all the more so for what remains unknown, despite the unprecedented scale and intensity of risk assessment efforts by federal and state agencies, city officials, university scientists, and citizens groups. As I've wrestled to collect and make sense of my case materials in the years since the disaster, I've come to better appreciate the significant and lasting power that the absence of knowledge can hold over decision-making about residents' soil remediation options, the city's recovery plans, or changing patterns of land use. These experiences and observations inform the arguments I build here. In the course of the essay I will turn frequently to the case material to illustrate ideas, support claims and push against conventional wisdom. For now, some background will help orient readers otherwise unfamiliar with the basic features of the case.

Hurricane Katrina hit southeast Louisiana on 29 August 2005, triggering a systemic failure of the federal hurricane levee protection system surrounding the City of New Orleans. At peak volume, 131 billion gallons of salt water inundated eighty percent of the city's land area (Campanella 2007, Smith and Rowland 2007). As the flood waters gradually receded, a layer of chemical-laced sediment blanketed the city (Nelson and Leclair 2006). In response, a dozen state and federal regulatory organizations led by the U.S. Environmental Protection Agency (EPA) and the Louisiana Department of Environmental Quality (LDEQ) began a year-long effort to characterize the nature and type of contamination in residential areas of four flood-impacted parishes. Targeting 195 different contaminants, project scientists collected more than 1,800 samples and conducted more than 400,000 chemical analyses, using the test results to calculate short- and long-term human health risk. Nearly a year after beginning the assessment project, EPA (U.S. EPA 2006) released a final summary report concluding that "the sediments left behind by the flooding from the hurricanes are not expected to cause adverse health impacts to individuals returning to New Orleans." The risk assessment project was a bright light in the otherwise dark confusion of Katrina's aftermath; politicians, business leaders, and residents alike celebrated the headline news that "EPA Declares N.O. Safe" (Brown 2006).

Asking "is the city safe?" as regulatory scientists, business leaders and city officials have done directs analytical attention to what is known about soil contamination and environmental risk. My study takes a different approach, instead asking "what remains unknown and why is that?" This question trains one's investigative focus in a different direction, toward the ways in which the absence of knowledge organizes and structures regulatory practice, including scientific and political constructions of risk. As I pursue this alternative line of inquiry, I am finding considerable evidence that what became known about soil quality in post-Katrina New Orleans depended in large part on what remained unknown. Before I turn to showing that the lessons from New Orleans have broad relevance to STS's understanding of ignorance, I examine the stakes for STS in underdeveloping theories of ignorance, and consider some of the opportunities, challenges, and methodological problems that contribute to this situation.

What is at stake for STS?

The social study of ignorance is marked by a long, but somewhat tortured intellectual history. As a topic of social analysis, ignorance has weathered prolonged periods of scholarly inattention punctuated occasionally by flurries of short-lived interest (for reviews see Smithson 1989 and Gross 2007). We are in the midst of one of those bursts now. Interest in the non-production of knowledge as an anthropological, historical, and sociological problem is on the rise, as illustrated by a number of recent conference panels, books, edited volumes, and special issues in academic journals addressing the topic (Bauchspies and Croissant 2014, Gross 2010, High et al. 2012, McGoey 2012a, Proctor and Schiebinger 2008, Sullivan and Tuana 2007) as well as the first *Handbook of Ignorance*, planned for publication in 2015 (Gross and McGoey forthcoming 2015). This work is highly diverse in theoretical orientation, scope, and topical focus, as illustrated in this Handbook in chapters by Hecht, Kinchy, Mayer and colleagues, and Waidzunas, each of which offers particular renderings of ignorance and related themes. These chapters collectively demonstrate the viability of ignorance as a research domain deserving sustained and thoughtful inquiry.

This current burst of attention notwithstanding, an accounting of STS epistemic practices reveals a wildly lop-sided ledger. The field has persistently emphasized the production, growth, and circulation of scientific knowledge over its non-production, decline, or sequestration. By giving such disproportionate attention to knowledge production, most STS scholars have in effect collectively followed the methodologically suspect path of sampling on a rather small class of dependent variables: facts, artifacts, and other socially recognized products of scientific work. I believe there are significant costs associated with this chronic imbalance.

One example of such costs is that we have not cultivated a tradition of systematically tracking the consequences of the non-production of knowledge. Government agencies such as the National Science Foundation meticulously track the value added to knowledge as it feeds technological development and economic growth, but these agencies do not generate comparable statistics tracing the potential value lost to scientific inactivity or failure. In the absence of such comparative statistical data, we are left to rely mostly on anecdotal data from scientists whose grants are not funded, innovations are not patented, or ideas are not picked up and who offer subjective projections of the broader potential losses to science stemming from their experience with failure.[3]

This is clearly inadequate. To the extent that our collective work neglects to study failure, we limit a deeper, relational understanding of the social dynamics of science. As a result, today STS lacks a systematic theory and analysis of the structural conditions of scientific work that inhibit knowledge production practices and subsequently shape epistemic cultures through those absences. Instead, theoretical attention in STS generally remains focused on achievements – variously won, but won nonetheless. Over time, this imbalance has the further consequence of narrowing possibilities for scholarly engagement in public debate on questions involving, for example, sciences' role in creating new forms of inequality.

For example, the racial and economic structure of the population of flooded New Orleans households mirrored the racial and economic structure of the city as a whole – African American and white households were impacted by flooding at levels roughly proportionate to their share of the total population (Campanella 2007)[4]. In contrast, the sampling and testing strategies employed by regulatory agency scientists distributed knowledge of soil conditions unevenly across that same flood zone. The uneven sampling produced areas represented officially by more or less information about soil quality, but also excluded some neighborhoods entirely from the risk assessment process (Frickel and Vincent 2011)[5]. Teasing apart the social and environmental consequences of this spatial inequality are difficult, but my research with collaborators has found that, in contrast to what theories of environmental justice might lead us to expect, knowledge produced from testing sediment and soil samples was concentrated in neighborhoods whose residents were disproportionately African American and lower-income. Wealthier and whiter neighborhoods that flooded received disproportionately less regulatory attention overall (for some of the details see Frickel et al. 2009). An added dimension of this complexity is that areas that were excluded from sampling altogether collectively represented a widely diverse sub-set of New Orleans neighborhoods, poor and wealthy, black and white, and in between. Indeed, some of the city's most economically and racially diverse neighborhoods were completely bypassed by regulatory efforts to produce place-specific knowledge of soil conditions and environmental risk (Frickel and Vincent 2011).

The larger point is that while the absence of knowledge clearly advantages some groups and disadvantages others, in New Orleans this dimensional axis runs orthogonally through race and class, complicating standard notions of racial and class bias. Just as the flood created a new form of environmental inequality (flooded vs. not flooded), regulatory agencies' efforts to assess risk within the flood zone created a new form of epistemic inequality (sampled vs. not sampled).

So, while it can be extremely difficult to measure systematically, the absence of knowledge of soil contaminant levels in certain neighborhoods undoubtedly limited residents' ability to contest official practices and decisions – illustrating Harding's (2006) claim that knowledge inequalities within science can reflect and deepen existing and new structural and cultural inequalities in the broader society.

In this way, ignoring ignorance can be counter-productive. It constrains opportunities for theory building and comparative analysis and fails to capture the full range of social practices that lead to fields of knowledge and non-knowledge. In the remainder of this chapter, I describe some of the opportunities for conceptual elaboration to redress this problem and some of the challenges accompanying them, beginning with some observations on the ontological ambiguities that can unintentionally hamper efforts to pin down ignorance as an object of analysis.

Ignorance as challenge and opportunity

Ontological ambiguities

An informative difference between studies of knowledge and studies of ignorance is that the former exhibits an ontological clarity that the latter often lacks. Revisiting the early constructivist studies in the sociology of scientific knowledge (SSK) such as Collins' *Changing Order* (1985) or Latour and Woolgar's *Laboratory Life* (1979), one finds clear statements of what these theorists understand scientific knowledge actually to be. We may argue with their theories about the social nature of scientific knowledge, but few would question whether scientific knowledge exists. By contrast, the literature on ignorance exhibits far less confidence in the ontological status of its subject.

This ontological ambiguity is perhaps best illustrated with reference to uncertainty, a concept that is often described as roughly synonymous with ignorance or as representing a particular type of ignorance. There are two related difficulties here. One difficulty is that uncertainty can be temporary, as when potential knowledge has not yet been produced, but it can also be endemic, as with stochastic processes that are inherently open and can only be understood through probabilistic knowledge. The former is not yet known and is thus ontologically ambiguous; the latter is unknowable and thus has no ontology (Rescher 2009, cited in Croissant 2014). A second ontological difficulty is that most STS studies that are about uncertainty actually investigate how scientific and policy decision-making advances (or does not advance) under conditions of limited knowledge (for example, Jamieson 1996, Shackley and Wynne 1996). To the extent this literature treats social and political constructions of uncertainty, its primary focus remains on what becomes known; that is, how knowledge of uncertainty is stabilized (Hoffman-Reim and Wynne 2002). By and large, this literature does not directly confront the problem of ignorance, theorizing what remains unknown and why that is. As a result, and somewhat paradoxically, studies of uncertainty generally do not resolve the ontological ambiguity of their subject. The STS literature on risk, described by Van Loon (2002: 2) as a virtual object that is forever "becoming-real," faces similar challenges.

More broadly, difficulties with the ontological status of non-knowledge has led to a strong tendency – one might even say a preoccupation – toward abstract philosophical discussions about what non-knowledge precisely is and what forms it takes, with many authors proposing different taxonomic systems for organizing a growing family of concepts. The result has been a miniature cascade of categorization schemes. Bernstein's (2009a, 2009b) detailed parsing of "non-knowledge" into sub-classes of ignorance, stupidity, error, and unreason is one example among many. These efforts at categorical refinement have given rise to a confusing number of more specific terms, including functional ignorance (Moore and Tumin 1949), specified ignorance (Merton 1987), specified known ignorance (Böschen et al. 2010), meta-ignorance (Smithson 1989), and ignorance of ignorance (Ravetz 1993). As Matthias Gross (2007: 744) has observed, many of these efforts "are certainly well thought through, but they rarely lead to clarification" in part because they give "little or no attention to or links with concrete examples or data."

To various degrees then, studies of uncertainty-making and the various taxonomic exercises on offer tend to side-step the central problem which, again, is to explain how and why certain kinds of scientific knowledge are not created, or certified or sent into circulation. Referring to the New Orleans case, for example, we might ask not how scientists constructed uncertainty surrounding environmental risk, but what role the absence of knowledge played in securing the EPA's claim that sediment left in Katrina's wake posed little or no risk to returning residents.

267

To answer that question in as comprehensive manner as possible, we would need to identify the institutional mechanisms that generated an architecture of ignorance as the question of risk moved from the domain of policy, to regulatory practice, to its uptake in civil society. For guidance in meeting that goal, I turn next to important opportunities emerging from scholarly work that explicitly positions ignorance as a historical and social problem to be investigated in the context of established social theory.

Epistemological opportunities

Concerned more with their topic's epistemology and social dynamics than with its uneasy ontology, a growing stream of empirical studies have generated two distinct but related claims about ignorance as it relates to the culture and social organization of science (for example, Hess 2007, Kleinman and Suryanarayanan 2013, Sullivan and Tuana 2007, Proctor and Schiebinger 2008, Gross 2010). The first claim is that ignorance is inherent in technoscientific practice, not as a simple byproduct or passive outcome of scientific work but as "an active production" in itself, one that is complexly bound up with knowledge, and thus also with politics and power (Tuana 2008: 109). The second, related claim is that ignorance, like knowledge, is shaped by and reflective of broader relations of power. Thus, for Tuana and others, studying what is not known is important because it "has the potential to reveal the role of power in the construction of what is known and provide a lens for the political values at work in our knowledge practices" (2008: 109–110).

My research in post-Katrina New Orleans lends empirical support to both of these claims. There, power is inscribed and enacted through environmental assessment protocols that limited a priori the risk-related questions regulators were legally required to answer and thus limited the kinds of questions regulators in fact asked. In turn, questions regulators asked shaped what came to be known and not known. In this way, the *power to not know* also shaped officials' decisions to follow those standard protocols, even though they were developed years earlier to deal with industrial accidents involving known types and quantities of contaminants – disasters to be sure, but nothing like the urban-scale complexities that confounded meaningful knowledge production following the 2005 hurricane and flood.

We can also trace power through the organization of environmental testing data and the blind spots such organization creates. In New Orleans, test data generated from soil and sediment samples was delivered to the public in two main forms. In one form the data was condensed into brief occasional reports containing general summaries of environmental conditions around the city but almost no place-specific information that might have been of practical use to city residents. The testing data were also made available online in raw form, completely disaggregated in a dataset containing hundreds of thousands of records but without any summary or interpretation. This approach established EPA's claim to institutional transparency, but it also rendered the data uninterpretable by the very public that had demanded complete access to that information. These dual moves to simultaneously over-aggregate and completely disaggregate the test data helped consolidate institutional power within the regulatory agencies. It did so on one hand by masking the ignorance of experts with a nearly impenetrable "wall" of data and, on the other, spotlighting the knowledge of experts in ways that silenced citizen concerns and helped maintain the publics' dependence on agency experts for meaningful information about contaminant levels and their spatial distribution.

If, as these examples suggest, power informs what is not known as well as what is known, ignorance studies presents STS with some important theoretical opportunities that should not be missed. But exploiting opportunities for conceptual development will require figuring out how to do empirical research when there is literally no "there" there.

Accessing the unknown: methodological considerations

I believe that the most important issues we face in the study of ignorance are methodological (Frickel 2014). How do we most effectively gain access to the unknown? How do we study what does not exist? We can gain some initial methodological traction on this problem by considering how studies of ignorance simultaneously extend and challenge the logic of analytical symmetry, a bedrock principle of SSK.

Symmetry and asymmetry

In *Knowledge and Social Imagery* (1976), David Bloor famously insisted that SSK produce symmetrical explanations of true and false beliefs. Latour and Callon (1992) extended the argument for symmetry to humans and non-humans. Ignorance studies offer another opportunity to extend the logic of symmetry again, this time to the presence and absence of knowledge. Such a program would seek to explain not only why some knowledge counts but also why some knowledge doesn't exist to be counted. At the same time, studies of ignorance present a challenge to Bloor's requirement that SSK also produce deeply empirical accounts. As Croissant (2014: 26) has recently suggested, the non-production of knowledge often yields little direct data and so studying ignorance in a systematic way will require that we "articulate methodological parameters necessary for studying things that aren't there."

One such condition may be a renewed reliance on inference. Early geneticists inferred knowledge of the structure and function of genes through studies of genetic mutations (Frickel 2004); physicists do the same in searching for quarks or bosons (Knorr Cetina 1999). STSers can do it too. Studying the effects of ignorance can cast indirect light back onto the structures that produce it. As Croissant (2013) suggests, this may necessarily substitute for a strictly symmetrical sociology of scientific knowledge, when direct empirical data is non-existent or inaccessible. But in arguing for the value of inference in studying absences, she also rightly cautions against overreliance on counterfactuals, imputing intentionality to actors' inaction, and committing Type II errors (false negatives) which, in this case, would involve assuming that the absence of evidence confirms the existence of an absence.

At present I see no clear navigational route through these hazards of method and logic, but in my opinion that's as it should be. They are part of the new terrains of ignorance and we will need to learn as we go, open-eyed and nimble-footed to avoid missteps when possible. If nothing else, relying more on inference may help nurture a move away from the strict empiricism that characterizes so much of STS research, with its close focus on practice and performativity (for example, MacKenzie et al. 2007, Pickering 1994), and help spark renewed interest in some older concepts such as norms, interests, rules and logics – concepts that may prove useful as elements of an institutional theory of ignorance (see Hess and Frickel 2014).

The challenges that non-knowledge poses to Bloor's strict prescription for a symmetrical SSK extend to STS theory more generally, helping to illuminate some of the limitations of existing theories and approaches. For example, micro-sociological replication studies, such as those pioneered by Collins (1985), have proven extremely useful for understanding processes of consensus formation among scientists but will not go far in helping us understand how absences and silences structure what is not replicated or transmitted. Similarly, we will need to learn more than "how to follow scientists and engineers through society" (Latour 1987) if we are to understand the inactivity of those people, things, and places that slip through the webbing of technoscientific actor-networks. As I discuss next, these challenges are reinforced by an epistemic culture within STS that strongly privileges agency and intentional action over structure and unanticipated consequences (but see Frickel and Moore 2005, Kleinman 2003).

Strategic vs. normative ignorance

As a field, STS maintains a methodologically individualist orientation to research that privileges agency and intentionality over structure in accounts about science and knowledge. This focus on actors as entrepreneurial change agents is replicated in ignorance studies as well. It can be seen most clearly in work that theorizes ignorance as "strategic" (McGoey 2012b), purposefully orchestrated by actors (most often individuals) with interest-driven goals and as a product of the intentional exercise of power. In these accounts, ignorance comes about most often through secrecy (Rappert 2014), censorship (Galison 2008), deceit and suppression (Proctor 1995, Markowitz and Rosner 2002), denial (McGoey 2012a), and doubt (Oreskes and Conway 2010, Michaels and Monforton 2005).

Each of these studies is highly insightful. But taken together such studies advance a conceptualization of ignorance that is overly narrow in two senses. This conceptualization is empirically narrow in that it misses ignorance resulting from the *unintended* consequences of social action. But it is also theoretically narrow, resting on the nominally functionalist assumption, pace Merton (1973), that ignorance derives from deviant science, i.e., assertions of fact that are politically motivated or otherwise self-serving. This assumption leads logically to the troubling conclusion that more public transparency or scientific autonomy (depending of the source of interest) will reduce ignorance by rendering science less deviant. While this may be an appropriate conclusion to draw in specific cases where abuses of political and economic power distort scientific research or where abuses of scientific power distort policy formation and implementation, a robust theory of ignorance in science requires an analytical framework that can accommodate a broader range of processes and outcomes.

The specific challenge here is to theorize ignorance not only as resulting from strategic goal-oriented action, but also as a product of the structural pressures, institutional arrangements and normative cultures that order everyday scientific practice and decision-making. Two important steps toward this broader understanding have emerged from studies of undone science. Hess (2007) has shown how the political economy of scientific fields operates to promote research agendas that are tied to funding opportunities and academic tenure requirements. Over time, research agendas pattern scientific behavior, decision-making and resource distribution to produce pockets of undone science that become taken for granted by researchers working in the field. Kleinman and Suryanarayanan (2013) deepen this argument by examining the ways in which disciplinary cultures, or "epistemic forms," privilege certain ways of knowing over others. Epistemic forms passively reinforce existing areas of undone science but can also be deployed to actively resist non-conformist methods or approaches that would see undone science get done[6]. Together, the studies by Hess and Kleinman and Suryanarayanan illustrate some of the conceptual benefits of moving beyond an actor-oriented, strategic view of ignorance to consider how the institutional structure of scientific fields produce and manage ignorance as a regular consequence of scientific business-as-usual. The New Orleans case provides additional evidence for the importance of institutional structure in understanding not just how ignorance is initially produced, but also how it is reproduced over time.

To better understand the institutionalization of ignorance in EPA's Katrina response project, Michelle Edwards and I studied how the Agency utilized human toxicity values, risk standards, and risk assessment protocols in relation to sediment and soil test results (Frickel and Edwards, 2014). "Human toxicity values" are numerical values that describe dose-response relationships for toxic substances and can be expressed for carcinogenic or non-carcinogenic effects. Risk standards use human toxicity values and other factors (for example, assumptions about an exposed person's weight, age, and gender) to calculate the upper limits of "acceptable risk." Risk assessment protocols, in turn, base risk management decisions (for example, whether or not to pursue additional testing or soil remediation) on whether identified contaminant levels exceed risk standards for a particular contaminant.

From September 2005 through July 2006, EPA scientists identified 141 contaminants in New Orleans soil and flood sediment. We traced this knowledge through the three interconnected policy frameworks of human toxicity values, risk standards, and risk assessment protocols to document how different ways of producing ignorance combine to structure decision-making about risk and safety. In part, we found that the carcinogenicity studies that regulatory scientists rely on to calculate lifetime cancer risk for more than 70% of these 141 chemical substances simply do not exist[7]. This undone science makes it impossible to develop risk standards using direct evidence. Instead, regulatory scientists created risk standards through a bureaucratic process of extrapolation – substituting knowledge of the effects of contaminants for which background studies do exist for knowledge of the effects of other contaminants for which background studies do not exist. Once this sleight of hand is accomplished, risk standards generated from extrapolated carcinogenicity values – rendered as numerical values indicating lifetime cancer risk – become indistinguishable from risk standards that are directly generated from actual studies. This is one way that ignorance is simultaneously produced, hidden from view, and institutionalized as meaningful regulatory science.

Another notable finding from this study is that in many ways the risk assessment project seems largely to have been conducted "off the shelf." In building their risk assessment, for the most part, regulators relied on existing sampling procedures, testing methods (some several decades old), quality assurance protocols, and management options rules. The unprecedented nature of the regulatory response lies in the quantity of the tests it produced, not in any substantial methodological or analytical innovations developed to meet the specific requirements of this one-of-a-kind catastrophe. This finding inversely complements Mayer and colleagues' analysis (Chapter 24 of this book) of the "sniff test" and the barriers to innovation encountered by FDA officials in the context of public controversy surrounding seafood safety testing following the BP oil spill in 2010. In different ways, these two studies seem to suggest that the institutional production of ignorance, and the public's unintentional or unwitting validation of that ignorance, is not necessarily contained by discrete episodes of scientific controversy, but instead can be replicated across different regulatory bureaucracies in response to different kinds of crisis. Indeed, a general conclusion emerging from analyses of different disasters – judging from this book's chapters dealing with oil spills, nuclear reactor accidents, shale gas development, and catastrophic flooding – is that these events produce "knowledge vacuums" and thus greatly intensify societal need for new knowledge (Frickel and Vincent 2010). The mad rush from all directions to regain some semblance of epistemic order (as well as social and political order) creates conditions that render the production of ignorance distinctly visible and therefore accessible to empirical investigation by those willing to sift through the rubble.

Four ways forward

From a growing stream of empirical scholarship I find four emerging strategies for studying ignorance. Each of these approaches is useful in different ways. The currently most popular strategy is to focus on *knowledge sequestration*, or the ways in which existing knowledge is prevented from circulating. Studies adopting this approach address the problem of ignorance in relative terms: the knowledge exists or once existed, but is kept hidden, made inaccessible, is lost or becomes forgotten[8]. The sequestration of knowledge can be intentional or unintentional. Proctor's study of the politics of cancer research illustrates the intentional production of ignorance resulting from the tobacco industry's purposive efforts to keep knowledge of the dangers of smoking confidential (Proctor 1995). This focus on purposive action is echoed in Oreskes and Conway's work on climate change (2010). An example of the unintentional loss of knowledge is Wylie's (2008) analysis of how archeological evidence regularly goes missing as a result of accepted disciplinary methods for archiving and accessing materials gathered from earlier field studies. The advantage of analyzing sequestration is that it can offer explanations derived from direct empirical data, providing one is lucky enough or persistent enough to find it given the systematic non-collection of this kind of data noted earlier.

A second strategy is to study social action in the context of acknowledged ignorance. When actors understand themselves to be operating from a position of not-knowing, they plan for and talk about what they do not know and they act accordingly. This *inferential approach* is illustrated in studies by Matthias Gross (2010) and Linsey McGoey (2012b). Gross introduces the concept of "surprise" to show how knowledge of ignorance can be used productively when it is antic-ipated and incorporated into planning and ecological design. Similarly, McGoey shows how regulatory disputes are shaped by efforts of litigants and regulators to cultivate ignorance about the effects of pharmaceutical drugs. In both examples, the lack of knowledge operates positively, as an organizational resource for pressing claims and guiding action. Studies that follow this strategy gain leverage on ignorance by accessing indirect data on its effects.

A third strategy takes historical processes seriously and examines ignorance as an emergent process structured temporally by degrees and types of *selective attention*. David Hess's (2007) examination of "alternative pathways" in science takes this approach. His work describes how over time areas of knowledge production can atrophy from chronic inattention that generates pockets of "undone science" (see also Frickel et al. 2010). Selective attention, and thus, igno-rance, can also be created by the structural exclusion of women, minorities, and people affected by industrial technoscience, as we've known for some time (Epstein 1998, Fortun 2001, Schiebinger 2000), or by the ontological assumptions that lead researchers to ask overly narrow questions of their research subjects concerning, for example, environmental mutagens (Frickel 2004), gay men (Waidzuna, Chapter 3 of this book) or human brains (Jordon-Young 2010).

A fourth strategy, emerging in my recent work in New Orleans, combines elements of the other three strategies to identify and explain how "knowledge gaps" are produced and institu-tionalized over time, space, and across knowledge domains. I have defined knowledge gaps as "organizationally circumscribed domains of unrealized knowledge" (Frickel and Vincent 2011: 12). I use this concept to study the ways regulatory scientists, city officials, and residents iden-tify, think about, and address problems of environmental risk. To trace these linked processes and their consequences empirically, I have focused my investigation on three domains where knowledge gaps are created or reproduced: in the field and laboratory where regulatory experts collected and analyzed soil and sediment samples; "upstream" in the policy frameworks and protocols that governed official risk assessment practices; and "downstream" in civil society responses to the short-lived and often frustrated contestation of official risk claims by environ-mental activists, neighborhood groups, and some city officials. Just as knowledge circulates through expert, political, and lay communities, knowledge gaps are also carried from one domain to another and combine in mutually reinforcing ways to create architectures of igno-rance that extend across scientific, regulatory, and civil society fields. This approach suggests the utility of comparative and historical research designs that allow scholars to systematically inves-tigate temporal and spatial processes of ignorance production and their institutionalization within and among different social domains.

These four emerging approaches – studying knowledge sequestration, using inferential logics, selective attention, and tracing the way that ignorance flows through different institutional domains – are starting points, not time-worn recipes. They are also not mutually exclusive, and I anticipate important insights will come from combining these strategies in different ways.

Conclusion: terra incognita

The study of ignorance promises not only to extend existing STS theory in interesting directions, but to recalibrate it toward different planes of investigation and understanding. In New Orleans, the consequences of ignorance generated by post-disaster risk assessment are real and lasting. When the EPA issued its final report in August 2006, responsibility for soil remediation effectively shifted from federal and state government to city government, which was economically broke and functionally broken, and to individual homeowners, who have faced strong market disincentives in conducting environmental testing on their own dime. While there was a groundswell of local interest in soil remediation in the first year or two following the flood, today – now nine years into the recovery – it's fair to say that a truly unique opportunity to detoxify the contaminated soils of a once-empty American city has long since passed.

Federally-funded housing development projects are required by law to conduct environmental assessments. But these and the scattered soil remediation projects undertaken by environmental organizations and civic groups are too few and too small to make a substantive difference given the widespread nature of soil contamination in this and many other historic cities. A lasting irony has been that the neighborhoods where investments in soil sampling and testing were most heavily concentrated have been among the slowest neighborhoods to repopulate. By contrast, many of the neighborhoods that have repopulated and seem to be thriving culturally and economically are also among those neighborhoods distinctive for the absence of official knowledge of the environmental quality of their soils. These knowledge gaps matter. They invisibly shape the texture of daily life in a city where intense local concern over soil quality and remediation following the flood has more recently taken a back seat to the issue of "food security" and a burgeoning urban farming and community gardening movement that is literally taking root in backyards and vacant lots across the city, potentially transforming contaminated urban soils into someone's supper. In this context, the intellectual opportunities and challenges presented to STS by the non-production of knowledge are complicated by a sense of moral urgency and paradox: to understand what we do not know.

Notes

1 This chapter benefitted from suggestions by Gabrielle Hecht, Daniel L. Kleinman and Kelly Moore, editors. London: Routledge. Small portions of this essay are developed from material previously published or forthcoming in several articles and book chapters. These sources are attributed in the text.

2 Scholars have used various terms to describe the non-production of knowledge. These terms include absence, agnotology, knowledge gaps, ignorance, non-knowledge, negative knowledge, silence, and undone science, among others. In this essay I use "ignorance" as a general covering term.

3 Kelly Moore deserves credit for this observation.

4 Today the population of Orleans Parish is smaller and, although it remains a majority black city, has proportionately fewer African American and low income residents (U.S. Census 2010).

5 Abby Kinchy and colleagues have used similar methods to track the spatial distribution of knowledge investments in water quality monitoring in the Marcellus shale formation of central Pennsylvania (Kinchy et al. 2013; Kinchy Chapter 14 of this book).

6 For a similar argument, framed in terms of overcoming the disciplinary challenges of interdisciplinary research in biomedical settings, see Albert et al. 2008.

7 This gap is also disturbingly large for non-carcinogenic (i.e., toxicological) effects, ranging from 37%–89% for different measures (Frickel and Edwards 2014).

8   Studies of manufacturing doubt also target circulation, but rather than keeping knowledge contained, this involves putting new knowledge into circulation that casts suspicion on extant understandings of, for example, climate change or the health effects of tobacco or coffee.

References

Albert, M., Laferge, S., Hodges, B.D., Regehr, G., and Lingard, L. 2008. "Biomedical Scientists' Perception of the Social Sciences in Health Research." *Social Science & Medicine* 66: 2520–2531.

Bauchspeis, W.K. and Croissant, J.L. eds. 2014. "Absences," a special issue of *Social Epistemology*, 28(1)

Bernstein, J.H. 2009a. "Nonknowledge: The Bibliographical Organization of Ignorance, Stupidity, Error, and Unreason: Part One." *Knowledge Organization* 36(1): 17–29.

Bernstein, J.H. 2009b. "Nonknowledge: The Bibliographical Organization of Ignorance, Stupidity, Error, And Unreason: Part Two." *Knowledge Organization* 36(4): 249–260.

Bloor, D. 1976. *Knowledge and Social Imagery.* Chicago: University of Chicago Press.

Böschen, S., Kastenhofer, K., Rust, I., Soentgen, J., and Wehling, P. 2010. "Scientific Nonknowledge and Its Political Dynamics: The Cases of Agri-Biotechnology and Mobile Phoning." *Science, Technology & Human Values* 35: 783–811.

Brown, M. 2006. "Final EPA Report Deems N.O. Safe." *The Times-Picayune*, [online] August 19. Available at: *www.nola.com/news/t-p/frontpage/index.ssf?/base/news-6/1155971580163240.xml&coll=1>* (accessed 28 September 2010).

Campanella, R. 2007. "An Ethnic Geography of New Orleans." *Journal of American History* 94(3): 704–716.

Clarke, A.E. and Fujimura, J.H. eds. 1992. *The Right Tools for the Job.* Princeton: Princeton University Press.

Collins, H.M. 1985. *Changing Order: Replication and Induction in Scientific Practice.* Chicago, IL: University of Chicago Press.

Croissant, J.L. 2014. "Agnotology: Ignorance and Absence, or Towards a Sociology of Things That Aren't There." *Social Epistemology*, 28(1): 4–25.

Epstein, S. 1998. *Impure Science: AIDS, Activism, and the Politics of Knowledge.* Berkeley, CA: University of California Press.

Fortun, K. 2001. *Advocacy after Bhopal: Environmentalism, Disaster, New Global Orders.* Chicago: University of Chicago Press.

Frickel, S. 2004. *Chemical Consequences: Environmental Mutagens, Scientist Activism, and the Rise of Genetic Toxicology.* New Brunswick: Rutgers University Press.

Frickel, S. 2014. "Absences: Methodological Note about Nothing, in Particular." *Social Epistemology* 28(1): 86–95.

Frickel, S. and Edwards, M. 2014. "Untangling Ignorance in Environmental Risk Assessment." In *Powerless Science? The Making of the Toxic World in the 20th Century*, eds. N. Jas and S. Boudia. London, Berghahn Books, 215–233.

Frickel, S. and Gross, N. 2005. "A General Theory of Scientific/Intellectual Movements." *American Sociological Review* 70: 204–232.

Frickel, S. and Moore, K. eds. 2005. *The New Political Sociology of Science: Institutions, Networks, and Power.* Madison, WI: University of Wisconsin Press.

Frickel, S. and Vincent, M.B. 2007. "Katrina, Contamination, and the Unintended Organization of Ignorance." *Technology in Society* 29: 181–188.

Frickel, S. and Vincent, M.B. 2010. "Disaster Science: Between Calamity and Recovery," *Items & Issues*, Social Science Research Council (Sept. 15), *http://itemsandissues.ssrc.org/disaster-science-between-calamity-and-recovery.*

Frickel, S. and Vincent, M.B. 2011. "Katrina's Contamination: Regulatory Knowledge Gaps in the Making and Unmaking of Environmental Contention." In *Dynamics of Disaster: Lessons in Risk, Response, and Recovery*, eds. R. A. Dowty and B. L. Allen. London: Earthscan, 11–28.

Frickel, S., Campanella, R., and Vincent, M.B. 2009. "Mapping Knowledge Investments in the Aftermath of Hurricane Katrina: A New Approach for Assessing Regulatory Agency Responses to Environmental Disaster." *Environmental Science & Policy* 12(2): 119–133.

Frickel, S., Gibbon, S., Howard, J., Kempner, J., Ottinger, G., and Hess, D. 2010. "Undone Science: Charting Social Movement and Civil Society Challenges to Research Agenda Setting." *Science, Technology & Human Values* 35(4): 444–473 (lead article).

Galison, P. 2008. "Removing Knowledge: The Logic of Modern Censorship." In *Agnotology: The Making and Unmaking of Ignorance*, eds. R. N. Proctor and L. Schiebinger, pp. 37–54. Stanford, CA: Stanford University Press.

Gross, M. 2007. "The Unknown in Process: Dynamic Connections of Ignorance, Non-Knowledge and Related Concepts." *Current Sociology* 55(5): 742–759.

Gross, M. 2010. *Ignorance and Surprise: Science, Society and Ecological Design*. Cambridge, MA: MIT Press.

Gross, M. and McGoey, L. Forthcoming 2015. *The Handbook of Ignorance*, London: Routledge.

Harding, S. 2006. "Two Influential Theories of Ignorance and Philosophers' Interests in Ignoring Them." *Hypatia* 21(3): 20–36.

Hecht, G. 2009. "Africa and the Nuclear World: Labor, Occupational Health, and the Transnational Production of Uranium." *Comparative Studies in Society and History* 51(4): 896–926.

Hess, D.J. 2007. *Alternative Pathways in Science and Industry: Activism, Innovation, and the Environment in an Era of Globalization*. Cambridge: MIT Press.

Hess, D.J. 2011. "Bourdieu and Science Studies: Toward a Reflexive Sociology." *Minerva* 49(3): 333–348.

Hess, D.J. and Frickel, S. 2014. "Fields of Knowledge and Theory Traditions in the Sociology of Science". *Political Power and Social Theory* 27: forthcoming.

High, C., Kelly, A.H., and Mair, J, eds. 2012. *The Anthropology of Ignorance: An Ethnographic Approach*. New York and Basingstoke: Palgrave Macmillan.

Hoffman-Reim, H and Wynne, B. 2002. "In Risk Assessment, One Has to Admit Ignorance." *Nature* 416(14 March): 123.

Jamieson, D. 1996. "Scientific Uncertainty and the Political Process." *Annals of the American Academy of Political and Social Science* 545(May): 35–43.

Jordon-Young, R. 2010. *Brain Storm: The Flaws in the Science of Sex Differences*. Cambridge, MA: Harvard University Press.

Kinchy, A., Parks, S., and Jalbert, K. 2013. "Fractured Knowledge: Mapping the Gaps in Public and Private Water Monitoring Efforts in Areas Affected by Shale Gas Development." Unpublished manuscript.

Kleinman, D.L. 2003. *Impure Cultures: University Biology and the World of Commerce*. Madison, WI: University of Wisconsin Press.

Kleinman, D.L. and Suryanarayanan, S. 2013. "Dying Bees and the Social Production of Ignorance." *Science, Technology and Human Values* 38(4): 492–517.

Knorr-Cetina, K. 1999. *Epistemic Cultures: How the Sciences Make Knowledge*. Chicago: University of Chicago Press.

Kuhn, T. 1962. *The Structure of Scientific Revolutions*. Chicago: University of Chicago Press.

Latour, B. 1987. *Science in Action: How to Follow Scientists and Engineers through Society*. Cambridge, MA: Harvard University Press.

Latour, B. and Callon, M. 1992. "Don't Throw the Baby out with the Bath School! A Reply to Collins and Yearley." In *Science as Practice and Culture*, eds. A. Pickering. Chicago: University of Chicago Press, 343–368.

Latour, B. and Woolgar, S. 1979. *Laboratory Life: The Social Construction of Scientific Facts*. Cambridge, MA: Harvard University Press.

MacKenzie, D., Muniesa, F., and Siu, L. 2007. eds. *Do Economists Make Markets? On the Performativity of Economics*. Princeton, NJ: Princeton University Press.

Markowitz, G. and D. Rosner 2002. *Deceit and Denial: The Deadly Politics of Industrial Pollution*. Berkeley: University of California Press.

McGoey, L., ed. 2012a. "Strategic Unknowns: Toward a Sociology of Ignorance." Special issue of *Economy and Society* 41(1).

McGoey, L. 2012b. "The Logic of Strategic Ignorance." *British Journal of Sociology* 63(3): 553–576.

Merton, R.K. 1973. *The Sociology of Science*. Chicago: University of Chicago Press.

Merton, R.K. 1987. "Three Fragments from a Sociologist's Notebook: Establishing the Phenomenon, Specified Ignorance, and Strategic Research Materials." *Annual Review of Sociology* 13: 1–28.

Michaels, D. and Monforton, C. 2005. "Manufacturing Uncertainty: Contested Science and Protection of the Public's Health and Environment." *American Journal of Public Health* 95(S1): S39–S48.

Moore, W.E. and Tumin, M.M. 1949. "Some Social Functions of Ignorance." *American Sociological Review* 14(6): 787–795.

Nelson, S.A. and Leclair, S.F. 2006. "Katrina's Unique Splay of Deposits in a New Orleans Neighborhood." *GSA Today* 16: 4–9.

Oreskes, N. and Conway, E.M. 2010. *Merchants of Doubt*. New York. Bloomsbury Press.

Pickering, A. ed. 1992. *Science as Practice and Culture*. Chicago: University of Chicago Press.

Pickering, A. 1994. "After Representation: Science Studies in the Performative Idiom." *PSA: Proceedings of the Biennial Meeting of the Philosophy of Science Association.* Volume Two: *Symposia and Invited Papers (1994)*: 413–419.

Proctor, R.N. 1995. *Cancer Wars.* New York: Basic Books.

Proctor, R.N. and Schiebinger, L. eds. 2008. *Agnotology: The Making and Unmaking of Ignorance.* Stanford, CA: Stanford University Press.

Rappert, B. 2014. "Present Absences: Hauntings and Whirlwinds in –graphy." *Social Epistemology* 28(1): 41–55.

Ravetz, J.R. 1993. "The Sin of Science: Ignorance of Ignorance." *Knowledge: Creation, Diffusion, Utilization* 15(2): 157–165.

Rescher, N. 2009. *Ignorance: On the Wider Implications of Deficient Knowledge.* Pittsburgh, PA: University of Pittsburgh Press.

Schiebinger, L. 2000. *Has Feminism Changed Science?* Cambridge, MA: Harvard University Press.

Shackley, S. and Wynne, B. 1996. "Representing Uncertainty in Global Climate Change Science and Policy: Boundary-Ordering Devices and Authority." *Science, Technology, & Human Values* 21(3): 275–302.

Shapin, S. 1995. "Here and Everywhere: Sociology of Scientific Knowledge." *Annual Review of Sociology* 21: 289–321.

Smith, J. and Rowland, J. 2007. "Temporal Analysis of Floodwater Volumes in New Orleans after Hurricane Katrina." In *Science and the Storms: The USGS Response to the Hurricanes of 2005*, eds. G.S. Farris, G.J. Smith, M.P. Crane, C.R. Demas, L.L. Robbins, and D.L. Lavoie. U.S. Geological Survey Circular 1306, 57–61.

Smithson, M. 1989. *Ignorance and Uncertainty: Emerging Paradigms.* New York: Springer-Verlag.

Sullivan, S. and Tuana, N. eds. 2007. *Race and Epistemologies of Ignorance.* Albany, NY: SUNY Press.

Tuana, N. 2008. "Coming to Understand: Orgasm and the Epistemology of Ignorance." In *Agnotology: the Making and Unmaking of Ignorance,* eds. R.N. Proctor and L. Schiebinger. Stanford, CA: Stanford University Press, 108–145.

U.S. Environmental Protection Agency 2006. "Summary Results of Sediment Sampling Conducted by the Environmental Protection Agency in Response to Hurricanes Katrina and Rita." August, 17. Available at *www.epa.gov/katrina/testresults/sediments/summary.html* (accessed September, 28 2010).

Van Loon, J. 2002. *Risk and Technological Culture: Towards a Sociology of Virulence.* London: Routledge.

Wylie, A. 2008. "Mapping Ignorance in Archaeology: The Advantages of Historical Hindsight." In *Agnotology: the Making and Unmaking of Ignorance,* eds. R.N. Proctor and L. Schiebinger. Stanford, CA: Stanford University Press, 183–205.

16

Political Ideology and the Green-Energy Transition in the United States

David J. Hess

VANDERBILT UNIVERSITY

The problem of environmental limits is one of the most important political challenges facing modern societies, because failure to address the limits could lead to catastrophic disasters. The most pressing environmental limit is anthropogenic greenhouse gases, but there are many others, including the destruction of habitats, loss of freshwater resources, and persistent chemical pollutants in the environment. Because markets do not adequately internalize long-term environmental costs, it is necessary for public policy to guide the redesign of large technological systems (LTSs) – such as electricity, transportation, and food production – so that they are more sustainable. However, the definition of the goal (the design of systems that are in some sense more sustainable) and the optimal pace of reform are highly contested politically. Thus, the study of the transitions of LTSs requires an approach that can interpret and understand the political processes involved. This chapter will outline the approach that I have been developing for the study of the sustainability transition, apply it to the case of the politics of the green-energy transition in the U.S. during the Obama administration, and discuss some general implications for theory development in science and technology studies (STS). The implications will include the importance of scale and geographical unevenness in the framework for the study of sustainability transitions.

Conceptual background

Contemporary STS research on technological change can be divided into two main traditions. First, constructivist accounts, such as the social construction of technology and actor-network theory, draw attention to the role of actors in negotiating changes in sociotechnical systems (for example, Bijker et al. 1987). From this body of work a set of influential and useful concepts has emerged for the study of technological change, including interpretive flexibility, social negotiation, enrollment, obligatory points of passage, and closure or stabilization. The second main approach, the study of large technological systems (LTSs), emphasizes the development and transition of those systems, the imbrications of social and material processes, and the factors that affect system change and stasis (Geels 2005, Grin et al. 2010, Hughes 1983).

The approach adopted here builds on and extends both the agency-based and transition frameworks by beginning with field theory and drawing attention especially to the role of political ideology in technological change. Social fields are networks of actors (both individuals and organizations) who share a common definition of what is at stake but have different viewpoints about what the outcomes of action in the field should be (Bourdieu 2005, Fligstein and McAdam 2012). Actors engage in relations of cooperation and conflict to achieve dominance for a particular vision of what the field should be and who should dominate it. For the study of technological change, the most relevant social fields are the political field (characterized by conflicts over the control of legislatures and government administrations in order to influence policy outcomes), the scientific field (where conflicts over priorities or agendas in a research field that affects technological innovation and evaluation are central), and the industrial field (in which conflicts over market position and the dominance of one type of design or product over another are fundamental; see also our discussion in Moore et al. 2011). These fields are cross-cut by differentials of power and actors associated with more and less privileged groups both within fields and across fields. Thus, conflicts of class, race, gender, and inequality are essential for analyzing fields. Furthermore, the fields themselves can be analyzed at variable scale from the microsocial to the global.

Whereas functionalist approaches to the similar meso-sociological concept of institutions emphasized unity and coherence under relatively stabilized systems of norms and rewards (Merton 1973), field analysis draws attention to relations of cooperation and conflict, the importance of strategies, and the role of power in the sense of the differential capacity to influence outcomes. Although it is possible for completely egalitarian fields to exist, in general social fields tend to be characterized by at least some agents in relatively subordinate positions, and they sometimes see themselves as challengers to "incumbents" in dominant positions. For example, in the scientific field there are conflicts between the dominant networks with their mainstream research programs and challengers who often occupy less powerful institutional positions (for example, Brown 2007, Frickel and Gross 2005). Challengers may also be aligned across social fields; for example, some scientists, often in subordinate positions in their own research fields, may form alliances with social movements that have identified "undone science," that is, systematically underfunded areas of research that may be of broad potential benefit (Frickel et al. 2010, Hess 2011).

One of the weaknesses of current formulations of field theory is that the concept of culture tends to be narrowly understood in terms such as "habitus" and "social skill." The Geertzian approach to culture as models of and for action is more open, but Geertz, like Merton for the study of institutions, emphasized the coherence and integration of cultural systems rather than their contested aspects (Geertz 1973). Thus, one needs a method that emphasizes culture as only partially shared, sometimes unconscious, and sometimes contested. When applying this approach to the political field, I have focused on tensions among political ideologies, just as in the scientific field one might study tensions among research programs or paradigms. This approach is similar to one emerging in the sociology of science and organizations that examines the role of competing institutional logics (for example, Berman 2012). Although this approach has proven very valuable for studying contemporary change in science and society, the institutional logics perspective tends to have a slightly different analytical focus from the one adopted here. Institutional logics are generally understood as relations among cultural systems rooted in diverse social institutions such as religion, the market, the corporation, the state, professions, the family, and community (Thornton et al. 2012). Individuals and organizations may bring together such logics for strategic purposes, such as when J.C. Penny used a religious institutional logic to counteract customers' attractions to the family logic of the small business (ibid.). Instead, in this study the focus is more on cultural systems that emerge as part of the social relations of the field and are defined by mutual opposition, a level of analysis that one might term "field logics." In the political field these logics appear as an "ideological field," a dimension of the political field that involves contestation over the systems of legitimate political principles that shape and are shaped by political discourse and policy. We might think of the ideologies as cultural systems, but reanchored in a field sociology that eschews the integrationist assumptions of Geertzian culturalism and Parsonian functionalism.

In the United States and most other industrialized Western countries, the central ideological conflict in the political field is between some variant of social liberalism (often called "social democracy" in Europe) and an opposing configuration of neoliberalism. The former is associated with the view that the government should exercise a relatively strong redistributive and regulatory function with respect to the market and that relatively high levels of government intervention in the economy are justified in order to serve collective goals such as health, environmental protection, and social fairness. In its strongest forms, social liberalism forms a continuum with socialism, which advocates increased government ownership of crucial industrial sectors such as health care, energy, communication, and transit. In contrast, neoliberal ideology articulates the position that markets should be enabled wherever possible, that they should replace policy wherever possible, that their regulation should be minimized, and that nonprofit and for-profit organizations are the best sites for solving the social problems associated with redistribution. In its strong form the view involves market fundamentalism, which advocates widespread deregulation, dismantling of welfare-state protections, and privatization of public assets.

In studies of the green-energy transition in the U.S., I have shown that a second ideological tension has also played an important role in policy disputes (Hess 2012a). Developmentalism involves support for government intervention in the economy in order to nurture and protect local or domestic industries in the face of global competition. In the nineteenth century, the U.S. used protectionist and industrial policies to build up its industrial base and to protect it from European competition, and in the twentieth century such policies were common in the newly industrializing countries that practiced import-substituting industrialization. Developmentalist policies and ideology also continued in the U.S. throughout the twentieth century at the state government level in economic development policies and programs. Just as there is a tension between neoliberalism and social liberalism, so developmentalism is sometimes challenged by a less influential form, localism, which focuses on the development of the locally owned small-business sector, often through an import-substitution rhetoric and strategy.

Before using the concept of ideology in the study of the politics of LTS design and transitions, two qualifications are necessary. Ideologies are ideal types. Sometimes specific statements by political actors and specific policies approximate those ideal types, but politics is the art of compromise, and consequently ideologies often appear as compromise formations, in which elements of disparate ideologies are overlaid and syncretized. Furthermore, the explicit debates over political ideology and the changes in policies, organizational routines, and everyday practices that coincide with those debates interact with deeper transformations of cultural practices (Rose et al. 2006). For example, the debates over social liberalism and neoliberalism contribute to a doxa of a diffuse cultural logic of responsibility, with differences between a social sense of responsibility and an alternative, enterprising sense. Likewise, the tensions in economic development ideologies create a doxa of place-based identity, with tensions between a protective and globalist sense. This point will not be developed here, but it is important to flag it in order to avoid misunderstandings about the relationship between ideologies, which are relatively explicit and self-conscious systems of meaning that orient strategy and identity in a field, and the more implicit cultural logics that become embedded in a wide range of practices across diverse social fields.

The remainder of the chapter will show how these concepts apply to the politics of the green-energy transition in the United States. Consistent with the goals and format of this volume, the analysis will include a discussion of the relationship between institutionalization and disruption of technology transitions, political values and material culture, and scalar and spatial dynamics of the politics governing the transitions.

Disruptions

In the environmental policy field in the U.S., as has occurred in many other policy fields in many countries, there was a long-term transition from command-and-control regulation to a second generation of policies that relied more on market mechanisms (Mazmanian and Kraft 1999). In turn, these changes were part of the broader neoliberalization of the political field that began during the 1970s and at first involved regulatory roll-back (Harvey 2005, Peck 2010). By the 1990s, growing scientific knowledge of the effects of greenhouse gases on global warming led to increasing calls for policy reforms from environmentalists and allied political leaders, but these policies were in conflict with both the interests of the fossil-fuel industry and with neoliberal ideology, because a response to climate change required extensive government regulation of industrial processes (Klein 2011). The Byrd-Hagel Resolution of the U.S. Senate, which passed with a 95–0 vote, blocked the ratification of the Kyoto Protocol. The text of the resolution also voiced the view that restrictions on greenhouse gases would harm the country's economic viability and that developing countries should also join such agreements. In terms of underlying ideologies, the resolution combined the neoliberal view that environmental regulation was potentially harmful to markets with the developmentalist view that the U.S. should defend itself against treaties that would be prejudicial to domestic industries (U.S. Senate 1997). This view was embraced by the Republican president George W. Bush (2000–2008), who opposed all major initiatives toward a green-energy transition.

The election of President Obama in 2008 disrupted the stasis in the federal government's environmental policy field, because his promise to create five million green jobs was a central goal in the first years of his administration. The frank acknowledgement of climate change and its linkage to job creation and business development was a substantial change from the previous administration. In the 2008 election, the president drew on support from a diverse coalition that included the Blue-Green Alliance of labor and environmental organizations, which provided some financial support and on-the-ground voter turn-out for Democratic Party candidates. Although portions of the green jobs policies were consistent with social liberalism (such as the programs that supported weatherization of low-income households and job creation for relatively unskilled workers), Obama's rhetoric and programs drew on developmentalist ideology to frame green-energy policies as instruments of energy independence, job creation, domestic manufacturing revival, and business development. Thus, environmental reform was linked to the economic crisis and the goal of job creation through green industrial development. The first legislative victory associated with the transition, the American Recovery and Reinvestment Act of 2009, included support for a wide range of green technologies and green jobs programs, but the administration also adjusted annual budgets in favor of renewable energy and energy efficiency. In 2009 the House of Representatives passed the American Clean Energy and Security Act (HR 2454), a sweeping law that would have created potentially millions of green jobs by bringing a cap-and-trade system of emissions regulation to the country, a national renewable portfolio standard for electricity (20 percent by 2020), and a national energy-efficiency standard.

However, by 2010 the disruption was itself disrupted. Political conservatives joined with wealthy donors associated with the fossil-fuel industry to mount a successful campaign in the U.S. Senate to block the corresponding bill, the Clean Energy Jobs and America Power Act (S. 1733). Senator John Kerry framed the bill not in social liberal terms as an environmental initiative but instead in defensive developmentalist language as "a bill for billions of dollars to create the next generation of jobs, and a bill to end America's addiction to foreign oil" (Lieberman 2010). Although lobbyists for the renewable energy industries and advocates from the Blue-Green alliance worked to support the legislation, they had significantly lower spending capacity than the fossil-fuel industry (Open Secrets 2013). Spending by wealthy donors associated with the fossil-fuel industry also had an effect on the mid-term elections of 2010. Of the 100 newly elected members of Congress in 2010, 94 made some kind of anti-green pledge, such as signing the "No Climate Tax Pledge" of the fossil-fuel funded organization Americans for Prosperity (Johnson 2010).

Many of the newly elected members of Congress and some of the existing members also endorsed skepticism and denialism of climate science, a position that was especially prominent in the Tea Party movement within the Republican Party. An opinion poll of self-identified Democrats, Republicans, independents, and Tea Party advocates showed that the first three groups believed that global warming was happening and supported at least a modest renewable portfolio standard law and global cap-and-trade treaties, whereas Tea Party supporters held the opposite view and even opposed the federal government mandate to shift to fluorescent light bulbs (Leiserowitz et al. 2011). Originally promoted by fossil-fuel companies, the denialist movement spread to include conservative news media and conservative foundations and think tanks (Goldenberg 2012, Jacques et al. 2008). Once elected, anti-green Republicans held hearings on climate science and invited climate deniers to testify. In October, 2011, the House Science, Space, and Technology committee issued a strategy letter that targeted climate science funding across a wide range of government agencies, from the Department of Energy to the National Aeronautics and Space Administration, the National Oceanic and Atmospheric Administration, and the National Science Foundation (Climate Science Watch 2011, U.S. House of Representatives 2011). Although the Senate, which was controlled by the Democratic Party, blocked many of the proposed budget cuts, the attack on climate science funding had a chilling effect on federal government agencies.

The situation of a disrupted green-energy transition in the U.S. contrasts with that of some countries in Europe and even Asia, where fossil-fuel firms have a weaker position with respect to governments, and the industry in those countries has tended to engage in an energy diversification strategy and accept government initiatives toward a green-energy transition. In the U.S., there is an interfield configuration that facilitates industrial influence on the political field in general and enables a network of wealthy donors associated with the fossil-fuel industry to neutralize green-transition policies. The configuration has affected the field relationship between science and the state, which in the energy-environment policy field is altered from the traditional pattern found in other advanced, industrialized countries and even in many other policy fields in the U.S. Although the dominant networks of the climate research field (indeed, almost all researchers in the field with any symbolic capital) agree that greenhouse gases represent a significant threat to the stability of the global climate and that action is needed, they have little capacity to affect the policy process (Hess 2014). The traditional understanding of the expertise-policy process is no longer a helpful guide, because the capacity for scientists to present their knowledge as apolitical and neutral with respect to the political field's ideologies is lost. An epistemic rift emerges as the science of global warming becomes equated with social liberalism, the road to serfdom, and European-style strangulation of the economy through state directed intervention.

After the electoral defeat of the Democrats in 2010 and the purging of many moderate (and often pro-environment) Republicans during the same election, the president backed away from the five-million green jobs frame and, to a large degree, from any new green-energy initiatives in the Congress (Roberts and Kincaid 2012). The 2012 election revealed a slate of Republican presidential contenders who were either outright climate-science deniers or opposed to the need for climate mitigation policies. However, the president also remained mostly silent on the issue and embraced an "all of the above" strategy on energy policy. As a result of the lack of support for green industrial policy in the U.S., the level of investment in green technology is lower than that of other G-20 countries on a per capita basis. In absolute figures China's government investment in clean tech surpassed that of the U.S. in 2009, and according to an analysis by Pew Charitable Trusts, Asian countries were poised to become global leaders in clean technology by 2040 (PEW Charitable Trusts 2010a, 2010b). American companies are dominant in the software-intensive smart-grid industry, but during the 2000s they lost their prominent position in several areas of manufacturing, such as solar photovoltaics and batteries.

With policy initiatives blocked in Congress, the president shifted to administrative measures, such as voluntary fuel-efficiency constraints with the automotive industry, green-energy goals for the military, and the regulation of carbon-dioxide under existing pollution laws. In response to the Obama administration's policy of greening the government and the military, by the summer of 2012 Republicans were also attacking the military's green energy policy (Abramson 2012). Although Democrats retained control of the White House and Senate in the 2012 election, the political opportunity structure for green-transition policies at the federal government level remained closed except through administrative measures.

Technological differences and political positions

The previous discussion presents a first-level narrative of the politics of the green-energy transition in the U.S.: it became caught up in partisan disputes that had a left-right polarity associated with neoliberal and social liberal positions. To overcome the opposition, Democrats adopted strategies that drew on developmentalist ideology, repackaged social liberal regulations through market-oriented policy instruments such as cap-and-trade policy, and developed political coalitions in favor of specific types of green technologies.

The process of building and maintaining support for green-transition policies resulted in the disaggregation of the broad category of "green-energy technology." The different types of green technology became associated with different political programs and ideologies. For example, the weatherization programs and "green collar" jobs that Democrats advocated were addressed to the needs of low-income, urban constituencies, because the programs created jobs and generated savings on home heating and air conditioning costs, and some programs involved unions such as Laborers International Union of North America. This element of the Obama administration programs was closest to the ideal of a "green New Deal" that some members of the labor-environmentalist coalition envisioned in 2009. Other programs, such as high-speed rail and support for wind farms, were popular with unions associated with manufacturing, including the United Steelworkers, which played a leading role in the Blue-Green Alliance (Hess 2012a).

Unions also endorsed developmentalist policies especially for trade-related issues. They drew attention to the loss of manufacturing jobs and the need for a more defensive posture with respect to the aggressive developmentalism of trading partners, especially China (see Chen, Chapter 7 in this book). One response from the administration was the "Buy American" program of the American Reinvestment and Recovery Act, and another was the administration's decision to investigate a trade complaint launched by the United Steelworkers, which claimed that China had manipulated trade agreements and significantly damaged green manufacturing in the U.S. (United Steelworkers 2010). In 2012 the administration responded to additional complaints from the solar and wind industries by initiating tariffs of up to 250% on imported photovoltaic panels and similar tariffs on wind turbine towers (Cardwell 2012, International Trade Commission 2012). By developing domestic procurement goals and adopting a more aggressive stance toward trading partners, the administration was able to shift developmentalist rhetoric to the side of green-transition policies (unlike the framing in the Kyoto Protocol debate, when developmentalism worked against such policies). Although the administration utilized developmentalist ideology rather than social liberal rationales, the shift was still opposed on neoliberal grounds. For example, Republican Congressman Fred Upton used the bankruptcy of Solyndra, a solar manufacturing firm that had been supported with a government loan guarantee, to "question whether the government is qualified to act as a venture capitalist, picking winners and losers in speculative ventures and shelling out billions of taxpayer dollars to keep them afloat" (Wald and Savage 2011). The president responded by vowing not to surrender the U.S. solar industry to China (Condon 2011).

In summary, a defensive developmentalist ideology served to parry ongoing attacks on green industrial policy: the Obama administration defended its support for green-transition policies by arguing that they were opportunities for job creation and business development rather than burdensome regulations or inappropriate interventions in the economy. But Democrats also attempted to clothe their demand policies, such as carbon regulation, in market instruments such as carbon trading, thus providing some armor for the policy initiatives against the attacks of market fundamentalists. By creating new markets, advocates of the green energy transition also encouraged divisions in the industrial field between the fossil-fuel industry and financial capital, which found new investment opportunities in green technology and green financial products. Here, government programs oriented toward research and development, such as the Advanced Research Project Agency-Energy (ARPA-E 2012), provided support to cutting-edge, high-technology projects, which in turn created opportunities for venture capital.

Other policies were consistent with localist movements, that is, movements that supported increased levels of local ownership and control. For example, property-assessed clean energy financing provided a financial mechanism for owners of homes and small businesses to invest in rooftop solar energy without long-term liquidity risks (Hess 2012a, 2013). Monthly repayment could be set up to be roughly equal to monthly energy savings, and at the end of the repayment period, ownership of energy production was partially transferred to the building owners. However, the associations were not necessarily stable. For example, by 2010 rooftop solar energy was receiving billions of dollars of investment from technology companies and the financial services industry, which developed third-party finance agreements that enabled homeowners to benefit from rooftop solar but restricted long-term local ownership (Hess 2013). Thus, a localist form of developmentalism, associated with rooftop ownership financed by local governments, was shifting into a mainstream form of developmentalism, associated with third-party ownership and financing from the technology and financial sectors.

Some of the programs – such as rooftop solarization, home weatherization, and even ARPA-E funding – were less controversial than others, such as high-speed rail (Williams 2011). The latter became a particularly focused target of attack by Republican governors, who framed it not as an economic development strategy but as an example of social liberals' profligacy and corruption by special interests (meaning the unions). In short, green technology was not a single, coherent category. Rather, it came to acquire political meanings because of linkages between types of green technology and political constituencies: weatherization with low-income groups and traditional Democratic Party urban constituencies; high-technology development with venture capital and the innovation industries; rooftop solar with localism and later with the financial industry; high-speed rail and wind with steel manufacturers, the construction industry, and unions.[1]

Uneven transitions and scalar dynamics

At this point the analysis of the green-energy transition has shown two types of unevenness: over time across political administrations and Congressional majorities, and across different types of green technology, which are associated with different political constituencies. Another source of unevenness involves scale, specifically the difference between politics and policy outcomes at the federal and state levels. When one switches attention to the state-government level, the history of green-transition policies in the U.S. becomes more complicated. Whereas at the federal government level there is a brief period of support at the beginning of the first Obama administration, especially during 2009, but it is blocked in Congress by the end of 2010, at the state-government level, there is a much more variegated pattern, with a relatively steady and unbroken transition in some states. Even during the years of 2000–2008, when the Republican administration of George W. Bush blocked almost all green-energy transition proposals (other than some modifications included in omnibus energy bills), there were significant developments in many state governments. Approximately half of all state governments, often in concert with large cities, developed a suite of green industrial policies that included both demand-side measures such as renewable portfolio standards and supply-side policies such as support for business development (DSIRE 2013, Hess 2012a). Furthermore, the developments were often bipartisan and in some cases even supported by Republican governors. Many of the policies were continuous with general economic development programs in the states, where state governments expanded their support for high-tech industrial clusters to include "clean tech." Thus, developmentalist ideology, which according to Eisinger (1986) has tended to flounder at the federal government level because of sectional rivalries, has always been more widely accepted at the state government level, where there are recognized industrial strengths and factor endowments that make industrial policy more bipartisan. Furthermore, dislocations from deindustrialization tend to force Republicans and Democrats alike to endorse strong economic development programs. This section will consider variations across industries and scale related to the green industries in the U.S. Specifically, I will contrast the transition in two industrial fields, biofuels and electricity, and the transition at the federal- and state-government levels.

The development of biofuels is treated here as a "green" technology, although it is highly controversial from the perspectives of land use, social justice, and climate mitigation. In the U.S. biofuels are associated with inefficient corn-based ethanol, and the extensive use of corn for ethanol production has driven up food prices and led to environmental degradation (Farrell et al. 2006, Runge 2010). Next-generation cellulosic ethanol and algae-based biodiesel may be more environmentally benign than corn- and soy-based biofuels, but they still rely on heavy use of natural resources, including fossil fuels and water. Notwithstanding the shortcomings of biofuels from an environmental perspective, they have received relatively high levels of bipartisan support in the U.S. Although the corn-belt states of the Upper Midwest have been the strongest proponents, states in all regions have agricultural industries and have supported biofuel production. Policy on biofuels tends to avoid the sectional rivalries that have often been the downfall of other industrial policies in the U.S. at the federal-government level. Because of the widespread geographical support for the industry, at the national government level there is a renewable fuels standard. After 2010 bipartisan support for biofuels in the federal government eroded to some degree; for example, in 2012 Congress ended the biofuels subsidy and tariff, and it also placed limits on the cost of biofuels consumed by the military (Pear 2012). However, the underlying support for the industry in the form of a national renewable fuels standard remained in place.

In contrast with biofuels, in the U.S. there is no national renewable electricity standard, and, as noted above, the attempt to institute one in 2010 led to bitter partisan divisions. However, by the 2000s, some state governments were enacting renewable electricity standards, energy-efficiency goals, and in some cases system benefits charges (which funded green transition policies). In the Northeast, states also developed a regional cap-and-trade system starting in 2008, and California launched a similar program in 2012. Groups of Midwestern and Western states, together with Canadian provinces, also developed plans for the regulation of greenhouse gases. Overall, Democratic Party leaders in city and state governments were stronger supporters of green-energy policies than Republican leaders, but several Republican governors were also supportive. Generally, the states that have been the strongest supporters of a renewable electricity standard are on the West Coast and in the Northeast and to some degree the Midwest. Those states tend to have lower levels of employment in the coal, natural gas, and oil industries, and they also tend to be Democratic Party strongholds (Hess 2012a).

The backlash against green-energy policy at the federal government level can also be seen at the state-government level. In a multivariate analysis of 6000 votes by state legislators on green-energy legislation, my collaborator and I showed that several of the significant predictors of negative Republican Party support for green-energy legislation, including the shift from President Bush to President Obama, were indicative of a backlash against the Democratic Party (Coley and Hess 2012). The election of Republican governors and legislators at the state government level in 2010 also led to reversals on green-energy policy reforms. For example, New Jersey withdrew from the Regional Greenhouse Gas Initiative, and other states backed away from previous initiatives to move toward cap-and-trade legislation. Republican legislators in Colorado, Michigan, Montana, Ohio, Washington, and West Virginia also attempted to reverse renewable portfolio standard (RPS) laws (Hess 2012a). Attacks by Republican legislators on unions also weakened one of the significant coalition partners for green-energy reforms. However, in states that retained Democratic Party control over the governor's office and state legislature, there was continued deepening of green-energy policy reforms. Thus, even with the Tea Party mobilization, there were dozens of green-energy reform laws passed in state legislatures in 2011 and 2012. Although most of the laws were passed in "blue" (Democrat-controlled states), some of the laws were passed in states with Republican governors but with growing and strong green-energy industries. Likewise, our analysis of state-government economic development plans and programs indicates that Republican governors were often supportive of clean-technology industries as part of the portfolio of support for economic development (Hess and Mai 2014). Thus, at the state government level developmentalist concerns can be associated with a more bipartisan approach, at least for clean technology industrial policy.

Opportunities for bipartisanship at the state government level were also related to the ideological valence associated with proposed laws (Coley and Hess 2012). Increases in renewable portfolio standards became increasingly controversial after 2009, because Republican opponents could frame them as a tax that households and businesses could ill afford during an economic recession. Thus, Democrats could be portrayed as spendthrift social liberals, but the traditional defense of Democrats, that the policies helped mitigate social inequality, was neutralized, because the "taxes" had a disproportionate effect on lower-income homes due to the relatively high percentage of energy expenditures in the household budgets of lower-quintile homes. In contrast, enabling laws that allowed markets to operate more efficiently in favor of green-energy initiatives in the private sector, such as laws in support of property-assessed clean-energy bonds, tended to achieve higher levels of bipartisan support.

In states where Democrats were firmly in power, such as California, New York, and Oregon, there was considerable legislative and executive activity even after 2010. Under Governor Jerry Brown, California continued its green transition record with dozens of new laws that the governor signed (although he vetoed some, too), and in New York Democratic Governor Andrew Cuomo supported green-transition legislation, openly discussed the reality of climate change, and argued that government should to respond to it (Vielkind 2012). Thus, the narrative that one encounters at the federal government level – the rise-and-fall of a green-energy transition with the corralling of climate science – is only partially accurate at the state government level. Particularly on the West Coast and in the Northeast, Democratic Party leaders continued to deepen green transition policies, to support clean-technology industrial clusters, and to argue for the importance of mitigation and adaptation to climate change. In this sense, significant parts of the U.S. looked more like large portions of Europe.

The variegated pattern is consistent not only with partisan divisions between red (Republican) and blue (Democratic) states; the ratio of "clean-energy" jobs to oil-and-gas jobs is higher in states with strong green-transition policies. For example, the ratio is above 1.0 in states that are leaders in the green-energy policy field: 2.1 for California, 2.5 for Massachusetts, 1.5 for New York, and 7.5 for Oregon. In contrast, the most laggard states for green-energy policy tend to have very strong fossil-fuel industries, such as Louisiana (.13 ratio) and Wyoming (.06).[2] The pattern suggests that some states have reached a tipping point in which the green-energy economy has become more economically powerful than the fossil-fuel industry. For example, in Iowa, where there are developed political constituencies for the state's wind and biofuels industries (and a 2.3 ratio of clean-energy to oil-and-gas jobs), the Republican governor elected in 2010 continued to support those industries (Eilperin 2012, Zimmerman 2011). Furthermore, the trend is toward a gradual transition in relative industrial strength toward more clean-energy jobs, including in "red" states that are generally considered strongholds of anti-green and pro-Republican support (Muro et al. 2011).

As the energy industry undergoes a transition from domination by fossil-fuel firms and jobs to one dominated by green-energy and related jobs, another important change may occur. The financial industry, such as venture capital, has increasingly become aligned with the rising industry, and it has become more willing to invest in political campaigns that support green-transition policies. For example, in a significant clash in California during 2010, out-of-state oil-and-gas firms, including one associated with Charles and David Koch, supported a ballot proposition that would have indefinitely delayed the implementation of the state's cap-and-trade regime. However, a coalition of labor, environmental, ethnic minority, and clean-tech organizations put together a winning campaign. The campaign framed the ballot proposition as a ploy by "out-of-state" oil interests that wanted to make California's air dirtier, thereby playing a chord of defensive developmentalism against the fossil-fuel industries' austerity message. In addition to the clash of messaging, the coalition outspent the fossil-fuel industry by a two-to-one ratio, with much funding coming from the wealthy donors associated with the state's venture capital and financial services industries (Hess 2012a, National Institute on Money in State Politics 2012).

Conclusion

The problem of explaining why there is a failed or stalled green-energy transition in the U.S. (at the federal government level and in many but not all states) requires a different kind of theory of science, technology, and society than has been typical in STS up to this point. This is not to say that existing conceptual frameworks are irrelevant. Certainly, the idea of coalition politics, which is central to understanding political processes in countries with electoral democracies, involves actors who build networks and negotiate compromises. Likewise, there are long-term transitions underway in LTSs from current configurations based on high levels of fossil-fuel consumption to new, post-carbon configurations. Those transitions require government policy to protect emergent technological "niches," such as solar energy, from market conditions until the new technologies achieve the scale and cost-competitiveness that enable them to compete in markets. Thus, transition dynamics are also important ingredients of a conceptual framework.

However, the study of the politics of the green-energy transition in the U.S. requires a broader framework in order to address the problem of a blocked political opportunity structure. Specifically, I have drawn attention to ideology, field structure, industrial and political power, and scale. In the U.S., the political field is much more subject to industrial influence than in some other countries, due high levels of private spending on political campaigns and a two-party system that tends to absorb and dampen the political extremes that are found in parliamentary systems. Although the American political system has elections, free speech, and other features generally associated with democratic processes, it functions, especially at the federal government level, more as a corporatocracy. Powerful industrial interests have a tremendous capacity to slow or block even extensive political coalitions, as we have seen in the case of the green-energy policies that failed at the federal government level in 2010, and in this case the fossil-fuel industry has tended to build alliances with political conservatives. However, the industrial field itself is divided, and as the case of the ballot proposition in California indicates, support from a countervailing industry such as finance can be politically crucial. The finance and technology sectors see opportunities in the green transition, and they tend to support a more developmentalist agenda, which when combined with the social liberal agenda of unions and urban constituencies, can be the basis of successful green-transition coalitions for Democratic Party strategists.

As a continental society with a federal government, a blocked political opportunity structure at the federal government level does not necessarily entail similar blockages at other scales of government, and industrial power that is effective at one level may be less so at another level. After 2010 conservative groups such as the American Legislative Exchange Council targeted green-energy policies in state legislatures, but after two years they could not demonstrate much success in their goal of achieving a roll-back of state-level renewable portfolio standards (Anderson 2011). Furthermore, there are interactions between the state and federal level: experiments in Chicago and California provided models for the Obama administration's green jobs policies, and continued developments in the Northeast and on the West Coast provided a refuge for the further development of green-transition policies after the political opportunity structure closed down at the federal government level in 2010.

In this situation of an uneven transition with a powerful industrial counter-mobilization, scientific research associated with the green-energy transition has lost much of its political effectiveness. The broad corralling of social liberalism, which took place over decades through the gradual construction of conservative think tanks and media (Barley 2010), has been extended to the scientific field. Traditional deference of policymakers to technical expertise, which scientists carefully protect through boundary work and boundary organizations (Guston 2001), has given way to a politicization of climate science and green-energy research. In the crudest forms, climate-related policy reforms are viewed as corrupt give-aways to special interests that are tainted by an agenda of social liberalism. Here, mappings from the left and the right can converge on meanings of specific technologies, such as weatherization for the poor and high-speed rail for the steelworkers.

More generally, the analysis of ideologies as contrasting cultural systems in political fields also has benefits for the historical sociology of contemporary science, technology, and industry. Although the transition in the political field in the U.S. (and in many other countries) during the 1980s and 1990s from the dominance of social liberalism to the dominance of neoliberalism is well recognized, the analysis of ascendant political ideologies such as localism and mainstream developmentalism (both with historical antecedents) enables the study of the historical changes in the political field to escape from a pendulum swing view of history. In the pendulum swing view, the future can become a return to social liberalism, just as a dialectical view of history awaits a revolutionary transformation to socialism. In contrast, the analysis developed here draws attention to the slow rise of mainstream developmentalism in the political field and its ideological foil, localism. The capacity for developmentalism to become dominant in the political field is closely connected with the declining relative position of the country in the global economy. The invigoration of industrial policy and defensiveness in trade relations that have emerged since 2000 with respect to green-tech and manufacturing are sites where one can track in some detail these underlying shifts. The interweaving of the green transition in technology with the global economic transition in economic primacy is likely to favor those who advocate more developmentalist positions in the political field. If this diagnosis is correct, the future trend of the political field in the U.S. is more likely to look like the protectionist past of the nineteenth century than the free trade past of the twentieth century. Likewise, rather than the social liberal emphasis on the redistributive and market corrective functions of the state and rather than the neoliberal emphasis on shifting political decision-making to the wisdom of markets, one will find increasing dominance of advocates of an interventionist state but with a razor-sharp focus on making and keeping good jobs and domestic industries in an increasingly innovative and competitive global economy. It remains to be seen how the global hegemon will handle its relative decline. Of the many possible future scenarios, a new developmentalism, especially a greenish one, may turn out to be a relatively benign historical outcome.

Notes

1 For a more detailed discussion of the ideological valences of specific forms of green technology, see Hess 2012b.
2 The values are calculated from comparable employment estimates from the Independent Petroleum Association of America 2009 and Pew Charitable Trusts 2009. Retail petroleum (for example, filling station) jobs are not counted, and coal industry jobs are not included. The correlation for all fifty states of the jobs ratio with a clean-energy policy index that I developed was .38, p <.01.

References

Abramson, L. 2012. "Military's Green Energy Criticized by Congress." *National Public Radio*, July 5. *www.npr.org/2012/07/05/156325905/militarys-green-energy-criticized-by-congress*.

Anderson, D. 2011. "Koch-Funded Group Mounts Cut-and-Paste Attack on Regional Climate Initiatives." *Grist*, Mar. 16. *http://grist.org/climate-policy/2011-03-16-koch-group-alec-cut-paste-attack-regional-climate-initiatives/full*.

ARPA-E (Advanced Research Projects Agency-Energy). 2012. "Programs Main Overview." *http://arpa-e.energy.gov/ProgramsProjects/Programs.aspx*.

Barley, S. 2010. "Building an Institutional Field to Corral a Government: A Case to Set an Agenda for Organization Studies." *Organizational Studies* 31(6): 777–805.

Berman, B. 2012. "Explaining the Move toward the Market in Academic Science: How Institutional Logics Can Change without Institutional Entrepreneurs." *Theory and Society* 41: 261–99.

Bijker, W., Hughes, T., and Pinch, T. eds. 1987. *The Social Construction of Technological Systems*. Cambridge, MA: MIT Press.

Bourdieu, P. 2005. *The Social Structures of the Economy*. Malden, MA: Polity Press.

Brown, P. 2007. *Toxic Exposures: Contested Illnesses and the Environmental Health Movement*. New York: Columbia University Press.

Cardwell, D. 2012. "U.S. Raises Tariffs on Chinese Wind Turbine Manufacturers." *New York Times*, July 28, B4.

Climate Science Watch. 2011. "House Science Committee Republicans Aim to Slash Climate and Sustainable Energy Programs." Oct. 25. *www.climatesciencewatch.org/2011/10/25/house-science-committee-republicans-aim-to-slash-climate-and-sustainable-energy-programs*.

Coley, J., and Hess, D.J.. 2012. "Green Energy Laws and Republican Legislators in the United States." *Energy Policy* 48(1): 576–583.

Condon, S. 2011. "Obama on Solyndra: I Won't Surrender to China on Green Tech," *CBS News*, Oct. 6. *www.cbsnews.com/8301-503544_162-20116750-503544.html*.

DSIRE (Database of State Incentives for Renewables and Energy Efficiency). 2013. "Rules, Regulations, and Policies for Renewable Energy." *www.dsireusa.org/summarytables/rrpre.cfm*.

Eilperin, J. 2012. "Unusual Coalitions Clash over Wind Power Tax Credit." *Washington Post*, Sept. 20. *http://articles.washingtonpost.com/2012-09-20/national/35495033_1_wind-energy-wind-power-wind-project*.

Eisinger, P. 1986. *The Rise of the Entrepreneurial State: State and Local Economic Development Policy in the United States*. Madison: University of Wisconsin Press.

Farrell, A., Plevin, R., Turner, B., Jones, A., O'Hare, M., and Kammen, D. 2006. "Ethanol Can Contribute to Energy and Environmental Goals?" *Science* 311 (5760): 506–508.

Fligstein, N., and McAdam, D. 2012. *A Theory of Fields*. Oxford, UK: Oxford University Press.

Frickel, S. and Gross, N. 2005. "A General Theory of Scientific/Intellectual Movements." *American Sociological Review* 70(2): 204–32.

Frickel, S., Gibbon, S., Howard, J., Kempner, J., Ottinger, G., and Hess, D. 2010. "Undone Science: Social Movement Challenges to Dominant Scientific Practice." *Science, Technology, and Human Values* 35(4): 444–473.

Geels, F. 2005. "The Dynamics of Transitions in Socio-technical Systems: A Multilevel Analysis of the Transition Pathway from Horse-drawn Carriages to Automobiles (1860–1930)." *Technology Analysis and Strategic Management* 17(4): 445–476.

Geertz, C. 1973. *The Interpretation of Cultures*. New York: Basic Books.

Goldenberg, S. 2012. "Conservative Think Tanks Step up Attacks against Obama's Clean Energy Strategy." *The Guardian*, Feb. 7. *http://m.guardian.co.uk/environment/2012/may/08/conservative-thinktanks-obama-energy-plans?cat=environment&type=article*.

Grin, J., Rotmans, J., and Schot, J. 2010. *Transitions to Sustainable Development: New Directions in the Study of Long Term Transformative Change*. New York: Routledge.

Guston, D. 2001. "Boundary Organizations in Environmental Policy and Science: An Introduction," *Science, Technology & Human Values* 26(4): 399–408.

Harvey, D. 2005. *A Brief History of Neoliberalism*. Oxford: Oxford University Press.

Hess, D.J. 2011. "To Tell the Truth: On Scientific Counterpublics." *Public Understanding of Science* 20(5): 627–641.

Hess, D.J. 2012a. *Good Green Jobs in a Global Economy*. Cambridge, MA: MIT Press.

Hess, D.J. 2012b. "The Green Transition, Neoliberalism, and the Technosciences." In *Neoliberalism and Technosciences: Criticial Assessments*, edited by L. Pellozzoni and M. Ylönen, 209–230. Northampton, MA: Edward Elgar.

Hess, D.J. 2013. "Industrial Fields and Countervailing Power: The Transformation of Distributed Solar Energy in the United States." *Global Environmental Change*, 847–855.

Hess, D.J. 2014. "When Green Became Blue: Epistemic Rift and the Corralling of Climate Science." In *Fields of Knowledge: Science, Politics, and Publics in the Neoliberal Age*, ed. by S. Frickel and D. Hess. Under review.

Hess, D.J., and Mai, Q. 2014. "Political Parties and Clean Technology: An Analysis of State Government Economic Development Programs in the U.S." Under review.

Hughes, T. 1983. *Networks of Power: Electrification in Western Society, 1880–1930*. Baltimore, MD: Johns Hopkins University Press.

Independent Petroleum Association of America. 2009. "IPAA Oil and Gas Producers in Your State: 2008–2009." *www.ipaa.org/wp-content/uploads/downloads/2012/01/2008-2009IPAAOPI.pdf.*

International Trade Commission. 2012. "Fact Sheet: Commerce Finds Dumping and Subsidization of Crystalline Silicon Photovoltaic Cells, Whether or Not Assembled into Modules from the People's Republic of China." October 10. *http://ia.ita.doc.gov/download/factsheets/factsheet_prc-solar-cells-ad-cvd-finals-20121010.pdf.*

Jacques, P., Dunlap, R., and Freeman, M. 2008. "The Organization of Denial: Conservative Think Tanks and Environmental Skepticism." *Environmental Politics* 17(3): 349–385.

Johnson, B. 2010. "The Climate Zombie Caucus of the 112th Congress." *Think Progress*, Nov. 19. *http://thinkprogress.org/climate-zombie-caucus.*

Klein, N. 2011. Keynote Lecture at the Annual Conference of the Business Alliance for Local Living Economies, Bellingham, WA. *http://vimeo.com/25839058* (accessed October 11, 2012).

Leiserowitz, A., Maibach, E., Roser-Renouf, C., and Hmielowski. J. 2011. "Politics and Global Warming: Democrats, Republicans, Independents, and the Tea Party." Yale University and George Mason University. New Haven, CT: Yale Project on Climate Change Communications. *http://environment.yale.edu/climate/news/PoliticsGlobalWarming2011.*

Lieberman, J. 2010. "Kerry, Lieberman – American Power Act Will Secure America's Energy, Climate Future." May 12. *www.lieberman.senate.gov/index.cfm/news-events/news/2010/5/kerry-lieberman-american-power-act-bill-will-secure-americas-energy-climate-future.*

Mazmanian, D. and Kraft, M. 1999. *Towards Sustainable Communities: Transitions and Transformations in Environmental Policy.* Cambridge, MA: MIT Press.

Merton, R. 1973. *The Sociology of Science.* Chicago: Unviersity of Chicago Press.

Moore, K., Kleinman, D.L., Hess, D., and Frickel, S. 2011. "Science and Neoliberal Globalization: A Political Sociological Approach." *Theory and Society* 40(5): 505–532.

Muro, M., Rothwell, J., and Saha. D. 2011. *Sizing the Clean Economy: A National and Regional Green Jobs Assessment.* Washington, DC: The Brookings Institution.

National Institute on Money in State Politics. 2012. "PROPOSITION 23: Suspends Certain Air Pollution Control Laws Until Unemployment Drops Below Specified Level." *www.followthemoney.org/database/StateGlance/ballot.phtml?m=716.*

Open Secrets. 2013. "Lobbying: Energy and Natural Resources." *www.opensecrets.org/lobby/indus.php?id=E&year=a.*

Pear, R. 2012. "After Three Decades, Tax Credit for Ethanol Expires." *New York Times*, Jan. 1. *www.nytimes.com/2012/01/02/business/energy-environment/after-three-decades-federal-tax-credit-for-ethanol-expires.html.*

Peck, J. 2010. *Constructions of Neoliberal Reason.* New York: Oxford University Press.

Pew Charitable Trusts. 2009. *The Clean Energy Economy: Repowering Jobs, Business, and Investment across America.* Washington, DC: Pew Charitable Trusts.

Pew Charitable Trusts. 2010a. *Global Clean Power: A $2.3 Trillion Opportunity.* Washington, DC: Pew Charitable Trusts.

Pew Charitable Trusts. 2010b. *Who's Winning the Clean-Energy Race: Growth, Competition, and Opportunity in the World's Largest Economies.* Washington, DC: Pew Charitable Trusts.

Roberts, J.T. and Kincaid, G. 2012. "No Talk but Some Walk: Obama Administration Rhetoric on Climate Change and International Climate Spending," paper presented at the Annual Meeting of the American Sociological Association, Denver, 2012.

Rose, N., O'Malley, P., and Valverde, M. 2006. "Governmentality." *Annual Review of Law and Social Science* 2: 83–104.

Runge, C. 2010. "The Case against Biofuels: Probing Ethanol's Hidden Costs." *Yale Environment 360*, March 11. *http://e360.yale.edu/feature/the_case_against_biofuels_probing_ethanols_hidden_costs/2251.*

Thornton, P., Ocasio, W., and Lounsburgy. M. 2012. *The Institutional Logics Perspective: A New Approach to Culture, Structure, and Process.* New York: Oxford University Press.

United Steelworkers. 2010. "United Steelworkers' Section 301 Petition Demonstrates China's Green Technology Violate WTO Rules." *http://assets.usw.org/releases/misc/section-301.pdf.*

U.S. House of Representatives. 2011. "Discretionary Spending Recommendations Offered by the Following Members of the House Committee on Science, Space, and Technology: Rep. Ralph M. Hall, Rep. F. James Sensenbrenner, Rep. Lamar Smith, Rep. Judy Biggert, Rep. W. Todd Akin, Rep. Michael McCaul, Rep. Steven Palazzo, Rep. Andy Harris, and Rep. Randy Hultgren." October 14, *http://science.house.gov/sites/republicans.science.house.gov/files/Letter_to_Joint_Select_Committee.pdf.*

U.S. Senate. 1997. "S. Res. 98." Report 105-54. *www.gpo.gov/fdsys/pkg/BILLS-105sres98ats/pdf/BILLS-105sres98ats.pdf.*

Vielkind, J. 2012. "Cuomo – "Climate Change is a Reality… We Are Vulnerable." *Albany Times-Union,* Capitol Confidential blog, Oct. 31. *http://blog.timesunion.com/capitol/archives/162798/cuomo-climate-change-is-a-reality-we-are-vulnerable.*

Wald, M. and Savage, C. 2011. "Furor over Loans to Failed Solar Firm." *New York Times,* Sept. 15: A25.

Williams, T. 2011. "Florida's Governors Rejects High-Speed Rail Line, Fearing Costs to Consumers." *New York Times,* Feb. 16. *www.nytimes.com/2011/02/17/us/17rail.html.*

Zimmerman, C. 2011. "Iowa Governor Signs Bill to Boost Ethanol and Biodiesel." *Ethanol Report,* May 26. *http://domesticfuel.com/2011/05/26/iowa-governor-signs-bill-to-boost-ethanol-and-biodiesel.*

17

Risk State

Nuclear politics in an age of ignorance

Sulfikar Amir

NANYANG TECHNOLOGICAL UNIVERSITY

Introduction

In his seminal work on risk society, Ulrich Beck writes, "In advanced modernity the social production of wealth is systematically accompanied by the social production of risk" (Beck 1992: 19). While Beck's proposition accurately describes risk relations in advanced modernity, a compelling question arises: how should we characterize the social production of risk in less advanced modernity in which the majority of the world's population lives? Does Beck's statement imply that less advanced modernity faces less risk? To modify Beck, I argue that in less advanced modernity the social production of wealth is systematically surpassed by the social production of risk. To put this differently, as the result of structural factors, the introduction of complex technologies in the global south leads risk to grow faster than wealth in that part of the world.

Exploring technology and society in the non-western world, this chapter focuses on risk and modernity, a recurring theme in science and technology studies (STS) and sociological scholarship. As researchers have shown in myriad of case studies (for example, Jasanoff 1986, Wynne 1989, Kleinman et al. 2005), risk is deeply embedded in the foundation of modernity in which technology constitutes the preeminent causal agent of risk production and distribution. Most studies of the interplay between technology and risk, however, are situated in technologically advanced societies. There is a relative paucity of discussion of how technological risk looms especially large in less advanced modernity, countries marked by vulnerable institutions and weakly enforced regulations. This chapter is intended to fill this lacuna by giving an account of peculiar circumstance where less advanced modernity poses greater risk.

In advanced modernity, as noted by sociologists (for example, Douglas and Wildavsky 1983, Luhmann 1993, Giddens 1990), risk emanates primarily from the material constitution of a cornucopia of advanced technological products. The predicament is rooted in high degrees of uncertainty in which the inability to predict and to the control side effects of modern technological systems inevitably entails rampant production of risk (Power 2007, Callon et al. 2001). This is a paradox of scientific knowledge: it has severe limits while playing a key role in wealth generation (Wallerstein 2004, Bijker et al. 2009). As a result, the problem of risk is always regarded as a problem of knowledge, because it is in this domain where uncertainty lurks behind fascinating advancements of technology. Thus, modern societies are constantly faced with the challenge of overcoming uncertainties that characterize technological culture (Van Loon 2002).

In less advanced modernity, the picture is likely to be very different. Technology-related risk is an equally problematic issue, but greater risk emerges from another source. Although the uncertainty of technological knowledge remains a critical source of risk, the potential hazard from the utilization of high technology is prompted less by the limits of knowledge than by *precariousness*, *incompetence*, and *fragility* with regard to the institutional (in)capacity that plagues multiple layers of institutions in less advanced modern societies. To understand this requires taking modernity as a cultural system built not only on materiality that defines how one society achieves techno-economic progress, but more importantly on the arrangement of social and political institutions that govern everyday life. Hence, the underlying factor that renders a modernity advanced or less advanced rests on the institutional structure and culture that underpins social stability and order in contemporary society.

It is through this particular lens that this chapter aims to illuminate how technological risks tend to accrue in less advanced modernity as a result of institutional predicaments. Toward this end, I focus on a controversy surrounding the interaction between nuclear power and society in which the institutions of less advanced modern society appears to be the primary agent of risk production. Since the early twenty-first century, nuclear power has arisen as a potential alternative to fossil fuel; it has been touted as one of the best sources of climate friendly energy for humankind. Despite the Fukushima nuclear meltdown in 2011, the global obsession with nuclear power has not substantially diminished (Ramana 2013).

In less advanced societies of the developing world, the controversy over nuclear risk voiced by global anti-nuclear movements did not render nuclear power less attractive. Commitment to nuclear power production has remained firm, in particular among emerging economies that are obsessed with the superiority of nuclear power and desperately need to secure energy supply. Situating nuclear power in less advanced modernity thus requires an examination of the nexus between risk, technology, and institutions, for the three are inextricably interwoven in the formation of modernity.

My argument is that the state is the central institution that *escalates* risk production in less advanced societies. These countries are typically governed by what I term "risk states." The structure and culture of risk states – or more specifically, the discrepancy between state capacity and the nature of risk embedded in high technology – create a situation in which these states are prone to augment the likelihood of hazards. To delineate how the interplay between the state and technological risk unfolds, the empirical setting of risk state in this chapter is situated in Indonesia, the fourth most populous country in the world. Like many developing states, Indonesia is strongly tempted to have nuclear power capacity in order to provide sufficient energy for industrial and economic growth. While the effort to produce nuclear power has been halted by strong resistance from the grassroots groups in the country, the Indonesian state remains unflinchingly adamant about introducing nuclear power to the nation. I elucidate the socio-political implications of such an ambitious venture, underlining what I call "institutionalized ignorance" as a puzzling trait of a risk state in coping with uncertainty and complexity of high technology.

A tale of an emerging economy

If one travels across the so-called global south, a cluster of less economically advanced countries, it is not too difficult to find states obsessed with the perceived benefits of sophisticated technologies and inclined to turn a blind eye to the potential hazards. Indonesia is such a risk state, a country eagerly pursuing nuclear power production. A detailed analysis of the Indonesian risk state allows me to illuminate the main features of my argument.

Situated in the equatorial line between Asia and Australia, Indonesia is an archipelagic nation made up by more than 13,000 islands. Despite the fact that Indonesia is a predominantly Muslim country, and it is the largest Muslim population in the world, the social, cultural, and political dynamics of this country are largely shaped by multiculturalism and pluralistic ideologies, which are rooted in the history of polities in the archipelago long before the Europeans came to Southeast Asia (Ricklefs 2001).

During the formative years of the nation under the leadership of the first president, Sukarno (1945–1967), Indonesia underwent a turbulent period marked by strong ideological contestation between different political groups. A dramatic transition occurred in 1965 after a failed coup organized by what is known as the 30 September Movement; this historic moment was followed by massive annihilation of the communists.[1] The bloody event led to the rise of Suharto as the strongman of the New Order regime.

For the next three decades, the New Order ruled the country with an authoritarian system strongly supported by the military (Vatikiotis 1993). The striking feature of the New Order regime lies in developmental visions that sought to modernize the Indonesian economy and society by drawing on the western prescriptions of industrialization and high technology development. Relying heavily on oil revenues, the New Order government built the country's economic infrastructures and gradually improved the socio-economic wellbeing of Indonesian people (Hill 2000). Despite some criticism addressed to the New Order's culture of corruption and nepotism, Indonesia's economic achievements received praise and admiration from international observers in the middle of 1990s. Indonesia's rapid industrial transformation earned the country the label of "new Asian tiger" (Hill 1997).

But the Asian financial crisis severely struck Indonesia in 1997–1998, and turned the country upside down. Among the Southeast Asian countries affected by the financial crisis, Indonesia was the worst; the crisis spread from the financial and economic sectors to the political system, shaking up the stability of the New Order authoritarianism. In May 1998, President Suharto tendered his resignation, marking the abrupt collapse of the New Order government. A transition government was formed to bring Indonesia to a new political era, namely, liberal democracy.

Post-New Order Indonesia embarked on a different set of economic and political rules. Political reforms were implemented across a broad range of institutions from the central government in Jakarta all the way down to district administrations throughout the country. It was a massive overhaul that attempted to replace the New Order authoritarian structure with a new democratic system. The one-party system prevalent in the New Order was replaced by a multi-party system in which the central government is now composed of several political parties that bring different interests and agendas to the state politics. In contrast with the political system in the New Order era where the parliament was no more than a rubber stamp to Suharto's idiosyncratic decisions, the post-New Order era reflects the growing power of the parliament to monitor the executive branch in making and executing policy. Within this new institutional setting, all developmental programs and projects proposed by the executive branch of the government now must acquire approval from the legislative branch. The fundamental changes in the Indonesian political system are by no means without consequences. Some studies have shown that the new political system resulting from democratization and decentralization have had positive as well as negative impacts on the ways in which public policies are conceived and executed (see, for example, Aspinall and Mietzner 2010). Corruption and inefficiency are among grave problems Indonesia is still suffering from, not to mention political contestations that affect the quality of government. The bottom line is that the Indonesian state continues to lack institutional capacities despite political reforms that had been undertaken at multiple levels.

Along with the political liberalization of the post-New Order political system, came economic liberalization. Suharto's administration was in fact friendly to market-oriented economics. The beginning of the Suharto rule was marked by the opening of Indonesian economy to foreign capital, deemed necessary to stimulate growth for domestic markets. Thus neoliberal policies were not new to Indonesia when they spread in the post-Suharto era. Yet the New Order government was modeled, to some extent, after the developmental state in which economic management is not entirely organized around market mechanisms. The role of the state remained pivotal in ensuring the provision of basic public needs, such as education, health and food supply, and infrastructure (Thee 2012).

The prominence of the state in Indonesia's political economy gradually dissolved as a result of the Asian financial crisis at the turn of the twenty-first century. To cope with the plunging exchange values of the rupiah, the Suharto government accepted a conditional loan from the International Monetary Fund (IMF). This is the moment when a set of neoliberal policies were reluctantly implemented by the Suharto administration and its successor. For the next five years, the Indonesian economy was brutally liberalized following the Washington Consensus's prescription of "structural adjustments." Consequently, as the role of the state declined, the management of many public sectors was taken over by the markets, which accommodated the interests of the private corporations and multinationals. During this period, the majority of Indonesians lived in dire economic circumstances and the middle class shrunk dramatically (Hill and Shiraishi 2007).

Twelve years after the painful crisis that dramatically changed Indonesia, the country was back on track. Despite a few problems at local levels, many praised the flourishing of democracy in Indonesia (Ananta et al. 2005). Although poverty and unemployment remain relatively high, numerous economic observers have been amazed by how Indonesia has managed to improve the economy so successfully; it has changed from a country desperately relying on foreign aid into one of the largest emerging economies in the world (Geiger 2011). Between 2001 and 2012, Indonesia's GDP jumped from US$160 billion to US$850 billion and it is predicted to reach US$1.1 trillion by 2013, making Indonesia eligible for the G20 membership.

Despite the 2008 global economic downturn that affected developed economies in Europe and North America, Indonesia was among a very few countries able to keep constant growth, thanks to the large proportion of domestic consumption that contributed to GDP. According to a well-known consulting company, Indonesia's growth trend is likely to continue, and by 2030 even surpass two major industrialized countries, Germany and the United Kingdom.[2]

Another indicator worth noting is the growing size of middle-class population. Rapid economic revitalization has allowed millions of Indonesians to enjoy economic benefits, particularly in the urban cities. This large group of citizens constitutes the new middle class whose consumption plays a major role in maintaining stable growth. One estimate shows that by 2030 there will be 170 million people in the consuming class, which is more than a half of Indonesia's projected population. This rosy prediction depicts Indonesia as an economic powerhouse that will play a strategic role for the global economy in a near future.[3]

Like other emerging economies, Indonesia is facing one common problem. The rise of middle class and the ongoing reindustrialization will continue only if sufficient supply of energy is provided to feed the myriad of economic activities in the manufacturing and financial sectors. For many years, energy production in Indonesia depended heavily on the country's oil resources. In fact, oil was the main source of revenue that the New Order exploited to finance national development. But this is no longer the dominant source of energy. Indonesia's oil reserves are rapidly depleting, and by 2005, Indonesia had become a net oil importer. The combination of a short supply of energy and the soaring energy demands to fuel an industrial- and consumption-based economy means that Indonesia faces the threat of an energy crisis. This poses an uneasy challenge to the state. It is in this looming crisis of energy that nuclear power enters the picture, along with a new politics of risk.

Nuclear in the state

While the early twenty-first century global trend toward nuclear power prompted Indonesia to embrace nuclear energy, it was the nuclear technocrats at the National Agency of Nuclear Power (BATAN) who picked up this momentum and mobilized it in the national energy policy discourse. To analyze nuclear politics in Indonesia thus requires understanding the history of its nuclear program since the 1950s, for this historical and institutional development forged Indonesia's strong penchant for adopting nuclear power at the beginning of the twenty- first century (Amir 2010a).

The aspiration to develop nuclear power capacity first appeared in the early years of post- independence, when President Sukarno felt worried about the radioactive fallout resulting from a series of thermonuclear tests conducted by the United States in the Pacific Ocean. Sukarno's concerns led to the formation of the Commission of Radioactivity Research in 1954. Although the commission was formed to assess the risks of nuclear fallout, commission members ended up developing an interest in acquiring nuclear power, as this technology was deemed strategic for Indonesia's economic development. In 1959 the Sukarno administration officially launched the Institute of Atomic Energy (LTA), the first governmental agency meant to build nuclear capacity in research and development.

In spite of Sukarno's inclination toward the eastern bloc, Indonesia managed to secure a US$350,000 grant from the U.S. *Atoms for Peace Program* designed by the Eisenhower adminis- tration to supply basic information and equipment of nuclear technology to research institutions throughout the world while containing nuclear proliferation. This financial support was spent in the installation of Indonesia's first nuclear reactor in Bandung, a 250-kilowatt Triga Mark II, mainly used to produce isotopes and to facilitate neutron physics experiments. As the Institute of Atomic Energy (LTA) grew in importance in the middle of the 1960s, it was renamed BATAN. At this point, Sukarno expressed interest in building an atomic bomb, and hoped to expand the capability of BATAN toward this end. This hope was short-lived, however, due to limited resources and the unstable political situation at that time. Sukarno lost power in the aftermath of the 1965 coup, and his ambition to turn Indonesia into a nuclear-weapon state vanished.

When Suharto came to power after the coup, Indonesia's nuclear program was reoriented toward civilian use. The New Order's friendly relations with the United States played a key role in focusing the nuclear program around non-military purposes. In 1970 Indonesia signed a non-proliferation treaty with the International Atomic Energy Agency (IAEA) and fully rati- fied it in 1978. As a result of the massive overhaul of the central bureaucracy by the New Order regime, BATAN was relocated under the Ministry of Research and Technology, headed by Suharto's closest aide, B.J. Habibie. During this time, the agency was given the task of regulat- ing and promoting nuclear research and development. BATAN's regulatory function ended when the parliament passed the 1997 Nuclear Power Act, resulting in the creation of the National Nuclear Regulatory Committee (BAPETEN).

Although no longer a ministerial-level agency, BATAN underwent a remarkable expansion during the New Order period. With a considerable amount of financial support from the New Order government, BATAN was able to develop capacities for conducting research and applying nuclear science and technology for non-military uses. To complement the first reactor in Bandung, another smaller research reactor was constructed in Yogyakarta in 1979. The Yogyakarta reactor had a capacity of 100-KW, and was primarily used to develop accelerator technology, for material processing, and for operator training. In 1987, Indonesia's nuclear program moved one step further when the G.W. Siwabessy research reactor began operation. This German-made 30-MW multi-purpose reactor is located in Serpong, on the outskirts of Jakarta, and is part of the Center for Science and Technology Development (PUSPIPTEK). The Siwabessy reactor was a highly visible materialization of the New Order's interest in nuclear science and technology, and remains the largest nuclear reactor in Southeast Asia to date.

BATAN is one of the largest governmental organizations in Indonesia. The agency employs highly educated and well-trained researchers and engineers, and it has grown rapidly since its founding. For the past thirty years, it has succeeded in applying nuclear science and technology to produce isotopes for a variety of non-military uses, from agriculture, medicine, to industry. However, the majority of BATAN high officials believed from the outset that the original mission of this agency was to develop nuclear power. The original plan for nuclear power development was initiated in 1972 when BATAN collaborated with the state-owned utility company, PLN, to form the Joint Preparatory Committee for Nuclear Power Construction under the auspices of the International Atomic Energy Agency (IAEA). For a few years, this committee carried out a series of studies to find a suitable location for a nuclear power station in Java. In 1980, BATAN submitted a proposal to construct Indonesia's first nuclear power station. But Suharto turned down the proposal on the grounds that nuclear energy was not economically viable.

It was at the end of the 1980s when the New Order began to approve the possibility of adopting nuclear energy. With the help of Japanese Mitsubishi's subsidiary New Japan Engineering Consultant (NEWJEC), BATAN conducted a feasibility study to construct a nuclear power station in the Muria peninsula of Central Java. In 1994, the study resulted in a proposal that suggested that the government construct twelve 600-MW reactors. But once again, the proposal had not been approved. When the Asian crisis came to cripple Indonesia's economy in 1997, the nuclear power agenda became more unlikely to be materialized.

The failure of nuclear power development in Indonesia under the New Order authoritarianism merits attention. One may be intrigued about why the construction of nuclear power plants was never materialized, given that the authoritarian politics of the Suharto regime provided favorable environments to realize a controversial technology without significant public resistance. Plenty of risky large-scale development projects, including dams, highways, and chemical plants, were effortlessly executed because there was little possibility of resistance to the New Order state. This development did not take place with regard to BATAN-proposed nuclear power plants. Instead, the New Order government deliberately delayed the project, not because of public protest, but more likely due to the political economy of oil that very much directed Indonesia's energy policy from the 1970s throughout the early 1990s. During this period, Indonesia's abundant oil resources rendered nuclear power development less urgent than it was in South Korea and Japan for instance. Despite the failure to accomplish its original goal, BATAN achieved incredible progress in terms of the development of human resources and scientific research over the course of its thirty-year existence. Its contributions to industrial and medical sectors are also not trivial. It has established a reputation as a respected nuclear institution in the region. It is this set of institutional arrangements that has kept Indonesia's focus on nuclear power for years, outlasting the New Order regime itself.

After being idle for a few years in the aftermath of the economic crisis, the Indonesian nuclear power program was revived in 2006. It was preceded by a decision of the Ministry of Energy and Mineral Resources (ESDM) to include nuclear power in the 2004 National Energy Policy. This was the first time nuclear power appeared in the formulation of energy policy, rather than as part of science and technology policy as it had been earlier. BATAN certainly played a significant role in advocating nuclear power as a strategic source for the energy supply system. The nuclear revival was in fact prompted by an in-depth study called the Comprehensive Assessment of Different Energy Sources for Electricity Generation in Indonesia (CADES), a scientific study on the different scenarios of energy input. Financially and technically supported by IAEA, the study was meant "to assess comprehensively the potential contributions of various energy options to the optimal long-term development of Indonesia's energy supply and demand consistent with sustainable development" (cited in Amir 2010a). Although the CADES report assessed a variety of potential energy sources, nuclear power was disproportionally emphasized. The report concluded that Indonesia must harness nuclear energy as a means to sustain the national energy supply for two reasons: first, nuclear power was considered environmentally friendly, and second, the cost was seen as more competitive than conventional resources. The report's strong recommendation for nuclear power clearly represented the technocratic logic that sought to direct the decision-making processes of the state. This proved to be effective: the Ministry of Energy and Mineral Resources (ESDM) adopted the recommendation of CADES in the updated national energy strategy, paving the path for nuclear power to become a significant source of energy in Indonesia. Two years later, President Yudhoyono signed the Presidential Decree No. 5, which laid out the grand plan for Indonesia's energy mix by 2025. This composition of national energy supply included a total of 2% nuclear energy, a small – but sufficient – amount to get the nuclear program materialized once and for all (Amir 2010b).

What is strikingly different in the present nuclear power program, compared with what transpired in the past, is the interplay between democratic politics and nuclear politics. Despite deep concern by a small group in parliament and civil society groups, the Indonesian parliament was generally enthusiastic about the proposed construction of nuclear power stations. Provided that the parliament serves as representative of the people, the parliament's striking inclination toward approving the nuclear agenda gave nuclear technocrats a sense of legitimacy in pushing to bring nuclearization closer to reality. There are two reasons why most politicians in the parliament were firmly convinced that nuclear power was indispensable. First, the sense of an energy crisis was widely shared among politicians, which also tended to hold an optimist view of the potential of nuclear energy to help address the emerging crisis. Second, possessing nuclear capacity was regarded as politically strategic for Indonesia. Many members of the parliament regarded nuclear power as a source of respect and esteem from neighboring countries. The bottom line here is that nuclear power in Indonesia is driven by a need for energy security and status, not necessarily in terms of military interest. As other writers have noted, the ability to be seen as technologically self-sufficient and independent is a common feature in nuclear politics of the state (Massey 1988, Jasper 1990). The question is how this nuclear politics intersects with public concern over its risks, in a new democratic environment.

The specter of risk

Risk is one of the most noticeable factors in technology-society relations. It is defined, in strictly technocratic terms, as the probability that hazards and dangers will occur. As noted earlier, risk is a prominent discourse in modern cultures that are characterized by deep penetration of scientific technology in everyday life. In sociological studies of risk, risk emanates from a wide spectrum of sources in social, cultural, and political spheres, and has two critical aspects. First, risk is always perceived, and it is the perception of risk that informs collective actions and responses toward technological objects (Douglas and Wildavsky 1983, Slovic 2000). Second, in many modern societies, concerns are often expressed about the production and distribution of risk and vulnerability (Beck 1992, Perrow 1999). This dimension of risk calls for an investigation that seeks to unpack how complex technological systems spread risks across societies in ways that increase our vulnerability to accidents and disasters. The bottom line is that the locus of risk and technology lies in *uncertainty*, a condition that characterizes virtually every form of modern technology due to the complexity of the world and the limits of knowledge.

Perhaps there is no more controversial technology than nuclear power. Ever since nuclear power was fashioned out of the same technology that produced the atomic bomb, it has faced social and political challenges (Welsh 2000, Sovacool and Valentine 2012). The root cause is the risk, which can turn into long-term social and environmental damages if it is inadequately controlled. So tremendous is the risk embedded in nuclear energy systems that it requires a centralized structure of control, which renders this particular technology highly political and authoritarian in nature (Winner 1986). The sixty-year history of nuclear power, as such, is replete with tensions, conflicts, lack of transparency, and minor and major accidents, culminating in nuclear disasters in three different social and political contexts, namely, Three Mile Island, Chernobyl, and Fukushima. Gabrielle Hecht's (2012) seminal work on nuclear materials reveals the accumulated troubles plaguing nuclear power production that vividly exhibit the vulnerability of this particular technology (see also Hecht's work on nuclear workers, Chapter 20 in this book). This damaged image notwithstanding, nuclear power has continued to grow in popularity and continues to be seen as viable energy option for any state.

The expansion of nuclearity that we are seeing in many parts of the world today has been challenged at multiple levels by civil society organizations that deem this technology to be one of the greatest dangers to democracy and civil society (Flam 1994). In an emerging democracy like Indonesia, public opinion matters a great deal and has significant impacts on how policies are made, chosen, and deliberated. The nuclear power program is not an exception. Thus, despite the growing currency of nuclear power in post-New Order Indonesia, the technocratic power that created and promoted the nuclear power program has faced a considerable challenge from civil society groups, which have held very different views about how nuclear power might possibly harm people. Given the contention over nuclear power in other democracies, it comes as no surprise that the Indonesian nuclear revival in 2006 triggered massive public opposition strongly organized by environmental groups and manifested in mass organizations. An open political atmosphere, along with the free flow of information, allowed anti-nuclear activists to mobilize support from various sectors of civil society, who questioned the government's seemingly malicious plan to bring nuclear power into the disaster-prone country.

These public expressions are a direct result of the democratization Indonesia has undergone since the 1998 *reformasi*, the dramatic political change that opened up new possibilities for civic engagement in public affairs. A set of political transformations as part of the democratic transition provided a venue for concerned citizens to openly question the logic, assumption, and vision that guided Jakarta's strong support of nuclear power. The way different citizen groups managed to work together to scrutinize techno-economic calculations justifying the production of nuclear energy was remarkable. A body of scientific information on the consequences of nuclear power collectively gathered by civil society groups formed what Sheila Jasanoff (2005) terms "civic epistemology" whereby the social, cultural, and environmental risk of the presence of nuclear power stations in Java was broadly assessed. This public assessment of nuclear risk informed the efforts of the anti-nuclear movement to convince the general public that nuclear energy is not an option for the country. This shared concern over nuclear risk resulted in a strong alliance of anti-nuclear groups that include environmental NGOs, students, intellectuals, anti-nuclear scientists and engineers, and even religious groups (Amir 2009). This opposition has formed an extended network of movement groups that spans from Jakarta all the way down to Jepara, Central Java, the proposed site for Indonesia's first nuclear power station.

The massive opposition to the nuclear power program caused a few years of delay in the construction of the first nuclear power plant. This might be regarded as a victory for the anti-nuclear movement in Indonesia, since it was able to put on hold the attempt of BATAN to realize its old ambition. The question of how the proposed nuclear power plants will be kept safe remains the predominant element of the controversy. But there is something peculiar about the discourse of nuclear risk in Indonesia as voiced by anti-nuclear groups. Every group may have different emphasis on which risk factor that concerns them. However, there is one poignant factor that appears to raise more apprehension to all anti-nuclear groups as well as the public in general: the likelihood of nuclear accident and disaster because of the lack of institutional capacity and competence of the state for handling high-risk technology.

The problem of state capacity has loomed large since the collapse of the New Order authoritarianism. The demise of Suharto's power had a significant impact on the way the state carried out its duties, a common shift found in post-authoritarian regimes. The development of a multi-party system and decentralized politics were part and parcel of what caused the weakening of the state capacity to ensure public safety and security. This lethargic condition was clearly indicated in a series of events in which the state miserably failed to quickly respond to an array of recurring crisis situations such as terrorism, natural disasters, and infrastructure accidents. In recent years the Indonesian public transport authority has had a notoriously high record of deadly accidents such as airplane crashes, train derailments, and ferry sinkings. Even worse is the terrible failure to mitigate the Sidoarjo mudflow disaster in eastern Java, which displaced thousands of Sidoarjo residents and wasted millions of dollars on impotent measures aimed at ending the flow of hot mud.

The inability of the state to properly manage this large scope of public safety and security problems prompted many Indonesians to believe that the state was utterly unable to oversee, let alone operate, nuclear power plants, which come with built-in systemic risk.[4] The bottom line is that the Indonesian state has weak institutional capacity due to inefficiency, lack of transparency and coordination, and widespread corruption plaguing virtually every layer of the state bureaucracy.

Although nuclear technocrats refer to thirty years of experience in operating nuclear research reactors in BATAN facilities as a proof of competence, the anti-nuclear groups argue that safe utilization of nuclear power necessarily requires institutional capacity and competence more than just the ability to operate research reactors. In other words, the credentials of BATAN in research and development is not considered reassuring because the scope of nuclear power production is far beyond research-oriented activities, where the agency is seen as competent (see Amir 2009).

As Mary Douglas (1985) has pointed out, risk is always perceived whether it is defined by lay people or by technical experts, and perception is a cognitive impulse largely influenced by social and cultural circumstances. Hence, the prolonged weak condition of the state in Indonesia strongly affected the way the public looked at nuclear risk as a fundamentally institutional problem. The lived experience of individual citizens in going through a long list of public safety troubles and disasters due to the ineptitude of the state to perform its duties across public sectors led to a widely shared perception that the state was producing life-threatening risk in the proposed nuclear power development. A big contradiction appears in this picture: while the state supported an aggressive campaign to promote nuclear power as a safe system of electricity generation, at the same time it repeatedly showed massive failures in protecting citizens from dangers and hazards.

The gist of the risk problem lies in declining trust of the public to the state. In this light, what concerns the public was less the technical risks of nuclear power than the institutional risk caused by the visible incompetence of the state even in performing its mundane duties. This means that threat to public safety does not emerge from the technology itself but from the state. In other words, the risk is embodied in the institutional precariousness of the state.

Institutionalized ignorance

In the aftermath of Fukushima nuclear disaster, many believe that nuclear energy will abruptly decline and soon be replaced by other options for renewable energy. This is the case in several countries, where the government is determined to bring nuclear power into a gradual phase-out. However, a great number of other countries instead express confidence in nuclear energy. In the post-Fukushima world, this may seem awkward. As noted above, this is what has unfolded in Indonesia, one of those countries where government officials are eager to embark on the nuclear power program despite what transpired at Fukushima.

The organized resistance drawn by the anti-nuclear groups may have delayed the introduction of nuclear energy but it did not terminate the entire program. On the contrary, the nuclear power program remained in progress, though at a slower pace. Strong protest and criticism from the anti-nuclear movements entailed only changes in the scheduling and siting of the reactors that BATAN had proposed to the government. Since the Muria Peninsula of Central Java was no longer politically suitable to host the construction of Indonesia's first nuclear power plant due to formidable refusal in the area, BATAN decided to relocate the site to another location. Indonesia's vast geography and the low level of opposition in some areas gave nuclear technocrats many options for siting nuclear power plants.

Bangka Island, a province outside of Java, was eventually designated as the new location. In the new site, BATAN effortlessly acquired approval from the provincial government, which was enthusiastic about nuclear power. Lack of electricity led the Bangka Island government along with its citizens to believe that the construction of nuclear power plants would benefit the area.[5] This enthusiasm was bolstered by financial support from the National Development Planning Agency (BAPPENAS). This money was specifically allocated for preparation measures, including so-called "socialization" aiming to increase a positive image of nuclear energy in the public.[6]

The eagerness of a country like Indonesia to go nuclear can be explained from an amalgamation of three major factors – political, symbolic, and economic. The first factor is the Indonesian "technopolitical regime" (Hecht 1999) consisting of technological practices, networks of individual researchers and technocrats, and pro-nuclear politicians in political party. The nuclear technopolitical regime in Indonesia is concentrated in BATAN, but is tightly connected to universities and domestic and foreign/international research institutions and agencies. Reinforced by a socio-technical imaginary (Jasanoff and Kim 2009) upon which nuclear politics bases its *raison d'état*, the technopolitical regime was able to survive political changes and managed to carry its vision on over generations. The longevity and staying power of BATAN has meant that despite strong disapproval of many groups of civil society, the impetus for going nuclear has remained firm.

Hand-in-hand with the role of the technopolitical regime in pushing nuclear power forward in Indonesia are symbolic factors. In a post-colonial state, modern high technology embodies not only material and physical properties but also symbolic meanings which the state seeks to produce as a quintessential part of national identity. The magnificence of technological artifact such as the nuclear reactor engenders what Itty Abraham (1998) aptly describes as a fetish in which the material and physical superiority of high technology is deeply embraced by the state, animating the post-colonial imagination. This is acutely exhibited in the discourse of nuclear power circulated among state bureaucrats, technocrats, and pro-nuclear politicians. In this discourse, the construction of nuclear power is regarded as a testimony to the state's capability for conquering superpower technology.

Inextricably intertwined with the political and symbolic forces is the economic rationale. The proven ability of nuclear power to provide large-scale production of electricity, as shown for years in several countries most notably France, Japan, and South Korea, simply justifies the value of the technology to boost economic growth. The bottom line is the interest in growth. For a country that is undergoing an unprecedented economic boom, growth serves as the ultimate goal that informs the Indonesian government in making decisions. As an emerging economy, growth is almost everything to Indonesia because it is the yardstick that defines the wellbeing of the country. After years of economic malaise in the aftermath of the Asian crisis, the return of growth is certainly seen as a kind of historic momentum not to waste; it is regarded as an imperative of socio-economic progress that must be maintained, if not increased. It is the profound interest in growth that becomes the underlying logic pushing the venture to go nuclear, for a steady supply of energy is the backbone in every effort to keep the economy from stalling. In a modern economy, sustainable production of energy allows industries and economic activities in all sectors to run and to stay productive, as well as to power homes, hospitals, schools, and public facilities. The bitter experience of economic crisis taught Indonesia an important lesson, which is that growth is what matters in maintaining the legitimacy of the state and that growth depends on high and stable levels of energy production. The close association of nuclear power with growth entails a conviction that nuclear power is inevitably needed in order for the state to be able to keep the economy afloat for a long run.

A combination of the aforementioned factors creates a penchant to overlook the fact that nuclear risk is likely to increase because of the lack of institutional capacity in the state. It is a situation whereby interests outweigh concerns, and the perceived necessity prevails over potential consequences. Furthermore, ignorance becomes the norm of governance that clouds good and clear judgment. The pursuit of one specific goal dismisses all other scenarios that may give the same desired outcome at the end. This is not to stay that the technocratic logic that drives the nuclearization process did not acknowledge risk potentials at all. Risk analysis is indeed part of BATAN's preparation measures to ensure that Indonesia's venture into nuclear energy would be safe and secure. The problem is that risk in the large-scale undertaking is defined mostly as a set of technical issues that simply demand technical solutions. This has resulted in a set of recommendations in which safety, security, and safeguards become the jargon, and the approach emphasizes a narrow range of factors.

The ignorance lies in the technocratic epistemology that prompted this mode of risk analysis. In this context, the broader spectrum of social and institutional risks remains largely *ignored*. As technocratic analysis of risk is strongly focused on potential hazards that come from technical problems, it fails to acknowledge weak institutional capacity of the state, which is the reason for public outcry and is directly relevant to nuclear risk. If critically scrutinized, there seems to be a discrepancy between what nuclear technocrats consider risk and the structure of the institutions under which the nuclear energy project was implemented. Nuclear technocrats were oblivious to the myriad of institutional problems – such as corruption, lack of coordination, and bureaucratic inefficiency – which could gravely affect the operation of nuclear power. BATAN officials appeared to ignore the fact that the successful operation of any technological system is underpinned by a network of transparent, well-integrated and well-governed organizations and infrastructures.

In this light, the ignorance of institutional risk plaguing Indonesia's nuclear power program was a result of the failure of technocratic thinking in which the adamant desire for nuclear power eclipsed the ability to comprehend the complex nature and high-level uncertainty of nuclear risk. The spread of blind enthusiasm among decision-makers in the parliament as well as the central government as illustrated above added to ignorance of the risk. These individuals were supposed to examine each and every consideration when making a decision that would affect society at large, but have failed to do so.

The propensity to downplay the social and institutional risk of nuclear power production produced what I refer to as *institutionalized ignorance*. Ignorance is institutionalized when an array of legal instruments such as laws, decrees, and regulations are conceived following narrow-minded assessments of risk that aim to legitimize the undertaking of large-scaled technological projects with a high degree of risk. It signifies a paradox in contemporary culture in which the growth of knowledge expands the realm of ignorance (Gross 2010). *Ignorance implies intentional actions to eschew the responsibility for having a comprehensive view of risk in assessing highly uncertain technological adventures.* The use of technical jargon blended with ostensibly nationalist interests cement the institutionalization of the ignorance of risk.

The persistence of Indonesia's nuclear program is precisely rooted in institutionalized ignorance, for it exploits institutional authority in pushing forward the construction of nuclear plants amid social protests against the suitability of the high-risk technological enterprise. There are several legal products that empirically exemplify institutionalized ignorance in the nuclear power program. The strongest is "the 2007 Law of Long-Term Planning of National Development, 2005–2025." The Indonesian parliament passed the law as a response to the need to have a macro long-term development planning, a technocratic measure inherited from the New Order's developmental technocracy. What nuclear advocates found profoundly gratifying is that the law explicitly states that Indonesia should begin operating nuclear power plants between 2015 and 2019 as part of the national energy system. The law is the first of its kind to specifically oblige the central government to implement the nuclearization agenda. However, there is no mention in the law of how nuclear risk should be governed. It provides "constitutional" legitimacy to turn BATAN's proposal into reality; consequently, it also renders critics and actions against nuclear power "unconstitutional." With this law, nuclear technocrats could argue that the government could be the subject of impeachment if it failed to accomplish this long-term goal.[7]

The prevalence of institutionalized ignorance has considerable implications for democracy. As noted before, democratization had allowed Indonesian people to express their opinion and concern over the state's technocratic efforts to introduce high-risk technology in society. However, limitations mark how democracy works. It does not always produce benevolent decisions, for democracy is part of politics and power struggle. In a young democracy like Indonesia, decision-making remains heavily influenced by powerful groups in which technocratic thinking often times serves to justify political economic interests of the state and its allies. This is illustrated in the nuclear power program in which the law that legitimatized nuclear power was a product of the democratic system but contradictory to public concerns. Rather than create an open atmosphere that allows comprehensive public examination of high-risk technology, the Indonesian democracy provides a venue for institutionalizing ignorance with far-reaching consequences. Democracy seems to have failed to govern risk as a political problem. Certainly this is not the end of the story. What remains to be seen is whether the ongoing democratic transition in Indonesia could turn back the political pendulum and switch ignorance into awareness.

Conclusion

As STS scholars have noted, the outcomes of technology-society interaction vary across societies. A multitude of technical, social, and political factors play key roles in determining how society responds to the introduction of new technology and how modern technology is able to keep its grand promise (Bijker et al. 1987, Bijker and Law 1992, MacKenzie and Wajcman 1999). This chapter situates the technology-society relation in the discourse of risk, focusing specifically on the role of nuclear power in the emerging economy and democracy of Indonesia.

In a time when emerging economies are increasingly preoccupied with technological accomplishment, the pursuit of high technology seems appealing to the state because it promises to bring improved well-being and higher prestige. This is the main reason why states in the global south are strongly eager to undertake efforts to create large-scale technological projects in which the vocabulary of risk seems to disappear from the developmental nomenclature.

Nuclear power is not the only technology that bears great risk. However, the risk of nuclear power is believed by some to exceed that of other large scale technologies because it has a broad spectrum of impacts across spatial, biological, and temporal dimensions. Hence, nuclear risk is a matter of scale, spread, and temporality of destruction. This causes nuclear power to remain highly controversial in contemporary societies despite its contribution to substantial socioeconomic progress in some countries, such as Japan, South Korea, and France.

As pointed out in this chapter, there are multiple dimensions of risk in nuclear power development. Each of these dimensions plays out differently in different socio-political contexts. Indonesia's nuclear power program highlights one particular dimension of risk: institutional risk. Beyond the technical hazards of nuclear power, the development of nuclear power in Indonesia occurs in an environment of a state with limited capacities. This has been among the concerns of the critics of nuclear development in Indonesia.

The confrontation between nuclear technocrats and civil society groups in the nuclear power controversy underlines one sociological aspect that characterizes technology-society relations, namely, trust. A state with fragile capacities does not engender trust among citizens. This constitutes a fundamental problem in less advanced modernity and certainly in Indonesia. Moreover, in an environment of institutional risk and ignorance, where the state is not adequately equipped with organizational competence and genuine commitment to the well-being of society, the social production of risk may exceed social production of wealth.

One last note: the logic of the state, albeit democratic, is not necessarily congruent with people's concerns. For the state possesses its own interests, which govern how it makes decisions and take actions. The failure of democracy to control this logic results in ignorance of the state, channeled through institutional regulations. The institutionalization of ignorance can have far-reaching implications because it obstructs the possibility for social control of high-risk technology. In the end, the state becomes the embodiment of social production of risk.

Notes

1   There is an ample collection of works investigating this tragic event. The most noted, yet controversial, account is the so-called "Cornel Paper" written by Anderson and McVey (1971). For a more recent study, see Roosa (2006).

2   See Oberman et al. (2012).
3   Ibid.
4   The empirical material of this observation can be found in Amir (2010a).
5   See "Nuclear Power Plants to be Built in Bangka Belitung, Governor Says." *The Jakarta Post*. July 06 2011.
6   A budget of approximately $50 million was allocated by BAPPENAS to facilitate the preparation of basic infrastructure documents such as siting studies, environmental impact analysis, etc.
7   A statement from former minister of research and technology exemplifies this argument.

References

Abraham, I. 1998. *The Making of Indian Atomic Bomb: Science, Secrecy and the Postcolonial State*. London: Zed Books.

Amir, S. 2009. "Challenging Nuclear: Antinuclear Movements in Postauthoritarian Indonesia." *East Asian Science, Technology and Society* 3: 343–366.

Amir, S. 2010a. "The State and the Reactor: Nuclear Politics in Post-Suharto Indonesia." *Indonesia* 89: 101–147.

Amir, S. 2010b. "Nuclear Revival in Post-Suharto Indonesia." *Asian Survey* 50: 265–286.

Amir, S. 2012. *The Technological State in Indonesia: The Co-Constitution of High Technology and Authoritarian Politics*. London: Routledge.

Ananta, A., Arifin, E.N., and Suryadinata, L. 2005. *Emerging Democracy in Indonesia*. Singapore: Institute of Southeast Asian Studies.

Anderson, B. and MacVey, R. 1971. *A Preliminary Analysis of October 1, 1965 Coup in Indonesia*. Ithaca, NY: Cornell Southeast Asia Program.

Aspinall, E. and Mietzner, M. 2010. *Problems of Democratization in Indonesia: Elections, Institution, and Society*. Singapore: ISEAS.

Beck, U. 1992. *Risk Society: Towards a New Modernity*. London: Sage Publications.

Beck, U. 2007. *World at Risk*. London: Polity Press.

Bijker, W. and Law, J. 1992. *Shaping Technology/Building Society: Studies in Sociotechnical Change*. Cambridge, MA: MIT Press.

Bijker, W., Bal, B., and Hendriks, R. 2009. *The Paradox of Scientific Authority: The Role of Scientific Advice in Democracies*. Cambridge, MA: MIT Press.

Bijker, W., Hughes, T., and Pinch, T. 1987. *The Social Construction of Technological Systems: New Directions in the Sociology and History of Technology*. Cambridge, MA: MIT Press.

Callon, M., Lascoumes, P., and Barthe, Y. 2001. *Acting in an Uncertain World: An Essay on Technical Democracy*. Cambridge, MA: MIT Press.

Douglas, M. 1985. *Risk Acceptability According to the Social Sciences*. New York: The Russell Sage Foundation.

Douglas, M. and Wildavsky, A. 1983. *Risk and Culture: An Essay on the Selection of Technological and Environmental Dangers*. California: University of California Press.

Flam, H. 1994. *States and Anti-Nuclear Movements*. Edinburgh: Edinburgh University Press.

Geiger, T. 2011. *The Indonesia Competitiveness Report 2011: Sustaining the Growth Momentum*. Geneva: World Economic Forum.

Giddens, A. 1990. *The Consequences of Modernity*. California: University of California Press.

Gross, M. 2010. *Ignorance and Surprise: Science, Society, and Ecological Design*. Cambridge, MA: MIT Press.

Hecht, G. 1999. *The Radiance of France: Nuclear Power and National Identity After World War II*. Cambridge, MA: MIT Press.

Hecht, G. 2012. *Being Nuclear: Africans and the Global Uranium Trade*. Cambridge, MA: MIT Press.

Hill, H. 1997. *Indonesia's Industrial Transformation*. Singapore: ISEAS.

Hill, H. 2000. *The Indonesian Economy*. Cambridge: Cambridge University Press.

Hill, H. and Shiraishi, T. 2007. "Indonesia after the Asian Crisis." *Asian Economic Policy Review* 2: 123–141.

Jasanoff, S. 1986. *Risk Management and Political Culture: A Comparative Analysis of Science in the Policy Context*. New York: The Russell Sage Foundation.

Jasanoff, S. 2004. *States of Knowledge: The Co-production of Science and Social Order*. London: Routledge.

Jasanoff, S. 2005. *Designs on Nature: Science and Democracy in Europe and the United States*. Princeton: Princeton University Press.

Jasanoff, S. and Kim, S.H. 2009. "Containing the Atom: Sociotechnical Imaginaries and Nuclear Power in the United States and South Korea." *Minerva* 47: 119–146.

Jasper, J.M. 1990. *Nuclear Politics: Energy and the State in the United States, Sweden and France.* Princeton: Princeton University Press.

Kleinman, D.L., Kinchy, A.J., Handelsman, J. 2005. *Controversies in Science and Technology: From Maize to Menopause.* Madison, WI: University of Wisconsin Press.

Luhmann, N. 1993. *Risk: A Sociological Theory.* Berlin: Walter de Gruyter.

MacKenzie, D. and Wajcman, J. 1999. *The Social Shaping of Technology.* Buckingham: Open University Press.

Massey, A. 1988. *Technocrats and Nuclear Politics.* London: Avebury Gower.

Oberman, R., Dobbs, R., Budiman, A., Thompson, F., and Rosse, M. 2012. *The Archipelago Economy: Unleashing Indonesia's Potential.* New York: The McKinsey Global Institute.

Perrow, C. 1999. *Normal Accidents: Living with High Risk Technologies.* Princeton: Princeton University Press.

Power, M. 2007. *Organized Uncertainty: Designing a World of Risk Management.* Oxford: Oxford University Press.

Ramana, M.V. 2013. "Nuclear Policy Responses to Fukushima: Exit, Voice, and Loyalty." *Bulletin of the Atomic Scientists* 69: 66–76.

Ricklefs, M.C. 2001. *A History of Modern Indonesia Since c.1200.* Stanford: Stanford University Press.

Roosa, J. 2006. *Pretext for Mass Murder: The September 30th Movement and Suharto's Coup d'état in Indonesia.* Madison: The University of Wisconsin Press.

Slovic, P. 2000. *The Perception of Risk.* London: Routledge.

Sovacool, B. and Valentine, S. 2012. *The National Politics of Nuclear Power.* London: Routledge.

Thee, K.W. 2012. *Indonesia's Economy Since Independence.* Singapore: ISEAS.

Van Loon, J. 2002. *Risk and Technological Culture: Towards a Sociology of Virulence.* London: Routledge.

Vatikiotis, M. 1993. *Indonesian Politics Under Suharto: Order, Development, and Pressure for Change.* London: Routledge.

Wallerstein, I. 2004. *The Uncertainties of Knowledge.* Philadelphia: Temple University Press.

Welsh, I. 2000. *Mobilising Modernity: The Nuclear Moment.* London: Routledge.

Winner, L. 1986. *The Whale and the Reactor: A Search for Limits in an Age of High Technology.* Chicago: University of Chicago Press.

Wynne, B. 1989. "Sheep Farming After Chernobyl: A Case Study in Communicating Scientific Information." *Environment* 31: 11–39.

18

From River to Border

The Jordan between empire and nation-state

Samer Alatout

COMMUNITY AND ENVIRONMENTAL SOCIOLOGY, THE NELSON INSTITUTE FOR ENVIRONMENTAL STUDIES

More than a decade ago, Lee and Roth (2001) asked the question "How can ditch and drain become a healthy creek" in order to demonstrate the exhaustive work performed by an environmental non-governmental organization (NGO) to turn a heavily polluted "ditch" into a "healthy creek." In this chapter, I am asking the same question, only in reverse: how did a historically significant, extremely healthy river become a ditch?

At present the lower Jordan River is nothing more than a trickle! The reason is that while Israel has been diverting the upper Jordan from Lake Tiberias in the north to the Negev Desert in the south using its National Water Carrier since the mid-1960s, Jordan, at the same time, has been diverting the river's largest tributary, the Yarmuk River, into its East Ghor Canal (see Figure 18.2). The result is not only that the Jordan River has almost disappeared in its lower stretches but also that the Dead Sea has been shrinking at an alarming rate, to the degree that large, threatening sinkholes have been appearing in its vicinity (see Figure 18.1). So, the river does not invoke the positive emotions familiar to us from a great number of travelogues. It is a disappointment for Christian pilgrims who get extremely disappointed upon reaching the point where Christ is said to have been baptized, and for environmentalists (Palestinian, Jordanian, Israeli, and international) who are troubled by the looming ecological disaster. Some use the current state of the Jordan River to argue on behalf of the river and its restoration, others look for technical fixes like the proposed Red Sea–Dead Sea Canal, which would arguably replenish the Dead Sea, and yet others view the river's current state as an expected and acceptable result of progress. But why and how was the Jordan River turned into a ditch? What were the forces of history that worked their magic against nature, if not even against God, and seem to have won? How did that happen despite the reverence that the Jordan River has in people's imaginations?

My brief answer to these questions is that the Jordan River had the misfortune of becoming a *border*, initially in the politics of empire (1840s–1923) and subsequently in the politics of nation-states (1948 until the present).[1] By turning the river into a border, water became a territorial object. Empires, however, cared about the river as a border in order to demarcate spheres of territorial influence and not in order to use its waters in development, at least not chiefly or essentially. The consequence of turning the river into a border and water into a territorial

Figure 18.1 Dead Sea sinkholes

Source: Image taken by Sean Pavone. http://www.123rf.com/photo_12449778_sink-holes-near-the-dead-sea-in-ein-
gedi-israel.html, accessed May 25, 2013.

object led the nation-states of the late-1940s and early-1950s to use the very sovereign claims over territory to redefine water as a matter of national security and as a resource that could and would be used in order to secure the nation-states. It is this history of turning the Jordan River into a border, water into a territorial object, and its utilization in the discourse of national security that might explain how the river became a ditch.

Empires, boundaries, and territorial borders

The year 1993 might not have registered for many of us. After all, the year was relatively quiet and at a measurable distance from arguably some of the most monumental events of the second half of the twentieth century: the fall of the Soviet Union in 1989 and the U.S.–Iraq war in 1990–1991. However, the year is significant for the purposes of this chapter for it was then that both Bruno Latour and Edward Said each published a significant book *We Have Never Been Modern* by Latour and *Culture and Imperialism* by Said[2]. Both emphasized a process by which human beings have tended to separate, demarcate, isolate, distinguish, and draw boundaries.

For Latour, this process is about the construction of distinctions not only between objects (nature, politics, and discourse), but also between modes and strategies of knowing and speaking about those objects (biophysical sciences, social sciences, and language). In a word, for Latour, modernity is one continuous engagement with differentiation: one science from the other, one form of knowledge from the other, and one object from the other. Said, on the other hand, is interested in studying empire and new forms of imperialism.[3] He focuses on the profoundly different and incommensurable narratives through which colonizers and colonized construct their colonial experiences, understand empire and imperialism. In the process, he argues, the identities of the colonizer and the colonized are grounded in assumptions of radical difference.

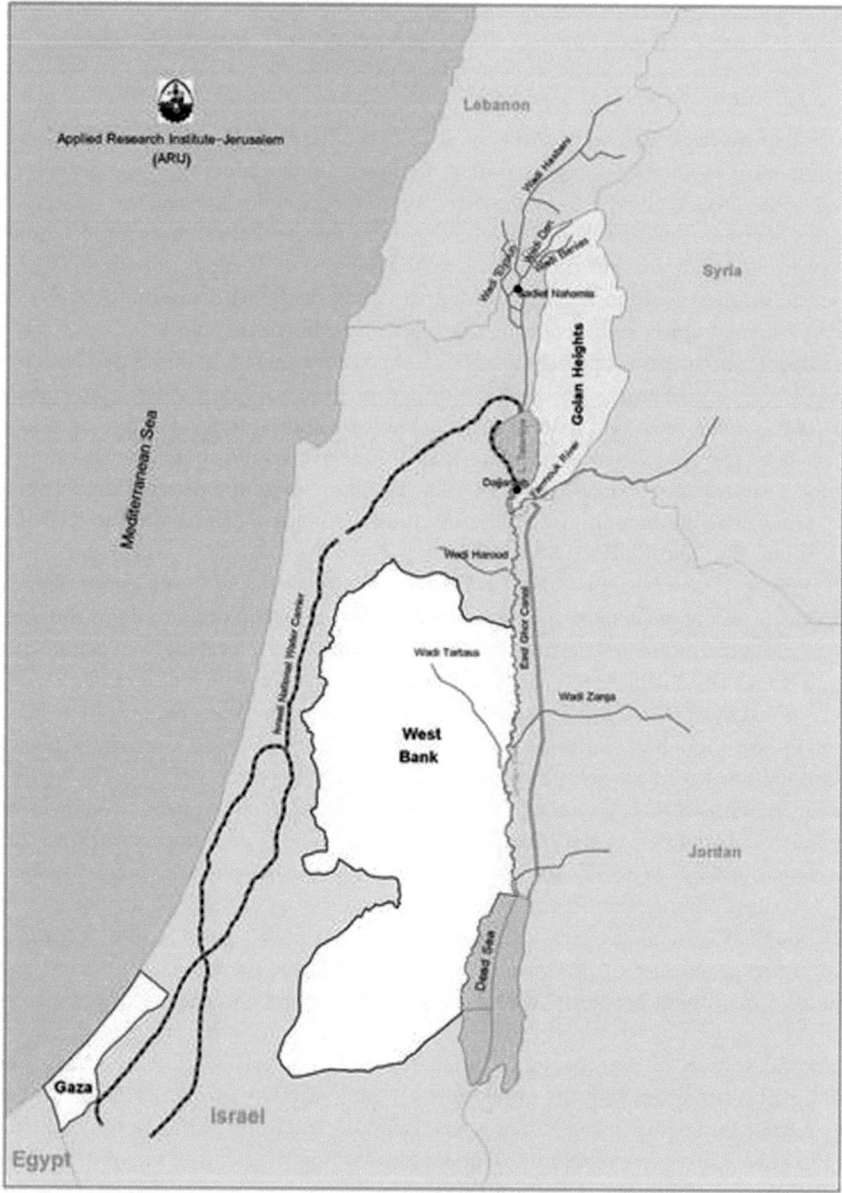

Figure 18.2 East Ghor Canal and the National Water Carrier

Source: Applied Research Institute of Jerusalem www.arij.org/publications(7)/Papers/2005/Roots%20of%20
Water%20Conflict%20in%20the%20Eastern%20Mediterranean.pdf, accessed May 30, 2013.

Latour and Said, however, insist that treating distinctions as if they reflect real and concrete identities and objects, misses a great deal of what takes place outside formal and disciplinary practices. They use similar terms, "mediation" in the case of Latour, and "hybridization" in Said's analyses, to illuminate what seem to be the underlying realities of the world. For Latour "mediation" continuously links the natural, the social, and the linguistic. For Said, "hybridization" links the colonized and the colonizer, empire and its subjects. Processes of mediation and hybridization never abate; they are continually taking place, even as we make the distinctions ever more stark.

Despite certain overlaps, Said and Latour speak to two different audiences. Said's work has been most influential in colonial and postcolonial studies; it underscores the role of imperialism in initiating the process of hybridization where the colonizer and colonized, the imperialist and the subject of empire are thrown into a web of relations that neither can fully avoid. Latour, on the other hand, builds on and contributes to science and technology studies (STS) by emphasizing the seamless modern networks of humans and non-humans and the difficulty of disconnecting a natural object from its life as a political and a linguistic object.[4]

What is valuable about these approaches to STS and postcolonial scholarship is that both are acutely aware of the contradictory, yet closely connected, processes of "purification" and "mediation." However, for the most part, science and technology studies neglects imperialism as a necessary context for knowledge production and modernity, and postcolonial studies, commonly, takes knowledge production at face value. By attending to the relationship between imperialism, knowledge production, and modernity, I hope to make a contribution to STS and postcolonial studies (see also Philip 1998, 2003).

I read the history of water technoscience and politics in historic Palestine *as if empire mattered* and *as if imperialism and colonialism were a significant* part of that history. In order to do so I suggest reading both Said and Latour in Thomas Gieryn's (1983 and 1999) vocabulary of boundary-making. Read from that vantage point, Said and Latour seem to be in search of boundaries, not as reflections of real distinctions, but as the result of a politics of boundary maintenance. They are interested in how those boundaries are accomplished, what relations and strategies of power achieve them, and what sorts of inclusions and exclusions they enable.

While all three scholars share a symbolic notion of boundaries and boundary-making, none of them pays focused attention to the spatial-territorial implications of boundary-making, i.e., to resultant, actual, and real *borders*. I focus attention on this particular phenomenon of *territorialization*: (1) territorial borders as, in part, the effects of boundary-making in science and in politics; (2) boundary-making in technoscience and politics as it translates into territorial borders and creates a new set of dynamics that fortify and shape practices of inclusion and exclusion; and (3) territorial borders and how they make apparent the politics of boundary-making.[5]

In the second section of the chapter, I revisit the issue of the historical boundaries of Palestine and their relationship with the Jordan River. I demonstrate how the notion that there is a *necessary* relationship between the Jordan River, Palestine, and its territoriality has a history, and that is the history of the encounter between empire and the river during the mid- to late-1800s – including the progressive definition of Palestine as the sole object of interest for the Zionist colonial project. During that period, the Jordan River was *made* into a local, Palestinian river. I argue that turning the Jordan from a local, Palestinian river into the eastern border of Palestine was an achievement of empire and under imperial pressures and concerns; the river basically became a border between empires, especially France and Britain that had a specific understanding of the role of rivers in demarcating the territorial states (1914–1923).[6] In the process of creating the geohydrological map of Palestine, the Jordan River slowly became a border between nation-states (1923 to the mid-1960s). The strategic significance of the river-as-border was not limited to its concrete demarcation of different territorialities, which was its main function in the encounter with Empire, but also in the role it played in turning water itself into a territorial object that is subject to the sovereign-security-developmental needs of emerging nation-states. It is this *border effect*, i.e., reconceptualizing water as a territorial-developmental-security object, which led to its diversion from the river bed through the Israeli National Carrier and the East Ghor Canal, thus turning the river into a ditch. In the conclusion I pose a few questions about the territorialization of boundary making in science.

The Jordan River and Palestine: before empire

To understand the increased political, scientific, and economic interest in Palestine, its boundaries, and its relationship with the Jordan River during the late-1800s and early-1900s, it is necessary to place that interest in both the context of the imperialist project of dividing the weakening Ottoman Empire and the Zionist project of settling Palestine and turning it into a Jewish state.

Since Herodotus (450 B.C.), there have been almost 2,500 years of writings about Palestine, and yet at the turn of the twentieth century, Palestine's territorial boundaries were still unclear. While Palestine initially meant the narrow edge of the eastern Mediterranean shore where the Philistines lived and after whom the place was named, Palestine's administrative boundaries have depended on the political-administrative needs of the empires that controlled the area. Historical maps clearly demonstrate how untenable Palestine's changing boundaries have been and that there is no necessary correspondence between Palestine as a concept and any specific territory marked as Palestine.

Westerners' interest in Palestine's boundaries and the Jordan River started in the mid-1800s. It was initially motivated by the continuous weakening of the Ottoman Empire which led the western powers (including the United Kingdom, France, and Italy) and their eastern counterparts (mainly Russia) to begin discussions over dividing the Empire, especially its Asiatic regions. There is no doubt that biblical narratives made Palestine especially important to all of these powers, but there was also an increasing thirst for the commercial and strategic wealth the area might provide.

The first modern exploration of Palestine was undertaken by the United States. The U.S. Congress dispatched and funded an expedition of the Jordan River, its sources, and the Dead Sea in 1847. The Congressional mandate was partly couched in biblical terms (finding archaeological connections with biblical narratives),[7] but also in commercial, political-strategic, and techno-scientific terms as well.[8] The results of that exploration can be found in two publications, one of which is an official document by the Department of the Navy, *Official Report of the United States Expedition to Explore the Dead Sea and the River Jordan* (1851), and the other is a personal account of the expedition by its commander W.F. Lynch (1849), *Narrative of the United States' Expedition to the River Jordan and the Dead Sea*. What seems clear from this record is that the Jordan River at the time was not thought of as a territorial border and, even when it progressively defined jurisdictional authority, the river did not matter for everyday life or even everyday governing.[9]

In addition to its biblical, political-strategic, scientific, and economic significance, there was an imperial motivation behind this expedition, as well. Lynch describes how important it was to raise "the American flag ... in Palestine," for the first time in areas "without consular precincts" (1849: 119). There was a distinct patriotic fervor in the way he described the American role in the "regeneration" of what he characterizes as "a hapless people" (1849: 119). He goes on to depict Palestine's native population in terms familiar to us from other imperial adventures. "In Syria," he argues, "we may look for more unadulterated specimens of the Muslim character than in the capital of the empire" (1849: 124). He portrays these "unadulterated" people, the "Arabs" of Acre, in unflattering terms: "thievery," "wretched," "mobs," and "notorious thieves" (1849: 125–126).[10]

This mix of biblical, commercial, imperial-strategic, and technoscientific interests undergirded every expedition thereafter. Some of the well-known explorations are those of the British Palestine Exploration Fund (PEF) (1881), *The Survey of Western Palestine*, and the later report (PEF 1889), *The Survey of Eastern Palestine*. It is quite difficult, if not impossible, to think of these more recent explorations as separate from or unrelated to increasing Zionist interest in settling Palestine. The influence of pre-Zionist thought of turning Palestine into a Jewish state is obvious in most of the documents produced by the PEF. For example, in 1875, in the midst of exploring Palestine, the Earl of Shaftesbury exclaimed in the annual meeting: "Let us not delay to … prepare… [Palestine] for the return of its ancient possessors…" (PEF 1875: 115–116).

The PEF went to great lengths to establish the notion of Palestine as an empty or almost empty space that was ripe for Jewish settlement. Their survey of western Palestine (PEF 1881: 331–362), pointing to differences in height, color, and customs of the different sectors of Palestinian native population, argued that the people of Palestine were composed of multiple races and *did not and will not constitute a national* community. This was in contradistinction to the Jewish people who, according to the PEF, constituted a national community despite their diaspora. This type of argument was often used, by the PEF and later on by the different imperial powers and the Zionist Organization, to legitimize the calls to turn Palestine into a Jewish state.

These arguments did not go unchallenged, and it is in these criticisms that we can discern the politics of Palestine's future, long before the Zionist movement was even established. Among others, Selah Merrill, who served as the American Consul General in Jerusalem, (1882–1885, 1891–1893, and 1898–1907) and was a well-known archaeologist, theologist and Hebrew scholar, presented a substantial critique of interpretations like those of the PEF's. Merrill argued that the natives of Palestine did, in fact, constitute a national community, and he suggested that many of the archaeological sites that were claimed to be of biblical relevance were actually not.

These disputes were framed, in part, as disagreements over what constituted legitimate scientific methodologies, findings, and scientific authority. The symbolic boundaries demarcating legitimate archaeological and other science from non-science (these were the actors' categories), the biblical-archaeological and territorial mapping of Palestine, as well as the human-scientific mapping of its population, cannot be separated from the widespread debates during the 1880s about turning Palestine into a Jewish state, the territoriality of which was already being debated and would prove crucial for turning the Jordan River into a border a few years later.[11]

Now in terms of the relationship between east and west of the Jordan River, all of these writers agreed on one matter: there were no clear territorial boundaries dividing the administrative units of the Syrian possessions of the Ottoman Empire. There was also a consensus that the everyday relationships (commercial, cultural, and political) between east and west of the river were extensive. In other words, the Jordan River did not function as a territorial boundary in practice, but as a body of water across which residents engaged in relations of all sorts and the banks of which were settled, at times, by the same families and relatives on the two different sides.[12]

The reading of the multiple connections between east and west of the Jordan River, all the surveys, maps, and travels that reflected or attempted to construct the spatiality of Palestine turned the Jordan into a local river – a local, Palestinian river to be exact – across which extensive interchange occurred. The popularization of the view that the Jordan River was a Palestinian river cannot be isolated from the activities of the British and American Zionists (including the surveys of the British and the American Palestine Exploration Funds). Nor can it be isolated from the Zionist Organization, established in 1897, which, at least since its 1904 conference, focused on turning Palestine into a Jewish state. The political vision of the Zionist Organization was indeed as faithful and as maximalist in terms of territory as that of the new archaeological knowledge of Palestine, i.e., turning the whole of Palestine, on both sides of the Jordan River, into a Jewish state.

This is probably the first attempt to turn Palestine into a political-territorial project. It was indeed under cover of empire (its military and its archaeology) that this project gained momentum. But, as will become clear in the next section, this project conflicted with relations of empire and had to be tamped down.

Jordan River as border: times of empire

The archaeological (by scientific explorations) and political (by Zionists – see Figure 18.3) construction of Palestine (1840s–1910s) as an area that extends from the Mediterranean in the west to what was then called the Syrian Desert in the east, thus including both banks of the Jordan River, was successful to a large degree (see Figure 18.2). In the multiple negotiations between the British, the French, the Russians, the Zionists, and the Arabs, Palestine always included both banks of the Jordan River, even as late as 1922. This construction of the territoriality of Palestine, however, was short-lived. While the mandate documents of April 1922 truncated the boundaries of Palestine in the north (they placed the Litani River in Lebanon and split the sources of the Jordan River among Lebanon, Syria, and Palestine), an amendment to the Palestine Mandate in September 1922 created the new country of Transjordan and thus separated the east and west sides of the river. Turning the Jordan River into a border was a matter of empire, as I demonstrate further. It was in the management of imperial relations between the French and the British, as well as between the British and the Arab and Zionist forces, that the Jordan River found itself having to demarcate territories and identities.[13]

A number of promises about the future of Palestine that were made by the British during World War I conflicted. For example, in 1915 Palestine was promised to the Emir of Mecca, Sharif Hussein, as part of the imagined Arab State in exchange for his support against the Ottomans.[14] In 1916 a different mapping of the area was agreed upon with the French in what came to be known the Sykes-Picot Agreement (see Figure 18.3). Here, part of Palestine was to be under British influence with Jerusalem as an international area. In 1917 the British promised to use its imperial powers to help the Zionist Organization establish a Jewish National Home in Palestine.[15] By 1918, when the Ottoman armies were decidedly defeated, the British had to sort out those conflicting promises. Much has been written about these promises, which, rightfully places most of the current conflicts in the Middle East at the feet of the British and French governments and their imperial adventures in the area during this period.

In any case, in the initial phases after World War I, both east and west of the river were placed under the British Mandate of Palestine. So, the Jordan River was, as most of the mapping of the 1800s indicated, a Palestinian river. British authorities themselves differed on whether to keep west and east of the river under the administration of Palestine or to split them apart. It is clear, however, that most of the British civil servants with Zionist connections, including the High Commissioner of Palestine, Herbert Samuel, were in favor of keeping east and west of the Jordan under one administration and of continued Jewish immigration and settlement on the east side of the river.[16]

The borders resulting from the Palestine Mandate reflect imperial interests as interpreted and negotiated by the civil servants of the time. Even when these different constituencies had agendas that might not be chiefly concerned with empire, everyone had to do the detailed and tedious work of articulating those interests, whatever they may be, (pro- or anti-Zionist, pro-

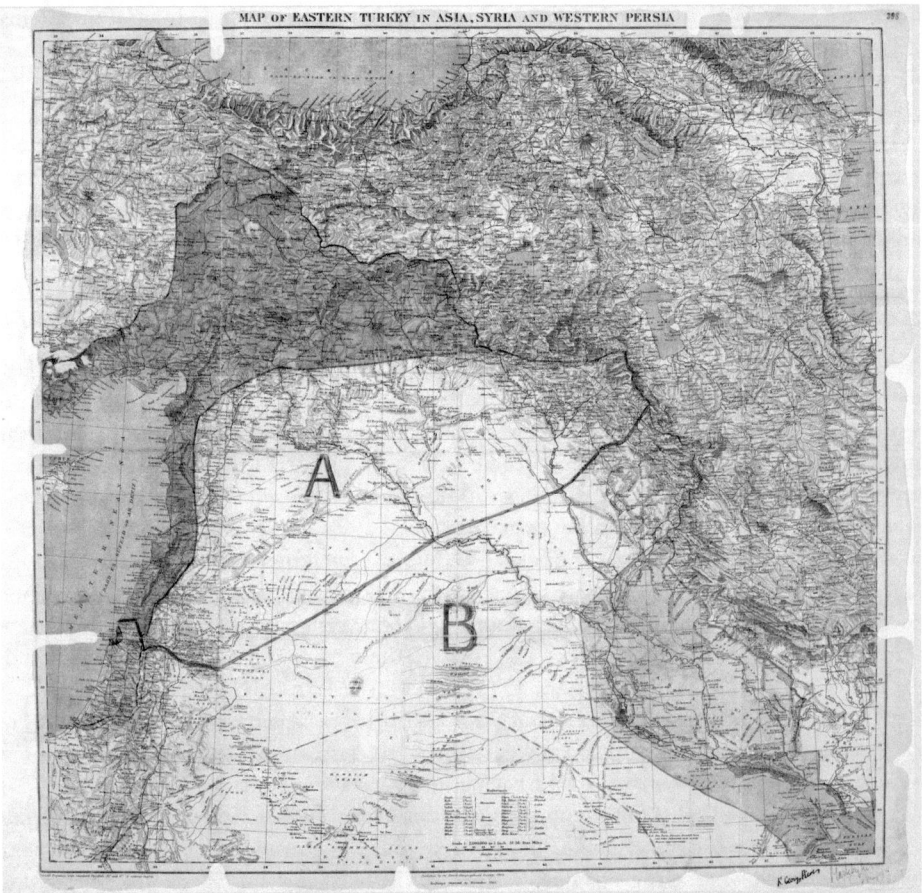

Figure 18.3   The Sykes-Picot Agreement, 1916

or anti-Arab, in favor of stronger or weaker collaboration with the French) with those of empire. We can see this in correspondence (1914–1922) between the War Office, the Office of the Secretary of the Colonies, the Office of the High Commissioner of Palestine after 1918, as well as in correspondence with the French on the boundaries of the territories under French and British influence. Two salient points emerge from this correspondence: (1) there was extensive discussion of whether to keep east of Palestine part of the Mandate of Palestine or to separate that territory once and for all; (2) there was extensive discussion of the degree to which the British should insist, against the wishes of the French, on including the sources of the Jordan River in the north in the territory of the Palestine Mandate.

Two brief examples demonstrate that interests of empire were paramount in British positions on the role of the Jordan River as a border during this period (1914–1923). The first example shows that Palestine in and for itself was unimportant to British politics. It was important only in the field of imperial relations. The second illustrates that while the British used an extended definition of Palestine, their agreement with dividing it into two mandates came out of a concern for the protection of British imperial interests in Arabia and the Gulf.

Scholars of the Middle East agree that the Report of the Committee of Imperial Defence on Asiatic Turkey (from now on De Bunsen 1915) became one of the most important strategic documents for how the British conducted negotiations throughout the war and after – it became a source of legitimacy that was referenced by different British authorities in favor of one or another solution to the Palestine issue. It is also true that this report demonstrates the fact that Palestine, in and for itself, was an unimportant element in British politics. The Committee from the very beginning acknowledges that British interests in Asiatic Turkey are part of the contemporary configuration of imperial struggles and that they cannot be seen in isolation from the interests of other imperial powers. British interests, in other words, "are *circumscribed by those of other powers* ... any attempt to formulate them must as far as possible be made to fit in with the known or understood aspirations of those who are our *Allies* today, but may be our *competitors* tomorrow" (De Bunsen 1915: 2, emphasis added).

In a very frank discussion about adding territories to British control, the committee acknowledged that there has to be a balance between "the prospective advantages to the British Empire" by dividing Turkey among the Allies on the one hand and the "inviolable increase of Imperial responsibility," on the other. The De Bunsen Report continues: "Our Empire is wide enough already, and our task is to consolidate the possessions we already have, to make firm and lasting the position we already hold, and to pass on to those who come after an inheritance that stands four-square to the world" (De Bunsen 1915: 2).

In setting the priorities of the British Empire, the committee was clear that Palestine was secondary to British interests and that it mattered only inasmuch as it contributed to British consolidation of their control over the Arabian/Persian Gulf and Iraq. It gave the highest priority to "the final recognition and consolidation of our position in the Persian Gulf," to the "prevention of discrimination of all kinds against our trade throughout the territories now belonging to Turkey," and to the "security for the development of undertakings in which we are interested, such as oil production, river navigation, and construction of irrigation works" (De Bunsen 1915: 3). They even explicitly put the "question of Palestine and the Holy Places of Christendom" on a list of lower priorities and set it "aside" as it is one of those "questions which could be discussed and settled in concert with other Powers" (De Bunsen 1915: 3). They go on to emphasize that any scheme used to divide the Asiatic possessions of the Ottoman Empire should try to minimize territorial conflict between the Powers (at the time France, Russia, and Italy) (De Bunsen 1915: 4-6).

In another secret and more recent document dated 9 April 1918, *Handbook on Northern Palestine and Southern Syria*, the War Office clearly states Palestine's relationship with the Jordan River: there is a western Palestine and an eastern Palestine that are strongly connected. The British at the time treated the river as a local body of water that runs through Palestine itself, not as a border that demarcates two different political or administrative units. For the British, the struggle between 1915 and 1922 was to manage the multiple pressures placed on them while at the same time protecting their main interest in the area: building a secure railway between the Gulf and the Mediterranean for oil trade and military needs. Their distrust of the French as a permanent ally prevented them from using another railway that already existed between Iraq and the Mediterranean, but one that went through the French Mandate area of Syria.

Pressures from the Emir of Hedjaz (Hussein Ben Ali) who insisted on the establishment of an Arab State that would include all the Arab provinces of the Ottoman Empire, the Zionist maximalist interest in Palestine on both banks of the river, and the French interest in protecting their Syrian and Lebanese Mandates from the encroachment of Arab nationalists under the leadership of Emir Faisal (one of the sons of Emir Hussein) led to, in the end, the decision of the British to use the Jordan River as a border. In addition, the northern boundaries of Palestine (including the northern borders of Transjordan) were also drawn with specific reference to the Jordan River and its sources.

Ultimately, the interests of empire and the multiple negotiations waged by the British led to the September 1922 conclusion, in which the terms of the Mandate were formally changed: the Jordan River became Palestine's eastern border, and only one of its sources in the north, the Dan River, was located in the territory of Palestine. This outcome altered the political geography of the area. It not only structured space in such a way that limited Jewish immigration to western Palestine; it also structured the future possibilities of what could be imagined as the geohydrological system of Palestine and future partitioning plans.

From empire to nation-states: discovering and then extending the geohydrological system of Palestine

In this section I focus attention on two incidents in which Zionist research on the "geohydrological system of Palestine," under the new political geography of the area, created new territorial opportunities for the partitioning of Palestine. The first case, in the late 1930s, is limited to research on the boundaries of groundwater resources and their correspondence to the political boundaries of the proposed Jewish state under the partition plan of 1937. The second case deals with the extension of the boundaries of the hydrological system of Palestine in the 1940s to include the Jordan River and how that extension corresponded to the territorial framework of the partition plan of 1947. Turning the Jordan River into a ditch was a direct result of this hydrological research and its territorial effects on the emerging states.

The geohydrological construction of national space

The Zionist Organization was forced to rethink its strategy in Palestine, after its imaginary as presented in the Paris Conference of 1919 (see Figure 18.4) failed. While the Zionists argued that the Jordan River was a crucial resource for the development of Palestine, the fact that the river became a border and not a local Palestinian river meant that its use was subject to international law and agreements. So, the most that Zionists were able to secure regarding the Jordan River was a 70-year concession from the British and the French for the use of the Jordan/Yarmuk system to produce hydropower. Beyond that, the use of the river for irrigation was limited to its basin as was customary in river management.

Since the establishment of the Mandate in 1923, much of Zionist politics went into defining what establishing a "Jewish national home" in Palestine might mean and what the shape of Palestine in that framework would be. Despite the multiple interpretations of what a Jewish national home in Palestine could look like, its meaning progressively undermined *cultural* interpretations (Palestine as a center of Jewish culture) and favored, especially since the late-1930s, the *political* interpretation of a Jewish national home, i.e., establishing a sovereign, Jewish nation-state.[17] This shift coincided with many events that took place during the 1930s: the consolidation of power in favor of Labor Zionism within the political institutions of the Jewish community of Palestine and the ascendance of David Ben-Gurion to a leadership position; the perceived threat of the emergence of strong Palestinian national aspirations in the early and mid-1930s; and the Peel Commission's report suggesting the partitioning of Palestine and the consequent White Paper of 1939 that was perceived as anti-Zionist. Only at this stage did Zionist water research and policy institutions begin treating, and in the process constructing, Palestine as a unified national space.[18]

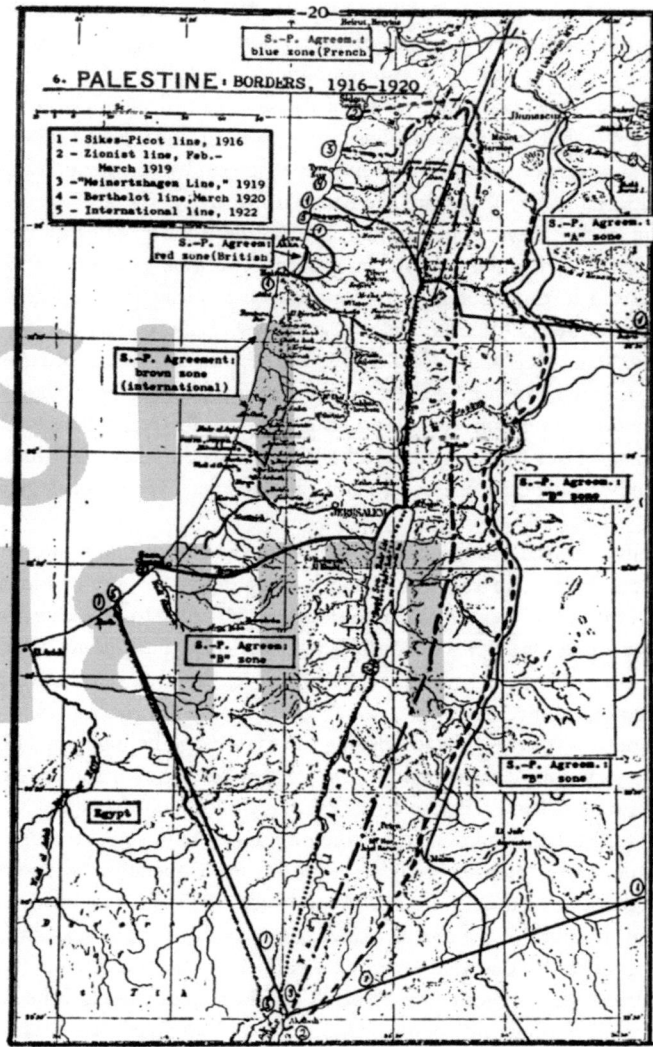

Figure 18.4   The Zionist proposal for the 1919 Paris Conference

The shift in favor of perceiving Palestine's water resources in terms of one national geohydrological system enabled and solidified the perception of Palestine as a national space. More significantly, however, perceptions of the national space extended as much as, and followed closely, the perceived geohydrological mapping of Palestine. What water experts constructed as the geohydrological system of Palestine had direct effects on what was imagined as the boundaries of the nation–state itself.

Geohydrological research shifted in two ways that are significant for our understanding of the role of the Jordan River in constructing Palestine. The first shift is in the importance given to groundwater research. In the 1920s and early-1930s, research on water resources figured only to the extent needed in the process of studying the geological structure of a particular area of

Palestine – for example, for the purposes of mining. During the late 1930s, coinciding with the emergence of water as a political resource in the struggle over Palestine's absorptive capacity, i.e., Palestine's ability to support millions of new Jewish immigrants, and Palestine's partitioning plans, research on groundwater resources in the whole of Palestine gained significance as a separate field of study. The second shift was from viewing groundwater resources and subterranean aquifers as separate and discrete to the perception of groundwater resources as constitutive of one geohydrological system that spans the national space.

The main water consultant of the Zionist organization in Palestine was Dr. Leo Picard of the Hebrew University. In 1931, Picard published his dissertation, *Geological Researches in the Judean Desert*, which he submitted to the Geological Department of the Imperial College of Science in London. As suggested by its title, the study was not especially concerned with Palestine as a contiguous space. It was regional in dimension and was carried out in order to investigate phosphate and bituminous limestone deposits in the area surrounding *Nabi Musa*.[19] This area had been the destination of pilgrims for a long time, not only for worship, but also for the collection of the aromatic black "Moses-stone." The Zionist-owned Palestine Mining Syndicate was interested in mineral deposits in the area and thus initiated Picard's geological survey.

What stands out as especially significant in this study is that water was not yet an important concern in geological research. Indeed, in the three-page section on water supply, Picard deploys the already known information about water resources *only* in order to elucidate the geologic structure of the area under study, *not* in order to tell us something about water resources themselves. The situation changed soon thereafter, however. In later studies, water became the main object of geological investigations.

Even when Picard moved on to the study of water during the mid–1930s, he intended only to study certain water resources within certain regions. One of his best known studies was concerned with the Western Emek (Plain of Esdraelon).[20] His conclusion was that water was scarcely available in that part of the country and that the geological structure was not conducive to groundwater aquifer formation.[21] Another study of Picard's, in cooperation with M. Avnimelech, was also regional in character (Picard and Avnimelech 1937). It was, however, concerned with the geological structure of the coastal plain and not with its water resources.

It was in the second half of the 1930s that water became a political resource in the conflict over immigration policies (Alatout 2009). From 1936, Picard's collaborator Avnimelech began collecting information about wells throughout Palestine (Picard 1940). Moreover, with Picard's support, Avnimelech also established an archive for that purpose at the Hebrew University in Jerusalem.

In 1940 Picard published his first work on the hydrology of Palestine as a unified space.[22] For the first time, a systemic study of Palestine's groundwater resources was available. In his study, Picard used the detailed archives in order to reconstruct Palestine in geohydrological terms. He divided Palestine into seventeen geohydrological regions arrived at from information gathered about different drillings throughout the country. Although different regions had different sub-soil geological structures, Picard was able to connect them all through the conclusion that, in Palestine, the Turonian is the main aquiferous and water-holding stratum. What is most significant about this study is the fact that for the first time Palestinian water resources were perceived as one explainable system and, also for the first time, a geohydrological map was introduced in order to convey the national character of Palestine's resources (see Figure 18.5). In short, if it were possible to view Palestine in terms of a single unified geohydrological system, it was also possible to view Palestine as a single nation, a nation defined in geohydrological terms.

Figure 18.5 The hydrological system of Palestine as depicted by Picard, 1940

The fact that Picard's study came at a time when Zionist land and water policies began to perceive Palestine in national terms is not the only example of political influence on water research. Picard's conception of the geohydrological system of Palestine was limited by Jewish settlement: he focused on areas where Jewish immigrants settled and where Zionist settlement institutions focused their attention. Two omissions in Picard's study are especially telling. Firstly, his perception of Palestine's water resources was limited to groundwater aquifers and, in consequence, excluded the Jordan River system. The significance of this exclusion is that it went hand in hand with the then-most influential legal doctrine concerning the use of the river: the Jordan River had to be utilized within its basin and could not be diverted to irrigate arable lands far from its immediate vicinity.[23] Hence, riverine water resources could not be used to justify a region-wide finding of water abundance, an argument that Zionists insisted on as a way to justify opening up Palestine's borders for millions of new immigrants (Alatout 2009). The second example is the exclusion of the Negev desert from the discussion of Palestine's geohydrological system (see Figure 18.5).

The Peel Commission's Partition Plan of 1937 proposed to divide Palestine into a Jewish state and an Arab state. The territorial correspondence between the planned Jewish state and the boundaries of the geohydrological system of Palestine was not coincidental. It makes sense that some might argue that the water system as imagined by Picard might not have been the most crucial issue in the partition plan; indeed, some would argue that it was the patterns of settlement of Jewish immigrants that determined the territoriality of the proposed Jewish state. That is true to a degree. However, what is missing in those arguments is the mutually supportive function of settlement and water research. It is also true that settlement policies, especially in the mid-1930s, focused on areas that have easily accessible water resources. In other words, settlement policies, water research and the construction of a new conception of a Palestinian geohydrological system, and the territoriality of the proposed state were all bound together in a seamless web of cause and effect.

Boundaries between what is science and non-science, between the availability of abundant water resources or the lack thereof, translated into territorially-focused research on the boundaries of the geohydrological system of Palestine, and into a proposal for a territorial state that followed the same borders (see Figure 18.6). As can be seen in Figure 18.6, despite the different scales, the political boundaries of the imagined nation-state and the geohydrological boundaries of water resources overlap almost perfectly. Effectively, the geohydrological boundaries served to legitimize the proposed political boundaries of the imagined nation-state and vice versa. Indeed, when the Zionists accepted the Peel Partition Plan of Palestine, they were keenly aware that population distribution in the area gave cover for the plan's territorial proposal. But Picard's geohydrological research provided subsoil structural legitimacy for those same territorial boundaries of the proposed state. It is no exaggeration to say that this system of aquifers helped naturalize the political boundaries of the newly imagined nation-state.

The importance of this example, despite its exclusive focus on groundwater resources, is that it underscores a number of crucial arguments advanced here: (1) boundaries of what is science and non-science (are those the real or imagined hydrological boundaries of Palestine and are these aquifers indeed connected?) translate into divergent territorialities; and (2) that the boundaries of the hydrological system in scientific discourse have clear effects on what is imagined as the territorial boundaries of the nation-state. As I demonstrate further, by constructing the criteria of a hydrological system of Palestine, the grounds for expanding that system to include the Jordan River were paved.

Extending the boundaries of the nation: the Jordan River and the hydro-technical construction of national space

Zionist politics of the early 1940s, which focused on expanding its territorial interest beyond the Peel Partition Plan of 1937, played a great role in constructing an expanded vision of the hydrological system of Palestine, which, in turn, legitimized its new territorial vision. Thanks

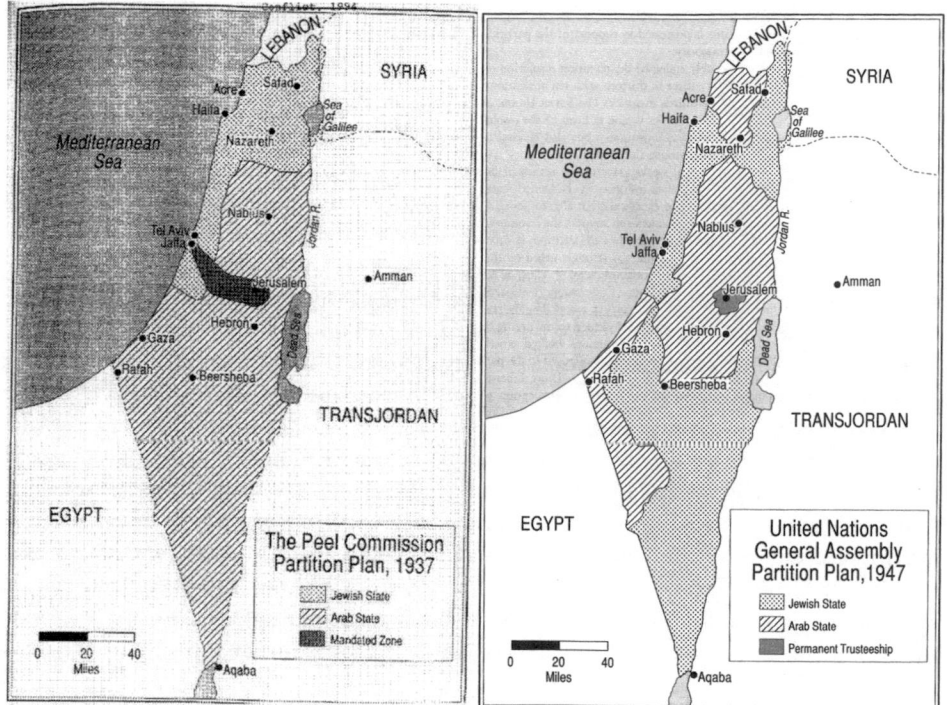

Figure 18.6 (a) Peel Commission Partition Plan, 1937; (b) The United Nations' General Assembly's Partition Plan, 1947

to Walter Lowdermilk, a renowned American soil conservation expert, the geohydrological construction of the national space preceded and shaped the political construction of the nation-state. Lowdermilk, who worked at the United States Department of Agriculture (USDA), was sponsored by the Mandate Government to research the construction of an airport. Given his membership in the mainly Christian pro-Zionist American Palestine Committee, the Zionist Agency seized upon this opportunity and provided Lowdermilk with extensive assistance during his visit. His interest, however, immediately turned to water and irrigation possibilities under the influence of the Jewish Agency and his personal friendship with Simha Blass, a water consultant for the Jewish Agency in Palestine.[24] This was particularly useful for the Zionist Organization in light of a study published by the government of Transjordan (Ionides 1939) that argued that the Jordan River was, indeed, needed for the irrigation of both sides of the river through east and west bank canals and that there was too little water to allow for any Jewish immigration into Palestine.

In 1944 Lowdermilk published the results of his research in *Palestine: Land of Promise*. The book became one of the most influential studies of Palestine's water availability and use during the Mandate period.[25] In it, Lowdermilk championed a project that would divert the Jordan River from the north to the Negev desert in the south, and dubbed it (in a recognizable allusion to the U.S. Tennessee Valley Authority) the Jordan Valley Authority.[26] The assumption here was that while Palestine's water was abundant it was badly distributed, geographically as well as temporally. In order to have access to agricultural water throughout the year, a whole system

321

of canals, aqueducts, cisterns, and reservoirs was needed. The project, Lowdermilk argued, could support four million more immigrants upon completion.[27]

Although the study was not detailed, and perhaps partly because of that, Lowdermilk deployed a multitude of rhetorical strategies in order to support his, and the Zionist, claim about Palestine's absorptive capacity. First, he stressed his status in the field of soil conservation as well as his affiliation with the USDA and other institutions like the TVA and the United States Bureau of Reclamation as sources of technical legitimacy. Second, he deployed the Bible as a legitimate source of knowledge of Palestine's history and population, and in the process legitimized a Judeo-Christian conception of Palestine. Third, he used a progressivist rhetoric and developmental discourse that fell on the side of the modern rather than the pre-modern, the farmer rather than the nomad, and the Jew rather than the Arab.

In order to enlist as many powerful experts as possible in support of his claim about Palestine's absorptive capacity, especially from the United States and Britain, and to legitimize his own research, Lowdermilk drew heavily on his institutional affiliations, including that with the TVA. Lowdermilk's was only the first successful attempt at using U.S. experts as an institutional support for Zionist, and later Israeli, water policies.[28] The TVA was a powerful source of technical legitimacy, especially when Zionist, and later Israeli, projects had to be presented in international forums in the United States, Britain or the United Nations. As we have seen, Lowdermilk went so far as to call his project the Jordan Valley Authority (JVA),[29] drawing an overt parallel between the TVA and the JVA. In sum, Lowdermilk's project resulted in a strong and long-lasting institutional relationship between the TVA and the Zionist and Israeli policy-making bodies.

Lowdermilk's use of the Bible as one of the main resources for legitimate knowledge of water and population history in Palestine was certainly calculated. In the forward of Lowdermilk's book, Sir John Russell stressed the fact that both he and Lowdermilk were of the Christian faith: "Like myself, he [Lowdermilk] is not Jewish, and can view the enterprise quite dispassionately" (Russell in Lowdermilk 1944: 7).[30] The deployment of the Bible and Christianity in understanding Palestine's history, while appearing politically disinterested, reflected a Judeo-Christian view of Palestine in which a Jewish presence was seen as legitimate. This Judeo-Christian conception of Palestine also subscribed to Zionist rhetorical strategies that used the Bible as a source of legitimacy. One of the biblical verses which Lowdermilk, as well as Ben-Gurion and other Zionists, often marshaled is the following: "Behold, the Lord thy God giveth thee a good land, a land of water brooks and fountains that spring out of the valleys and depths, a land of wheat and barley, of vines, figs and pomegranates, of olive oil and honey, a land in which thou shalt eat bread without scarceness, thou shalt not lack anything in it."[31] Based partly on this description, Lowdermilk concluded "that the Land of Israel was capable of supporting and actually did support at least twice as many inhabitants as at present" (24).[32]

Given this Judeo-Christian conception of Palestine as the land of abundance, one question experts had to engage concerned the factors that brought about the changed conditions of Palestine – apparent desertification and decrease in population. Specifically, the question was whether these new conditions of scarcity were induced by climatic changes that had become more or less permanent. Many experts asserted that the climate of Palestine had changed drastically, especially during the seventh century.[33] This theory, however, meant the acceptance of British assertions of Palestine's limited absorptive capacity, at least as far as water and land resources were concerned. Lowdermilk took a static view instead, and insisted that Palestine's climate had not undergone such changes and that the climate had remained constant since biblical times. He argued that the reason for the transformation of the country from a prosperous, populous, and predominantly agricultural haven into a desert should be found in the people who inhabited the land, not in its climate. Arab nomads who invaded Palestine during the seventh century were the reason for its decay, he reasoned. Rather than maintaining their villages, cities, and agricultural systems, they moved from one place to another, invading one agricultural community after the other and bringing ruin to those communities.[34]

Lowdermilk wove together four tightly connected strands of argument: that a Judeo-Christian conception of an abundant Palestine was legitimate; that the climate of Palestine continued to guarantee possibilities of abundance; that the reason for the apparent scarcity was the "primitive Arab" inhabitants of Palestine; and that, thus, Jewish presence in Palestine is politically and historically legitimate (biblical arguments), technically possible (conditions of abundance), as well as necessary (for a new modernized Palestine).

The potential effects of the development of the Jordan Valley Authority, metaphoric as well as material, were immediately seized upon by Zionists and Zionist supporters in the United States. Emanuel Neumann, a radical Zionist and a member of the Zionist World Executive since 1931, encouraged the publication of Lowdermilk's findings for some time before their actual publication.[35] He also took it upon himself to enlist the cooperation of David Lilienthal, then the head of the TVA.[36] Lilienthal, in turn, secured the cooperation of many of the TVA technical cadre. For example, Colonel Theodore Parker, Chief Engineer of the TVA, designed the necessary studies and recommended many of the technical experts, including James Hays, who played a major part in the realization of the project.

Even before the publication of Lowdermilk's study, and after reviewing one of the drafts, Neumann constituted a Commission on Palestine Survey, which he chaired. James Hays, who was a Project Manager of the TVA at Bristol, Tennessee, and an irrigation and power engineer, became the Commission's Chief Engineer and worked closely with a volunteer Engineering Consulting Board.[37] In mid-1945, Hays produced his preliminary report and later, in 1947, produced a more comprehensive one under the title, *The TVA on the Jordan*.

This project functioned on more than one political level. Lowdermilk's argument in 1939 that Palestine would be capable of absorbing many more millions of Jewish immigrants than the British argued, was partly based on his conceptualization of a diversion scheme of the Jordan River from the north of Palestine to the fertile lands of the Negev Desert in the south (see Figure 18.7). In 1942, the Biltmore program radicalized the Zionist position regarding the partition of Palestine. After the Zionist Organization agreed to the Peel Partition Plan in 1937, following the geohydrological maps produced by Picard, the Biltmore conference reneged and produced a new plan in which the whole of Palestine was to become a Jewish state. Lowdermilk's proposal, hydrologically and developmentally, supported this position.

In the years to come, the Lowdermilk-Hays project proved important in the sense that the Zionist position during the United Nations partition negotiations insisted on the inclusion of the Negev and the Western Galilee within the boundaries of the Jewish state (see Figure 18.8). Both were essential to the project. In his forward to the Hays' volume, Neumann declared that:

> Those who had been responsible for working out the details of the United Nations partition plan, were familiar with the basic aspects of the Lowdermilk-Hays project and took it largely into account in drawing the boundaries of the new states.
>
> *Hays 1948: xv–xvi*

He went on to demonstrate how negotiations at the United Nations not only adopted the partition plan of the United Nations Special Committee on Palestine (UNSCOP), but also corrected and improved the original UNSCOP proposals by including in the Jewish State area

Figure 18.7 The Jordan River Project of Walter Lowdermilk, 1944

the Sahl el Battauf depression in Galilee – a natural reservoir essential to the execution of the project. The Jewish State was thus provided with far-reaching possibilities for utilizing the most vital natural resource of the country for large-scale irrigation, agricultural colonization and hydro-electric development.

Hays 1948: xvi[38]

The Lowdermilk-Hays project not only facilitated the emergence of the Negev as part of the Zionist imaginary of Palestine, but also legitimized that emergence in technical terms. The empty fertile Negev irrigated by the Jordan waters would indeed become the promised land of abundance. As Lowdermilk laid out his vision:

> In no other place in the world is there the setting, the drive, and the possibility of demonstrating how the decline of misused and damaged lands may be preserved by the production of abundance through devotion and love of the land, and full and scientific use of the resources of land and water, power and minerals. Such full and scientific use of resources is made particularly feasible by the fact that the valley of the Jordan River and the coastal plain of Palestine offer a combination of natural features that set the stage for one of the most unique and far-reaching reclamation projects on earth, comparable to the Tennessee Valley Authority of the United States in scope and function.
>
> *Hays 1948: vi*

Comparing the maps of the United Nations Partition Plan (Figure 18.8) and the Lowdermilk-Hays project (Figure 18.7) demonstrates how the territoriality of Israel followed closely that of the hydrological system as imagined and extended by Lowdermilk-Hays.

It is clear, then, that the scientific investigation of water in Mandate Palestine was shaped by the territorial outcomes of the imperial encounter and, most significantly, from the fact that the Jordan River was turned into a border. As a border the river made the territorial possessions of empires easier to manage and resulted in a peaceful coexistence between the French and the

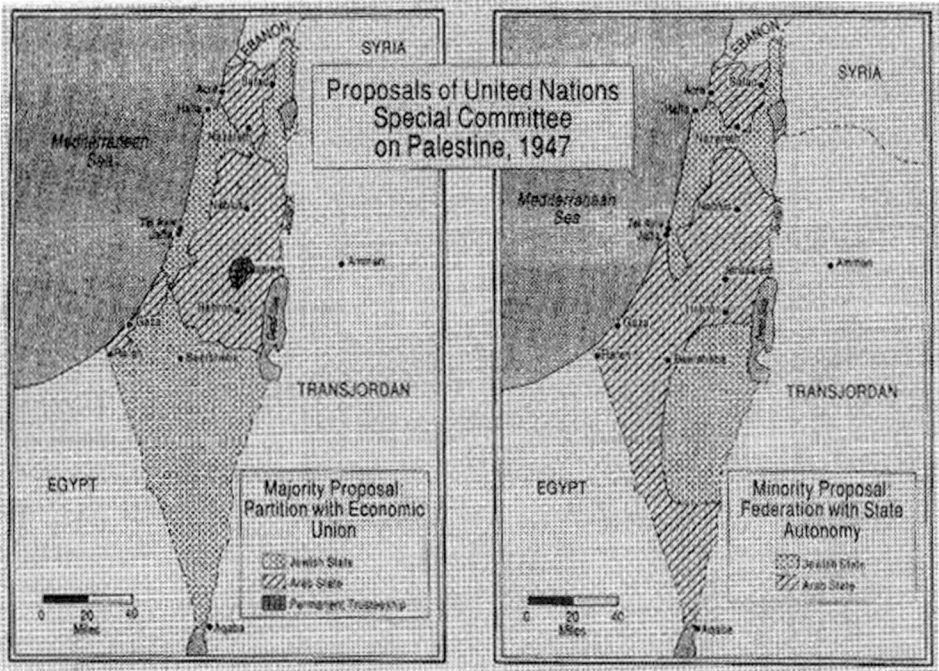

Figure 18.8 The United Nations' General Assembly's partition plan, 1947, Tessler.

British, at least for a while. But at the same time, it threw the river into the middle of new territorial arrangements and newly constructed identities, the bases of which were the new emerging nation-states. This is precisely the hydropolitics that led to the use of the river's water in diversion schemes and to turning the river into a ditch.

Conclusion: Water as a Territorial Object

The encounter with imperialism turned the Jordan River into a border. The various empires that engaged in Palestine at the turn of the twentieth century left the Jordan River underutilized for the most part: (1) because of the conclusions of hydrological studies of the river as unsuitable for irrigation; and (2) so that it would not become a source of friction between the contending empires. However, this political-geographic event, turning the river into a border, invested the river with new political meanings and turned its water into a *territorial object*, subject to new discourses of empire, in the initial stages, and sovereignty and national security, later on during the 1940s and 1950s.

Presumably, the ultimate test of sovereignty is the protection of the territorial integrity of the state and the state's ability to exercise its political-territorial power. That is often understood as the management of the flows along and across a state's borders. In Israel, this conception of national security was strongly linked to an open policy of Jewish immigration, settlement of empty spaces, and the use of agriculture as an anchoring practice for what it means to be a Jewish subject and a citizen of the state. That is why the National Water Carrier and the diversion of the Jordan River were essential during the 1960s. But this means that for the Jordan River to play its function as a border, i.e., to play its role in securing the state, it had to be emptied of its water: *it is only by emptying the Jordan River bed that the security of the Israeli state was possible* – at least that has been the dominant belief in Israeli policy and governance institutions.

Even though the national security discourse was not as powerful a driver in Jordan as it was in Israel during the 1950s, there were a number of political and economic calculations (including the settlement of Palestinian refugees along the banks of the river, a general developmental discourse that stressed the utilization of rivers, and a politics of dilution of Palestinian identity) that were used to legitimize the diversion of the Yarmuk River into the East Ghor Canal. Even though it technically did not divert the water out of the river's basin like the Israeli project, the ultimate effect was very similar: return flow can hardly make up for the loss of water used in irrigation.

The boundaries between appropriate and inappropriate archaeological and hydrological knowledge, between knowledge that is scientific and that which is not, defined the territorial boundaries of Palestine. In the initial stages of the encounter with western research and encroachment, these contested scientific claims were essential in defining the boundaries of Palestine in such a way that the Jordan River was thought to run through it. It was the encounter with imperialism that reframed knowledge of the archaeology and hydrology of Palestine in such a way that produced the Jordan River as a border, as the eastern limit of Palestine. Further geohydrological research, or I should say the very boundaries between what is and is not a legitimate hydrological research, defined the territoriality of Palestine in different ways at different times (1930s–1940s).

Going back to the main argument of the paper that engages Bruno Latour and Edward Said, I would like to emphasize two arguments stressed to different degrees throughout this chapter: (1) Politics of empire is important in understanding modernity and its institutions, including and especially that of technoscience. In this sense, the lenses we use to understand modernity (technoscience) need to be enriched with an understanding of the politics of empire if only to understand our present more fully; (2) Our understanding of empire should take seriously the notion that empire is itself a product of systems of knowledge, and scientific knowledge in particular. I argue in this chapter that a reorientation toward empire in STS, in contradistinction with modernity, might open up productive engagements with the institutions of modernity and with the present.

In this chapter I chose to focus attention on the territorial implications of boundary making in technoscience, i.e., its *border effects*. I also limited my study to two particularly large scales of political organization where the importance of territory and borders is hardly ever challenged: empires and nation-states. In addition, I used as my example geohydrological systems (river and groundwater aquifers) that lend themselves easily to territorial demarcations – even though this latter point might be a judgment in hindsight and not a foregone conclusion. At this point, I am left wondering what these *border effects* of boundary making in technoscience might look like at different scales of ecological/biological and political organization.

I conclude by pointing to a number of questions that I think become empirically and theoretically compelling here: does the process of boundary making in science *always* have territorial implications, is it always spatialized, and is it always productive of borders? Does the fact that a border is necessarily occupied by two or more sovereignties (regardless of the scale of those sovereignties, be they individuals, communities, or states) mean that either conflict *will* ensue or that the object itself will be torn apart – utilized in a sovereign-territorial framework that is ultimately about security? It seems to me that interrogating these questions is especially important for environmentally significant objects like rivers.

Notes

1 When I speak of empire, I do not mean to imply the interests of any individual imperial power. Rather, I am referring to a relational system in which different individual Powers (French, British, Russian, Italian, Ottoman, etc.) responded to challenges and desires in a field of struggle with other individual and collective powers.

2 I would not want to spend too much time on a side conversation, but I'd like to suggest, in part as a thought exercise, that these books in particular were written in the shadow of the events I mentioned.

3 Said defines imperialism as the process by which empire was made possible and differentiates that from colonialism, which is the process of conquering a territory and vacating it for the purposes of settlement.

4 While this is not crucial for the discussion here, it should be noted that while the power of hybridization lies in direct human action for Said (trade, military, and cultural expansion under imperialism), sometimes Latour seems to imply that mediation is in the *nature of things*, i.e., despite the wishes of the humans to demarcate and differentiate.

5 Science studies scholars who focus on social worlds are interested in understanding what might make a (boundary) object solid enough to sustain an identity across social worlds, but flexible enough to be subject to articulation or translation across these very social worlds (Star and Griesemer 1989, Fujimura 1987 and 1992). Instead, in this paper I am limiting my interest in boundary objects to the degree that they themselves become borders, and territorial borders to be exact.

6 The northern border of Palestine was also decided with the Jordan River in mind. Although the river itself did not constitute the border, its main sources were instrumental in the negotiations. I do not deal with that border in this paper, but it is important to note that at the end of the negotiations and when the territoriality of Palestine was more or less stable by 1923, the Litani River was placed fully in Lebanon. The three headwaters of the Jordan were divided and the Hasbani River was placed in Lebanon, the Banias in Syria, and the Dan in Palestine. For a detailed treatment of the negotiations over the northern borders see Garfinkle 1994.

7 Reflecting on that, the leader of the expedition, William Lynch, tells one Palestinian "notable" about how the expedition was sent "from a far distant but powerful country to solve a scientific question…

[that] with solutions of scientific questions, we hoped to convince the incredulous that Moses was a true prophet" (133). For an excellent contribution to the politics of archaeology and state building in Israel, see Abu El-Haj (2002).

8   The strategic (secret) nature of the mission is obvious in Lynch's mandate to Dr. Anderson, the expedition's geologist: "you are to make no communication, verbal or otherwise, of the labours or results thereof, of yourself or any member pertaining to it, save to myself officially, until relieved from the obligation by the Hon. Secretary of the Navy." On the possible commercial benefits of this mission, he also instructs his geologist to obtain "mineralogical specimens," to ascertain "if the surrounding regions be volcanic," collecting and studying the soil "on the eastern shore especially, as formed by disintegration, and the nature of vegetation as connected with it." He goes on to request that "the soil in which grapes of such extraordinary size are said to grow should be collected for analysis, to ascertain if the chemical composition has any influence on the size of the fruit." Moreover, "specimens of mud from various parts of the sea, river and lake, should be collected and placed in air-tight vessels.... It is most important to ascertain whether birds live on the shores, or fish within the depths, of the Dead Sea." Last, but "not less, to note carefully every stream and fissure, their direction and their depth, and to ascertain, if possible, whether the former are perpetual, or only temporary, torrents" (137–138).

9   One example makes the vagueness of the river as a jurisdictional border clear. Lynch approached the Mushir (governor) of Beirut for a permission to survey the east bank of the river. The Mushir, according to Lynch (1849: 114), was "uncertain whether the eastern side of the Jordan was included in his jurisdiction or in that of the Pasha of Damascus." Even the maps showing the area under his jurisdiction were inconclusive.

10 Lynch described the Arabs of Palestine in terms that continue and solidify racist assumptions in the U.S. and in most empire's relations with overseas *natives*. For example, speaking about a welcoming Sheikh, Lynch argues that "But for his costumes, he would, in our country, pass for a genteel negro, of the cross between the mulatto and the black" (149). On another occasion, Lynch describes a dinner scene in a tent: "What a patriarchal scene! Seated upon their mats and cushions within, we looked out upon the fire, around which were gathered groups of this *wild* people, who continually reminded us of *our Indians*" (150). These points of reference are not coincidental or innocent; they underscore the terms with which Lynch related to the Palestinian population.

11 One example can illustrate the fact that these scientific controversies were extremely relevant to political positions over Palestine. Merrill for his part was a staunch critic of turning Palestine into a Jewish state and argued strongly against its implications for the area and for peace. On one occasion, Merrill was very critical of the Blackstone Memorial (1891), in which more than 400 Christian authorities lobbied the United States' president to "restore" Palestine to the Jews: "Why shall not the powers which under the treaty of Berlin, in 1878, gave Bulgaria to the Bulgarians and Servia to the Servians now give Palestine back to the Jews? These provinces, as well as Roumania, Montenegro and Greece, were wrested from the Turks and given to their natural owners. Does not Palestine as rightfully belong to the Jews?"

12 As we saw from Lynch (1849), the position of the Jordan River as a territorial border was unclear at best. We know the following for a fact. Most of the area known to us as Syria, Lebanon, and historic Palestine (the area in Figure 18.2) was known as the Province of Syria and was divided into a number of administrative districts reporting directly to Damascus (except for the district of Jerusalem that reported directly to Constantinople). However, beginning in the 1850s the Ottoman Empire instituted a new set of regulations that came to be known collectively as *Reorganization* (*Tanzimat*). Those changed so much of life in the area. In part justified through a new conception of modernization needed to face the encroachment of Western Powers and in part an attempt to reduce the effects of the new pan-Arab national identity that was on the rise in Greater Syria itself during the late 1800s, the Ottoman Empire divided the province of Syria into a number of provinces and districts. These gradual changes culminated in splitting Syria into a number of provinces in 1888. On the eve of WWI the area looked administratively very different (Figure 18.2): there were 3 major provinces, Aleppo, Syria, and Beirut, each divided into a number of districts (Aleppo contained the districts of Aleppo, Urfa, and Marsh; Beirut was divided into the districts of Nablus, Acre, Beirut, Tripoli, and Latakia; Syria was divided into the districts of Hama, Damascus, Hauran, and Ma'an). At the time of World War I those territorial administrative distinctions were new and minimal in terms of how people conducted their lives.

13 I need to reemphasize here that despite the importance of the Jordan River resources for imperial politics and for defining the northern boundaries of Palestine, I will be focusing exclusively on its eastern boundaries in this chapter.

14 The Hussein-McMahon correspondence did not produce a map of the area, only a textual under-standing. However, because the correspondence mentioned Palestine and never specified western or eastern Palestine, as of 1939, the British government acknowledged that the promise included all of Palestine.

15 The Balfour Declaration did not come with a map either. However, Winston Churchill's White Paper of 1922 made it clear that the British interpretation of the promise is that the Jewish National Home would be established in Palestine and never in the whole of Palestine. That was the basis of banning Jewish settlement on territories east of the Jordan after 1922.

16 As a matter of fact, there was an immediate attempt at settling the east bank of the river by Jewish immigrants.

17 The first official call for the establishment of a Jewish state in Palestine came in the 1942 Biltmore program. This was the most important document to come out of the Extraordinary Zionist Conference, held by American Zionists at the Biltmore Hotel in New York City in May of 1942. It was addressed by David Ben-Gurion.

18 Historian Baruch Kimmerling (1983) points to the fact that "contiguity" in land settlement became a very important concept in Zionist land settlement policy at the same period.

19 Arabic name for the area meaning "Prophet Moses."

20 This was the first study to investigate water as such on a regional level (Picard 1936).

21 It is important to mention that not only were such studies of water resources regional in dimension, but also that they were not careful about the implications of their findings. Water was yet to acquire political significance, which happened about the same time. Because of its conclusion, this study played a role in the hands of the Peel Commission. During one of the sessions, the Commissioners referred to this particular study in order to delegitimize Zionist perceptions of water abundance and in defense of Mandate experts' perceptions of scarce water resources. However, at a later stage, Picard qualified his conclusion of the scarcity of water in Western Emmek (Picard 1940).

22 Picard (1940: 1) reasons that the impetus for this research is the practical problems stemming from the need for water during wartime, but, also, in order to complete "the systemic development of … boring archives which existed now for a number of years."

23 See Russell's introduction of Lowdermilk 1946.

24 Lowdermilk's involvement in Palestine mixed his strong Zionist ideological underpinnings of the importance of restoring Jewish national life in the promised land of Palestine and his love for Zionist settlement as a form of "social experiment."

25 During interviews with Israeli water scientists and engineers who worked during the 1940s and 1950s, Lowdermilk and his book were described in glowing terms. In an old interview, Hillel Shuval, who was a professor of Hydrology at the Hebrew University, described Lowdermilk's book as "visionary." He went on to describe the effects the book had on his own life. Shuval was a college student when the book was published and it led to his decision to focus on studying water engineering and work-ing in the water sector upon immigration to Palestine, interview 28 July 1997.

26 As we will see later on this study was followed by another study by James Hays (1948) which attempted the operationalization of the Lowdermilk study.

27 Lowdermilk was of course interested in the project of Jewish colonization of Palestine as part and parcel of his theological connection to the Zionist project.

28 The Zionist Organization used the TVA extensively. The TVA was perceived as an important instru-ment of development and progress throughout the world at the time. Ironically, its connection with the Zionist project in Palestine might stem from the historical linkage often made between labor Zionism and socialist, anti-imperialist strategies around the world.

29 One of the most influential Zionist studies of water resources, which was turned into Israel's First National Plan, was based on Lowdermilk's project. It was illustratively titled: *The TVA on the Jordan*. See James Hays (1948).

30 Russell, "Forward," in Lowdermilk, *Palestine: Land of Promise*, p 7.

31 Quoted in ibid. p. 24 from Deut.Vii. 7–9. Lowdermilk also repeats this passage in many of his writ-ings during the forties and fifties.

32 His conservative estimate in this paragraph (that Palestine supported double the population) was meant as a worst scenario condition. His general assumption in the book is that Palestine could support four million more immigrants.

33 For an excellent review of this debate, see Arie Issar (1990).

34 Lowdermilk (1944: 52). Notice the exoneration of the Hebrew as well as Western cultures from the

ruin that befell Palestine. In the end, the demarcation lines between the Arabs on the one hand and Western and Hebrew cultures on the other are redrawn in many fashions throughout the book. They are drawn between the settled and the nomad, the civilized and the uncivilized, the industrious and the lazy, the progressive and the traditional. For example, in the process of enumerating the various powers that were in control of Palestine from time to time, Lowdermilk retained the concept of "invaders" for the Arab forces only. After mentioning the Pharaohs, Barak and Sisera, Gideon, Saul and Jonathan, Lowdermilk goes on: "then came Xerxes, Sennacherib, Alexander the Great, the Romans under Titus, the Arab *invaders*, the Crusaders under Richard the Lion-Hearted, Saladin, the Turks, Napoleon" (55). Emphasis added.

35 Neumann, who became a member of the Zionist executive since 1931, was one of the Zionist leaders who, along with David Ben-Gurion and other Zionist radicals, toppled the reign of Chaim Weizmann in the World Zionist Organization in the early 1930s. The main reason was the fact that Weizmann was a gradualist-Zionist – believing that achieving Zionist objectives should be gradual and without unnecessary alienation of the British Mandate. However, the immediate reason for his defeat in the Zionist movement was an interview with a newspaper in which he argued that Zionists never had the intention of reconstituting Palestine with a Jewish majority. Other Zionists found that a retreat from the objectives of the "Jewish National Home." See Laqueur 1997, *A History of Zionism*.

36 In 1946 Lilienthal moved on to become the first Chairman of the Nuclear Energy Commission.

37 The members of this board were well-known experts and continued to act as consultants to the Zionist movement and the state of Israel during the 1940s and 1950s. Some of those experts, whose inclusion was to provide technical legitimacy were the following: Dr. Abel Wolman of Johns Hopkins University, served as the Chairman of the National Water Resources Board of the United States; Harry Bashore, United States' previous Commissioner of Reclamation; Theodore Parker, who after his death was replaced by C. Blee, Chief Engineer of the TVA; lastly, John Savage, at one time Chief Designing Engineer of the Bureau of reclamation.

38 The Battauf reservoir which was meant to be the main reservoir in the project was abandoned in 1953 because it was proven to leak to a large degree. It was assumed to have the potential of storing more than 1,000 mcm.

References

Abu El-Haj, N. 2002. *Facts on the Ground: Archaeological Practice and Territorial Self-Fashioning in Israeli Society*. Chicago, IL: University of Chicago Press.

Adams, W. 1874. *The Jordan Valley, and the Dead Sea*. London: T. Nelson and Sons.

Alatout, S. 2006. "Towards a Bio-Territorial Conception of Power: Territory, Population, and Environmental Narratives in Palestine and Israel." *Political Geography* 25: 601–21.

Alatout, S. 2008. "'States' of Scarcity: Water, Space, and Identity Politics in Israel, 1948–1959." *Environment and Planning D: Society and Space* 26: 959–982.

Alatout, S. 2009. "Bringing Abundance into Environmental Politics: Constructing a Zionist Network of Water Abundance, Immigration, and Colonization." *Social Studies of Science* 39: 363–394.

Blake, G.S. and Goldschmidt, M.J. 1947. *Geology and Water Resources of Palestine*. Jerusalem, Palestine: Government of Palestine.

Cleveland, W. 2000. *A History of the Modern Middle East*. Boulder, CO: Westview Press.

Condor, C.R. 1889. *The Survey of Eastern Palestine: Memoirs of the Topography, Orography, Hydrography, Archaeology*. London: The Committee of the Palestine Exploration Fund.

De Bunsen Committee. 1915. *Committee of Imperial Defence: Asiatic Turkey, Report of a Committee*. British National Archives, CAB 42/3/12.

Department of the Navy. 1851. *Official Report of the United States Expedition to Explore the Dead Sea and the River Jordan*. DC: National Observatory. Written by W.F. Lynch.

Fujimura, J. 1987. "Constructing Do-able Problems in Cancer Research: Articulating Alignment." *Social Studies of Science* 17: 257–293.

Fujimura, J. 1992. "Crafting Science: Standardized Packages, Boundary Objects, and Translation." In *Science as Practice and Culture*, edited by A. Pickering, 168–211. Chicago: University of Chicago Press.

G. S. (I.). 1918. *Handbook on Northern Palestine and Southern Syria*. London: Government Printer.

Garfinkle, A. 1994. *War, Water, and Negotiation in the Middle East: The Case of the Palestinian-Syrian Border, 1916–1923*. Tel Aviv: Tel Aviv University.

Gelvin, J. 2005. *Israel Palestine Conflict: One Hundred Years of War*. Cambridge, UK: Cambridge University Press.

Gieryn, T. 1983. "Boundary-Work and the Demarcation of Science from Non-Science: Strains and Interests in Professional Ideologies of Scientists and on boundary." *American Sociological Review* 48: 781–795.

Gieryn, T. 1999. *Cultural Boundaries of Science: Credibility on the Line*. Chicago, IL: University of Chicago Press.

Government of Palestine. 1947. *Memorandum on the Water Resources of Palestine*, presented by the Government of Palestine to the United Nations Special Committee on Palestine in July, 1947. Jerusalem: Government Printer.

Great Britain. 1937. *Palestine Royal Commission: Minutes of Evidence Heard at Public Sessions*. London, UK: H.M. Stationery Office.

Great Britain. 1946. *A Survey of Palestine: Prepared in December 1945 and January 1946 for the Information of the Anglo-American Committee of Inquiry*. Palestine: Government Printer.

Gruenbaum, L. 1946. *Outlines of a Development Plan for Jewish Palestine*. Jerusalem, Palestine: Weiss Press.

Hays, J. 1948. *The T.V.A. on the Jordan: Proposals for Irrigation and Hydroelectric Development in Palestine*. Washington, DC: Commission on Palestine Surveys.

Ionides, M.G. 1939. *Report on the Water Resources of Transjordan and Their Development: Geology, Soils, and Minerals*. London: Crown Agents for the Colonies.

Issar, A. 1990. *Water Shall Flow from the Rock: Hydrology and Climate in the Lands of the Bible*. Heidelberg, Germany: Springer-Varleg.

Kimmerling, B. 1983. *Zionism and Territory: The Socio-Territorial Dimensions of Zionist Politics*. Berkeley, CA: Institute of International Studies.

Kimmerling, B. and Migdal, J. 1993. *Palestinians: The Making of a People*. New York, NY: The Free Press.

Latour, B. 1993. *We Have Never Been Modern*. Cambridge, MA: Harvard University Press.

Laqueur, W. 1997. *A History of Zionism: From the French Revolution to the Establishment of the State of Israel*. New York, NY: Fine Communications.

Lee, S. and Roth, W. 2001. "How Ditch and Drain Become a Healthy Creek: Re-Presentations, Translation and Agency During the Re/Design of a Watershed." *Social Studies of Science* 31: 315–356.

Lowdermilk, W. 1944. *Palestine: Land of Promise*. New York, NY: Harpers & Brothers Publishers.

Lynch, W.F. 1849. *Report of the Secretary of the Navy, with a Report Made by Lieutenant W. F. Lynch of an Examination of the Dead Sea*. DC: Government Publishers.

Merrill, S. 1883. *East of the Jordan: A Record of Travel and Observation in the Countries of Moab Gilead and Bashan during the Years 1875–1877*. New York: Charles Scribner's Sons.

Palestine Exploration Fund. 1875. *Quarterly Statement for 1875*. London: The Committee on Palestine Exploration Fund.

Palestine Exploration Fund. 1881. *The Survey of Western Palestine: Special Papers on Topography, Archaeology, Manners and Customs, ETC*. London: The Committee of Palestine Exploration Fund.

Palestine Exploration Funds. 1889. *The Survey of Eastern Palestine: Memoirs of the Topography, Orography, Hydrography, Archaeology, Etc*. London: The Committee of Palestine Exploration Fund.

Philip, K. 1998. "English Mud: Toward a Critical Cultural Studies of Colonial Science." *Cultural Studies* 12: 300–331.

Philip, K. 2003. *Civilizing Nature: Race, Resources, and Modernity in Colonial South India*. Rutgers, NJ: Rutgers University Press.

Picard, L. 1931. *Geological Researches in the Judean Desert*. Jerusalem: Goldberg's Press.

Picard, L. 1936. *Conditions of Underground-Water in the Western Emmek (Plain of Esdraelon)*. 1936. Bulletin of the Geology Department, Hebrew University, Jerusalem, Vol. I, No. I.

Picard, L. 1940. *Groundwater in Palestine*. Bulletin of the Geology Department of the Hebrew University, Jerusalem, Vol. 3, No. 1.

Picard, L. and Avnimelech, M. 1937. *On the Geology of the Coastal Plain*. Bulletin of the Geology Department of the Hebrew University, Jerusalem, Vol. I, No. 4.

Star, S. and Griesemer, R. 1989. "Institutional Ecologies, 'Translations', and Boundary Objects: Amateurs and Professionals in Berkeley's Museum of Vertebrate Zoology, 1907–1939." *Social Studies of Science* 19: 387–420.

Twain, M. 1869. *The Innocents Abroad or the New Pilgrims' Progress*. Hartford, CT: The American Publishing Company.

19

State-Environment Relationality

Organic engines and governance regimes[1]

Patrick Carroll and Nathaniel Freiburger

UNIVERSITY OF CALIFORNIA, DAVIS

Introduction

This chapter deals with the historical development of what we term the "enviro-state." The term describes a set of discursive, organizational, and material relationalities between science and governance that have developed since the scientific revolution. We use the term "state-environment relationalities" to capture the complexity of the connections. In so doing, we open up analytical space beyond the narrow confines of the language of "the relationship" between "the state" and "the environment." The latter language obscures complex relationalities, giving the impression that the state is a macro actor that stands in a relationship with something *external* to it called "the environment." The enviro-state is generated in practices of environing, meaning surrounding and penetrating. These practices stem from the networking of science and government around boundary objects (Star and Griesemer 1989) such as land, water, lithium, and other "natural resources," and result in deep entanglements between what is considered as nature, infrastructure, and governance. Regimes of governance that include both state agents and technoscientists give modern states the specific character of being technoscientific. It is important to note that regimes of "governance" are *not* confined to official government. To the extent that official government permits "private" (what is private is constituted by law) agents to govern, those private agents can form part of a regime of governance (see Bakker 2007 for the conceptualization of governance in terms of who gets to make decisions and how). Similarly, to the extent that non-governmental organizations participate in political decisions, they too are part of the regime of governance. Of course, agency (understood here as decision-making power) within regimes of governance will vary both temporally and spatially.

Taking our cue from Richard White (1995), who described the Columbia River as an "organic machine," we conceptualize the enviro-state in a similar manner. But the enviro/technoscientific state is not simply a hybrid of nature and culture and a cyborg-like formation. It is an engine in motion. It is in process. It is not that it merely *is* a particular type of entity, but that it is generative, a driving force, a constant movement "forward" in the sense of "economic development," human population growth, and resource exploitation. Thus, we describe the enviro-state as an "organic engine." In order to develop an ontology of the enviro-state as something assembled from discourse, practices, and materialities, we maintain that it is best conceived as a "thing." To conceive of the enviro-state as a "thing" is not to conceive of it as an "object" out there. Despite the tendency to equate the word "object" with "thing," the two are far from equivalent (Latour 2005, Ingold 2010). For instance, one can go to a thing, but not to an object. That is because the oldest meaning of a thing is a gathering, particularly a political assembly where decisions are made. We adopt the term because it evokes the enviro-state as a regime of governance, but equally because it evokes it as an assemblage. We only add that it is a material assemblage comprised of both humans and non-humans.

We build our argument on analytic resources derived from Foucaultian governmentality studies (Burchell et al. 1991, Rose and Miller 1992, Foucault 2004, Dean 1999) and Actor Network Theory – ANT (Callon 1986, Law 1992, Latour 1996, Law and Hassard 1999). How we do so will become clear in our theoretical discussion of state-environment relationality and through our empirical cases. For now, we wish to simply point out how our theoretical argument diverges from the dominant forms of these literatures. First, governmentality studies developed in opposition to state theory and as a result has neglected developing an ontology of the state (see Carroll 2009). While we agree that the state is an effect of myriad micro level actions, what Foucault termed "micro-physics," we hold that the ontology of the state is fundamentally material. Hence the conceptualization of the state as a thing. Indeed, we think it is somewhat ironic that a theory that pays attention to the material at the micro level ends up with a rather nebulous image of what the state is. Second, we do not flatten the analysis into "discourse," sometimes rendered as "discursive practices." Despite the invocation of "practice" and "technologies," both tend to be subsumed into discourse. Governmentalities appear at base to be, precisely, mentalities. They are "rationalities of government" (Barry et al. 1996). We maintain a strong analytic distinction between discourse, practice, and materiality (practice being the nexus of the discursive and the material).

With respect to ANT, we do not adopt a "generalized symmetry" that homogenizes, through the concept of "actants," human and non-human actors. While ANT has done much to demonstrate that humans are not the only entities that can act, we maintain that it is not necessary to collapse all actors into the single category of "actant" to recognize this. We retain the term "agent" to characterize those actors whose action involves choice, design, purpose, goals, etc. This particular formulation of agency views it as a special kind of action. It is unique to *living* human beings. A storm has no design behind it, it has no goals and involves no choices. To say both humans and non-humans have agency obscures this critical difference. We believe it is important to maintain the distinction in terms of the politics of technoscience. ANT is often criticized for lacking a critical edge, and for neglecting less powerful humans (Haraway 1992). While we do not have a normative political agenda in this chapter, maintaining the conventional understanding of agency allows for political critique in terms of accountability in the sense of accounting for which humans get to meaningfully exercise agency and which do not. Still, we also maintain that storms can act, and that they can materially resist human goals. Andrew Pickering (1999) has acknowledged this, yet he continues to ascribe the agency captured by our concept of "agents" to non-humans which clearly do not have goals or make decisions. We reserve the term "force" for action characteristic of phenomena like storms. We further distinguish a third actor, collectively understood as material culture. Material culture does not have agency in the sense of a capacity to make choices or have goals. However, it does embody designs and goals. Hence material culture has agency once removed from its human creators. For this reason, we reserve the term "actant" for material culture. That term signifies a degree of material agency.

Thus we have three distinct actors: humans, who are also agents; actants, which embody design and thereby agency once removed; and forces (more precisely, if somewhat cumbersomely, "non-designed materialities"), which have no design or purpose of any kind, but nonetheless can act with powerful effect. The particular arrangements of agents, actants, and forces allow each empirically specific capacities to act. We do not see the action of each of our actors as mutually exclusive, in the modernist sense. For instance, the flow of water in a valley or a salt flat only becomes a "flood" in relation to human presence. A flood is a "problem," and problems only exist in relation to human designs. It is nonetheless importantly caused by the action of forces such as storms.

We also depart from classical ANT in stressing the concept of "material relationality" over linguistic metaphors such as "translation" and "semiotics" to deal with particular, empirical arrangements of agents, actants, and forces. We are aware that ANT is described as a material relational theory, but in a particular context ANT took a linguistic turn. The concept of "material semiotics" was deployed to bring in non-humans without relying, as Collins and Yearley (1992) argued was required, on natural scientists' *accounts* of non-humans (in doing so one would return to a fully humanist analysis). In addition to the ways that our theoretical tools diverge from many contemporary uses of ANT, we employ a now neglected term, the concept of "punctualization." Though crafted within earlier versions of ANT (Law 1992), this concept has largely fallen by the wayside (Latour 2007), which is somewhat odd, since Law gave it so much importance (see below). It refers to the process whereby something is black-boxed, the process through which heterogeneous actor-networks are rendered as free-standing singularities, often as actors. The concept is fundamental to our critique of the notion of the state as a single macro actor, and our alternative ontology of it as an actor-network. To quote Law:

> This, then, is the core of the actor-network approach: a concern with how actors and organizations mobilize, juxtapose, and hold together the bits and pieces out of which they are composed; how they are sometimes able to prevent those bits and pieces from following their own inclinations and making off; and how they manage, as a result, to conceal for a time the process of translation [we would say articulation] itself and so turn a network from a heterogeneous set of bits and pieces each with its own inclinations, into something that passes as a punctualized actor.
>
> *Law 1992: 386*

State-environment relationality

In this section we briefly address state theory, then discuss how and when an equivalence developed between "nature" and "*the* environment," and conclude with a discussion of state-environment relationality and how we propose to conceive of it. The character of "the state" is highly contested and the literature on "it" is vast and varied. It is beyond the scope of this chapter to rehearse those debates (see Steinmetz 1999, Jessop 2001, Adams et al. 2005, Carroll 2009, Passoth and Rowland 2010). As indicated above, we employ a cultural analytic of the state, informed by Foucault inspired governmentality studies and theoretical resources in ANT. Rather than taking the distinction between society (or economy) and the state as a given, Mitchell (1991) set a Foucaultian agenda to reveal the means through which "the uncertain yet powerful distinction between state and society is produced" (78). In a similar vein we do not take the distinction between state and environment as given. Rather we historicize both "*the* state" and "*the* environment." We see the state as an unfolding process, as an "effect" of actions rather than a macro actor in its own right (see Mitchell on the "state effect" in Steinmetz 1999).

We argue that "the environment" comes into existence in the mid-twentieth century as part and parcel of the emergence of the modern "environmental movement." In both cases, the ongoing iteration of the macro distinctions in social science actually helps create, rather than merely describe, some reality (as a general point see Law 2004; more specific arguments to this effect have been made about economics and "the economy" as in Callon 2007, MacKenzie 2006, MacKenzie et al. 2008, MacKenzie and Millo 2003, Mitchell 2005). Treating the state as an effect avoids its reification as a concrete superhuman-like actor. By treating the environment as an emergent reality, we are better able to see how state and environment are deeply entangled to the point where the boundary between them, like that between state and society, simply will not hold.

Work informed by ANT has critiqued the idea of the state as a macro singular actor. Carroll, for instance, has argued that states not only do not act, they "cannot act" (Carroll 2006: 19, see Meyer 1999 for an early undeveloped note on this). When looked at from a micro and historical perspective, states appear not as actors, but as complex assemblages of ideas/discourses, practices/organizations, and materialities. Human actors certainly act and speak in the name of the state, and these we designate as "state agents." States, however, are literally made from the material forms of land, bodies, built environment, infrastructure, water, and other "natural resources." The assemblage of these elements gives states the character of heterogeneous actor-networks (Passoth and Rowland 2010).

With respect to the environment, the new "environmental movement" achieved something quite spectacular, something that centrally informed the macro conceptualization of the state/environment binary: it managed to create an equivalence between "nature" and "*the* environment." The term environment is widely used as an adjective: the political environment, the institutional environment, the built environment, etc., but when one uses the term "the" environment it is most immediately understood to mean "nature" or some element of it, such as air, water, forest, mountains, etc. The nature-environment synonym has been fully institutionalized in the strong sociological sense: it is taken for granted, naturalized, both popularly and in science. But like everything else, it has a history.

The *Oxford English Dictionary on Historical Principles* (OED) (1933) is always the best place to start when trying to understand meanings of particular words in the English language. The word "environment" derives from Middle French, the earliest documented meanings being "the action of circumnavigating, encompassing, or surrounding; the state of being encompassed or surrounded." In this sense it was often used in the military context of being surrounded. By the eighteenth century it could also mean the "area surrounding a place or a thing; the environs, surroundings, physical context." Early travelogues used it in the sense of "picturesque environment," and this may have set the stage for the use of the word to describe geological formations. For instance, the United States Geological Survey (USGS), spoke of the "geologic environment" (1903) of oil pools on the Gulf Coast (see also McGee 1896 for an early usage of "environment"). This usage resonates with OED's definition at 2(b), where environment refers to the "physical surroundings or conditions in which a person or organism lives, develops, etc., or in which a thing exists; the external conditions in general affecting the life, existence, or properties of an organism or object." Interestingly, the cited text of this use is from Herbert Spencer in 1855, where he divided "the environment" of a plant into two halves, soil and air (today "environment" would, in the context of the movement, refer to soil, air, *and* plant). A subsequent usage arises in his treatise on morals: "The organism is continually adapted to its environment." One can find occasional attempts to draw equivalence between environment and nature at the turn of the twentieth century. For instance, in a piece from 1896 entitled "Influence of Environment upon Human Industries or Arts," one occasionally finds the term "nature or environment" (Mason 1896). This meaning only begins to gain wider coinage by the mid-twentieth century. The OED, using a source from 1948, defines it as the "natural world," specifically in relation to "man's impact on the environment." This sense, or the use of the word at all, is noticeably absent from an extensive collection of classic writings on conservation and preservation, published around the turn of the twentieth century (Stradling 2004). In this respect the contrast with classic writings of the new environmental movement (1968–1982) is striking (Stradling 2012).

The conceptual coupling – nature as environment and human impacts upon it – is the central framework of the new environmental movement of the 1960s. In one of the seminal texts of that movement, Rachel Carson protested the "most alarming of all man's assaults on the environment is the contamination of air, earth, rivers, and sea with dangerous and even lethal materials" (1962: 6). From the 1960s onward the term "environment" would punctualize an array of new distinct groups with different emphases, foci, strategies, etc., into the single reference point of "the environmental movement." However, and this is critical, by the time the modern environmental movement was cementing the nature-environment synonym, the "nature" in question had already been fundamentally transformed by human artifice. As we seek to show, that transformation is inextricably bound up with modern technoscientific state formation. Once we historicize the nature-environment synonym, we see how deploying "the environment" as an analytic category is problematic. It is especially inadequate when set in an interactional but fundamentally separated relationship with a macro entity called "the state." Work inspired by Foucault, sometimes called "green" or "environmental governmentality," replaces analysis of state and environment as distinct categories with a focus on the "complex networks of people, local communities and global organizations that are able to secure, however temporarily, the right disposition of things" (Whitehead 2008: 426). The analytical shift is one toward entanglements, relinquishing dualistic boundaries that cannot hold (Haraway 1985). We are not engaged in "environmental governmentality," which like all governmentality studies tends to abandon analysis of "the state" entirely. However, our analytic frame is similar. We aim to provide evidence of state-environment entanglement, an entanglement that results from the practices of environing. The latter is effected by the networking of science and government, which results in shifting regimes of governance and the construction of the modern state as an organic engine.

We now turn to our cases. The first case is that of Ireland, c. 1650–1900. Here we locate the beginning of the practices of techno-governmental environing in the context of the scientific revolution. We then turn to the case of California, c. 1850–2000, where we focus on reclamation and the emergence of water as a key boundary object between science and governance. We conclude with the case of Bolivia, a case that affords a view of newly emergent science-government relationalities around the object of lithium in the *Salar de Uyuni* (salt flat). The cases involve different regimes of governance – regimes that are constantly in process over time, sometimes shifting violently – but the entanglement of state-environment in all three reveals that each shares the ontology of being technoscientific organic engines. Also, while they have scaler variability in many (sometimes counter-intuitive) respects, we believe that each speaks to an entity that can justifiably be characterized as an "enviro-state." For instance, while the case of Bolivia is focused on the localized region of an isolated salt flat, its significance lies in the way it reveals how lithium is serving as a boundary object between science and government in the context of visions of a new regime of governance that the Bolivians call a "pluri-national state." They also differ in chronological scale, but we argue that the same points hold, that the cases collectively illustrate our argument about modern technoscientific states in terms of organic engines, regimes of governance, and the practices of environing.

Ireland

The Irish case illustrates aspects of the new experimental science which can be identified in the processes of environing in the subsequent cases of California and Bolivia. Francis Bacon famously hailed how the new experimental science would usher in the "empire of man over nature." Edgar Zilsel (1942) was perhaps the first to identify the critical development that explains how such a new empire – or state – might be possible. Zilsel noted that the key to the scientific revolution was the "breakdown of the social barrier" between the mechanical arts and more purely intellectual pursuits. The result was what Carroll-Burke (2001) has called "engine science," an inherently powerful form of inquiry in which technologies are immanent, particularly the forms of scopes, meters, graphs, and chambers (for example, pumps).

Engine science targeted the material world through the practice of experiment, and had an affinity with new forms of political and economic knowledge, which were equally experimental. In this context William Petty is a key human agent, causing a revolution in the discourse and practice of what today we call "political economy." Petty argued that science and politics were "instruments" for developing wealth in the service of aggrandizing the state. His political economy was expressed in what he called "political anatomy," "political medicine," and "political arithmetic" (Carroll 2006). Petty's first rule for governing a state effectively was procurement of a map, and he produced an extremely accurate one for Ireland. A modern map is a powerful actant, a constitutive element of a modern regime of governance.

The case of Ireland exemplifies the character of the new science as Ireland was explicitly viewed as ground for experiment. Petty is again key. While in the nineteenth century William Nassau Senior could remark about how "experiments are made in that country and pushed to their extreme consequences," Petty was the first to promote and attempt such experiments. He was not only a founding member of the Royal Society of London, but co-founded a sister organization in Ireland, the Royal Dublin Society. He used his large estate in Kerry as his own focused experiment, often in communication with William Penn who was doing the same thing in what is now Pennsylvania. Petty was always referred to as a guiding light for later efforts, such as the extensive geological surveying, and the remarkable Ordnance Survey. Petty even conducted the first census of the country. In the mid-nineteenth century the most exhaustive census ever was conducted in Ireland. It was a "sociometer" of unheard of dimensions.

The new science led to the scoping of Ireland, its graphing, metering, and eventual transformation. The work carried out on the material (land, bodies, built environment, etc.) of the new science generated a techno-territory and a new enviro-state. Land is a particularly illustrative boundary object. This is evident through the process of reclamation of marshes and bogs, deforestation, drainage and the "rectification" of rivers. Fields were created, surrounded by drainage ditches, and cleared of rocks and other impediments. "Artificial grasses" developed in the Royal Dublin Society were planted across the island. In the nineteenth century an island-wide arterial drainage system was built by the government through the department of public works. The country was utterly engineered throughout the eighteenth and nineteenth centuries and gradually incorporated into a new regime of governance inextricably bound up with techno (or engine) science. A process of environing occurred, a process of surrounding and penetrating that which surrounds, constituting a new state *as* a new "environment."

This new environment, from well-ordered homes, to manicured lawns and gardens, to clean streets, to reordered rural spaces, reshaped rivers and coastlines, is fully "stateized" (Painter 2006). The nexus of science and government is often *represented* in terms of discrete realms of the state, society, the economy, and the environment. The latter three are seen as given, while the former is viewed as an actor that intervenes in them, regulates them, puts demands upon them, etc. (Mol and Buttel 2002). However, we maintain that the ontology of the state appears as an assemblage of many bits and pieces that are discursive, organizational, and material. That is, "the state" appears more like a heterogeneous actor-network than a singular punctualized actor. It is also in this respect that we see the emergence of the state as an organic engine, a thing relentlessly driven forward, first in the name of improvement, then progress and economic development. Since before the twentieth century there was no "environmental movement," let alone "preservation movement," and no concept or language of nature *as* environment, there were no actors who could counter or shape the transformation of the country or the regime of governance it was subjected to according to those terms. For all intents and purposes, "the environment" did not exist.

California

As in the Irish case, in California one of the major boundary objects between science and government was land. Once again reclamation was at the top of the government's agenda. In 1850, when California joined the union, the federal government granted the states possession of what was described as "swamp and overflowed" lands, with the only condition that the land be "reclaimed." The government acted immediately, claiming the state contained six to ten million acres of swampland and requesting that the federal government permit the state surveyor-general to establish the actual acreage (McDougall 1852: 15). Much of the swampland was located in the Sacramento Valley and that was where most of the focus was until the twentieth century. Similar to Ireland, the goal of the government was economic development and the extraction of revenue, but the construction of an effective regime of governance for that purpose would be no easy task. Reclamation in California would require construction of a complex set of variously articulated and resisting relationalities among agents, actants, and forces. Intermittent years of extremely heavy rains combined with steep and fast flowing rivers draining into the Sacramento River, a river surrounded by low-lying basins, confronted human designs with unpredictable and almost intractable non-designed materialities. The regime of governance initially conceived for reclamation was passive and permissive in terms of the role of formal government. Land would simply be granted to homesteaders for a nominal fee with the expectation that they reclaim their own patch of land. But as the work of the surveyor-general gradually provided government some insight into the character of the valley, a shift occurred toward a regime of governance in which state government agents, in the form of a Board of Swampland Commissioners, would attempt a short-lived effort to coordinate the landowners' reclamation works. For a variety of reasons, many farmers resisted the government efforts (Carroll 2012). The rivers had not yet been metered and the scoping conducted was of limited value. State agents, with few effective actants for allies, were largely powerless in the face of suspicious landowners.

This was the context for a radical shift in the regime of governance in which the state commissioners board was abolished, and authority over reclamation vested in county government. Apart from the county surveyors (now ex-officio engineers) processing applications, landowners were once again left to govern their own land as they saw fit. Large landholders began building works without much concern for the forces at work in the valley as a whole, or for the consequences for other landowners. The result was that landholders up and down the river built levees that forced "flood" waters onto the land of others. The courts were already becoming important actors in the regime of governance in California by crafting, case by case, who had rights to the use of water – the legislature had not passed any legislation on the question. But in this context, as suit and counter-suit were filed, the agency of a single judge had profound consequences. Judge H.W. Hurburt declared that "self-protection" was "the first law of nature" and as such landowners could not be restricted in their attempts to protect their own properties (Kelley 1998: 94). The regime of governance weighted heavily toward the rights of individual landholders was secured in court and a levee battle was unleashed, reaching crisis levels with landowners destroying each other's levees. The levees were "thick with politics" (Bijker 2007). They became critical actants that mangled (Pickering 1999) the goals and designs of human agents, turning farmer against farmer. State actors mobilized the landholders' levees as powerful allies supporting the case that only the authority of state agents, with knowledge of the entire valley, could engineer effective protection. Non-humans can act, but they cannot speak. In the context of the levee battles, state agents were most advantageously positioned to speak for the levees, and thus align these actants with their own designs.

As the century progressed, the state engineers gained more and more allies in the form of meters and graphs. The graphs permitted them to more forcefully argue for legislation that would create a regime of governance in which formal government would be key. The problem of reclamation began to be viewed in relation to problems such as hydraulic mining and inland navigation, as well as a host of others that put state-wide designs central to the process of environing, and the construction of the state as an organic engine. In 1911, a critical human agent in the Army Corps of Engineers (USACE) by the name of Thomas H. Jackson argued that problems of reclamation, hydraulic mining, and navigation were "inseparably connected" and needed to be "considered under one general project" (Jackson Report 1911: 4). This conclusion shifted decision-making power within the regime of governance away from landholders and toward state actors. It also signaled a shift in which land as a boundary object was being displaced and subsumed into that of water. A range of issues, including reclamation, flood control, mining debris, navigation, forestry management, drinking supplies, industrial development, power generation, salinity control, and irrigation, were punctualized as a single "water problem." In 1915 the governor convened a "state water conference" at which all these issues were both gathered together discursively and graphically (*Report, State Water Conference*, State of California 1916). But the punctualization of "water problems" and the apparent centralization of the issues in state government masked the complexity involved in both these moves. Rather than a centralized actor-state taking over the singular problem of water, the number of distributed actors requiring articulation and assemblage multiplied. Rather than punctualizing the state as a coherent actor, the state became an even more complicated *thing*.

Nonetheless, in 1911 state actors moved boldly with a plan for a valley-wide flood control infrastructure. It involved a huge floodway, numerous weirs and pumps, and hundreds of miles of levee. A new state reclamation board with extensive "police powers" was created. Major government projects, impacting many people, typically involve coercive measures. ANT tends not to pay as much attention to powers of coercion as it does to practices of enrolling allies. Governmentality studies, due to a focus on the liberal rationalities of government, and its lack of attention to the state, has also neglected the importance of the continued growth of police governmentality in the nineteenth and twentieth centuries. In California, those landholders who could not be enrolled were compelled into alignment by the police power of state government. It took more than a quarter of a century to build the flood control infrastructure and it cost about $100 million. The result was that the Sacramento Valley was utterly transformed and incorporated into a new regime of governance in which the role of state agents and actors was strengthened. At the southern reaches of the river, in the delta region, two huge steam dredges scooped out as much material as was excavated for the Panama Canal, so that the river could handle the outflow from the bypass system. This helped protect an area of about half a million acres that had been farmed increasingly from the 1860s. About 1100 miles of levees were eventually built in the estuary, creating islands that were farmed and giving it the character of a delta with a thousand miles of navigable waterways. The rich peat soil of the delta, when exposed to oxygen, caused microbial action to result in subsidence, so that the islands are perpetually sinking, many far below sea level. The thick tules that covered the area had been clear-cut. By the early-twentieth century, the delta was "essentially a 'man-made' landscape" (Thomson 1961: 4). But as with the Sacramento Valley as a whole, the delta region was not only transformed by artifice and infrastructure, it was incorporated into complicated governance regimes, held together by district, local, state, and federal organizations, as much as by levees, pumps, meters, etc. No one sought to stop the transformative action. While a new "preserva-tionist" movement had emerged alongside the conservationist movement, its leaders were inspired by a spiritualism derived from what were considered the most majestic aspects of "nature," such as Yosemite and Hetch Hetchy valleys. John Muir fought bitterly against the damming of the Hetch Hetchy Valley, but cared little about the draining of the "swamps" of the Sacramento Valley. Certain aspects of "nature" were to be preserved in national parks, but there was little conceptualization of nature as "the environment." So as in the case of Ireland, the regime of governance and actor-network built around flood control and reclamation had no "environmental" actors within it. That would have to wait until the mid-to-late-twentieth century.

While portions of the San Joaquin Valley were also subject to inundation by water, it was much more arid than the Sacramento. In this context the invention of the centrifugal pump became another critical actant that inserted state and federal government more formally into the regime of governance in the Central Valley. This time the main issue was less one of drainage and more one of irrigation. Pumping from the San Joaquin Valley aquifer expanded rapidly in the early-twentieth century. By the 1920s the aquifer was already seriously depleted and in some places farms were already being abandoned. This spurred on the development of the Central Valley Project (CVP). Though the plan, developed in the 1920s and approved by state legislation in 1933, envisioned flood control through head water dams, storage for dry years, power generation, and salinity control in the delta, it was centrally designed to save San Joaquin Valley agriculture due to the loss of ground water. The project's infrastructure is comprised of sixteen reservoirs, thirty-nine pumping plants, two pumping generating plants and seven power plants, and has seven million acre-feet of water delivery capacity per year within a storage capacity of thirteen and a half million acre-feet (State of California Water Project Atlas 1999: 10). It is a technoscientific assemblage of massive proportions, an assemblage that further propelled the formation of the state as an organic engine. It also expanded federal government power within the regime of governance now environing the entire Great Central Valley. Once again there was no environmental movement at the time, and thus there was no resistance to the project in terms of protecting "nature."

State government embarked on an equally ambitious plan in the form of the State Water Project (SWP), begun in the 1950s just as the CVP was being completed. The SWP remains the largest infrastructural project ever built by any state government in the union. The infra-structure consists of 32 storage facilities with a capacity of almost six million acre feet, more than 650 miles of canals, aqueducts, and pipelines, 5 power plants, 17 pumping plants, and 139 pumping units. The humble pump, and the mundane valves that make it work, had come a long way in 300 years. The CVP and SWP utterly transformed the San Joaquin Valley. A new "envi-ronment" was "built," and the entire Central Valley, from the Sacramento Valley in the north to the delta in the middle and the San Joaquin in the south, bears little resemblance to its "natural" state. The Sacramento River had become, to use Karen O'Neil's term, a "river by design" (O'Neil 2006). It should be noted here that such material infrastructure critically constitutes government organization, as well as it being the product of government action. The two are co-productive of each other (Jasanoff 2004). Every major infrastructure project intrinsically involved the construction of government organization. With respect to the flood bypass system, it is the State Reclamation Board and its subordinates. Though the Reclamation Bureau existed prior to the CVP, its organization was expanded and involved different kinds of expert knowledge. With respect to water issues generally, state government organization transitioned from the Swampland Commissioners, to the State Engineer's office, to the Division of Water Resources within the Department of Pubic Works. Finally, with the commencement of construction of the SWP, the Division of Water Resources became a department in its own right.

The vastness of the works, the profundity of the transformation, the extent of the science and governance network, but most of all the pace of the transformation make California an exemplary case of environing, of the process of surrounding or constituting that which surrounds. The case exemplifies the formation of the enviro-state. The state of California would be unrecognizable absent the water infrastructure with its army of technoscientists, government officials, material culture of infrastructure and other actors. We see how "the state" and "the environment" are inextricably entangled. No clear boundary can be drawn between them because the "environment" now exists *within* a regime of governance. California has been made into an organic engine where the "natural" and "artificial" are lashed together, and designed for relentless "growth," and the growth in this case is truly phenomenal: in little more than a century and a half the state's population has grown from a few thousand settlers (plus about 150,000 American Indians), to about 40 million people (the Indian population was rapidly decimated, reduced to 30,000 by the 1870s), and its level of productive economy is in the top ten worldwide.

This observation allows us to come full circle in terms of our ontology of state-environment relationality. The new environmental movement that emerged from the 1960s, at the heart of which was the nature-environment synonym, confronted a world that cannot be captured by an analytic that takes the environment as a given, as nature, and as bounded and separate from "the state." This is not to say that the separation of state and environment, the nature-environment synonym, and the conceptualization of the state as a macro actor, cannot serve effective political action. On the contrary, the SWP was well underway before the emergence of an effective environmental movement, but once that movement emerged it immediately engaged with all the issues that concerned it as they related to water. The Sierra Club was no longer restricted to protecting majestic nature as in the preservationist era. For instance, in the early 1970s the organization published a blistering critique of Californian water politics under the title *The Water Hustlers* (Boyle et al. 1971). While the movement was too late to stop the SWP, it did play a major role in stopping the final part of the project, the plan for a "peripheral canal" to move water around the delta on its way to the San Joaquin Valley and southern California (though Governor Brown is currently pushing ahead with a plan for giant tunnels to do the same thing). The emergence of "environmental actors" subsequently leads to significant changes in the regime of governance in the Central Valley. For instance, the Endangered Species Act resulted in restrictions on the amount of water that can be pumped south, and nothing in the way of major earthworks can be undertaken without extensive environmental review. In this context we can see the emergence of a fully formed enviro-state, one in which state-environment relationality is not only deeply entangled, but in which environmental actors have shifted the regime of governance on the basis of ecological principles, and through the construction of the nature-environment synonym.

Bolivia

In our final case study, that of Bolivia, water also figures centrally in the environing process of the enviro-state. It is not water as "natural resource" that we illustrate here. Rather, water figures into the arrangement of agents, actants, and forces constituting the possibilities for another type of resource: lithium. The case is intriguing because it addresses how lithium is contemporaneously becoming an object of technoscience and a new imagined state described as "pluri-national," and how the region of the *Salar de Uyuni* (Uyuni Salt Flat) is reconfigured through the arrangement of agents, actors, and forces in changing regimes of governance. The Irish case, by contrast, presented a broad view of the nature of the scientific revolution and its deployment in Ireland, highlighting the importance of land as a boundary object between technoscience and government there. The California case also focused on the boundary object of land, but showed how it became subsumed into the boundary object of water. It showed how, through the boundary object of water, massive infrastructure was engineered into the landscape, and new regimes of governance established.

Lithium describes a heterogeneous set of discursive, organizational, and material relationalities involved in the environing processes immanent in the composition of an enviro-state. Here we follow the force of "flooding" of the *Salar de Uyuni* to illustrate the way that technoscience and government work jointly to reconfigure the region of the salt flat. This occurs through the interweaving of scientific and political understandings (agents), engineering products (actants), and the phenomenon of flooding (forces) in relation to "lithium" or a "lithium resource," which acts as a boundary object that ties them together. The entrance of lithium into the interplay of scientific understandings (geo-scientific studies), engineering products (i.e., infrastructure), and non-designed material processes (inundation) reconfigures the *Salar* into a relational entity that cannot be reduced to a state/environment relation in the disentanglement/purification practices of traditional boundary conventions. In comparison to the Irish and Californian cases, the Bolivian case illustrates a transformation of a governance regime currently under way. This transformation is articulated within the discourse of establishing the "pluri-national state" (see Albó and Barrios 2006 for a review of the pluri-national state idea/project). Most, if not all, of the treatments of the pluri-national state prioritize the juridical and legal/constitutional dimensions, to the neglect of the material dimensions of this new type of state. The ways that it is being built into and out of the environment, in the sense of environing, is subsumed in an analysis of the appropriate discursive/legal/economic constitution of the state (Radhuber 2012, Santos 2007, Tapia 2007). Constitutional principles are certainly important. However, environing processes – as a matter of material constitution and one that is relatively neglected – is one of the principal sites of the politics of the pluri-national state (one only needs to contrast the transcontinental highway set to be built through the Indigenous Territory and National Park, Isiboro Secure and the lithium production plant in the *Salar de Uyuni* to see how these environing processes produce distinct political effects). Instead of concentrating on the pluri-national state *idea*, here we focus our attention on one case of environing in the *material constitution* of the pluri-national state. This section illustrates the entanglements of agents, actants and forces brought about by lithium's role in the reconfiguration of the region of the *Salar de Uyuni* in the environing processes linked to the prior regime of governance, and those entailed in a transformation of that regime.

Geology, geochemistry, and sedimentology

The Andes mountains continue to fascinate geologists, particularly in terms of their origins. However, beginning in the mid-to-late-1960s and picking up in the early-to-mid-1970s, interest in the Andes began to focus not on the origin of the mountain range, or its ability to support tectonic theories of the evolution of the Earth, but on phenomena within the Andes (such as drainage basins). As part of this shifting focus, the region of the *Salar de Uyuni* became interesting not only to geologists, but also sedimentologists, geochemists, chemists, and various engineering disciplines. As research amassed on the drainage basins of the Andes, the number of studies "directed at the nature and potential of evaporitic deposits and their actual state" began to grow (Arduz Tomianovic 1988: 10). The list of agents involved in this research is long, and together they constitute part of a governance regime aimed at resource identification and concession. Private companies such as the Foote Mineral Company, foreign state agencies such as the United States Geological Service (USGS) and the U.S. National Aeronautics and Space Administration (NASA), as well as the Bolivan Geological Service were the major agents involved.

In the early seventies Bolivian geologists worked with NASA to develop minerals exploration technology via satellite, a program called the Earth Resources Technology Satellite (ERTS). The research intended to provide "new information on the relationship and regional distribution of volcanic centers and effusive products in the *Altiplano*" (High Plains of the Andes). According to Brockmann "interest in such rocks is high because they appear to be the source of salt, borax, sodium sulfate and sodium carbonate, and other salts of potential economic value" (1974: 564). Satellite images (graphing) became critical allies to teams of geologists and their future sampling programs, given that they provided useful hydrologic information on this closed basin and specifically on "flood" patterns.

In 1977 a team of geologists from the USGS had identified the evaporite basins of the Andes as not only interesting in their implications for geology, but also as sources of "salts of potential economic value." They arrived that year to "verify" the existence of lithium resources that had been mentioned in a conference in 1976 that proposed that these evaporite basins might be of particular interest. To aid them, a friend (the Bolivian geologist, Carlos Brockmann) employed in NASA's satellite imagery project generously "furnished ERTS data about flooding of the *Salar de Uyuni*" which was used in a field investigation in 1976 that included "reconnaissance of the regional geology and hydrology, the study of geomorphology and mineralogy of the salar crusts, and the sampling of salt crust and near-surface brines" (Ericksen et al. 1977: 8, 3). A report on the research concluded that *Salar de Uyuni* had formed under conditions favorable to the accumulation of lithium-rich brines. One of those conditions was annual flooding. "Lithium values in brines of *Salar de Uyuni* … are due, at least in part … to wind-generated currents in surface brines during periods of flooding…" (Ericksen 1977: 34). Over a decade later another geologist interested in evaporite basins affirmed the lithium/flooding connection: "…the (flood) water evaporates rapidly and only the dissolved components added by the river remain. This very restricted area can be considered, with regard to K, Li, Mg, and B [Potassium, Lithium, Magnesium and Boron] as a playa lake" (Risacher and Fritz 1991: 223).

Contrast this description of the *salar* as a "playa lake," rich in mineral deposits, that developed in the arrangement of geo-scientific studies, satellite technology and core sampling, and the non-designed material force of flooding to another description of the *salar* by colonial officials:

> It is of the most unspeakable solitude, and abandonment that one could imagine … one doesn't encounter even the smallest village of people, and even less the possibility of one, given that one cannot see even a single plot where they could even plant any species of seeds, because it is thus, the sterility … that does not even permit a single shrub, nor a single straw, that instills fear in those who cross it, suspicious of perishing of hunger, and what is more, from thirst for the lack of coming onto any sources of water…
>
> *Priest of the Parish of Llica in 1791, quoted in Andrade 2009: 75*

While desolation figures prominently in these descriptions, they subsequently point to the most consequential (for colonial officials) characteristic of the *salar*: the dangerous condition produced by its flooding. This treacherous situation is recounted in tales (of colonial officials) of lives lost when flooding turns the ten-thousand square kilometer surface into a reflection of the sky and its crossing impossible:

> In the rainy season it cannot be crossed, not even by the indians, given that the whole of this area floods and appears as a sea, and it has been widely heard that a priest from Lípez, his church, and some mestizos, perished on that route.
>
> *from correspondence in the Archbishop's of Las Charcas archives, 1840,*
> *quoted in Andrade 2009: 76*

In the colonial context when the *Salar* floods, it is just that: a flood, as an act of nature, or for the Spanish colonizers, perhaps, an act of God. The region is understood primarily in terms of danger, in particular the potential loss of human life (that of colonial officials). In the colonial regime of governance, flooding is just one more part of the "sterility" of the *salar*. But, beginning in the 1970s the arrangement of agents, actants, and this force reconfigure the *salar*. In the wake of geological studies, notably high concentrations of lithium *values* became linked to "flooding," and inundation is no longer a simple matter of the sterility of the "natural" environment of the *salar*. Inundation ceases to be a force (of nature) independent of human designs. Geologists (from both American and Bolivian state agencies and the two largest privately held lithium companies), and geological practices, as well as satellite technology and analytical chemistry techniques, had already begun to reconfigure the region through the practices of environing in terms of surrounding or encompassing. The new combination of agents (geo-scientists, prospectors of private firms, state agents), actants (ERTS satellite imagery, sampling techniques (gas powered drilling), new knowledge of the selective precipitation of components of the brine), and forces (radiation, winds, precipitation) began to obligate practitioners to enter a space of agreement on the significance of inundation as a factor in the production of a lithium resource. "Flooding" is reconfigured by the activities carried out with respect to the boundary object "evaporite resources," namely, lithium. Still, the phenomenon of flooding was of interest to a particularly limited group, one that nonetheless constituted a regime of governance based on resource concessions and security of contracts tied to them. That is, until the distribution of agents, actants, and forces is reconfigured again in 2009 with the placement of extensive infrastructure on the surface of the *salar* under the National Direction of Evaporite Resources (GNRE). Here, the flooding of the salt flat becomes something other than the link between inundation and lithium values in the evaporite resource deposits of the *salar*. This infrastructure was the means through which the *Salar de Uyuni* was subjected to a more penetrating environing than before.

Infrastructure

In 2009, at the First International Forum on the Industrialization of Lithium and Other Evaporitic Resources, in the city of La Paz, Bolivia, representatives from the GNRE described to an international audience at the Central Bank of Bolivia infrastructure projects that had been completed and ones currently being built. These projects are centered in the *salar* region as it is linked to lithium mining, and after the nationalization of gas resources in Bolivia they constitute one of the economic pillars of the pluri-national state: a pillar based not on resource concessions (as in the regime of governance of the 1970s to the 1990s), but on government led technoscientific development of a pluri-national state. The articulations of the lithium project by agents from the regional farmer's union (United Regional Federation of Laborers and Farmers of the Southern Altiplano–FRUTCAS) and the Bolivian president, Evo Morales, signal a reconfiguration of the regime of governance in the environing of the Salar: both maintain that the project will take place under the direction and control of Bolivian (non-governmental and state) agents. It is to be one hundred percent Bolivian (state) owned and controlled. The infrastructure constituting the project includes: asphalt highways from Uyuni to Huancarani and Potosí to Uyuni; supply of electricity to the region of more than 30 megawatts; gas duct to Uyuni and Rio Grande (most likely from Tarija); augmenting the current capacity of the gas duct to Potosí; construction and extension of the railway to the lithium pilot plant; freshwater catchment and the treatment of water with high salinity; radio bases in two municipalities near the lithium project; and the construction and putting into orbit the telecommunications satellite "Tupak Katari" (Corporación Minera de Bolivia 2009). In addition to this list, two projects that are central to our analysis here include a potassium chloride production plant and a lithium carbonate plant, both of which are built in the southeast corner of salar, which experiences the highest levels of inundation and had already been identified as the area with the highest lithium concentrations in the 1970s.

The lithium concentrations/inundation link is no longer the most important one with respect to flooding. The value of lithium related infrastructure and the potential damage to it by periods of inundation are brought together as a potential concern in periods of flooding. "Uyuni, at 12,000 feet above sea level, is a region largely forgotten by previous governments. There are a few secondary roads and an old rail line, both of which wash out in periodic flooding. Virtually everything will have to be built from scratch and skilled personnel will need to be recruited from outside the area" (Mares 2010: 15). Many critics of the project question whether the massive infrastructure to be built would be functional due to regular inundation of the *salar*. "Seasonal flooding of the salt flats slows the evaporative process in the pools relative to the evaporation rate at competing sites (40 percent of the rate at Atacama) and thus contributes to higher costs" (Mares 2010: 15). The inundation-lithium-infrastructue relation has become "a very serious problem not only for the aspirations of ... Potosí, but for all the expectations that lithium has created" (Proyecto de Litio 2011). In the particular arrangement of agents, actants and forces entailed in the environing processes in the region means that technoscientists, government officials, and local inhabitants are forced to consider flooding in terms of the actual and potential infrastructure destroyed by it. Flooding has economic costs and implication for economic competitiveness.

By 2010, the nineteen million dollars invested by the GNRE, through a loan from the Central Bank of Bolivia, has been materialized and spatialized on the surface of the *Salar de Uyuni* in the form of two processing plants and some three hundred plus hectares of dikes built from compacted salt and lined with impermeable geotextile. This situates the force of inundation in terms of the potential damage to the evaporation pools, roads, vehicles, processing plants, as well as the slowing of the evaporation process, dilution of the brines, etc. "Flooding" becomes effectively an arrangement of agents (engineers and officials from the GNRE), actants (sampling programs and built infrastructure), and the force of inundation. In turn, the landscape of the salar, its lithium, and its flooding become inseparable from human constructions and material infrastructures.

Where's the flood?

It is the boundary object of lithium that begins to draw agents together in the interpretation of the "force" of inundation in the *salar*. First, it is geologists who, through the practices of measuring, sampling, and monitoring the conditions of the evaporite processes of Uyuni, begin to situate inundation in terms of identifying the "causes" of concentrations of lithium in particular places (which began as a concern of sedimentologists, and geochemists). Second, and some three decades later, the event of the "flooding" of the *salar* comes under a different arrangement, one that includes a threat to the state's construction of a lithium project and the cost of the destroyed infrastructure tied to Bolivia's "natural resource." Inundation as a "force" of nature, is no longer just this raw fact, if indeed it ever was. That there is no mere "natural fact" that simply appears as the capriciousness of the forces of nature, or non-designed materialities, is a particularly potent illustration of the ontology of the enviro-state. All "real raw facts" come under a meaning given to them by their entrance into the complex relationality of environment and state, where their meaning is presented in a metric of their effects, and often in the register of costs, and caught in the interplay of particular arrangements of agents, actors and forces. The environing of the region is ongoing, resulting in greater and greater entanglement of "the environment" and "the state." The technoscientific dimensions, and the networking of technoscience and governance around a boundary object, are the same as in the cases of Ireland and California. But in Boliva the techno-governmental aspects of the regime of governance are being articulated with the vision of a new pluri-national state, one that might differ substantially from modern, liberal states in terms of the conception and distribution of resources, rights, and territory.

Conclusion

We have argued against the idea that the state is an actor that stands apart from the environment. A state that acts relative to the environment makes sense if one views the state as a regime with a head in the sense of Hobbes' Leviathan. But with the development of the technoscientific state over the last three centuries we see the emergence of an array of relationalities that constitute regimes of governance, and which cut across what are considered state and society, state and economy, and state and environment. Various boundary objects such as land, infrastructure, people/bodies, water, forests, lithium, and others serve as nodes around which these relationalities are formed, relationalities that are discursive, organizational, and material.

The image of the state one is left with after our analysis is a gathering, an assemblage, an actor-network, in short a "thing." It is a thing assembled from human and non-human, from the natural and artificial, though it confounds these very distinctions. To use Haraway's term, it is a cyborg (1985). The point is not that it simply *is* a cyborg, but that its fundamental characteristic is a generative one. It is designed for relentless growth, hence the concept of an organic engine. This is the ontology – or one of them – of the technoscientific state. It is always in process – it is never fixed in the sense of a "matter of fact." It is always contested and agitated, always a "matter of concern" (Latour 2004, 2008). At the heart of that concern, and its process of contestation, is the effort of various agents to establish particular accounts of the state, of nature, of the environment, of water, of lithium (and so on) over and above others. That is, human agents attempt to establish these complex things as simple matters of fact, though the facts of the matter vary with the aims of such actors. In this sense everyone so engaged, including the authors of this paper, participates in ontological politics.

Though the cases of Ireland, California, and Bolivia are in many respects very different, we argue that in every case one can discern a process of environing, of surrounding and penetrating that which surrounds. The result is what we call the "enviro-state," a concept that moves beyond the regulatory state idea embodied in the notion of an "environmental state" (Mol and Buttel 2002). The latter conceptualization maintains the macro distinction between an actor-state and an external environment (often synonymous with "nature"). While this conceptualization has, and can continue to facilitate actions to create new regimes of governance aimed at protecting a conceived external environment, it fails to recognize how the very successes of such efforts result in even more complex relationalities and entanglements that evade its conceptual grasp.

We believe our argument is generalizable because it is rooted in an understanding of the character of the new science that developed in the seventeenth century, and has been similarly mobilized and networked within modern political regimes. The new science was characterized by powerful technologies of inquiry that permitted material engagements with a nature conceived as external to culture: it was "engine science." But the new science was itself a new set of cultural values that aimed at creating an "empire of man over nature." As such, nature became more rather than less political. A new ontological politics was born that both separated nature from "man" and reconceived nature as something to be known and exploited. The invention of political economy, the development and use of political "instruments" to expand wealth and aggrandize "the state," was part and parcel of the new science. It was never simply about knowing nature. It was about controlling and exploiting "it." But in the process regimes of governance were created that undermine any simple distinction between the state on the one hand and the environment on the other. In an important sense, at least in the context of the modern technoscientific state, the state is the environment, and the environment the state.

We would like to finish with a simple example of the apparent contradictions that follow from the formation of the enviro-state: the way environmental protection has been integrated into existing engineered systems. In the case of California, wildlife refuges have been created within the engineered flood plain. For instance, in the middle of the Sutter section of the Sacramento Valley flood bypass one finds an 11,000 acre "wildlife habitat refuge." The Sacramento National Wildlife Refuge Complex website explains that 90% of California's wetlands are gone, and that new wetlands "cannot be created naturally" (alongside a picture of a bulldozer in a "marsh"). At another point it states that the wetland in all five of the complex's refuges are "almost entirely manmade." Thus the scientists/officials of government are themselves aware of the inconsistencies in the idea that a "wildlife" area is built into an engineered landscape that is held together by governing discourses and organizations. Eco-feminist Carolyn Merchant provocatively argued that the scientific revolution brought about the "death of nature." When discourse and theory reduce "the environment" to "nature," are we also forced to speak of the death of *the* environment? The alternative, which we are arguing for, is to recognize that "the environment" is hybrid, built, and governed. There is no outside of it, and it has no existence outside the complexes of human agents, actants, forces, and the regimes of governance and environing processes from which it has been engineered.

Note

1   Patrick Carroll's research on California was conducted with support from the National Science Foundation (#SES0646982). Nathaniel Freiburger's research in Bolivia was conducted with support from the National Science Foundation (#SES1058330) and the Wenner Gren Foundation. Additional historical research on lithium was conducted with support from the Chemical Heritage Foundation. The authors are grateful for this support. We would also like to thank the editors of this collection for their constructive feedback on earlier drafts.

References

Adams, J., Clemens, E.S, and Orloff, A.S. (Eds.). 2005. *Remaking Modernity: Politics, History, and Sociology* Durham: Duke University Press.

Albó, X. and Barrios, F.X. 2006. "Por una Bolivia Plurinacional e Intercultural con Autonomías," in *Informe Nacional Sobre Desarrollo Humano en Bolivia*. La Paz, Bolivia: United Nations Development Program.

Andrade, C. 2009. "Bolivia y sus Rutas Etno-ecoturísticas en los Llip'is de Potosí: Historia del Salar de Uyuni." Unpublished Manuscript. Sucre, Bolivia.

Arduz Tomianovic, M. 1988. "Investigaciones sobre Los Salares del Altiplano Boliviano, Su Historia Resumida." *Ciencia* 1 (1): 10–17.

Bakker, K. 2007. "The 'Commons'Versus the 'Commodity': Alter-globalization, Anti-privatization and the Human Right to Water in the Global South." *Antipode* 39(3): 430–455.

Barry, A., Osborne, T., and Rose, N. (Eds.). 1996. *Foucault and Political Reason: Liberalism, Neo-liberalism, and Rationalities of Government*. Chicago: The University of Chicago Press.

Bijker, W.E. 2007. "Dikes and Dams, Thick with Politics." *Isis* 98: 109–123.

Boyle, R.H., Graves, J., and Watkins, T.H. 1971. *The Water Hustlers*. New York: Sierra Club.

Brockmann, C.B. 1974. "Earth Resources Technology Satellite Data Collection Project, ERTS-I Bolivia." Pp. 559–577 in *Third Earth Resources Technology Satellite-1 Symposium*, edited by Stanley C. Freden, Enrico P. Mercanti, and Margaret A Becker. NASA Goddard Space Flight Center, Washington, DC: NASA.

Burchell, G., Gordon, C., and Miller, P. (Eds.). 1991. *The Foucault Effect: Studies in Governmentality*. Chicago: The University of Chicago Press.

Callon, M. 1986. "The Sociology of an Actor-Network: The Case of the Electric Vehicle," in *Mapping the Dynamics of Science and Technology: Sociology of Science in the Real World*, edited by Michel Callon, John Law, and Arie Rip. New York: Macmillan.

Callon, M. 2007. "An Essay on the Growing Contribution of Economic Markets to the Proliferation of the Social." *Theory, Culture & Society* 24(7–8): 139–163.

Carroll, P. 2006. *Science, Culture, and Modern State Formations*. Berkeley, CA: University of California Press.

Carroll, P. 2009. "Articulating Theories of States and State Formation." *Journal of Historical Sociology* 22(4): 553–603.

Carroll, P. 2012. "Water and Technoscientific State Formation in California" *Social Studies of Science* 42(4): 489–516.

Carroll-Burke, P. 2001. "Tools, Instruments, and Engines: Getting a Handle on the Specificity of Engine Science." *Social Studies of Science* 31(4): 593–626.

Carson, R. 1962. *Silent Spring*. New York: Haughton Mifflin Company.

Collins, H.M. and Yearly, S. 1992. "Epistemological Chicken," in *Science as Practice and Culture*, edited by Andrew Pickering. Chicago: The University of Chicago Press.

Corporación Minera de Bolivia, Dirección Nacional de Recursos Evaporíticos de Bolivia. 2009. "Proyecto Planta Industrial de Carbonato de Litio y de los Productos Industriales Derivados com son el Cloruro de Potasio, Sulfato de Potasio, Ácido Bórico Contenidos en el Salar de Uyuni." Presentation at the *Primer Foro Internacional de Ciencia y Tecnología de Litio y Otros Recursos Evaporíticos*. La Paz, Bolivia.

Dean, M. 1999. *Governmentality: Power and Rule in Modern Society*. Thousand Oaks: Sage Publications.

Ericksen, G.E., Vine, J.D. and Raul Ballón, A. 1977. "Lithium-Rich Brines at Salar de Uyuni and Nearby Salars in Southwestern Bolivia." Washington, DC: United States Printing Office.

Foucault, M. 2004. *Security, Territory, Population*. New York: Palgrave Macmillan.

Haraway, D. 1985. "A Manifesto for Cyborgs: Science, Technology and Socialist Feminism in the 1980's," in *The Haraway Reader*, edited by Donna Haraway. New York: Routledge.

Haraway, D. 1992. "The Promise of Monsters: A Regenerative Politics for Inappropriate/d Others," in *Cultural Studies*, edited by Lawrence Grossberg, Cary Nelson, and Paula Treichler. New York: Routledge.

Hayes, C.W. and William Kennedy. 1903. *Oil Fields of the Texas-Louisiana Gulf Coastal Plain*. Department of the Interior. Washington: Government Printing Office.

Ingold, T. 2010. "Bringing Things to Life: Creative Entanglements in a World of Materials," *Realities Working Papers* 15. (*www.manchester.ac.uk/realities*).

Jackson Report, in U.S. House Doc 81, 62nd Congress, 1st Session 1911.

Jasanoff, S. 2004. *States of Knowledge: The Co-production of Science and Social Order*. New York, NY: Routledge.

Jessop, B. 2001. "Bringing the State Back In (Yet Again): Reviews, Revisions, Rejections, and Redirections." *International Review of Sociology* 11(2): 149–173.

Kelley, R. 1998. *Battling the Inland Sea: Floods, Public Policy, and the Sacramento Valley*. Berkeley: University of California Press.

Latour, B. 1996. "On Actor-Network Theory: A few Clarifications." *Soziale Welt.* 47: 369–381.

Latour, B. 2004. "Why Has Critique Run Out of Steam? From Matters of Fact to Matters of Concern," *Critical Inquiry* 30 (Winter): 225–248.

Latour, B. 2005. "From Realpolitik to Dingpolitik: Or, How to Make Things Public," pp. 1–31 in *Making Things Public: Atmospheres of Democracy*, edited by Bruno Latour and P. Weibel. Cambridge, MA: MIT Press.

Latour, B. 2007. *Reassembling the Social: An Introduction to Actor-Network Theory.* Oxford: Oxford University Press.

Latour, B. 2008. *What Is the Style of Matters of Concern.* Amsterdam: Van Gorcum.

Law, J. 1992. "Notes on the Theory of the Actor-Network: Ordering, Strategy, and Heterogeneity," *Systems Practice* 5(4): 379–393.

Law, J. 2004. *After Method: Mess in Social Science Research.* New York: Routledge

Law, J. and Hassard, J. (Eds.). 1999. *Actor Network Theory and After.* Malden, MA: Blackwell Publishers.

MacKenzie, D. 2006. *An Engine, Not a Camera: How Financial Models Shape Markets.* Cambridge, MA: The MIT Press.

MacKenzie, D. and Millo, Y. 2003. "Constructing a Market, Performing Theory: The Historical Sociology of a Financial Derivatives Exchange," *American Journal of Sociology* 109(1): 107–145.

MacKenzie, D., Muniesa, F., and Siu, L. (Eds.). 2008. *Do Economists Make Markets? On the Performativity of Economics.* Princeton: Princeton University Press.

Mares, D.R. 2010. "Lithium in Bolivia: Can Resource Nationalism Deliver for Bolivians and the World?" in *Energy Market Consequences of an Emerging U.S. Carbon Management Policy.* James A. Baker III Institute for Public Policy of Rice University.

McDougall, J. 1852. Governors Address to State Assembly, 1852. *Journal of the Fourth Session of the Legislature of the State of California.* San Francisco: State of California.

McGee, W.J. 1896. "The Relation of Institutions to Environment," pp. 701–712 in *Annual Report of the Board of Regents of the Smithsonian Institution, Showing the Operations, Expenditures, and Condition of the Institution.* Washington, DC: Government Printing Office.

Mason, O.T. 1896. "Influence of Environment Upon Human Industries or Arts," *Annual Report of the Board of Regents of the Smithsonian Institution, Showing the Operations, Expenditures, and Condition of the Institution.* Washington: Government Printing Office.

Meyer, J. 1999. "The Changing Cultural Content of the Nation-State: A World Society Perspective," in *State/Culture: State Formation after the Cultural Turn*, edited by George Steinmetz. Ithaca and London: Cornell University Press.

Mitchell, T. 1991. "The Limits of the State: Beyond the Statist Approaches and Their Critics," *American Political Science Review* 85(1): 77–96.

Mitchell, T. 1999. "Society, Economy, and the State Effect," in *State/Culture: State-Formation after the Cultural Turn*, edited by George Steinmetz. Ithaca: Cornell University Press.

Mitchell, T. 2005. "The Work of Economics: How a Discipline Makes Its World." *European Journal of Sociology* 46(2): 297–320.

Mol, A.P.J. and Buttel, F.H. (Eds.). 2002. *The Environmental State Under Pressure.* Oxford: JAI.

Oxford English Dictionary. 1933. *The Oxford English Dictionary, Being a Corrected Re-Issue with an Introduction, Supplement, and Bibliography of A New English Dictionary on Historical Principals, founded mainly on the materials collected by The Philological Society and edited by James A.H. Murray, Henry Bradley, W.A. Craigie, [and] C.T. Onions.* Oxford: Clarendon Press.

O'Neil, K.M. (2006) *Rivers by Design: State Power and the Origins of U.S. Flood Control.* Durham, NC: Duke University Press.

Painter, J. 2006. "Prosaic Geographies of Stateness," *Political Geography* 25(7): 752–774.

Passoth, J.-H. and Rowland, N.J. 2010. "Actor-Network State: Integrating Actor-Network Theory and State Theory," *International Sociology* 25(6): 818–841

Proyecto de Litio a Punto de Ahogarse 2011, March 25. *La Patria*, retrieved from *www.hidrocarburos bolivia.com/bolivia-mainmenu-117/mineria-siderurgia/41208- proyecto-del-litio-a-punto-de-ahogarse.html.*

Pickering, A. 1999. "The Mangle of Practice: Agency and Emergence in the Sociology of Science," pp. 372–393 in *The Science Studies Reader*, edited by Mario Biagioli. New York: Routledge.

Radhuber, I. 2012. "Indigenous Struggles for a Plurinational State: An Analysis of Indigenous Rights and Competences in Bolivia." *Journal of Latin American Geography* 11(2): 167–193.

Risacher, F. and Fritz, B. 1991. "Quaternary Geochemical Evolution of the Salars of Uyuni and Coipasa, Central Altiplano, Bolivia." *Chemical Geology* 90: 211–231.

Rose, N. and Miller, P. 1992. "Political Power beyond the State: Problematics of Government." *British Journal of Sociology* 43(2): 173–205.

Santos, B.S. 2007. "La Reinvención del Estado y el Estado Plurinacional." Pp. 1–66. Cochabamba Alianza Interinstitucional CENDA, CEJIS, CEDIB.

State of California 1916. *Report, State Water Problems Conference.* Sacramento.

State of California 1999. *California State Water Project Atlas.* Sacramento: Department of Water Resources.

Star, S.L. and Griesemer, J.R. 1989. "Institutional Ecology, 'Translations' and Boundary Objects: Amateurs and Professionals in Berkeley's Museum of Vertebrate Zoology, 1907–39." *Social Studies of Science* 19(3): 387–420.

Steinmetz, G. (Ed.) 1999. *State/Culture: State-Formation after the Cultural Turn.* Ithaca: Cornell University Press.

Stradling, D. (Ed.). 2004. *Conservation in the Progressive Era: Classic Texts.* Seattle, WA: University of Washington Press.

Stradling, D. (Ed.). 2012. *The Environmental Moment, 1968–1982.* Seattle, WA: University of Washington Press.

Tapia, L. 2007. "Una Reflexión sobre la Idea de Estado Plurinacional." *OSAL (Buenos Aires: CLACSO).* Año VIII (22): 47–63.

Thompson, J. 1961. "How the Sacramento-San Joaquin Delta Was Settled," Sacramento, CA: County of Sacramento, Superintendent of Schools.

White, R. 1995. *The Organic Machine: The Remaking of the Columbia River.* New York: Hill and Wang.

Whitehead, M. 2008. "Cold Monsters and Ecological Leviathans: Reflections on the Relationships between States and Environments." *Geography Compass* 2(2): 414–432.

Zilsel, E. 1942. "The Sociological Roots of Science." *American Journal of Sociology* 47: 544–552.

Part V
Technoscience as Work

20

Invisible Production and the Production of Invisibility

Cleaning, maintenance, and mining in the nuclear sector[1]

Gabrielle Hecht

UNIVERSITY OF MICHIGAN

Large-scale disasters have become privileged sites for STS analysis. Bhopal, Chernobyl, the Challenger, Katrina: such catastrophic events reveal the fissures of our technopolitical regimes, the fallacy of the nature-culture divide, and the power relations that shape and are enacted by our infrastructures. They offer particularly propitious places for exploring scalar shifts and slippages. Disasters transport us from O-rings and reactor instrumentation panels to structural secrecy in sociotechnical systems, and from those systems to problems of planetary pollution and human security. Along the way, we trace the tendrils of sociotechnical systems in order to disrupt notions of center and periphery; we hop through geographic and temporal scales; we see social dynamics and technopolitical relationships that might otherwise remain invisible. In short, disasters allow us to observe "the world in a machine."[2]

So too with the hydrogen explosions at the three Fukushima Daiichi nuclear power plants in March 2011, which launched one of the largest disasters in industrial history. Nearly two years after the explosions, radioactive fish with levels of contamination up to 2500 times the legal limit offered evidence of ongoing radioactive leaks. Conservative estimates judge that "cleaning up" the mess – to the extent that this is even possible – will take four decades and cost more than $125 billion. Along the way, thousands of workers will be exposed annually to levels of radiation well in excess of 20 milliSieverts, the maximum limit recommended by the International Commission for Radiological Protection (ICRP) for normal working conditions.

At one level, of course, working conditions in the three devastated reactors are anything but "normal." By definition, states of emergency entail a suspension of the ordinary, a breach of normal rules. In the nuclear sector, this eventuality has been codified: ICRP recommendations allow for higher exposures during post-accident recovery operations. At least half of Chernobyl's roughly 700,000 "liquidators" were exposed to 100 milliSieverts of radiation, and many received far higher doses.[3] Drawing legitimacy from such precedents, immediately after the accidents the Japanese government raised exposure limits for workers to 250 milliSieverts, and for the general public from 1 to 20 milliSieverts. Widespread outrage ensued: citizens and experts loudly denounced the fact that infants were being permitted radiation exposures equivalent to the ICRP maximum for industry workers.

Outrage rapidly focused on the dysfunctions that had enabled the reactors accidents. The Japanese public and its political representatives instinctively understood the observation that disaster experts have made time and again: states of emergency may be exceptional, but they also reveal the ordinary (mal)functioning of a society and its institutions. The 2012 Japanese parliamentary report on the reactor accidents, for example, locates their roots not in the earthquake and tsunami, but rather in the social, political, and technological relationships that structured Japan's nuclear industry.[4]

This chapter takes the aftermath of the Fukushima accidents as the starting point for an exploration of work and workers commonly considered marginal to technoscientific enterprise, and proceeds through a series of spatial and temporal shifts. We begin with the contract workers hired to "clean up" the exceptional mess in Japan. We then move back in time and across oceans, to consider the subcontractors hired to conduct ordinary reactor maintenance and refueling in Japan, France, and elsewhere. Maintenance is the unseen, decidedly unspectacular work essential to keep any technological assemblage working – work so invisible and unglamorous that most scholars avoid studying it, preferring instead to focus on acts of creation and construction. Without these workers, sociotechnical systems could not function: they may be socially marginal, but they are technopolitically central to the production of nuclear power (and all other industries).

In nuclear and other systems, subcontracting has consequences for occupational health, as well as for transnational knowledge production (about the effects of low-level radiation exposure in the nuclear case). Contemplating these consequences, in turn, takes us to another apparently peripheral part of the global nuclear industry: uranium production. After a quick comparative consideration of knowledge production about the dangers of radon exposure in mines, we land in Gabon. The final empirical section of the chapter examines labor and occupational hazards there, including the efforts of Gabonese mineworkers to make themselves and their illnesses visible on the global technoscientific stage.

(In)visibility is a central theme of this chapter. It seeks to make visible labor that – by virtue of its unexciting nature or (apparently) peripheral location – often remains hidden in our accounts of technoscientific work: ordinary maintenance, African mining. The chapter also considers how the hazards of this labor are rendered invisible. Although radiation lends itself particularly well to exploring this theme, social science work on occupational and environmental illness has shown that all contaminants require complex, technoscientific infrastructures to become visible and actionable.[5] Instruments, labor relations, scientific disciplines, expert controversy, and lay knowledge combine to create what Michelle Murphy has called regimes of perceptibility – assemblages of social and technical things that make certain hazards and health effects visible, and leave others invisible.[6] While this chapter confines itself to the nuclear industry, therefore, many of its themes – contract labor, the transnational distribution of danger, the regimes by which industrial hazards become (im)perceptible – resonate for other technoscientific endeavors, and other sociotechnical systems.

The Fukushima cleanup

At this writing, more than 24,000 men have engaged in cleanup and decontamination work following the meltdowns of the three reactors at Fukushima Daiichi. The vast majority of these people sign up out of sheer economic necessity. They are subcontract employees, recruited through nationwide temp agencies, among local residents rendered unemployed by the disaster, and among the thousands of day laborers who eke out an existence in the notorious slums of Japanese cities. As one contract worker observed, "If [day laborers] refuse, where will they get another job? ... I don't know anyone who is doing this for Japan. Most of them need the money."[7]

Over the course of the three years following the accidents, media reports began to focus on the murky hiring practices that recruited workers for the cleanup. The longstanding relationships between *yakuza* (organized crime) groups and the nuclear industry received special attention. Suzuki Tomohiko, a journalist who went undercover as a cleanup worker, reported that the Tokyo Electric Power Company (TEPCO) had explicitly asked a *yakuza*-affiliated recruitment company for "men who [were] expendable."[8] *Yakuza* leaders, in turn, saw the nuclear industry as safer and a more reliable source of revenue than drugs. In the words of one boss interviewed by Suzuki,

> Nukes are a cash cow for us. A steady source of income. Once they're up and running, they just keep on giving. We can get by on only the one line of work. Never have to dirty our hands with meth. Because I really hate drugs. I'd rather wear construction boots than get into the pharmacy business. From your perspective in the broader society, nukes come with a whole treasure chest of taboos, but that's exactly what makes them such a horn of plenty for us in the underworld.[9]

In May 2012, the Fukushima prefecture police arrested a *yakuza* boss for dispatching gang members to the cleanup site. According to Suzuki's report, however, the role of *yakuza* in nuclear operations went much deeper than simple recruitment: they had been present at some plants from their inception, for example, by arranging power company payoffs to local fisheries in exchange for their support in reactor siting decisions.

A survey of cleanup workers conducted by TEPCO in late 2012 found that almost half had been hired under circumstances that violated Japanese labor laws. Not all of these recruitments involved organized crime. But the murky hiring practices left workers vulnerable to abuse. Some never received written contracts. Others had their allowances – allocated for living expenses, or for dangerous work – siphoned off by their employers. One of the most common violations involved disguising a worker's actual employer: a man might be hired by one subcontractor, only to receive his instructions and pay from another.

One effect of such practices was to dilute – or simply eliminate – responsibility for worker safety. Who ensured that workers did not receive excess radiation exposures: the company with which they signed a contract (assuming a contract even existed), or the company that issued their instructions? Who kept track of the total dose accumulated by any given worker? One quarter of the surveyed workers did not receive reports of their radiation exposures from their employers.

Clearly, then, there was a huge gap between principle and practice. In principle, cleanup workers – regardless of their employer – were issued with protective clothing and dosimeters. These were checked at the end of each shift. When a temp worker reached his exposure limit, he was to be assigned to a different post. More commonly, however, temp workers simply lost their jobs when they reached the limit. No surprise, then, that some occasionally chose to leave their dosimeters in a corner in order to prolong their employment. Subcontracting companies, meanwhile, shifted the economic pressures that they experienced onto their employees. In one documented case, a supervisor ordered his team to make lead-lined cases for their radiation detectors, so that these would register lower doses.

The physical environment onsite only aggravated such problems. In the two weeks following the accidents, some 40 percent of these cleanup workers did not wear radiation monitors, because most of the 5000 devices that could have been made available were washed away by the tsunami. Two years later, many parts of the reactors remained impenetrable.

The devastation caused by the earthquake, tsunami, and hydrogen explosions continues to make the work environment unpredictable. Considerable engineering effort has gone into designing robots that can operate in highly radioactive environments. So far, however, even the best model can only be used for reconnaissance: if it falls, it needs human help to get up.

It's tempting to dismiss these working conditions as the unfortunate but inevitable fallout of an extraordinary event. The working environment at Fukushima Daiichi certainly has many unique aspects. But as the small handful of scholars and activists who have studied labor in the nuclear industry have shown, the social and technological relationships that shape that environment have a long history.[10] This history must be apprehended on several scales simultaneously, from the apparently "global" scale at which radiation exposure limits are negotiated to the local practices that enact (or violate) those limits.

A short history of a "global" number

A few months after the reactor accidents, the Japanese government brought radiation exposure limits for nuclear industry workers back down to the "normal" level of 50 milliSieverts per year. The change was in part a response to public outrage, and in part a means of claiming that the radiation situation was "under control." Even this "normal" limit, however, was controversial. Let's take a quick look at its history.

The International Commission for Radiological Protection is an international non-governmental organization that dates back to 1928, when it was founded by physicists and radiologists seeking to define limits for their own occupational exposures. After World War II, the ICRP's membership grew, its aims broadened, and it began issuing recommendations on permissible radiation doses in all manner of occupations.[11] Other institutions with interests in radiation protection included the International Labor Organization (ILO), the International Atomic Energy Agency (IAEA), and the United Nations Scientific Committee on the Effects of Atomic Radiation (UNSCEAR), which was tasked with collating and analyzing all available data on the biological and environmental effects of ionizing radiation. In principle, the division of labor had UNSCEAR collating scientific data, the ICRP articulating the "fundamental philosophy" of radiation protection in the form of quantitative and qualitative recommendations, and the IAEA and the ILO developing "codes of practice."[12] As is the case with most institutions that claim "global" legitimacy, however, all these organizations lacked enforcement powers. They produced knowledge and prescriptions that claimed universal purview, but only national authorities were empowered to translate recommendations and codes into legal limits and regulatory structures.

In the 1970s, the ICRP enunciated a principle meant to guide work practices in all nuclear facilities: ALARA, the recommendation that at all times doses be kept As Low As Reasonably Achievable. ALARA was conceived as a guideline for policymakers in the face of ongoing scientific controversy. By the 1970s, most international experts subscribed to the linear no-threshold hypothesis, which held that there was no such thing as a "safe" dose of radiation: all exposure had some harmful biological effect. But some still clung to the threshold hypothesis, which held that radiation exposures below a certain level had no discernible effect upon human organisms (and which nuclear operators found much more palatable). In light of mounting evidence, the ICRP adopted the linear no-threshold model as its working hypothesis. But it did not go so far as recommending that all nuclear operations cease.

Instead, the ICRP sought a way to resolve the fundamental tension between the linear no-threshold model and industrial use of nuclear energy. Noting that no industry operated under perfect safety conditions, the ICRP proposed ALARA as a means of keeping the hazards of nuclear work comparable "to those that are accepted in most other industrial or scientific occupations with a high standard of safety."[13] Maximum limits ensured that no worker would receive radiation exposures known to have "deterministic effects" (in other words, exposure levels which demonstrably caused immediate health problems). The limits also sought to minimize the "stochastic effects" of low-level exposure (in other words, radiation levels which sometimes, in some people, induced cancers decades after the exposures). Because radiation affects tissues differently – reproductive or blood-forming organs are more susceptible to radiation damage than muscle, for example – different organs were assigned separate exposure limits. The cumulative, whole-body dose limit was set at 50 milliSieverts.

Acknowledging that there was no safe threshold, the ICRP stressed that its recommended limits were "boundary conditions for the justification and optimization of procedures *rather than ... values that should be used for purposes of planning and design*."[14] In other words, employers shouldn't *plan* for workers to absorb maximum permissible doses. The existing limit should be the *outer* boundary, the limit of acceptability. The "as low as" part of the ALARA principle enjoined the industry to keep exposures well under the limit at all times. At the same time, the "reasonably achievable" part of ALARA offered a way to calculate the amount of money spent on radiological protection: the cost of preventing deaths from exposure should compare to that spent per life "saved" in other industries.

In 1990, a review of new research on the biological effects of low-dose exposures prompted the ICRP to lower its whole-body occupational limit from 50 milliSieverts to 20 milliSieverts. Previously, the ICRP had considered only fatal cancers and two generations of hereditary effects in its calculation. The new limit took non-fatal cancers into account.

The ICRP has never had regulatory power, however. It can only produce recommendations. Nations set their own limits. These do not necessarily match ICRP guidelines (for reasons ranging from controversy over the science to industry pressure). Thus, more than two decades after the ICRP lowered its recommended limit to 20 milliSieverts, the annual threshold for U.S. and Japanese radiation workers remains at 50 milliSieverts.

Ordinary maintenance

Specific limits aside, most national nuclear regulatory agencies officially embrace the ALARA principle. "Reasonably achievable," however, is an elastic concept, one that inherently blends cost and risk mitigation. The ICRP itself calls for balancing the minimization of average exposures against total expense, and encourages power plant operators to apply "optimization" principles to radiation protection. Subcontracting reactor maintenance offers operators one route to such optimization.

Electric utilities seek maximum technological control over their operations, to ensure that power plants run smoothly, customers get their electricity, and profits get made. But reactors are extremely complex installations. Even during normal operations, circumstances cannot be predicted and planned to the last detail. As Constance Perin and Pierre Fournier show in their ethnographies of work in U.S. and French nuclear reactors, small glitches can have big consequences. For example, an unlabeled circuit breaker can lead an exhausted technician to cause a reactor circuit to trip, costing the electric utility millions of dollars in lost production. When unexpected events occur, the culture of control that undergirds reactor design and operation can impede improvisation, undermining the human capacity to react in real time. Sometimes, this leads to the very accidents that control procedures are designed to prevent.[15] Reactor maintenance operations are designed, among other things, to minimize such unexpected events.

Ordinarily, reactors need to be shut down every twelve to twenty-four months for refueling and maintenance. These shutdown periods are known as outages. During these times, spent fuel is removed from the core and new fuel is added. Outages also offer crucial opportunities to inspect, clean, and repair valves, pipes, steam generators, electrical systems, control panels, and other reactor components.

Radiation affects inert materials as well as biological organisms. The older the reactor, the more corroded and fragile its components, and the more radioactivity they emit. Maintenance thus gets more onerous, time-consuming, costly, and dangerous over time. In the "hottest" parts of a reactor, even wearing the most effective possible protection gear, an employee can potentially absorb a significant proportion of his yearly dose (whether that be 50 or 20 milliSieverts) in a few minutes of maintenance work. In addition, reactors go offline when they are shut down for maintenance. Today, reactor outages cost well over $1 million a day. There are thus strong incentives to get through maintenance procedures quickly.

The subcontracting system that currently governs clean up at the Fukushima plants was originally conceived in the early 1970s, as a means of managing the labor requirements and radiation exposures imposed by reactor outages.[16] Utilities hire subcontractors, who divide maintenance operations into smaller units, for which they in turn hire other subcontractors, who hire others … and so on, for a total of eight levels of subcontracting. The hierarchy serves to spread radiation exposure over a large number of workers, and thereby comply with regulatory limits on the maximum exposures of individuals. Operations begin with workers at the bottom of the hierarchy, who are sent to decontaminate work sites in the "hot zone." They scrub instruments, pipes, floors, walls: basically, any place or piece of equipment that will require repair or intervention. This decontamination work gives the skilled employees who perform the actual equipment maintenance more time to do their jobs. In principle, this "management by dose"[17] ensures that no one individual receives more than the mandated annual maximum.

Since the beginning of Japan's nuclear industry, the bottom levels of the subcontracting hierarchy have been populated by unskilled, temporary workers. In a brief study published in 1986, Yuki Tanaka reported the early involvement of *yakuza* syndicates in labor recruitment: "In the worst cases, Yakuza members use[d] intimidation to get workers to go to nuclear power plants in order to make up the numbers required during regular inspections." Whether or not they had been recruited by gangs, those at the bottom of the hierarchy were all "unskilled and comparatively older workers." Tanaka described this population as follows:

> There are ex-miners who lost their jobs at coal mines because of the drastic change in the government energy policy, day laborers from Kamagasaki and Sanya, discriminated against *buraku* people (similar to untouchables), farmers away from their homes during the slack season, and local retired workers.[18]

Day laborers thus served as "radiation fodder." Subcontractors worked them to the limit of their allowable exposure, then let them go (until the next time). Employees in the upper echelons of the subcontracting hierarchy, by contrast, were generally technicians and skilled workers whose companies specialized in nuclear power plant maintenance. These were salaried employees, and over the years they built up considerable expertise in reactor maintenance. In some sense they too served as radiation fodder; subcontracting at all levels minimized the exposures of full-time utility workers.

Efforts to make Japanese subcontractors' working conditions visible began in the late 1970s, with the publication of two memoirs written by men who had worked at different levels of the system. Both described the onerous physical constraints imposed by working in highly radioactive environments. Both made clear that all too often, workers chose – or were pressured – to remove protective equipment or bypass safety procedures in order to speed up their jobs. As sociologist Paul Jobin recounts in his recent work on Japanese reactor maintenance, efforts to unionize these workers were spearheaded by photographer-activist Higuchi Ken'ichi, whose powerful photos captured the experience of those at the bottom of the hierarchy.[19] Those efforts met with no success.

Jobin reports that since 1991, Japanese Labor Standards Offices (under the authority of the Ministry of Health and Labor) have granted compensation for radiation-induced cancers or leukemias to no more than six workers, or, in posthumous cases, their families (eight others died of acute radiation after the accidents at Tokaimura in 1999, and Mihama in 2004). Some of these cases benefited from extensive publicity through court litigation and support from civil society organizations. In other cases, families insisted on keeping their names secret because they "feared opprobrium from the company or the community… it's not well viewed to be the parent of an 'irradiated' [person]."[20]

Although these compensation cases seem to offer hope, their number is tiny compared to the total number of Japanese nuclear power plant workers, which topped 80,000 in 2009 according to the figures cited by Jobin. Nearly 90 percent of all labor in Japanese nuclear power plants since the late 1980s has been subcontracted. During any *one* job, subcontracted workers have received two to three times the annual radiation dose absorbed by utility employees. That's assuming that dosages were recorded honestly. Mr. Yokota – a decontamination worker who subsequently headed a small radiation protection company, which catered to subcontractors, and then fell prey to cancer – gave Jobin a step-by-step description of how he had contributed to falsifying records. In the course of one interview, Mr. Yokota produced the "no anomaly" stamp he'd used to fake medical reports in cases where the annual occupational health visit had revealed abnormal blood results signaling the possible beginning of cancer or leukemia.

The use of temporary workers to manage reactor outages is by no means limited to Japan. In the U.S., workers who dive inside highly radioactive reactor steam generators during maintenance outages are known as "jumpers," "glow boys," or "sponges." A single intervention can expose them to one-quarter of their yearly allowable dose: after four jobs, they are "cooked out of work" for the year. Some are temporary workers; others are employees of firms that specialize in outage services.

Similar approaches guide maintenance throughout the European nuclear industry. France – which depends on nuclear power for at least three-quarters of its electricity every year – adopted the Japanese system in the late 1980s. In her vivid account of French "nuclear servitude" (a term used by the industry itself), sociologist Annie Thébaud-Mony found that subcontracted labor accounted for 80 percent of total radiation exposure in French nuclear plants. Invoking the rem (Roentgen equivalent man), an older unit of radiation exposure still used in everyday speech, French reactor maintenance workers drily refer to themselves as "rem beasts" or "rem meat." Many live a nomadic life, moving from reactor to reactor all around the country. They rarely make more than minimum wage, so although they receive (minimal) housing allowances, many prefer to maximize their revenues by living out of their vehicles. For years, their temporary status excluded them from the powerful labor unions to which most French utility workers belong.

Invisibility and ignorance

A major consequence of the subcontractor employment system is invisibility – of worker exposures, of occupational diseases, of the true collective dose generated by nuclear power plants, and of maintenance failures. Let's briefly address each of these elements in turn.

1   **Greater, and unrecorded, exposures.** As we saw for some of the Fukushima cleanup
    crew, workers sometimes see a short-term financial incentive in abandoning their dosime-
    ters for certain jobs, so that their radiation exposures are not officially recorded. This
    prolongs their employment, but it also increases their doses. While the practice is currently
    most prevalent among temporary workers, historically it has not been unique to this popu-
    lation. In the early decades of the French nuclear program, for example, abandoning one's
    dosimeter could be a sign of dedication and virility.[21]

2   **Illnesses don't become "occupational disease."** Subcontract workers are often
    dubbed nuclear gypsies (in Japan) or nomads (in France) because they move around from
    workplace to workplace, living out of trailers. This intense mobility makes it very difficult
    to maintain accurate annual or lifetime exposure data. Many severe health problems thus
    never get recorded as "occupational disease." Workers rarely benefit from compensation,
    because their diseases cannot be linked to past exposures in ways that are scientifically or
    legally persuasive. Kept out of labor unions, they do not have access to structures that
    might help them change these conditions.

3   **Collective dose.** Each year, nuclear plants report the sum total of doses absorbed by all
    their employees, a figure known as the "collective dose" of a plant. This figure is used as an
    indicator of the overall working conditions at the plant, and contributes to assessments of
    its overall safety record. But utilities don't include the exposures of subcontracted workers
    in their data.[22] That, in turn, means that data for any given nuclear power plant vastly
    under-reports the *true* collective dose (i.e., the total exposure received by the sum of both
    utility and subcontract workers).

4   **Maintenance failures.** Unlike full-time plant employees, even the most skilled contract
    workers do not have daily experience operating a reactor, or daily contact with its equip-
    ment and instrumentation. The low social status of contract workers, furthermore, makes
    it difficult for them to report irregularities they might notice. This situation has been linked
    to maintenance failures in both France and Japan.[23]

These layers of invisibility compound each other, and contribute to the invisibility of the labor
itself. This matters not simply for individual workers and their struggles, but also for "global"
scientific knowledge production. For example, an international epidemiological study on the
relation between radiation exposure and cancer conducted between 1990 and 2005, which
surveyed 400,000 nuclear industry workers in 15 countries, didn't include subcontracted maintenance workers. The majority of people working in French and Japanese nuclear power plants,
which provided two of the largest population samples, were therefore excluded from the
study.[24]

Local invisibilities thus scale up into global ones. Indeed, the absence of maintenance workers from the data feeding scientific knowledge production on radiation effects offers a textbook
case of what historian Robert Proctor calls *agnotology*: the "conscious, unconscious, and structural production of ignorance." Proctor and others have shown that ignorance about
occupational and environmental toxins is not always a mere absence of knowledge; it can be
actively produced, sometimes by the very same processes that produce knowledge. As sociologist Scott Frickel argues in this volume and elsewhere, the resulting "institutionalized
ignorance" produces structures of "non-knowledge," akin to Murphy's regimes of (im)percep-
tibility. Taken together, structures of knowledge and "non-knowledge" operate dialectically to
shape possibilities for social and political action.[25]

Uranium production and the mechanisms of invisibility

Reactors form the technopolitical core of today's nuclear industry, its economic *raison d'être*. They are fueled by uranium, whose fission produces both electricity and radioactivity. While uranium lies at the material core of the nuclear power industry, its production is widely seen as lying on the industry's technopolitical periphery. Consider uranium mining in Africa. During the Cold War, six African countries – South Africa, Namibia, Gabon, Madagascar, Niger, and Congo – together provided between 20 and 50 percent of the capitalist world's uranium. Much like the maintenance employees who keep reactors running, the African workers who toiled in these mines have been largely absent from the scientific studies (and historical narratives) of the "nuclear age."

It's tempting to explain these absences by invoking colonialism, global inequalities, and capitalist exploitation: that, at least, is how the (small number of) NGOs that have attended to African uranium have framed their explanations.[26] There's no question that patterns of inequality contribute in fundamental ways to the broad-scale invisibility of African uranium workers and their exposures. But invoking grand social forces does not actually *explain* very much, not the least because inequalities were configured differently in colonial and postcolonial Gabon, for example, than in apartheid South Africa. We must understand not just the invisibility, but also its ongoing consequences for the production of knowledge and ignorance, and for the lives and health of workers. Even a quick overview of the mechanisms of invisibility requires us to invoke multiple spatial, temporal, and analytic scales.

Rendering radiation perceptible

Consider one of the primary occupational hazards for uranium miners: radiation exposure. Uranium naturally decays into radon, a radioactive gas that in turn decays into a variety of other radioactive substances known as radon daughters. These decay processes release radioactive alpha particles, which miners inhale. The alpha particles remain lodged in their lungs, and some (but not all) miners develop lung cancer anywhere from ten to thirty years after exposure. High-grade mines can also present high levels of gamma radiation, exposure to which can lead to leukemia and other types of cancer, as well as produce genetic effects that carry forward to future generations.

At one level this etiology might seem straightforward, especially considering that lung disease among workers in uranium-rich mines has been documented since the sixteenth century, when Paracelsus and Agricola attributed lung ailments among miners in the Schneeberg district of Saxony to the inhalation of metallic vapors. After the symptoms were linked to malignant tumors in the late nineteenth century, "Schneeberg lung cancer" was listed as an occupational disease. In the 1920s, researchers began to suspect inhaled radon as the trigger. One study published in a U.S. journal in 1937 reported that lung cancer had killed some 30 percent of autopsied uranium miners. In 1940, the new Nazi administration in Karlsbad issued regulations governing radiation exposure in the uranium industry.

After World War II, however, when superpower enthusiasm for atomic bombs sparked a uranium boom, the dangers of radon exposure quickly became invisible in the U.S. A small handful of scientists in the U.S. Atomic Energy Commission (AEC) and the Public Health Service (PHS) did express concern about radon levels in the uranium mines of the Colorado Plateau. But the AEC refused to monitor radon or radiation levels in the mines, despite being the sole legal buyer of the ores they produced. AEC administrators insisted that the mines were not special, nuclear workplaces, but rather ordinary worksites that could be regulated by state-level mining bureaus. State bureaus, however, had no radiation expertise. The mines were run privately – sometimes on a shoestring, by speculators. The PHS launched a massive study of radon exposure on the mines, but operators would only let project staff take readings if they kept their true purpose hidden from the miners. This meant that workers were not given portable dosimeters; instead, PHS scientists placed radiation detectors in mine shafts and calculated average exposures from these instrument readings. These averages were then correlated with evidence from medical records and clinic visits. By 1963, the PHS study offered epidemiological evidence for excess lung cancer among miners, who still knew nothing of this occupational danger. The AEC refused to increase the price it paid for ore in order to fund radiological protection in the mines; the number of excess cancers reached into the hundreds. Nevertheless, for decades the mining industry – and the AEC – insisted that these cancers came from smoking, or from the negative synergistic effects of smoking and mining. Not until 1990, did the U.S. government officially recognize the health hazards of uranium mining on the Colorado Plateau.[27]

In France, regimes of perceptibility were constructed differently. Unlike the U.S., where uranium mines were privately run, in France most uranium mines were operated directly by the state via its Commissariat à l'Énergie Atomique (CEA). Whereas the U.S. AEC refused to monitor radiation in mines, CEA radiation protection experts monitored radiation and radon levels in French uranium mines from the 1950s onward. Workers were issued portable dosimeters, whose readings were supplemented by ambient instruments. In principle, CEA radiation protection officers could order mine superintendents to remove workers with excess exposures.[28] All told, they developed extensive infrastructures to make radiation in French mines perceptible, and to generate employment records that included individual radiation exposures.

Making *radon* perceptible, however, is not the same thing as making *illness* visible. The CEA's expertise lay in health physics and other aspects of radiation exposure and protection. Using animal experimentation, for example, CEA experts had proved to their own satisfaction that radon exposure caused cancer independently of negative synergistic effects with tobacco consumption. Assuming that animal experimentation (and related research on how different organs responded to radiation) provided definitive guidelines for setting exposure thresholds, they worked to keep worker exposures under those thresholds and produced extensive data sets on radiation levels in French mines, but for decades they failed to do any epidemiological follow-up. The question of whether exposure levels under the official threshold might have health consequences could not be answered with French data.

Americans constructed an epidemiological regime of perceptibility, while the French constructed a dosimetric one. Both produced knowledge, but both also produced ignorance. Although these experts read each others' work, debated in conferences, and deliberated together in ICRP and IAEA committees, the knowledge generated in one country did not readily travel to the other.[29] In fact, the knowledge-generation mechanisms put in place by the CEA in France even had trouble reaching its mines in Africa.

Knowledge, ignorance, and social action

In order to supply France's growing nuclear power program, the CEA and its affiliates operated uranium mines in southern Madagascar during the 1950s and 1960s, eastern Gabon from the 1950s until the 1990s, and northern Niger from the 1970s to the present day. Although radon and radiation levels were measured in some of these mines, the data were used for managerial purposes rather than for health monitoring. An examination of radiation monitoring at the uranium mine in Mounana, Gabon illustrates the significance of this distinction. It also demonstrates how the dialectical – and transnational – production of knowledge and ignorance shapes the possibilities for social action.[30]

From 1961, Mounana's first year of operation, mineworkers wore dosimetric film-badges that measured their exposure to gamma radiation. These were collected monthly, and sent to France for processing. Radon levels were measured using ambient dosimetry: instruments were scattered around the mine shafts, and monthly averages recorded for each shaft. Thus, every month managers could produce a report on overall radiation levels. When these reports showed that a worker had exceeded his yearly allowance of radiation – which, depending on the levels, could occur in just a few months – he would (in principle) be moved to another post. In 1970, a new site director – frustrated by the expense of maintaining shafts under France's maximum permissible levels – realized that the International Labor Organization used a different formula for calculating allowable thresholds than the French authorities. Applying the ILO formula to Mounana's environment led to a three-fold increase in the site's maximum permissible exposure. Overnight, the number of "over-exposed" workers dropped to zero, without their actual exposures changing.

Mounana miners were never informed about their precise readings – neither before nor after this change. Nor were radiation levels transmitted to the site's medical service. Radiation exposure, in other words, never became part of a worker's medical file. Nor did the radiation readings become part of the data sets which French radiation protection experts took so much pride in.

Such absences multiplied across spatiotemporal scales. Whether in the 1960s or the 1990s, no Gabonese (or Nigériens, or Malagasies) attended the endless meetings where the ICRP or IAEA derived limits, set standards, and developed codes of practice. In the early 1990s, an international group of epidemiologists reanalyzed data from the eleven *existing* studies of radon and lung cancer risk that covered miners in Australia, Canada, China, Czechoslovakia, France, Sweden, and the U.S. African exposures were excluded from reanalysis because they'd never existed *as data* in the first place.[31] These systemic invisibilities thus penetrated, in deep and lasting ways, the efforts to produce universally applicable prescriptions and place-less knowledge.

Nevertheless, it would be a mistake to portray Gabonese uranium workers as helpless. After the Mounana mine shut down in 1999, workers and area residents grew increasingly suspicious about their health. Inspired by reports of Aghirin'man, an NGO concerned with occupational and environmental illness in active Nigérien uranium mines, a group of Mounana residents formed the *Collectif des anciens travailleurs miniers de Comuf* (CATRAM) in 2005. CATRAM demanded a health and environmental monitoring program along with a fund to disburse medical compensation claims from Areva, the French nuclear fuel cycle corporation that inherited responsibility for the mine's legacy in 2001. One member noted that Gabonese workers "did not, during their entire careers at the Mounana uranium mine, benefit from the attentive medical surveillance reserved for their expatriate colleagues. During their leaves in France, the latter systematically underwent hematology examinations and cancer screening."[32] Sick Mounana residents, by contrast, frequently did not know whether they had cancer or a different disease.

As the Gabonese soon learned, screening had not immunized French expatriates against disease. The same year CATRAM formed in Gabon, a former French resident of Mounana formed her own advocacy group to seek compensation for expatriate workers who had since contracted cancer. She, in turn, enlisted three other French NGOs – with legal, radiation, and medical expertise respectively – to investigate the health and environmental legacy of uranium mining in Africa. French, Gabonese, and Nigérien organizations joined forces, and sent teams to Gabon and Niger. The Mounana team took independent environmental readings and surveyed nearly 500 former COMUF employees about their health and work experience.[33]

Survey responses echoed narratives I heard in the course of my own historical research. The vast majority of workers reported no formal training on radiation or radon-related risks. At best, they learned about risks by word of mouth from other workers. Employees were not required to wear protective gear, and all work clothing was washed at home. As one former employee reported, "we were so unaware of the risks that we smoked and ate at the workplace, and since we never wore protective gloves, we ate and inhaled whatever was on our hands and in the air, [including] after maintenance operation[s] that left yellowcake powder suspended in the air."[34] Employees did not receive reports of their radiation exposures. Everyone agreed that the Gabonese state had done nothing to monitor working conditions or occupational health. One former medical doctor testified that company clinicians had no training in uranium-related occupational health, and that the company's radiation protection division consistently refused to provide dosimetric readings to the medical division.

Absent the ability to conduct full medical examinations, the NGO team's assessment of health outcomes couldn't be conclusive. One clear pattern emerged, however. Half of the surveyed Gabonese workers reported pulmonary distress, which also appeared (at least anecdotally) to affect their families disproportionately. Workers did testify to satisfactory medical attention during employment – even though, the report added parenthetically, "the doctor didn't have nuclear expertise and the nature of the examinations wouldn't have detected internal contamination." After the site closed, Gabonese "felt abandoned" at the very moment that their health problems had become more severe.[35] Back in France, eleven of the seventeen expatriates surveyed suffered from cancer.

In Paris in April 2007, the NGOs released their reports on Mounana and on Areva's mining operations in Niger. Areva responded by promising to install "health observatories" in both places.[36] "Observatories" were a far cry from remediation and compensation. Discussions between Areva and the NGOs dragged on. A pair of documentaries shown on French television in 2009 raised the stakes. The first documented radioactive contamination produced by uranium mining in France itself; the second presented a visual accounting of the problems that NGOs had identified in Gabon and Niger.

In June 2009, the medical and legal French NGOs announced that they'd reached an "unprecedented" agreement with Areva to form a "pluralist group" of experts to oversee the health observatories set up by Areva in Gabon, Niger, and elsewhere. The committee's tasks included defining protocols for data collection, analyzing the results obtained by all the health observatories, and making proposals to improve occupational and environmental health at the sites. Expressing skepticism about the prospect of prompt, effective remediation, however, the radiation-focused NGO declined to participate in a process that could legitimate Areva as a responsible corporate actor without producing real change. While the agreement laid out procedures for ensuring balance between Areva and the NGOs in selecting committee members and charting goals, no one sought to include local representatives on the pluralist committee or the health observatories, oversights particularly striking given how little Gabonese (or Nigérien) mineworkers trusted their governments.

In the two years following these arrangements, dozens of former mineworkers have presented themselves to the health observatory at Mounana. The doctor in charge of examining them, however, has declared that she couldn't reach any conclusions because she didn't have national databases – most notably, cancer registries – against which to compare the health of former workers. The African health observatories have yet to produce a single instance of compensation. (In France, only two former expatriate employees have obtained compensation, in both cases via court litigation.) In late 2012, the two French NGOs pulled out of the arrangements, declaring them a charade.

In the forty years of Mounana's operation, radiation exposure data traveled no further than managerial reports. This absence produced ignorance across time and space: exposure records did not go to the workers, or to their doctors, or to scientists producing knowledge of the relation between radon exposure and lung cancer, or to the international organizations producing radiation exposure standards. Workers understood that ignorance was being actively produced. Not until the mine shut down, however, could they find the transnational allies required to mount a competing regime of perceptibility. Resources and timing, however, meant that this regime was more successful at identifying ignorance than at producing actionable knowledge.

Conclusion: scales of invisibility

Reactors are the most visible part of the nuclear power system, never more so than when they explode or meltdown. Catastrophes such as the one at Fukushima also reveal dimensions of the system that often remain invisible. Exploring one invisible dimension can lead to others. Thus this chapter has moved from cleanup at Fukushima to reactor maintenance more generally, and from vulnerable reactor cores to the often-invisible production of their constituent element, uranium. These geographical and temporal shifts have led us through shifting analytic scales: from workplace and monitoring practices, to the production of scientific papers and international regulations, and back again to workplaces.

STS scholarship that attends to transnational dynamics often does so in order to examine how technoscience travels. To be sure, we've seen some of that in this chapter. But the shifts we've followed also show us that knowledge, technologies, and practices don't *necessarily* travel; they don't even necessarily follow the experts and institutions that produce them. The result is not just invisibility, but the active production of ignorance, via the absence of hundreds of thousands of maintenance workers and uranium miners from investigations on the relationship between radiation exposure and illness.

Such patterns are by no means unique to the nuclear industry, as scholars who explore occupational and environmental hazards in other domains have demonstrated. Furthermore – as their work also shows, and as we've seen here – combating workplace hazard is not merely a matter of producing more knowledge, of filling the ignorance gap. Making hazards visible is only a first (and often contested) step: even if stakeholders achieve a rough working consensus that a substance is hazardous, controversies continue over the degree of toxicity, the establishment of working limits, the cost of adhering to limits, the means of enforcement, and the standards of proof required for compensating ill workers. These controversies, in turn, are also shaped by wider fields of power, which can include not just entities typically included in STS analyses (such as international organizations and multinational corporations), but also those that aren't (such as organized crime and corrupt governments). Knowledge, ignorance, and the dynamics between them play out in all fields, and at all these scales.

Notes

1 Portions of this essay previously appeared in Gabrielle Hecht, "Nuclear Janitors: Contract Workers at the Fukushima Reactors and Beyond," *The Asia-Pacific Journal*, Vol 11, Issue 1, No. 2, January 14, 2012; other portions draw upon Gabrielle Hecht, *Being Nuclear: Africans and the Global Uranium Trade* (MIT Press and Wits University Press, 2012). Used with permission. Full source citations are in the originals.
2 The phrase is from Edwards 2000. Disasters have given rise to scholarship in many different disciplines. For an overview of the origins of disaster expertise in the U.S., see Knowles 2011. A foundational text

on the ordinary problems of complex technological systems is Perrow 1999; two classic case studies are Vaughan 1997 and Fortun 2001.

3   Figures for Chernobyl are endlessly controversial: the number of liquidators, their exposures, and the number of deaths and illnesses resulting from the accident have been subject to debate for years. For STS treatments of the Chernobyl accident, see Petryna 2002, Kuchinskaya 2011, and Schmid 2014.

4   The National Diet of Japan, *The Official Report of Fukushima Nuclear Accident Independent Investigation Commission* (Executive Summary), 2012.

5   Frickel 2004, Markowitz and Rosner 2002, Mitman, Murphy, and Sellers 2004, Proctor 2012, Rosner and Markowitz 1991, Sellers 2004, Sellers and Melling 2011.

6   Murphy 2006.

7   Quoted in Cordula Meyer, "Fukushima Workers Risk Radiation to Feed Families," Spiegel Online International, Sept. 21, 2011.

8   Suzuki Tomohiko, *The Yakuza and Nuclear Power: An Undercover Report from Fukushima Daiichi* (Tokyo: Bungeishunju Ltd. 2011), translation excerpt at *www.booksfromjapan.jp/publications/item/1176-the-yakuza-and-nuclear-power-an-undercover-report-from-fukushima-daiichi*, accessed 13 December 2011.

9   Ibid.

10   Caufield 1990, Fournier 2012, Hecht 2009, 2012, Jobin 2012, Perin 2005, Petryna 2002, Thébaud-Mony 2011.

11   For an insider history of radiological standards in the U.S., see Walker 2000.

12   Boudia 2007, 2008.

13   "ICRP Publication 22. Implications of Commission Recommendations that Doses be kept as Low as Readily Achievable. A report of ICRP Committee 4." Pergamon Press for the ICRP, 1973), 3.

14   B. Lindell, "Basic Concepts and Assumptions behind the new ICRP recommendations," in IAEA, *Application of the Dose Limitation System for Radiation Protection: Practical Implications*. Proceedings of a Topical Seminar on the Practical Implications of the ICRP Recommendations (1977) and the Revised IAEA Basic Standards for Radiation Protection, Vienna, 5–9 March 1979, 3. Emphasis mine.

15   Perin 2005, Fournier 2012.

16   Jobin 2011.

17   The term comes from Thébaud-Mony, 2011.

18   Tanaka 1997.

19   Higuchi and the workers he followed are featured in the 1995 BBC film "Nuclear Ginza," available for viewing at *www.youtube.com/watch?v=W_qb0uAc1dg*, accessed 13 December 2012.

20   Jobin 2012, p. 100.

21   Fournier 2012, Hecht 2009.

22   For France, see Thébaud-Mony 2011, for Japan, see Tanaka 1997.

23   Including for Fukushima: The National Diet of Japan, *The Official Report of Fukushima Nuclear Accident Independent Investigation Commission* (Executive Summary), 2012.

24   Cardis, E. et al., "Risk of cancer after low doses of ionising radiation: Retrospective cohort study in 15 countries," *British Medical Journal,* 331 (7508, July 9, 2005): 77, cited in Thébaud-Mony, op. cit., xxii–xxv.

25   Frickel 2008.

26   For example, Greenpeace International 2010. *Left in the Dust: Areva's Radioactive Legacy in the Desert Towns of Niger.*

27   For details, see Hecht 2012, chapter 6.

28   Robert Avril et al., "Measures adopted in French uranium mines to ensure protection of personnel against the hazards of radioactivity," *Proceedings of the Second United Nations International Conference on the Peaceful Uses of Atomic Energy, Held in Geneva, 1–13 September 1958.* Vol. 21: *Health and Safety: Dosimetry and Standards* (Geneva: United Nations, 1985): 63.

29   For details, see Hecht 2012, chapter 6.

30   This section is taken from Hecht 2012, chapter 7. See the original for complete archival citations.

31   Jay H. Lubin et. al., "Radon and Lung Cancer Risk: A Joint Analysis of 11 Underground Miners Studies," NIH report 94-3644 (National Institutes of Health, 1994). I use "African" deliberately here – this study also excluded South Africa, Namibia, and Congo.

32   Jules Mbombe Samaki, "Memorandum sur la nécéssité de la prise en compte de la Veille sanitaire et du dédommagement des anciens travailleurs miniers," Libreville, 25 avril 2005. Private, anonymous communication.

33   Samira Daoud and Jean-Pierre Getti, "Areva au Gabon: Rapport d'enquête sur la situation des

travailleurs de la COMUF, filiale gabonaise du groupe Areva-Cogéma." Sherpa, 4 avril 2007. *www.asso-sherpa.org.*

34 Samira Daoud and Jean-Pierre Getti, "Areva au Gabon: Rapport d'enquête sur la situation des travailleurs de la COMUF, filiale gabonaise du groupe Areva-Cogéma." Sherpa, 4 avril 2007, p. 7.

35 Ibid., 24.

36 "L'observatoire de Mounana," *L'union,* 1 juin 2007.

References

Boudia, S. 2007. "Global Regulation: Controlling and Accepting Radioactivity Risks," *History and Technology,* 23:4 389–406.

Boudia, S. 2008. "Sur les dynamiques de constitution des systèmes d'expertise scientifique. La naissance du système d'évaluation et de régulation des risques des rayonnementsionisants," *Genèses,* Vol. 70: 26–44.

Caufield, C. 1990. *Multiple Exposures: Chronicles of the Radiation Age.* Chicago: University of Chicago Press.

Edwards, P.N. 2000. "The World in a Machine: Origins and Impacts of Early Computerized Global Systems Models" in *Systems, Experts, and Computers: The Systems Approach in Management and Engineering, World War II and After,* eds. T.P. Hughes and A.C. Hughes. Cambridge, MA: MIT Press, 221–254.

Fortun, K. 2001. *Advocacy after Bhopal: Environmentalism, Disaster, New Global Orders.* Chicago: University of Chicago Press.

Fournier, P. 2012. *Travailler dans le nucléaire: enquête au coeur d'un site à risques.* Paris: Armand Colin.

Frickel, S. 2004. *Chemical Consequences: Environmental Mutagens, Scientist Activism, and the Rise of Genetic Toxicology.* Rutgers: Rutgers University Press.

Frickel, S. 2008. "On Missing New Orleans: Lost Knowledge and Knowledge Gaps in an Urban Hazardscape," *Environmental History* 13 (October): 643–650.

Hecht, G. 2009 *The Radiance of France: Nuclear Power and National Identity after World War II.* Cambridge, MA: MIT Press, new edition.

Hecht, G. 2012. *Being Nuclear: Africans and the Global Uranium Trade.* Cambridge, MA and Johannesburg, South Africa: MIT Press and Wits University Press.

Jobin, P. 2011. "Dying for TEPCO? Fukushima's Nuclear Contract Workers," *The Asia-Pacific Journal,* Vol. 9, Issue 18 No 3, May 2, 2011.

Jobin, P. 2012. "Fukushima ou la radioprotection, retour sur un terrain interrompu," in *Santé au travail: approches critiques,* edited by A. Thébaud-Mony, Jobin, P., Daubas-Letourneux, V., Frigul, N., 83–104. Paris: La Découverte.

Knowles, S.G. 2011. *The Disaster Experts: Mastering Risk in Modern America.* Philadelphia: University of Pennsylvania Press.

Kuchinskaya, O. 2011. "Articulating the signs of danger: Lay experiences of post-Chernobyl radiation risks and effects," *Public Understanding of Science* May Vol. 20, No. 3: 405–421.

Markowitz, G. and Rosner, D. 2002. *Deceit and Denial: The Deadly Politics of Industrial Pollution.* Berkeley: University of California Press.

Mitman, G., Murphy, M., and Sellers, C., eds. 2004. *Landscapes of Exposure: Knowledge and Illness in Modern Environments. Osiris,* Vol. 19.

Murphy, M. 2006. *Sick Building Syndrome and the Problem of Uncertainty: Environmental Politics, Technoscience, and Women Workers.* Durham, NC: Duke University Press.

Perin, C. 2005. *Shouldering Risks: The Culture of Control in the Nuclear Power Industry.* Princeton: Princeton University Press.

Perrow, C. 1999. *Normal Accidents: Living with High-Risk Technologies.* Princeton: Princeton University Press, updated edition.

Petryna, A. 2002. *Life Exposed: Biological Citizens after Chernobyl.* Princeton: Princeton University Press.

Proctor, R. 2012. *Golden Holocaust: Origins of the Cigarette Catastrophe and the Case for Abolition.* Berkeley: University of California Press.

Rosner, D. and Markowitz, G. 1991. *Deadly Dust: Silicosis and the Politics of Occupational Disease in Twentieth-Century America.* Princeton: Princeton University Press.

Schmid, S. 2014. *Producing Power: The Pre-Chernobyl Origins of the Soviet Nuclear Industry.* Cambridge, MA: MIT Press.

Sellers, C. 2004. *Hazards of the Job: From Industrial Disease to Environmental Health Science.* Chapel Hill: University of North Carolina Press.

Sellers, C. and Melling, J. eds., 2011. *Dangerous Trade: Histories of Industrial Hazard across a Globalizing World*. Philadelphia, PA: Temple University Press.

Tanaka, Y. 1997. "Nuclear Power Plant Gypsies in High-Tech Society," *The Other Japan: Conflict, Compromise and Resistance since 1945*, edited by Joe Moore, 251–271. M.E.Sharpe. Amityville, NY.

Thébaud-Mony, A. 2011. *Nuclear Servitude: Subcontracting and Health in the French Civil Nuclear Industry*. Armonk, NY: Baywood Publishing Company.

Vaughan, D. 1997. The Challenger Launch Decision: Risky Technology, Culture, and Deviance at NASA. Chicago: University of Chicago Press.

Walker, S.J. 2000. *Permissible Dose: A History of Radiation Protection in the Twentieth Century*. Berkeley: University of California Press.

21

Social Scientists and Humanists in the Health Research Field

A clash of epistemic habitus[1,2]

Mathieu Albert and Elise Paradis

UNIVERSITY OF TORONTO AND UNIVERSITY OF CALIFORNIA SAN FRANCISCO

Introduction

Over the course of the late twentieth century, academic science has evolved from a logic of science for its own sake where the search for truth had intrinsic value (Friedland and Alford 1991) to a logic where science has become increasingly evaluated on the basis of its economic value and societal usefulness (Gumport 2000, Popp Berman 2012a, 2012b, Slaughter and Leslie 1997). The view of science that underpins this vision is one where scientists tackle so-called "real-world problems" and find solutions that benefit society in varied ways, what Gibbons and his colleagues call "Mode 2" knowledge production (1994) and Ramirez (2006, 2010) has characterized as the shift toward the socially useful university: one that behaves like a rational actor and shows its importance for the broader society.

Canada has followed these global trends. Since World War II, Canada's science policy agenda has emphasized collaborative research, in the hopes of increasing the utility and use of academic knowledge. A cluster of new policy and funding initiatives based on a market logic were implemented to facilitate interdisciplinary research, accelerate collaboration and commercialize academic research (Albert and McGuire 2014, Cameron 2004, Fisher et al. 2001, 2005, Polster 2002, Snowdon 2005).

The value and contribution of interdisciplinarity to the knowledge production exercise is now taken for granted, and only rarely contested (among the few critical perspectives on interdisciplinarity see Cooper 2012, Moore 2011, Hoffman 2011, Jacobs and Frickel 2009, Laberge et al. 2009, Weingart 2000). Proponents believe it is a means to maximize innovation and economic growth and see it, in its ideal or idealized form, as a proven way to generate "better" research and better solutions[3] (for example, see Committee on Facilitating Interdisciplinary Research 2004, Frodeman et al. 2010, Hadorn et al. 2008, Hall et al. 2012). Better research is thought to arise from interdisciplinarity when a plurality of approaches are brought to the study of a "problem" by a diverse set of researchers brought together in research teams, centres, departments or faculties. One of the assumptions made by this model is that researchers from all disciplines will *equally* contribute to the research design and participate in the study of the problem, but our research suggests that several structural barriers limit social scientists' and humanists' ability to be full contributors in the health research field. Indeed, these barriers make it impossible for their excellence to be recognized, and their epistemes to enrich and transform health research.

In the health research field, interdisciplinarity is also increasingly valued and can be seen in the transformation of both research funding and at the level of medical school faculty. Many funding agencies have developed programs specifically to intensify interdisciplinary research, and some were created for this specific purpose. For example, in 2000, the Canadian Institutes of Health Research (CIHR) was established with an express mandate to forge a health research agenda across disciplines (Government of Canada 2000). Seven years later, the U.S. National Institutes of Health created nine interdisciplinary research consortia "as a means of integrating aspects of different disciplines to address health challenges that have been resistant to traditional research approaches" (NIH 2007). Within medical schools in the United States and Canada, the number of faculty with PhDs has grown impressively (from 21,932 in 1997 to 30,363 in 2008, a 38% growth in 11 years), showing a growing commitment to research and a diversification of their staff. Within clinical departments in particular, the number of PhDs grew by 50%, from 11,479 to 17,182 during the same period (AAMC 1998, 2009).

In this chapter, we would like to trouble what we believe is the embellished story of interdisciplinarity. It is not fully demonstrated, we argue, that interdisciplinary research finds holistic solutions to "real," complex "problems" through the equal contributions of scholars across a range of disciplines. We focus on the untold story of social scientists and humanists who work in medical schools to show how interdisciplinarity has mostly resulted in these scholars' adaptation to the rules of the health research field dominated by the biomedical sciences, rather than in a transformation of the health research field to be inclusive of their different epistemic habitus. We use neo-institutional theory and Pierre Bourdieu's social theory to show how the discourse of interdisciplinarity is decoupled from – i.e., does not fit – its actual practice, and how the interdisciplinary health research field creates new power hierarchies or reproduces old ones among scientific disciplines.

Using a broad range of data (institutional, financial and interview data), we question the feasibility and expected outcomes of interdisciplinarity by showing how the different disciplines that discursively constitute the interdisciplinary health research field (biomedical sciences, health services, epidemiology/public health, the social sciences and humanities) actually hold different levels of legitimacy and thus different scientific authority to define and lead research agendas. In the Canadian context, this disparity manifests itself through a broad range of symbolic and organizational acts of domination. We organize these data using the different types of decoupling outlined by Bromley and Powell (2012), making visible the gaps among the policy, practices and purported outcomes of interdisciplinarity. First, we focus on the underrepresentation of social scientists and humanists on various decisional and advisory committees in the country's largest interdisciplinary health research funding agency. Second, we explore the *financial decoupling* faced by social scientists and humanists, which limits their ability to fully participate in the health research enterprise, disseminate their work and network with other health research colleagues. Third, we discuss the everyday professional experiences of social scientists and humanists working in medicine to show how the practice of interdisciplinarity sometimes gets in the way of doing better science.

Two major aspects of our study distinguish it from previous research on interdisciplinarity. First, many studies have favored ethnographic and phenomenological approaches without consideration of the structural aspects of research environments and the power relations that result. Several of these studies have as an objective the improvement of the interdisciplinary collaborative process through the identification of the elements enabling or limiting collaboration (see, for example,

Dewulf et al. 2007, Jeffrey 2003, Lau and Pasquini 2004, Lélé and Norgaard 2005, Maasen 2000). In contrast, our study is rooted in the principle that research environments are inherently structured, such that the question of power must necessarily be taken into consideration if we are to understand relationships among disciplines, and the organization of scientific work. Our study thus moves away from interactionist approaches that focus solely on visible interactions among actors (as though they were living in a cultural and structural vacuum) and is inspired by Pierre Bourdieu's (1987) "constructivist structuralist" approach. According to this approach, actors are embedded in a social universe where social and symbolic structures, which predate their own entry into this universe, influence actions and social relations. Actors contribute to the reproduction of these structures or transform them based on their own practices and on the power they hold within this structure.

A second distinctive aspect of this study is the environment it is concerned with: faculties of medicine. Several studies of interdisciplinarity have focused on emerging or temporary interdisciplinary teams, for example the creation and functioning of new teams or interdisciplinary research centers (for example, Jeffrey 2003, Stokols et al. 2003). Because of their recent development, these environments are typically only partly institutionalized. The power relations among disciplines, while present (MacMynowski 2007, Williams et al. 2002), are not fully cemented into an established social order. In contrast, faculties of medicine are highly institutionalized and hierarchical organizations where various structural mechanisms (standardized evaluation criteria, policies governing the supervision of graduate students based on funding held by scientists, temporary and non-tenured faculty positions, etc.) maintain and reproduce the social order. Social science and humanities researchers who join a faculty of medicine thus enter a material and symbolic space that was stratified prior to their entry. Consequently, the challenges they face and their work experiences are likely to be different from those they would face in a context where norms are still emerging.

Theoretical framework

To make sense of this situation – the gap between the discourse on interdisciplinarity and its actual practice, and the perpetuation of the power hierarchy it creates in the academic context – we turn to neo-institutional theory, Bourdieu's concepts of doxa and to our own Bourdieu-inspired concept of epistemic habitus.

Neo-institutionalist scholars study, among other things, the way new cultural ideas, such as human rights, education for all, science for development, emerge, diffuse, and become taken for granted (Emirbayer and Johnson 2008, Finnemore 1993, Powell and DiMaggio 1991, Meyer et al. 1997). Interdisciplinarity is one such cultural idea: it emerged after World War II and was crystallized in a 1972 OECD report titled *Interdisciplinarity: Problems of Teaching and Research in Universities* (Klein 1990). Interdisciplinarity was adopted more broadly as a research ideal by science policy-makers during the following decades, and integrated with Innovation System policies (see Albert and Laberge 2007, Shariff 2006). While academics have built distinctions between *inter-*, *multi-*, and *trans-*disciplinary research (for example, Rosenfield 1992), interdisciplinarity is the most frequently used term in health research (Paradis and Reeves 2012), and is often used as an umbrella term that includes the other subtypes of research by both academics and lay people. Thus, we use interdisciplinary here to denote the broader, more inclusive version of cross-disciplinary interaction.

In their classic paper, Meyer and Rowan (1977) offer an explanation for the gap between formal policy and actual practices. They theorize modern organizations as the "dramatic enactments of the rationalized myths pervading modern societies, rather than as units involved in exchange – no matter how complex – with their environments" (1977: 346). Societal beliefs are seen to have a concrete impact on both individual and organizational practices. For organizations such as universities, environmental or institutional pressures come from codified law, from soft laws such as standards, ratings, rankings, rights-based claims, general social or professional norms, and from the need to secure legitimacy (Bromley and Powell 2012, Ramirez 2006, 2010). Decoupling or loose coupling can occur when institutional discourses and demands suggest a course of action that is not closely aligned with organizational goals and everyday practices (Bromley and Powell 2012).

Bromley and Powell (2012) distinguish between two different "types" of decoupling: policy/practice decoupling and means/ends decoupling (see Figure 21.1, adapted from Bromley and Powell). Policy/practice decoupling explains the gap between policies and their enactment. It can highlight whether and why policies are unimplemented or routinely violated, and the lacking fit between policies and their implementation. Means/ends decoupling explains the gap between practices and their purported outcomes. It can highlight how changing practices does not yield the expected outcomes, and how ultimately new policies may fail to achieve their goals. Means/ends decoupling happens when formal structures have real organizational consequences, work activities are altered, and policies are implemented and evaluated, but scant evidence exists to show that these activities are linked to organizational effectiveness or outcomes (Bromley and Powell 2012: 14).

In the case of interdisciplinarity, we see "policy" as the governmental initiatives to promote interdisciplinarity, as well as the content of these policies (i.e., discourse) that frame interdisciplinarity as a privileged way of finding "real-world" solutions to "real-world" problems; "practices" as the daily enactment of interdisciplinarity (collaborative research/problem-solving, multi-disciplinary evaluation of research activities, etc.); and putative "outcomes" as increased knowledge production and "better," more innovative and economically-generative research (for example, increased number of patents, science-inspired technology, economic growth).

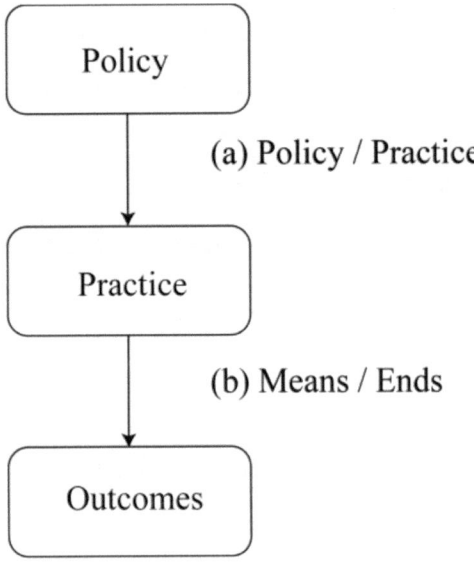

Figure 21.1 Two types of decoupling
Source: Adapted from Bromley and Powell (2012).

While the concept of decoupling is useful in highlighting the discrepancy between the call for interdisciplinarity from policymakers and its actual practice, it does not help us understand how this gap may be generated and how it may unintentionally serve as a means for reproducing the hierarchy among disciplines. Pierre Bourdieu's concept of doxa (1987) and our concept of epistemic habitus, inspired by Bourdieu's concept of *disciplinary* habitus (2004), may help understanding these processes. Doxa is for Bourdieu "a set of fundamental beliefs which does not even need to be asserted in the form of an explicit, self-conscious dogma" (Bourdieu 2000: 16). Doxa thus refers to the taken-for-granted assumptions or shared beliefs within a field. In *Raisons Pratiques* (1994), Bourdieu adds a political dimension to this definition of doxa by specifying that it is the particular worldview of the dominant group that is imposed onto all members of the group and perceived as universal. In this sense, doxa could be seen as an arbitrary viewpoint made natural and according to which social actors model their actions.

Building on Bourdieu's concept of *disciplinary* habitus, the concept of *epistemic* habitus (Albert et al. 2014) emphasizes that scientific practices are the result of a socialization process. The concept of *epistemic* habitus retains from Bourdieu's original concept the notion that scientists internalise a system of schemes of perceptions, judgments and practices through their academic training and professional experience. This internalization provides scientists with a *practical sense* that orients their actions in accordance with the field's doxa (Bourdieu 2004). Unlike Bourdieu, we focus on the epistemic schemes of thought acquired by scientists through their academic education, research activities, and career paths, regardless of their current disciplinary and/or departmental affiliations. Epistemic habitus varies within the interdisciplinary health research field and, we argue, contributes to the decoupling of goals and outcomes and of means and ends.

Methods

We used two methods of data collection: document and policy analysis and semi-structured interviews. The purpose of the document and policy analysis was to explore whether policy/practice decoupling occurs in the health research field. We used a wide array of documents produced by the Canadian Institutes of Health Research (CIHR), including annual reports (2000–2011), international review panel reports (2006, 2011), responses from the CIHR to the international review panel reports (2006, 2011), discussion documents on funding (2012) and other documents posted on the funding agency website. To explore the financial impediments faced by social scientists and humanities scholars working in faculties of medicine, we collected data about average annual granting amounts at the Social Sciences and Humanities Research Council of Canada (SSHRCC) and at CIHR, the costs related to various academic activities such as conferences registration, publication in clinical journals, and supervision of graduate medical students. We compared these costs with those incurred to engage in social scientific and humanities-based academic activities such as conferences.

To investigate means/ends decoupling, we conducted semi-structured interviews with twenty social scientists and humanists working in nine faculties of medicine across Canada. Inclusion criteria were: holding a doctoral degree from a social science or a humanities department (for example, anthropology, sociology, human geography, education, history) and having held a primary appointment in a faculty of medicine for at least two years. To increase the likelihood that respondents' epistemic habitus was characteristic of "traditional" social scientists' and humanities scholars' schemes of thought and scholarly practices, we excluded faculty members who had received their training in departments or programs such as nursing, epidemiology, statistics, and health promotion. Faculty members who had MD degrees with additional training in qualitative research or epidemiology were also excluded. This strategy was developed to target individuals who had internalized the logic of one of the social science (sociology, anthropology, etc.) or humanities (history, philosophy, literature, etc.) fields through their academic training and acquisition of epistemic habitus, rather than that of medical fields such as epidemiology or nursing. In more practical terms, we targeted participants whose epistemic habitus values scholarly practices such as theory-based research, critical social science and the publishing of books, book chapters and extensive peer-reviewed articles (8000 words or more, which is standard in the social sciences and the humanities). Resume and publications of all potential participants were examined before their inclusion in our sample. In-person and phone-based interviews lasted between 60 and 90 minutes and were audio-recorded with the participants' consent. Follow-up interviews were conducted when further clarification was needed. The interview data were analyzed using thematic content analysis.

Results

The next section illustrates decoupling between discourses and practice in CIHR leadership committees by illuminating the representation imbalance between social scientists and humanists and biomedical scientists.

Decoupling at the Canadian Institutes for Health Research

In 2000, the Government of Canada decided to replace the Medical Research Council of Canada with the Canadian Institutes for Health Research (Government of Canada 2000) to promote interdisciplinary research on a wide range of determinants of health rather than research restricted to a more traditional biological focus. The Canadian Institutes of Health Research Act stated the goal of the agency as follows:

> The objective of the CIHR is to excel, according to internationally accepted standards of scientific excellence, in the creation of new knowledge and its translation into improved health for Canadians, more effective health services and products and a strengthened Canadian health care system, by (...) encouraging *interdisciplinary*, integrative health research (...) that include bio-medical research, clinical research, research respecting health systems, health services, the health of populations, *societal and cultural dimensions of health* and environmental influences on health, and other research as required.
>
> *CIHR Act 2000: 3–4; our emphasis*

While the CIHR Act does not explicitly mention "social sciences" and "humanities," it has been widely, if not unanimously, understood by social scientists and humanities scholars (and by the broader Canadian scientific community) that the inclusion of the "societal and cultural dimensions of health" in CHIR's sphere meant the inclusion of traditional disciplinary-trained social scientists and humanities scholars into the realm of health research (see Graham et al. 2011, Plamondon 2002). This is also the Social Science and Humanities Research Council's (SSHRCC's) interpretation of CIHR's mandate, set in SSHRCC's 2009 decision to stop funding health-related social science and humanities research projects and redirect them to CIHR. This decision has had structural consequences, which we discuss later.

With a budget close to one billion dollars for 2012–2013, CIHR is the largest funding agency for health research in Canada. Therefore, decisions made by leadership committees regarding issues such as research priorities, strategic development and budget allocation have a significant impact. In light of the interdisciplinary mandate given to CIHR by the federal government, one would assume governance committees would include representation from all disciplines and research sectors, including members of the social sciences and humanities community. To assess academic representation, two of CIHR's executive committees, the Governing Council and the Science Council, and two International Review Panels instigated by CIHR were evaluated. As Figure 21.2 shows, the leadership space is almost entirely populated by scholars with biomedical backgrounds.

According to the CIHR Act (2000, updated version, March 18, 2013), CIHR's Governing Council consists of no more than eighteen Canadians who are appointed by the Governor in Council to renewable three-year terms (CIHR Act 2000). The Governing Council is an advisory body to the President of the CIHR. Its role is to oversee the direction and management of CIHR, to develop strategic directions, goals and policies; evaluate the agency's overall performance; and approve the budget (CIHR 2012). As a formal matter, "Council members [are supposed to] represent a wide range of backgrounds and disciplines, reflecting CIHR's broad mandate and vision" (CIHR 2011a: 72). From 2002 to September 2012, this "wide range" of members included: 31 biomedical scientists, 14 individuals with no academic appointments[4], 2 epidemiologists/population health scientists, 1 health services researcher and 2 social sciences or humanities scholars (see Figure 21.2). In total, since the creation of the CIHR in 2000, 50 individuals have served on the Governing Council; only 2 of those were from the social science and humanities community.

The second executive committee we examined is the Science Council. This committee consists of the CIHR President, the Chief Scientific Officer, the Vice-President Knowledge Translation & Public Outreach, the Executive Vice-President, the Director of Ethics, the Chief Financial Officer and thirteen Scientific Directors of the CIHR Institutes.[5] The mandate of the Science Council is to provide "scientific leadership and advice to Governing Council on health research and knowledge translation priorities and strategies," and to recommend "investment strategies in accordance with CIHR's 5-year Strategic Plan" (CIHR 2011b).

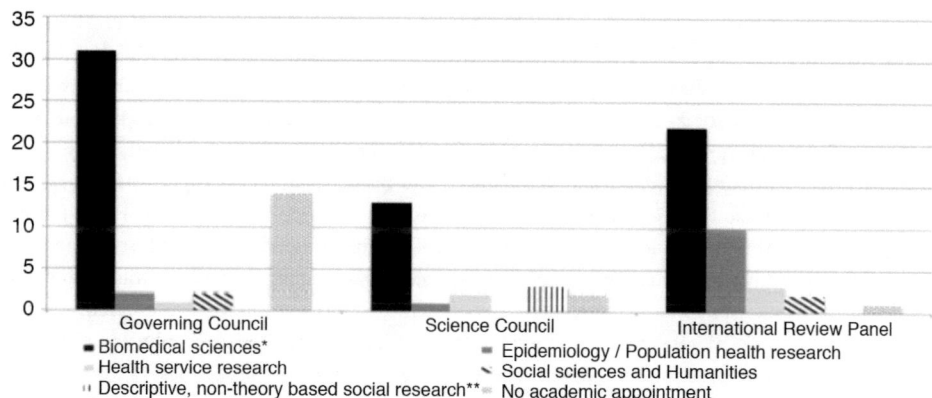

Figure 21.2 Number of members with specific background, by committee

Notes: * Includes basic and clinical research.
    ** Includes scholars using qualitative or quantitative methodological tools such as interviews, focus groups, observation, survey, without engaging in any scientific or conceptual social analysis. Typically, this type of research is conducted by researchers who have no formal training in any of the social science disciplines.

It also approves funding decisions for "all funding opportunities and initiatives," recommends "effective and efficient allocation of research funds" to the Governing Council and oversees "policies, programs and processes which enable delivery of CIHR's strategic plan." In November 2011, members of the Science Council included: 13 biomedical scientists, 2 individuals with no academic appointments[6], 2 health services researchers, 1 epidemiologist/population health scientist, 3 researchers involved with descriptive, non-theory based social research (excluded as per our definition in the Methods section earlier; also see Figure 21.2 for inclusion criteria), and zero social sciences or humanities scholars (see Figure 21.2).

Finally, we examined two external committees mandated by the CIHR's Governing Council in 2006 and 2011 to undertake a review of CIHR. The specific role of these International Review Committees was to evaluate the internal structures and performance of CIHR (CIHR June 2006), to assess CIHR's effectiveness in fulfilling its mandate as outlined in the CIHR Act, and to suggest how CIHR might more effectively achieve its mandate (CIHR June 2011). Again, scientific representation on those two committees was not evenly distributed across disciplines: altogether members of the two review panels included 22 biomedical scientists, 10 epidemiologist/population health scientists, 3 health services researchers, 2 social sciences or humanists and 1 committee member with no academic appointment (see Figure 21.2).

In sum, across the four committees we assessed, biomedical scientists dominated representation with about 60% of votes; epidemiologists had between 4% and 26% of votes; health services research between 2% and 10%; and social scientists and humanities researchers controlled between 0% and 5% of votes. Meanwhile, individuals without any academic appointments possessed between 3% and 29% of votes.

While CIHR was created with the mandate to foster interdisciplinary research and support the entire spectrum of health-related research (Government of Canada 2000), our data show that this call for inclusiveness has not materialized in CIHR's leadership committees. We interpret this gap as an occurrence of decoupling between the rhetorical push for interdisciplinarity and its practice at CIHR governance level.

Building on the large body of work on symbolic boundaries (Bourdieu 1984, 2004, Gieryn 1999, Lamont and Molnar 2002) and work on the cultural dimension of disciplines and scientific practices (Albert et al. 2008, 2009, Becher and Trowler 2001, Knorr-Cetina 1999), we argue that the social scientists and humanities scholars representation deficit on CIHR committees is likely to induce decisions – skewed by biomedical scientists epistemic habitus – that will maintain the current social and doxic order of the health research field. An example of this can be seen in the recommendations made by the 2011 International Review Panel. While many of the recommendations were intended to bolster the biomedical research community, no one addressed issues related to social science and humanities research. The Review Panel recommended that CIHR "provide sufficient funding for randomized controlled trials," to pay "particular attention to clinical investigators who must balance clinical service obligations with research," "to establish Centres of Excellence of Clinical and Translational Research," and to "catalyze new areas of research…, including the domains of mathematics, physics, computer and materials sciences, bioinformatics and certain engineering disciplines such as bioengineering." Recommendations were also made "to facilitate the development of a national bioinformatics strategy" and to explore "areas such as ecology, operations research or the study of complexity" (CIHR 2011c). The CIHR Governing Council accepted these recommendations (CIHR 2011c).

We believe it is unlikely that this striking oversight of the social sciences and humanities is the result of intentional acts of discrimination or the conscious performance of boundary-work.

Given the scholarship on professional socialization, it seems more likely to reflect biomedical scientists' taken-for-granted assumptions about how interdisciplinary research in health should develop. As members of a socio-cultural community (i.e., epistemic community), biomedical scientists envision the world and act accordingly through the cognitive categories they have internalized, i.e., through their epistemic habitus (Albert et al.2014).The enactment of biomedical scientists' taken-for-granted assumptions through the Review Panel recommendations and CIHR Governing Council concurring response is likely to further legitimize the current power hierarchies among disciplines in health research and reinforce the institutional arrangements privileging biomedical science.

Financial decoupling: access, inclusion, and exclusion

Another manifestation of the decoupling between policy and practice can be seen in the gap between the pro-interdisciplinary inclusiveness discourse and the financial barriers social scientists and humanists confront in their efforts to fully participate in the health research enterprise. Financial decoupling, we argue, is a consequence of the discordance between the field's logic (its dominant doxa) and the epistemic habitus of researchers who joined the health research field after prolonged disciplinary socialization in the social sciences or humanities. In the current organizational and symbolic order of academia, the financial resources necessary to function in the health research field surpass the resources social scientists and humanities scholars are accustomed to having and actually need to carry out their research projects.

Social scientists and humanities scholars in medicine in Canada have until 2009 applied to and obtained grants from the Social Sciences and Humanities Research Council of Canada (SSHRCC). SSHRCC grants are awarded based on the accepted rules of the games in the social sciences and humanities, where research requires few or no instruments or laboratories, and is simultaneously less intensive in human capital. Indeed, in the social sciences or humanities, a faculty member typically hires a few modestly paid graduate students to help do the work; in contrast, biomedical science projects often include project managers, lab managers, several research assistants and graduate students. Consequently, the sums awarded by SSHRCC are typically much smaller than those awarded by CIHR, its equivalent funding body in medicine, such that the funds allocated to social scientists and humanities scholars in health research are typically much smaller than those of their clinical or biomedical colleagues.

Looking at data from each body's recent annual reports, we get insight into the size of this funding gap. In fiscal year 2010–2011, SSHRCC awarded $19M in funding across 370 health and related life sciences and technologies grants[7], for an average of $51,351 per grant (SSHRCC 2011). In contrast, in fiscal year 2010–2011, CIHR funded 14,139 researchers with $753M (CIHR 2011a). First-year awarded amounts averaged $134,000, but these grants are often multi-year grants, which potentially close to doubles or triples the amount per grant; furthermore, on average, CIHR awardees hold $162,000 of funding annually. Unfortunately, the data are not perfectly comparable, but the lowest possible ratio of the (underestimated) CIHR funding to SSHRCC funding in health is 2.6:1, and could be as high as 5:1. Success in the interdisciplinary health research domain defined by CIHR thus implies at least 2.6 times more grant support for researchers than provided by SSHRCC on average in health and related life sciences and technologies. While we don't dispute the fact that biomedical and clinical research currently costs more than social science and humanities research, we argue that this higher cost is the result of a symbolic struggle over which research is important and which is not, or less, and thus should not be taken for granted. There could be another symbolic order – another doxa – where the value, legitimacy and primacy of biomedical science are lower than those of the social sciences and humanities (Frank and Gabler 2006). For this to happen, the balance of power in the health research domain would need to change dramatically. As noted by Bourdieu (1988, see also Wacquant 1993), however, this change would be dependent on similarly dramatic changes in the field of power.

The distribution of money across the discursively constructed "themes" of the interdisciplinary medical research domain funded by CIHR also provides revealing information about interdisciplinary power dynamics. In its 2010–2011 annual report, CIHR indicates the following spending: $475M for biomedical research; $58M for health systems/services research; $129M for clinical research; and $91M for social, cultural, environmental and population health (including epidemiology). A further $213M was awarded to research projects that did not specify a theme. In proportion to the full budget, then, biomedical and clinical researchers reap 63% of all funding; among theme-assigned projects, 80%, in contrast with 9% and 12% respectively for social, cultural, environmental and population health research. Although we do not have data on what percentage of applications were successful across areas of research, it is clear that the interdisciplinary field of medicine is dominated by biomedical and clinical science.

Indirect support for this claim can be seen in the recommendations made by the CIHR International Review Panels, as we have seen earlier, to increase the level of resources allocated to clinical and biomedical sciences, despite the fact that they already receive the greatest share of CIHR funding (CIHR 2011c). Further support can also be found in the fact that only 18% of the interdisciplinary peer-review committees at the CIHR (10 of 54) include panelists with *some* expertise in social science and/or humanities (Albert et al. 2009). Those committees include, for example, the Health Policy & Systems Management Research committee, the Humanities, Law, Ethics & Society in Health committee, and the Health Services Evaluation & Interventions Research committee. Biomedical research committees (such as Randomized Controlled Trials and Microbiology and Infectious Diseases) constitute the majority of committees at CIHR and do not count panelists with expertise in the social sciences and humanities.

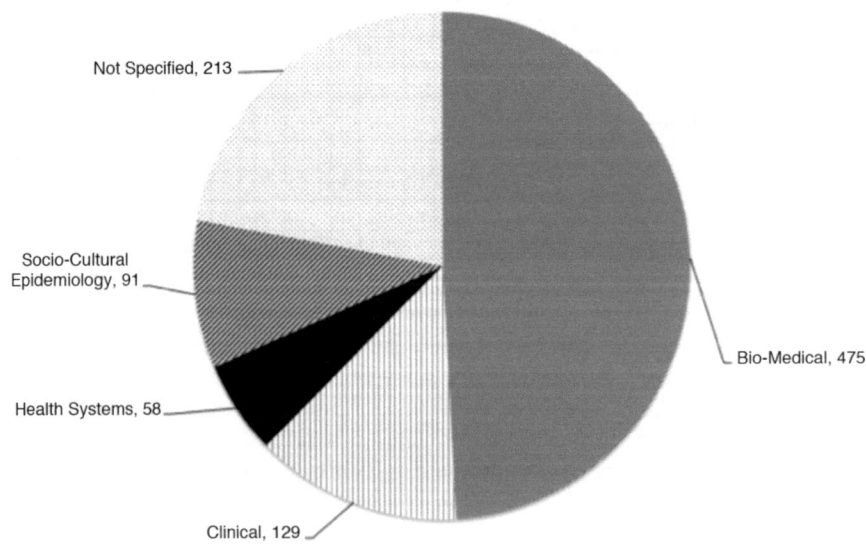

Figure 21.3 CIHR funding allocation per research domain for fiscal year 2010–2011 ($ millions)

One could argue that the recent Canadian reforms that have forced social scientists and humanities researchers to apply to CIHR instead of SSHRCC may benefit them. Indeed, they can now theoretically obtain 2.6 times more money for the same research. The reality of professional/epistemic habitus suggests a different outcome, however. Researchers internalize the requirements, or doxa, of their field through exposure, and by the time they graduate from their PhD program, they have been socialized into one set of expectations defined importantly by their own advisors' habitus and understanding of field norms. As such, even if it is possible for any researchers to do as we did – compare average grant amounts across granting agencies – and infer that future grants should be revised and matched to the new agency's funding levels (i.e., at least 2.6 times higher), the likelihood that a researcher would do so is absurdly small, especially in the first few years after the sea change for two main reasons. First, unless one has been exposed to granting practices in biomedical research (probably not the case for most social science and humanities scholars), he or she may not envision a potential systemic funding discrepancy. Second, the research problems he or she chooses to investigate and how he or she will actually study them (for example, as an individual researcher or as a team leader with multiple graduate students and research assistants) are dictated by his or her field of inquiry and epistemic habitus. Externally dictated changes in funding availability are unlikely to transform the grant-writing and research practices of social scientists and humanities researchers in the face of such a newfound largesse.

The size of grants in the social sciences and humanities has serious consequences for those who try to compete in the field of the biomedical sciences. For example, if they want to be seen as productive by colleagues from an altogether different background, social scientists and humanities scholars need to publish in clinical journals and attend clinical conferences. The *Canadian Medical Association Journal*, the leading clinical journal in Canada and fifth international in terms of impact factor, recently announced changes to its publishing policy: beginning in January 2013, it will charge fees of $2,750 per article (CMAJ email to EP). CMAJ cites the need to provide free access to research: an "author pays" model, instead of a "reader pays" model (Fletcher 2012). We know of no social science or humanities journal that asks fees for publication.

Similarly, the registration fees for clinical and clinical-type conferences such as medical education conferences also contrast starkly with those of social sciences and humanities conferences. For example, the early registration fee for the Association of American Medical Colleges' 2012 annual conference was US$600; early registration fee for members of the Association for Medical Education conference in Europe was €498 (US$652); early registration fee for the 2012 Canadian Conference on Medical Education was CND$695 (US$702); and a British Medical Journal conference (2013) on evidence-based medicine cost £495 (US$797). In contrast, the American Sociological Association 2013 conference non-member early registration fee is US$360, the Modern Language Association 2013 conference early registration non-member fee is US$270, the American Historical Association non-member early registration fee is US$212, and the American Anthropology Association non-member early registration fee for the 2012 conference is US$383. The significantly higher registration fees for clinical and clinical-type conferences may operate as a structural obstacle for social scientists and humanists who attempt to share their work with the medical world and gain visibility (and eventually status) within this community, or require a conscious effort to write conference and publishing fees as line items in grants.

Another contributor to academic success is a scholar's ability to recruit and retain talented graduate students. In a Canadian Faculty of Medicine (Institute of Medical Science 2008), faculty members need to secure a minimum of $25,000 per year for two years for a Master student (total = $50,000) and $27,000 per year for five years for a PhD student (total =

$135,000). When students are unable to capture external funding, it is the supervisor's responsibility to provide financial support out of his or her grant. The barrier is then two-dimensional – financial and symbolic – and hinders social scientists' and humanists' ability to build their reputations as graduate student supervisors. As noted earlier, the 2.6:1 ratio of CIHR to SSHRCC means that, as a proportion, students are more expensive to social scientists and humanists than they are to biomedical researchers. The same master's student annual salary constitutes 49% of the average SSHRCC health-related grant, but only 19% of the average first-year CIHR grant. To make matters worse, many medical students and other health research graduate students are told that the social sciences and humanities are a distraction from their other professional goals.

The gap between the financial resources that social scientists and humanists typically can access and the cost associated with academic life in the health research field may constitute a serious obstacle to the acquisition of scientific legitimacy within this field. Decoupling here plays out as discordance between the inclusive interdisciplinary discourse and the resources needed to actually achieve the inclusiveness of social scientists and humanists in medical schools. This financial decoupling is an example of policy/practice decoupling: while in theory, interdisciplinary policies foster the diversification of research, the absence of mechanisms to achieve financial inclusiveness has the effect to further reinforce pre-existing epistemic hierarchies.

We will now turn to the social scientists and humanists themselves and explore whether or not they experience discordance between faculties of medicine's expectation with regard to productivity and their own disciplinary-learned expectations, i.e., with their own epistemic habitus. We will focus on evaluation criteria; a central element of academic life.

Decoupled academic experience of social scientists and humanists in faculties of medicine

Interview data suggest that most social scientists and humanists see themselves as misfits or outsiders in their work environment. Researchers mention evaluation criteria to assess productivity as one of the key reasons for that alienation. These criteria are perceived as incongruous with those they have been accustomed to in their domain for several reasons. For some, the concrete ways to measure productivity in their clinical department comes as a cultural shock. The two following quotes represent well the overall description made by participants:

> The metrics I'm measured on are very simplistic, and it can be boiled down to one word: volume. The key thing that people here value is lots of stuff, so lots of publications, lots of grants, lots of presentations, lots of students, so it's about a volume game.
>
> *SSH20*

> The dominant definition of excellence equates productivity with quality, to some extent. Well, I'm not sure that there's any consistent understanding of quality other than the impact factor of the journal that you might publish in, or receiving peer-reviewed grants. So I think predominantly it becomes a quantitative assessment of how many grants you're pulling in and how many papers you're publishing.
>
> *SSH07*

As these quotes suggest, the social scientists and humanists we interviewed feel that the predominant criteria according to which they are assessed is the number of papers they write and grants they receive. Do they have the perception that the content and quality of their work is taken into account by their assessors? Again, here are some representative quotes:

> One of the things that bothers me immensely about the particular environment that I work in is that one's academic excellence is very much dictated by number of publications regardless of the quality of the publications.
>
> *SSH03*

> I had a paper published in the *American Journal of Sociology* two years ago. This journal is one of the top journals in North American sociology, but here it's irrelevant. Irrelevant. My colleagues, biostatisticians and epidemiologists, they just don't have a clue, not a clue.
>
> *SSH15*

For most of our participants, being in medicine implies something similar to moving to another country, a country with its own rules, expectations, value system, and legitimate strategies to establish reputation. For all participants in our study, adaptation was necessary. For some, adaptation was successful, for others it failed. Let's look first at a quote illustrating success. The first quote is from a full professor who has been in a faculty of medicine for thirty years:

> To be fair, for all my criticisms and so on, I have become a full professor, and I have been respected, and I am sought after. I've succeeded, right? So they can't have been that hard on my social science, because I did write social science. I didn't really compromise.
>
> *SSH05*

However, when this participant provided contextual details, the flavor of his experience changed:

> Because my career took place at a time when I could survive – and I did survive, and I did thrive – I have never felt as though I couldn't end up saying what I wanted to say, but that's largely to do with the time I've been here. I don't think it's the same anymore, I think it's increasingly getting harder to do social science research. ... You can't go anywhere with sociological imagination anymore, and if as a student you're going to get out of here in a decent amount of time, if you're going to get funding, if you're going to get off to a decent start, you have to better just align yourself with where things are at now. There are very few people that are willing to take a risk and do a really social science-y kind of thing.
>
> *SSH05*

The second quote from this participant suggests that despite the rise of the interdisciplinary discourse in Canada, it has become harder for social scientists and humanists to have a satisfying career in medicine over time. This increased difficulty illuminates Bromley and Powell (2012) means/ends type of decoupling: the discourse of interdisciplinarity has led to the growing presence of social scientists and humanists in the health research field as well as to the transformation of the governmental funding mechanisms, but has not delivered on the purported benefits of interdisciplinarity. Indeed, the push for interdisciplinarity has made it harder for social scientists and humanists to be evaluated by their peers and to be successful according to the norms they were socialized into: their epistemic habitus. While engaging in interdisciplinary research, social scientists' and humanists' inputs are often devalued, and the space for true multi-disciplinary research has shrunk with the growing cross-disciplinary competition for funds at CIHR. Social scientists and humanists are separate, and unequal. Consequently, several have had to compromise or dilute their work to fit the dominant publication model in medicine, that is articles in the range of 3,000–4,000 words, characteristically without theoretical grounding or substantive literature review and discussion.

The following two junior social scientists are currently experiencing these constraints and therefore do not seem to have the same positive experience that their senior colleague had. These two quotes are representative of many participants' dissatisfaction with their work environment and own career as it developed within this environment:

> I plateaued and now what happened to me is, I don't give a shit, so I'm kind of coasting downward in terms of trajectory. I'm not in an upward swing, because in a way I've kind of just given up. I find myself in a game that I can't stand. I can't stand the basic struggle for legitimation, I'm really repulsed by what these criteria to be legitimate in this context are. I can hardly wait to retire. I just want out.
>
> *SSH15*

> I need to think about where I'm going from here. Obviously I need to look into other work. I'm demoralized. I sat down with my supervisor a couple months ago and said, 'I'm really demoralized.'
>
> *SSH02*

We can draw a provisional conclusion from our interview data: although interdisciplinarity is high on the agenda of the Canadian health research funding agencies and faculties of medicine, it is not clear to what extent social scientists and humanists can contribute to interdisciplinary dialogues given how much standards of excellence vary between disciplines and the continued asymmetry in financial means, access to graduate students, and so forth. Their legitimacy and scientific authority are not recognized by the dominant epistemic habitus in the health research field. As a result, many such scholars have had to learn to play the scientific game according to rules decided *by* biomedical scientists *for* biomedical scientists in order to be successful. This move has diluted their own standards of excellence and led them to become a different type of scholar (i.e., to acquire a new epistemic habitus). The experience of biomedical scientists has not been so tainted by the standards of the social sciences and humanities. Adaptation has been mostly unidirectional (on the issue of unidirectional adaptation see medical anthropologists' account of their experience working in interdisciplinary health research teams, Barrett 1997, Foster 1987, Lambert and McKevitt 2002, Napolitano and Jones 2006, Williams et al. 2002).

Conclusion

Decoupling and power relationships in medical schools and in the broader health research field can take multiples forms and manifest themselves in various sites. In this chapter we have focused on a limited set of manifestations of decoupling, concentrating on the variety of epistemic habitus represented on CIHR leadership committees and on financial decoupling because they make visible the discrepancy between the calls for interdisciplinarity and the structural constraints social scientists and humanists face in faculties of medicine and health research more broadly. Furthermore, as we have shown with our interview data, these structural constraints are discordant with the lavishness of the discourse about the collaborative and the spirit of openness inherent to interdisciplinarity. To be successful and build the scientific legitimacy and authority recognized by the dominant biomedical epistemic habitus, many social scientists and humanists have had to adapt to a new social structure. Several among those who were unable to engage in such adaptation faced discouragement and disillusion with the game of science.

Neo-institutional theory has typically focused on the mechanisms of compliance and harmonization within organizations – through normative, mimetic and coercive forces (DiMaggio and Powell 1983). While we recognize the importance of these forces, the story we are telling here highlights the importance of habitus in slowing down or thwarting change and adaptation among individuals. Indeed, habitus is developed through a socialization process that is field-specific, and is therefore embodied, durable, unconscious, and idiosyncratic. Social scientists and humanists in medicine have entered a field that has been structured without them, whose rules and power structures are unknown to them. Through our documented and policy analysis, we have shown the structural forces at play in the interdisciplinary health research field, and the power relations they re/produce. With our interviews, we documented social scientists' and humanists' experience of dissonance as it specifically relates to their epistemic habitus: their attempts to adapt to the doxa of their new field, and their frustrations with rules of the game that according to several, disadvantages them. In their chapter on gender in the knowledge economy, our colleagues Vardi and Smith-Doerr (Chapter 22, this book) have highlighted similar processes: how "gendered organizations" and the gendered structure of competition negatively impact women. The potential of the knowledge economy to reshape work practices and institutionalized hierarchies has not materialized into equal opportunities for women.

In their case, as in ours, decoupling does not necessarily result from an intentional or conscious act of discrimination against or belittling of a group: women or social scientists and humanists. Instead, we argue that in many cases it is an unforeseen and undocumented consequence of the doxa of the health research field, that is the field-specific taken-for-granted and unformulated assumptions about the natural order of things within the biomedical and clinically dominated research field. Taken-for-granted assumptions get materialized in structural arrangements that legitimize and facilitate some practices while delegitimizing and discouraging others. In this sense, structural arrangements are also structuring forces shaping actors' practices. Whether it is intentional or not, the decoupling of pro-interdisciplinarity discourse and current social structures puts social scientists and humanists in the uneasy position of "misfits." The institutional setting within which they find themselves (the reward system, the standard knowledge production practices, the lab structure of their working environment, the pressure to capture large grants, work in teams, and publish in high impact factor journals, etc.) is often in dissonance with their epistemic habitus. Trying to advance their career within this system structurally confines social scientists and humanists to a position of inferiority before the biomedical and clinical scientists, as the system has not been created to espouse their own academic practices, but rather those of the more powerful group.

Another potential impact of the dissonance between the health research field logic and social scientists' and humanists' epistemic habitus is what David Hess has called "undone science" (2007). In the context of this chapter, "undone science" is the health-related social science or humanities research that could be done but is not because of the structural constraints exerted by this logic upon social scientists and humanities scholars. Social scientists' and humanists' embeddedness in the health research field orients and constrains their research. It allows them to ask certain types of questions (How can we change people's behavior to make them eat better?), but makes it less likely they will ask others (How would a better redistribution of resources across society benefit population health?). The fact that SSHRCC stopped funding health-related research in 2009 has meant that social scientists and humanists' work is not evaluated by their peers, but rather by scientists whose epistemic habitus may clash with theirs. The consequent dilution, transformation, or partial disappearance of health research inspired by the social science and humanities epistemes is thus practically unavoidable in the current Canadian context. As made clear by Bak (Chapter 23, this book), the focus on a specific set of research questions – be they inspired by utilitarian and nationalistic goals or by a specific scientific episteme – has important impacts not only on the type of research that gets funded, but also on the work lives of scientists and their satisfaction with their jobs.

If health research funding agencies and faculties of medicine are serious about fostering interdisciplinary or multi-disciplinary departments and institutes in which social scientists and humanists will thrive, they need to make room for different research practices. Otherwise, interdisciplinarity in health research may result in being an empty mantra with the real effect of subordinating non-clinicians and non-basic scientists to the rules of the powerful.

To our knowledge, our research is the first to look at interdisciplinarity within an institutional context and to take into consideration institutionalized power relationships. Existing work has been mostly preoccupied with scientists' interactions within research teams and research centers rather than with cross-disciplinary long-term relationships within a structured organizational field. To strengthen and nuance our preliminary findings and conclusions, further research should adopt a longitudinal approach to see if social scientists and humanists adapt or can change the structure of the field and create conditions where they can join interdisciplinary endeavors without having to leave their epistemic habitus in the cloakroom. Similarly, further research could compare the experiences of social sciences and humanities scholars across countries, but also across departments.

Notes

1   Authors listed in alphabetical order.
2   This research was supported by the Canadian Institutes of Health Research, grant # KTE-72140 and the Social Science Research Council of Canada. The authors wish to thank Kelly Moore and Daniel Kleinman for their insightful feedback on several versions of this chapter.
3   Epistemologically, one may argue that this position is rooted in naïve empiricism. Such arguments assume that the nature of reality is such that its intrinsic features can only be grasped by interdisciplinary approaches. This realist epistemological position implies an objective reality that can be revealed with the proper tool (see Weingart 2000), rather than a reality that is constructed and thus always elusive and amenable to various scientific (re)constructions.
4   Individuals with no academic appointments include representatives from the ministry of health, the business and pharmaceutical sectors and the health management system sector.
5   CIHR is structured around thirteen virtual geographically distributed institutes that each supports health research in biomedical, clinical, health systems and services and population health. Scientists funded by CIHR become members of one or more institutes. Directors of those institutes are considered scientific leaders in their respective domain.
6   Individuals with no academic appointments include members of the CIHR management team.
7   Although SSHRCC has stopped funding health-related projects in 2009, these data represent amounts received by ongoing projects funded before or in 2009, as well as projects focused on health-related technologies, broadly understood.

References

Albert, M. and Laberge, S. 2007. "The legitimation and dissemination processes of the innovation system approach: the case of the Canadian and Québec science and technology policy." *Science, Technology & Human Values* 32(2): 221–249.
Albert, M. and McGuire, W. 2014. "Understanding change in academic knowledge production in a neoliberal era." In *Fields of Knowledge: Science, Politics and Publics in the Neoliberal Age,* edited by S. Frickel and D. Hess. Bingley, UK: Emeralds Books.

Albert, M., Laberge, S., and Hodges, B.D. 2009. "Boundary work in the health research field: Biomedical and clinician scientists' perceptions of social science research." *Minerva* 47(2): 171–194.

Albert, M., Laberge, S., and Hodges, B.D. 2014. "Who wants to collaborate with social scientists? Biomedical and clinical scientists' perceptions of social science". In *Collaboration Across Health Research and Medical Care: Healthy Collaboration*, edited by B. Penders, N. Vermeulen, and J. Parker. London UK, Burlington VT: Ashgate Publishers.

Albert, M., Laberge, S., Hodges, B.D., Regehr, G., and Lingard, L. 2008. "Biomedical scientists' perception of social science in health research." *Social Science & Medicine.* 66(12): 2520–2531.

Association of American Medical Colleges (AAMC) 1998. *AAMC Data Book: Statistical Information Related to Medical Education.* Washington, DC: AAMC.

Association of American Medical Colleges (AAMC) 2009. *AAMC Data Book: Medical Schools and Teaching Hospitals by the Numbers.* Washington, DC: AAMC.

Barrett, B. 1997. "Identity, ideology and inequality: Methodologies in medical anthropology, Guatemala 1950–1995." *Social Science & Medicine* 44(5): 579–587.

Becher, T. and Trowler, P.R. 2001. *Academic Tribes and Territories.* Buckingham: The Society for Research into Higher Education and Open University Press.

Bourdieu, P. 1984 [1979]. *Distinction: A Social Critique of the Judgment of Taste* (trans. Nice, R.). London. Routledge.

Bourdieu, P. 1987. *Choses dites.* Paris: Minuit.

Bourdieu, P. 1988 [1984]. *Homo Academicus* (trans. Collier P.). Cambridge: Polity Press.

Bourdieu, P. 1994. *Raisons pratiques.* Paris: Seuil.

Bourdieu, P. 2000 [1997]. *Pascalian Meditations* (trans: Nice R.). Stanford: Stanford University Press.

Bourdieu, P. 2004 [2001]. *Science of Science and Reflexivity* (trans: Nice R.). Cambridge: Polity Press.

Bromley, P., and Powell, W.W. 2012. "From smoke and mirrors to walking the talk: Decoupling in the contemporary world." *The Academy of Management Annals* 6(1): 483–530.

Cameron, D.M. 2004. *Collaborative Federalism and Post-secondary Education: Be Careful What You Wish For.* Kingston, ON: Queen's University. John Deutsch Institute for the Study of Economic Policy.

Canadian Institute of Health Research 2006. *International Review Panel Report 2000–2005.* Ottawa.

Canadian Institutes of Health Research Act 2000, last amended, June 29, 2012. Government of Canada: Ottawa.

Canadian Institutes of Health Research 2011a. *Moving Forward: Annual Report 2010–2011.* Ottawa, ON, Canada. (accessed June 25, 2013) *www.cihr-irsc.gc.ca/e/documents/cihr_annual_report_2010-11_e.pdf.*

Canadian Institutes of Health Research 2011b. *Science Council.* (accessed June 25, 2013) *www.cihr-irsc.gc.ca/e/34008.html.*

Canadian Institutes of Health Research 2011c. *CIHR Response and Action Plan – 2011 International Review Panel Recommendations. Introduction and Overall Response.* (accessed June 25, 2013) *www.cihr-irsc.gc.ca/e/44567.html.*

Canadian Institute of Health Research 2011. International Review Panel Report 2005–2010. Ottawa.

Canadian Institutes of Health Research 2012. *Role of Governing Council,* (accessed June 25, 2013) *www.cihr-irsc.gc.ca/e/38103.html.*

Committee on Facilitating Interdisciplinary Research, Committee on Science, Engineering, and Public Policy, National Academy of Sciences, National Academy of Engineering, Institute of Medicine 2004. *Facilitating Interdisciplinary Research.* Washington, DC: The National Academies Press.

Cooper, G. 2012. "A disciplinary matter: Critical sociology, academic governance and interdisciplinarity." *Sociology* 47(1): 74–89.

Dewulf, A., François, G., Pahl-Wostl, C., and Taillieu, T. 2007. "A framing approach to cross-disciplinary research collaboration: Experiences from a large scale research project on adaptive water management." *Ecology and Society* 12(2): 14. *www.ecologyandsociety.org/vol12/iss2/art14.*

DiMaggio, P.J. and Powell, W.W. 1983. "The iron cage revisited: Institutional isomorphism and collective rationality in organizational fields." *American Sociological Review* 48(2), 147–160.

Emirbayer, M. and Johnson, V. 2008. "Bourdieu and organizational analysis." *Theory & Society.* 37: 1–44.

Finnemore, M. 1993. "International organizations as teachers of norms: The United Nations Educational, Scientific, and Cutural Organization and science policy." *International Organization* 47(4): 565–597.

Fisher, D., Atkinson-Grosjean, J., and House, D. 2001. "Changes in academy/industry/state relations in Canada: The creation and development of the networks of centres of excellence." *Minerva* 39(3): 299–325.

Fisher, D., Rubenson, K., Clift, R., Lee, J., MacIvor, M., Meredith, J., Shanahan, T., Jones, G., Trottier, C., and Bernatchez, J. 2005. *Canadian Federal Policy and Post-Secondary Education: Case Report.* New York: Alliance for International Higher Education Policy Studies.

Fletcher, J. 2012. "Research for all." *Canadian Medical Association Journal* 184(5): 1657.

Foster, G.M. 1987. "World health organization behavioral science research: Problems and prospects." *Social Science & Medicine* 24(9): 709–717.

Frank, D.J. and Gabler, J. 2006. *Reconstructing the University: Worldwide Shifts in Academia in the 20th Century.* Stanford: Stanford University Press.

Friedland, R. and Alford, R.R. 1991. "Bringing society back in: Symbols, practices, and institutional contradictions." In *The New Institutionalism in Organizational Analysis*, edited by W.W. Powell and P.J. DiMaggio, 232–262. Chicago: The University of Chicago Press.

Froderman, R., Thompson Klein, J., and Mitchman, C. 2010. *The Oxford Handbook of Interdisciplinarity.* Oxford: Oxford University Press.

Gibbons, M., Limoges, C., Nowotny, H., Schwartzman, S., Scott, P., and Trow, M. 1994. *The New Production of Knowledge: The Dynamics of Science and Research in Contemporary Societies.* Thousand Oaks, CA: Sage.

Gieryn, T.F. 1999. *Cultural boundaries of science. Credibility on the line.* Chicago: University of Chicago Press.

Government of Canada 2000. Canadian Institutes of Health Research Act. Ottawa. (accessed June 25, 2013) *http://laws-lois.justice.gc.ca/eng/acts/C-18.1/FullText.html.*

Graham, J., Adelson, N., Fortin, S., Bibeau, G., Lock, M., Hyde, S., Macdonald, M.H., Olazabal, I., Stephenson, P., and Waldram, J. 2011. "A manifesto: The end of medical anthropology in Canada?" *University Affairs* March: 37.

Gumport, P.J. 2000. "Academic restructuring: Organizational change and institutional imperatives," *Higher Education* 39(1): 67–91.

Hadorn, G.H., Hoffmann-Riem, H., Biber-Klemm, S., Grossenbacher-Mansuy, W., Joye, D., Pohl, C., Wiesmann, U., and Zemp, E. 2008. *Handbook of Interdisciplinary Research.* Dordrecht: Springer.

Hall, K.L., Vogel, A.L., Stipelman, B.A., Stokols, D., Morgan, G., and Gehlert, S. 2012. "A four-phase model of transdisciplinary team-based research: goals, team processes, and strategies." *Translational Behavioral Medicine* 2(4): 415–430.

Hess, D.J. 2007. *Alternatives Pathways in Science and Industry: Activism, Innovation, and the Environment in an Era of Globalization.* Cambridge, MA: MIT Press.

Hoffman, S.G. 2011. "The new tools of the science trade: contested knowledge production and the conceptual vocabularies of academic capitalism." *Social Anthropology* 19(4): 439–462.

Institute of Medical Science 2008. University of Toronto (accessed June 25, 2013) *www.ims.utoronto.ca/current/funding/policy.htm.*

Jacobs, J.A. and Frickel, S. 2009. "Interdisciplinarity: A critical assessment." *Annual Review of Sociology* 35: 43–65.

Jeffrey, P. 2003. "Smoothing the waters: Observations on the process of cross-disciplinary research collaboration." *Social Studies of Science* 33(4), 539–562.

Klein, J.T. 1990. *Interdisciplinarity: History, Theory, and Practice.* Detroit: Wayne State University Press.

Knorr-Cetina, K. 1999. *Epistemic Cultures. How the Sciences Make Knowledge.* Cambridge, MA: Harvard University Press.

Laberge, S., Albert, M., and Hodges, B.D. 2009. "Perspectives of clinician and biomedical scientists on interdisciplinary health research." *Canadian Medical Association Journal* 181(11): 797–803.

Lambert, H. and McKevitt, C. 2002. "Anthropology in health research: From qualitative methods to multidisciplinary." *British Medical Journal* 325: 210–213.

Lamont, M. and Molnar, V. 2002. "The study of boundaries in the social sciences." *Annual Review of Sociology* 28: 167–195.

Lau, L. and Pasquini, M. 2004. "Meeting grounds: Perceiving and defining interdisciplinarity across the arts, social sciences and sciences." *Interdisciplinary Science Reviews* 29(1): 49–64.

Lélé, S. and Norgaard, R.B. 2005. "Practicing interdisciplinarity." *Bioscience* 55(11): 967–975.

Maasen, S. 2000. "Inducing interdisciplinarity: Irresistible infliction? The example of a research group at the Center for Interdisciplinary Research (ZiF), Bielfeld, Germany." In *Practising Interdisciplinarity*, edited by P. Weingart and N. Stehr, 173–193. Toronto: University of Toronto Press.

MacMynowski, D.P. 2007. "Pausing at the brink of interdisciplinarity: Power and knowledge at the meeting of social and biophysical science." *Ecology and Society* 12(1): 1–14.

Meyer, J.W. and Rowan, B. 1977. "Institutionalized organizations: Formal structures as myth and ceremony." *American Journal of Sociology* 83(2): 340–363.

Meyer, J.W., Boli, J., Thomas, G.M., and Ramirez, F.O. 1997. "World society and the nation state." *American Journal of Sociology* 103(1): 144–181.

Moore, R. 2011. "Making the break: Disciplines and interdisciplinarity." In *Disciplinarity Functional Linguistic and Sociological Perspectives*, edited by F. Christie and K. Maton, 87–105. London: Continuum.

Napolitano, D.A. and Jones, C.O.H. 2006. "Who needs 'pukka' anthropologists? A study of the perceptions of the use of anthropology in tropical public health research." *Tropical Medicine and International Health* 11(8): 1264–1275.

National Institutes of Health. 2007. *NIH Launches Interdisciplinary Research Consortia.* Washington, DC: The Institutes. (accessed June 25, 2013) *www.nih.gov/news/pr/sep2007/od-06.htm*.

Organization for Economic Cooperation and Development (OECD) 1972. *Interdisciplinarity: Problems of Teaching and Research in Universities.* Paris: Organization for Economic Cooperation and Development.

Paradis, E. and Reeves, S. 2012. "Key trends in interprofessional research: A macrosociological analysis from 1970 to 2010." *Journal of Interprofessional Care* 27(2): 113–122.

Plamondon, R. 2002. *Transforming Health Research in Canada: The Making of the Canadian Institutes for Health Research.* Ottawa: Public Policy Forum.

Polster, C. 2002. "A break from the past: Impacts and implications of the Canada foundation for Innovation and the Canada Research Chairs initiative." *Canadian Review of Sociology and Anthropology* 39(3). 275–299.

Popp Berman, E. 2012a. "Explaining the move toward the market in U.S. academic science: How institutional logics can change without institutional entrepreneurs." *Theory and Society* 41(3): 261–299.

Popp Berman, E. 2012b. *Creating the Market University: How Academic Science Became an Economic Engine.* Princeton: Princeton University Press.

Powell, W.W. and DiMaggio, P.J. 1991. *The New Institutionalism in Organizational Analysis.* Chicago: University of Chicago Press.

Ramirez, F.O. 2006. "The rationalization of the university." In *Transnational governance: Institutional dynamics of regulation*, edited by M.L. Djelic, and K. Sahlin-Anderson, 225–244. Cambridge: Cambridge University Press.

Ramirez, F.O. 2010. "Accounting for excellence: Transforming universities into organizational actors." In *Higher Education, Policy, and the Global Competition Phenomenon*, edited by V. Rust, L. Portnoi, and S. Bagely, 43–58. London: Palgrave.

Rosenfield, P.L. 1992. "The potential of transdisciplinary research for sustaining and extending linkages between the health and social sciences." *Social Science & Medicine* 35(11), 1343–1357.

Shariff, N. 2006 "Emergence and development of the National Innovation Systems concept." *Research Policy* 35(5): 745–766.

Slaughter, S. and Leslie, L.L. 1997. *Academic Capitalism: Politics, Policies and the Entrepreneurial University.* Baltimore and London: Johns Hopkins University Press.

Snowdon, K. 2005. *Without a Road Map: Government Funding and Regulation of Canadian Universities and Colleges.* Research Report W | 31 Work Network. Canadian Policy Research Networks (accessed June 25, 2013) *http://rcrpp.ca/documents/40781_en.pdf*.

Social Sciences and Humanities Research Council of Canada 2011. *Talent, Insight, Connection.* 2010–2011 Annual Report. Ottawa, ON, Canada (accessed June 25, 2013) *www.sshrc-crsh.gc.ca/about-au_sujet/publications/SSHRC_Annual_Report_2010 11_e.pdf*.

Stokols, D., Fuqua, J., Gress, J., Harvey, R., Phillips, K., Baezconde-Garbanati, L. et al. 2003. "Evaluating transdisciplinary science." *Nicotine & Tobacco Research* 5 supp. 1: S21–S39.

Wacquant, L.J.D. 1993. "From ruling class to field of power. An interview with Pierre Bourdieu on *La noblesse d'État*." *Theory, Culture & Society* 10: 19–44.

Weingart, P. 2000. "Interdisciplinarity: The paradoxical discourse." In *Practising Interdisciplinarity*, edited by P. Weingart and N. Stehr, 25–42. Toronto: University of Toronto Press.

Williams, H.A., Jones, C., Alilio, M., Zimicki, S., Azevedo, I., Nyamongo, I., Sommerfeld, J., Meek, S., Diop, S., Bloland, P.B., and Greenwood, B. 2002. "The contribution of social science research to malaria prevention and control." *Bulletin of the World Health Organization* 80(3): 251–252.

22

Women in the Knowledge Economy

Understanding gender inequality through the lens of collaboration

Itai Vardi and Laurel Smith-Doerr[1]

BOSTON UNIVERSITY AND UNIVERSITY OF MASSACHUSETTS AT AMHERST

Introduction

This chapter explores issues of gender inequality in the knowledge economy. Though the definition of "knowledge economy" may vary significantly and is itself gendered (Walby 2011), the increasingly blurred lines separating private firms and universities (Kleinman and Vallas 2001) call for expanding the analytical perspective. In our analysis we thus adopt a broad definition of "knowledge economy" that includes science-intensive industries, such as information technology, biotechnology, and nanotechnology (Moore et al. 2011: 510), as well as traditional knowledge producing organizations in the academic and non-profit research sectors. Because U.S. for-profit firms employ 59% of workers whose highest degree is in science and engineering (NSF 2012), we feel it is critical to examine gender inequality in for-profit settings as well as in academia where it is most often studied. The knowledge economy is increasingly central to comprehending broad social processes underway in science and technology, and it thus also holds an important key to understanding gender dynamics.

Within science, new general patterns of work would seem to favor the entrance of women into the knowledge economy. These include changes that have expanded opportunities and increased the number of locations for work in science: industry jobs are valued by young scientists over academic ones (Smith-Doerr 2005, Vallas and Kleinman 2007), new scientific posts are created, like the proliferation of postdoctoral positions (for example, Frehill and Lee 2013), and new forms of organization, such as flatter network forms (for example, Powell 1990, DiMaggio 2001), flourish. One might expect that women would do well in this new knowledge economy. In the uncertain global economy, traditionally masculine blue-collar manufacturing jobs have shifted from affluent countries to poor countries. Thus, many men are doing worse than their fathers, especially men without advanced education to enter the knowledge economy in the first world. At the same time, women in the U.S. are earning degrees at higher rates than men at all levels and in many fields. An expanding knowledge economy would suggest that we are well placed for entering a new, more equal era for women in science. Indeed, while women in the natural sciences and engineering still face formidable barriers, evidence amassed in the past three decades in the U.S. seems to point to several positive and dramatic changes. The rise in women's overall education and labor market participation (for example, Coontz 2011), the closing of degree attainment gaps by gender in most scientific fields (for example, NSF 2011), and the introduction of new flexible work patterns (for example, Smith 2001) all suggest that the knowledge economy has created new possibilities for women and men to create and work in more equitable and open contexts.

In this chapter, we seek to evaluate the processes of gender (in)equality in the knowledge economy. In what ways have women benefited from the rise of the new industrial sector? How have efforts for challenging gender inequality in science and technology been stymied or reversed? While growing research on the intersection of gender and other social identities (race, ethnicity, and nationality) informs our understanding of women in the knowledge economy, here we take a new approach by focusing on *collaboration*. Not only is collaboration burgeoning in the academy (as evidenced by increasing co-authorship patterns, for example, Wuchty et al. 2007), it also provides the foundation for new knowledge industries (Powell et al. 1996, Powell and Snellman 2004). Collaboration, as a work pattern characteristic of the knowledge economy, provides a potentially interesting social location and process in which to observe gender relations at the intersection of individual efforts and organizational contexts. If collaboration is a key form of work, how is collaboration related to longstanding problems like gender inequality in science? The collaborative model of science contrasts with the traditional model of a single principal investigator (PI) who focuses on gaining fame for being the first to discover or invent something (Merton 1973).

In this chapter we explore both the benefits and disadvantages of collaboration for women. Can collaboration provide for new equitable settings, or are the lucrative incentives for collaboration squeezing out opportunities for women? Do collaborative efforts shape women's access to resources? How may collaboration disrupt or re-institutionalize old gender orders? Are the career risks of collaboration distributed equally between men and women? Is collaboration itself gendered, so that some collaborators are more valued than others?

Overall, we argue that the interrelationship between the two phenomena – gender (in)equality and collaboration – is less than straightforward. Collaboration may disrupt gender inequality in some contexts, and institutionalize it in others. While the larger structural contexts of flexibility and collaboration in the new knowledge economy may open up new opportunities for women in science, the interactional contexts of collaboration on the micro level may yet reinforce gender inequalities. In this chapter, we use preliminary interview data from our ongoing research on collaboration in the chemical sciences to illustrate our points about the ways collaboration may serve to expand or contract gender inequality in knowledge-intensive work settings. Before presenting our argument, however, we briefly review several broader trends of gender (in)equality in the knowledge economy, focusing on approaches that privilege individual rights, whether through formal legislation or changing norms about women's roles in paid work.

Shifts in the institutionalization and disruption of gender inequality in the knowledge economy

The first wave of the women's movement at the turn of the last century opened higher education to women, who entered college in significant numbers by the 1970s. The overall shift has been dramatic: today women have overtaken men in enrollment and passing exams in schools and universities throughout the global North (Walby 2011). Women's share of the advanced degrees in science and engineering fields rose during the end of the last century as well. By 2009, the U.S. reached gender parity in granting PhDs across all fields (NSF 2012). In some fields, most notably the biological sciences, there is gender parity in doctoral recipients. But other fields that feed into the knowledge economy have seen very slow gender integration. In engineering, which is an important training ground for the for-profit sector of knowledge workers, women are still only 25% of the new PhDs (NSF 2012).

Table 22.1 outlines some important general trends in the institutionalization and disruption of gender (in)equality that have affected women in the knowledge economy. We define institutionalization as the formalization of social relationships into similar looking structures or patterns and the routinization of practices into norms or laws. In contrast, we see disruption as a process of social change that involves overt or tacit interruptions of established patterns and structures. We also recognize that the relationship between the two processes – institutionalization and disruption – is often dialectic: the disturbance of established patterns may result in new relationships and structures, or ultimately reinforce the existing social forms. In mapping the literature, we distinguish between scholarly approaches to studying institutions that examine the more readily apparent individual level of women's rights (for example, Rossiter 2012) and those institutional approaches that examine the less visible organizational level of gendered structures and practices in the knowledge economy (for example, Huber 1974, Smith-Doerr 2011). At both the level of individual women's rights and the level of organizational and institutional change our focus is on the industrialized world, where knowledge economies have been established. At the individual level we note scholarship on both formal legal and informal cultural shifts in the history of gender relations. At times, institutionalization of inequality is disrupted by social change that shifts the direction of social structures or practices. Rules or norms that excluded women from higher education or the best jobs in science and engineering through the first half of the twentieth century were increasingly disrupted by the second wave of the women's movement that gathered momentum in the 1970s. Women scientists challenged advertisements for scientific positions that openly posted different starting salaries for male and female job candidates and pursued equal pay in legal forums as well, primarily through working to pass the Equal Pay Act and pressuring government officials to implement Equal Employment Opportunity (EEO) rules for gender after 1972 (Rossiter 1998).

Table 22.1 Institutionalization and disruption cycle in gender inequality (particularly in the U.S.): Historical and analytic trends

		Institutionalization	Disruption
Individual level (more apparent)	Explicit formal rules.	Policies excluding females from education, employment.	Women's movements challenge and change rules/ EEO laws.
	Explicit informal norms.	Gender/family roles, intersected by race.	Normative shifts and political economy's structural changes increase paid labor.
Organizational level (less visible)	Subtle practices.	Gendered organizations/ hierarchies.	New forms of organization may disrupt.
	Career incentives.	Competition.	Collaborative practices may create opportunities.

We know from sociological studies of EEO implementation, however, that legislation to extend rights does not necessarily mean equitable outcomes (Edelman 1992, Dobbin 2009). EEO compliance often takes the form of window dressing, like the *pro forma* "diversity training" requirements that do not change daily behavior in organizations. Psychological level studies of implicit bias suggest that the institutionalization of equal pay in laws may be undermined by our shared preferences for men and masculinity (Valian 1999, Moss-Racusin et al. 2012). What the historical data on formal legal recourse show is that there is a tendency for equality gains and attempts to institutionalize them to be disrupted by a return to the status quo of masculine bias. Given that legislative and other rule changes have mixed effects, we turn to an investigation of the effects of more informal, normative changes in gender relations that affect the workplace and their contributions to gender (in)equality.

Normative changes, particularly the shift away from expecting women to remain in the private sphere of the home, have accompanied women's entrance into the paid workforce. Women now make up nearly half of all of the workers in the U.S. labor market and are expected to work outside the home, but the expectations for unpaid domestic labor among women remain formidable (Padavic and Reskin 2002). Women scientists and engineers do more than their share of the unpaid labor at home (Misra et al. 2012). A handful of studies of scientists and engineers in knowledge economy industries show that women express greater concern about work-family balance than men do (Herman et al. 2012, Eaton 2003). From the literature, we see evidence of two conflicting normative expectations for women in the knowledge economy: to work (all hours) in the lab, and to do (most) of the work at home. Perhaps the first expectation is not surprising: since the 1960s, the liberal feminist idea that women should engage in paid work is widely accepted as normative. Highly educated women are often drawn to jobs in which they can produce knowledge, and individual women in professional jobs express high levels of satisfaction with their work (Bielby and Bielby 1989). Despite the extensive movement of women into demanding professional jobs, powerful norms continue to dictate that women do a disproportionate amount of labor in the home (Hochschild 2012).

The gendering of work is further compounded by considering the various ascribed social identities that women must negotiate in the work environment. These social identities include citizenship, and race and ethnicity. White women and native citizens of industrialized countries are advantaged compared to women of color and immigrants. Paid work often requires proof of citizenship or special visas, especially in the United States, so there are extra barriers to entering the knowledge economy for women who come from countries where few opportunities to work in scientific fields exist (Zippel and Frehill 2011). Even when women are citizens of an industrialized country, the biases against women of color create durable inequalities. Glenn's (2002) scholarship on the historical intersection of race and gender in U.S. citizenship demonstrates how particular stereotypes were attached to African American women, Asian American women, and Latinas, restricting their work opportunities and calling the "legitimacy" of their citizenship into question. While the study of the intersection of race and gender is growing, fewer studies of the intersectionality of race and gender in *science* have been conducted. But consider a couple of these exceptional studies, which begin to advance our understanding of intersectional perspectives on gender inequality. In Ong's investigation of academic settings, women faculty of color often report experiencing greater marginalization than do white women or men of color (Ong 2005). In Branch's (2011) study of computer science, African American women entering computing found themselves marginalized and limited to occupational ghettos such as data entry.

Computing and information technology presents an interesting case for contemplating the relation between women's roles and changing cultural and structural patterns in the knowledge economy. Donato (1990) characterizes computer programming as a case of an occupation that was feminized at first when it was seen as "clerical work." In the U.S., the first computer programmers were white women who ran ballistics trajectories on the ENIAC computer during World War II. Later, when men discovered the power and value of computing, women were replaced and computing became a male domain. But women entered again: Wright and Jacobs (1994) chart the gender integration of computing during the 1980s. When occupations, like computer systems analyst, were growing rapidly, women (especially white and Asian women) did move into the field. But unlike other gender integrating fields, computing did not become devalued and neither did women move into occupational ghettos; pay remained high even as women's proportion of some computing jobs meant that the field seemed to be re-feminizing.

By the late 1990s, there was a precipitous decline in women's participation in computing. Despite the high growth in internet-related information technology (IT) occupations, women were disappearing from computing. The evidence for women's share of computer science bachelor's degrees is telling: women went from 37% of the computer science majors in 1985 to only 18% by 2009. A study of computer science majors at Carnegie Mellon University, where women drop out of computer science (CS) at twice the rate that men do, sheds some light on how women became discouraged from completing computer science majors (Margolis and Fisher 2001). As CS majors became increasingly associated with the stereotype that "someone who is myopically obsessed with computers is a perfect fit" for computer science (Margolis and Fisher 2001: 65), together with unengaged teaching styles and large classes, women were disproportionately driven away from the major. While some men were also driven away from CS by the single-mindedness of programming culture and large introductory "weed-out" classes, women were disproportionately likely to leave the major at Carnegie Mellon and to report dissatisfaction with these aspects of the CS major in the study. Moreover, contrary to the stereotype that the male "geeks" are better suited for CS, the evidence from Margolis and Fisher's (2001) study demonstrated that obsession and unlimited programming hours does *not* translate into better performance in the major.

In response to results from this study and others on CS majors, Harvey Mudd College (another engineering-focused school in the U.S.) has made CS introductory courses smaller and trained faculty to change the culture of competitive "one-up-manship" in the classes (Alvarado et al. 2012). Since these programmatic changes were introduced, the percentage of CS majors at Harvey Mudd who are women has risen from 10% to 40% in less than a decade (Klawe 2011). The case of women in computing speaks to the importance of both structural opportunities and cultural biases to gender equality in the knowledge economy. When new occupations open up in high-tech fields, women do enter. Policies and interventions can change the interaction contexts in the classroom and thus disrupt the gendered sorting mechanisms.

Institutionalization and disruption at the meso- and macro-levels

While thus far we have focused on the more visible historical-legal changes and shifts in cultural attitudes, it is also clear that there remains wide gender inequality in the knowledge economy that does not seem to be fully explained by how individuals encounter laws or norms. Beyond laws and norms, a focus on the less visible organizational contexts of inequality (particularly how collaboration unfolds) in the knowledge economy is key to understanding the institutionalization and disruption of inequality in these science-intensive industries. These less apparent organizational contexts of inequality are represented in the lower half of Table 22.1.

At the macro-level of economic life, the coming of the knowledge economy disrupted the gender order characteristic of the traditional economy. Fueled by global processes of capital movements and de-industrialization (Bell 1973), the knowledge economy enhances the importance of human capital (Becker 1964, Kleinman and Vallas 2001) while promoting new organizational forms of work (Castells 1996, Leadbetter 2000, Powell 1990). Whereas older industrial forms of production and consumption rested on fixed or "harder" forms of capital – typically dominated by managerial cultures of pyramidal hierarchy, accumulation through competition, and technological mastery – the new economy's reliance on human, "fluid" or "flexible" forms of capital offered women an opportunity to reduce gender inequality.

Women in several knowledge-intensive fields are more likely to be in positions of greater authority than in manufacturing or service industries (Smith-Doerr 2004, Walby et al. 2007). Despite these gains, salient forms of gender inequality persist, as older forms of power and differentiation become reinstated in the economic sphere. As the segregation of women in education and the workforce endures, women's increasing overall levels of human capital do not necessarily translate into the kinds of capital required to succeed in knowledge-intensive fields (Tam 1997, Tomaskovic-Devey and Skaggs 2002).

Most labor practices in the knowledge economy occur within organizations. The concept of "gendered organization" provides a useful framework for understanding both the durability and instability of gender orders in knowledge economy workplaces. According to this theory, most contemporary work organizations are structurally gendered (Acker 1990, Ridgeway 2001, Britton 2000). By patterning divisions of labor, symbols, interactions, professional identities, and structural logics in terms of a *distinction* between male and female, masculine and feminine, work organizations systematically disadvantage women, their bodies, and their sexuality (Acker 1990). Echoing the two dimensions of the gendered organization concept, it is useful to discuss the specific effects of gendered organizations in the knowledge economy in terms of *organizational cultures*, and *organizational forms*.

In terms of organizational cultures, processes of gendering operate in the knowledge economy by imbuing organizations with values, dispositions, identities, and social expectations that have real consequences for men and women. The deep historical association of science and technology with dominant forms of masculinity is well documented (Cockburn 1985, Oldenziel 1999, Wajcman 1991), as is its ideological links to broader systems of masculine power and control (Schwartz-Cowan 1979, Fox Keller 1983, Faulkner 2000, 2001). The persistence of a masculine culture in science and engineering has been described as "tenacious" (Sappleton and Takruri-Rizk 2008: 298). The central norms of science – rationality, efficiency, and the rigid adherence to rules and order – define the typical ideals for success in the field, including decisions about hiring and promotion. Traditionally masculine behaviors such as competitiveness, individualism, and self-promotion are tied here to the concept of the "ideal worker" (Miller 2004). Importantly, women experience the consequences of this culture differently than men, as their respective role conflicts and structural constraints diverge (Evetts 1998). The inherent instability of the "new economy," with its increased job insecurity, demands for "flexibility," fragmented careers, and ever-shifting skill sets (Bauman 2000, Sennett 1998), exacerbate these difficulties for women. Among other things, women are more likely to be the "trailing spouse" in heterosexual marriages, and if moving ahead in the knowledge economy means moving geographically across long distances, women are disadvantaged relative to men. Such gender differentiation, however, often occurs through highly subtle dynamics that may be difficult to detect. Benschop and Doorewaard (1998) call these dynamics the "gendered subtext of organizations." Dana Britton's (2010: 9–10) research on gender inequality in promotions to full professor in academic settings captures the operation of the gendered subtext of organizations. One of her interviewees notes that it is not formal constraints that hold women back. Instead, women are "just not one of the guys, so they're not going to be promoted."

Organizational forms also play a pivotal role in determining the degree of gendering within the workplace. While bureaucratic-hierarchical logics that reduce subjectivity in personnel decisions may prove beneficial for some women in science and technology (Bielby 2000, Baron et al. 2007, Guthrie and Roth 1998, Fox 2001), the newer networked, horizontal structures typical of the knowledge economy have been shown to reconfigure gendered organizations into more equitable environments (Smith-Doerr 2004, Kalev 2009, Leadbetter 2008). This flatter, network form of organization is, as we have pointed out above, characterized by collaborative relationships. When characterized by pooling and exchanging resources, collaboration requires malleable structures that ease the connections between workers of differential status. It is these flexible forms that afford women new opportunities to reconfigure power relations in science. Collaboration is thus associated with the network form of organization more than with hierarchical forms of organization.

But such flattening and its consequences are not a universal trend in knowledge-intensive settings: in academia, increasing marketization and bureaucratization means that knowledge production is carried out under the reemergence of rigidly hierarchal, rule-based forms of organization (Moore et al. 2011, Kleinman and Vallas 2001). Longer working hours and multiplying task demands in the past two decades (research, teaching, mentoring, service, and now entrepreneurship) have had adverse effects on women faculty members (Acker and Feuerverger 1996, Jacobs and Winslow 2004, Fox and Xiao 2013), who unlike men often negotiate the tensions between academic careers and parenthood by sacrificing research, the strongest determinant of promotion (Misra et al. 2012). The pressures of domestic responsibilities, disproportionately experienced by women, often contravene the prevalent norm of working extremely long hours (Sappleton and Takruri-Rizk 2008). For example, in a study of the University of California (UC) faculty, women with children reported spending the least amount of time on professional work like research and the most on caregiving, when compared to men in any family status and to women without children (Mason and Goulden 2004). Even though the mothers in the UC study did professional work 51 hours per week on average, their average fell short by a full day's work – 8–9 hours per week – compared to their childless colleagues (Mason and Goulden 2004: 99).

Social capital accumulation is also gendered. In industrial contexts, women often find it difficult to enter key corporate and social networks because of the long working hours many firms expect (Roberts and Ayre 2002). Further, professional networks may create webs of restricted access (Bourdieu and Wacquant 1992), as certain forms of informal interaction in science and technology fields actually enhance cultures of exclusivity such as "old boy clubs." If exclusionary practices in the old economy rested largely on gendered socialization into dominant sex roles (Kanter 1977), newer forms of segregation are predicated on gendered knowledge and practices in ways that leave women outside of power circles. For example, in her study of men and women in various medical roles in healthcare, Lindsay (2008) found that women were excluded from crucial professional networks by the frequent cultural association between complex technical competence and masculinity. Women performing such tasks as anesthesia or suturing were often met with surprise or resistance, unlike their male counterparts. Similarly, Kellogg (2011) found that surgical residents who enjoy high status cultivate a culture of "machismo" (what she calls "Iron Men") that defines strong professional ability by linking mastery of surgery with radical individualism and competition.

The case of collaboration

As discussed above, the knowledge economy has accentuated the importance of alternative modes of work, not only in terms of how organizations are structured, but also in how workers interact with each other. How are these different forms of labor associated with gender relations and structures? To explore this question, we take a closer look at one such instance: collaboration. While collaboration is by no means a new phenomenon in scientific research (Beaver and Rosen 1978, 1979a, 1979b), the centrality of collaborative work has increased significantly in recent decades (Wray 2006). What this means is that although the relationship between looser and less hierarchical work structures and collaboration is strong, it is not historically necessary. Scholarship has noted that as problems become more complex, detail-specific, and global in scope, they cannot be properly grasped by any single scientist, field, or institution; this complexity has been associated with an increase in collaboration (Shrum et al. 2007).

Although seemingly simple, "collaboration" itself can prove to be quite an elusive concept. As Katz and Martin (1997) have shown, problems in pinpointing the meaning of collaboration include proper measurement, discerning the various levels of cooperation, and evaluating collaborative costs. While most working definitions in the literature describe collaboration as involving a team consisting of two or more individuals working together to advance a common goal (for example, Amabile 2001, Melin and Persson 1996), we prefer a broader sociological conceptualization. We understand collaboration as a *form of social interaction in professional contexts that involves both the sharing and withholding of knowledge, resources, and technologies*. We recognize that research collaboration describes relationships between individuals and between organizations (for example, Bozemen et al. 2012).

Ostensibly, collaboration in knowledge-intensive fields possesses the capacity to benefit women by undermining gendered organization logics and traditional divisions of labor. In a famous case, Marie Curie's success can be attributed to the kind of collaboration she developed with her husband and his scientific team, which stressed both group efforts and individual achievements (Pycior 1993). Collaboration does not mean that individual scientists produce less; in fact, scientific collaboration is linked to greater productivity for individual scientists (Zuckerman 1967, Melin 2000). There is productivity at the organizational level as well: for instance, companies with higher levels of collaborative R & D are more likely to go public (Powell et al. 1996). Collaboration is also associated with "preferential attachment" (Merton 1968) that can lead to high status, promotions, and increased funding (Drori et al. 2003, Sonnert and Holton 1996). One advantage in collaborative environments is the opportunity to choose and shift between partners (Smith-Doerr 2004). Through collaboration's close affinity to interdisciplinary research, cooperative work increases women's social capital by exposure to associated disciplines, circles of expertise, and support networks (Pfirman and Balsam 2004). Women in initial stages of acculturation into the IT profession, for example, have pointed to collaboration as an important feature that attracts their commitment to the professional culture (Guzman and Stanton 2008).

Still, the overall picture is complex. While collaboration has been associated with increased output and promotions, not *all* collaboration has been found to enhance productivity (Lee and Bozeman 2005, Katz and Martin 1997). Problems with the assignment of authorship credit are not uncommon (Wray 2006), and conflict can dampen productivity. Collaborative outcomes and impacts are far from universal.

As the motivations for collaborating are divergent and numerous, the reasons for shared work affect the dynamics and outcomes of collaborative projects. Collaborative activities often are focused on technology development, software, or patents rather than publications. Thus, Bozeman and colleagues (2012) distinguish between collaboration that produces increments in knowledge and that which produces increments in wealth. In each collaborative relationship, different stakeholders place divergent values on various outputs, affecting the process of collaboration and its perceived effectiveness (Siegel et al. 2003). D'Este and Perkmann's (2011) diverse list of academic scientists' motivations to enter into collaboration with industry highlights these tensions, ranging from commercializing to learning. Complicating matters, these diverse motivations are hard to disentangle from one another. At the organizational level, Kleinman and Vallas (2001) also note the converging and overlapping motivations for science as universities increasingly focus on profit and industrial firms' focus on collaborative science.

In our own research, we are finding that chemical scientists employ narratives about motivations for collaboration that acknowledge the tensions in collaborative work, including the time and funding pressures that often shape collaboration and even organizations in the knowledge economy. One male scientist whom we interviewed discussed how his experience with the "flexibility" of the biotechnology industry during his postdoc led him straight into the academy:

> At the time [I interviewed as a postdoc], two of the big departments [in the firm] were receiving the bulk of their research funds, one was [A], one was [B]. I was there at the time… [Dept. A was] … developing this drug called [X], which turned out to be a really big thing, if you have this specific type of cancer. The [B] department was not putting what they call "candidates" into the pipeline, so they were spending a lot of money, but they weren't coming up with good drug targets. So, when I got there [to the company] a week later, they decided to terminate the entire [B] department, so it was about a hundred people that got fired. Um, probably about 15 or 20 of them were actually running labs. We're talking about people with PhDs, you know, running labs.

In this case, the flexibility of the firm led to shutting down a large portion of the organization and to layoffs. Perhaps this scientist's reaction would have been different if he had been working in the successful department with the "really big" drug coming out. But his observation of others who lost their jobs in the for-profit world led him to believe that financial motivations for scientific collaboration would not permit him the autonomy he wanted:

> I recognized at the time that it's a financial decision. You can't spend 250 million dollars on a department that's not going to make targets. But at the same time I decided… I really enjoy the liberty of you know, really being able to explore.

Not only do motivations for scientific collaboration differ, but collaboration is also not always beneficial for participants. Conflicting interests and the different positions of researchers in social structures mean that teamwork, even among colleagues, is not inherently equitable. While research on the "dark side of collaboration" (Bozeman et al. 2012) has focused mainly on the dilemmas of authorship (for example, Wainwright et al. 2006), ethics and socio-political dynamics (Shrum et al. 2001, 2007), student exploitation (Slaughter et al. 2002, Baldini 2008), and conflict of interests (for example, McCrary et al. 2000), much less in-depth work has been done on gender problems in collaboration. Is there, then, a way to reach some general conclusions about the various gendered effects of collaboration in the knowledge economy? We suggest that several key dimensions for understanding collaboration can serve as a basis for a systematic analysis of its potential for disruption or institutionalization of gender patterns. These include *teamwork dynamics, social and human capital imbalances*, and *collaboration outside of one's field*.

First, network organizations are characterized by both inter-organizational collaboration and teamwork within firms (Smith-Doerr and Powell 2005). Working in more horizontally organized collaborative relationships with other colleagues provides a way of overcoming or mitigating the traditional disadvantages women scientists face when working as individuals, namely the marginalization of their research areas, smaller and less-well-equipped labs, and a general exclusion from institutional power circles (MIT 2002, Roos and Gatta 2006). Pooling, sharing, and exchanging resources thus become not only a means of increasing productivity but a group strategy for circumventing social obstacles. Collaboration, however, may pose different kinds of dilemmas for women employed in network organizations. Working in teams frequently requires one to convince colleagues by emphatically arguing a case. As research on industrial corporations and engineering firms has shown, being perceived as too forceful may be detrimental to women's promotion while passivity could be seen as a weakness and confirm men's preexisting conceptions of women's diminished abilities (Kanter 1977, Evetts 1998).

In our research, we are finding a similar Catch 22 in collaboration, particularly for women scientists. Women's careers are unlikely to advance if they do not collaborate, but collaboration can be a tricky kind of relationship in negotiating credit, division of labor, and responsibility – all of which are processes that may be subtly gendered. The comments of a very senior woman industry scientist (who had academic experience) on how to negotiate collaboration in one's career imply this "damned if you do, damned if you don't" problem. If you collaborate too much, you may be suspect and exposed to non-beneficial collaboration, but if you do not collaborate, you have no career. In the interview, she noted the potential problems with collaboration, such as women not getting enough credit, but then went on to talk about the need for collaboration, if done carefully, in the following excerpt:

> It is important for everyone to find someone in order to be sharing information about your cohort – whatever stage of career you are in. You have to be careful about who you talk to and think about: is the information good for people to know or not? You don't want to discourage young women, for example. So you have to think about what kind of discussion you can have with a particular group. The other question is, do you have a choice about sharing information or is it demanded? Privacy and not prying is important for a healthy climate.

This scientist's comments note that both sharing and withholding information is part of collaboration. She also hints that walking the fine line of doing enough collaboration, but being careful about with whom you collaborate, may be particularly difficult for young women scientists. Still, her message is that mentors should *not* discourage women from collaboration, because it is crucial in every career stage.

Second, social and human capital imbalances should be considered. Organic models of interaction and group development, characteristic of research collaboration, may indeed offer women new avenues for professional growth in the knowledge economy. But the *ad hoc* process of decision making in academia, coupled with a lack of written rules and procedures governing collaborative practices and responsibilities, also accentuates bench scientists' reliance on the lead principal investigators (PIs), who often have little time or interest to pursue issues "outside of the science." In looser structures in academe, where PIs and group supervisors have final authority in both substantive work and administrative issues but have little training as managers, women who are less assertive experience frustrations of not being "in the know" about promotions and professional opportunities (Roth and Sonnert 2010). As the recognized leaders of all academic projects coming out of their labs, PIs tend to receive the credit for the productivity from their lab. Because on average men are more likely to hold positions as PIs and women are more likely to be in supportive roles such as graduate students in labs across many scientific fields, there is a kind of gendered Matthew Effect. Those scientists who are more well-known receive greater credit for the collaborative work (Zuckerman 1988) and those scientists are commonly men (Rossiter 2012 calls this the "Matilda Effect").

Collaboration can also disrupt the traditional connections between social capital and human capital, opening up new opportunities. By underscoring the assets of knowledge and expertise, collaboration may elevate women's overall status in scientific fields. It can also benefit women's individual careers. Take, for example, the story a young woman PhD student told in an interview about how collaboration with someone outside of her university has helped to launch her career:

> I was looking forward to a new project, and talking to [the PI] about it. I had some ideas… she [the PI] talked to [outside collaborator] and he said he was really interested in the project, he wanted to look at the [data with a particular instrument], but he doesn't know how to use [the instrument]. So [the PI] brought up the project and I thought it was really interesting, and she gave me his email, I contacted him, and we began.

Next, the PhD student describes their co-authored publication, her first one, as a "good" experience. She goes on to describe the importance of working closely in this collaboration:

> Well, in the case of [the outside collaborator], he came here to visit, which I think was really important, even though we're not very far along in the project. It's hard to just communicate with somebody by email, but when he came here we were able to throw out ideas and really get to learn what the strengths of the other person, you know, are, and what do they know, and running dialog for a long time. So I think that was important.

This is one example of many that we heard about the benefits of collaboration to individual careers.

Furthermore, in light of the growing significance of occupations that transcend particular organizational cultures (such as IT, where workers in high-tech regions like Silicon Valley may often switch jobs), collaboration offers women alternative networks for cooperation and development that sidestep the constraints posed by particular gendered organizations (Guzman and Stanton 2008). Scientific organizations that encourage personal connections between researchers, but that do so without the transparency, collective rewards, and flexibility that have been found to advantage women scientists' careers in network organizations (Smith-Doerr 2004), may also disadvantage women. Research has shown that men tend to collaborate with more powerful figures (who are usually men), based on network homophily (Ridgeway and Smith-Lovin 1999). Homophily, where "birds of a feather flock together," is a social process by which similar people are more likely to form social ties. The human capital from skill training and career-relevant experience is often garnered through informal ties, but because of gender homophily in informal networks women may disproportionately lack access to key information regarding opportunities for developing their human capital, and for promotion and rewards (Padavic and Reskin 2002). Furthermore, men tend to choose collaborators based on instrumentality and previous experiences, whereas women are less likely to choose collaborators based on whether or not the collaborator has skills complementary to their own (Bozeman and Gaughan 2011). Women may pursue shared research not so much as a strategic calculation of pooling various forms of human capital but for other reasons such as a common work ethic or service in mentoring. At least one study found that women academic scientists at all faculty ranks collaborate with other women in higher percentages than male academics in the same ranks (Bozeman and Corley 2004). Collaboration also provides women with a sense of community, a support network, and an opportunity to relate to others (Pfirman and Balsam 2004).

Third, collaboration with non-researchers or others outside one's field increasingly becomes a crucial component for successfully pursuing team projects. Scientists, however, may develop ambivalent attitudes towards collaborators "outside" of science, in part due to the latter's liminal position (as "outside insiders"). Albert et al. (2008), for example, note the ambivalence with which biomedical scientists treat the social scientists with whom they are supposed to collaborate. In Chapter 21 of this book, Albert and Paradis further demonstrate that social scientists and humanists are the collaborators who must adjust to the epistemic culture of medicine, rather than biomedical collaborators adjusting to incorporate others in interdisciplinary collaboration. More research on whether and how these "outsider" dynamics of collaboration are gendered is needed. Furthermore, a strong sense of group achievement runs the risk of omitting individual women's important contributions to team discoveries – an elision that has characterized much of the history of women in science (Lohan 2000, Wajcman 1991). The lack of formal rules guiding collaborative work results in unarticulated expectations from PIs in academia, which creates frustration among marginal actors (for example, women in administrative support, human resources, and other areas not immediately linked to the scientific mission) who complain about lack of guidance and feedback as well as restricted access to important information on career advancement (Roth and Sonnert 2010). Women collaborators who come from under valued disciplines or subfields may face double suspicion stemming from a sense of superiority over marginalized others. For example, professional women such as product writers and communication specialists have been found to be especially vulnerable to gendered and classed stereotyping while working in collaborative science teams (Brady 2003).

Even among scientific colleagues within the same field, women may be less valued as collaborators. One of our interview respondents from industry argued that evaluation biases against women in scientific careers, evident from the research literature, are also present in collaboration:

> Men and women are perceived differently. Even the space you have is affected by gender [referring to the famous 1999 MIT report on lab size inequalities]. In collaboration, a man and a woman might not ask for things the same way and there are issues with inequity in how they are heard. These are small things rather than overt.

She went on to discuss the differences in collaboration prospects by field, and the importance of women's presence in visible numbers for ease of collaboration:

> In the biological sciences, people collaborate with each other. Women can collaborate with other women, because there are a lot of them. In the harder sciences, there are still issues.

Interestingly, this scientist points to the homophily assumption, that women will collaborate with other women. Her remarks sum up some of the relevant themes on collaboration we are drawing from the interviews with more than 70 chemical scientists: motivations for collaboration matter; women experience a "damned if you do, damned if you don't" perspective on collaboration, and while both men and women clearly articulate the benefits of collaboration, women's contributions to collaborative work may remain undervalued. These seemingly contradictory themes underline our basic argument: collaboration sometimes expands and sometimes contracts gender inequality in science. The next step (which is beyond the scope of this chapter) is to better understand the contexts and conditions where collaboration can disrupt gender inequality.

Conclusion

In this brief essay on women in the knowledge economy, we have argued that the institutionalization and disruption of social forces affecting gender equality can be viewed both historically as a cyclical, chronological process, and analytically as a set of processes occurring at different levels of social action (see Table 22.1). At the level of individuals, and considering the rights of women to work in the knowledge economy, we see historical shifts from the institutionalized exclusion of women from education and paid employment to women's movements that fought for legal changes. These legally instituted changes, however, can be circumvented by subtle, implicit biases toward masculinity that are shared by both men and women. At the organizational level, the shifts are often less visible, being located inside firms or universities rather than discussed in court cases or media representations of women. The gendered organization of science still favors dominant masculinity and hierarchies led by men in ways that usually go unnoticed. New forms of organization (such as the network form that appears in some knowledge economy sectors like biotechnology), however, may disrupt traditional masculine power structures. Collaboration, now a staple feature of scientific knowledge production, allows for an alternative sociality that may undercut cultural and structural patterns of discrimination. The individual competition in classic academic science careers may also be challenged by more collaborative practices.

To better grasp the various links between new forms of knowledge labor and gender inequality, more research is essential. While these links between knowledge production and gender inequality are quite diverse and perhaps even contradictory, we delineate below three broad areas where research on collaboration can inform those concerned with the status of women in the knowledge economy. First, more work on for-profit scientific contexts is crucial not only for discerning the effects of collaborative processes on gendered organizations and institutions operating under commercial logics, but also as a key comparative context to examine related dynamics in academia. As collaborative ties between universities and private industry firms continue to intensify through the sharing of personnel, materials, and technologies, we must understand how collaboration may lead to institutionalization or disruption of gender equality. Collaboration across academia and industry combines market interests and basic research motivations, raising new critical questions about the relationship between gender inequality and motivations for scientific collaboration. We need to know more about whether there is a gendered evaluation bias of collaborators, whether the beneficial outcomes of collaboration are gendered, and whether the motivation for collaboration is related to unequal outcomes in collaboration.

Second, more work should concentrate on the connections between collaboration and gender in the context of increasing tension between the globalization of scientific research and localizing trends in knowledge production. On the one hand, general processes such as outsourcing, migrant knowledge workers, and the increasing precariousness of work suggest that both women and men in high-tech science and engineering face global pressures for increased mobility and non-standard work patterns. On the other hand, national and ethnic identities still play an important part in shaping the possible gendered effects of geographic relocation, job insecurity, and shifting temporal commitments on collaboration. Chapter 23 in this book, by Hee-Je Bak, demonstrates the need for such a nuanced analysis. Bak shows that while embracing a strong commitment to a commercial perspective on the value of scientific knowledge, R&D in South Korea is centrally directed by the government. Although many Korean scientists recognize the value of basic research and intellectual curiosity, public and private pursuits of technoscientific knowledge are merged under an ethos of global competition. What this case tells us is that we need to know more, not only about the dark side of collaboration and its gendered consequences, but also about the different national contexts in which collaboration acquires positive meanings.

Finally, the general acceptance of collaboration in the knowledge economy raises interesting research questions for women in science policy and science education. Given that collaboration is a form of sociality that is both organic and institutionally patterned, can policy interventions create incentives to make collaborative work more equitable? If so, what would these programs look like? In order to facilitate collaboration, issues that should be raised in the policy creation process include: whether people can be professionally trained to collaborate in equitable ways; to what extent power can be neutralized in collaborative settings; and how values such as individuality, competition, and economic rationality can be reconciled with collective, non-instrumental, and educational forms of collaboration.

Note

1   Corresponding author: Laurel Smith-Doerr, Department of Sociology, University of Massachusetts, Amherst, MA 01002; Lsmithdoerr@soc.umass.edu. We would like to acknowledge support from NSF grant 1064121, and our colleagues Jen Croissant, Angela Stoutenburgh, and Claire Duggan. All opinions are those of the authors and do not necessarily reflect the views of the National Science Foundation.

References

Acker, J. 1990. "Hierarchies, Jobs, Bodies: A Theory of Gendered Organizations." *Gender & Society* 4(2): 139–158.

Acker, S. and Feuerverger, G. 1996. "Doing Good and Feeling Bad: The Work of Women University Teachers." *Cambridge Journal of Education* 26(3): 401–422.

Albert, M., Laberge, S., Hodges, B., Regehr, G., and Lingard, L. 2008. "Biomedical Scientists' Perception of the Social Sciences in Health Research." *Social Science & Medicine* 66(12): 2520–2531.

Alvarado, C., Dodds, Z., and Libeskind-Hadas, R. 2012. "Increasing Women's Participation in Computing at Harvey Mudd College." *ACM Inroads* 3(4): 55–64.

Amabile, T., Patterson, C., Mueller, J., Wojcik, T., Odomirok, P., and Marsh, M. 2001. "Academic-Practitioner Collaboration in Management Research: A Case of Cross-Profession Collaboration." *Academy of Management Journal* 44(2): 418–431.

Baldini, N. 2008. "Negative Effects of University Patenting: Myths and Grounded Evidence." *Scientometrics* 75(2): 289–311.

Baron, J., Hannan, M., Hsu, G., and Koçak, Ö. 2007. "In the Company of Women: Gender Inequality and the Logic of Bureaucracy in Start-Up Firms." *Work and Occupations* 34(1): 35–66.

Bauman, Z. 2000. *Liquid Modernity*. Cambridge, UK: Polity Press.

Beaver, D. and Rosen, R. 1978. "Studies in Scientific Collaboration: Part I – The Professional Origins of Scientific Co-authorship." *Scientometrics* 1(1): 65–84.

Beaver, D. and Rosen, R. 1979a. "Studies in Scientific Collaboration: Part II – Scientific Co-authorship, Research Productivity, and Visibility in the French Scientific Elite, 1799–1830." *Scientometrics* 1(2): 133–149.

Beaver, D. and Rosen, R. 1979b. "Studies in Scientific Collaboration: Part III – Professionalization and the Natural History of Modern Scientific Co-authorship." *Scientometrics* 1(3): 231–245.

Becker, G. 1964. *Human Capital: A Theoretical and Empirical Analysis with Special Reference to Education*. Chicago, IL: Chicago University Press.

Bell, D. 1973. *The Coming of the Post-industrial Revolution*. New York, NY: Basic Books.

Benschop, Y. and Doorewaard, H. 1998. "Six of One and Half a Dozen of the Other: The Gender Subtext of Taylorism and Team-based Work." *Gender, Work and Organization* 5(1): 5–18.

Bielby, W. 2000. "Minimizing Workplace Gender and Racial Bias." *Contemporary Sociology* 29(1): 120–129.

Bielby, W. and Bielby, D. 1989. "Family Ties: Balancing Work and Family Commitments in Dual Earner Households." *American Sociological Review* 54(5): 776–789.

Bourdieu, P. and Wacquant, L. 1992. *An Invitation to Reflexive Sociology*. Chicago, IL: Chicago University Press.

Bozeman, B. and Corley, E. 2004. "Scientists' Collaboration Strategies: Implications for Scientific and Technical Human Capital." *Research Policy* 33(4): 599–616.

Bozeman, B. and Gaughan, M. 2011. "How Do Men and Women Differ in Research Collaborations? An Analysis of the Collaborative Motives and Strategies of Academic Researchers." *Research Policy* 40(10): 1393–1402.

Bozeman, B., Fay, D., and Slade, C. 2012. "Research Collaboration in Universities and Academic Entrepreneurship: The-State-of-the-Art." *Journal of Technology Transfer* 38(1): 1–67.

Brady, A. 2003. "Interrupting Gender as Usual: *Metis* goes to Work." *Women's Studies* 32(2): 211–233.

Branch, E. 2011. *Opportunity Denied: Limiting Black Women to Devalued Work*. New Brunswick, NJ: Rutgers University Press.

Britton, D.M. 2000. "The Epistemology of the Gendered Organization." *Gender & Society* 14(3): 418–434.

Britton, D.M. 2010. "Engendering the University through Policy and Practice: Barriers to Promotion to Full Professor for Women in the Science, Engineering and Math Disciplines." In *Gender Change in Academia: Re-mapping the Fields of Work, Knowledge, and Politics from a Gender Perspective*, edited by B. Riegraf, B. Aulenbacher, E. Kirsch-Auwärter, and U. Müller, 15–26. Weisbaden: VS Verlag.

Castells, M. 1996. *The Information Age, Volume 1: The Rise of the Network Society*. Oxford, UK: Blackwell.

Cockburn, C. 1985. *Machinery of Dominance*. London, UK: Pluto Press.

Coontz, S. 2011. *Strange Stirring: The Feminine Mystique and American Women at the Dawn of the 1960s*. New York, NY: Basic books.

D'Este, P. and M. Perkmann. 2011. "Why Do Academics Engage with Industry? The Entrepreneurial University and Individual Motivations." *Journal of Technology Transfer* 36(3): 316–339.

DiMaggio, P. 2001. "The Futures of Business Organization and Paradoxes of Change." In *The Twenty-First Century Firm: Changing Economic Organization in International Perspective*, edited by P. DiMaggio, 210–243. Princeton, NJ: Princeton University Press.

Dobbin, F. 2009. *Inventing Equal Opportunity*. Princeton, NJ: Princeton University Press.

Donato, K. 1990. "Programming for Change? The Growing Demand for Women Systems Analysts." In *Job Queues, Gender Queues: Explaining Women's Inroads into Male Occupations*, edited by B. Reskin and P. Roos, 167–182. Philadelphia, PA: Temple University Press.

Drori, G., Meyer, J., Ramirez, F., and Schofer, E. 2003. *Science in the Modern World Polity: Institutionalization and Globalization*. Stanford, CA: Stanford University Press.

Eaton, S. 2003. "If You Can Use Them: Flexibility Policies, Organizational Commitment, and Perceived Performance." *Industrial Relations* 42(2): 145–167.

Edelman, L. 1992. "Legal Ambiguity and Symbolic Structures: Organizational Mediation of Civil Rights Law." *American Journal of Sociology* 97(6): 1531–1576.

Evetts, J. 1998. "Managing the Technology But Not the Organization: Women and Career in Engineering." *Women in Management Review* 13(8): 283–290.

Faulkner, W. 2000. "Dualisms, Hierarchies and Gender in Engineering." *Social Studies of Science* 30(5): 759–792.

Faulkner, W. 2001. "The Technology Question in Feminism: A View from Feminist Technology Studies." *Women's Studies International Forum* 24: 79–95.

Fox, M.F. 2001. "Women, Science and Academia: Graduate Education and Careers." *Gender & Society* 15: 654–666.

Fox, M.F. and W. Xiao. 2013. "Perceived Chances for Promotion among Women Associate Professors in Computing: Individual, Departmental, and Entrepreneurial Factors." *Journal of Technology Transfer* 38: 135–152.

Fox Keller, E. 1983. *A Feeling for the Organism: The Life and Work of Barbara McClintock*. New York, NY: W. H. Freeman.

Frehill, L. and M. Lee. 2013. "The Role of Foreign-Born Women in the Expansion of the Biomedical Postdoc Workforce." Paper presented at the Eastern Sociological Society meetings, Boston, MA.

Glenn, E.N. 2002. *Unequal Freedom: How Race and Gender Shaped American Citizenship and Labor*. Cambridge, MA: Harvard University Press.

Guthrie, D. and Roth, L. 1999. "The State, Courts, and Equal Opportunities for Female CEOs in U.S. Organizations: Specifying Institutional Mechanisms." *Social Forces* 78(2): 511–542.

Guzman, I. and Stanton, J. 2008. "Women's Adaptation to the IT Culture." *Women's Studies* 37(3): 202–228.

Herman, C., Lewis, S., and Humbert, A. 2013. "Women Scientists and Engineers in European Companies: Putting Motherhood under the Microscope." *Gender, Work and Organization* 20: 467–478.

Hochschild, A. 2012. *The Outsourced Self: Intimate Life in Market Times*. New York: Metropolitan Press.

Huber, J. 1974. "Forget This Career Stuff." *Chronicle of Higher Education*, May 13 issue.

Jacobs, J. and Winslow, S. 2004. "The Academic Life Course, Time Pressures and Gender Inequality." *Community, Work & Family* 7(2): 143–161.

Kalev, A. 2009. "Cracking the Glass Cages? Restructuring and Ascriptive Inequality at Work." *American Journal of Sociology* 114(6): 1591–1643.

Kanter. R.M. 1977. *Men and Women of the Corporation*. New York, NY: Basic Books.

Katz, J. and Martin, B. 1997. "What is Research Collaboration?" *Research Policy* 26(1): 1–18.

Kellogg, K. 2011. *Challenging Operations: Medical Reform and Resistance in Surgery*. Chicago, IL: University of Chicago Press.

Klawe, M. 2011. "Increasing the Participation of Females in Computing Careers." *Journal of Computing Sciences in Colleges* 27(1): 98–100.

Kleinman, D. and Vallas, S. 2001. "Science, Capitalism, and the Rise of the 'Knowledge Worker': The Changing Structure of Knowledge Production in the United States." *Theory & Society* 30(4): 451–492.

Leadbeater, C. 2000. *Living on Thin Air: The New Economy*. London, UK: Penguin.

Lee, S. and Bozeman, B. 2005. "The Impact of Research Collaboration on Scientific Productivity." *Social Studies of Science* 35(5): 673–702.

Lindsay, S. 2008. "The Care-tech Link: An Examination of Gender, Care and Technical Work in Healthcare Labor." *Gender, Work & Organization* 15(4): 333–352.

Lohan, M. 2000. "Constructive Tensions in Feminist Technology Studies." *Social Studies of Science* 30(6): 895–916.

Margolis, J. and Fisher, A. 2001. *Unlocking the Clubhouse: Women in Computing*. Cambridge, MA: MIT Press.

Mason, M. and Goulden, M. 2004. "Marriage and Baby Blues: Redefining Gender Equity in the Academy." *Annals of the American Academy of Political and Social Science* 596(1): 86–103.

Massachusetts Institute of Technology. 2002. "A Study on the Status of Women Faculty in Science at MIT." *Report of the School of Science, Committees on the Status of Women Faculty*. Cambridge, MA: MIT Press.

McCrary, S., Anderson, C., Jakovljevic, J., Khan, T., McCullough, L., Wray, N., and Brody, B. 2000. "A National Survey of Policies on Disclosure of Conflicts of Interest in Biomedical Research." *New England Journal of Medicine* 343(22): 1621–1626.

Melin, G. 2000. "Pragmatism and Self-organization: Research Collaboration on the Individual Level." *Research Policy* 29(1): 31–40.

Melin, G. and Persson, O. 1996. "Studying Research Collaboration Using Co-authorships." *Scientometrics* 36(3): 363–377.

Merton, R. 1968. "The Matthew Effect in Science." *Science* 159(3810): 56–63.

Merton, R. 1973. *The Sociology of Science: Theoretical and Empirical Investigations*. Chicago, IL: University of Chicago Press.

Miller, G. 2004. "Frontier Masculinity in the Oil Industry: The Experience of Women Engineers." *Gender, Work and Organization* 11 (1): 47–73.

Misra, J., Lundquist, J., and Templer, A. 2012. "Gender, Work Time, and Care Responsibilities among Faculty." *Sociological Forum* 27(2): 300–323.

Moore, K., Kleinman, D., Hess, D., and Frickel, S. 2011. "Science and Neoliberal Globalization: A Political Sociological Approach." *Theory and Society* 40(5): 505–532.

Moss-Racusin, C., Dovidio, J., Brescoll, V., Graham, M., and Handelsman, J. 2012. "Science Faculty's Subtle Gender Biases Favor Male Students." *Proceedings of the National Academy of Sciences* 109(41): 16474–16479.

National Science Foundation. 2011. Survey of Earned Doctorates. *www.nsf.gov/statistics/sed/2011/start.cfm*. Washington, DC: National Science Board.

National Science Foundation. 2012. *Science and Engineering Indicators*. Washington, DC: National Science Board.

Oldenziel, R. 1999. *Making Technology Masculine: Men, Women and Machines in America 1870–1945*. Amsterdam, NL: Amsterdam University Press.

Ong, M. 2005. "Body Projects of Young Women of Color in Physics: Intersections of Gender, Race, and Science." *Social Problems* 52(4): 593–617.

Padavic, I. and Reskin, B. 2002. *Women and Men at Work*. 2nd edition. Thousand Oaks, CA: Pine Forge Press.

Pfirman, S. and Balsam, P. 2004. "Women and Interdisciplinary Science: Promise and Peril." *Women, Work, and the Academy.* Barnard Center for Research on Women (9–10 December). *www.barnard.edu/bcrw/womenandwork/pfirman.htm.*

Powell, W. 1990. "Neither Market nor Hierarchy: Network Forms of Organization." *Research in Organizational Behavior* 12: 295–336.

Powell, W. and Snellman, K. 2004. "The Knowledge Economy." *Annual Review of Sociology* 30: 199–220.

Powell, W., Koput, K., and Smith-Doerr, L. 1996. "Interorganizational Collaboration and the Locus of Innovation: Networks of Learning in Biotechnology." *Administrative Science Quarterly* 41(1): 116–145.

Pycior, H. 1993. "Reaping the Benefits of Collaboration While Avoiding Its Pitfalls: Marie Curie's Rise to Scientific Prominence." *Social Studies of Science* 23(2): 301–323.

Ridgeway, C.L. 2001. "Gender, Status, and Leadership." *Journal of Social Justice* 57(4): 637–655.

Ridgeway, C.L. and Smith-Lovin, L. 1999. "The Gender System and Interaction." *Annual Review of Sociology* 25: 191–216.

Roberts, P. and Ayre, M. 2002. "Did She Jump or Was She Pushed? A Study of Women's Retention in the Engineering Workforce." *International Journal of Engineering Education* 18: 415–421.

Roos, P. and Gatta, M. 2009. "Gender (In)Equity in the Academy: Subtle Mechanisms and the Production of Inequality." *Research in Social Stratification and Mobility* 27(3): 177–200.

Rossiter, M. 1998. *Before Affirmative Action: Women Scientists in America, 1940–1972.* Baltimore, MD: Johns Hopkins University Press.

Rossiter, M. 2012. *Women Scientists in America: Forging a New World since 1972.* Baltimore, MD: Johns Hopkins University Press.

Roth, W. and Sonnert, G. 2010. "The Costs and Benefits of 'Red Tape': Anti-bureaucratic Structure and Gender Inequity in a Science Research Organization." *Social Studies of Science* 41(3): 385–411.

Sappleton, N. and Takruri-Rizk, H. 2008. "The Gender Subtext of Science, Engineering & Technology (SET) Organizations: A Review and Critique." *Women's Studies* 37: 284–316.

Schwarz-Cowan, R. 1979. "From Virginia Dare to Virginia Slims: Women and Technology in American Life." *Technology and Culture* 20(1):51–63.

Sennett, R. 1998. *The Corrosion of Character: The Personal Consequences of Work in the New Capitalism.* New York, NY: Norton.

Shrum, W., Chompalov, I., and Genuth, J. 2001. "Trust, Conflict and Performance in Scientific Collaborations." *Social Studies of Science* 31(5): 681–730.

Shrum, W., Chompalov, I., and Genuth, J. 2007. *Structures of Scientific Collaboration.* Cambridge, MA: MIT Press.

Siegel, D., Waldman, D., and Link, A. 2003. "Assessing the Impact of Organizational Practices on the Relative Productivity of University Technology Transfer Offices: An Exploratory Study." *Research Policy* 32(1): 27–48.

Slaughter, S., Campbell, T., Folleman, M., and Morgan, E. 2002. "The 'Traffic' in Graduate Students: Graduate Students as Tokens of Exchange Between Academe and Industry." *Science, Technology and Human Values* 27(2): 282–313.

Smith, V. 2001. *Crossing the Great Divide: Worker Risk and Opportunity in the New Economy.* Ithaca, NY: Cornell University Press.

Smith-Doerr, L. 2004. *Women's Work: Gender Equity vs. Hierarchy in the Life Sciences.* Boulder, CO: Lynne Rienner Press.

Smith-Doerr, L. 2005. "Institutionalizing the Network Form: How Life Scientists Legitimate Work in the Biotechnology Industry." *Sociological Forum* 20(2): 271–299.

Smith-Doerr, L. 2011. "Contexts of Equity: Thinking about Organizational and Technoscience Contexts for Gender Equity in Biotechnology and Nanotechnology." In *Nanotechnology and the Challenges of Equity, Equality and Development: The Yearbook of Nanotechnology in Society 2,* edited by S. Cozzens and J. Wetmore, 3–22. New York: Springer.

Smith-Doerr, L. and Powell, W. 2005. "Networks in Economic Life." In *The Handbook of Economic Sociology,* 2d ed., edited by N. Smelser and R. Swedberg, 368–402. Princeton: Princeton University Press and Russell Sage Foundation.

Sonnert, G. and Holton, G.1996. "Career Patterns of Women and Men in the Sciences." *American Scientist* 84(1): 63–71.

Tam, T. 1997. "Sex Segregation and Occupational Gender Inequality in the United States: Devaluation or Specialized Training?" *American Journal of Sociology* 102(6): 1652–1692.

Tomaskovic-Devey, D. and Skaggs, S. 2002. "Sex Segregation, Labor Process Organization, and Gender

Earnings Inequality." *American Journal of Sociology* 108(1): 102–128.

Valian, V. 1999. *Why So Slow? The Advancement of Women*. Cambridge, MA: MIT Press.

Vallas, S. and Kleinman, D. 2007. "Contradiction, Convergence and the Knowledge Economy: The Confluence of Academic and Commercial Biotechnology. *Socio-Economic Review* 6(2): 283–311.

Wainwright, S., Williams, C., Michael, M., Farsides, B., and Cribb, A. 2006. "Ethical Boundary-work in the Embryonic Stem Cell Laboratory." *Sociology of Health & Illness* 28(6): 732–748.

Wajcman, J. 1991. *Feminism Confronts Technology*. Cambridge, UK: Polity Press.

Walby, S. 2011. "Is the Knowledge Society Gendered?" *Gender, Work and Organization* 18(1): 1–29.

Walby, S., Gottfried, H., Gottschall, K., and Osawa, M. eds. 2007. *Gendering the Knowledge Economy: Comparative Perspectives*. Basingstoke, UK: Palgrave Macmillan.

Wray, K. 2006. "Scientific Authorship in the Age of Collaborative Research." *Studies in History and Philosophy of Science Part A* 37(3): 505–514.

Wright, R. and Jacobs, J. 1994. "Male Flight from Computer Work: A New Look at Occupational Resegregation and Ghettoization." *American Sociological Review* 59(4): 511–536.

Wuchty, S., Jones, B., and Uzzi, B. 2007. "The Increasing Dominance of Teams in Production of Knowledge." *Science* 316(5827): 1036–1039.

Zippel, K. and Frehill, L. 2011. "Gender and International Collaborations of Academic Scientists and Engineers: Findings from the Survey of Doctorate Recipients, 2006." *Journal of the Washington Academy of Sciences* 97(1): 49–69.

Zuckerman, H. 1967. "Nobel Laureates in Science: Patterns of Productivity, Collaboration, and Authorship." *American Sociological Review* 32(3): 391–403.

Zuckerman, H. 1988. "The Sociology of Science." In *Handbook of Sociology*, edited by N. Smelser, 511–574. Newbury Park: Sage.

23

The Utilitarian View of Science and the Norms and Practices of Korean Scientists[1]

Hee-Je Bak

KYUNG HEE UNIVERSITY, KOREA

The utilitarian view of science and national interest in science

Although science is commonly viewed as universal and an international ethos prevails among scientists, scientific work and careers are frequently linked to a nation and to utilitarian goals. This is not surprising, given the fact that major sources of funding for research and development (R&D) are overwhelmingly national. In particular, major national governments have viewed science and technology as national resources and attempted to mobilize them in the interest of each nation. Since the 1960s, the effect of science and technology on national economies has been a growing concern. A number of nation-states have made efforts to bolster economic competitiveness by promoting scientific advance. Science and technology policies in most countries have thus focused on how to better harness R&D and position strategic technologies for industrial competitiveness (Branscomb and Florida 1999, Elzinga and Jamison 1995, Ergas 1987, Okimoto 1989).

Regardless of the common goal, actual policies depend heavily on the political and social conditions in each country. In the U.S., the government's involvement in the economy tends to be viewed with suspicion, and how much and in which way the state should be involved in scientific enterprise for economic development has been controversial (Branscomb and Florida 1999, Kleinman 1995). In contrast, countries where the central government has acted aggressively for economic development have been more likely to be involved deeply in scientific enterprise (Choi 1996, Evans 1995).

This chapter discusses the ways in which the Korean government has mobilized science and technology for Korea's economic development and in so doing, how its utilitarian view of science has affected the norms and practices of Korean scientists. As an exemplar of developmental states, South Korea has long been known for its active role in economic development. The efforts of the Korean government to mobilize scientific and technological resources to move its economy beyond labor-intensive industries are not especially well known outside the country. In early 1960s, the Korean government set its first national plan for science and technology promotion, brought a number of Korean scientists back from abroad, and set up state-funded research institutes to assist local firms, which were struggling to enter knowledge-based industries. Since the 1980s, it has also pushed university research toward applied and developmental research by distributing government fund in these directions. In this chapter, I argue that the Korean government's efforts to mobilize science and technology for economic development not only had a distinctive impact on the institutional development of Korean science but also encouraged Korean scientists to embrace the utilitarian values of science that view scientific research as a tool for industrialization and nationalism in science that equates scientific advance with national progress.

The utilitarian view and nationalism in science have been associated with Korean scientists' adherence to "post-academic" norms and practices of science (Ziman 2000). The promotion of science and technology for economic development often went hand-in-hand with an emphasis on commercializing scientific research results. Science and technology studies (STS) research on privatization and commercialization of science in the U.S. has pointed to the ways that university-industry ties undermine academic norms of science, such as open sharing of research results and encourage post-academic norms of science, and the pursuit of applicability of research and intellectual property rights (Bok 1982, Croissant and Smith-Doerr 2008, Krimsky 2003, Rosenzweig 1985). However, in the Korean case it has been the role of government rather than university-industry relationships that has prompted movement away from academic and toward post-academic norms. In the U.S., for instance, academic scientists have become increasingly concerned with intellectual property rights (a post-academic orientation) as their research has connected them with private firms. By contrast, in Korea, the government has played a principal role in promoting the interest of academic scientists in applicability of their research and intellectual property rights. The Korean government has prompted scientists to identify the pursuit of applicability of their research with their contribution to national progress, which, in turn, helped them accept the norm of the pursuits for intellectual property rights and even the norm making secrecy among scientists acceptable.

In the following pages, I first discuss the critical role of the Korean government in industrial transformation of the country and the mobilization of scientific research for industrialization. Then, I describe the norms and practices of Korean scientists in terms of disinterestedness and communality and discuss the influence of the Korean government on scientists' norms and practices.

The central role of the Korean government in economic development

After thirty-five years of Japanese occupation and the Korean War, South Korea was economically devastated. As late as 1960, its economy was barely sustained by economic aid and military assistance from the U.S. However, after just one generation, South Korea transformed itself from one of the poorest countries in the world to one of the world's economic powerhouses, especially in high technology. A number of Korean companies, such as Samsung, LG, and Hyundai, are now known as world leaders in semiconductors, mobile phones, electronic displays, chemical cells, shipbuilding and the auto industry.

Scholars have been interested in and attempted to account for this exceptional success story of late industrialization. Among many explanatory factors, the major role of the state has often been emphasized (Amsden 1989, Evans 1995). The Korean government has planned and orchestrated economic development of the country since the mid-1960s, although the growth of private companies and the globalization process have been reducing the power of the state significantly in recent years.

Many observers considered the first five-year plan for national economic development launched in 1962 as the beginning of economic development in Korea (Kim and Leslie 1998).

From that time, economic development in Korea had clear ideological motivations and implications. First of all, after seizing power by a military coup in May 1961, President Park Chung Hee and his colleagues tried to legitimize the military coup with the rapid economic development of the country. Furthermore, the Korean government presented the nation not just as being better off economically than in prior years, but also as a rapidly modernizing country easily surpassing its nemesis, North Korea (Kim and Leslie 1998). Economic development and industrialization was promoted as a national calling and even the raison d'être of the nation, and science was portrayed as a major contributor to the country's progress.

Through the series of national economic development plans, the Korean government envisioned three steps of economic development. The first was to develop export sectors in labor-intensive light assembly industries, such as textiles, by capitalizing mainly on its comparative advantage in labor costs. The next step was to develop capital-intensive heavy industries, such as steel. The third was to develop knowledge-intensive high technology industries, such as electronics. To accomplish this, the Korean government nurtured the country's research capabilities and linked them to local industries, in some cases taking over R&D for private firms (Evans 1995, Kim and Leslie 1998). The vision of transforming its economy into a knowledge intensive one finally came true when Korea became the world leader in DRAM (dynamic random access memory) semiconductors in the 1990s.

National mobilization of scientific resources for economic development

The Korean government has long acknowledged the critical role of scientific research in the Korean economy's attempts to advance in areas such as chemicals and high technologies (Amsden 1989, Choi 1996, Evans 1995). In this context, in modern Korea, scientists and scientific research have been treated explicitly as national resources to be mobilized for the national goal of economic development (Kim 2010). The following briefly describes three cases that demonstrate how the Korean government mobilized science and technology for its national goals for utilitarian purposes.

National planning for mobilizing science for economy

Since the 1960s, the Korean government has promoted economic development of the country through the five-year plans for national economic development. Science and technology policies in Korea were integrated with the economic development plans. Therefore, the Economic Planning Board, instead of Department of Education, was responsible for the first five-year plan for technology promotion, the first official science and technology policy in Korea, which was initiated in 1962. The subtitle of the plan for technology promotion was "Supplement to the first five-year plan for national economic development," and the primary goal of it was securing technical personnel needed for economic development (Jeon 1982, Song 2007). The name of the plan changed to the plan for science and technology promotion in 1966 with the introduction of the second five-year plan. The close tie between the plan for science and technology promotion and the plan for national economic development continued into the 1990s. As late as 1993, the general plan for national science and technology promotion was included as a subsection of the five-year plan for new economy (Song 2007).

The Economic Planning Board was also responsible for the development of human resources. It prepared for the first "Five-year plan for human resources development (1962–1966)" in accordance with the five-year plan for national economic development. One of the main goals of the plan was to coordinate efforts to reach the total university enrollment quota in science and engineering fields. Despite the opposition of the Department of Education, it was determined that the Economic Planning Board set the enrollment quota in science and engineering fields (Jeon 1982: 122–131). These examples demonstrate the nature of the Korean government's interest in science. South Korea integrated its national science and technology policy explicitly into its national economic plan and thus set the stage for Korean scientists' adherence to post-academic norms.

State-funded research institutes for industrial needs

Another important aspect of the Korean government's economic development strategy was the establishment of public research institutions. In 1966, the Korea Institute of Science and Technology (KIST) was founded in the interest of economic development. Making critical contributions to the growth of the heavy chemicals industries in Korea in the 1960s and 1970s, it became known as the most successful model of a research institute for economic development in the developing world (Kim and Leslie 1998). Unlike other research institutes in developing countries at that time that supported the conduct of basic research, KIST aimed to assist Korean industries by solving problems of direct interest to them. That is, KIST was set up and operated to assist domestic industry, importing appropriate technologies and applying them in actual production and supporting research aimed at solving problems of concern to local firms (Kim and Leslie 1998, Moon 2006, Park et al. 2001).

KIST made important contributions to the growth of the shipbuilding, steel, chemical, and electronic industries in Korea. The success of KIST led the Korean government to set up subsequent state-funded research institutes dedicated to research relevant to specific knowledge-intensive industries, including the Korea Institute of Machinery and Materials, the Korea Institute of Chemical Technology, and the Electronics and Telecommunications Research Institutes. In 1970, these state-funded institutes accounted for 58.51% of total national R&D expenditures. As the Korean economy has developed, this proportion has declined to about 12.67% in 2010 and the proportion of total national R&D expenditures industry is responsible for has increased from 12.56% to 74.80% during the same period.

When the Korean government set up KIST and other public research institutes, one of the daunting tasks it faced was recruiting qualified researchers. The Korean government tried to bring the many Korean scientists working abroad back home. Toward this end, the Korean government provided not only financial incentives, but also portrayed these scientists as national heroes dedicated to creating national economic prosperity. It has been underscored through the media that these scientists were persuaded to come back home because of their sense of mission: they should help the country make economic progress through industrialization (Choi 1995). With such general support by the government, KIST researchers enjoyed a high social status (Moon 2006).

Shaping university science's movement toward applied research

The way the Korean government supported academic science also reveals its utilitarian view of science. In the 1960s and 1970s, the Korean government viewed universities primarily as educational institutions and provided minimal support for academic research. The proportion of university R&D expenditure to total national R&D expenditure remained less than 10% until 1977 when the Korean Science Foundation was established, signaling a leap in the government's financial support for university research.

One of the reasons for the Korean government's initial half-hearted support for university research was science policymakers' mistrust of university scientists. From the Korean government's utilitarian view of science, what Korea needed was applied research, which would be able to assist Korean industry directly. In the eyes of many bureaucrats, however, university professors resided in an ivory tower and conducted research for personal curiosity irrelevant to national needs (Jeon 1982). This mistrust explains the Korean government's establishment of free-standing state-funded research institutes and also a new graduate school specifically for educating scientists and engineers to aid industry – Korea Advanced Institute of Science (later, Korea Advanced Institute of Science and Technology, KAIST). KAIST's environment supported the utilitarian purpose of science for industry and national defense. Most student theses dealt with practical questions related to Korean industry and national defense, and about 70% of KAIST students entered industry and state-funded research institutes after getting their degrees (Park et al. 2001).

The increasing prominence of industry-funded science prompted discussion over how to coordinate the best public and private research. In the wake of those discussions, the Korean government began to support large scale, long-term national R&D projects for fundamental technologies at state-funded research institutes. At the same time, it attempted to take advantage of the potential of university research capacities and increased its support in that area (Hwang and Yun 2003). From 1985 to 2010, university R&D expenditure increased 12.67 times, even after controlling for inflation. Government funding was responsible mainly for the increased research in universities, so that the proportion of government support among the university R&D expenditure increased from 8.8% in 1987 to 72.4% in 2010. Without question, the Korean government became the major supporter of university research. In 2010, private companies were the second largest supporter of university research, accounting for slightly over 10% of total university R&D expenditure.

Through the policy judgment of "selecting and concentrating," the Korean government's support for scientific research has concentrated on a few areas that are believed to have high potential for commercialization. Consequently, as government support for university research has increased remarkably, so has the proportion of that research devoted to application and development. Korean university scientists have thus become increasingly concerned with the possibility of commercializing their research and directing it toward applied and developmental ends. Although the Korean government's support for basic research increased rapidly, its support for applied and developmental research increased even faster. The proportion of applied and developmental research in university R&D expenditure increased from 26.92% and 7.22% in 1983 to 32.81% and 31.20% in 2010, respectively, while the proportion of basic research decreased from 65.86% to 35.34% during the same period. In contrast, university research in major industrial countries, such as the U.S., Japan, Germany, or France, continued to focus on basic research: in terms of R&D expenditure, the proportion of basic research in university research in these countries has remained high, from around 50% in Japan to around 90% in France, and relatively stable during the same period (Bak 2006).

Culture of Korean scientists: embracing the utilitarian view of science

The economic development motivation for government-supported scientific research in modern Korea has inhibited the institutionalization of pure Mertonian-type scientific norms among Korean scientists. Efforts by the Korean government to mobilize science to support economic development encouraged scientists to embrace a utilitarian view of science, which Ziman has suggested is central to the "post-academic" norms of science (Ziman 2000). In this section, I first demonstrate that post-academic norms of science are more prominent than pure

Mertonian-type norms among scientists in the Korean public sector. Then I will examine how these scientists embrace the utilitarian view of science, focusing on the role of the state and the nationalist view of science.

Post-academic norms over Mertonian norms of science

Many analysts have suggested that a broad transformation in the institutional structure of science is underway (Berman 2011, Croissant and Smith-Doerr 2008, Etzkowiz 1989, Etzkowiz et al. 1998, Gibbons et al. 1994, Kleinman 2003, Slaughter and Rhodes 1996, 2004, Slaughter and Leslie 1997, Vallas and Kleinman 2008, Ziman 2000, see also Vardi and Smith-Doerr, Chapter 22 of this volume). According to some of these analysts, these structural changes have been accompanied by changes in the culture of science. A number of these studies, which focus on western societies, have suggested that the norms and practices of industrial science, including valuing industrially applicable knowledge and using of secrecy in the interest of protecting intellectual property rights, have been integrated into academic science where the ideal norms of science, such as communality and disinterestedness, had previously been dominant. Moreover, this literature has also reported that despite some integration, there has been a growing tension between the academic and industrial scientific cultures in university settings (for example, Blumenthal et al. 1996, 1997, Krimsky 2003, Ziman 2000).

In Korea, while there is some tension between the two scientific cultures, the norms and practices of industrial science have been prominent even among scientists in the public sector. Korean scientists have embraced the utilitarian view of science, since the earliest efforts to use science to bolster economic development. The results of "the Survey of the Norms and Values of the Korean Scientific Community (hereafter Scientific Norms Survey)" conducted in 2006 demonstrate that the majority of Korean scientists in the public sector embrace the utilitarian view of science. This web-based survey was carried out by the Research Institutes of Information Society at Kyung Hee University with the support of the Korea Research Foundation. Using the stratified random sampling method, the survey was administered to 435 professors in 16 universities and 252 researchers in seven state-funded research institutes.

The Scientific Norms Survey is composed of two components. The first includes an array of questions about scientists' motivations for becoming scientists. The second focuses on scientists' commitment to traditional academic norms of openness and communality. Putting these two portions of the survey together points at once to Korean scientists' simultaneous commitment to academic and post-academic norms and their unambiguous commitment to the use of science for utilitarian ends.

The results of the Scientific Norms Survey suggest that, before becoming scientists, Korean scientists had a traditional Mertonian image of science. The survey asked scientists to reveal their major motivations for choosing their career in science by selecting 2 from 8 possible options. Among them, "To fulfill intellectual desires (61.3%)" was followed by "Autonomy of research (43.6%)." At the same time, 27.6% of respondents indicated that "Contributions to the national advancement" was one of their two motivations for following a scientific career and 15.9% indicated that "Contribution to the welfare of humankind (15.9%)" was important. For scientists who were over 60 years old, those motivated by advancing the national interest was higher than for the sample as a whole at 34.1%. Thus, in response to this question, the profile of respondents suggests a mix of academic motivations (autonomy, etc.) and a more utilitarian orientation (a commitment to national advancement and human welfare).

Even while the majority of scientists were motivated by curiosity and work autonomy, traditional motivations, their perspectives regarding actual scientific practices were far from the traditional ideology of science. The Scientific Norms Survey includes a wide range of questions about scientific norms. Among them, the question items measuring the norms of disinterestedness and communality are particularly relevant to this study, since these two norms are implicitly against the utilitarian view of science.

Disinterestedness was defined by Robert Merton as the principle that scientific activities be independent of scientists' individual prejudice (Merton 1973); however, its meaning later extended to include independence from the influence of external interests (Gaston 1978). Thus, disinterestedness typically refers to a principle that scientific activities should be guided only by the pursuit of truth and be independent of the religious and political beliefs, economic gain, or the popularity of a research topic.

Merton also identified communality as a norm of science, which has been frequently discussed in relation to commercialization of science. The norm of communality calls for scientists to communicate their findings publicly, reflecting the fact that scientific research relies upon previous scientific work accumulated by other scientists. Therefore, the norm of communality certainly conflicts with the culture of commerce in which privatization is used to generate profit and research secrecy is a common practice. Indeed, as the commercialization process has increasingly penetrated into U.S. academic science, there has been growing concern about secrecy and delayed publication among scientists (Bok 2003, Grobstein 1985, Krimsky 2003). In particular, intellectual property rights have been criticized for discouraging information exchanges and thus slowing scientific advance. It has been reported that scientists hold their findings for quite some time due to intellectual property rights concerns and contractual agreements with industrial research funders (Blumental et al. 1996, 1997).

In the Korean survey, one item directly measured respondents' perceptions of the emphasis on the applicability of science, which may be viewed as being against the norm of disinterestedness. For the statement that "Scientists must pay attention to the applicability (the potential of commercialization) of their research," positive responses were overwhelming. About two thirds agreed (agree: 52.3%, strongly agree: 15.8%) with the statement, while only 6.8% disagreed (disagree: 5.6%, strongly disagree: 1.2%), with about 25% of responses in the neutral category. In contrast, much smaller portions of respondents agreed with the statement that "A research topic should be selected (by scientists) only based upon intellectual curiosity and scientific consideration." About 40% of respondents agreed (agree: 31.7%, strongly agree: 9.8%), while about 38% disagreed (disagree: 29.3%, strongly disagree: 7.3%) and 22% of respondents exhibited a neutral position.

The results from the survey also suggest that Korean scientists generally view bestowing intellectual property rights on scientists as legitimate. Most responses to the statement "Bestowing intellectual property rights for new scientific findings is a legitimate reward to scientists and funding organizations," were favorable. The majority of Korean scientists, 61.5% of respondents, agreed with the statement and 21.3% strongly agreed. Fewer than 4% disagreed with the statement, and 13.5% provided neutral responses. In contrast, respondents revealed a lukewarm support for the statement "Since scientific findings are the common good, all the scientific findings should be released openly without restriction." Fewer than half of the respondents agreed (34.9% "agree" and 12.0% "strongly agree"), while about 30% of respondents disagreed, with about 25% providing neutral answers.

The most striking finding concerns the respondents' attitude toward secrecy in research activities. The survey presented scientists the following proposition: "If necessary, the publication of a research article can be delayed more than six months to secure intellectual property rights." The period of six months has been used as the boundary between traditional academic research practice and secrecy in previous research (Blumenthal et al. 1996, 1997). In Korea, scientists widely viewed substantial publication delays as legitimate: 61.1% of respondents agreed with the above statement and 16.5% strongly agreed. Only 6.0% answered "disagree" and 0.9% "strongly disagree." 15.5% took a neutral position.

In sum, the Scientific Norms Survey shows that while characteristics traditionally associated with academic science motivated many Korean scientists to choose their career in science, the majority of Korean scientists in practice expressed the utilitarian view of science by emphasizing industrially applicable research and the legitimacy of secrecy in the interest of intellectual property protection. While there are ways in which survey respondents' answers could be seen as contradictory, I would suggest that overall post-academic norms have a stronger hold on the Korean scientific community than pure Mertonian norms.

The role of the state and the nationalist view of science

This section discusses how Korean scientists in the public sphere came to embrace the utilitarian view of science and, in so doing, highlights the critical role of the state and the nationalist view of science in shaping norms of science. For this purpose, I rely on the results of in-depth interviews I conducted with thirty-two Korean university scientists between 2003 and 2005. The purposive sampling method was used to make sure that interviewees have as many diverse characteristics as possible. A total 12 physicists and 20 biologists in six universities were interviewed. 14 interviewees were professors, 4 associate professors, and 14 assistant professors at the point of interview. Each interview took about 1.5 to 2 hours.

In the interviews, few scientists contended that scientists should be interested in only scientific values when selecting research topics. Almost all interviewees pointed out the prospects for funding was one of the most important factors in selecting a research topic. They tended to acknowledge that the interest from outside the scientific community narrowly defined has influenced their process of selecting research topics, and at the same time, they asserted that the influence of the Korean government in distributing research funds was more important in topic selection than was direct connection with industry. The following remark from a young physics professor conducting research on an electronic display highlights the effect of the state:

> Our country has maintained a "selecting and concentrating" policy. Under the policy, the state demands specifying concentrated research areas from among selected areas. Then, [when] what I want to research is … not included in the selected areas … I have to ponder between what I want to research into and what the state demands, because if I conducted the research, then, I might face problem in funding. Research requires funding! … Let's say a project can get funding more easily. Then, the center of gravity tends to move toward it. A display is one of the selected areas [by the state].

This interview and many others suggest that governmental funding for research has shaped research practices of Korean scientists in the public sector. Indeed, almost all interviewees confirmed that it is quite common that scientists change their research topics from their original interest to one with better funding opportunity. "There are many researchers around who changed their research topics [due to funding opportunities]" said a biologist. Another interviewee remarked that funding pushes scientists toward applied research. A cytochemistry researcher argued "In the past, molecular biologists tended to conduct pure molecular biological research. But because it [getting funding] didn't work well, they have increasingly cooperated with agricultural researchers."

An especially interesting finding is the way Korean scientists justified the post-academic norms of science. Selecting research topics for funding opportunities may be viewed as a behavior against the norm of disinterestedness. However, in interviews, few respondents pointed to this tension. On the contrary, Korean scientists in the public sector tended to emphasize the applicability of selected research topics in terms of the contribution to national progress: since the government would allocate the funding for national needs, selecting research topics for the funding opportunity means working for the national need rather than following personal interest. Similarly, even those who conduct mainly basic research emphasized applicability in selecting research topics. The basic scientists certainly were more likely to argue that more support for basic research would be needed for scientific advance. However, they based their argument on the claim of applicability, suggesting that basic research would generate scientific findings on which industrial application would be carried out later. Indeed, more than half of basic scientists interviewed mentioned that science policies emphasizing industrial applicability would be inevitable, given Korea's economic status.

Also, in the interviews consistent with the survey results discussed above, most scientists regarded intellectual property rights as rightful rewards for scientific endeavor. That said, there was some disagreement between scientists doing basic and those doing applied research. While applied scientists tended to have a positive view of intellectual property protection, a few of the scientists I interviewed, especially those who conduct basic research, expressed their concern about slower information flow due to intellectual property rights considerations. Still, when their research was potentially profitable, many scientists, including basic researchers, ranked intellectual property rights as more crucial than publishing a journal article and even accepted secrecy as legitimate. According to a theoretical physicist,

> We cannot prevent research from moving toward commercialization. ... Scientists who conduct pure research like me do not need to pay attention to it [a patent]. ... Applied research should go for patents. I don't feel uncomfortable with the fact that a certain scientific finding is not being published [and is instead protected as intellectual property].

Such responses might not be surprising, given that the majority of Korean scientists tend to share a strong utilitarian view of science. An interviewee reported that her research team decided not to publish their research on the rice genome in *The Plant Cell* because in the review process, the journal asked authors to release the database without guaranteeing their ownership rights:

> Rice is a crop with a lot of economic potential. So the request of offering it [the database] completely is, we think, against the interest of the Korean government who funded this research. In fact, the genome database of Arabidopsis [Arabidopsis thaliana] is free of charge. But it is not a cash crop but a crop for basic science. Then, can we apply the same principle to cash crops? .... If we release the database and fail to obtain patents, then what we do in Korea is merely informatization – creating a database. We worried about losing economic opportunities.

This example demonstrates graphically the view common among Korean scientists, as made clear in the survey data, that economic interest and secrecy are more important than the norm of communality. Especially interesting in this case is the way the interviewee justified her research team's decision to give up publishing their findings by connecting the benefits of patenting their research outcome to national economic interest instead of scientists' individual rewards. She emphasized that not just economic opportunity but national interest was at stake.

Noting that the Korean government supported their research, she also claimed that her research team was very willing to provide their data to other Korean scientists upon request. In doing so, she identified intellectual property rights of her research with national interest and, thus, justified her research team's decision to give up publication.

Linking intellectual property rights to national interest was quite common in the interviews I conducted. Indeed, national wealth was one of the most frequently used words when interviewees talked about intellectual property rights. A physicist noted, for example, that "If we applied well, having intellectual property rights may yield wealth to the nation." Moreover, appealing to nationalism in science, some interviewees advocated secrecy more bluntly. As a young plant biologist remarked when talking about intellectual property rights: "I am thinking over national wealth that we may be able to obtain through secrecy rather than worrying about it."

Another notable finding from the interviews relates to the ways in which scientists became interested in intellectual property rights. Interviewees tended to mention the influence of the Korean government rather than the commercialization of science itself or the pull of the private sector. The literature on commercialization of U.S. academic science has pointed out that the culture of commerce stemming from industry-university ties was a primary factor in explaining academic scientists' acceptance of intellectual property rights and secrecy (Blumenthal et al. 1996, 1997, Bok 1982, Krimsky 2003). That is, academic scientists started to consider intellectual property rights because of a contract with a private firm. In Korea's case, however, the government instead of industry appears to have motivated the increased interest in intellectual property rights and secrecy among public sector scientists.

As discussed earlier, science has been mobilized nationally by the Korean government for economic development. In doing so, Korean scientists have come to embrace the utilitarian view of science, viewing it as a tool for national industrialization, and have come to identify science with national advance. Historical experience may account for scientists' tendency to connect the economic benefits that result from scientific research to national interest rather than individual rewards. By helping scientists see patenting research outcomes, industrialization, and national interest as identical, the nationalistic view of science may also have helped scientists accept the norm of the pursuits for intellectual property rights and the practice of secrecy without much resistance.

Conclusion: the state and culture of science

During the middle of the twentieth century, economic development was a major government priority in South Korea. Korean policymakers viewed scientific research as a crucial tool for the economic development of the nation and mobilized scientists toward that end. The government's utilitarian view of science was revealed clearly in its national plan for science and technology promotion, which accompanied the national plan for economic development. This utilitarian view led the Korean government to be reluctant to support scientific research in higher education prior the mid-1980s, viewing the basic research orientation of university professors as unlikely to aid economic development. Instead of supporting university scientists, the Korean government established mission-oriented, state-funded research institutes to conduct research to assist local industries. Even when the Korean government began to support university research seriously, it invested primarily in applied and developmental research.

The strong utilitarian view of science promoted by the Korean state created a distinctive research environment for Korean scientists, and, in turn, had significant effects on the norms and research practices of scientists in universities and public research institutes. In Korea, the assigned role of scientists has been to produce instrumental knowledge for industrialization, and the state has distributed rewards with the aim of realizing this objective. The distribution of government funds for applied and developmental research in universities appears to have contributed much to scientists' perceptions of the purpose of scientific research. The nationalist view of science and the state-guided commercialization of science in Korea also helped legitimatize scientists' pursuit for intellectual property rights, the emphasis by scientists on the applications of scientific research, and even secrecy, because each of these was seen as helping to realize national, rather than personal, interests.

The STS literature has discussed different types of scientific norms and practices and revealed tensions especially among academic and industrial research norms. Focusing on the ramification of growing academy-industry ties, the existing literature has focused on the ways through which the culture of commerce has transformed the Mertonian-type norms and practices of academic science (Etzkowitz 1989, Etzkowitz et al. 1998, Hackett 1990, Kleinman 2003, Krimsky 2003, Owen-Smith and Powell 2001, Slaughter and Leslie 1997, Slaughter and Rhodes 2004, Ziman 2000). In particular, scholars in the U.S. have expressed widespread concern about how university-industry relations (UIRs) have undermined academic norms. They suggest that UIRs threaten and, indeed, are changing the normative landscape of science. These researchers have suggested that increased connections between academia and industry have been and are threatening norms of openness and free exchange (Bok 1982, Croissant and Smith-Doerr 2008, Krimsky 2003, Rosenzweig 1985).

In Korea, however, the Mertonian-type norms have never had the priority many analysts see in the pre-1980s United States, even among scientists in the public sector. On the contrary, Korean government efforts to put science in the service of national economic development have prompted scientists to accept secrecy and less-than-open sharing of research results in the interest of national economic advance. Well before analysts in the U.S. worried about how UIRs were shaping the norms of science, Korean government policies were shaping the norms of science and doing so in a way that deviates from the Mertonian ideal and in a way analysts see as more consistent with the values associated with the incursion of industry into academic science.

One might argue that the nationalist view of science is merely rhetoric used by scientists to justify their pursuit of individual interests in their research (Gilbert and Mulkay 1984, Mulkay 1976). Indeed, some scientists in Korea have mobilized rhetorical framings such as "the fundamental technology of Korea" and "Science has no borderline, but scientists have their fatherland" intentionally to justify their research and to obtain support from the state. It is not easy, therefore, to precisely determine the extent to which nationalism in science is used rhetorically for personal advantage and to what extent it is an internalized norm to which Korean scientists are truly committed. It is worth noting, however, that many basic researchers in Korea whose research has little to do with economic development expressed belief in the value of intellectual property protection and secrecy in pursuit of national economic development.

In sum, the experience of Korea suggests that, in developmental states, the state may wield greater influence on the norms and practices of scientists in the public sector than does the spread of industry directly into academia. Perceiving science as a resource and a tool for national economic competitiveness, the state can promote the utilitarian view of science not only through how it distributes support for scientific research but also through the new organizational systems it creates. Furthermore, unlike industry that represents private interests, the state is often seen to represent the public good, which helps scientists justify post-academic norms and practices in terms of national interests. Put differently in Korea, a quintessential developmental state, scientists may dismiss traditional academic norms because they are viewed as being inconsistent with national well-being. In all, we see a very different set of forces in Korea transforming the norms of science than we find in the U.S. and other western countries.

Note

1   I appreciate thoughtful comments on the earlier version of this chapter from Daniel L. Kleinman and Kelly Moore. Writing of this essay was supported, in part, by Korean Research Foundation (KRF-2010-330-B00169).

References

Amsden, A.H. 1989. *Asia's Next Giant: South Korea and Late Modernization*. New York, NY: Oxford University Press.

Bak, H.-J. 2006. "Characteristics and Transformation of Korean Academic Science." *Korean Journal of Social Theory* 30: 213–244 [in Korean].

Berman, E.P. 2011. *Creating the Market University: How Academic Science Became an Economic Engine*. Princeton, NJ: Princeton University Press.

Blumenthal, D., Campbell, E., Causino, N., and Louis, K. 1996. "Participation of Life Science Faculty in Research Relationships with Industry: Extent and Effects." *New England Journal of Medicine* 335: 1734–1739.

Blumenthal, D., Campbell, E., Anderson, M., Causino, N., and Louis, K.S. 1997. "Withholding Research Results in Academic Life Science: Evidence from a National Survey of Faculty." *Journal of American Medical Association* 277(15): 1224–1228.

Bok, D. 2003. *University in the Marketplace*. Princeton, NJ: Princeton University Press.

Bok, S. 1982. "Secrecy and Openness in Science: Ethical Considerations." *Science, Technology & Human Values* 7(1): 32–41.

Branscomb, L.M. and Florida, R. 1999. "Challenges to Technology Policy in a Changing World Economy." In *Investing in Innovation*, edited by L.M. Branscomb and J.H. Keller, 3–39. Cambridge, MA: MIT Press.

Choi, H.-S. 1995. *Research Institutes Where the Light Never Goes Out*. Seoul: Chosun Ilbo-Sa [in Korean].

Choi, Y.-H. 1996. "The Path to Modernization." In *Korea at the Turing Point: Innovation-based Strategies for Development*, edited by L.M. Branscomb and Y.-H. Choi, 13–28. Westport, CT: Praeger.

Croissant, J.L. and Smith-Doerr, L. 2008. "Organizational Contexts of Science: Boundaries and Relationships between University and Industry." In *The Handbook of Science and Technology Studies* (3rd Edition), edited by E.J. Hackett, O. Amsterdamska, M. Lynch, and J. Wajcman, 691–718. Cambridge, MA: The MIT Press.

Elzinga, A. and Jamison, A. 1995. "Changing Policy Agendas in Science and Technology." In *Handbook of Science and Technology Studies* (2nd Edition), edited by S. Jasanoff, G. Markle, J. Petersen, and T. Pinch, 533–553. Thousand Oak, CA: Sage.

Ergas, H. 1987. "Does Technology Policy Matter?" In *Technology and Global Industry: Companies and Nations in the World Economy*, edited by B.R. Guile and H. Brook. Washington, DC: National Academy Press: 191–245.

Etzkowitz, H. 1989. "Entrepreneurial Science in the Academy: A Case of the Transformation of Norms." *Social Problems* 36(1): 14–29.

Etzkowiz, H., Webster, A., and Healey, P. eds. 1998. *Capitalizing Knowledge*. Albany: SUNY Press.

Evans, P. 1995. *Embedded Autonomy: States & Industrial Transformation*. Princeton, NJ: Princeton University Press.

Gaston, J. 1978. *The Reward System in British and American Science*. New York: A Wiley-Interscience Publication.

Gibbons, M., Limoges, C., Nowotny, H., Schwartzman, S., Scott, P., and Trow, M. 1994. *The New Production of Knowledge: The Dynamics of Science and Research in Contemporary Societies*. Thousand Oaks, CA: Sage.

Gilbert, N. and Mulkay, M. 1984. *Opening Pandora's Box*, Cambridge: Cambridge University Press.

Grobstein, C. 1985. "Biotechnology and Open University Science." *Science, Technology & Human Values* 10(2): 55–63.

Hackett, E.J. 1990. "Science as a Vocation in the 1990s." *Journal of Higher Education* 61(3): 241–279.

Hwang, H. and Yun, J. 2003. "Transformation of Knowledge Production Structure in Universities." In *Social History of Knowledge Transformation*, edited by Society for Korean Social History 291–318. Seoul: Munhakgua Jisung-sa [in Korean].

Jeon, S.-K. 1982. *Science and Technology Policy in Korea: A Testimony of a Policy-maker*. Seoul: Jeungwoo-sa [in Korean].

Kim, D. 2010. "Formation of National Organization of Scientists: Creation of the Korean Federation of

Science and Technology Societies." In *The Korean Scientific Community*, edited by Kim, H.-S. et al. 127–145. Seoul, Korea: KungRee [in Korean].

Kim, D.-W. and Leslie, S.W. 1998. "Wining Markets or Winning Nobel Prizes? KAIST and the Challenges of Late Industrialization." *Osiris* 13: 154–185.

Kleinman, D.L. 1995. *Politics on the Endless Frontier: Postwar Research Policy in the United States*. Durham, NC: Duke University Press.

Kleinman, D.L. 2003. *Impure Cultures: University Biology and the World of Commerce*. Madison, WI: University of Wisconsin Press.

Krimsky, S. 2003. *Science in the Private Interest*. Lanham, MD: Rowman & Littlefield.

Merton, R.K. 1973. *The Sociology of Science: Theoretical and Empirical Investigations*. Chicago, IL: Chicago University Press.

Moon, M.Y. 2006. *Early History of KIST, 1966–1980: From Contract Research to National Project Research*. Seoul National University (unpublished Ph.D. thesis) [in Korean].

Mulkay, M. 1976. "Norms and Ideology in Science." *Social Science Information* 15(4/5): 637–656.

Okimoto, D.L. 1989. *Between MITI and the Market: Japanese Industrial Policy for High Technology*. Stanford, CA: Stanford University Press.

Owen-Smith, J. and Powell, W.W. 2001. "Careers and Contradictions: Faculty Responses to the Transformation of Knowledge and Its Uses in the Life Science." *Research in the Sociology of Work* 10: 109–140.

Park, S., Shin, D., and Oh, D. 2001. *100 Years of Our Science*. Seoul: Hyunam-sa. [in Korean].

Rosenzweig, R.M. 1985. "Research as Intellectual Property: Influences within the University." *Science, Technology & Human Values* 10(2): 41–48.

Slaughter, S. and Leslie, L.L. 1997. *Academic Capitalism: Politics, Policies, and the Entrepreneurial University*. Baltimore, MD: Johns Hopkins University Press.

Slaughter, S. and Rhodes, G. 1996. "The Emergence of a Competitiveness Research and Development Policy Coalition and the Commercialization of Academic Science and Technology." *Science, Technology & Human Values* 21(4): 303–339.

Slaughter, S., and Rhodes, G. 2004. *Academic Capitalism and the New Economy*. Baltimore, MD: Johns Hopkins University Press.

Song, S. 2007. "A Content Analysis on the S&T Comprehensive Plans in Korea: Focusing on Five-Year Plans." *Journal of Science & Technology Studies* 7(1): 117–150 [in Korean].

Vallas, S.P. and Kleinman, D.L. 2008. "Contradiction, Convergence and the Knowledge Economy: The Confluence of Academic and Commercial Biotechnology." *Socio-Economic Review* 6(2): 283–311.

Ziman, J. 2000. *Real Science: What It Is, and What It Means*. Cambridge: Cambridge University Press.

24

Science as Comfort

The strategic use of science in post-disaster settings

Brian Mayer, Kelly Bergstrand and Katrina Running

UNIVERSITY OF ARIZONA, UNIVERSITY OF ARIZONA AND IDAHO STATE UNIVERSITY

The recent catastrophic disasters of the twenty-first century have fueled a growing body of literature in science and technology studies (STS) on the relationship between science and the social dynamics of disaster prevention, management, and recovery. From the accidental release of nearly five million barrels of oil from the explosion of the Deepwater Horizon oil rig in 2010 to the nuclear meltdown at the Fukushima Nuclear Power Plant in 2011, the need for improving our understanding of how technoscience is implicated in the way political institutions prepare for and deal with disasters has never been clearer. As Fortun and Frickel (2012) note, "disaster has been a blind spot in STS." It remains to be seen whether our existing understandings of science and society are applicable to the unique and often chaotic conditions surrounding disasters. Thus, while there are some studies on the impact of regulatory responses to disasters (for example, Frickel et al. 2009), as well as citizen participation in science post-disaster (for example, McCormick 2012), STS would benefit from greater understanding of the use of science and technology in the recovery from disruptive events. Toward this end, we examine the use of seafood testing procedures following the Deepwater Horizon oil spill. We find that beyond simply generating knowledge, the seafood testing program served multiple purposes, from attempting to boost consumer confidence in Gulf seafood to making science accessible to seafood workers. In short, science was used as an institutional tool to reduce uncertainty generated by the disaster for both consumers and producers of seafood. Despite these aims, the testing garnered widespread negative reactions in the media, reflecting a culture of public distrust of government that had emerged in the aftermath of the disaster.

In crisis settings, the credibility and legitimacy of science and technology become especially contested as various stakeholders compete for the regulatory and cultural authority to define the scope of the problem and design potential strategies for management and recovery (Clarke 1991, Fortun and Frickel 2012, Freudenburg 1997). A similar phenomenon can be seen in development contexts in which stakeholders contest the costs and benefits of embracing risky new technologies (Amir, Chapter 17 of this book). In extreme situations, including environmental disasters in economically developed nations and rapid economic growth in emerging nations, when new forms of science and technology must be rapidly created and deployed, existing rules and standards governing the use of technoscience, as well as established cultural scripts reinforcing their legitimacy, may be lacking. In the absence of preexisting norms and rules, modern science and technologies have the potential to exacerbate anxiety by asking the public to have faith in unproven measures. Thus, in the process of institutionalizing unfamiliar approaches to technoscience it is important for scientists and regulators to be aware that the uncertain context in which new or modified forms of science are introduced can lead to contestation of their credibility and legitimacy.

We examine challenges to the legitimacy of new science in response to a disaster, specifically the 2010 Deepwater Horizon oil spill in the Gulf of Mexico. This oil spill was the largest in the history of the United States, releasing 200 million gallons of oil and causing environmental harm to ecosystems and economic disruption in coastal communities dependent on seafood and tourism industries. In the wake of this catastrophe, governmental agencies and scientists implemented sensory testing, in which participants used their sense of smell and taste to determine the presence of petroleum-based oil in seafood. This attempt to institutionalize the science of sniff testing as a legitimate disaster response was met with both public ridicule and enthusiasm by various stakeholders. It thus provides a useful case study for assessing both the contestation of novel scientific practices in a disaster situation and the strategic use of science in managing public responses to disasters. One of the primary purposes of the testing was to calm the fears of consumers wary of eating Gulf seafood due to possible oil-related contamination. For communities along the Gulf coast who were already struggling with oil cleanup, lost tourism, and fishery closures, a nationwide boycott of Gulf seafood would be economically disastrous. Sensory testing provided a highly visible and easy to understand form of seafood inspection accessible to the general public, accompanied by government-approved messages that the seafood was safe to eat.

The deployment of sensory testing also included community outreach programs that trained people working in the seafood industry to use their noses to detect oil in seafood. Seafood workers in coastal communities lived with daily uncertainty about what the ultimate effects of the oil spill would be, such as whether they would lose their jobs due to fishing restrictions or perceived contamination of seafood resulting in lackluster sales. The sniff test science empowered people to use their own skill sets and recover some amount of control in an uncertain situation through an easy-to-learn scientific procedure (Otwell 2012). In these two ways – reassuring consumers and providing tools to seafood workers – the use of sensory testing after the Deepwater Horizon oil spill serves as an example of "comfort science" wherein scientific processes are undertaken with the goal of reassuring a concerned public in the aftermath of environmental disruption.

The changing role of science in a culture of competing claims, risk and uncertainty

The public's trust in the credibility and legitimacy of science has undergone substantial change in the past few decades (Beck 1992, Giddens 1991, Moore 2008). During industrialization, technoscientific advancements elevated the role of science in managing complex new technologies and led the public to increasingly trust scientific endeavors and scientific authority (Barber 1990, Parsons 1962). As a result, the period from industrialization to the middle of the twentieth century was marked by a widely held cultural view of science as objective, noncontroversial, and generally beneficial for society. This perspective on science has been challenged in recent years, however, due to public debates between scientists on the moral and political implications of their work, the media's tendency to highlight scientific disagreement in its coverage of science-related issues, and legal battles in the wake of industrial disasters (Button

2010, Fortun 2001, Moore 2008). Moreover, recent studies have demonstrated growing cleavages between social groups and their level of trust in science and technology (Gauchat 2012, McCright and Dunlap 2011, Mooney 2005). Highly visible and contentious scientific debates such as those concerning climate change present a divided scientific community, providing an opening for lay audiences to superimpose other social partitions such as political ideology or religion to make sense of the debate. Indeed, political ideology and religiosity (Gauchat 2012) seem to be driving the public further apart in their perception of whether scientific and technological advancements represent objective and beneficial sources of information.

The public's skepticism toward science has also been aggravated by technoscientific disasters, such as nuclear reactor meltdowns and the discovery of chemical pollutants in our air, food, and water (Carson 1962, Fortun and Frickel 2012). These events have led to a variety of negative and unintended consequences for society and individuals' health, generating existential insecurity regarding the level of trust the public should place in science (Beck 1992, Giddens 1991). Beck (1992) argues that the catastrophic risks brought about through modern industrialization are attributed by the public to science gone wrong and a lack of practical management in the scientific endeavor, which contributes to backlash against science. The public's understanding and acceptance of science may be contingent on trust in the institutions conducting the science (Bucchi and Neresini 2008, Wynne 1995). Elsewhere in this volume, Amir (Chapter 17) similarly argues that weak and poorly developed political institutions can undermine public trust in science, especially in cases where the state is seen as unable to effectively oversee the implementation of risky technologies. In the case of science in post-disaster settings, especially when the disaster is human-caused, competing estimates of scope and a sense of doubt about science's ability to keep us safe further exacerbate the perception of bias and corruption within the scientific process.

Crisis situations also present a number of distinctive circumstances that intensify the ambivalence many in the modern public feels toward scientific authority. When crises occur, demands for rapid responses often require novel or reimagined implementations of technoscientific fixes – solutions that potentially introduce new fears or uncertainty into the public sphere (Fortun and Frickel 2012). The chaos in the aftermath of crises makes it difficult for the public to feel sufficiently informed, thereby taking away their ability to participate in the decision-making process which has become increasingly desired as ultimate scientific authority is questioned (Moore 2008). Furthermore, the rapid deployment of technoscientific solutions often challenges existing rules and regulations, requiring a temporary relinquishment of procedural requirements and regulatory oversight and casting even more doubt on the infallibility of scientific recommendations.

The public's reaction to the scientific tests implemented in the wake of the Deepwater Horizon oil spill to assess the safety of potentially oil-tainted seafood is an example of modern science's multiple and contested roles at a moment of particular risk and uncertainty. With competing claims over the size and impacts of the oil spill broadcast daily over multiple forms of media, the general public was faced with a cacophony of opinions and contradictory data on which to base decisions concerning their consumption of seafood. The U.S. National Oceanic and Atmospheric Administration (NOAA) and the U.S. Food and Drug Administration (FDA) attempted to utilize the historical legitimacy of sensory analysis to promote the validity of the sniff tests with some success. The implementation of this particular form of science, and the challenges with which it was met, reveals much about how regulatory science is used and perceived in a post-disaster setting.

In addition to possible public contestation, scientific processes in disaster contexts face the challenge of providing functions that extend beyond simply collecting data and expanding scientific knowledge. In the highly uncertain climate generated by disasters, governmental agencies and other institutions may strive to reduce anxiety and reassure the public, avoiding the damage that could ensue from a panicked population. One such strategy is to use "comfort science" where science is used to reduce uncertainty about disaster-related hazards and potentially restore faith in economic or political institutions. In the aftermath of the Deepwater Horizon oil spill, we see these additional functions of science being used by governmental agencies to bolster consumer confidence in the safety of seafood harvested in the Gulf of Mexico. Such efforts attempted to thwart boycotts of Gulf seafood that threatened the collapse of an economy already made vulnerable by declining tourism in the region.

In order for comfort science to be effective, it has to be communicated to and accepted by everyday people, and one strategy for accomplishing this is to increase public participation in scientific processes. By making science accessible and comprehensible to lay audiences, this could potentially aid in getting individuals to accept, rather than ignore or reject, scientific findings that work to avert unrest and maintain stability in the crisis setting. As Wynne notes, "even a technically literate person may reject or ignore scientific information as useless in the absence of the necessary social opportunity, power, or resources to use it" (1995: 363). Additionally, engaging the public in scientific procedures and forums may serve to increase their sense of being a stakeholder in the process, which in turn may quell negative reactions. Some institutions may also seek public participation as a way to gain public legitimation and avoid controversy, particularly in regard to sensitive scientific issues (Bucchi and Neresini 2008). In the case of the Deepwater Horizon oil spill, we see such strategies in action where sensory analysis may have been used strategically to reassure consumers that seafood was safe, frame science in a way the public could understand, and integrate public participation through community trainings in "sniff test" methods.

The challenges of a technoscientific disaster: the Deepwater Horizon oil spill

On April 20, 2010, the Deepwater Horizon drilling rig located fifty miles off the coast of Louisiana in the Gulf of Mexico exploded, killing eleven men working on the platform. Initial attempts to stop the spill were unsuccessful, and the well released an estimated 185 to 205 million gallons of crude oil before it was capped nearly three months after the explosion. Uncertainty was rampant throughout the crisis, from competing estimates of the amount of escaped oil to widely varying results in the models predicting its spread across the Gulf of Mexico. Estimates of the spill were generated by scientists from BP, the U.S. Coast Guard (USCG), the National Oceanic and Atmospheric Administration (NOAA), the Department of Energy (DOE), the U.S. Geological Survey (USGS), and a broad spectrum of university-affiliated and independent scientists. Estimates of the total number of gallons released each day of the 87-day event ranged from 42,000 in the early weeks of the spill to 2.6 million gallons a day when the well was finally capped.

Fisheries were closed where oil was sighted and coastal response efforts were mobilized according to rough estimates of which shores were predicted to be impacted by oil. Responsibility for the oil spill was contentious, with a complex web of organizational actors sharing responsibility for the spill. BP, which owned the majority of rights to drill in the Macondo Prospect where the well was located, contracted with Transocean to lease the Deepwater Horizon oil rig and subcontracted with Halliburton for cement work around the well. Regulatory responses were equally complicated, with BP seemingly taking the lead in the response effort, supervised by the USCG, informed by NOAA, occasionally in contact with the

Environmental Protection Agency (EPA), and in regular communication with a multitude of state and local agencies from the five states along the Gulf of Mexico.

The economic impact of the Deepwater Horizon oil spill was immediately felt by the seafood industry. From fisheries closures in Texas and Louisiana in 2010 to the continued tarnishing of the reputation of seafood caught throughout the Gulf of Mexico, the production of seafood across the region has not rebounded as of 2012 despite a return in tourism. Consumers questioned the safety of seafood caught in the Gulf of Mexico and were no doubt worried and confused by the competing estimates of the spread of oil and the biological harm it might be causing. Flying high above the heads of attendees at seafood festivals in 2010 and 2011 in locations as far from the Gulf of Mexico as New York and Chicago, banners proudly declared "No Gulf Seafood Sold Here!" Menus in seafood restaurants in Las Vegas and Denver informed customers that no items served from their kitchens contained Gulf caught shrimp or oysters. The perception that Gulf of Mexico seafood was contaminated spread quickly across the country and became a major scientific concern for the agencies charged with monitoring the safety and quality of the American food supply (Danielson 2011, Jervis 2010, Jonsson 2010).

In response to widespread concern about potentially tainted seafood, NOAA and the FDA increased their testing of Gulf seafood, hoping to restore public confidence that regulatory science was being performed in a neutral and objective manner. The contentiousness of other scientific assessments involved in the oil spill, such as early misleading estimates of the amount of oil being released, created an incentive for regulatory agencies to make science accessible to the public. Further, there was a pragmatic need for a faster seafood screening technique, since sending each sample to a lab for analytic chemical testing could be both time consuming and expensive. Thus, in order to accentuate the testing using standardized laboratory procedures, NOAA and the FDA also implemented supplemental olfactory sensory perception tests, or what came to be known as the "sniff test," whereby trained professionals would use their olfactory senses to determine whether or not seafood was tainted with oil. This olfactory test was used to generate credibility for all federal testing; the high degree of visibility of the oil in the Gulf needed to be met with an equally visible scientific practice (Stein 2012). For the media, the use of the sniff test raised as many questions as it answered, including whether the professionally trained human nose could deliver credible results. Our assessment of the use of the sniff test to detect tainted seafood reveals a major disconnect between the expectations of the regulatory scientists and what the public perceived as junk science.

The science behind the "sniff test": sensory evaluation in theory and practice

Although the reactions of the general public to sniff tests in response to the Deepwater Horizon oil spill largely indicated surprise and unfamiliarity with sensory-based testing, the standardized use of human beings' sensory abilities to evaluate the quality of products is far from a new phenomenon. Historically, buyers of many products would test samples off shiploads to assess the quality of goods, leading to grading standards in wine, tea, coffee, butter, and meat. Some of these standards persist to modern day (Meilgaard et al. 2007). In the early 1900s, professional tasters were used in a variety of industries, and companies have long employed experts for assessing product quality, such as perfumers, flavorists, brewmasters, winemakers and coffee or tea tasters (Meilgaard et al. 2007; Stone et al. 2012).

In the mid-twentieth century, sensory analysis became more formalized and scientifically rigorous. In the 1940s, Scandinavians developed the "triangle test" which looked at three samples in which only two were from the same batch, and participants had to identify the third dissimilar sample; this method was used in breweries to screen judges for their beer evaluation abilities (Helm and Trolle 1946, Lawless and Heymann 2010). At this time, industry consultants also developed a "Flavor Profile" method which offered a more reliable way of describing sensory attributes of a product and moved away from expertise housed in the individual to the use of consensus-based decisions by a panel of approximately six experts (Moskowitz 1993, Stone et al. 2012). In the 1950s and 1960s, academics began publishing books on sensory evaluation, and the study of sensory analysis was introduced by several university programs (Jellinek 1985). In the 1970s, researchers developed Quantitative Descriptive Analysis, which documented food sensory characteristics in reliable, precise and reproducible ways, and became widely adopted by companies and research laboratories alike (Moskowitz 1993, Stone et al. 1974).

Today, "sensory evaluation has emerged as a distinct, recognized scientific specialty" and is defined as a scientific discipline used to evoke, measure, analyze and interpret responses to characteristics of products as perceived by the senses of sight, smell, touch, taste and hearing (Stone et al. 2012: 13, 15). It is described as a "quantitative science" which calls for controlled testing conditions to collect numerical data analyzed through statistical methods and interpreted properly by informed professionals (Lawless and Heymann 2010). Scientists continue to improve the methodology of sensory testing, with several journals devoted to the topic, and sensory tests are being used in a variety of fields, including the food and beverage industry, as well as in creating personal hygiene products, detecting environmental odors, chemical testing and the diagnosis of illness (Meilgaard et al. 2007). Indeed, sensory evaluation procedures and analysis techniques have been developed for everything from canned chicken (Lyon 1980) to soy sauce (Jeong et al. 2004).

There are a number of advanced instruments for evaluating food, including "electronic noses," but human beings' senses continue to play an important role because human sensory perceptions provide odor and taste cues, as well as holistic impressions, that can be missed by instruments (Jellinek 1985). Moreover, the human nose has the potential to detect some odors at very low concentrations. Meilgaard et al. (2007) notes that the most sensitive gas chromatographic method can detect approximately 10^9 molecules per milliliter, and there are numerous odor substances where the human nose is ten to a hundred times more sensitive than gas chromatographs. Interestingly, experts in smell, such as perfumers and wine tasters, gain their talents not necessarily through a highly sensitive nose, but through training and cognitive skills that allow them to make full use of the sensory information available to them.

Sensory tests are particularly advantageous following an oil spill as they can be done quickly and can evaluate large numbers of fish and shellfish, which helps target appropriate samples for chemical analyses (Davis et al. 2002). The NOAA Seafood Inspection Program has performed sensory analysis to detect oil contamination in multiple oil spills, including the 1989 Exxon Valdez spill in Alaska, the 1996 spill in Rhode Island, the 1999 spill in Coos Bay, Oregon, and multiple oil spills in the San Francisco area (NOAA 2011). In fact, sniff tests have been described as the "gold standard" to detect tainted seafood (Schmit 2010).

Given the importance of sniff tests, regulatory agencies have developed scientific guidelines for conducting sensory evaluation in an oil spill. Almost ten years before the Deepwater Horizon oil spill, NOAA published a technical memorandum, "Guidance on Sensory Testing and Monitoring of Seafood for Presence of Petroleum Taint Following an Oil Spill" (Reilly and York 2001). This document summarizes the procedures and scientific principles that underpin sensory evaluations of seafood in response to oil spills. First, rigorous methods are used for collecting seafood samples. Statistical sampling methods can help to determine how many species should be collected from an area. Ideally, control samples will be taken from areas not yet affected by the oil spill. Second, the testing process involves several important components. While assessors can perform well in the field, laboratory sites are preferable given the ability to create a controlled environment and to avoid interference from other parties, such as the media or industry representatives. At the heart of sensory testing is the use of trained individuals, or assessors, who evaluate the seafood and who are selected based on: (1) Sensory acuity and the ability to describe perceptions analytically; (2) Potential for developing an analytic capability which includes training in test procedures, recognizing and identifying sensory attributes, and refining sensitivity and memory to provide precise, consistent, standardized reproducible sensory measurements; and who are (3) Monitored to ensure effective performance, such as by including known clear or spiked samples (Reilly and York 2001). Assessors vary from being trained to perform a specific task to being experts with a high degree of sensory ability, training and experience with sensory tests. Ideally, these would be seafood product experts from government agencies who have years of training and experience that they can bring to evaluate seafood tainted from an oil spill.

For the sample testing, assessors are given raw tissue portions of organisms, standardize the distance from their nose to the samples, and take two or three short shallow sniffs, termed "bunny sniffs" (Reilly and York 2001). Odors are perceived through a region at the top of the inside of the nose; a sniff is more effective than breathing for drawing air to this area (Jellinek 1985). Assessors then take a break of at least one minute between samples and "cleanse their nose," such as by sniffing the backs of the hand or arm or the headspace over a glass of clean water. If there is a taint in the odor, the sample fails and the assessors take no further action. If nothing is detected, assessors repeat the odor sniffing process for a cooked sample of the seafood, and if it passes this test as well, it proceeds to flavor testing. To evaluate the flavor, assessors place a standardized amount of the sample in their mouth and expectorate, cleansing their mouths between samples (Reilly and York 2001).

The NOAA guidelines offer a portrait of what sensory testing entails in response to an oil spill, but what actually happens on the ground when mobilizing to respond to an unexpected disaster? In response to the oil spill, multiple agencies coordinated to develop sampling protocols that would lead to the reopening of fisheries in both state and federal waters, and these protocols called for both sensory analysis and chemical testing to determine whether oil contaminants were present in seafood. As suggested by the sensory test guidelines, control samples of uncontaminated fish and shellfish were collected to serve as reference samples for sensory tests; in fact, at the University of Florida, so many samples of uncontaminated fish were collected that they surpassed existing freezer space (Houck 2010). A prominent source of sensory testing was an expert panel based at a NOAA lab in Pascagoula, Mississippi, where seven expert sensory assessors (whose identities were kept secret to avoid being targeted if fishing grounds remained closed) sniffed as many as 36 samples a day (Mui and Fahrenthold 2010). The experts were able to smell oil diluted to one part per million, which is forty times more sensitive than the average smelling ability (Drogin 2010). If three of the seven experts fail a sample, then the area it comes from remains closed to fishing (Severson 2010). Samples passed by the expert panel were sent to a NOAA laboratory in Seattle where they were evaluated using gas chromatographs, mass spectrometers and analytic chemistry to look for polycyclic aromatic hydrocarbons, a harmful component of oil (Flatow 2011). If such contaminants were found at harmful levels, then the area remained closed.

In addition to the expert panel, NOAA officials also trained people, often state and local inspectors, to work in other locations, such as docks, seafood processors, and restaurants (Skoloff 2010). These trainings were added as an extra precaution to prevent contaminated seafood from entering markets, particularly in cases where seafood was suspected of being caught in areas closed to fishing due to the oil spill. One newspaper account describes the experiences of Gary Lopinto, a seafood program manager for Louisiana's health department and one of approximately sixty seafood safety workers trained at the NOAA lab:

> At the NOAA lab, Lopinto was taught to sniff cucumbers, watermelon or even canned corn to clear his nostrils. But this is the real world. Amid the din of heavy machinery grading and sorting oysters at the Motivatit processing plant, Lopinto neutralized his nose by smelling his sleeve. Then he scooped a jiggly oyster out of its shell and held it up to his moustachioed face. He sniffed. "If there's any detection of oil, you're gonna got a nasal sensation ... or maybe a little gas smell," he said. "This, believe it or not, smells like corn to me."
>
> *Mui and Fahrenthold 2010*

In addition to NOAA, several other organizations were involved in training and inspecting seafood. The Mississippi State University Coastal Research and Extension Center in Biloxi trained about sixty seafood workers to use their sense of smell to recognize seafood tainted by oil by exposing them to a variety of scents that helped build a sensory memory to recognize and distinguish odors (Templeton 2010). Additionally, a University of Florida program trained inspectors to detect oil in shrimp, oysters, and crabs (Houck 2010).

Individuals, especially those with ties to either the seafood industry or the network of "olfactory professionals" employed in the sniff testing process, were for the most part reassuring about the effectiveness of sniff tests to detect even small amounts of potentially harmful oil and chemical dispersants. Steve Otwell, Seafood Specialist and Director of the Aquatic Foods Product Lab in Gainesville, Florida expressed confidence in the sniff test explaining, "It just turns out that the human nose is an effective tool to measure concentrations at that level of food safety" (Houck 2010). Steve Wilson, chief quality officer for NOAA's seafood inspection program, conveyed similar confidence, estimating that the sniffers are accurate 80% of the time (Mui and Fahrenthold 2010). Even those with no obvious self-interest in touting the seafood as safe reported high levels of confidence in the government's testing procedure: Ralph Portier, an Environmental Sciences Professor at Louisiana State University reportedly remarked, "this is probably the safest seafood entering the U.S. market right now" (Jervis 2010).

In sum, the sensory analysis conducted by federal and state agencies employed established scientific guidelines developed in the field of sensory evaluation over many years to identify seafood contaminated by oil. However, despite the fact that the sniff tests universally deemed Gulf seafood safe, and the results of supplemental chemical testing indicated that no oil contaminated samples were ever detected (Taylor 2012), many commercial fishermen in the Gulf of Mexico as well as consumers throughout the United States doubted the scientific credibility of the agencies employing the sensory and chemical testing. For many, during this time of heightened uncertainty, the use of sensory assessment represented an effort to appease seafood workers and consumers more than a scientifically credible tool for evaluating food safety.

Contested science: reactions to sensory tests

When reports of the government's use of sensory analysis for testing Gulf seafood hit the public airwaves and newsprint, reactions ranged from cautiously optimistic to downright suspicious. In particular, the use of "sniff tests" to find contaminated seafood was described by some media commentators and members of the general public as "unscientific," "crazy" and "ridiculous" (Roberts 2010, Peterka 2010, Skoloff 2010). Ron Kendall, Director of the Institute of

Environmental and Human Health at Texas Tech and a specialist on the environmental impact of oil spills, expressed the opinion that the smell test was inadequate and more precise testing and analysis would be required. "Everyone is racing around to give an answer before we have really done the science," he warned (Severson 2010). Similarly, there appeared to be public skepticism regarding the safety of seafood. A spokesperson for the Louisiana Seafood Promotion and Marketing Board noted, "What we have is the results of testing. So there are folks who choose not to believe the results of those tests and that's a challenge for us" (Smith 2010). An article in *Marketwatch* echoed this sentiment: "Although dozens of government agencies at federal, state and local levels have tested Gulf seafood and water composition, the amount of public doubt still remains high" (Ngai 2010). In interviews, laypeople expressed doubts or misunderstandings about the sniff tests with comments such as, "They're going to smell it? No way. How they gonna know? I ain't eating any of it. I don't trust the nose" (Skoloff 2010).

The purported lack of science behind testing for the presence of the widely used chemical dispersant, Corexit, was another frequently cited reason for concern. Despite the FDA's assurances that their studies showed the dispersant was unlikely to be harmful because it did not accumulate in fish (FDA 2010), during the immediate aftermath of the BP spill, when media outlets were running numerous stories on the government's testing procedures, there was no chemical or tissue test for the presence of dispersants in seafood. Lisa Suatoni, a senior scientist with the Natural Resources Defense Council, told the *New York Times* that the lack of government specifics about the testing techniques for dispersants resulted in troubling questions about the scientific basis by which states were making decisions about the status of fisheries (Quinlann 2011). The sole reliance on sniff tests to find contamination in seafood by Corexit also raised concerns because, according to Kevin Kleinow, Professor of Aquatic Toxicology at Louisiana State University, "a number of surfactants that are used in dispersants have very little odor" (Dearen and Bluestein 2010). Ultimately, in response to public pressure, the FDA and NOAA did develop chemical tests for Corexit. In their testing for Corexit, NOAA reported that all samples tested below FDA safety limits and that over 99 percent of the thousands of samples tested did not have detectable residue (Schwaab et al. 2011).

Another general objection to government monitoring and testing was a lack of trust in the government's credibility after the role it played in early, misleadingly low projections of the amount of oil being spilled after the Deepwater Horizon accident. Chuck Hopkinson, Director of the Georgia Sea Grant program at the University of Georgia, said in an interview that the government had lost credibility by putting out deceptive information early in the disaster (Smith 2010). CNN reporter John Roberts also cast doubt on the legitimacy of the government's assurances by reminding people that government officials who had declared the air at Ground Zero safe for clean-up workers had "intentionally misled" those workers and the general public about health effects of exposure to airborne hazardous substances (Roberts 2010). Even employees of NOAA, such as marine habitat specialist Kris Benson, agreed "wholeheartedly that the public perception of the federal response is not good," and noted that the "credibility issue has been very disheartening" (quoted in Smith 2010).

This loss of trust in the government was exacerbated by many citizens' perceptions that BP was actually running the show, or that sniff tests were being favored due to their speed and cost-effectiveness rather than their accuracy. In an interview with Keith Olbermann, Spike Lee, who made a documentary about the Deepwater Horizon oil spill, expressed the former view, accusing the United States government of echoing BP-created public statements in the wake of the spill. Some of the officials interviewed may have unwittingly contributed to public doubt by framing sensory analysis in terms of budget practicalities. For example, Gerald Wojtala, Director of the International Food Protection Training Institute, which is funded by the FDA and the

W.K. Kellogg Foundation and works with regulatory agencies and public health officials to maintain federal food safety standards, was quoted as explaining,

> There are a lot of sophisticated tests, but when you think about it, do you want to run a test that takes seven days and costs thousands of dollars? This [sniff test] saves a lot of time and money, and it puts more eyes and noses at different points in the system.
>
> *quoted in Skoloff 2010*

Thus, multiple factors combined to contest the legitimacy of scientific practices used in the aftermath of the Deepwater Horizon oil spill. The public generally reacted negatively to the idea that people could smell oil contamination at levels as low as one part per million in seafood, leading to skepticism of the scientific rigor behind these methods. Additionally, people were uncomfortable with the initial use of sensory testing as the only means of detecting Corexit, as well as with the standards set by the FDA for defining tolerable risk. This skepticism was compounded by distrust in government authorities due to initial misleading statements about the amount of oil being released in the spill and BP's role in cleaning it up. These examples are indicative of the potential problems that unfamiliar scientific methods can face when implemented in times of uncertainty and anxiety that emerge post-disasters, and these cases illuminate the processes of contestation of science that can occur in unexpected disaster situations.

Comfort science: sensory analysis testing to reassure the public

Following an initial period of uncertainty and skepticism regarding the extent of the spill and potential for contamination of the food chain, a concerted effort by economic, regulatory, and political actors emerged to assure the public and promote consumer confidence that Gulf seafood was safe. White House Executive Chef Cris Comerford even joined the campaign, traveling to Louisiana on a "fish fact-finding mission" and concluding "This is very, very good seafood. All the scientists are doing everything they can to ensure whatever comes to the market is good for public consumption. It tastes good and it's safe. What more do you need to know?" (quoted in O'Neil 2010). Kevin Griffis, of the U.S. Department of Commerce summed up the basic sentiment well: "The message we're delivering is simple: The seafood in your grocery store or local restaurant is safe to eat, and that goes for the seafood harvested from the Gulf" (quoted in Skoloff 2010).

Governmental regulatory agencies were well aware of consumer skepticism of seafood safety and tailored strategies to alleviate such concerns. For instance, a joint statement by officials at NOAA, the FDA, and the Louisiana Department of Health and Hospitals states in regard to seafood testing: "The results of the tests, all publically available, should help Americans buy Gulf seafood with confidence: the seafood has consistently tested 100 to 1,000 times lower than the safety thresholds established by the FDA for the residues of oil contamination" (Schwaab et. al 2011). By making the seafood testing results publically available and appealing to seafood consumers, these agencies undertook specific strategies to use science to increase consumer confidence. The report then continues: "Scientists expected seafood would metabolize and excrete dispersant and that it was unlikely to be taken up by seafood in large quantities, but to support consumer confidence, NOAA and FDA worked to develop a chemical test to detect traces of the dispersant in fish tissue" (Schwaab et al. 2011). The governmental regulatory agencies are clearly indicating that they did not develop dispersant tests to gather additional information or expand scientific knowledge; rather this was done to increase consumer confidence in seafood safety. Thus, by their own admission, government agencies practiced comfort science, using scientific procedures with the primary purpose of reassuring a concerned public in the aftermath of a disaster situation and with the goal of preventing further damage, in this case an economic boycott of Gulf seafood.

Because comfort science is, by its nature, designed for lay audiences, there is an incentive to make the science accessible and comprehensible to non-experts such that easily-understood methods may translate more effectively into behavior modifications. Sensory analysis is particularly effective in this respect. The practice of trained smelling, unlike using complicated equipment such as gas chromatographs, is a skill accessible to the public and relatable to everyday experiences of using olfactory senses to gather information and make decisions. As Dr. John Stein, the Deputy Science Director of the Food Inspection Program at NOAA states in an interview about sensory testing: "If you think about your ability to detect something in your refrigerator, it has an off odor, you can detect it at very, very low levels" (quoted in Hansen 2010). Thus, sensory analysis – which seafood workers and residents could learn to do themselves – framed the science behind oil detection in seafood in a way that was accessible and could be utilized by the general public, thus theoretically empowering and reassuring them. Government and academic institutions also encouraged public participation in outreach programs that trained participants in sensory analysis methods. Although many citizens were skeptical of expert undertaken sniff tests, these seafood oil detection programs, which taught related skills to seafood workers and other residents, had widespread appeal, spreading by word of mouth and attracting fishermen, environmentalists and other officials who viewed it as an opportunity to learn more and gain some measure of control over the situation (Otwell 2012).

The outreach programs reflect a strategy of using public participation to gain legitimacy and reduce contestation (Bucchi and Neresini 2008). However, the sensory analysis trainings also served the important purpose of making the science practical to participants by providing tangible benefits to those most affected by the oil spill: seafood workers. Residents affected by the BP oil spill experienced relatively high levels of psychological stress (Gill et al. 2012), and seafood workers, in particular, were economically hit by the disaster. Thus, individuals involved in the seafood industry had much to gain from comfort science, both to reassure the seafood consumers keeping the industry afloat and to provide seafood workers some relief from the stress of the disaster. The sensory analysis outreach trainings, by providing a skill for detecting oil in seafood, did appear to provide some measure of control in the uncertain climate generated by the disaster. Steve Otwell describes the trainings as "partially entertainment and partially comfort" and states: "The thing was that people appreciated someone was actually trying to help. Not saying that the sky was falling" (2012).

In interviews, seafood workers emphasized that the outreach programs provided a way of dealing with the effects of the oil spill, with interviewees saying they participated because they wanted "new tools to help them keep their businesses going" given the hardships of the oil spill and believed that the trainings would help them "guarantee the safety of their products" (Templeton 2010). While the outreach sensory analysis programs may have provided some reassurance, as well as new skills, to participants, they also served to tap into a valuable resource: the everyday knowledge, skills and practical expertise of seafood workers. One benefit of public engagement in science is accessing the everyday knowledge of participants. As Bucchi and Neresini (2008: 451) state, "[l]ay knowledge is not an impoverished or quantitatively inferior version of expert knowledge; it is qualitatively different." People who live and work in arenas affected by technological disasters may have additional insight into the effects of these disasters. For example, Wynne (1989) notes that governmental experts failed to make use of the lay knowledge of British farmers in identifying sources of radiation and managing affected sheep flocks. Similarly, those who work in the seafood industry in the Gulf coast are well-situated to notice changes to fish and shellfish that could result from an oil spill. Steve Otwell describes these skills as commonplace: "Those working in this industry use their noses in their day-to-day business. Most fisherman and seafood producers can tell by the way the air smells if it is low or high tide" (quoted in Templeton 2010).

Scientific challenges in the uncertain climate of disasters

In the era of uncertainty following the Deepwater Horizon oil spill, sensory tests served the pragmatic function of providing quick and inexpensive testing for harmful chemicals. However, the easy-to-understand mechanics of the test were also geared toward gaining acceptance among everyday people and calming fears about Gulf seafood – a strategy which, according to media portrayals, proved to be largely unsuccessful. A second strategy of community outreach programs in sensory analysis training was more warmly received, with the trainings described as a way to reassure the fishers, the consumers, and regulators that all was well in the production of Gulf Coast seafood (Otwell 2012). The community trainings were also a way to bring people who were most affected by the oil spill, seafood workers who live and work daily on or near the waters of the Gulf Coast, into the scientific process, providing them with some level of comfort and control in a difficult time. Thus, both the general sensory analysis testing and the community extension programs served as examples of comfort science in that they sought to do more than systematic data collection; they aimed to reassure seafood consumers and workers alike in a disaster setting.

When disasters occur, especially technoscientific disasters with origins in human activities, regulatory authority and institutional norms of credible scientific processes can change in the interest of managing the public's reaction. In the wake of the Deepwater Horizon oil spill, the worst technoscientific disaster in recent U.S. history, there was substantial pressure on those managing the disaster to maintain order and reduce the damages to economic sectors that were already taking a hit, especially the seafood and tourism industries in the Gulf. In response to this pressure, a loose coalition of regulatory actors implemented a fast, low-cost method of testing Gulf seafood for safety: the sniff test. The sniff test might have served to reassure the public that purchasing Gulf seafood was safe and thus helped minimize damages to the seafood industry and increase trust in the post-disaster scientific response of governmental agencies. However, while the sniff test did confer some benefits, such as increasing public engagement, it also was heavily criticized as lacking scientific rigor. The broadcast images of public officials sniffing seafood for oil came across as almost comical, and rather than being accepted as a legitimate scientific endeavor, the sniff tests were derided as ineffective. Thus, sensory analysis, a scientific practice not well-known to the public, was contested in regard to its legitimacy and credibility in the stressful climate of the Deepwater Horizon oil spill where people were worried about their health and safety.

The Deepwater Horizon oil spill presents an interesting clash of contestation and comfort themes in the reactions to, and functions of, regulatory policies and practices following large-scale environmental disruptions. This case study illustrates the complex role that science plays in such tumultuous and uncertain times. Public agencies and other scientific actors may try to make use of comfort science to maintain stability and prevent panic by developing relatively simple methods. However, successfully implementing "comfort science" can be particularly challenging in technoscientific disaster settings, when the public may have lost faith in the institutions and procedures that allowed for the disaster to occur in the first place. Our findings indicate that, indeed, there was considerable skepticism directed toward the agencies engaged in managing the disaster's aftermath – including cynicism that BP was playing too large a role. It is also possible that the public's trust in science decreases after technoscientific disasters because people interpret these tragedies as evidence that scientific advancement is too risky, thus calling into question the favorability of science and technology in general. In light of these findings, future disaster response efforts should keep in mind the loss of trust in science and technology that these types of disasters evoke, and Disaster Science and Technology Studies research should continue to identify the critical elements of an effective and widely supported response strategy to technoscientific disasters.

References

Barber, B. 1990. *Social Studies of Science*. New Brunswick, NJ: Transaction Publishers.

Beck, U. 1992. *Risk Society: Towards a New Modernity*. London, UK: Sage.

Bucchi M. and Neresini, F. 2008. "Science and Public Participation." Pp. 449–472 in *The Handbook of Science and Technology Studies*, third edition, edited by E.J. Hackett, O. Amsterdamska, M. Lynch, and J. Wajcman. Cambridge, MA: MIT Press.

Button, G. 2010. *Disaster Culture: Knowledge and Uncertainty in the Wake of Human and Environmental Catastrophe*. Walnut Creek, CA: Left Coast Press.

Carson, R. 1962. *Silent Spring*. NY: Houghton Mifflin.

Clarke, L. 1991. *Acceptable Risk? Making Decisions in a Toxic Environment*. Berkeley, CA: University of California Press.

Danielson, E. 2011. "A Study of the Economic Impact of the Deepwater Horizon Oil Spill." Report prepared for Greater New Orleans, Inc. Retrieved June 20, 2013 (*http://gnoinc.org/wp-content/uploads/Economic_Impact_Study_Part_III_-_Public_Perception_FINAL.pdf*).

Davis, H.K., Moffat, C.F., and Shepherd, N.J.. 2002. "Experimental Tainting of Marine Fish by Three Chemically Dispersed Petroleum Products, with Comparisons to the Braer Oil Spill." *Spill Science & Technology Bulletin* 7: 257–278.

Dearen, J. and Bluestein, G. 2010. "Gulf Seafood Declared Safe: Fishermen Not So Sure." *The Associated Press*, August 2. Retrieved October 9, 2012 (*www.organicconsumers.org/articles/article_21346.cfm*).

Drogin, B. 2010. "Sniffing Out the Oil." *Los Angeles Times*, July 13. Retrieved Septemeber 14, 2012 (*http://articles.latimes.com/2010/jul/13/nation/la-na-fish-sniffers-20100713*).

FDA (Food and Drug Administration). 2010. "Protocol for Interpretation and Use of Sensory Testing and Analytical Chemistry Results for Re-Opening Oil-Impacted Areas Closed to Seafood Harvesting Due to the Deepwater Horizon Oil Spill." Retrieved October 11, 2012 (*www.fda.gov/food/ucm217601.htm*).

Flatow, I. 2011. "Assessing the Health of the Gulf, Post-spill." *National Public Radio*, January 21. Retrieved September 28, 2012 (*www.npr.org/2011/01/21/133117322/assessing-the-health-of-the-gulf-post-spill*).

Fortun, K. 2001. *Advocacy after Bhopal: Environmentalism, Disaster, New Global Orders*. Chicago, IL: University of Chicago Press.

Fortun, K. and Frickel, S.2012. "Making a Case for Disaster Science and Technology Studies." STS Forum on Fukushima, February.

Freudenburg, W.R.1997. "Contamination, Corrosion, and the Social Order: An Overview?" *Current Sociology* 45: 19–40.

Frickel, S., Campanella, R., and Vincent, M.B.. 2009. "Mapping Knowledge Investments in the Aftermath of Hurricane Katrina: A New Approach for Assessing Regulatory Agency Responses to Environmental Disaster." *Environmental Science & Policy* 12: 119–133.

Gauchat, G. 2012. "The Politicization of Science in the Public Sphere: A Study of Public Trust in Science in the U.S., 1974–2010." *American Sociological Review* 77: 167–187.

Giddens, A. 1991. *Modernity and Self-Identity: Self and Society in the Late Modern Age*. Stanford, CA: Stanford University Press.

Gill, D.A, Picou, J.S., and Ritchie, L.A. 2012. "The Exxon Valdez and BP Oil Spills: A Comparison of Initial Social and Psychological Impacts." *American Behavioral Scientist* 56: 3–23.

Hansen, L. 2010. "Gulf Fish Cordoned Off for Seafood Sniffers' Inspection." *Weekend Edition Sunday of National Public Radio*, July 18. Retrieved September 23, 2012. (*www.npr.org/templates/story/story.php?storyId=128600385*).

Helm, E. and Trolle, B. 1946. "Selection of a Taste Panel." *Wallerstein Lab Communications* 9: 181–194.

Houck, J. 2010. "UF Set to Train Fish Inspectors to Sniff Oil." *Tampa Tribune*, May 24. Retrieved October 9, 2012 (*http://duke1.tbo.com/content/2010/may/24/na-uf-set-to-train-fish-inspectors-to-sniff-oil/*).

Jellinek, G. 1985. *Sensory Evaluation of Food: Theory and Practice*. Chichester, England: Ellis Horwood.

Jeong, S.Y., Chung, S.J., Suh, D.S., Suh, B.C., and Kim, K.O. 2004. "Developing a Descriptive Analysis Procedure for Evaluating the Sensory Characteristics of Soy Sauce." *Journal of Food Science* 69: 319–325.

Jervis, R. 2010. "First Catch from the Gulf: Is the Seafood Safe?; Concerns Remain as Waters Open after Spill." *USA Today*, August 24. Retrieved October 9, 2012 (*http://usatoday30.usatoday.com/printedition/news/20100824/1afirstcatch24_cv.art.htm*).

Jonsson, P. 2010. "Gulf Oil Spill: Can Region Keep its Seafood on America's Dinner Tables?" *Christian Science Monitor*, September 24.

Lawless, H. and Heymann, H. 2010. *Sensory Evaluation of Food: Principles and Practices*. New York: Springer.

Lyon, B.G. 1980. "Sensory Profiling of Canned Boned Chicken: Sensory Evaluation Procedures and Data Analysis." *Journal of Food Science* 45: 1341–1346.

McCormick, S. 2012. "After the Cap: Risk Assessment, Citizen Science, and Disaster Recovery." *Ecology and Society* 17: 31.

McCright, A. and Dunlap, R. 2011. "The Politicization of Climate Science and Polarization in the American Public's Views of Global Warming, 2001–2010." *Sociological Quarterly* 52: 155–194.

Meilgaard, M.C., Civille, G.V., and Carr, B.T. 2007. *Sensory Evaluation Techniques*, fourth edition. Boca Raton, FL: Taylor & Francis Group.

Mooney, C. 2005. *The Republican War on Science*. New York: Basic Books.

Moore, K. 2008. *Disrupting Science: Social Movements, American Scientists, and the Politics of the Military, 1945–1975*. Princeton, NJ: Princeton University Press.

Moskowitz, H.R. 1993. "Sensory Analysis Procedures and Viewpoints: Intellectual History, Current Debates, Future Outlooks." *Journal of Sensory Studies* 8: 241–256.

Mui, Y.Q. and Fahrenthold, D.A. 2010. "Gulf Seafood Must Pass the Smell Test: Government Trains Inspectors to Sniff Out Contaminated Catch." *The Washington Post*, July 13. Retrieved October 9, 2012 (*www.nola.com/news/gulf-oil-spill/index.ssf/2010/06/trained_noses_to_sniff_out_gul.html*).

Ngai, C. 2010. "Gulf Seafood Declared Safe, But Doubts Linger." *Marketwatch*, November 5. Retrieved October 9, 2012 (*http://articles.marketwatch.com/2010-11-05/industries/30767012_1_gulf-seafood-samples-deepwater-horizon-oil-spill*).

NOAA. 2011. "Seafood Safety Factsheet." July, 21. Retrieved September 15, 2012 (*http://docs.lib.noaa.gov/noaa_documents/DWH_IR/reports/Seafood_safety/Seafood_safety_FACT_SHEET.pdf*).

O'Neil, C. 2010. "Officials, Chefs Tout Gulf Seafood Safety." *Atlanta Journal-Constitution*, September 22. Retrieved October 9, 2012 (*www.ajc.com/news/lifestyles/health/officials-chefs-tout-gulf-seafood-safety/nQkNL/*).

Olbermann, K. and Robinson, E. 2010. "COUNTDOWN for August 17, 2010." *MSNBC*, August 17.

Otwell, S. 2012. Personal Interview, August 30.

Parsons, T. 1962. "The Institutionalization of Scientific Investigation." Pp. 7–15 in *The Sociology of Science*, edited by B. Barber and W. Hirsch. London: Free Press.

Peterka, A. 2010. "Gulf Spill: Groups Skeptical of Federal Seafood-Safety Testing." *Greenwire, Public Health* 10(9), December 17.

Quinlann, P. 2011. "More Questions Than Answers on Dispersants a Year After Gulf Spill." *New York Times*, April 22.

Reilly, T.I. and York, R.K. 2001. *Guidance on Sensory Testing and Monitoring of Seafood for Presence of Petroleum Taint Following an Oil Spill*. NOAA Technical Memorandum NOS OR&R 9. Seattle, WA: National Oceanic and Atmospheric Administration.

Roberts, J. 2010. "Testing Used to Examine Seafood in Gulf; Is Gulf Seafood Safe?" *CNN*, August 6.

Rotkin-Ellman, M., Wong, K.K., and Soloman, G.M. 2011. "Seafood Contamination after the BP Gulf Oil Spill and Risks to Vulnerable Populations: A Critique of the FDA Risk Assessment." *Environmental Health Perspectives* 120(2): 157–161.

Schmit, J. 2010. "Protecting Gulf Seafood from Oil Spill; No-Fishing Zone Spreads." *USA Today*, June 4, p. 3B.

Schwaab, E., Kraemer, D., and Guidry, J. 2011. "Consumers Can Be Confident in the Safety of Gulf Seafood." Food and Drug Administration. Retrieved September 21, 2012 (*www.fda.gov/Food/FoodSafety/Product-SpecificInformation/Seafood/ucm251969.htm*).

Severson, K. 2010. "As Oil and Fear Spread, Gulf Fishing Rules Tighten." *The New York Times*, July 14. Retrieved October 9, 2012 (*http://query.nytimes.com/gst/fullpage.html?res=9D04E5DC143DF937A25754C0A9669D8B63&pagewanted=all*).

Skoloff, B. 2010. "Trained Noses to Sniff Out Gulf Seafood for Oil." *The Associated Press,* June 7. Retrieved September 21, 2012 (*www.nola.com/news/gulf-oil-spill/index.ssf/2010/06/trained_noses_to_sniff_out_gul.html*).

Smith, B. 2010. "Experts Disagree Whether Gulf Seafood Is Safe to Eat." *The Associated Press State & Local Wire,* November 19. Retrieved October 9, 2012 (*www.nola.com/news/gulf-oil-spill/index.ssf/2010/11/experts_disagree_whether_gulf.html*).

Stein, J. 2012. Personal Interview, September 12.

Stone, H., Bleibaum, R., and Thomas, H. 2012. *Sensory Evaluation Practices*, fourth edition. Waltham, MA: Elsevier.

Stone, H., Sidel, J.L., Oliver, S., Woolsey, A., and Singleton, R.C. 1974. "Sensory Evaluation by Quantitative Description Analysis." *Food Technology* 28: 24–33.

Taylor, M.R. 2012. "Gulf Seafood Is Safe to Eat after Oil Spill." U.S. Food and Drug Administration. Retrieved June 20, 2013 (*https://blogs.fda.gov/fdavoice/?tag=gulf-seafood*).

Templeton, K. 2010. "Seafood Workers Sniff for Safety." Mississippi State University, Office of Agricultural Communications, September 2, 2010. Retrieved September 23, 2012 (*http://msucares.com/news/print/fcenews/fce10/100902.html*).

Wynne, B. 1989. "Sheep Farming after Chernobyl: A Case Study in Communicating Scientific Information." *Environment Magazine* 3(12): 10–39.

Wynne, B. 1995. "Public Understanding of Science. Pp. 361–388 in *The Handbook of Science and Technology Studies,* edited by S. Jasanoff, G.E. Markle, J.C. Petersen, and T. Pinch Thousand Oaks, CA: Sage.

Part VI
Rules and Standards

25

Declarative Bodies

Bureaucracy, ethics, and science-in-the-making[1]

Laura Stark

VANDERBILT UNIVERSITY

I. Introduction

Perhaps the greatest flaw in Max Weber's thinking was that he sincerely believed he was right. Weber was the twentieth century's preeminent theorist of bureaucracy, and his description of a paradoxical modern world has guided science and technology studies (STS) scholars since the field's founding. The framework Weber created was immediately influential, as apparent in social theorist Robert Merton's decision in 1936 to write his Ph.D. dissertation on the rise of modern science as a companion to Weber's book on the rise of modern capitalism, *The Protestant Ethic and the Spirit of Capitalism* (2008; Merton 1938). Since then, Weber's framework has remained influential in STS. Steven Shapin's *The Scientific Life* (2010), for example, is best read as a twenty-first-century homage to Weber's "Science as a Vocation" (Weber 2004).

Weber (1864–1920) both theorized and experienced a world permeated by large-scale administrative organizations – both public agencies and private firms – which he thought grew out of the peculiar variety of capitalism that cropped up in Europe around the time of the Protestant Reformation. For Weber, the defining feature of modern capitalism was not free exchange, but the odd fact that people *saved*, in the financial sense, as evidence that they *were saved*, in the religious sense. En masse, Europeans cultivated the self-discipline, restraint, and asceticism that allowed them to accumulate wealth that they poured back into family businesses, eventually shifting financial exchange outside of the household, where it had originated (Weber 2008). In Weber's vision of history, businesses grew in scale and in number along with people's wealth. Governments expanded apace, bringing taxation and the rule of law to give a democratic anchor to the global capitalist economy. Private associations emerged to protect financial interests of the well-educated professional classes that already enjoyed political advantage. The upshot for Weber was that in an effort to manage their size and scale, these growing businesses, associations, and governments set up large apparatuses staffed by permanent employees simply to administrate the organization – apparatuses he called bureaucracies.

In both private firms and public agencies, this historically and geographically unique variant of capitalism prompted a world that was – and remained – highly rule-bound, or in Weber's terms, "rational." For him, "rational" did not mean logical, but indicated, in the context of organizations, that their bureaucracies were governed by agreed-upon, seemingly impersonal, rules. Weber was puzzled and impressed by the suasion of rationality. Even when moderns disagree with the actions prescribed by impersonal rules, Weber observed, they typically follow the rules nonetheless, without being compelled by brute force. Weber had a name for this power to prompt others to obey without physical violence: "legitimate domination" (Weber 1978, see especially volume 1, chapter 3). By Weber's description, there were a variety of forms of domination – both legitimate and illegitimate – but the rational authority of seemingly impersonal, shared rules was the moral force that drove modernity.

The ascetic sensibility, which sixteenth-century Calvinists cultivated to assure themselves they had a secure place in the kingdom of heaven, had been transposed, according to Weber, into a secular, capitalist world that was itself the product of self-restrained Calvinists. Science was indispensable to the impersonal administration of states and private enterprises in this environment:

> A rational, systematic, and specialized pursuit of science, with trained and specialized personnel, has only existed in the West in a sense at all approaching its present dominant place in our culture. Above all is this true of the trained official, the pillar of both the modern State and of the economic life of the West.
>
> *15, Protestant Ethic, Parsons translation*

The work practices and ways of reasoning endemic to science played handmaiden to a rational world because it made everything *in theory* knowable, predictable, and fair – even human beings. An austere work ethic (indeed, the Protestant ethic) and the economic practices that went along with it (for example, bookkeeping, investing) had been transposed onto a secular economic system. This combination created a new human experience: one in which moderns dispatched the tasks at hand according to protocol and established procedure without caring to ask to what ends their work would be put. Those who exemplify Weber's scientist-bureaucrat in the present day include nuclear scientists, environmental policy makers and other specialists working on small assignments directed by large organizations to accomplish end goals with which they may or may not agree – if only they stopped to think about it (Espeland 1998, Gusterson 1996, Thorpe 2006). Messy moral questions were translated into technocratic inquiries precisely to provide easy answers to important existential questions about how we should live, individually and collectively (Espeland 1998, Evans 2002). The result, Weber lamented, was a modern amorality for which he blamed and pitied both scientistic thinking and scientists themselves.

How wrong he was. This chapter argues that, in some settings, scientists meaningfully debate the ultimate ends of scientific work within the structures of bureaucracy and yet without devolving to rote, technocratic rule following. To be sure, Weber and scholars following his tradition have described two ways that scientists have participated in modern bureaucracies: as insiders acting as the classic science-bureaucrat Weber bemoaned and as outsiders resisting bureaucracy. First, scholars have shown how as insiders scientists have taken jobs in the service of large organizations, often as full-time employees who follow rules and protocols precisely without using personal judgment to consider whether their actions were right or wrong. If scientists in these roles used their judgment and discretion to make decisions, they had breached the rules designed to keep them in check. Second, scholars have shown, in the spirit of Weber, modern conditions prompt their own subversion through science. Particularly after World War II, large-scale funding of science by governments and industry served to train and employ the scientists who ultimately worked to dismantle bureaucratic capitalist imperatives through activist organizations (Moore 2008, Thorpe 2006, Wisnioski 2012).

These two approaches are accurate, but they offer only a partial description of the place and practices of scientists in bureaucratic administration. This chapter widens the scope of STS and extends Weber's theory of bureaucracy to correct his narrow vision of how scientists enact bureaucracy. The chapter explains how modern organizations have not only limited scientists' roles to technocratic cogs or engaged them as antagonists, but also spawned a new kind of bureaucratic structure that empowers science experts (and others) to use their discretion, rather than to avoid idiosyncratic judgments, to make decisions about other people and the collective good – so long as they work together. I call these new groups within bureaucracies "declarative bodies" (Stark 2012).

This chapter sets out the three defining features of declarative bodies (Section II) and uses two case studies to explain the two broad tasks declarative bodies carry out for organizations. The first case illustrates a declarative body assigned the task of making rules: the American Psychological Association's Cook Committee, formed in the late 1960s to write a new section on research on human beings for the association's code of ethics (Section III). The second case examines a type of declarative body formed to apply established rules – namely Institutional Review Boards (IRBs) – by examining the inner workings of IRBs at three universities in the present day (Section IV). In the chapter's penultimate section (V), I consider how the analytic category of "declarative bodies" opens new research questions that STS scholars can explore in other areas as well.

At the broadest level, declarative bodies have direct, specific effects on the work of individual scientist-scholars by requiring precise changes to how they carry out and express their findings. Scientist-scholars obey the authority of declarative bodies because of the symbolic and material resources at stake – whether grant money, article publication, career advancement, or project approval. Consequently, declarative bodies actively alter the process of knowledge-making and the products of knowledge itself. Not only do declarative bodies set broad boundaries around what is acceptable or unacceptable, good or bad; the work of declarative bodies is, quite directly and immediately, integrated within the practices of knowledge-making taking place in labs, fields, museums, and clinics.

II. Three defining features of declarative bodies

Public agencies and private businesses purporting to operate fairly or democratically face a persistent challenge: their legitimacy depends on the appearance that they govern according to impersonal, equitable rules of law, and yet they rely on select groups of elites to create and interpret these rules (Weber 1978: 979–991, Turner 2001). In order to manage this challenge, science comes into play in two ways in bureaucracies. First, organizations draw on scientists[2] *to establish rules*, regulations, and policies through a process that Weber called "lawmaking." Second, organizations rely on scientists *to apply existing rules*, regulations and policies to specific cases, a process that Weber (awkwardly) named "lawfinding" (1978: 653–654).

The roles of scientists themselves in carrying out these two tasks have changed dramatically in the past century. To make decisions in their name, many large organizations came to rely on groups composed of scientists who are expert in content areas (for example, anthropology), rather than individuals who are trained in fields of administration itself (for example, human resources). Specialists in science, medicine, and other fields of knowledge have been incorporated in growing numbers into the process of writing and applying rules on behalf of governments and private corporations – whether profitable or charitable – as members of committees that I call "declarative bodies" because their declarations in speech and text ("revise," "amend," "move") alter the workaday practice of other scientists. In their policies, organizations officially authorize members of declarative bodies to use their discretion in making decisions rather than to avoid judgment through rigid application of rules. Often, for example, academics serve national and global organizations as members of groups that decide how public grant money should get allocated (Brenneis 1994, Guetzkow et al. 2004, Lamont 2010), what research practices are ethically acceptable (Bosk and DeVries 2004, Evans 2002, Hurlbut 2010), and how environmental and health policies should be enforced (Benson 2010, Burnett 2012, Imber 1986, Jasanoff 2007).

Declarative bodies have three charateristics in common that suggest how they work and why they are useful for organizations. First, the knowledge experts who serve on these deliberative bodies for organizations act as bureaucrats only fleetingly and for a particular decision-making purpose. Anthropologist Donald Brenneis (1994: 25) uses the term "nonce bureaucrats" – bureaucrats only here and for the time being – to describe knowledge experts who interpret rules to meet specific and temporary bureaucratic needs, such as evaluating grant proposals, reviewing clinical trial data, deciding abortion requests, or revising school curricula. Participants typically serve for a few hours per week or month, and most likely would not identify themselves as bureaucrats. Although they carry out the work of bureaucratic administration, these academics, clinicians, lawyers, and others learn to embody "bureaucratic selves" temporarily in situations where they do the work. The implication is that "bureaucracy" at least since the late twentieth century might best be thought of as a discursive setting, in which speakers adopt the language and logic appropriate to the task at hand, rather than a physical location that is clearly demarcated and solely dedicated to organizations' administration.

Second, knowledge specialists are urged to use discretion when they make decisions. For regular full-time bureaucrats working on their own, use of discretion would be an abuse of power – and thresholds, standards, and other forms of quantitative rules are aimed at staving off this threat (for example, Riles 2011). For knowledge experts, however, their mandate is precisely to use their discretion to create and interpret qualitative standards. Scientists' license to use their judgment has been built into rational – that is to say, rule-bound – decision-making rules. In sum, scientists, clinicians, humanists, and other knowledge specialists are valuable to organizations not for their ability to forego personal judgment. To the contrary: knowledge workers are free to use their uncommon knowledge to apply general rules to particular cases. Putting the first two features of declarative bodies together and in contrast to Weber's ideal type bureaucracy, it is fair to say that the value of nonce bureaucrats is their apparent ability to use discretion soundly, not to avoid it.

Third, the necessary condition for sound discretionary decisions is that choices are made among a group of experts acting as a single unit, a legal body though not one human body (Koch 1997: 47). The aim is to create a "decision-making community" whose ultimate decisions can be respected and considered legitimate because they include "different personalities, agendas, value systems, types of expertise and experiences, etc. ... often by design" (8–9). Fair decisions within a democratically-operating organization would ideally appear beyond the whims of an individual, even a highly-educated one, and as a result legitimate discretionary decisions have to be made by a group, or "body," of multiple experts. The consequence is that organizations empower experts to use discretion only to the extent they can make decisions as a committee.

Taken together, the cases show that declarative bodies consider competing *forms* of scientific evidence to write and interpret rules. The cases that follow display the techniques that declarative bodies use to make decisions and, consequently, to affect scientific knowledge production.

III. Case 1: lawmaking and APA's ethics committee

At the end of the 1960s, thousands of American psychologists wrote to their professional organ-ization, the American Psychological Association (APA), and described research on human beings that they thought might be unethical. They were completing a survey from a group of six leading experts in their discipline, whom the heads of the organization had empowered to write an ethics code setting standards for how psychologists should treat their research partic-ipants (Stark 2010). This section focuses on the Cook Committee's work on the question of whether – or how – psychologists could use deception, which was one of six research practices that psychologists were specifically asked about in the survey. Ultimately, the committee agreed that deception was an acceptable research practice and they wrote this rule into their final prod-uct, *Ethical Principles in the Conduct of Research with Human Participants* (1973), which was added to the APA's existing ethics code the same year.

This case study examines the Cook Committee as an example of a declarative body to illus-trate the three features of declarative bodies: experts serve temporarily for a specific decision-making purpose, are authorized to use their discretion, and have the imperative to reach a unified decision as a single social actor. The section also explains how members of declarative bodies draw on competing forms of evidence and how they resolve conflicts stem-ming from the rival implications of these competing forms. In the case of the Cook Committee, its six members had at hand the results of a recent and massive survey of American psychologists. These survey results were intended to be a source of (seemingly) objective evidence that I demarcate "matters of fact," and as such were meant to serve as the basis for the ethics code. However, committee members also drew heavily on their first-hand knowledge as researchers in writing the ethics code – a form of evidence I call "professional experience." I give additional examples of these two competing forms of evidence and spell out the implica-tions for declarative bodies in the second case, which looks at the inner workings of IRBs in the process of rule-finding and that I describe in the later section. The case of rule-making by the APA ethics committee that follows here demonstrates that experts who constitute declara-tive bodies draw on their professional experiences to establish seemingly impersonal, universal rules that distinguish good from bad, right from wrong. The judgments of Cook Committee members based on their training and track record as practicing scientists largely explains the policies that the entire organization ultimately adopted.

Temporary bureaucrats

In 1966, the officers of the American Psychological Association established a declarative body, formally named Ad hoc Committee on Ethical Standards in Psychological Research, to write an update to its existing code of ethics that addressed how researchers could treat their research participants. Rather than assign administrative staff to the project, APA officers selected six esteemed university faculty members from across the United States to constitute the commit-tee. For APA leaders, it was important that the Cook Committee anchor the new ethics guidelines to empirical evidence, and so the committee members began their work by collect-ing data. This was a massive administrative undertaking, and resulted in a membership survey of 19,000 members between 1968 and 1970. After the Committee received the last survey response, they worked for the next year and a half individually and collectively at regular writ-ing retreats to compose the *Principles*.

Discretion

The Cook Committee took the questionnaire responses very seriously, but the meanings and implications of the incidents were infinitely flexible. The Committee's first writing retreat in Boulder in February 1970 suggested the troubles the members would have in arriving at common interpretations of the questionnaires. "A considerable portion of the two days," chair Stuart Cook recorded, "was devoted to discussion of ethical problems and proposed principles in order to give the Committee members a feel for the difficulties in arriving at positions on which we could agree."[3]

One strategy that Committee members used in drafting the ethics rules was to insert guidance grounded on their experiences if they felt that their first-hand knowledge as researchers exemplified a basic principle that the survey responses suggested but did not actually articulate. I call this form of evidence "professional experience." Committee member Gregory Kimble, for example, was assigned to write the first iteration of the deception principles in preparation for the February meeting, and two out of the fourteen incidents were his own stories. By his own account, Kimble "edited every incident heavily and," as he described it, "ended up writing my own revisions in every case."[4]

Given his training and experience, it might make sense that Kimble elaborated on deception in his section drafts, if it were not for the reams of survey responses he had intended to marshal. Kimble considered himself an experimental psychologist, and he built his career testing theories of learning. The political popularity of stimulus-response theory had reached its zenith in the early 1960s, and while it retained a foothold in the discipline, its practitioners eventually found that policy makers were losing interest in the field's services (to control prisoners, mental patients, children) when the theory encroached on democratic ideals. To be sure, Kimble felt that deception should not to be used cavalierly. But the practice created a "serious conflict … between honesty on the one hand and the scientific value of the data to be obtained on the other." Kimble felt that psychologists could, and indeed *should*, manage the conflict between honest practice and valid data by anchoring decisions to a utilitarian calculus. "There is a positive correlation," Kimble argued, "between the importance of the problem (to the discipline not just the investigator) and the degree of deception that can be justified."[5]

Depending on their training and research, however, scientists can embody and indeed value different elements of their first-hand knowledge, which they can nonetheless warrant as "professional experience." In the case of the APA ethics committee, this issue came to the fore because each of the committee members was assigned the task of seeking advice on the code from a set of senior psychologists and leaders of subfields within psychology. Committee member Brewster Smith interviewed Alberta Siegel, a prominent psychologist of child development and former editor of the journal *Child Development* (1964–1968), and Diana Baumrind, also a child psychologist. Smith's interviews with the women made him increasingly stalwart in his conviction that deception was indefensible under any circumstances, in contrast with the position of fellow committee members such as Kimble and also McGuire. Siegel, who by that time had read a working draft of the committee's principles, told Smith that the document was "unduly permissive of deception." Smith told his chairman Stuart Cook: "I think her case is a serious one and warrants reconsideration of our initial chapter."[6] One month later Smith spoke and later corresponded with Baumrind. Her position, as Smith heard it, was that "behavioral research is contributing to the moral ills of society and the influence is a direct one." To Smith's mind, she was not an alarmist: "I think we need to take this more seriously than we have in our present report," Smith told Cook. "Generally, I think her letter is worth a very careful reading and recommend that it be circulated to the committee."

The interviews with Siegel and Baumrind were a catalyst for Smith, and in their wake he advocated aggressively against deception. "[S]peaking for myself (MBS), I think we have to highlight deception-manipulation as raising a different order of issues than the others that we deal with," he explained to Cook. "I think our document can be radically faulted as it stands unless we make this distinction."[7] In sum, the interviews with Siegel and Baumrind gave Smith the impetus and justification to urge that deception be treated with an absolutist approach rather than a utilitarian approach in the first published draft of the *Principles*.

In practice, the Committee members collapsed their roles as editors and authors. As they tidied members' survey responses into one coherent statement of psychologists' views, the committee members drew in their own autobiographical vignettes.

Consensus

In the summer of 1970, the committee members met in La Jolla, California to pull together a unified version of the *Principles* from the sections that they had each written independently. When they tacked together the separate sections, however, they found that the resulting document was full of contradictions. For example, four of the six committee members had all written about the ethics of deception within their assigned sections formally divided into stress, informed consent, invasion of privacy, and deception proper. Troublingly, their guidelines about deception in each of these sections were at odds with one another.

As a solution, the Committee members decided to include in the introduction of the document (which had yet to be written) an overarching statement about how psychologists should make ethics decisions in general. Having agreed to circulate suggestions, committee member William McGuire proposed that the Committee manage their text's "internal contradictions" by encouraging psychologists to "utilize a cost-benefit calculus," but other members resisted because they believed that some practices were wrong in any circumstances regardless of the benefits.[8] For some, deception appeared to call for a utilitarian approach, which was the position that lying was acceptable if, as McGuire put it, the benefits outweighed the costs. On the other side, deception appeared to warrant a "deontological" approach (to use Smith's term), which was the view that people have a few basic rights and freedoms that should not be violated in any circumstances. The members had come to their views based on their own professional experiences, which they each regarded as equally valid, and while their conflict over deception was mostly collegial, it was also the source of the quandary: how would they resolve a disagreement based on equally valid but ultimately incommensurable professional experience?

In effect, committee members' divergent perspectives undermined their own authority to write the ethics principles in the document that set them out. When the first full draft of the *Principles* was published in 1971, it betrayed the division among Cook Committee members over the ethics of deception. Uncannily, Cook anticipated the reaction that APA members would indeed have to the draft *Principles*:

> From one side it is seen as a panic-bred renunciation of our most effective research practices – an underhanded blow to the development of a psychological science badly needed by society. From the other it is viewed as a mealy-mouthed whitewash of reprehensible research behavior – a guide book that might well be subtitled 'How to justify taking advantage of research subjects.'[9]

Their fifth iteration was published in July 1971 in the association's newsletter, *The APA Monitor*, in order to get feedback from APA members who would have to vote to pass the *Principles* as an addendum to their ethics code. The Committee intended this draft to be essentially the final version but the draft embodied their divergent views – their lack of consensus, and thus of authority to act as "the committee." In light of their split opinions, the APA Board of Scientific Affairs saluted the Committee's two-year writing effort – and promptly sent the men back to the drawing board.

Using a different form of evidence altogether, critics of the draft argued that the Cook Committee had drafted rules about the use of deception that were at odds with the views of everyday psychologists, which the Cook Committee might have been aware of, critics charged, had the committee been good, objective empiricists and attended to the matters-of-fact of the survey results. This criticism reveals the importance of discretion for members of declarative bodies and shows that psychologists recognized many forms of "empirical evidence": the committee's first-hand knowledge and disembodied data. The crux of the committee's internal differences centered on the various implications of one form of evidence (namely, different members' professional experiences); but the committee's disagreements with workaday psychologists whom they claimed to represent turned on a disagreement over which *form* of evidence – matters-of-fact or professional experience – were most persuasive.

Committee members took up both problems when they recognized that they had to settle their own disagreements because the committee only had legitimate authority, to use Weber's terms, if the committee members composed rules in a single voice. Throughout the fall of 1971 and spring of 1972, the Cook Committee assessed responses to the first draft of the *Principles*, and came to agree with their professional experiences as psychologists through their training and eventual research. Cook and McGuire were particularly disposed, by the demands of their own studies and experiences, to remain open to deception practices. As Cook explained, "When I was a subject, I expected to be deceived."[10] McGuire, who did research on attitude change himself, stressed that deception was acceptable in part because it was impossible for people to make truly autonomous decisions. What does it mean to get consent from research participants, he asked,

> when one considers also that the individual's choices are not capricious nor independent of contemporary and earlier influences from his environment? What does it mean to say that the person should be free from coercion to participate, at the same time as we say that people's decisions must be motivated or that they are affected by forces that act on them?[11]

Cook explained the Committee's experiences to APA members in the May 1972 issue of the *Monitor*, which accompanied the second published draft of the *Principles*: "Reactions to the first draft brought to light a number of disagreements among psychologists that, in turn, affect reactions to proposed ethical principles." Consequently, members of the Cook Committee tuned their opinions and decided: "Openness and honesty are essential characteristics of the relationship between investigator and research participant. When the methodological requirements of a study necessitate concealment or deception, the investigator is required to ensure the participant's understanding of the reasons for this action and to restore the quality of the relationship with the investigator."[12] These fifty words comprising one principle on deception replaced the tens of pages detailing and debating seven principles addressing deception in the first draft of the rules. The practice of deception was no longer described as unethical, nor was its use defined as a violation of subjects' rights and psychologists' standards. Rather, deception was neutralized in such a way that the ethical question was not whether the practice had been used, but whether the appropriate techniques for using deception had been deployed, a determination that shaped the possibilities for American psychologists' research practices and the resulting knowledge they could produce.

By rendering a decision as a single social actor, the Cook Committee touched knowledge-making practices of thousands of American psychologists because the rules they were authorized to set altered the daily motions and embodied experiences of scientists at the time and ever since. The case demonstrates that within bureaucracies *legitimate* "empirical evidence" takes multiple forms, including experiential knowledge and not only the objective matters-of-fact. Several members of the Cook Committee used deception in their own research, and they ultimately used their discretion and authority as members of a declarative body to write principles that corresponded to their professional experience. Their decision on deception may have been a convenient answer to the question of whether and how psychologists could use the practice in research, but it also fit well with their own experiences borne from decades of training and research that six specific psychologists had conducted first hand.

IV. Case 2: lawfinding and university IRBs

Institutional Review Boards are state-mandated groups that enact human subjects regulations at most American hospitals, universities, and research institutions (Stark 2012). This section draws on my long-term ethnographic observations and audio recordings of the meetings of three IRBs from 2003 to 2004. Created by federal regulation in 1974, these deliberative groups have the authority to decide whether researchers at institutions that receive public money intend to treat their "human subjects" appropriately. Today, most universities, hospitals, and research institutions in the United States and abroad maintain at least one IRB or contract with a commercial IRB.

As a companion to the previous case study of the APA ethics committee between 1966 and 1973, this section considers Institutional Review Boards as an example of a declarative body in the present-day. With this additional case, I again document the three features of declarative bodies (knowledge experts act as bureaucrats fleetingly, use discretion, and make decisions acting as a single unit) and then elaborate on the features of decision-making processes within declarative bodies introduced in the case of the APA ethics committee. Specifically, this section explores how knowledge experts, each with a distinct claim to expertise, reach the consensus necessary in bureaucratic settings to speak as a single social actor and thereby ground their claim to legitimate authority.

Temporary bureaucrats

IRBs are an example of a form of state decision making that gained ground within the U.S. federal government starting in the late 1960s. During the Johnson Administration, the U.S. Congress passed an unprecedented number (130) of federal statutes that dramatically expanded the scope of federal programs (Levitan and Taggart 1976, McKenzie and Weisbrot 2008). These programs were overwhelmingly aimed at regulating the natural environment and human health and, consequently, specialists in science, medicine, and other fields of knowledge were incorporated into the process of writing and applying regulations (Katznelson 1996: 28; Kerwin 2003: 15).

Federal human subjects regulations were part of that effort. The administrative rules were passed in 1974 through the National Research Act. The regulations required that universities, hospitals and other institutions create review boards to evaluate all research on people under their auspices if the institution received public money (which most did: McCarthy 2008). Human subjects regulations and other protective efforts placed decision making outside of civic control and into the hands of formally-defined experts.

In 1965, NIH Director James Shannon made the first formal move to institute IRBs when he asserted that NIH-sponsored research "should be discussed with a peer group" that is, peers of the investigator, "for a basis of sound judgment." Eventually, the U.S. Surgeon General approved Shannon's 1965 recommendation, and informed major universities, hospitals, and professional associations that to receive federal money for research, they would have to "provide prior review of the judgment of the principal investigator or program director by a committee of his institutional associates."[13] As originally imagined and subsequently required, these committees of peers reviewed risks and benefits of scientists' proposed research, and assessed "the appropriateness of the methods used to secure informed consent."[14] The mandate for peer evaluation of research is the backdrop for the present-day requirement that IRBs include experts in the topics under review by the committees.

The assumption that experts should use discretion was built into the structure of IRBs. When he required peer review of federally-funded research starting in 1966, the Surgeon General did not circulate guiding principles but wrote that "the wisdom and sound professional judgment of you and your staff will determine what constitutes the rights and welfare of human subjects in research, what constitutes informed consent, and what constitutes the risks and potential medical benefits of a particular investigation."[15] The standard material of ethics codes – risks, benefits, and consent – were not defined in advance but were to be, according to the Surgeon General, whatever a local committee of experts and others deemed appropriate.

Discretion

The IRB members I observed reached decisions not by applying formulaic rules, but by settling who among them had the best justification for his or her recommendation. Using their own pragmatic and case-specific knowledge, board members arrived at decisions by determining the best evidence for a good course of action within their specific, local settings. The evidence that IRB members used took a variety of forms, but here I point out the important differences in the two forms that I introduced in the previous section on the APA ethics committee: matters-of-fact and professional experience. Although opinions base on people's first-hand experience are necessarily subjective and idiosyncratic, experts' embodied knowledge gives them value in bureaucratic settings and grounds their most persuasive form of evidence.

To be sure, board members often presented their views with seemingly objective findings during IRB meetings, often in the form of numbers, like statistics. This form of evidence tends to be authoritative because it is not associated with any one person, and instead can take a material form, for example, in a diagram, table, or image. I call warrants based on seemingly disembodied, objective evidence matters-of-fact. What I mean to designate with matters-of-fact is information that is shared specialized knowledge: the information that one could learn by studying academic books, reading scholarly journals, or hearing lectures in a given field (Collins and Evans 2007).

One example comes from my interview with a faculty member at the teaching hospital where I observed. During our conversation after an IRB meeting, I asked Kimberly, a young pediatrician, why she had urged board members to approve a hotly contested study without modifications, which some of her colleagues had strongly resisted. Here is how she explained her position:

> I mean, there were recent articles in the New England Journal that [said] the chance of actually having major toxicity from the trial is less than five percent and toxic death is less than one percent, like point five percent. And the chances of getting, having clinical

benefit, true response, is about ten percent. And stabilization of disease was like thirty to forty percent. So, actually the chances are not so bad in that stage in disease, that you will get benefits.

C6: 331

Kimberly immediately drew on matters-of-fact to justify her opinion: response rates from a premier medical journal. Interesting, Kimberly continued:

> There are certain people that I think make judgments based on sort of visceral feelings about things. And then there are other people that will listen to evidence-based arguments. And I'm probably a mix of both. I have definite/ I come to the table with definite visceral responses to things… But, in the end if you can give me a good scientific basis, that's what's going to sway me the most either way.

C6: 331

In addition to emphasizing the importance of matters-of-fact, or what she called "evidence-based arguments," Kimberly also described a second form of evidence that IRB members used in making decisions – and which she herself indicated that she used since she comes to the table "with definite visceral responses to things."[16] I call this second kind of evidence "professional experience." The cultural authority of knowledge-producing professions – the sciences, medicine, and the humanities, for example – depends on the notion that practitioners acquire unique and valuable skills through their training and professional experience. These experiences, moreover, are generally thought to translate into rare abilities to judge the quality, veracity, or ethics of knowledge outside of research settings.

As an example of "professional experience," consider one case in which members of the IRB were reviewing a study on arthritis in elderly people. They were recruiting arthritic men via courses offered through the National Arthritis Foundation. As a result, the investigators were able to tap into a population for which a good deal of information had already been collected about their condition. The primary IRB reviewer, Dr. Endicott, had given it a positive review, and the study seemed poised to sail through to full-board approval. Before the vote for approval, however, another board member, an exercise physiologist named Ulrich, argued for changes to the protocol that would seriously decrease study enrollment:

> **Ulrich (exercise physiologist):** With reference to my background, there's a health questionnaire (used in) other studies, and it does a good job of referring to: How do you feel (at different levels of exertion)?

NG, July: 350

Here, Ulrich specifically linked his opinion to what he called his "background." Then he continued by recounting one of his own experiences as a researcher:

> **Ulrich (continuing):** I'll give you an example: my first year here in 1982, we had someone on a treadmill who had a spinal cord injury with support, and the student came up and said 'excuse me, but I don't think we should be testing this subject. He has an aortic aneurysm.'

NG, July: 350

Other board members laughed. The laughter suggests how untoward Ulrich's anecdote sounded: he had encouraged a person with a serious heart condition to exercise. The debate continued, but in the end Ulrich's warrant based on his professional experience swayed members. Most influentially, Ulrich had persuaded Dr. Endicott, the main IRB member charged with presenting this study to the board. This analysis is one example of my observation that IRB members tended to defer to recommendations of fellow members who articulated experience as a knowledge producer that was most similar to the investigator. In some instances, this board member recommended harsher restriction on the investigator's work based on his own experience, and other times he urged the group to loosen constraints. Regardless of the direction of change, though, it was the similarity of the expert's professional experience with the case at hand that appeared compelling, rather than the relative authority of an expert's discipline in some seemingly objective hierarchy (Mallard et al. 2009).

Consensus

Different forms of evidence – matters-of-fact versus professional experience – can also yield conflicting implications. In instances when members of declarative bodies asserted incompatible views based on different forms of evidence, professional experience typically won out and won swiftly. Sustained disagreement threatened to undermine the ability of the declarative body to act as a unit, which would unravel its basis of authority.

The fundamental way in which IRB members reached consensus on how rules applied in specific cases was to settle who had expertise on the matter at hand. The group deliberations that I observed were essentially referenda on who had the most salient knowledge and thus whose advice should be followed. IRBs also had decision-making habits, which they used when they processed studies that felt familiar and routine. However, when the proper course of action was either ambiguous or contested, board members reached consensus by deciding whose recommendation was based on the most relevant knowledge, which all members could endorse as the view of "the IRB."

Declarative bodies in action

IRB members work to build consensus about how a given research protocol should be changed (or "modified" in the parlance of regulation). Technically, IRB members vote to decide how a researcher must alter her study. In all but two cases that I observed, IRB votes were unanimous, and in order to dissent, these members abstained from voting altogether, rather than cast a "no" vote. The pervasiveness of group consensus and the strictly symbolic function of voting can be seen as a product of the regulations (Coleman 2004, Guston 2006). Administrative law only authorizes "the IRB" to review and recommend changes to studies, and this feature of their structure disposes board members to reach unanimous decisions. Consequently, IRB members reach agreements by using their discretion to decide who has the best evidence on the matter at hand.

The value placed on scientists' rare, embodied experience is instrumental in modern bureaucracies. It allows experts to carry the authority they have derived from one setting to another, for example, from the bench to the bedside, the lab to the courtroom, the field to the review panel, and the armchair to the lectern. The authority vested in scientists' first-hand experience has prompted organizations' use of nonce bureaucrats and also grounded organizations' mandate that experts use discretion. As a result, science experts have created and enacted a vast range of postwar ethics policies that have also affected their daily work. In the case of IRBs, most American research universities, hospitals, and institutes in the United States had a group review procedure in place by 1970 because federal health agencies had made committee review a condition of funding.

As this section and the case of the APA ethics committee suggest, not all forms of evidence are equally persuasive within declarative bodies. Evidence based on experts' professional experience was most persuasive in the two cases because such evidence is based on embodied knowledge, which makes it a rare commodity. At the same time, claims based on professional experience refer to first-hand knowledge, making it nearly impossible to challenge on its own terms. These paradoxes open up questions about the process of decision making in declarative bodies and the factors that alter knowledge-making practices for scientists.

V. Declarative bodies: a research agenda

Small decisions have wide-reaching effects for how knowledge is made and by whom. The concept of declarative bodies promises to yield new approaches and answers to ongoing questions in STS: how do people make facts in the name of science?

1. Disparities in participation

In practice, some people who are members of declarative bodies participate more often and more influentially than others in decision-making discussions. This chapter suggests that the type of evidence that a speaker uses to justify her recommendations works to persuade or dissuade listeners of the merit of her view, and, consequently, of the wisdom of harnessing the group's authority to the recommendation.

Claims based on embodied knowledge are those that derive from first-hand experience (what I have referred to as "professional experience") – that allow us to say, I saw it with my own eyes. In sites of knowledge production, the authority of embodied knowledge is consistently persuasive enough that STS scholars have come to argue that it might offer an opportunity for more democratic decision-making practices in science and medicine. For example, sick people have leveled surprisingly persuasive claims to expertise – and thus authority – on their own illness (see, for example, Brown and Morello-Frosch's contribution to this volume, as well as Rabeharisoa and Callon 2002, Shapin 1995, Williams and Popay 2006). When non-credentialed people get access to sites of knowledge production, they are often thwarted because they do not have the manners of talk and behavior that signal their sensibilities are trained and thus trustworthy. For STS scholars, the implication is that we should explore how scientists and other knowledge specialists justify their views in bureaucratic settings. Advocates and scholars of "lay expertise" might reconsider the common assumption that non-experts' embodied experiences are a resource unique to them – which they might use persuasively if only they were included in deliberations.

Attention to declarative bodies enriches current efforts to explain decision outcomes in two ways. First, it adds to a fuller explanation of group decision making (and its disparities) by emphasizing how language strategies can build or break down consensus during the course of deliberations. Other factors certainly affect the decision-making process. The relationships that people build outside of meetings help to explain how groups achieve consensus in the official moment of deliberation (Lee 2007, Polletta 2002, Wilde 2007). It is also clear that decision outcomes are patterned on group members' demographics. But attention to forms of evidence reveals what it is about individuals' traits that make some members appear to have a knack for affecting decisions – namely, that the *types* of evidence that individuals have at their disposal are constrained by their life histories. By deferring to a members' professional experience – as opposed to personal stories or "evidence-based" knowledge – structures of authority are stabilized through group deliberation.

The members of declarative bodies who sway decisions most systematically, according to the case studies explored here, are those who justify their recommendations with a claim to research experience relevant to the case at hand. In the two cases explored in this chapter, group members recounted their research experiences as a means, in the absence of other routines for establishing credibility, to persuade colleagues that their advice was valid and worth following. Members who grounded their opinions on professional experience were the most persuasive but it is worth bearing in mind that "professional experience" is not a form of evidence available to all people members. Only science experts who achieved training as a relevant specialist could use this most persuasive justification. In short, not all forms of evidence are equally persuasive in bureaucratic settings, nor are they equally available to all potential participants.

2. Open and restricted settings

STS scholars might observe that the groups I call declarative bodies are sites of knowledge production, analogous to laboratories or libraries. As such, they open the possibility of exploring the place of expert judgment in science by focusing on discursive spaces rather than on individual actors. Declarative bodies enable comparative study of the range of sites in which knowledge is made.

Attention to declarative bodies allows scholars to compare how experts communicate officially versus when they are off the formal record in settings that are closed de facto (if not de jure) or are otherwise out of reach. The analytic of open and restricted offers a starting point for the urgent work of expanding conceptions of scientific settings and moves us past the problematic distinction of "public versus private." The distinction between open and restricted settings is based on people's relative access to spaces – not only physical settings but diffuse forums such as one-on-one talks and email exchanges. The concept allows scholars to carry out empirical studies that retain Weber's pragmatic suspicion that democratic ideals, such as transparency and freedom of information, conceal as much as they reveal (for an excellent model, see Riles 2011).

This is also important work given that governments, corporations, and other organizations are responsible for massive numbers of choices that affect people's everyday lives, and yet such decisions typically take place outside of public view. Declarative bodies often create and apply rules for organizations in settings in which non-experts have limited access and little involvement – in terms of both choosing the decision makers and scrutinizing their decisions. In settings where experts make decisions with little formal record, STS scholars might explore when first-hand, experiential, embodied knowledge might be most persuasive, as it was in the two cases examined in this chapter. When decisions are translated outside of these cloistered settings for public consumption, the grounds on which group members made decisions often drops away entirely or is scrubbed clean of personal idiosyncrasy.

This chapter advances research on science in the global context of large-scale organizations by suggesting that the ways in which experts interpret rules is contingent on their immediate audience. The cases here suggest that the techniques of persuasion that are most effective among experts in *restricted settings*, such as IRB meetings, have been shown to be ineffectual and often corrosive for experts who have been called to account for their views in *open settings*, for example, in congressional hearings (Thorpe and Shapin 2000), legislative debates (Porter 1995), and jury trials (Timmermans 2007).

VI. Conclusion

On November 7, 1917, Weber took the stage in Munich and delivered the address his student audience had paid him to produce: a lesson on science as a vocation. Where Marx in the nineteenth century had instructed, with ferocity and certainty, every man to change the world, Weber took the tact of a twentieth-century skeptic: read Tolstoy. Weber, conjuring the Russian author, asked his audience "Who if not science will answer the question: what then shall we do and how shall we organize our lives?" It was a riddling question, since Weber believed that science could not tell one how to live. Weber explained that he and others could make moral decisions as scientists so long as they avoided searching for moral guidance in the products of their scientific work, namely, matters of fact.

In his writing, Weber did not describe a space within the modern bureaucratic world in which scientists might create or deploy an ethics beyond the evidence of science that nonetheless took shape within the structures of bureaucracy. This chapter builds on Weber's theory and opens new areas of inquiry through analysis of declarative bodies. Drawing on two cases, I show how bureaucratic organizations have empowered expert groups I call declarative bodies to create and interpret rules. "Declarative bodies" demarcate groups of knowledge experts who are empowered by law to make decisions for public and private organizations. Contrary to Weber's explicit theory of bureaucracy, the new science-bureaucrats who comprise declarative bodies serve in this function temporarily and for a specific purpose and are encouraged to use rare, experiential knowledge to make judgments.

The say-so of these declarative bodies has tangible, material consequences. For example, the Federal Reserve Board decided which institutions American taxpayers would keep afloat in the 2008 financial crisis, and the Medicare Payment Advisory Commission tells doctors how much they will get paid for seeing low-income patients. Declarative bodies signal a particular and very recent understanding of the place of experts in bureaucratic administration.

At the broadest level, I argue that the process of knowledge-making, a topic that has anchored STS work since the 1980s' turn toward practice (Zammito 2004), has paid accurate but partial attention to the role of bureaucracy in knowledge production. Scholars have explored how bureaucratic ancillaries, such as standardization and routines (Heimer 2001, 2008, Timmermans and Berg 2003), have affected the process through which experts establish a workable definition of truth in scientific settings and how this knowledge has been applied in other settings. Scholars have also emphasized how scientists have pushed against bureaucratic organizations through personal and collective activism. As a complement, this chapter has shown that scientists also regularly serve temporarily in small groups in which they are authorized to make decisions using their judgment to write and apply rules, rather than by using precisely defined standards and measures.

Importantly, declarative bodies are integrated into the process of scientific knowledge-production because they actively revise, edit, and modify researchers' specific projects. The seemingly small and mundane changes that ethics review committees, editorial boards, and funding panels request researchers alter how they talk to their informants and patients, as well as how they move around their field sites and laboratories. Because of the material and symbolic resources at stake, researchers tend to comply with this new form of rational authority based on science by which their work is governed. The authority of declarative bodies can compel obedience from any researcher with an interest in a holding a budget that is bigger rather than smaller, getting research authorization sooner rather than later, or moving a reputation higher rather than lower.

Questions remain to be answered about how declarative bodies operate and these investigations are ripe for scholars in STS to take up because declarative bodies have direct effects on the practice of science. Although conventionally considered "outside" of science settings, this chapter has focused on how declarative bodies set the workaday motions of researchers and make consequential choices about studies in ways that affect what it is possible to know. The decisions of these expert groups are just words, and yet those words have the power to change the practice and the products of science.

Notes

1   I am grateful to Phil Brown, Daniel Kleinman, Kelly Moore, Gary Shaw, and Alistair Sponsel, each of whom gave excellent comments of precisely the sort that I needed at different stages of writing.
2   I use the term "scientist" to refer to knowledge-makers in all disciplines.
3   American Psychological Association archives, HS Library of Congress, Washington, DC, c427. Folder 10: "Minutes of meeting held Feb 20–21 1970."
4   427, f1, "preliminary draft" Feb 1970.
5   427, f1, "preliminary draft" Feb 1970.
6   c443, f5, "Interview with Alberta Siegel" 20 Oct 1970.
7   c443, f5. 20 Nov 1970.
8   c443, f7, McGuire to Cook Committee, 30 July 1970.
9   428 Folder 5: CESPR, Drafts, Published, First published in Monitor, 1971, p1–111. Cook's editorial to accompany the first published draft in the July 1971 *Monitor* read "…the differences among psychologists on research ethics are greater than any of us knew. No committee can resolve these differences – at least not in a way that will affect research behavior. Those of us who formulated the proposed principles contained in the insert brought to the task a wide range of viewpoints. Over time, we developed a common point of view – yet that point of view has already been characterized as inadequate and unwise."
10   Minutes from Rocky Mountain Psych Assoc, May 1971. APA, c431, f3.
11   McGuire "11am version of 1.5." 12 March 1972.
12   Second draft of the *Principles*. *APA Monitor* 3(5), May 1972.
13   Archives of the Office of NIH History, Bethesda, MD. Clinical Center Collection, Ethical, Moral and Legal Aspects, Folder 2, "Memo to the Heads of Institutions Conducting Research with Public Health Service Grants from the Surgeon General," 8 Feb 1966.
14   Ibid.
15   Ibid.
16   C6: 331.

References

Benson, E. 2010. *Wired Wilderness: Technologies of Tracking and the Making of Modern Wildlife*. Baltimore: Johns Hopkins University Press.

Bernstein, B. 2008. *Applied Studies towards a Sociology of Language*. 1st ed. London: Routledge.

Bosk, C.L. and DeVries, R.G. 2004. "Bureaucracies of Mass Deception: Institutional Review Boards and the Ethics of Ethnographic Research." *Annals of the American Academy of Political and Social Science* 595 (1) (September 1): 249–263.

Brenneis, D. 1994. "Discourse and Discipline at the National Research Council: A Bureaucratic Bildungsroman." *Cultural Anthropology* 9 (1) (February): 23–36.

Burnett, D.G. 2012. *The Sounding of the Whale: Science & Cetaceans in the Twentieth Century*. Chicago; London: University of Chicago Press.

Camic, C., Gorski, P., and Trubek, D. ed. 2005. *Max Weber's Economy and Society: A Critical Companion*. 1st ed. Stanford: Stanford University Press.

Carpenter, D.P. 2002. "Groups, the Media, Agency Waiting Costs, and FDA Drug Approval." *American Journal of Political Science* 46 (3): 490–505.

Coleman, C.H. 2004. "Rationalizing Risk Assessment in Human Subject Research." *Arizona Law Review* 46: 1.

Collins, H. and Evans, R. 2007. *Rethinking Expertise*. Chicago: University of Chicago Press.

Eliasoph, N. and Lichterman, P. 2003. "Culture in Interaction 1." *American Journal of Sociology* 108 (4) (January): 735.

Epstein, S. 1996. *Impure Science: AIDS, Activism, and the Politics of Knowledge*. Berkeley: University of California Press.

Epstein, S. 2007. *Inclusion: The Politics of Difference in Medical Research*. Chicago: University of Chicago Press.

Espeland, W.N. 1998. *The Struggle for Water: Politics, Rationality, and Identity in the American Southwest*. 1st ed. Chicago: University of Chicago Press.

Espeland, W.N. 2000. "Bureaucratizing Democracy, Democratizing Bureaucracy." *Law & Social Inquiry* 25 (4): 1077–1109.

Espeland, W.N. and Stevens, M.L. 1998. "Commensuration as a Social Process." *Annual Review of Sociology* 24 (1): 313–343.

Espeland, W.N. and Vannebo, B.I. 2007. "Accountability, Quantification, and Law." *Annual Review of Law and Social Science* 3: 21–43.

Evans, J.H. 2002. *Playing God?: Human Genetic Engineering and the Rationalization of Public Bioethical Debate*. 1st ed. Chicago: University of Chicago Press.

Evans, J.H. 2005. "Max Weber Meets the Belmont Report. Toward a Sociological Interpretation of Principlism." In *Belmont Revisited: Ethical Principles for Research with Human Subjects*, edited by James F. Childress, Eric M. Meslin, and Harold T. Shapiro, 228–243. Washington, DC: Georgetown University Press.

Guetzkow, J., Lamont, M., and Mallard, G. 2004. "What Is Originality in the Humanities and the Social Sciences?" *American Sociological Review* 69 (2) (April): 190–212.

Gusterson, H. 1996. *Nuclear Rites: A Weapons Laboratory at the End of the Cold War*. Berkeley: University of California Press.

Guston, D. 2006. "On Consensus and Voting in Science: From Asilomar to the National Toxicology Program." In *The New Political Sociology of Science*, edited by Kelly Moore and Scott Frickel, 378–404. Madison: University of Wisconsin Press.

Gutmann, A. 1980. *Liberal Equality*. Cambridge: Cambridge University Press.

Heimer, C.A. 2001. "Cases and Biographies: An Essay on Routinization and the Nature of Comparison." *Annual Review of Sociology* 27 (1): 47–76.

Heimer, C.A. 2008. "Thinking about How to Avoid Thought: Deep Norms, Shallow Rules, and the Structure of Attention." *Regulation & Governance* 2 (1): 30–47.

Hurlbut, J.B. 2010. "Experiments in Democracy: The Science, Politics and Ethics of Human Embryo Research in the United States, 1978–2007." Ph.D., United States, Massachusetts: Harvard University.

Imber, J. 1986. *Abortion and the Private Practice of Medicine*. New Haven: Yale University Press.

Jasanoff, S. 1998. *The Fifth Branch: Science Advisers as Policymakers*. Cambridge, MA: Harvard University Press.

Jasanoff, S. ed. 2004. *States of Knowledge: The Co-production of Science and the Social Order*. 1 New. Routledge.

Jasanoff, S. 2007. *Designs on Nature: Science and Democracy in Europe and the United States*. Princeton, NJ: Princeton University Press.

Katznelson, I. 1996. "Knowledge about What? Policy Intellectuals and the New Liberalism." In *States, Social Knowledge, and the Origins of Modern Social Policies*, edited by Theda Skocpol and Dietrich Rueschemeyer, 17–47. New York and Princeton, NJ: Russell Sage and Princeton University Press.

Kennedy, D. 2005. *The Disenchantment of Logically Formal Legal Rationality: Or Max Weber's Sociology in the Genealogy of the Contemporary Mode of Western Legal Thought*. Stanford: Stanford University Press. http://search.proquest.com.proxy.library.vanderbilt.edu/docview/60023355/13AC42907F92A8349CB/1?accountid=14816.

Kerwin, C. 2003. *Rulemaking: How Government Agencies Write Law and Make Policy*. 3rd ed. Washington, DC: Congressional Quarterly Press.

Kevles, D.J. 2000. *The Baltimore Case: A Trial of Politics, Science, and Character*. Reprint. New York: W. W. Norton & Company.

Koch, C.H. 1997. *Administrative Law and Practice*. 2nd ed. 3 vols. St. Paul, MN: West Pub. Co.

Lamont, M. 2010. *How Professors Think: Inside the Curious World of Academic Judgment*. Cambridge, MA: Harvard University Press.

Lawrence, C. and Shapin, S. ed. 1998. *Science Incarnate: Historical Embodiments of Natural Knowledge*. 1st ed. Chicago: University of Chicago Press.

Lee, C.W. 2007. "Is there a place for private conversation in public dialogue? Comparing stakeholder assessments of informal communication in collaborative regional planning". *American Journal of Sociology* 1 (July 2007): 41–96.

Levitan, S.A. and Taggart, R. 1976. "The Promise of Greatness: The Social Programs of the Last Decade and Their Major Achievements." Cambridge, MA: Harvard University Press.

Lipsky, M. 2010. *Street-Level Bureaucracy: Dilemmas of the Individual in Public Service, 30th Anniversary Expanded Edition*. 30 Anv Exp. Russell Sage Foundation.

Mackenzie, G.C. and Weisbrot, R. 2008. *The Liberal Hour: Washington and the Politics of Change in the 1960s*. New York: Penguin Press.

Mallard, G, Lamont, M., and Guetzkow. J. 2009. "Fairness as Appropriateness: Negotiating Epistemological Differences in Peer Review." *Science, Technology & Human Values* 34 (5) (September): 573–606.

McCarthy , C. 2008. "The Origins and Policies That Govern Institutional Review Boards." In *The Oxford Textbook of Clinical Research Ethics*. Edited by Emanuel, Ezekiel J. Oxford: Oxford University Press: 541–551.

Merton, R. K. 1938. "Science, Technology, and Society in Seventeenth Century England." *Osiris* 4: 360.

Moore, K. 2008. "Disrupting Science Social Movements, American Scientists, and the Politics of the Military, 1945–1975."

Petryna, A. 2009. *When Experiments Travel: Clinical Trials and the Global Search for Human Subjects*. Princeton, NJ: Princeton University Press.

Polletta, F. 2002 *Freedom Is an Endless Meeting: Democracy in American Social Movements*. Chicago: University of Chicago Press.

Porter, T.M. 1995. *Trust in Numbers*. Princeton, NJ: Princeton University Press.

Rabeharisoa, V. and Callon, M. 2002. "The Involvement of Patients' Associations in Research." *International Social Science Journal* 54 (171): 57–63.

Riles, A. ed. 2006. *Documents: Artifacts of Modern Knowledge*. 1st ed. Ann Arbor: University of Michigan Press.

Riles, A. 2011. *Collateral Knowledge: Legal Reasoning in the Global Financial Markets*. Chicago; London: University of Chicago Press.

Schneiderhan, E. and Khan, S. 2008. "Reasons and Inclusion: The Foundation of Deliberation." *Sociological Theory* 26 (1) (March): 1–24,100.

Shapin, S. 1995. *A Social History of Truth: Civility and Science in Seventeenth-Century England*. 1st ed. Chicago: University of Chicago Press.

Shapin, S. 2010. *The Scientific Life: A Moral History of a Late Modern Vocation*. Chicago: University of Chicago Press.

Shapin, S. and Schaffer, S. 2011. *Leviathan and the Air-Pump: Hobbes, Boyle, and the Experimental Life*. Princeton, NJ: Princeton University Press.

Stark, L. 2010. "The Science of Ethics: Deception, the Resilient Self, and the APA Code of Ethics, 1966–1973." *Journal of the History of the Behavioral Sciences* 46 (4): 337–370.

Stark, L. 2012. *Behind Closed Doors: IRBs and the Making of Ethical Research*. Chicago: University of Chicago Press.

Thorpe, C. 2006. *Oppenheimer: The Tragic Intellect*. Chicago: University of Chicago Press.

Thorpe, C. and Shapin, S. 2000. "Who Was J. Robert Oppenheimer? Charisma and Complex Organization." *Social Studies of Science* 30 (4) (August 1): 545–590.

Timmermans, S. 2007. *Postmortem: How Medical Examiners Explain Suspicious Deaths*. Chicago: University of Chicago Press.

Timmermans, S. and Berg, M. 2003. *The Gold Standard: The Challenge of Evidence-Based Medicine*. 1st ed. Philadelphia: Temple University Press.

Timmermans, S. and Epstein, S. 2010. "A World of Standards But Not a Standard World: Toward a Sociology of Standards and Standardization." *Annual Review of Sociology* 36 (1): 69–89.

Turner, S. 2001. "What Is the Problem with Experts?" *Social Studies of Science* 31 (1) (February 1): 123–149.

Waddell, C. 1989. "Reasonableness Versus Rationality in the Construction and Justification of Science Policy Decisions: The Case of the Cambridge Experimentation Review Board." *Science, Technology, & Human Values* 14 (1) (January 1): 7–25.

Weber, M. 1978. *Economy and Society: An Outline of Interpretive Sociology*, edited by Guenther Roth and Claus Wittich. Fourth ed. Berkeley: University of California Press.

Weber, M. 2001. *The Protestant Ethic and the Spirit of Capitalism*. Talcott Parsons, trans. London; New York: Routledge.

Weber, M. 2004. *The Vocation Lectures: Science As a Vocation, Politics As a Vocation*, edited by David S. Owen, Tracy B. Strong, and Rodney Livingstone. Indianapolis: Hackett Pub Co.

Weber, M. 2008. *The Protestant Ethic and the Spirit of Capitalism*, edited by Richard Swedberg. New York: W. W. Norton & Company.

Weber, M. 2010. *The Protestant Ethic and the Spirit of Capitalism*, edited by Stephen Kalberg. Revised. Oxford University Press, USA.

Wisnioski, M.H. 2012. *Engineers for Change: Competing Visions of Technology in 1960s America*. Cambridge, MA: MIT Press.

Zammito, J.H. 2004. *A Nice Derangement of Epistemes: Post-Positivism in the Study Of Science from Quine to Latour*. Chicago: University of Chicago Press.

26

Big Pharma and Big Medicine in the Global Environment

Anne E. Figert and Susan E. Bell

LOYOLA UNIVERSITY CHICAGO AND BOWDOIN COLLEGE

Introduction

In previous work we argued that sociologists need to expand our thinking about pharmaceuticalization, the process of understanding and/or treating social, behavioral, or bodily conditions with pharmaceuticals. The majority of sociological scholarship has investigated pharmaceuticalization as a primarily Western process and conceptualized it in modern terms (Bell and Figert 2010, 2012a, 2012b). In our view, the work of anthropologists and science and technology studies (STS) scholars who decenter the West as the starting point for research opens up new avenues for understanding the global dynamics of pharmaceuticalization. We have also argued in favor of adopting a postmodern theoretical lens which allows us to understand pharmaceuticalization both as a strategy of enhancement by individuals in resource-rich societies and as an exercise of power in resource-poor societies and to bring to light its multiple, multi-directional and at times apparently contradictory effects.

In this chapter we expand upon our previous work and focus on one essential part of the pharmaceuticalization process: global clinical trials and related ethical and research standards. We also consider the role of global clinical trials in reducing public health strategies from a broad array of disease prevention efforts to one seeking to improve the health of populations with pharmaceuticals. The issues we explore center upon the key research and ethical standards for global pharmaceutical development. We define, review and problematize the concept of ethical variability and show how it simultaneously upholds and disrupts Western ethical guidelines for human subjects research. In doing so, we show how global clinical trials contribute to the further pharmaceuticalization of public health worldwide with major implications for the lives of people globally. The degree and scope of how people interact with pharmaceuticals throughout the world is uneven because pharmaceuticalization and global clinical trials map onto global patterns of inequality. Some human bodies serve as research subjects whereas some bodies are pharmaceutical sales targets. Whereas some people in some areas of the world are (over)pharmaceuticalized, other people are (under)pharmaceuticalized. We conclude the chapter with a discussion of how and why STS perspectives on harmonization and variability in ethical and research standards shed light on the study of pharmaceuticalization and more broadly on the global dynamics of health inequality.

Conceptual framing: medicalization, biomedicalization, and pharmaceuticalization

Because studies of pharmaceuticalization have taken shape alongside the development of scholarship about medicalization and biomedicalization, we begin with a brief overview of these fields. The concept of medicalization was introduced to the medical sociology field in the 1970s to understand and look critically at "the involvement of medicine in the management of society" (Zola 1972: 488). Medicalization is now ubiquitously used in the social and medical sciences and has successfully moved into popular culture. One of the most influential definitions of medicalization comes from U.S. sociologist Peter Conrad (2005: 3) who declares that the essential meaning of the term is "*defining a problem in medical terms, usually as an illness or disorder, or using a medical intervention to treat it*". Medicalization explains a process of medical expansion in a modern society. It makes sense of how and why more and more conditions are defined and treated medically and increasingly pharmaceutically.

Current medicalization scholarship has refocused our analytic gaze from the power and authority of the medical profession to consider the active participation of patient/consumer/ users individually and collectively in medicalization processes (Brown and Zavestoski 2005, Crossley 2006), resistance to pharmaceuticals (Figert 2011, Williams et al. 2011), and the use of medical prescription drugs for non-medical purposes (Williams et al. 2008). It has also explored new "engines" of medicalization including the pharmaceutical industry (Conrad 2005) and technoscience (Clarke et al. 2003). Although medicalization is a capacious concept, it cannot fully capture the contemporary global dynamics of pharmaceuticals and technoscience. Thus scholars have introduced the concepts of biomedicalization, pharmaceuticalization, and pharmaceuticalization of public health, which are often more effective than medicalization alone in analyzing the nuances and complexities of the development, testing, expansion and distribution of pharmaceuticals in the world today (see Bell and Figert 2012a for a more extensive discussion).

Whereas the process of medicalization can be conceived of in modern terms of engineering, control, and rationalization, the process of biomedicalization can be conceived of in postmodern terms of networks, spirals, and complexity. Understanding both the definition and effects of biomedicalization helps to make sense of how and why more and more conditions are defined and treated medically and pharmaceutically in the twenty-first century. Biomedicalization, as established by Clarke and her colleagues, is a concept and analytic tool for identifying, disentangling, and explaining medical and pharmaceutical expansion in the twenty-first century. It captures "the increasingly complex, multisited, multidirectional processes of medicalization that today are being both extended and reconstituted through the emergent social forms and practices of a highly and increasingly technoscientific biomedicine" (Clarke et al. 2003, 162). One example of this is the application of screening technologies using molecular biomarkers that constitute new categories of people at risk and new opportunities for biomedical surveillance and intervention as well as self-monitoring and regimens of behavior change (Shostak 2010).

Although feminist scholars and activists in Women's Health Movements have looked critically at the development of the birth control pill and other reproductive technologies since the 1970s and social scientists have studied pharmaceuticals and the pharmaceutical industry for many years (for example, Boston Women's Health Book Collective 1973, Hartmann 1987, Gabe and Bury 1988), pharmaceuticalization as a unique term was introduced by anthropologists (Nichter 1989). A broadly accepted definition of pharmaceuticalization by sociologist John Abraham is "the process by which social, behavioral or bodily conditions are treated, or deemed to be in need of treatment/intervention, with pharmaceuticals by doctors, patients, or both" (Abraham 2010: 290). There are complex forces generating the expansion of pharmaceuticalization: Big Pharma's industry control over the science underpinning drug development and testing, skillful use of marketing, and "disease mongering"; physicians as prescribers, gatekeepers, and "developers of new medicines often in alliance with the industry"; affluent publics in consumer-oriented societies who use information technology and become "expert patients"; and governments and insurance companies.

Pharmaceuticalization scholarship builds on and has explicit ties with medicalization scholarship, but scholars generally agree that pharmaceuticalization can occur without medicalization, and vice versa. Studies of the pharmaceutical process, like those of the medicalization process, typically do so from the perspective of modern social theory. They draw inspiration and logic from the natural sciences, adopt an engineering mentality (Lock 2004), trace the drug development and approval process in terms of "countervailing powers" (Busfield 2006), and identify pharmaceuticalization as a search for control of behavioral, bodily, or social conditions (Bell and Figert 2012a).

In our work, we show that current research in anthropology provides a useful layer of understanding pharmaceuticalization (Bell and Figert 2012a). Whereas sociologists primarily study pharmaceuticalization by focusing upon power, economics and treatments in the West and the dynamics of the largest pharmaceutical companies (often called Big Pharma) and high-income nation states, anthropologists focus primarily upon the issues of pharmaceuticals in low or middle-income countries where the political economic systems are often post-colonial or post-communist. This focus has allowed anthropologists to conceptualize pharmaceuticalization differently and to examine political, economic and organizational dynamics that are less visible in the studies by sociologists.

The pharmaceuticalization of public health as outlined by Biehl and others suggests that there is both a political and an economic rationality to cutting back on disease prevention efforts in favor of a national pharmaceutical distribution policy (Biehl 2007, Whitmarsh 2008). From a neoliberal state perspective, it is cheaper and more efficient to diagnose and treat diseases pharmaceutically than to prevent them through traditional public health measures. The process of pharmaceuticalization and the policy of the pharmaceuticalization of public health are key factors in the expansion of the use of pharmaceuticals to treat medical and social problems. These two "strands" of pharmaceuticalization theory and research shed light on the uneven and unequal global processes of pharmaceuticalization. Thus, in the global North, pharmaceuticalization is primarily about expanding social and behavioral diagnostic categories and diagnoses, while in the global South, pharmaceuticalization is primarily about expanding access to medicines and public health or of increasing testing sites for pharmaceutical clinical trials. In the global South, citizens and non-governmental organizations (NGOs) – often in response to the HIV/AIDS epidemic – have mobilized and demanded access to certain drugs or treatments.

The pharmaceuticalization of public health scholarship also brings a postmodern framework to understanding pharmaceuticals today. We use the term "postmodernity" to refer to society based on information technology and characterized by interaction, contingency, fragmentation, volatility, and hybridity. In postmodern society, boundaries are blurred, such as between public/private, government/corporation, expert/lay, human/animal, and human/machine. Postmodern theory assumes that the political economic, cultural, organizational, and technoscientific trends and processes of pharmaceuticalization are complex and mutually constituted. The pharmaceuticalization of public health is manifest in macrostructural changes as well as in new personal identities, subjectivities, and configurations by seeking to connect global dynamics among states, non-governmental organizations (NGOs), pharmaceutical companies, and local communities (see Petryna et al. 2006, Clarke et al. 2010, Bell and Figert 2012a).

Much work remains to be done in exploring how pharmaceuticalization works globally, and we must recalibrate the balance between studies of the global North and global South (for example, Brazil (Biehl 2006, 2007), Barbados (Whitmarsh 2008), India (Sunder Rajan 2007, 2012), Thailand, Uganda and South Africa (Petty and Heimer 2011), and Poland (Petryna 2009)). The pharmaceuticalization of public health can be used to make sense of dynamics where states with less power and wealth define free access to pharmaceuticals as rights of citizenship (and a new subjectivity, pharmaceutical citizenship). It can make visible ways that pharmaceuticalization can contribute to the creation of new democratic tools for individuals, activist groups, and states. For example, participation in clinical trials of pharmaceuticals can be a strategy to gain access to drug treatments and medical care and accomplish what people believe to be in their best economic and medical interests. For societies with few resources, pharmaceuticalization can be a strategy for realizing the rights of citizens and improving population health. Defining rights of citizenship as access to pharmaceuticals creates new possibilities for entering into the grip of biomedical power, forecloses other approaches to improving population health and well-being, and contributes to the pharmaceuticalization of public health.

Standardization and global clinical trials

In this section, we show how the global expansion of clinical trials and the global standardization of research procedures and ethics foster pharmaceuticalization and the pharmaceuticalization of public health. Standardization is the process of constructing uniformities across space and time. These uniformities, created by multiple historically situated actors, are expressed in standards (Timmermans and Berg 2003). The standards, in turn, coordinate people and things in "new configurations" (Timmermans and Epstein 2010: 83). The standardization of both research procedures and ethics has facilitated the proliferation of global clinical trials, the portability of results and the global expansion of markets. These standards are expressed in and enforced by international organizations (for example, the International Conference on Harmonization of Technical Requirements for Registration of Pharmaceuticals for Human Use (ICH), the Agreement on Trade Related Aspects of Intellectual Property (TRIPS)), regulatory agents of the state (for example, U.S. Food and Drug Administration, Drug Controller-General of India), professional governance (for example, the Declaration of Helsinki), and codes and formal laws (for example, The Nuremberg Code, The Belmont Report). At the same time the development of global clinical trials has coordinated people and things into new configurations (for example, "Contract Research Organizations"), and subjectivities (for example, "treatment naïve populations," "treatment saturated populations" discussed later in this chapter). Increasingly, clinical pharmaceutical trials have been privatized by the development of a contract research industry (Fisher 2006, 2009).

While not all clinical trials are conducted with pharmaceuticals, many studies by anthropologists and sociologists focus on the expansion of global clinical pharmaceutical trials as primary components of pharmaceuticalization (for example, Petryna et al. 2006, Dumit 2012, Williams et al. 2008). In 2006, more than 2.4 million Americans participated in clinical trials (Dumit 2012: 18). Since the 1990s, the number of international subjects involved in clinical trials – including pharmaceutical trials – has grown substantially, from 4000 in 1995 to 400,000 in 1999 (Petryna 2006: 189). Until recently, much of the pharmaceutical and clinical research was conducted in the U.S. and Western Europe, but today it is likely to be conducted elsewhere. During February 2013, 29,623 clinical trials were actively recruiting study participants, and almost half of the trials (49%) were seeking subjects exclusively outside the United States (*www.clinicaltrials.gov* February 11, 2013).

There are multiple, multidirectional reasons that clinical trials are pushed and pulled globally. The time between first identification of an active agent with therapeutic potential and formal approval for marketing is 10–15 years and each new drug costs $897 million to develop (Busfield 2006, Petryna et al. 2006: 11).[1] Developing countries in particular are likely to have fewer regulations and a looser regulatory apparatus for enforcing ethical and research standards. They provide cheaper labor and lower infrastructure costs, reducing overall expenses of clinical trials by 30–50 percent (Sunder Rajan 2007: 72). In addition they reduce the time line for clinical testing by accelerating subject recruitment and improving the likelihood of showing drug effectiveness because their populations are more likely to be pharmaceutically or treatment naïve, that is to have little or no previous access to pharmaceuticals and no background medications at the time of the trial that might confound results.

Clinical trials are pushed and pulled globally to reduce time and expense both because in the U.S. and Western Europe patients and potential human subjects are increasingly skeptical of drug trials and because patients and physicians in Eastern Europe, Eurasia and the global South need the resources that pharmaceutical companies offer. The global expansion of clinical trials opens the possibility for individuals and communities to gain access to medicines otherwise unavailable to them (Biehl 2006, Nguyen 2005). The use of these pharmaceutically naïve subjects "creates efficient results, free of statistical noise" (Petryna 2007: 37).

Standardizing global research procedures

The standardization of clinical trials and research procedures and the actors involved in this standardization in the 1990s played an important role in the expansion of global clinical trials and pharmaceuticals. Global technical standards specifically for pharmaceutical research began to be institutionalized in the 1990s, exemplified by the development of Contract Research Organizations (CROs) and the International Conference on Harmonization (ICH). CROs create mobile clinical trial environments and trial results. CROs are private, for profit companies that implement and manage global clinical trials for large multinational pharmaceutical companies. Fisher (2009) reports that more than 75 percent of clinical drug trials in the U.S. are now conducted in the private sector. Since the U.S. is the largest pharmaceutical market in the world and by some recent estimates Big Pharma makes two-thirds of its profits in the United States, the process for pharmaceutical testing in the U.S. and its legitimation by the U.S. Federal Food and Drug Administration (FDA) guide how pharmaceutical companies conduct clinical trials outside as well as inside the U.S. (Harris 2013). There are four distinct phases of pharmaceutical testing for approval by the FDA. Each phase is designed to build upon the others and requires increasing numbers of participants. If a drug is determined to be safe in a small group of healthy volunteers (Phase 1 with 20–80 participants) and effective in treating the targeted condition (Phase 2 with 100–300 participants), the drug will move to Phase 3 which is characterized by a large group of participants (usually between 1,000 and 3,000 people) to confirm effectiveness and scrutinize any possible side effects. In some cases, a Phase 4 (also called "post-marketing") trial will be conducted. This phase is primarily observational and is non-experimental and may explore new uses or dosages, or use in new populations.

CROs work and serve as the "middlemen" or service providers (see Fisher 2009). In primarily low- and middle-income countries, CROs identify research sites, clinics, practitioners and recruit human subjects. International CRO's are important in the process of moving a drug from Phase 2 to Phase 3, because of the need for large numbers of participants. Their main source of revenue comes from conducting clinical trials efficiently and cost effectively (Petryna

2006: 38). CROs also help to ensure that clinical research complies with accepted technical standards and national and international ethical guidelines and thereby make "data from various international sites portable to and usable within the U.S. drug approval process" (Petryna 2011: 307).

From 1992 to 2004 the CRO market grew from $1 billion to $7 billion and by 2004 there were more than 1000 CROs worldwide. A recent survey of CROs found that pharmaceutical companies outsourced a wide variety of functions to CROs from design to site selection, study conduct, and medical writing (Getz and Vogel 2009). Sunder Rajan (2007) highlights the role that consulting firms, such as A.T. Kearney, play in helping Big Pharma find international testing sites. Kearney developed an attractiveness index for clinical trials (calculated by evaluating patient availability, cost efficiency, relevant expertise, regulatory conditions and national infrastructure) and determined that the most favorable pharmaceutical testing sites were China, India and Russia (Bailey et al. 2009: 57). Sunder Rajan (2012: 332) points out that unlike the pharmaceutical companies for which the locus of value lies in the valorized expansion of health, the locus of value for CROs is the valorized expansion of pharmaceutical clinical trials.

The development of the CRO industry occurred concurrently with and was fostered by the introduction of guidelines for clinical trials established in 1990 by the International Conference on Harmonization (ICH). The ICH was the product of international pharmaceutical regulators from the U.S., the European Union, Japan and the pharmaceutical industry (see *www.ich.org/about/vision.html*). In effect, the ICH established uniform research and technical requirements and standards such as randomized controlled trials (RCT) – in which subjects are randomly assigned to either a treatment group or a control group – in drug testing. At first the use of these standards "made clinical data from international research sites transferable and acceptable to regulatory bodies in" the major markets of Europe, Japan, and the United States (Petryna 2007: 30). Since 2007, the ICH has opened up its process and expanded its reach beyond these major markets. For example, representatives of drug regulatory agencies from Australia, Brazil, China, Chinese Taipei, Russia, India, Singapore, and South Korea have been invited to attend the ICH (*www.ich.org*). Together CROs and the ICH construct uniform standards for global clinical trials across time and space.

Standardizing global research ethics and ethical variability

The successful expansion of global pharmaceutical research depends on adhering to certain established international ethical standards. At the same time, global expansion fosters the transformation of these same ethical standards. Since the 1980s, the global dynamics of the pharmaceutical industry have played "an important role in shaping contexts in which ethical norms and delineations of human subjects are changing" (Petryna 2006: 34). In the common narrative of research ethics, ethical standards for clinical trials can be traced to the Nuremberg Code (1947) and the Declaration of Helsinki (1964), which provide both a moral framework and an explanation for how and why human subjects need and are protected in ordinary scientific and medical practice (Hoeyer 2009). In this section we argue that recent revisions to the Declaration of Helsinki have created a new configuration of unethical trials, that global ethical standards often produce the impetus, justification and tools for turning healthy populations into experimental subjects, and that a modernist frame of understanding cannot account for these effects.

There was no international statement differentiating between legal and illegal human experimentation until the Nuremberg Code (a set of ten points related to human experimentation targeting Nazi doctors and scientists) was established in 1947 during the World War II war crimes tribunal. The voluntary consent of the prospective human subject is the bedrock of the

Nuremberg Code. The Code requires that all unnecessary physical and mental suffering should be avoided, the degree of risk should never exceed the benefit which may derive from the tested drug or treatment and the research should be conducted by a scientifically-qualified person.[2] The Nuremberg Code continues to serve as a "blueprint for today's principles that ensure the rights of subjects in medical research" (Shuster 1997: 1436) although some scholars have convincingly argued that it was frequently ignored by scientists and physicians because it was really only for Nazi "barbarians" and not everyday scientists and physicians (Katz 1992, Rothman 1991, Hoeyer 2009).

The Declaration of Helsinki,[3] established in 1964 by the World Medical Association, seeks to guide physicians in research with human subjects, and leaves intact physicians' civil, criminal and ethical responsibilities under the laws of their own countries. Katz (1992) argues that in contrast to the Nuremberg Code, in the Declaration of Helsinki concerns over the advancement of science began to overshadow concerns over the integrity of persons. The Declaration has been amended regularly since 1964 but the most controversial amendments have to do with the issues of the use of placebos, international testing, informed consent, and access to treatment at a trial's conclusion. Effectively, by establishing that placebo trials are acceptable only when no proven treatment already exists, the 1996 and 2000 revisions to the Declaration created a new configuration of unethical clinical trials in the U.S. and other countries. In the 1996 revision to the Declaration, the idea of a placebo, an inert substance or one containing no medication, was introduced for the first time:

> In any medical study, every patient – including those of a control group, if any – should be assured of the best proven diagnostic and therapeutic method. This does not exclude the use of inert placebo in studies where no proven diagnostic or therapeutic method exists.
>
> *as quoted in Carlson et al. 2004: 698*

In contrast, the 2000 revision to this paragraph reads:

> The benefits, risks, burdens and effectiveness of a new method should be tested against those of the best current prophylactic, diagnostic, and therapeutic methods. This does not exclude the use of placebo, or no treatment, in studies where no proven prophylactic, diagnostic or therapeutic method exists
>
> *as quoted in Carlson et al. 2004: 700, emphasis in original*

The U.S. and the FDA have not recognized these recent amendments to the Declaration regarding the preference for testing new pharmaceuticals and vaccines against the best current methods instead of against a placebo, arguing that this would inhibit the development of good science and efficacious drugs (see Wolinsky 2006). Pharmaceutical trials funded by the U.S. government and its agencies continue to use placebo testing throughout the world.

International ethical guidelines for research involving human subjects – such as the 1996 and 2000 revisions to the Declaration of Helsinki – are being recast along with the movement of clinical trials globally. Revisions concerning the use of placebos in pharmaceutical research are directly related to a growing concern about international studies of maternal-fetal HIV transmission in developing countries (Carlson et al. 2004). In an article published in the *New England Journal of Medicine*, Lurie and Wolfe (1997) questioned why studies outside the U.S. sponsored by a U.S. government agency, the National Institutes of Health, used a placebo design even though there was already a known and effective treatment to prevent maternal-fetal HIV transmission available in the U.S. Supporters of the study design and implementation argued that "local cultural variables and deteriorated health services" made placebos acceptable and that it would be a "paternalistic imposition" for the U.S. to determine the appropriate design of research in regions of such poverty (Petryna 2007: 28–29). Similarly, local and national authorities in these regions argued that they should determine research conduct and treatment distribution.

As clinical pharmaceutical trials have become globalized, STS scholars and others have examined how enacting ethical standards internationally takes place. Petryna (2007) argues that "ethical variability" – the creation of local standards to recruit human subjects for clinical and pharmaceutical research – produces the conditions for the exploitation of "Third World subjects." Ethical variability legitimates the modification of ethical standards according to the local contexts of clinical trials. It has evolved as a tactic for weighing immediate health benefits or outcomes against protection and safety considerations and not as a strategy for being sensitive to those asked to enroll in clinical pharmaceutical trials (Petryna 2007, Farmer and Campos 2004).

More generally, participants adjust and use workarounds in implementing ethical standards. Drawing from their study of HIV treatment and clinical trials in the global North (two U.S. clinics) and South (one each in Thailand, Uganda and South Africa), Heimer and colleagues show that in all clinics, both North and South, "neither research subjects nor the recruitment and consent process actually live up to the ethical ideals as embodied in the institutions of informed consent" (Heimer 2012: 24). When researchers or state agencies try to implement any local or global standards, it is inevitable that the practices will include workarounds and adjustments.

Social scientists have observed repeatedly that the Nuremberg Code, Declaration of Helsinki, and similar ethical standards for research assume autonomy and choice of the individual "and downplay social and economic constraints on individual agency" (Marshall and Koenig 2004: 255; see also Fisher 2009). The result is that global ethical standards often provide the impetus, justification and tools for turning healthy populations into experimental subjects. As Angell warned in 1997:

> Research in the Third World looks relatively attractive as it becomes better funded and regulations at home become more restrictive. Despite the existence of codes requiring that human subjects receive at least the same protection abroad as at home, they are still honored partly in the breach. The fact remains that many studies are done in the Third World that simply could not be done in the countries sponsoring the work. Clinical trials have become a big business, with many of the same imperatives. ... Those of us in the research community need to redouble our commitment to the highest ethical standards, no matter where the research is conducted, and sponsoring agencies need to enforce those standards, not undercut them.
>
> *Angell 1997: 849*

To summarize, on the one hand the construction of a universal standard or ethical code of conduct for pharmaceutical clinical trials appears to be "good" or "just" because it is sensitive to imbalances of power and money. This interpretation would work within a modernist framework. On the other hand, treating all people as equal in a world characterized by inequality effectively serves to reinforce that inequality. The harmonization of ethical codes or standards for global clinical trials obfuscates the reproduction and exacerbation of global inequality. Furthermore, all global standards are practiced and implemented locally and thus entail local workarounds and adjustments in the field. The modernist frame, dominant in sociological accounts of pharmaceuticalization, cannot account for all of these practices and effects. A post-modern framework for understanding global clinical pharmaceutical trials helps the analyst move away from an either/or framing to understand that ethical variability is not always bad and standardization is not always good. Both variability and, as we discuss next, standardization, can produce different outcomes depending upon local settings and histories.

Disruptions to standardization?

In this section we show how standard sociological and modernist conceptual frameworks for understanding pharmaceuticalization are simply insufficient to explain the expansion of clinical trials to the global South and how in some respects those in low resourced countries benefit more by participation in these trials than those in high resourced countries. We address the question of how an understanding of standards and rules sheds light on the conceptualization and processes of global pharmaceuticalization and the pharmaceuticalization of public health, and we use work on the expansion of pharmaceutical clinical trials in India by Sunder Rajan to illustrate our argument (Sunder Rajan 2005, 2007, 2012).

As discussed above, with the expansion of the number of clinical trials, the need for human subjects increases, and trials are more and more likely to be conducted in the global South with the goal of producing portable results. One reason given for this is that these countries are seen as having fewer regulations and a looser regulatory apparatus (Petryna 2009). Since the 1990s, India has become incorporated into the globalized drug development sector. In his study of global pharmaceutical economies, Sunder Rajan (2007, 2012) contests the assumptions that ethical standards are "stricter" in the West. In 2005 India converted its guidelines for informed consent into laws (Schedule Y) and is now the only country in the world "where the violation of good clinical practice is a criminal rather than a civil offence" (Sunder Rajan 2007: 74). Indeed, Schedule Y focuses on ways of insuring informed consent from subjects who are poor and illiterate. In many ways, this means that local ethical standards in India are especially tight, and global harmonization could diminish the possibility of developing local standards such as these. On the surface, the informed consent process in India provides potential experimental subjects with the choice to freely participate or not participate in clinical pharmaceutical trials. Although subjects may freely give consent to participate in clinical trials, their access to pharmaceuticals ends along with the end of the trials. Experimental subjects are still exploited, or in Sunder Rajan's terms "merely risked" because for this population, clinical experimentation is not linked to the benefit of subsequent therapeutic access.

The harmonization of ethical standards provides the conditions for continued global pharmaceutical and economic inequality. In India, the apparatus of clinical trials simultaneously accepts Western bioethics standards of informed consent and rigorously applies them so its research results can travel. At the same time, its population will bear the burden without the benefit of research results. In its careful attempt to adhere to global (universal) standards, India creates conditions for the exploitation of Indian bodies (and by implication of Third World subjects more generally) because of the real economic rewards and the potential for further inclusion in the global pharmaceutical economy.

In the U.S. and most of the world, there is less attention to the ethics of how poverty and specific forms of indebtedness shape consent and decisions in pharmaceutically naïve populations[4] or whether the burden of research is balanced with tangible therapeutic benefits after completion of trials (Fisher 2009). Advocacy groups have learned to fight for access to pharmaceuticals for citizens in the global South (as they did successfully in Brazil) (see Biehl 2004, 2007) and for the importation of more affordable generic versions of medicines from foreign manufacturers, as Brazil, Argentina and South Africa have done for AIDS medicines (WHO Drug Information vs. 19, no. 3 2005, Access to Medicines).[5] One way that pharmaceutical companies respond to such activism and pressure is by providing therapeutic access through their compassionate use programs "which make the drugs tested in Phase 3 trials available to the sick volunteers for a fixed period of time after completion of the trial" (Sunder Rajan 2007: 79).

The dynamics of clinical pharmaceutical trials in India, Barbados and Brazil are representative of new forms of an international bioeconomy in which nations, the pharmaceutical industry and other corporate actors work to create global experimental sites. In this new phase of capitalism, clinical pharmaceutical trials establish places where experimental subject populations exchange human bodies for payment in the form of cash or access to treatment (Sunder Rajan 2006, Dumit 2012). In this context, the problem of the exploitation of third world "merely risked" subject populations is not the result of the harmonization of standards – either looser standards pulling clinical trials to the global South or tighter standards protecting experimental subjects in the global South – but reflects the reorganization and reconceptualization of global capital in relation to "life itself" since the 1980s.

New configurations

Although the global expansion of clinical trials works unevenly throughout the world, there is evidence that in some respects physical sites of the new bioeconomy such as health or pharmaceutical clinics in "poor" countries benefit more from these trials than do clinics in "rich" countries. Thus, in a study of clinical trials in countries at varying levels of development – the U.S., South Africa, Thailand and Uganda – Petty and Heimer (2011) and Heimer (2012) show that global HIV research can be more beneficial to countries in the global South than to the U.S. Clinics reconfigure their local practices of care and treatment to bring them in line with ICH standards: to produce accurate, complete, and verifiable study data and to ensure "that the rights and well-being of human subjects are protected" (Petty and Heimer 2011: 350). These reconfigurations include upgrading laboratory facilities to be able to do the complicated tests required by clinical trials and using laboratory equipment in study-specific ways to produce standardized results. The new configurations vary depending on clinics' existing resources, routines, and relationships. In poorer countries, where equipment is in short supply funders often pay to improve laboratory facilities so clinics can participate in research (Petty and Heimer 2011: 342). Once laboratory facilities are upgraded, clinics in poorer countries can employ them in both research and treatment. However, because materials are less easily repaired, replenished, or replaced, "the overall effect of altering the material environment in poorer countries is likely to be modest unless the flow of funds is very stable" (Petty and Heimer 2011: 357). By contrast, in richer countries, research-provided technologies duplicate already available medical equipment and doing the research has a less beneficial effect.

Clinical pharmaceutical trial participation can also reshape the clinics in ways that smooth the way for their later adoption of clinical research findings. The everyday actions and results of introducing new jobs, technologies, and standard operating procedures for clinical trials is as important to changing medical practice as is the influence of subsequent research results. Petty and Heimer (2011) identified three types of practices that are changed in the doing of clinical research. The introduction of new research-mandated tools alters the material environment, the introduction of new and/or retrained staff reorganizes staff relationships in the clinics, and the adoption of research practices changes clinic priorities. In other words, conducting clinical research is not just a means of testing new treatments that subsequently change medical practice.

The new routines for clinical trials change clinic practices so that new therapies will fit local conditions and can be translated into medical care (Petty and Heimer 2011).

In the process of conducting clinical trials, standardization of actions and practices for doing the research reshapes the clinics and gives further agency and sometimes bargaining power to the clinic staff to advocate for their patients.

Petty and Heimer (2011) found that in reconfiguring their local practices in order to participate in global clinical trials, clinics in the U.S. and the global South fostered a pharmaceutical approach to public health that ultimately necessitated and created reliance upon technoscience and biomedicine beyond money and supplies. More generally, participation in pharmaceutical clinical trials creates regimes of practice and enforces ways of thinking and action that focus on pharmaceutical solutions. It forecloses other ways of thinking about and treating public health problems. While providing certain kinds of benefits to resource poor countries, the pharmaceuticalization of public health projects (vaccines, pharmaceutical testing or treatments) excludes cheaper and more effective ways to treat the health of the general population. When clinics change how they work and think about the way to treat patients in adjusting to pharmaceutical trials, they narrow the gaze and focus to one that concludes pharmaceuticals are the ultimate solution to improving public health. This is especially problematic in states with fewer resources because it ultimately narrows the options to more technological and capital intensive solutions.

Conclusion

In this chapter we have argued that global research and ethical standards of pharmaceutical development – especially in global clinical trials – are institutionalized, disrupted and/or shaped by nation states and international bodies, local and global cultures, and multinational pharmaceutical firms. The standard pharmaceuticalization and modernist framework uses one of two possible narratives about why and how clinical trials and ethical practices have become standardized. In the first narrative, the institution of medicine in conjunction with international regulatory bodies successfully developed and adopted scientific and ethical frameworks for the conduct of clinical trials globally. The result is better, well-designed, portable, and ethical scientific research and pharmaceutical products. The second explanation suggests an alternative result, that global bodies are being exploited by the capitalistic expansion of pharmaceutical companies into the global South in the pursuit of cheaper trials and an undermedicalized surplus army of available bodies.

But neither of these modernist frames fully captures what is going on with global pharmaceutical trials. Ultimately, the outcomes of the clinical trial process internationally do not fit standard modernist narratives of either exploitation or the ethical advance of scientific research. A modern perspective on global clinical trials employs an either/or analysis. Global clinical trials can also be seen through a postmodern framework that captures the uneven and contradictory character of pharmaceutical trials occurring throughout the world. We show that bodies used in clinical trials may or may not ever benefit from pharmaceutical development and may or may not be exploited during and after the trials conclude. A postmodern perspective enables a more subtle analysis: pharmaceutical and clinical trial innovations are made possible by and at the same time foster major shifts in the global political economy. This ambiguity is especially apparent in the pharmaceuticalization process. Global pharmaceutical trials and ethical research standardization are complex, global, and multi-sited and involve remaking the technical, organizational, and institutional infrastructures of the life sciences and biomedicine. The pharmaceutical transformation of life and approach to public health is associated with a new, postmodern era in medicine and society more broadly.

To support our argument, we analyzed two cases: Sunder Rajan's study of clinical trials in India and Petty and Heimer's study of global clinical research in HIV clinics. Both of these cases explain how local circumstances help to make sense of pharmaceuticalization and the pharmaceuticalization of public health and both cases are better explained by a more postmodern than a modern frame. Sunder Rajan shows that the particular history of the pharmaceutical industry in India, Indian CROs, and labor exploitation are explanatory "forces" that have led to India's desire to be a location for clinical trials. In spite of the fact that most Indians may not benefit directly from pharmaceutical research, some poor and illiterate Indians do gain access to clinical trials after informed consent is carefully administered to them. Using Sunder Rajan's case of India, we conclude that a modernist explanation of either economic exploitation or benefit does not go far enough.

In the second case we show that the global expansion of clinical trials works unevenly throughout the world and further that in some respects clinics in "poorer" countries benefit more from these trials than do clinics in "richer" countries. For example, Petty and Heimer document how an unintended consequence of participating in global clinical HIV trials for those in poor countries is the reconfiguration of their organizational and medical practices. An additional consequence is the pharmaceuticalization of public health even though it may be a more expensive strategy. Through their participation in clinical trials, clinics create a regulatory, clinical and institutional apparatus that fosters a pharmaceutical approach to HIV. As they write: "…the costs of new pharmaceuticals can easily overwhelm the healthcare systems of poor countries, when investing in the lower-end of healthcare would surely be wiser" (Petty and Heimer 2011: 357). Public health becomes pharmaceuticalized with significantly different procedures and consequences. The contradictions, reversals, and production of new subjectivities such as pharmaceutical citizenship or reconfigured clinics are better explained by a postmodern than a modern theory of pharmaceuticals.

Finally, both of these cases show that while in theory the call for global ethical research standards appears to be a modern and scientific way forward, in reality the implementation of these standards is not "standard" and not always beneficial to clinics and patients in poor countries. Clinics or countries encourage and produce "workarounds" in their efforts to conform to standards. Distinctions such as ethical variability versus standardization – and the modernist assumptions and interpretations of their effects – fail to capture some of the surprising ways in which standards and variability shape the experiences of people in very different parts of the world who are part of a global pharmaceutical system, and thus modernist approaches do not help us fully comprehend the dynamics of global health inequality.

Notes

1 According to the Pharmaceutical Research and Manufacturers of America (PHRMA), only "one of every 10,000 potential medicines investigated by America's research-based pharmaceutical companies makes it through the research and development and is approved for patient use by the United States Food and Drug Administration" and on average it takes fifteen years of research and development and more than $800 million for each pharmaceutical that makes it to the market. (*www.phrma.org/innovation*) PhRMA "Innovation" Retrieved Sept. 12, 2007.

2 See *Trials of War Criminals before the Nuremberg Military Tribunals under Control Council Law* No. 10, Vol. 2, pp. 181–182. Washington, D.C.: U.S. Government Printing Office, 1949.

3 See the World Medical Association for the most current version of the Declaration *www.wma.net/en/10home/index.html*.

4 For example, in Mumbai, India, most of the subjects for clinical trials were recruited from among unemployed textile workers who had lost their jobs after the collapse of the textile industry in the 1980s and 1990s (Sunder Rajan 2005).

5   The Indian state has just begun to do this by issuing compulsory licenses for producing generic versions of patented medicines (see Harris, *New York Times* April 1, 2013).

References

Abraham, J. 2010. "The Sociological Concomitants of the Pharmaceutical Industry and Medications." In *Handbook of Medical Sociology*, edited by C. Bird, P. Conrad, A. M. Fremont, and S. Timmermans, 290–308. Nashville: Vanderbilt University Press.

Angell, M. 1997. "The Ethics of Clinical Research in the Third World." *New England Journal of Medicine-Unbound* Volume 337.12 (1997): 847–849.

Bailey, W., Cruickshank, C., and Sharma, N. 2009. "Make Your Move: Taking Clinical Trials to the Best Location." In *Executive Agenda*, A. T. Kearney: A. T. Kearney. *www.atkearney.com* retrieved January 13, 2013, 56–62.

Bell, S.E. and Figert, A.E. 2010. "Gender and the Medicalization of Health Care" In *Palgrave Handbook of Gender and Healthcare*, edited by E. Kuhlmann and E. Annandale, 107–122 City: Palgrave Macmillan.

Bell, S.E. and Figert, A.E. 2012a. "Medicalization and Pharmaceuticalization at the Intersections: Looking Backward, Sideways and Forward." *Social Science & Medicine* 75: 775–783.

Bell, S.E. and Figert, A.E. 2012b. "Starting to Turn Sideways to Move Forward in Medicalization and Pharmaceuticalization Studies: A Response to Williams et. al." *Social Science & Medicine* 75: 2131–2133.

Biehl, J. 2004. "The Activist State: Global Pharmaceuticals, AIDS, and Citizenship in Brazil." *Social Text* 22: 105–132.

Biehl, J. 2006. "Pharmaceutical Governance." In *Global Pharmaceuticals: Ethics, Markets, Practices*, edited by A. Petryna, A. Lakoff, and A. Kleinman, 206–239. Durham: Duke University Press.

Biehl, J. 2007. "Pharmaceuticalization: AIDS Treatment and Global Health Politics." *Anthropological Quarterly* 80: 1083–1126.

Boston Women's Health Book Collective. 1973. *Our Bodies, Ourselves: A Book by and for Women*. New York: Simon & Schuster.

Brown, P. and Zavestoski, S. 2005. *Social Movements in Health*. Oxford, UK: Blackwell Publishing.

Busfield, J. 2006. "Pills, Power, People: Sociological Understandings of the Pharmaceutical Industry." *Sociology* 40: 297–314.

Carlson, R.V., Boyd, K.M., and Webb, D.J. 2004. "The Revision of the Declaration of Helsinki: Past, Present, and Future." *British Journal of Clinical Pharmacology* 57: 695–713.

Clarke, A. Mamo, E.L., Fishman, J.R., Shim, J.K., and Fosket, J.R. 2003. "Biomedicalization: Technoscientific Transformations of Health, Illness, and U.S. Biomedicine." *American Sociological Review* 68: 161–194.

Clarke, A., Shim, J., Mamo, L., Fosket, J.R., and Fishman, J.R. 2010. *Biomedicalization: Technoscience Health and Illness in the U.S.* Durham, NC: Duke University Press.

Conrad, P. 2005. "The Shifting Engines of Medicalization." *Journal of Health and Social Behavior* 46: 3–14.

Crossley, N. 2006. *Contesting Psychiatry: Social Movements in Mental Health*. London, UK: Routledge.

Dumit, J. 2012. *Drugs for Life: How Pharmaceutical Companies Define Our Health*. Durham, NC: Duke University Press.

Farmer, P. and Campos, N.G. 2004. "New Malaise: Bioethics and Human Rights in the Global Era." *Journal of Law, Medicine & Ethics* 32: 243–251.

Figert, A.E. 2011. "The Consumer Turn in Medicalization: Future Directions with Historical Foundations." In *The Handbook of the Sociology of Health, Illness & Healing: Blueprint for the 21st Century*, edited by B. Pescosolido, J. Martin, J. Mcleod, and A. Rogers, 291–307. New York: Springer.

Fisher, J.A. 2006. "Co-ordinating 'Ethical' Clinical Trials: The Role of Research Coordinators in the Contract Research Industry." *Sociology of Health & Illness*, 28: 678–694.

Fisher, J.A. 2009. *Medical Research for Hire: The Political Economy of Pharmaceutical Clinical Trials*. New Brunswick, NJ: Rutgers University Press.

Gabe, J. and Bury, M. 1988. "Tranquillisers as a Social Problem." *Sociological Review* 36: 320–352.

Getz, K.A. and Vogel, J.R. 2009. "Successful Outsourcing: Tracking Global CRO Usage." *Applied Clinical Trials* 18: 42–50.

Harris, G. 2013. "Patent's Defeat in India Is Key Victory for Generic Drugs" *New York Times*, April 1.

Hartmann, B. 1987. *Reproductive Rights and Wrongs: The Global Politics of Population Control and Contraceptive Choice*. New York: Harper and Row.

Heimer, C.A. 2012. "Inert Facts and the Illusion of Knowledge: Strategic Uses of Ignorance in HIV Clinics." *Economy and Society* 41: 17–41.

Hoeyer, K. 2009. "Informed Consent: The Making of a Ubiquitous Rule in Medical Practice." *Organization* 16: 267–288.

Katz, J. 1992. "The Consent Principles of the Nuremberg Code: Its Significance for Then and Now", in G.J. Annas and M.A. Grodin (eds) *The Nazi Doctors and the Nuremberg Code*, pp. 227–39. Oxford: Oxford University Press.

Lock, M.M. 2004. "Medicalization and the Naturalization of Social Control." In *Encyclopedia of Medical Anthropology; Health and Illness in the World's Cultures*, edited by C. W. Ember and M. Ember, 116–124. New York: Kluwer Academic/Plenum.

Lurie, P and Wolfe, S.M., 1997. "Unethical Trials of Interventions to Reduce Perinatal Transmission of the Human Immunodeficiency Virus in Developing Countries." *New England Journal of Medicine* 337: 853–856.

Marshall, P. and Koenig, B. 2004. "Accounting for Culture in a Globalized Bioethics." *Journal of Law, Medicine & Ethics* 32: 252–266.

Nguyen, V.K. 2005. "Antiretroviral Globalism, Biopolitics, and Therapeutic Citizenship." In *Global Assemblages*, edited by A.O. and S.J. Collier, 124–144. Malden, MA: Blackwell Publishing.

Nichter, M. 1996 [1989]. "Pharmaceuticals, the Commodification of Health, and the Health Care-Medicine Use Transition." In *Anthropology and International Health: Asian Case Studies, Theory and Practice in Medical Anthropology and International Health*, edited by M. Nichter and M. Nichter, 265–326. Amsterdam: Gordon and Breach Publishers.

Petryna, A. 2006. "Gloalizing Human Subjects Research." In *Global Pharmaceuticals: Ethics, Markets, Practices*, edited by A. Petryna, A. Lakoff, and A. Kleinman, 33–60. Durham, NC: Duke University Press.

Petryna, A. 2007. "Clinical Trials Offshored: On Private Sector Science and Public Health." *BioSocieties* 2: 21–40.

Petryna, A. 2009. *When Experiments Travel: Clinical Trials and the Global Search for Human Subjects*. Princeton, NJ: Princeton University Press.

Petryna, A. 2011. "Pharmaceuticals and the Right to Health: Reclaiming Patients and the Evidence Base of New Drugs." *Anthropological Quarterly* 84: 305–330.

Petryna, A., Lakoff, A., and Kleinman, A. 2006. *Global Pharmaceuticals: Ethics, Markets, Practices*. Berkeley: University of California.

Petty, J. and Heimer, C.A. 2011. "Extending the Rails: How Research Reshapes Clinics." *Social Studies of Science* 41: 337–360.

Rothman, D. 1991. *Strangers at the Bedside: A History of How Law and Bioethics Transformed Medical Decision Making*. New York: Basic Books.

Shostak, S. 2010. "Marking Populations and Persons at Risk: Molecular Epidemiology and Environmental Health." In *Biomedicalization: Technoscience, Health, and Illness in the U.S.*, edited by A. E. Clarke, L. Mamo, J. R. Fosket, J. R. Fishman, and J. K. Shim, 242–262. Durham and London: Duke University Press.

Shuster, E. 1997. "Fifty Years Later: The Significance of the Nuremberg Code." *New England Journal of Medicine* 337: 1436–1440.

Sunder Rajan, K. 2005. "Subjects of Speculation: Emergent Life Sciences and Market Logics in the United States and India." *American Anthropologist* 107: 19–39.

Sunder Rajan, K. 2006. *Biocapital: The Constitution of Postgenomic Life*. Durham and London: Duke University Press.

Sunder Rajan, K. 2007. "Experimental Values: Indian Clinical Trials and Surplus Health." *New Left Review* 45: 67–88.

Sunder Rajan, K. 2012. "Pharmaceutical Crises and Questions of Value: Terrains and Logics of Global Therapeutic Politics." *South Atlantic Quarterly* 111: 321–346.

Timmermans, S. and Berg, M. 2003. *The Gold Standard: The Challenge of Evidence-Based Medicine and Standardization in Health Care*. Philadelphia: Temple University Press.

Timmermans, S. and Epstein, S. 2010. "A World of Standards but not a Standard World: Toward a Sociology of Standards and Standardization." *Annual Review of Sociology* 36: 69–89.

Whitmarsh, I. 2008. *Biomedical Ambiguity: Race, Asthma, and the Contested Meaning of Genetic Research in the Caribbean*. Ithaca and London: Cornell University Press.

WHO Drug Information. 2005. *Access to Medicines* 19(3).

Williams, S.J., Gabe, J., and Davis, P. 2008. "The Sociology of Pharmaceuticals: Progress and Prospects." *Sociology of Health & Illness* 30: 813–824.

Williams, S.J., Martin, P., and Gabe, J. 2011. "The Pharmaceuticalization of Society? A Framework for Analysis." *Sociology of Health & Illness* 33: 710–725.

Williams, S.J., Seale, C., Boden, S., Lowe, P., and Steinberg, D.L. 2008. "Waking Up to Sleepiness: Modafinil, the Media and the Pharmaceuticalisation of Everyday/night Life." *Sociology of Health & Illness* 30: 839–855.

Wolinsky, H. 2006. "The Battle of Helsinki." *EMBO Reports* 7: 670–672.

Zola, I. 1972. "Medicine as an Institution of Social Control." *Sociological Review* 20: 487–504.

27

On the Effects of e-Government on Political Institutions

Jane E. Fountain

UNIVERSITY OF MASSACHUSETTS AMHERST

Introduction

Research on e-government typically focuses on disruptive technologies and their presumed transformational effects on government. The internet and associated technologies are more than two decades old, and even cursory observation demonstrates that institutional change in government is often painstakingly slow. To theorize longer-term developments in e-government, an institutional perspective on e-government is sketched and illustrated in this chapter. An institutional approach invites one to examine interactions among people, technologies and structures over time and in political environments characterized in part by conflict over ideas, rights and resources to uncover mechanisms that contribute to stability and change.

To extend institutional perspectives to account for e-government, the chapter introduces the concept of a *digitally mediated institution* – that is, a government organization characterized by a high degree of digital infrastructure and widespread use of digital applications and tools. The chapter then sketches a partial review of various institutional mechanisms that underlie temporal features of institutional development including policy feedback, conventions, path dependence, and key dimensions of longer-term institutional development including timing, sequencing and more gradual patterns of change than are typically presented in disjunctive formulations. Selected concepts are then illustrated briefly through two case studies of state-level digitally mediated institutional stability and change focusing on Europe and the United States federal government. These cases highlight the influence of early events on subsequent paths of development, the importance of timing and sequencing, critical junctures and the ways in which policy entrepreneurs often appear as puzzlers exhibiting uncertainty but seeking to construct and employ appropriate logics. The chapter ends with a brief discussion of implications for science, technology and society.

In keeping with the major themes of this *Handbook*, the chapter seeks to shed light on how and why ideas, artifacts, and practices come to be institutionalized or disrupted in political institutions. Institutional perspectives connect micro-level processes with more macro-level organizational and societal systems. With respect to e-government, cultural values dominant in American and European politics – including democracy, strong association of technological development with progress and social betterment, citizen participation, and mistrust of central government, among other normative values – underlie many institutional reform initiatives.

What are digitally mediated institutions?

While much digital government research focuses on service provision and digital tools in governance, an institutional perspective invites examination of longer-term and deeper inter-relationships between the internet and state structure and behavior. I define the term, "digitally mediated institutions" as those political institutions that use a portfolio of digital information, systems and tools internally and across boundaries. Several dimensions of digitally mediated institutions differentiate them from other types of institutions. These dimensions include: *sunk costs* incurred in the development of large-scale socio-technical systems in public organizations; *rigidity* of many interfaces, systems architecture, code, and digital infrastructure; the *pressure* such systems exert on decision makers to re-engineer and re-structure to realize a return on invest-ment in cyberinfrastructure; and *network dynamics* including the strong tendency toward inter-operability – defined as the ability of multiple systems, applications, data, procedures and other rule regimes to work together, or to inter-operate – among organizational and inter-organizational actors in order to gain coordination benefits by leveraging digital information infrastructure.

Institutional development over time

Institutionalists have long been concerned with time and its role in institutional development. They have conceptualized states as institutional actors constrained by decisions and policies made in the past (Evans et al. 1985). They have traced the gradual evolution of institutional arrangements in part by demonstrating how small changes during periods of putative stasis may accumulate to yield transformative change (Thelen 2004). More recently, researchers have examined institutional developments in networked systems drawing out the particular features of networks, shared conventions and their role in emergence (Singh Grewal 2008).

Institutional theories complement rational choice models of institutional development by foregrounding boundedly rational, social constructionist action. In these pre-rational views, actors often are uncertain about the best course of action (or about their interests or prefer-ences). Relationships between political means and ends may be unclear. Calculative rationality fails to capture decision making in environments characterized by uncertainty and ambiguity. Moreover, actors do not simply calculate; they seek and employ logics of appropriateness in displays of legitimacy; they imitate successful models without understanding their underlying features. While such behavior may be strategic, it is not calculative as formulated in rational choice perspectives (DiMaggio and Powell 1983; Powell and DiMaggio 1991). Mimetic forces and a desire for legitimacy influence many institutional actors to adopt new technologies. Uncertainties surrounding emerging internet and social media use in e-government heighten pre-rational dispositions, producing an environment different from a traditional market or polit-ical setting.

Policy feedback

The institutional tradition in political science has long recognized that policies have politics. In a seminal study of tariff policies, E. E. Schattschneider in 1935 argued that "new policies create a new politics." Theda Skocpol and others (for example, Heclo 1974, Weir et al. 1988, Skocpol and Finegold 1995, Mettler and Soss 2004) have conceptualized policy feedback as a core dimension of state structure and capacity. Skocpol coined the term "policy feedback" to explain how "policies, once enacted, restructure subsequence political processes." Skocpol noted two types of feedback. New policies affect state capacity by restructuring or reinforcing administrative arrangements, and policies influence the capacity, goals and identity of social groups affecting interest group politics. Thus, timing and sequence are critical to the politics created by policies. Reviewing subsequent related research, Mettler and Soss (2004) traced three lines of policy feedback: First, policies influence the "political interactions of organized interests and policy makers"; second, public policies affect the "beliefs, preferences, and actions of diffuse mass publics"; and, third, "public policies affect the depth of democracy, the inclusiveness of citizenship, and the degree of societal solidarity" (2004: 60).

Researchers have defined institutions as bundles of rules or rule regimes. Public policies, including e-government policies, bundle similar rules. Pierson observes that "Most of the politically generated 'rules of the game' that directly help to shape the lives of citizens and organizations in modern societies are, in fact, public policies." He continues, "If policies as institutions matter for political scientists, it is because the influence of policies on social actors – on who they are, on what they want, on how and with whom they organize – is such that it changes the way these actors engage in politics" (2006: 116). Policy rule structures, once in place, shape preferences and influence channels of action available to political actors. The preferences of actors may shift around a policy structure, making subsequent changes in the structure not only inconvenient but also politically disadvantageous or logically implausible.

Conventions

Institutions tend to be highly stable. How does this stability come about and what sustains it? While deep sources of stability lie in normative ideas and values, researchers have increasingly described structural and processual mechanisms that also underlie stability. Among these are conventions. Conventions develop when actions are interdependent, when coordination is needed and when actors consent to a behavior, a process, or a standard in order to overcome coordination problems and share benefits. Conventions are rules that exhibit positive feedback as each actor develops and acts upon mutually reinforcing expectations that others will follow the convention. Thus, networks of actors who share conventions typically "lock in" agreements as they adapt.

Network forces powerfully affect conventions. As one researcher notes:

> The analysis of conventions is obviously relevant to a discussion of standards… However, the idea of network power focuses less on settled conventions than ones emerging due to a combination of extrinsic and intrinsic reasons. Therefore, it emphasizes the positive feedback dynamic central to the interdependent action that drives the adoption of one convention rather than another.

Path dependence

Long-term institutional developments are deeply influenced by their past. But what are the mechanisms of influence? Path dependent models tend to stress positive feedback loops. Specifically, when early events – possibly caused by accident or chance – influence subsequent decisions, a path is formed the retention of which may grow more attractive as its effects accumulate over time.

Institutionalists have demonstrated the explanatory usefulness and applications of path dependence and positive feedback frameworks across a range of social phenomena (Arthur 1994, David 1994, 2000, Pierson 2000, 2004). Path dependent processes are important to recognize because they counter the claim that rational action will correct inefficient paths. Path dependent processes may lead to unpredictable outcomes because of the strong effect of (sometimes accidental or aberrant) early events; inflexibility with respect to breaking out of a path once it has "locked in," "nonergodicity," meaning that early random events do not necessarily function as noise because of their potential future import, and the potential for producing inefficient paths because suboptimal solutions or arrangements may be reinforced and thus increasingly difficult to change. Understanding path dependent processes allows one to predict subsequent outcomes.

While theories of technology development and innovation have considered path dependence for some time, they have tended not to consider socio-technical path dependence *in the context of politics*, a context essential for study of e-government and, more generally, digitally mediated political institutions. Institutional behavior in political environments is characterized typically by *collective action* rather than by individual action, the structure and characteristics of which are significantly different from one another. The use of *authority* through formal institutional roles, public policies and legislation sets the rules of political behavior apart from those of markets, which operate through exchange. Unlike market-based behavior as portrayed in neo-classical models of choice, political actors routinely *adapt their expectations and behaviors* to political rules and policies because these rules define the constraint space for action.

Long-term institutional development

Historical institutionalists emphasize the importance of timing and sequence in political development highlighting the unfolding of events through time. Identifying *the specific mechanisms by which long-term effects occur* is necessary if comparisons across cases are to be made.

Institutionalists provide more powerful explanations of stability than of change. Indeed, conventions and path dependence provide accounts of increasing stability over time. A focus on stability presents a problem for students of institutional change. Some institutionalists have argued that institutions change only when external shocks force them to do so. A related line of research conceptualizes change as punctuated equilibrium. In this view, during punctuations, openness to innovation and change results in rapid developments, followed by institutional stability (see, for example, Krasner 1988). Still other lines of inquiry focus on a complex interplay between agency and structure in institutional development, noting possibilities for political entrepreneurs to intervene at "critical junctures" (Swidler 1986, Katznelson 2003, Thelen 2004, 2006, Orren and Skowronek 2004).

By contrast, other researchers argue for gradual yet transformative change over time (Thelen 2004, 2006, Grief and Laitin 2004, Clemens and Cook 1999). They theorize institutional change in part as *a process of mobilizing support* among political actors to develop, reinforce or revise institutional arrangements (Thelen 2004, Carpenter 2001). For example, Thelen argues that institutions themselves are the object of more or less continuous political contestation rather than stable arrangements that undergo renegotiation periodically. Still other researchers examine changing temporal patterns over long periods of time as a way of describing institutional change, in some cases emphasizing accumulation of small changes including technological change (Bell 1973) and, in other cases, conceptualizing thresholds that lead to critical periods in social movements (McAdam 1982).

Other mechanisms, or systematic explanations of causal factors, shape change over time. Thelen (2004) described layering as "the partial renegotiation" of institutional mechanisms or processes in situations when actors lack power or cognitive ability to comprehensively reconstitute a bundle of institutional dimensions (or rules). Similarly, Schickler, in a detailed study of institutional change in the U.S. Congress, used the term "disjointed pluralism" to conceptualize "institutions as multilayered historical composites that militate against any over-arching order in ... politics." This layering results in a sedimentation of rules, processes and other institutional arrangements that are "more haphazard than the product of some overarching plan" (2001: 15–18).

A stream of institutionalist accounts of stability and change examine the role of those actors who have lost political battles and, as a result of loss, emerge as catalysts for institutional change (see, for example, Clemens and Cook 1999, Clemens 2002, Thelen 2004). Others have high-lighted interactions among multiple institutions as precursors of institutional change often producing unanticipated results (Orren and Skowronek 2004) or among policy entrepreneurs whose political skills and network position allow them to articulate a new vision and to mobi-lize support for institutional change (Burt 1995, Padgett and Ansell 1993, Clemens and Cook 1999, Schickler 2001, Clemens 2006).

E-government as digitally mediated institutional development

Research on e-government can be enriched in explanatory power and validity by incorporat-ing institutional perspectives and extending them to account for the characteristics of digitally mediated institutions. Digitally mediated institutions and policies, of the type that abound in e-government, exhibit many dimensions of path dependence. These tendencies are directly related to the technological systems employed, thus increasing inertia and the probability of unanticipated, often suboptimal, outcomes as e-government systems and policies develop. Sunk costs are typically high for complex software and hardware systems, which are notoriously expensive to develop and maintain, making change potentially highly costly if policies change. In addition, these technical system dynamics are intertwined in complex ways with positive feedback and path dependence in politics.

The potential of networked systems lies, by definition, in their inter-operability. Thus, conventions – standards – are a prerequisite for shared benefits through coordination. Moreover, the attractiveness of interoperability extends beyond benefits to political actors to civil society – the users of such systems, citizens who may be able to gain access, information and trans-parency through such inter-operability. Digitally mediated institutions have intensified pressures to develop conventions.

The logics and complexity of digital systems influence their development. Digital systems tend to be opaque to non-technical decision makers. Among the implications of this statement are the effects on decision "quality" of lack of knowledge of system capacity. Recent current events concerning surveillance, privacy, data sharing and analytics demonstrate the general lack of knowledge of most political decision makers regarding the technical systems about which they develop policies. While institutionalists observe the use of logics beyond mere calculation and maximization among political actors, digitally mediated institutions also exhibit techno-logical logics, which stem from underlying norms and values within the profession of engineering and socialized through education and training into engineers, software specialists, and those who build and maintain such systems. Among the chief attributes of engineering logics are norms of efficiency, streamlining, "faster-better-cheaper," and a tendency toward stan-dardization and convergence as "efficient" solutions to coordination. This technical logic layers over the social and political logics that tend toward conventions as a means to overcome chal-lenges to coordination problems.

Digitally mediated institutions combine policy feedback and the dynamics of collective action and political mobilization with path dependence in technological systems thereby producing an additional layer of unexpected outcomes, dynamic and emergent network tendencies and greater complexity than institutional developments without digital infrastructure. Together, temporal mechanisms of institutional development and characteristics of digitally mediated institutions invite us to attend to longer-term, gradual developments that characterize most complex digital government. Digitally mediated institutional developments are more often characterized by long-term, gradual change than they are by disjunctive change, even when disjunctive technological innovation has taken place (Fountain 2001a). The two cases of digitally mediated institutional development that follow illustrate some of the ideas presented in the brief review above. The first case traces the promise and challenges of technology-enabled cross-agency collaboration in the U.S. federal government. The second case briefly examines the development of conventions and standards to protect intellectual property in Europe.

Enacting an institutional environment for cross-agency collaboration

In the early 1990s, policy makers in the Clinton Administration began building e-government to "transform" government institutions by leveraging information and communication technologies.[1] A key innovation was the "virtual agency," essentially a portal or "one-stop shop" containing all of the government's information and services organized by a specific subpopulation, for example, senior citizens, students or business. Government reformers made an explicit decision not to try to reorganize agencies and programs but to use virtual reorganization of information to streamline and improve services. Their preferences and strategy were influenced by institutionalized challenges to change. However, even efforts to build cross-agency information sharing on the internet were thwarted by deeply embedded layers of budget, oversight and other administrative processes that reinforced single-agency behavior and hindered coordination. During the Clinton Administration, disjunctive, technological change occasioned by the internet did not lead to disjunctive institutional change in the state. By contrast, technology often was enacted in ways that reinforced institutional norms and practices as well as bureaucratic politics leading to suboptimal outcomes (Fountain 2001a).

In spite of bureaucratic resistance, policy entrepreneurs continued to mobilize and work for innovation. For example, during the George W. Bush administration, twenty-five cross-agency e-government initiatives, originally termed the Quicksilver projects and carried forward from the Clinton Administration, were central. Policy makers forged new rules, including the establishment of the government's first Chief Information Officer and the Office of E-Government and Information Technology, in the U.S. Office of Management and Budget. Bush administration officials sought to consolidate information systems and streamline standard administrative functions such as travel, payroll, and authentication across the government. Cross-agency projects encompassed policy domains as diverse as disaster management, rulemaking, grants, benefits and loans.

The record of success for these projects varied. Some, including electronic rulemaking and grants management, succeeded as communities of practice among bureaucrats from across agencies developed shared, cross-agency conventions and standards. Others, including authentication, floundered due to political conflict or lack of convergent standards to coordinate activities. The combination of temporal mechanisms used across the Clinton and Bush administrations as part of their efforts to renegotiate norms of agency autonomy in order to leverage the benefits of networked governance included the development of new conventions, more or less constant efforts at renegotiation, and limited but important positive feedback as new legislation, rules, positions and understandings accumulated incrementally from 1993 to 2008.

During the first and second Obama administrations, mandates requiring cross-agency collaboration as a strategic imperative for improving government performance have grown in importance. As e-government innovations have matured into standard agency practice, demand for networked governance has been driven by calls for: solutions to pressing, complex policy problems that cross bureaucratic boundaries; cost savings and efficiency; reduction of duplication; and further leveraging of technology to enable agencies to share platforms, systems, applications and information.

The Government Performance and Results Modernization Act

The Government Performance and Results Modernization Act (GPRAMA) of 2010 (H.R. 2142) became law in 2011. It extends the Government Performance and Results Act (GPRA) of 1993 and requires stronger development of government-wide priority goals and greater use of cross-agency coordination. The law requires the Office of Management and Budget (OMB) to include cross-cutting, government-wide priority goals in its formulation of the annual performance plan, mandated originally under GPRA (Kamensky 2011). This instantiation of e-government innovations into formal legislation illustrates long-term institutional change in the federal government. Should one think of the legislation as a punctuation in previous equilibrium or a threshold reached through incremental accumulation of small changes or as a gradual transformation? These framings of events are essentially subjective; all exhibit validity. One can say with certainty that no external shock occasioned the legislation.

The GPRA Modernization Act clearly indicates that Congress endorses interagency activities. In stark contrast with traditional bureaucratic perspectives, GPRAMA makes clear that many strategies, priorities, and goals of the government lie *inherently* across agencies (U.S. Congress 2010). This shift in language and logics of appropriateness, encoded in legislation, evidence a formalization of new ideas, norms and practices.

President Obama's FY2012 budget named fourteen cross-agency priority goals, the first set of such goals in the nation's history. The projects grew out of existing administration priorities but respond directly to GPRAMA's requirements.

Outcome-Oriented Goals	Management Improvement Goals
• Exports • Entrepreneurship and Small Business • Broadband • Energy Efficiency • Veteran Career Readiness • STEM Education • Job Training	• Cybersecurity • Sustainability • Real Property • Improper Payments • Data Center Consolidation • Closing Skills Gaps • Strategic Sourcing

Figure 27.1 Cross-Agency Priority (CAP) goals

Institutional constraints on collaboration

Legal requirements for interagency collaboration are layered on an institutional environment still designed for agency autonomy. For example, several laws prohibit specific agencies from sharing data with other agencies to protect personal privacy or national security. In fact, many agencies still guard access to "their" data as part agency culture in spite of presidential directives on "open government" that require agencies to make data more accessible to the public and to share it with other agencies. Moreover, access to data remains problematic in spite of sunshine laws that require agencies to make information available to the public.

Legislation requires agencies to secure permission of Congress before developing shared interagency budgets for joint projects or operations. Most interagency project budgets entail complex memoranda of understanding that may require a year or more of negotiation. Still other laws and rules constrain development and use of shared budgets, operations and personnel. The appropriations process itself, a core function in government, is highly agency and program specific, making cross-agency projects difficult to develop and sustain.

Laws and regulations specify "the rules of the game" for departments and agencies that, in turn, shape the behavior of government officials. The structure of congressional committees and subcommittees fragments jurisdiction and oversight of cross-agency efforts (Radin 2012). Clearly, public policies as institutions circumscribe the environment for cross-agency collaboration in the federal government and specify many of the ways those collaborations will be designed and managed. Legal impediments can stymy forward motion of interagency working groups.

At least four broad types of institutional processes work against cross-agency collaboration: the vertical structure of bureaucracy, often called "stovepipes," which is the fundamental organizational form of the executive branch of government and three central governance processes – legislation, accountability, and budgeting.

By definition, bureaucracies have well-defined jurisdiction and authority relations ordered through a clear chain of command. Max Weber argued that bureaucracy was the only form of organization capable of coordination and control in industrializing societies (Weber [1922] 1978: ch. 11). While for the past thirty years or so, public managers and management experts have pursued the promise of e-government to forge more flexible, innovative and productive forms than traditional bureaucracy, the basic structure of bureaucracy persists – and with good reason (Fountain 2001b, Kettl 2006). Collaborative governance, networks across agencies and other cross-boundary arrangements have been layered over traditional bureaucratic organizations. They have not replaced them.

In recent years, legislators have mandated that agencies and programs cooperate to achieve public ends, but legislation often requires particular agency behavior without providing needed authority or resources. Thus, a sedimented cacophony of legislative rules simultaneously requires, incentivizes, prohibits and constrains cross-agency collaboration. Accountability is also problematic: accountability flows directly from the vertical structure of bureaucracy. A director is directly accountable to Congress for the actions of his or her agency.

Cross-agency collaboration blurs lines of authority and accountability. Public managers are challenged when asked to maintain vertical accountability in their agency activities while supporting "horizontal" or networked initiatives for which lines of accountability are less direct and clear. The budget process is organized to authorize and appropriate funds to individual departments for department-specific programs (Bardach and Lesser 1996, Allen et al. 2005). Shared resources form a significant source of cohesion for interagency collaboration, in part because they change the nature of the relationship from multiple exchanges to a shared system.

Researchers have found that the amount of resources shared by the group is one of the determinant factors for partnership effectiveness.

Although pockets of good practice have developed, institutional systems and policies to support interagency collaboration have lagged (Allen et al. 2005, Wilkins 2002, Fountain 2001a). Many agencies continue to define data and to implement procedures (including those for services to the public) in agency-specific ways in large part to focus on accountability to Congress. At the same time, standard administrative functions, such as grants administration, could be further harmonized across agencies to better serve the public. Although progress has been made, grants management, which still varies from one agency to another in spite of legislation that requires streamlining across agencies, is one of many examples where the traditional structure of accountability has hindered development of e-government (GAO 2013).

In spite of these challenges, for nearly thirty years policy entrepreneurs in the permanent senior civil service have mobilized, often with external interest groups and other stakeholders, and have accumulated practical experience over time with the development and governance of sustainable cross-agency operations, and this experience creates an environment conducive to the future development of e-government initiatives.

The two case studies that follow illustrate several of the concepts enumerated above. The first case, drawn from Europe, sketches the trajectory of a relatively new agency designed from the beginning with e-government in mind. The second case, based in the United States, depicts the constraints posed by history, culture, and the layering of legislation, practice and commitments over time.

Governance of trademarks in Europe

Created in 1993, the European Commission Office for the Harmonization of the Internal Market (OHIM) began with a mandate to strengthen the internal market of the European Union (EU) by working to lower and, when possible, remove barriers to "the free movement of goods and services" across Europe.[2] The legislation creating the agency, which became operational in 1996 and is based far from Brussels (the home of the European Union) in Alicante, Spain, also created the Community Trade Mark (CTM) and the Registered Community Design (RCD). A trademark or design registration from OHIM offers intellectual property protection for brand names and related images in all 27 EU member states. The CTM makes it possible to register once, pay one fee, and manage a trademark or design in one language. To make this vision of harmonization a reality would require digital data, processes and systems of the type central to e-government.

States traditionally have regulated intellectual property rights according to the theory that legal protection supports innovation and creativity as well as competition in market systems. In the EU, trademarks may be registered at several levels of governance: at the national level, through national offices within each member state, at the regional level in some instances (for example, through the Benelux – for Belgium, Netherlands and Luxembourg – Intellectual Property Office), at the European Community level through OHIM, and at the international level through the World Intellectual Property Office (WIPO) in Geneva. These partially nested rule regimes and institutions have been layered on one another as Europeanization and globalization have developed over time.

The primary governance and oversight bodies for OHIM – its Administrative Board and Budget Committee – were designed to reflect the negotiated compromises made to coordinate the interests of member states and the Commission. The Administrative Board consisted of representatives from each of the member states, each of whom wielded a vote on OHIM's policies. Administrative Board members largely came from the intellectual property (IP) offices of their home countries, rather than from relevant ministries. This resulted in parochialism, conflicts of interest and other tensions. The initial design of OHIM's governance bodies was meant explicitly to check "interference" from Brussels in the ability of OHIM to function autonomously. Oddly, the Commission was represented in OHIM's governance bodies but had no voting privileges.

As a new political institution, OHIM has affected national IP offices in complex ways, for example, by changing the opportunities for interest groups to influence intellectual property policies by introducing a new layer of governance and policy at the European level. In fact, the CTM was established to "Europeanize" many businesses in member states by making it easier for them to conduct business across national boundaries within Europe. The establishment of e-government practices and systems at OHIM challenged national IP offices to modernize and to improve their administrative operations. Most national offices have viewed the CTM as an institutional vehicle in competition with the national trademark, but over time, national office bureaucrats are realigning their expectations, preferences and activities to work effectively within the new politics created by this new policy.

Agency start conditions: early events set a path

From its operational beginning, civil servants within OHIM decided to develop a paperless office. The agency had 23,000 CTM applications on the first day they were made available in 1996. To their shock, OHIM's managers found that CTM applications during the first year would equal 43,000, overwhelming the operational and technical capacity of the agency even as strong demand legitimized the new and untested policy that gave rise to the CTM. An agency official observed: "National offices could fall back to paper if [their IT systems] failed. We did not have that possibility. We had no tradition to fall back on."[3]

From growth to productivity

OHIM launched its first website, OAMI-Online, in 1998 and began making documents available online. But the "paperless office" at that time provided only first-generation electronic sources of information and required staff to scan paper mail or faxes into digital form (although it soon became possible to import data sent via faxes directly into the system) and, throughout the trademark or design application examination process, to print, mail or FAX paper documents back to users or other entities.

By 2008, even *The Economist*, which has reported on the overwhelming failure rate of e-government projects, pronounced that "OHIM offers a streamlined, paperless operation and does much of its business online, keeping costs down and speeding up the processing of applications" (*Economist* 2008). What was the path by which this success was forged?

Wubbo de Boer, a Dutch lawyer and civil servant became the second president of OHIM in October 2000 bringing thirty years of experience and expertise to the job. The dynamic president and his managers developed a horizontal organization, building a senior management team without divisional separations which set out to focus intensively on the needs of what OHIM calls its "users," primarily large firms handling brand management and forming a set of powerful interests in the European economy. They created a Quality Management Department to devote sustained attention to analyzing and improving administrative processes. The president noted that the unit "created a point of reflection for many things to be said and thought that were not possible before: to do something that was fundamental." The autonomy and resources of the new agency allowed scope for re-imagining and structuring a new type of agency designed to leverage the internet.

The staff at OHIM, mostly European Commission civil servants, began to realize that the core strategy was shifting from building agency capacity through growth in staff to capacity building through productivity gains guided by simplification of processes and procedures, attention to user needs, careful measurement of performance and continued innovation using technology. In fact, at a crucial point, the president decided that staff growth would end, even as the volume of trademark applications continued to increase. This approach to building agency capacity prompted internal tensions that ultimately strained the entire European network of trademark institutions as OHIM continued to grow in CTM registrations and to make dramatic productivity gains by automating key steps in core tasks as part of its vision as an agency of the information age. This put it at odds with traditional civil service conventions.

During the early years of the agency, the recruitment of trademark and design examiners focused on lawyers and paralegals. But the skill mix required of examiners changed as e-business tools, to use the agency's term, and the use of large databases became embedded in the design of the examiners' work greatly reducing search costs and paperwork. Executives within OHIM worked assiduously to use internal staff mobility, in part through strong investments in training internal staff, to strengthen the skills of existing staff in order to reassign within the agency those whose jobs had been made obsolete by technology and who could master new skills required for the examiner positions. In 2001, to facilitate the operational transformation of the agency, the management team developed a policy enabling employees to receive twelve days of training a year, an unprecedented investment for an EC agency. In fact, mailroom personnel and other clerical workers were offered the opportunity and training to become examiners. OHIM also established generous, flexible telework policies for its employees.

In 2004, OHIM's managers established performance targets for each employee for each twelve-month period. By linking performance objectives to appraisal – and to the organizational culture – the notion of performance standards became salient throughout the organization though not without tension. This blending of neo-liberalism with informatization caused a paradoxical mixture of pride in performance and trepidation concerning job security. As we will see in the U.S. case study to follow, the complexities involved in linking performance objectives, appraisal, technological modernization and other dimensions that contribute to e-government are far more challenging when multiple agencies are involved in these developments. The success of OHIM is due in part to the authority granted to OHIM's agency executives to design a new agency. While the U.S. federal government and the European Commission may be rough analogues in terms of federalism, they are institutionally entirely different, not least because of their differing historical paths.

OHIM as a benchmark for Europe: the service charter

To create "external pressure" on the institution, OHIM conducted its first annual web survey of users in 2005 and published the results on its website in 2006. The agency surveys users annually, using a highly detailed instrument, and reports the results publicly with the explicit norm of transparency as key to public service and to pressure itself to closely monitor and improve performance. Based on user feedback, OHIM developed three primary service dimensions – timeliness, accuracy and accessibility – and began to analyze the work of examiners with a view to focusing their expertise on the core tasks of examination while assigning ancillary tasks, such as data entry and translation, to others. Building on their three service dimensions, OHIM's managers elaborated a series of quality standards for service dimension, drawing from user survey results.

In addition, the progressive introduction of more web-based information and e-business tools created a dynamic environment online for users with inevitable bumps in the road as new systems were developed, implemented and refined. The agency's focus on users and its commitment to transparency pressured OHIM's technology managers to build greater user participation into the design and development of new e-business tools. This alignment between continuous improvement in internal performance, through close communication with users, and mobilization of strong support from interested business groups formed a self-reinforcing cycle with strong path dependence, mobilization of interest groups and realignment of interests over time.

In 2008, Charlie McCreevy, European Commissioner for the Internal Market and Services, announced that "The Commission supports the ambition that (OHIM) should be the benchmark among industrial property offices, and targets for further improvement in the work of the Office are high" (OHIM Annual Report 2008). The service charter of OHIM, a set of performance targets expressed as commitments to users, and the performance standards within it, were used internally to suggest targets for individual employees and for units in order to measure their productivity and, in the aggregate, the agency's performance. The agency published on its website its actual performance against its service standards on a quarterly basis to promote transparency and accountability.

The agency was unusual among European political institutions because it possessed the financial means for substantial development projects and had invested approximately €30 million per year, or 20–25 percent of its budget, to build a "complete e-business service offering" in five years. (The agency's operating revenues consist primarily of the application fees.) As development of a digitally mediated institution continued, new tools, systems and databases gave rise to continued re-examination of work processes, first in the back office, for example, in routine, clerical tasks and, later, through simplification and streamlining of the core examination tasks. Moreover, by making its databases, search tools and other innovations accessible to the public and its users, the agency fostered substantial co-production of trademark and design registration. By 2010, OHIM was able to offer a comprehensive suite of e-business tools or "solutions" to its users.

The success of e-government requires not only technological developments but also a host of related changes in employee skills and work practices to align with organizational and administrative changes. While ample resources may make these developments more feasible, resources do not necessarily diminish the dislocation experienced by employees confronting rapid change. OHIM's senior management group had largely mandated administrative innovations on agency staff, producing tensions that could not be diminished solely through perks such as training and telework.

Ironically, given its commitment to measurement, only in November 2009 did OHIM implement its first employee survey. Some of the results were troubling; in fact, one manager characterized the response as a "staff protest vote in terms of the management policy." While the deep cultural shift in norms of work and productivity were applauded by some of OHIM's workers, the changes perplexed and angered other OHIM staff who wondered why an agency with a budget surplus and the highest productivity in Europe continued to push for higher performance levels. These internal tensions mirrored strains in the inter-institutional network of trademark agencies as well, and these would have to be negotiated as part of the interplay between institutional stability and change. Specifically, national IP offices were pushed to change due to advances at OHIM. While these tensions existed, they did not fundamentally inhibit the move to paperlessness and the accompanying increase in productivity. In fact, one of the paradoxes of the case is the near simultaneous mixture of pride and tension associated with advances at OHIM.

Using interoperability for European harmonization

As part of its role in a multi-level governance network, OHIM, initially under pressure from national trademark offices, developed a series of collaborative projects with national offices by which the European trademark system has been developing shared standards, shared platforms, shared classification systems, shared databases, shared tools and, through these inter-operability gains, shared understanding and a shared view of trademark and design in a federated system.

At the end of 2009, OHIM released its internal electronic file manager to national trademark offices through a free license. Subsequently, the agency made available a common trademark search engine tool to allow users to search for trademarks across the registers of WIPO, OHIM and EU national offices. Another tool shared with national trademark agencies provided the means for examiners to compare the classification databases of national offices online. Going further, OHIM worked with a group of national trademark agencies to produce a common database available in the twenty-two EU languages and for use by all IP offices. Still further, OHIM and national partners launched projects to create a single European platform for filing national, international and CTM applications through a single interface. Managers at OHIM and national agencies undertook to develop a pan-European web portal, which, OHIM claimed, would provide a central source for IP information within the EU.

All of these projects – and the significant institutional changes they would make possible – were due to convergence on conventions – shared technical standards and open source technologies in order to increase inter-operability within OHIM and, in turn, within the European system of national trademark and design offices. From 2003 to 2005, a group of technical experts in the trademark and design domains met four or five times each year to discuss and develop common standards, which would be necessary for harmonization of the internal market.

OHIM is widely considered the benchmark for trademark and design registration. Their experience and innovative capacity offered to national offices a set of important strategic and administrative practices, e-business tools, and other information resources that could be adopted or adapted to national settings. The cost savings to national IP offices of forgoing their own development of information systems was substantial. While performance standards and increasing productivity may have met more resistance, they became associated with e-governance through the institutional developments pursued by OHIM's managers. Opportunities for knowledge sharing among the national offices and with OHIM had made the vision of a European multi-level governance and administrative system for trademark and design operationally feasible. Although a thicket of legal, political and practical issues would require political negotiation and careful policy evaluation to harmonize, the technological systems and e-business tools required to run a multi-level, coordinated system were available for immediate use. While the layering of institutional arrangements is important, so is the layering of logics. In this case, computing logics, digitally mediated, are juxtaposed with logics of governance – subsidiarity, territoriality, and the shared understandings between states and the civil servants who inhabit public bureaucracies.

From the start, OHIM envisioned itself as a "paperless office." As a new agency, it had the scope to develop rules and arrangements that would forge and reinforce its e-government path. In fact, it is an unusual agency in its elaboration and synthesis of process management, analysis, training and technology development. Timing and sequencing are critical in OHIM's history. Begun just as the internet "revolution" is in full force, it had no legacy systems to change. In the larger European governance space, the agency was a first mover and, without any intention of doing so, developed systems that national IP offices could license and use with little modification thus creating a standard and fostering European conventions that allow inter-operability across the various levels of governance. Throughout its development, each new system and tool forged a path making subsequent information systems and practices easier to undertake and implement.

Conclusions

Both cases of digitally mediated institutional development presented here exhibited high degrees of uncertainty creating an environment of pre-rational choice and the use of logics of appropriateness. In contrast with the U.S. federal government, OHIM, and to some extent the new European Commission, had the distinct advantage of being "new" with new authorities and a new mission. Moreover, the development of inter-operability across the trademark and design registration policy domain in Europe benefited from operating within one specific policy domain. In the U.S. federal government, agencies focused on very different policy domains have attempted, with some success, to develop conventions and to overcome challenges to coordination. While impressive developments in e-government are found in the U.S. federal case, challenges to e-government developments across agencies are equally impressive in their tenacity.

The scale of the U.S. federal government dwarfs that of OHIM, thus increasing complexity. Moreover, the role of the U.S. Congress and its relationship to the executive agencies is quite different from that of the European Parliament and European Commission agencies. The U.S. Congress plays a much stronger role in legislation, appropriations and operations of agencies making change, including development of e-government, more challenging. Thus, the scope of the two cases differs, and the overall governance structures and history differ as well. In both cases, policy entrepreneurs – senior civil servants or officials with deep expertise, experience and long periods of engagement – forged communities of practice and searched out opportunities for movement. These entrepreneurs typically are skilled at mobilizing support among external stakeholders as well as those within government. These two cases offer portraits of longer-term institutional developments in different political systems. They are meant to display a range of mechanisms specifying temporal dimensions of institutional development and to highlight the ways in which digitally mediated institutions overlay and intensify institutional perspectives.

Ideas, artifacts and practices come to be institutionalized or disrupted through the actions of coalitions, through incremental redesign of operations and procedures with positive feedback and lock-in, at times, but also with the possibility that incremental changes will be reversed as new political regimes change paths. A key force for momentum is found in the fact that actors seek conventions to be able to engage in collective action. Digitally mediated institutions vividly tend toward conventions through development of standards for inter-operability.

In this chapter, I examine technoscience at the level of the state. State structure and capacity is built up from individuals, small groups, and communities of practice who puzzle over challenges, propose and develop quasi-solutions that require agreement, then develop policies and systems. In these cases, core ideas about e-government travel globally through professional networks. At the same time, state actors include those for whom the status quo represents considerable power. These actors counter some e-government developments with the force of highly stable institutionalized practices.

The U.S. and EU cases illustrate cultural values that emphasize democratic governance as a vehicle for modernity, speed and efficiency, using digital means whenever possible. So far, the result is a changing notion of boundaries, of agency autonomy, of federalism in Europe, and a highly imperfect but forward moving set of shared systems and processes in governance. The U.S. case in this chapter highlights the strength of early events and lock-in in path dependence. Agency centric institutions within the federal government, encoded in law and reinforced in agency–congressional relations, have protected agency autonomy and make e-government developments that would network agency capacity an ongoing challenge. The result is a mosaic that includes stunning innovation combined with equally impressive resistance to change.

In this chapter, I have sought to connect e-government with institutional mechanisms that describe and explain political stability and change in and across bureaucracies. Two cases drawn from complex political institutional developments in e-government over more than a decade illustrate interactions among actors, processes and new technologies as they unfold in institutional development. In these accounts, digital technologies are not leading to the demise of political institutions but are embedded in political conflicts and policy-making. In the case of the U.S. Government Performance and Results Modernization Act, a reconceptualization of the appropriate locus for policy-making, from single agency to networks of agencies, is a result of a series of gradual changes, only some of which are directly related to e-government. In the case of European trademark policies and practices, the use of shared information and standards has provided a strong platform on which competing interests have found a series of focal points to further cooperation amid contestation. This conceptualization of digitally mediated institutional development is meant to encourage more attention to the precise mechanisms and conceptualizations that describe and explain longer-term institutional developments and the influence of digital mediation in these processes.

Notes

1   For a more detailed account of this case, see Fountain 2013. This research was made possible by grants from the National Science Foundation, under grant numbers 0131923 and 0630239. The opinions, findings, conclusions and recommendations in this report are my own and do not necessarily reflect the views of the National Science Foundation.
2   This case is excised from a detailed study of the development of the European Commission Office for Harmonization of the Internal Market. See Fountain et al. 2010.
3   Quotations in this case study are drawn from interviews conducted by the author with OHIM managers and key stakeholders in 2009 and 2010.

References

Allen, B., Juillet, L., Paquet, G., and Roy, J. (2005) "E-Government as Collaborative Governance: Structural, Accountability and Cultural Reform," in *Practicing E-Government: A Global Perspective*, edited by Mehdi Khosrow-Pour, 1–15. New York: Idea Group.

Arthur, W.B. (1994) *Increasing Returns and Path Dependence in the Economy*. Ann Arbor: University of Michigan Press.

Bardach, E. and Lesser, C. (1996) "Accountability in Human Services Collaboratives – For What? and to Whom?" *Journal of Public Administration Research and Theory* 6, no. 2: 197–224.

Bell, D. (1973) *The Coming of Post-Industrial Society: A Venture in Social Forecasting*. New York: Basic Books.

Burt, R. (1995) *Structural Holes: The Social Structure of Competition*. Cambridge, MA: Harvard University Press.

Carpenter, D. (2001) *The Forging of Bureaucratic Autonomy: Reputations, Networks and Policy Innovation in Executive Agencies, 1862–1928*. Princeton: Princeton University Press).

Clemens, E. (2002) "Invention, Inovation, Proliferation: Explaining Organizational Genesis and Change," in *Social Structure and Organizations. Research in the Sociology of Organizations*, edited by Michael Lounsbury and Marc J. Ventresca. Amsterdam: Elsevier.

Clemens, E. (2006) "Lineages of the Rube Golberg State: Building and Blurring Public Programs," in *Rethinking Political Institutions: The Art of the State*, edited by Stephen Skowronek, Daniel Galvin, and Ian Shapiro. New York: New York University Press.

Clemens, E. and Cook, J.M. (1999) "Politics and Institutionalism: Explaining Durability and Change," *Annual Review of Sociology* 25.

David, P. (1994). "Why Are Institutions the 'Carriers of History'? Path Dependence and the Evolution of Conventions, Organizations, and Institutions," *Structural Change and Economic Dynamics*, 5, No. 2, pp. 205–220.

David, P. (2000). "Path Dependence, Its Critics, and the Quest for 'Historical Economics," in P. Garrouste and S. Ioannides, eds., *Evolution and Path Dependence in Economic Ideas: Past and Present*. Cheltenham, UK: Edward Elgar.

Dawes, S.S. and Prefontaine, L. (2003) "Understanding New Models of Collaboration for Delivering Government Service." *Communications of the ACM* 46, no. 1: 40–42.

DiMaggio, P. and Powell, W.W. (1983) "The Iron Cage Revisited: Institutional Isomorphism and Collective Rationality," *American Sociological Review*, 48: 147–160.

Economist (2008) "A money mountain," March 8–14, p. 73.

European Commission (2008) OHIM Annual Report.

Evans, P., Rueschemeyer, D., and Skocpol, T. (1985) *Bringing the State Back In*. Cambridge, UK: Cambridge University Press.

Fountain, J.E. (2001a). *Building the Virtual State: Information Technology and Institutional Change*. Washington, DC: Brookings Institution Press.

Fountain, J.E. (2001b) "Toward a Theory of Federal Bureaucracy in the 21st Century," in *Governance.com: Democracy in the Information Age*, edited by E. Kamarck and J. S. Nye, Jr. Washington, DC: Brookings Institution Press.

Fountain, J.E. (2013) "Implementing Cross-Agency Collaboration," IBM Center for the Business of Government.

Fountain, J.E., Galindo Dorado, R., and Rothstein, J. (2010) "OHIM: Creating a 21st Century Public Agency," Amherst, MA: National Center for Digital Government monograph.

Grief, A. and Laitin, D. (2004) "A Theory of Endogenous Institutional Change," *American Political Science Review* 98, no. 4.

Heclo, H. (1974) *Modern Social Politics in Britain and Sweden: From Relief to Income Maintenance*. New Haven: Yale University Press.

Kamensky, J.M. (2011) "GPRA Modernization Act of 2010 Explained," IBM Center for the Business of Government blog. January 6.

Katznelson, I. (2003) "Periodization and Preferences: Reflections on Purposive Action in Comparative Historical Social Science," in J. Mahoney and D. Rueschemeyer, eds., *Comparative Historical Analysis in the Social Sciences*. New York: Cambridge University Press.

Kettl, D. (2006) "Managing Boundaries: The Collaboration Imperative," *Public Administration Review* Volume 66, Issue Supplement s1, pages 10–19, December.

Krasner, S.D. (1988) "Sovereignty: An Institutional Perspective," *Comparative Political Studies* 21, no. 1: 66–94.

Lindblom, C. (1977) *Politics and Markets: The World's Political-Economic Systems*. New York: Basic Books.

March, J.G. and Olsen, J.P. (1989) *Rediscovering Institutions: The Organizational Basis of Politics*. New York: Free Press.

McAdam, D. (1982) *Political Process and the Development of Black Insurgency, 1930–1970*. Chicago, IL: University of Chicago Press.

Mettler, S. and Soss, J. (2004) "The Consequences of Public Policy for Democratic Citizenship: Bridging Policy Studies and Mass Politics," *Perspectives on Politics* 2, no. 1: 60.

Orren, K. and Skowronek, S. (2004) *The Search for American Political Development*. New York: Cambridge University Press.

Padgett, J.F. and Ansell, C.K. (1993) "Robust Action and the Rise of the Medici, 1400–1434," *American Journal of Sociology* 98: 1259–1319.

Pierson, P. (2000) "Increasing Returns, Path Dependence, and the Study of Politics," *American Political Science Review* 94, no 2: 251–267.

Pierson, P. (2004) *Politics in Time: History, Institutions, and Political Analysis*, Princeton: Princeton University Press.

Pierson, P. (2006) "Public Policies as Institutions," in *Rethinking Political Institutions: The Art of the State*, edited by Stephen Skowronek, Daniel Galvin, and Ian Shapiro, New York: New York University Press: 114–134.

Powell, W.W. and DiMaggio, P. (1991) *The New Institutionism in Organizational Analysis*. Chicago, IL: University of Chicago Press.

Radin, B. (2012) *Federal Management Reform in a World of Contradictions*. Washington, DC: Georgetown University Press.

Schattschneider, E.E. (1935) *Politics, Pressures and the Tariff: A Study of Free Private Enterprise in Pressure Politics, as Shown in the 1929–1930 Revision of the Tariff*. New York: Prentice-Hall.

Schickler, E. (2001) *Disjointed Pluralism: Institutional Innovation and the Development of the U.S. Congress*. Princeton: Princeton University Press.

Singh Grewal, D. (2008) *Network Power: The Social Dynamics of Globalization*. New Haven, CT: Yale University Press.

Skocpol, T. and Finegold, K. (1995) *State and Party in America's New Deal*. Madison, WI: University of Wisconsin Press.

Swidler, A. (1986) "Culture in Action: Symbols and Strategies," *American Sociological Review* 51: 273–286.

Thelen, K. (2004) *How Institutions Evolve: The Political Economy of Skills in Germany, Britain, the United States and Japan*. New York: Cambridge University Press.

Thelen, K. (2006) "The Evolution of Vocational Training in Germany," in *Rethinking Political Institutions: The Art of the State*, edited by Shapiro, Skowronek and Galvin. New York: New York University Press.

U.S. Congress. Senate Report 111-372. (2010) "GPRA Modernization Act of 2010: Report of the Committee on Homeland Security and Governmental Affairs, United States Senate, to accompany H. R. 2142." December 16.

U.S. General Accountability Office (GAO). GAO-13-383. (2013) "Grants Management: Improved Planning, Coordination, and Communication Needed to Strengthen Reform Efforts." May 23.

Weber, M. (1978 [1922]) *An Outline of Interpretive Sociology: Economy and Society*, 2 vols. ed. Guenther Roth and Claus Wittich. Berkeley, CA: University of California Press.

Weir, M., Shola Orloff, A., and Skocpol, T. (1988) *The Politics of Social Policy in the United States*. Princeton, NJ: Princeton University Press.

Wilkins, P. (2002) "Accountability and Joined Up Government." *Australian Journal of Public Administration* 61, no. 1: 114–119.

28

Science, Social Justice, and Post-Belmont Research Ethics

Implications for regulation and environmental health science

Rachel Morello-Frosch and Phil Brown

UNIVERSITY OF CALIFORNIA, BERKELEY AND NORTHEASTERN UNIVERSITY

Introduction

Research practices and regulation aimed at addressing environmental contamination have shifted dramatically in academic and policy arenas, due to challenges from social movements and innovative scientists who have collaborated directly with affected communities to conduct research linked to regulatory change. This shift has also catalyzed new initiatives from some federal and state agencies as well as foundations to support community-engaged science (Balazs and Frosch 2012). As science and technocratic decision-making increasingly shape policy and environmental regulation in the United States, some affected communities have marshaled their own scientific resources, often by engaging in research collaborations with academic partners. These community-academic collaborations entail direct public engagement in the scientific enterprise (including development of research questions, study protocol design, data collection, interpretation and dissemination of results) with an eye toward leveraging data to improve policy-making and protect public health. This form of community-engaged research pointedly resists forms of "scientization" in which decision-making is dominated by experts who work to ensure that debates over policy remain "objective" and divorced from their socioeconomic and political contexts (Morello-Frosch et al. 2011). Instead, community-engaged research eluci-dates the potential *and limitations* of medical science in solving persistent health problems that are socially and economically mediated. The experience of the environmental health move-ment, which is concerned with environmental causation of disease, demonstrates that lay pressure is critical to advancing environmental and occupational health.

In an environment of uncertainty and consequent contestation, the scientific data and analy-sis on which policy and regulation is based inevitably integrates "large doses of social and political judgment" (Jasanoff 1990, 229). Where judgment is a central part of data development and the data collected can affect communities, it makes ethical sense for communities to have a voice at the table. However, this not only creates institutional challenges to the scientific enterprise and the regulatory process, but also raises significant bioethical questions as

Institutional Review Boards (IRB) – the entities that oversee human subjects research in university and related settings – grapple with the governance of research projects that entail extensive engagement between study participants and researchers, the sharing of research results despite scientific uncertainties, and meeting the sometimes competing demands of community-level and individual-level research protections. We call these emerging tensions in human subjects' protection "post-Belmont research ethics."

In 1979, the Belmont Report established ethical principles for the use of human subjects in scientific research. Developed partly in response to the Tuskegee syphilis study[1], Belmont identified three basic principles, which are interpreted and applied by IRBs that oversee human subjects research. The first of these principles, "respect for persons," stresses that an individual's decision to become a research participant must be voluntary and calls for special protection for those who lack the capacity to make such a decision themselves (such as children). The second principle, "beneficence," calls on researchers to "do no harm" or barring that, to maximize the benefits of their research while reducing, as much as possible, risks to study participants. Finally, the principle of "justice" requires careful attention to the fair distribution of risks and benefits, calling on researchers to select study participants only "for reasons directly related to the problem being studied" and to vigilantly avoid the selection of subjects for "their easy availability, their compromised position, or their manipulability." Justice also requires that those who bear the risks of research should, whenever possible, be among the first to benefit from its insights (National Commission for the Protection of Human Subjects of Biomedical and Behavioral Research 1979).

Thus, IRBs' Belmont-driven approach to human subjects protection represents an *institutionalized* practice of research oversight. The highly bureaucratized nature of IRBs, which Stark (Chapter 25 of this volume) analyzes, has become a barrier to some of the protections that IRBs were intended to provide. While suitable for many biomedical applications, IRBs often strictly apply Belmont principles in ways that impede legitimate social science and community-based participatory research (CBPR). Indeed, formalized ethical protocols provide structured guidelines for research, but they do not fully address the uncertainties faced by researchers as they navigate new ethical terrains of community engaged science and the dynamic relationships between multiple parties within a research project (for example, individual study participants and their communities as well as researchers).

Our notion of post-Belmont ethics acknowledges the importance of formal ethical standards used by IRBs, while also recognizing that ethical research requires continued reflexivity, communication, and attention to *community* rights, not solely individual rights. In this approach, community harm and benefit is consistently taken into account, as is individual harm and benefit. Sharing information is emphasized, even in light of scientific uncertainties, which contrasts with the tendency of IRBs to withhold uncertain information for fear of unduly alarming study participants. In short, post-Belmont ethics *disrupt* the passive and paternalistic formulation of human subjects in favor of directly engaging study participants in the implementation and oversight of environmental health research.

As the icon of environmental contamination changes from drums of toxic waste to test tubes of human blood or breast milk, so too does the measurement of chemical contamination change from exposures outside the body to those inside it. Concerned citizens are just as likely to ask "is it in us?" as to ask what a nearby refinery emits from its smokestacks. Personal exposure research methods, including household air and dust sampling and biomonitoring of human tissues (such as blood, breast milk, or hair) have become important ways to examine the impact of chemical and other contaminants on humans. This expansion in how human exposures to environmental chemicals is measured, coupled with increased institutional support for community-engaged environmental health science more broadly, has opened the way for communities, scientists, and some regulatory officials to challenge established institutional norms in the realms of scientific methods and study protocol design; development of relevant research questions; the intersection of science, decision-making and policy advocacy; institutional review board (IRB) governance; and public participation in the scientific enterprise and the regulatory process.

Citizen, scientist and regulator challenges to established research norms and practices can all be seen through the lens of the *right-to-know* and *right-to-act* (and the inherent tensions between the two) approaches. The *research right-to-know* asserts the rights of study participants to know their results from exposure research on their homes and bodies, even if the implications for health effects and strategies for exposure reduction are uncertain. This orientation challenges established norms concerning IRB tendencies toward withholding data from study participants when health effects are not well understood. In all, the various confrontations with the established guiding framework of IRBs are part of a broad constellation of organizing and advocacy activities that provide communities with knowledge of contamination in their homes and bodies, while seeking to empower them to act in order to reduce or prevent future exposures.

Community-based participatory research (CBPR) in environmental health science is one mechanism through which citizens directly act on their right to know. CBPR has promoted changes in theories of disease causation and new lines of scientific inquiry that have helped to (re)shape scientific fact-making, particularly in regulatory science (Morello-Frosch et al. 2006; Morello-Frosch et al. 2011). This is exemplified in the cumulative impacts arena. Here, environmental justice advocates have long asserted that chemical-by-chemical and source-specific assessments of the health risks of environmental hazards are scientifically problematic because they do not reflect the cumulative impacts of multiple environmental and social stressors that are disproportionately faced by marginalized and vulnerable communities, and which may act additively or synergistically to harm health (Cal-EPA 2003; Sadd et al. 2011; Morello-Frosch and Shenassa 2006). This can involve chemical exposures from ubiquitous consumer products (for example, flame retardants or phthalates) and multiple chemical exposure that are place-based due to multiple emission sources in a geographic area. CBPR has helped advance the science of "cumulative impacts" by elevating the role of structural determinants and their associated social stressors in creating vulnerabilities to the adverse health effects of environmental hazards among people of color and the poor. Ultimately, this focus on cumulative impacts or the "double jeopardy" of environmental and social stressors is transforming how scientists study environmental health problems (Clougherty and Kubzansky 2009; NRC 2009; DeFur et al. 2007) and how regulators address them. We can think of CBPR as a reflection of a post-Belmont environment in which the rights of communities, not just individuals are taken seriously.

This chapter covers three elements of the research right-to-know and its transformative potential in the realms of environmental health science and regulation: contestations over cumulative impacts in regulatory science; CBPR strategies in personal exposure assessment science; and post-Belmont research ethics. We begin by discussing the role of communities in transforming regulatory science through its efforts to compel agencies to address cumulative impacts of environmental and social stressors, particularly in marginalized communities. We then examine ways in which CBPR is transforming scientific knowledge production with a focus on personal exposure assessment science and biomonitoring research. Next, we proceed with an examination of the ethical challenges of these collaborations, specifically whether and how exposure results are shared with study participants who want them. This issue in particular has raised ethical tensions between CBPR research collaborations and the IRBs that oversee the human subjects' protection of this work. We conclude with an exploration of reflexive research ethics – an approach to addressing some of the post–Belmont ethics questions emerging from CBPR personal exposure research. Reflexive research ethics provides a way for researchers to grapple with the *institutional* disruption of traditional research ethics, question their own ethical standpoints, and move toward a more integrated, and community-engaged perspective.

Contestations over cumulative impacts in regulatory science

The persistence of health disparities and environmental inequalities in the U.S. has placed environmental health science and regulation at a crossroads. Environmental justice advocates have long argued that their neighborhoods are beset by multiple environmental stressors, which include air and water pollution and substandard housing. Community leaders also contend that existing regulations fail to protect residents adequately because the regulations are focused narrowly on pollutants and their sources. Growing evidence shows that social stressors – including poverty, racial discrimination, malnutrition, and chronic health conditions – also disproportionately affect these communities (Adler and Rehkopf 2008). Research is beginning to show how the cumulative effects of social and environmental stressors can work in combination to produce health disparities (Clougherty and Kubzansky 2009; Morello-Frosch et al. 2010).

Environmental justice advocates have demanded that emerging scientific evidence on cumulative impacts be translated into valid and transparent tools for decision-making in environmental regulation and policy even as the science evolves (Los Angeles Collaborative for Environmental Justice and Health 2011; Cal-EPA 2003; Su et al. 2009). Current risk assessment practices address differential susceptibility for certain intrinsic biological factors (for example, age) by applying safety or default factors to protect biologically sensitive populations (such as children) in limited cases. Advocates have been working with regulatory scientists and academic researchers to ensure that the environmental risk assessment process better accounts for so-called extrinsic factors – including neighborhood poverty, unemployment, lack of food security, discrimination and other psychosocial stressors – that can contribute to the heightened vulnerability of disadvantaged communities (National Environmental Justice Advisory Council 2004; NRC 2009). One approach supported by advocates is to use cumulative impact screening to map, characterize, and target vulnerable communities for interventions that improve existing conditions and prevent future harm. Currently the regulatory burden of proof is placed on communities to demonstrate cumulative impacts, yet many disadvantaged neighborhoods may lack political clout or the capacity for civic engagement to push for regulatory action. The use of cumulative impact screening could remove this burden of proof from vulnerable communities and increase the likelihood that disadvantaged neighborhoods will receive focused regulatory attention. After nearly a decade of debate, negotiation and direct engagement in the regulatory process, advocates have moved some regulatory agencies at the federal, state, and local levels to incorporate elements of cumulative impacts such as those described above into assessment and planning procedures (EPA 2003; OEHHA 2007). Several agencies, such as the U.S. Environmental Protection Agency (EPA), are beginning to develop such tools to target enforcement and compliance activities nationally (EPA 2011), guide land use planning in California (CARB 2005), and inform regulatory programs at the California Air Resources Board (Pastor et al. 2010). These screening methods can help regulators and policy-makers target their efforts to remediate cumulative impacts, environmental inequities, and focus regulatory action at the neighborhood level. One key element of these screening approaches is the importance of engaging communities in method development, metric choices and scoring approaches as these evolve. CBPR has been used to ground-truth, or verify on the ground with direct observation, the results of one screening approach, the Environmental Justice Screening Method. Currently supported by the California Air Resources Board, developed with significant community input, and used in diverse regions across California, the Environmental Justice Screening Method uses roughly thirty environmental health, climate change, and social vulnerability measures to score and map neighborhoods based on three different dimensions: (1) proximity to hazards, (2) exposure to air pollution, (3) climate change vulnerability and (4) social and health vulnerability. These four scores are then added together in order to determine "cumulative impacts." The result is an easy-to-understand visual representation of which communities might require special consideration, such as targeted regulatory protection from further siting of emission sources, more compensatory resources, and additional participatory outreach (Morello-Frosch et al. 2012).

Biomonitoring and household exposure studies

Two case studies illustrate the ways in which community-engagement has transformed exposure assessment science and ethical norms in IRB governance and oversight of such projects. The first study, the Household Exposure Study (HES) involved a research collaboration between an independent research institute (Silent Spring Institute), a regional environmental justice advocacy organization (Communities for a Better Environment, CBE), and two academic institutions (the University of California, Berkeley and Brown University – the co-authors' current and former academic institutions) to measure indoor and outdoor levels of chemicals in a community bordering a major oil refinery in Richmond, California and in a nearby rural community in Bolinas, California that served as a regional comparison area. Protocols entailed sampling indoor and outdoor air and dust for pollutants from industrial emissions, transportation sources, and consumer products. HES partners collected air and dust samples from fifty homes and from nearby outdoor areas and tested these samples for more than 150 analytes, including, endocrine disrupting compounds as well as particulates, metals, polycyclic aromatic hydrocarbons (PAHs), ammonia, sulfates, and other pollutants originating from nearby industries, and which are commonly emitted from refineries (Brody et al. 2009). In the HES, community and academic partners worked collaboratively to develop innovative, transparent and scientifically valid communication materials to report back individual and aggregate sampling results to all participants who wanted them (Brody et al. 2007; Morello-Frosch et al. 2009).

The second study, known as Chemicals in Our Bodies (COB), is a collaboration between UC Berkeley, UC San Francisco, and the California Biomonitoring Program. Over ninety women seeking prenatal and delivery care at San Francisco General Hospital were recruited during their second trimester of pregnancy for a chemical biomonitoring study. Umbilical cord blood as well as maternal blood and urine were collected at delivery and tested for more than 100 chemical analytes. Women were also interviewed to assess potential chemical exposure sources in the home and workplace. Unlike the HES, the COB project did not apply CBPR methods. However, scientists engaged and worked closely with study participants to get their input on the development of report-back materials for those who wanted their results. Some of this work entailed qualitative interviews to assess what participants expected to learn from their involvement in COB. COB also involved direct usability testing, in which participants reviewed prototype report-back materials and gave feedback on elements that they viewed as problematic or difficult to understand. Providing chemical exposure results to study participants who wanted them in the COB project was motivated by a requirement under the California

Biomonitoring legislation (California Biomonitoring Program 2006). Although the levels of community engagement for the HES and COB were significantly different, both projects entailed working closely with study participants to ensure that results communication strategies were useful, meaningful and where possible, actionable.

As environmental justice CBPR projects study the sources and pathways of chemical exposures, they are also faced with the paucity of health effects data for many of the pollutants studied. This situation raises ethical and scientific challenges for whether and how to report results to study participants. In the context of CBPR, this means ensuring that exposure data are reported in ways that are meaningful and that elucidate potential strategies for individual or collective action to protect health. In general, participants tend to want their exposure results, which they often use as a tool for public health advocacy (Morello-Frosch et al. 2009). As a result the HES created bilingual materials (Spanish/English) including graphic displays for communicating aggregate and individual-level results, scientific uncertainties, and potential strategies for exposure reduction. Ultimately, the project found that its communication strategy for results return contributed to environmental health education and stimulated behavioral change and collective efforts to reduce exposures (Adams et al. 2011).

Despite the success of these communication efforts, this form of individualized report-back remains controversial: some university and public health department IRBs question whether the uncertainty regarding the health effects of contaminants or the lack of clear strategies to reduce exposures can cause undue worry and stress among study participants. Recent work suggests that this concern may not be warranted, as study participants generally want access to their personal data, even when the implications of contaminant exposures for future health effects are not well understood (Altman et al. 2008). IRBs are largely unfamiliar with CBPR research, reluctant to oversee community partners, and resistant to ongoing researcher-participant interaction. In resisting such interaction, IRBs can potentially violate the very principles of beneficence and justice that they are supposed to uphold. For example, some IRBs refuse to allow report-back of individual data to participants, which contradict the CBPR principles that guide a growing number of projects. Some allow for passive report-back protocols, whereby the burden is on participants to request data, rather than an active attempt by researchers to provide data in an accessible form (Brown et al. 2010). One IRB even went further by prohibiting research because it feared that if women knew that there were contaminants in their breast milk that they would forgo breastfeeding. This fear has not been borne out by any scientific evidence, however (Brody et al. 2009).

Perhaps more important from an ethics standpoint is not *whether* to provide study results to participants, but *how*. In the HES, collaborative partners agreed that individual-level report-back would be rolled out first, and then followed up by a community meeting where participants and other community members could hear about the aggregate air and dust sampling results. Many study participants attended the community meeting, bringing their results materials with them in order to share with others and ask questions. This extended the reach of the HES to broad audiences in order to leverage results to improve regulation and land-use decision-making. With the support of scientific partners, Communities for a Better Environment, along with some study participants, used data from the HES in their testimony before Richmond's Planning Commission to protest a conditional-use permit application by the nearby Chevron Refinery that would have expanded the facility's capacity to refine lower-grade crude oil and significantly increased pollutant emissions. The presentation of the HES results received significant media attention, as well as inquiries from the California Attorney General's Office. For the COB project, the strategy for developing and testing report-back materials with significant input from study participants from diverse racial/ethnic, socioeconomic and linguistic backgrounds will shape how California's Biomonitoring Program reports exposure results for future studies and surveillance programs it conducts across the state.

In general, by closely collaborating with social movement organizations, researchers can release results to the individual participants and aggregate results to the broader community before publishing them in scientific journals or discussing them with the media. For example, we have heard of community meetings that were closed to the media days or hours before scientific publications or press conferences, to ensure that the first to hear about results are those most directly affected by them. This example of timely dissemination of findings demonstrates one way of respecting the community's interest in co-directing the dissemination of study results while ensuring timely and productive publication for academic partners.

Institutionalizing and advancing the practice of individual report-back

Post-Belmont ethics embrace a CBPR perspective that promotes disseminating personal exposure results to study participants not only to communicate health information, but also to address disparities in access to knowledge that traditionally characterize "lay-expert" relationships (Sullivan et al. 2001). The CBPR approach must be strategic, however, since this framework raises potential conflicts of community versus individual right-to-know: the broad dissemination of biomonitoring results can adversely affect communities under study, even if the rights and confidentiality of individual study participants are protected. Indeed, communities exposed to toxic contaminants with significant health risks may be collectively or individually stigmatized (Morello-Frosch et al. 2009). Individual members of communities may be denied jobs or health or life insurance if they are associated with an "at risk" population. Collectively, a community perceived as "contaminated" may be passed over for programs or benefits, face stereotyping that affects the quality of health care, or suffer lost real estate values or financial liability for remediation (Weijer 1999). Finally, it is important to distinguish between a community's right-to-know and its right-to-act. This is particularly the case in the context of personal exposure research in occupational settings, where study participants may not necessarily be in work situations that enable them to take action to reduce exposures, either through the use of personal protective equipment or the reformulation of products used at the workplace (Senier et al. 2007; Pulido 1996). These challenges of report-back related to right-to-know and right-to-act can be proactively addressed if researchers purposefully develop protocols and communication strategies in partnership with study communities of a personal exposure project (Brown et al. 2010). Key to this process is a collective understanding about who represents the interests of study communities and how their issues can be effectively deliberated and incorporated into protocol development.

For example, the Navajo Nation maintains its own IRB to protect its people from research that would not directly help them (Sharp and Foster 2002). The Indian Health Service adds "respect for communities" to the Belmont principles and expects proposals to discuss whether there are tribal consultants who could be involved, whether there are community capacity-building benefits to the tribe and whether researchers will provide regular and timely community consultations. As federal funding increases for CBPR projects on environmental health such innovative approaches to addressing the ethical challenges of report-back are likely to become more common (Cordner et al. 2012; Hoover et al. 2012).

Finally, innovations in clinical ethics may have implications for results communication in environmental health research. As information technology makes medical records more accessible and patients call for greater transparency, their interest in reading their doctors' visit notes may increase. Some healthcare providers are experimenting with ways to invite patients to review doctor visit notes online with the goal of improving patient understanding of diverse indicators of their health status, fostering more productive communication, and promoting shared decision-making that leads to better health outcomes (Delbanco et al. 2010). Preliminary results from this work indicate that patients who had electronic access to doctors' notes reported feeling more informed and in control of their health care. The Open Notes experiment also promoted better medication adherence, and led to few privacy concerns, worry or confusion among patients. A significant portion of patients also reported sharing their doctors' notes with others (Delbanco et al. 2012). Digital results communication interfaces tested in the clinical setting could be adapted for applications to report back individual results to participants in personal exposure studies. However, such a strategy would require a community engaged approach to develop a digital interface that ensures respect for cultural and individual differences by providing options for receiving results, including views using text or graphs, in different languages, and aimed at diverse literacy levels.

Reflexive research ethics

In addition to the issues raised by individual-level report-back, critical ethical questions emerge regarding the dissemination and communication of exposure study results to the general public, particularly study communities. These questions require proactive reflection and evaluation of the ways in which research has beneficial or detrimental impacts on social movements and community partners. In our Household Exposure Study, we embarked on this process at several points throughout the multi-year research project. We used mutual assessment and evaluation methods (such as debriefings, anonymous evaluations, and short interviews) in formal and informal meetings to learn how partners experienced the academic-community collaboration as it evolved and whether it met diverse objectives, needs and goals. These ongoing reflexive exercises help to ensure that emerging ethical problems are uncovered and addressed in a timely way. This reflexive practice also guarantees a proactive, iterative process for institutionalizing report-back to individual study participants and the broader public, as outlined below.

Values and ethics determine what and how compounds should be measured in study participants and how results are disseminated. Throughout this scientific process, however, there are "moments of uncertainty," points at which scientists face ethical decisions, but lack formal ethics guidelines for emerging science, thus necessitating informal ethics which must constantly be formulated and reformulated (Cordner and Brown 2013). These moments are: (1) choosing research questions or methods, (2) interpreting scientific results, (3) communicating results to multiple publics, and (4) applying results for policy-making. Questions within each category are highlighted below:

1   *Choosing research questions or methods:* This moment of uncertainty leads to ethical tensions which can be unresolved if formal ethical guidelines lag behind the development of novel methods or do not adequately prepare researchers or practitioners to deal with the relevance of findings for non-scientific purposes. Ethical questions that can emerge regarding the production of policy-relevant research or the development of novel methods or scientific practices include: How should research questions be chosen? What methods are ethically appropriate? How should ethics of newly developed methods be evaluated? What are the diverse roles of scientific organizations, professional associations, government agencies, and IRBs? These research design and methods questions are ethical ones because researchers often fail to adequately consult communities in terms of understanding their needs and interests and how these may diverge from those of the researcher.

2   *Interpreting Scientific Results*: How should findings in one area of research be connected to findings in another area of research? How should findings be interpreted when the topic remains contested or inconclusive? How should researchers interpret the risk of a chemical they are studying when their research project evaluates only hazard or exposure but not both?

3   *Communicating Results to Multiple Publics*: General ethical questions that arise include: Who owns (or has the right to) community- or individual-level data? If results are to be shared with participants, how should results be presented and at what point along the research-dissemination timeline? Is it better to present the findings in technical terms or to simplify results to make them more comprehensible to lay audiences? How should results be tailored to specific audiences?

4   *Applying Results for Policy-Making*: Ethical questions related to this moment of uncertainty include: What is the appropriate involvement of researchers in applying their research findings to policy decisions? Should scientists be involved in public debates? How should uncertainty or lack of data inform health and environmental regulation? These are ethical questions since often researchers are discouraged from engaging in the realms of policy-making and practice, while community-engaged researchers view policy applications as integral to assessing the benefits of their work.

To address these questions, researchers must engage in what we term "reflexive research ethics," a self-conscious, interactive and iterative reflection upon researchers' relationships with research participants, relevant communities and principles of professional and scientific conduct (Cordner et al. 2012). Using reflexive research ethics, researchers need to engage in continued adjustment of research practices according to relational and reflexive understandings of individual and community level ethics. Reflexivity is not solely an individual researcher's endeavor, but rather a collective relationship between all actors in a research collaborative. In this process, researchers must move beyond viewing scientific ethics as static, individualized or one-size-fits-all standards or guidelines. Researchers need to understand research as a relational process that changes over time and always has social consequences.

Practicing reflexive research ethics requires researchers to identify and establish interactive discussions with multiple relevant actors, including research participants, local communities, academic disciplines, and people potentially impacted by research findings. Researchers must also identify norms and principles that govern their research; draw upon accumulated knowledge of how others have conceptualized, addressed, and reflected upon relevant ethical issues; assess ethical tensions which may arise from the prioritization of particular interests, publics or principles; and respond to emergent ethical tensions. Reflexive research ethics should govern all phases of the research process, including the *identification* of research questions and motivation; the *engagement* with community actors, social movements, knowledge institutions and other publics; the *production* of knowledge; the *interpretation or analysis* of data; the *presentation and dissemination* of research results; and the *use* of scientific knowledge (Cordner et al. 2012).

This is a logical accompaniment to community-based participatory research, which has a strong ethical direction to involve community partners in all those aspects of research. Community-based organizations have been the pioneers in seeking alternative ethical frameworks, including Native American tribal IRBs that require research to be relevant to tribes, and not simply aimed at advancing scientific knowledge (Arquette et al. 2002; Quigley 2006). Other related examples include community review boards which represent a specific community or litigant class (for example, residents affected by the Fernald nuclear weapons site (Gerhardstein and Brown 2005)) or broader interests of a neighborhood, which may involve many organizations (Grignon et al. 2008; Watkins et al. 2009; Bronx Health-Link and Campus-Community Partnerships for Health 2012).

Conclusion

Much like classic notions of scientific objectivity, research integrity is often understood as a commitment to adhering to established principles of research conduct in an unbiased fashion that focuses exclusively on the protection of individual study participants. Post-Belmont research ethics highlight the inherent tensions between individual and community rights and protections of research participants in the scientific enterprise. This is a form of disruption in how researchers, institutions, and IRBs view research subjects; the *passive subject* is replaced by the *active participant*. The post-Belmont perspective also acknowledges the perennial ethical challenge that participants' right-to-know their exposure results may not necessarily be connected with their right-to-act on those results by taking action to reduce chemical exposures. Most important, community-based participatory research methods demonstrate that research integrity and ethics are fluid, dynamic, value-laden and often contested guideposts that must be constantly and self-consciously reflected upon. This is what leads to the need for reflexive research ethics, a framework for ongoing evaluation, redefinition, and revision of forms of community engagement in the scientific enterprise. This constant evolution of community engagement in environmental health science inevitably raises conflicts between IRBs on one hand and CBPR researchers and their community partners on the other.

Moreover, increased lay involvement in science makes it necessary to have principles of democratic knowledge sharing that are robust, yet nimble. These principles include standards of measurement and risk assessment in regulatory science, addressing inequities in exposure to environmental hazards, and ethical and scientifically valid communication of results to participants. This type of communication is *expansive*, where Belmont protections have largely been *restrictive*. That expansiveness is a major force of disruption, which upsets the limited protections of traditional research ethics. The post-Belmont ethics we have posited provide a model for a broader integration of community-engaged research ethics that match the increased trend toward democratizing the scientific enterprise, the co-production of environmental health knowledge between communities and researchers, and the application of study results to support policy change.

Note

1 The Tuskegee Syphilis Study was an infamous clinical study conducted between 1932 and 1972 by the U.S. Public Health Service in collaboration with the Tuskegee Institute, to study the progression of untreated syphilis in rural African American men who thought they were receiving free health care from the U.S. government. Researchers enrolled a total of 600 impoverished sharecroppers from Macon County, Alabama; in exchange for participating in the study, the men were given free medical care, meals, and free burial insurance. They were never told they had syphilis, nor were they ever treated for it. According to the Centers for Disease Control, the men were told they were being treated for "bad blood." The 40-year study was controversial because researchers knowingly failed to treat patients appropriately after the 1940s validation of penicillin as an effective cure for the disease they were studying. Revelation of study failures by a whistleblower catalyzed changes in U.S. law and regulation on the protection of participants in clinical studies (Jones 1981; Reverby 2009).

References

Adams, C., Brown, P., Morello-Frosch, R., Brody, J., Rudel, R., Zota, A., Dunagan, S., Tovar, J., and Patton, S. 2011. "Disentangling the Exposure Experience: The Roles of Community Context and Report-back of Environmental Exposure Data." *Journal of Health and Social Behavior* 52(2): 180–196.

Adler, N. and Rehkopf, D. 2008. "U.S. Disparities in Health: Descriptions, Causes, and Mechanisms." *Annual Review of Public Health* 29: 235–52.

Altman, R., Brody, J., Rudel, R., Morello-Frosch, R., Brown, P., and Averick, M. 2008. "Pollution Comes Home and Pollution Gets Personal: Women's Experience of Household Toxic Exposure." *Journal of Health and Social Behavior* 49: 417–435.

Arquette, M., Cole, M., Cook, K., LaFrance, B., Peters, M., Ransom, J., Sargent, E., Smoke, V., and Stairs, A. 2002. "Holistic Risk-based Environmental Decision-making: A Native Perspective." *Environmental Health Perspectives* 110: 259–264.

Balazs C. and Morello-Frosch, R. 2012. "The Three R's: How Community Based Participatory Research Strengthens the Rigor, Relevance and Reach of Science." *Environmental Justice* 6(1): 9–16.

Brody, J., Morello-Frosch, R., Brown, P., Rudel, R. 2009. "Reporting Individual Results for Environmental Chemicals in Breastmilk in a Context That Supports Breastfeeding." *Breastfeeding Medicine* 4(2): 121–121.

Brody J., Morello-Frosch, R., Zota, A., Brown, P., Pérez, C., and Rudel, R. 2009. "Linking Exposure Assessment Science with Policy Objectives for Environmental Justice and Breast Cancer Advocacy: The Northern California Household Exposure Study." *American Journal of Public Health* 99: S600–S609.

Brody J., Morello-Frosch, R., Brown, P., Rudel, R., Altman, R., Frye, M., Osimo, C., Pérez, C., and Seryak, L. 2007. "Improving Disclosure and Consent: is it Safe? New Ethics for Reporting Personal Exposures to Environmental Chemicals." *American Journal of Public Health* 97(9): 1547–1554.

Bronx Health Link and Community-Campus Partnerships for Health. 2012. "Community IRBs and Research Review Boards: Shaping the Future of Community-Engaged Research." Albert Einstein College of Medicine, Bronx Health Link and Community-Campus Partnerships for Health. Available from *http://depts.washington.edu/ccph/pdf_files/Shaping_the_Future_of_CEnR.pdf* (accessed 11/26/2012).

Brown, P., Brody, J., Morello-Frosch, R., Tovar, J., Zota, A., Rudel, R. 2012 "Measuring the Success of Community Science: The Northern California Household Exposure Study." *Environmental Health Perspectives* 120: 326–331.

Brown, P., Morello-Frosch, R., Brody, J., Altman, R., Rudel, R., Senier, L., Pérez, C., and Simpson, R. 2010. "Institutional Review Board Challenges Related to Community-Based Participatory Research on Human Exposure to Environmental Toxins: A Case Study." *Environmental Health* 9: 39. Available from *www.ehjournal.net/content/9/1/39* (accessed 11/26/2012).

Cal-EPA Advisory Committee on Environmental Justice. (Cal/EPA) 2003. "Recommendations of the California Environmental Protection Agency Advisory Committee on Environmental Justice to the Cal/EPA Interagency Working Group on Environmental Justice, Final Report." Sacramento, CA. California Environmental Protection Agency. Available from *www.calepa.ca.gov/envjustice/documents/2003/7_11report.pdf* (accessed 11/26/2012).

California Air Resources Board (CARB). 2005. "Air Quality and Land Use Handbook: A Community Health Perspective." Sacramento, CA. California Air Resources Board.

California Biomonitoring Program. 2006. Senate Bill 1379. Chapter 8, Health and Safety Code Sections Health and Safety Code Sections 105440–105444. Available from *www.cdph.ca.gov/programs/Biomonitoring/Pages/default.aspx* (accessed 11/26/2012).

Clougherty, J. and Kubzansky, L. 2009. "A Framework for Examining Social Stress and Susceptibility in Air Pollution and Respiratory Health." *Environmental Health Perspectives* 117(9): 1351–1358.

Cordner, A., and Brown, P. 2013. "Moments of Uncertainty: Ethical Considerations and Emerging Contaminants" *Sociological Forum* 28(3): 469–494.

Cordner, A., Ciplet, D., Brown, P., and Morello-Frosch, R. 2012 "Reflexive Research Ethics for Environmental Health and Justice" *Social Movement Studies* 11: 161–176.

DeFur, P., Evans, G., Hubal, E., Kyle, A., Morello-Frosch, R., and Williams, D. 2007. "Vulnerability as a Function of Individual and Group Resources in Cumulative Risk Assessment." *Environmental Health Perspectives* 115: 817–824.

Delbanco, T., Walker, J., Darer, J., Emore, J., Feldman, H., Leveille, S., Ralston, J., Ross, S., Vodicka, E., and Weber, V. 2010. "Open Notes: Doctors and Patients Signing On." *Annals of Internal Medicine* 153: 121–125.

Delbanco, T., Walker, J., Bell, S., Darer, J., Elmore, J., Fraq, N., Feldman, H., Mejilla, R., Ngo, L., Ralston, J., Ross, S., Trivedi, N., Vodicka, E., and Leveille, S. 2012. "Inviting Patients to Read Their Doctors' Notes: A Quasi-experimental Study and a Look Ahead." *Annals of Internal Medicine* 157: 461–470.

Environmental Protection Agency (EPA). 2011. "The Environmental Justice Strategic Enforcement

Assessment Tool (EJSEAT)" Available from (*www.epa.gov/environmentaljustice/resources/policy/ej-seat.html* (accessed 11/26/2012).

Environmental Protection Agency (EPA). 2003. "Framework for Cumulative Risk Assessment." Environmental Protection Agency Document. Available from *www.epa.gov/raf/publications/pdfs/frmwrk_cum_risk_assmnt.pdf* (accessed 11/26/2012).

Gee, G., Payne-Sturges, D. 2004. "Environmental Health Disparities: A Framework Integrating Psychosocial and Environmental Concepts." *Environmental Health Perspectives* 112(17):1645–1653.

Gerhardstein, B., and Brown, P. 2005. "The Benefits of Community Medical Monitoring at Nuclear Weapons Production Sites: Lessons from Fernald." *Environmental Law Reporter* 35: 10530–10538.

Grignon, J., Wong, K., and Seifer, S. 2008. "Ensuring Community-Level Research Protections." Proceedings of the 2007 Educational Conference Call Series on Institutional Review Boards and Ethical Issues in Research. Seattle, WA: Community-Campus Partnerships for Health.

Hoover, E., Cook, K., Plain, R., Sanchez, K., Waghiyi, V., Miller, P., Dufault, R., Sislin, C., and Carpenter, D. 2012. "Indigenous Peoples of North America: Environmental Exposures and Reproductive Justice" *Environmental Health Perspectives.* Special Report. Available from *http://dx.doi.org/10.1289/ehp.1205422* (accessed 11/26/2012).

Jasanoff, S. 1987. "Contested Boundaries in Policy-relevant Science." *Social Studies of Science,* 17(2):195–230.

Jasanoff, S. 1990. "American Exceptionalism and the Political Acknowledgment of Risk." *Daedalus*: 61–81.

Jasanoff, S. 2004. *States of Knowledge: The Co-production of Science and Social Order.* New York: Routledge.

Jones, J. 1981. *Bad Blood: The Tuskegee Syphilis Experiment.* New York: Free Press.

Los Angeles Collaborative for Environmental Justice and Health. 2011. *Hidden Hazards: A Call to Action for Healthy, Livable Communities.* Los Angeles: Liberty Hill Foundation.

Morello-Frosch, R., and Shenassa, E. 2006. "The Environmental 'Riskscape' and Social Inequality: Implications for Explaining Maternal and Child Health Disparities." *Environmental Health Perspectives* 114(8): 1150–1153.

Morello-Frosch, R., Brown, P., and Zavestovski, S. 2011. "Environmental Justice, Contested Illnesses Struggles." In *Contested Illnesses: Citizens, Science and Health Social Movements.* Edited by P. Brown, R. Morello-Frosch and S. Zavestovski, pp. 3–14. Berkeley: University of California Press.

Morello-Frosch, R., Jesdale, B., Sadd, J., and Pastor, M. 2010. "Ambient Air Pollution Exposure and Full-Term Birth Weight in California." *Environmental Health* 9: 44. Available from *www.ehjournal.net/content/9/1/44* (accessed 11/26/2012).

Morello-Frosch, R., Brody, J., Brown, P., Altman, R., Rudel, R., and Pérez, C. 2009. "Toxic Ignorance and Right-to-Know in Biomonitoring Results Communication: A Survey of Scientists and Study Participants." *Environmental Health,* 8(1): 6. Available from *www.ehjournal.net/content/8/1/6* (accessed 11/26/2012).

Morello-Frosch, R., Pastor, M., Sadd, J., Pritchard, M., and Matsuoka, M. 2012. "Citizens, Science and Data Judo: Leveraging Secondary Data Analysis to Build a Community-Academic Collaborative." In *Environmental Justice in Southern California.* In *Methods for Community-Based Participatory Research for Health.* 2012. Edited by B. Israel, E. Eng, A. Schultz, E. Parker, and B. Josey pp. 547–580. New York, NY. Wiley and Sons Publishers.

Morello-Frosch, R., Zavestoski, S., Brown, P., McCormick, S., Mayer, B., and Gasior, R. 2006. "Social Movements in Health: Responses to and Shapers of a Changed Medical World." In *The New Political Sociology of Science: Institutions, Networks, and Power.* 2006. Edited by S. Frickel and S.K. Moore. Madison, WI: University of Wisconsin Press.

Morello-Frosch, R., Zuk, M., Jerrett, M., Shamasunder, B., Kyle, A. 2011. "Synthesizing the Science on Cumulative Impacts and Environmental Health Inequalities. Implications for Research and Policy-making." *Health Affairs* 30(5): 879–887.

National Commission for the Protection of Human Subjects of Biomedical and Behavioral Research. 1979. *The Belmont Report: Ethical Principles and Guidelines for the Protection of Human Subjects of Research.* Edited by National Institutes of Health. Bethesda, MD: Office of Human Subjects Research.

National Environmental Justice Advisory Council. 2004. "Ensuring risk reduction in communities with multiple stressors: environmental justice and cumulative risks/impacts." National Environmental Justice Advisory Council Cumulative Risks/Impacts Work Group. Available from *www.epa.gov/environmentaljustice/resources/publications/nejac/nejac-cum-risk-rpt-122104.pdf* (accessed 11/26/2012).

National Research Council (NRC) 2009 *Science and Decisions: Advancing Risk Assessment.* Washington, DC: National Academies Press.

Office of Environmental Health Hazard Assessment (OEHHA). 2007. "Environmental Justice Activities At OEHHA: Cumulative Impacts and Precautionary Approaches" California Environmental Protection Agency. Available from *http://oehha.ca.gov/ej/index.html* (accessed 11/26/2012).

Pastor, M., Jr., Morello-Frosch, R., and Sadd, J. 2010. "Air Pollution and Environmental Justice: Integrating Indicators of Cumulative Impact and Socio-economic Vulnerability into Regulatory Decision-making" Final Report for California Air Resources Board. Available from: *www.arb.ca.gov/research/apr/past/04-308.pdf* (accessed 11/26/2012).

Pulido, L. 1996. *Environmentalism and Economic Justice: Two Chicano Struggles in the Southwest.* Tuscon, AZ. University of Arizona Press.

Quigley, D. 2006. "A Review of Improved Ethical Practices in Environmental and Public Health Research: Case Examples from Native Communities." *Health Education and Behavior* 33: 130–147.

Reverby, S. 2009. *Examining Tuskegee: The Infamous Syphilis Study and its Legacy*, Chapel Hill: University of North Carolina Press.

Sadd, J., Pastor, M., Morello-Frosch, R., Scoggins, J., and Jesdale, B. 2011. "Playing it Safe: Assessing Cumulative Impact and Vulnerability through an Environmental Justice Screen Method in the South Coast Air Basin, California." *International Journal of Environmental Research and Public Health* 8: 1441–1459.

Senier, L., Mayer, B., Brown, P., and Morello-Frosch, R. 2007. "School Custodians and Green Cleaners: New Approaches to Labor-Environment Coalitions." *Organization and Environment* 20(3): 304–324.

Sharp, R. and Foster, M. 2002. "Community Involvement in the Ethical Review Of Genetic Research: Lessons from American Indian and Alaska Native Populations." *Environmental Health Perspectives* 110(S2): 145–148.

Su, J., Morello-Frosch, R., Jesdale, B., Kyle, A., and Jerrett, M. 2009 "An Index for Assessing Cumulative Environmental Hazard Inequalities to Socioeconomic and Racial-ethnic Measures with Application to Los Angeles County." *Environmental Science and Technology* 43(20): 7626–7634.

Sullivan, M., Kone, A., Senturia, K., Chrisman, N., Ciske, S., and Krieger, J. 2001. "Researcher and Researched–community Perspectives: Toward Bridging the Gap." *Health Education & Behavior* 28: 130–149.

Watkins, B., Shepard, P., and Corbin-Mark, C. (2009). "Completing the Circle: A Model for Effective Community Review of Environmental Health Research." *American Journal of Public Health* 99(S3): S567–S577.

Weijer, C. 1999. "Protecting Communities in Research: Philosophical and Pragmatic Challenges." *Cambridge Quarterly of Healthcare Ethics* 8: 501–513.

Wilson, S. 2009. "An Ecological Framework to Study and Address Environmental Justice and Community Health Issues." *Environmental Justice* 2(1): 15–23.

Index

Please note that page numbers relating to Notes have the letter 'n' following the page number, whilst those referring to Figures or Tables are in *italics*.

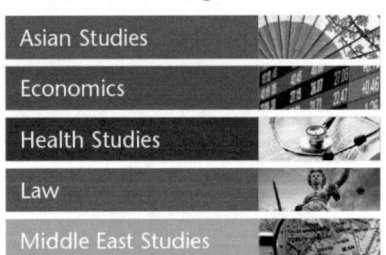
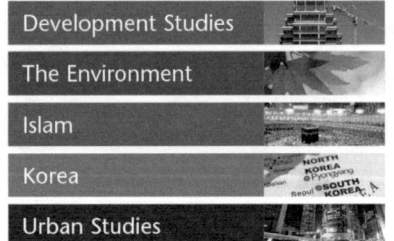